3rd International Conference on Data Mining, Communications and Information Technology (DMCIT 2019)

AA006530

Journal of Physics: Conference Series Volume 1284

Beijing, China
24 – 26 May 2019

ISBN: 978-1-5108-9567-6
ISSN: 1742-6588

Printed from e-media with permission by:

Curran Associates, Inc.
57 Morehouse Lane
Red Hook, NY 12571

Some format issues inherent in the e-media version may also appear in this print version.

Institute of Physics
Dirac House, Temple Back
Bristol BS1 6BE UK

Phone: 44 1 17 929 7481
Fax: 44 1 17 920 0979

techtracking@iop.org

Additional copies of this publication are available from:

Curran Associates, Inc.
57 Morehouse Lane
Red Hook, NY 12571 USA
Phone: 845-758-0400
Fax: 845-758-2633
Email: curran@proceedings.com
Web: www.proceedings.com

Table of contents

Volume 1284

2019 3rd International Conference on Data Mining, Communications and Information Technology (DMCIT 2019)

24–26 May 2019, Beijing, China

Accepted papers received: 26 June 2019
Published online: 22 August 2019

Preface

Papers

Automation and Innovative Manufacturing

IOP Conf. Series: Journal of Physics: Conf. Series **1284** (2019) 011001 doi:10.1088/1742-6596/1284/1/011001

PREFACE

DMCIT 2019 was held in Beijing, China, May 24-26, 2019. DMCIT 2019 was organized by Asia Pacific Institute of Science and Engineering. This conference aims to provide a platform for researchers and engineers to share their ideas, recent developments and successful practices in data mining, communication and the related topics in information technology. The participants of the conference were from almost every part of the world, with background of either academia or industry, even well-known enterprise. The success and prosperity of the conference is reflected high level of the papers received.

The proceedings are a compilation of the accepted papers and represent an interesting outcome of the conference. This volume covers 3 chapters: 1. Automation and Innovative Manufacturing; 2. Internet, Computing, Algorithms and Software; 3. Signal Processing and Telecommunications.

We would like to acknowledge all of those who supported DMCIT 2019. Each individual and institutional help were very important for the success of this conference. Especially we would like to thank the organizing committee for their valuable advices in the organization and helpful peer review of the papers.

We sincerely hope that DMCIT 2019 will be a forum for excellent discussions that will put forward new ideas and promote collaborative researches. We are sure that the proceedings will serve as an important research source of references and the knowledge, which will lead to not only scientific and engineering progress but also other new products and processes.

Anders Lindquist Norman C. Beaulieu

Prof. Anders Lindquist & Prof. Norman C. Beaulieu
Conference Committee Chair

IOP Conf. Series: Journal of Physics: Conf. Series **1284** (2019) 011002 doi:10.1088/1742-6596/1284/1/011002

COMMITTEES

Conference Committee Chair

Anders Lindquist, Shanghai Jiao Tong University, China

Norman C. Beaulieu, Beijing University of Posts and Telecommunications, China

Program Committee Chair

Wei Jia, University of Chinese Academy of Science, China

Publication Chair

Rung-Ching Chen, Chaoyang University of Technology, Taiwan, ROC

International Technical Committee

Ayman Aljarbouh, University of Grenoble Alpes, France

J. Amudhavel, Department of Computer Science & Engineering, KL University, India

Kwang Sik Chung, Korea National Open University, Korea

Dong HUANG, South China Agricultural University, China

M.A.JABBAR, Vardhaman college of Engineering Hyderabad, Telangana, India

Hongchun Qu, Chongqing University of Posts and Telecommunications, China

RajaniKanth Aluvalu, Vardhaman college of Engineering Hyderabad, Telangana, India

Ka-Chun Wong, City University of Hong Kong, Hong kong

Mustansar Ali Ghazanfar, University of Engineering and Technology, Taxila, Pakistan

Guan-Ling Lee, National Dong Hwa University, Taiwan

M. Awais Azam, University of Engineering and Technology, Taxila, Pakistan

IOP Conf. Series: Journal of Physics: Conf. Series **1284** (2019) 011003 doi:10.1088/1742-6596/1284/1/011003

Peer review statement

All papers published in this volume of *Journal of Physics: Conference Series* have been peer reviewed through processes administered by the proceedings Editors. Reviews were conducted by expert referees to the professional and scientific standards expected of a proceedings journal published by IOP Publishing.

A Feature Selection Method based on the Pearson's Correlation and Transformed Divergence Analysis

Ying Zhang[1,2], Lin Wang[1], Danyang Geng[1], Yunfei Ai[1], Wei Xia[1], Xuejiao Bai[1] and Shikai Sun[1]

[1] China Transport Telecommunications & Information Center, Beijing 100011, China
[2] Beijing Forestry University, Beijing 100083, China

august12@163.com

Abstract. An improved feature selection method has been presented, which is based on Transformed Divergence (TD) considering weights of classes and Pearson's correlation analysis. Using the improved method, this study evaluated several derived vegetation indices and texture measures based on Landsat-8 OLI data to determine their effect on improving the land cover classification separability in Jiangle county, Sanming city of Fujian province, China. The best vegetation indices combination was selected by the improved feature selection method and likewise, best textural combinations at different spatial resolution levels from multi-spectral bands or panchromatic band were obtained. The improved feature selection method found that a single feature could not maximize the separability of vegetation classes. When selecting four vegetation indices, the separability of vegetation classes can be maxmized significantly; and two textural measures were suitable for maxmizing the separability of vegetation classes. Overall, the result verifies that the feature selection method considering weights of classes and Pearson's correlation coefficient can select optimal features to maximize class separability.

1. Introduction

Land-cover classification using remotely sensed image has been being a hot study topic, which has attracted great attention [1, 2, 3, 4]. Landsat (TM, ETM+ and OLI) images are the most frequently used for classification and monitoring because of its free acquisition and high time phase [5]. The Landsat record has been extended because of the Landsat 8 Operational Land Imager (OLI). The OLI sensor has improved signal-to-noise performance and has been added new spectral bands in the blue and cirrus cloud-detection portion of the spectrum. Compared to the ETM+ panchromatic band, the OLI panchromatic band has a narrower wavelength range, which is more suitable to distinguish vegetation from non-vegetation area. Therefore, OLI data has greater potential for improving vegetation mapping.

Many vegetation indices and texture features based on remotely sensed images have been developed to improve classification performance in the area of thematic mapping [6, 7, 8, 9, 10, 11]. Using vegetation indices (VI) has many advantages relative to single original band, such as enhancing spectral differences for plants, attenuating atmospheric effect, soil background and shadow. Image textures are also an effective feature in improving the classification result. The common used texture measures are Gray Level Co-occurrence Matrix method (GLCM), which is extracted using different window sizes [12]. However, the application of a large volume of features inevitablely increases the computational time and the classification accuracy has little improvement. To select the suitable features, many researchers have developed a large number of feature selection methods to select optimal feature or features combination [13, 14, 15].

Selection of feature algorithm is an important hot topic in land-cover classification[12]. Many features have similar information and are not all useful for classification. To increase the separability of categories, it is necessary to use suitable method to identify the optimal features. Though most previous studies have used VI and texture measures for thematic mapping, the influence from sample amounts of different classes is not considered and the number of best features [13] used remains unclear for different features types. The objectives of this research are to develop a feature selection method for selecting optimal VIs and textural features in the study area. In addition, evaluate how many features are suitable for vegetation classification when using different feature types.

2. Data and processing

This study area is located in Jiangle county, FuJian province, Southeast China (see Figure 1). The study area has a central subtropical climate, which is characterized by abundant rainfall and sufficient heat. The annual average temperature is 19.47 °C and the average rainfall 1613.96 mm. The forest species are dominated by *pinus massoniana, cedarwood, schima superba* and *castanopsis sclerophylla.*

Four vegetation classes including coniferous forest, broad-leaved forest, mixed forest, farmland were located and marked during ground survey. In the front of fieldwork, we used Worldview-2 image obtained in December 2015 to select the candidate sample locations in the laboratory. A total of 261 field samples were surveyed and 2355 sample pixels were obtained. Table 1 displays the amounts of training and validation samples for each class.

The sensor data, Landsat-8 OLI, was used in this research. The OLI sensor collects data using 9 shortwave spectral bands and the multispectral bands have a spatial resolution of 30 m. The panchromatic band has a spatial resolution of 15 m. The OLI image used in the study was produced on 17 October 2014. The sun elevation angle and sun azimuth angle are 50.309 and 149.441°, respectively. A image-to-image geometric correction was performed for the OLI image and the correction error is 0.351 pixels. The image was resampled to 30-meter of spatial resolution. The C correction model was used for reducing topographic effects. Subsequently, FLAASH atmospheric correction were performed and then the reflectance was obtained.

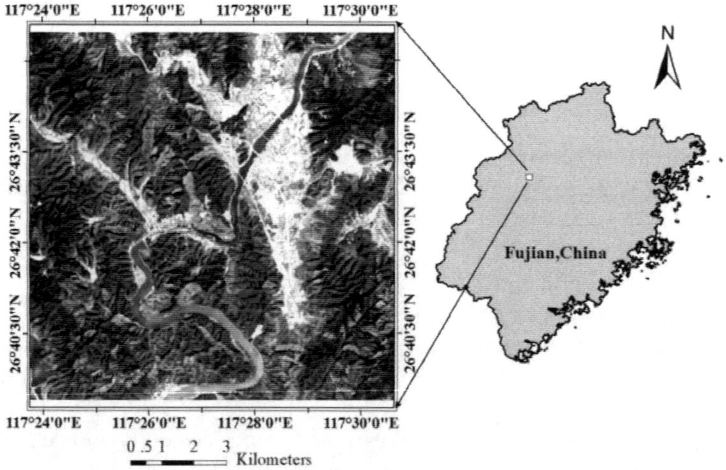

Figure 1. Study area - Jiangle county, Fujian province, China.

Table 1. Number of field samples, training and validation samples.

Types	Surveyed Fields	Training(Pixels)	Validation(Pixels)
Coniferous Forest	37	120	226
Broad-Leaved Forest	27	87	134
Mixed Forest	38	112	265
Farmland	43	126	298
Felling Area	23	66	134
Built-Up Area	36	132	187
Bare Land	28	87	144
Water	29	72	165
Total	261	802	1553

3. Method

3.1. General methodology

First, the VIs and textural measures were obtained using the reflectance image. And then, transformed divergence algorithm (TD) considering weights of classes and correlation analysis were used to select suitable vegetation indices and texture measures at different spatial resolution levels.

3.2. The improved feature selection method

Transformed Divergence (TD) distance is an accurately quantitative indicator for classes separability, which was proved in many studies [16]. However, TD distance does not consider the weights of different classes influenced by sample amounts of different classes. So, accurate separability of multi-classes necessarily takes into account inconsistent sample number for all classes. In the study, the TD_{bh} distance considering weights of sample is used and it is calculated by Equation (2).

$$TD_{ij} = 2(1-\exp(\frac{-D_{ij}}{8})) \tag{1}$$

$$TD_{bh} = \sum_{i=1}^{N}\sum_{j>i}^{N}\sqrt{p(w_i)\times p(w_j)\times TD_{ij}^{2}} \tag{2}$$

$$D_{ij} = \frac{1}{2}tr((C_i - C_j)(C^{-1}_i - C^{-1}_j)) + \frac{1}{2}tr((C^{-1}_i - C^{-1}_j)(\mu_i - \mu_j)(\mu_i - \mu_j)^T)$$

where i and j is the class, TD_{ij} is the transformed divergence distance of two classes, D_{ij} is the divergence distance of two classes, C_i and C_j are covariance matrixes of two classes, μ_i and μ_j are the mean vector, T is transpose of the matrices, N is the number of classes, $p(w_i)$ and $p(w_j)$ are prior probabilities of two classes i, j.

The prior probabilities can be calculated using sample data in Table 1.

The features with greater TD_{bh} values were firstly selected using the equation 2. Then the correlation analysis was performed using Pearson's correlation analysis. The features with high separability and low correlation were finally selected for classification. When two or more features were selected, the best feature combination is identified by the following equation (3).

$$BFC = \sum_{i=1}^{m} TD_{bhi} / \sum_{j=1}^{m} R_{ij} \tag{3}$$

Where TD_{bhi} is the TD value considering the prior probabilities of samples on the feature i, R_{ij} is the correlation coefficient between two features, and m is feature number.

3.3. Selection of vegetation indices

Primary purpose of feature selection is enhancing separability of vegetation types in this paper. Therefore, the separability of features was evaluated only using training samples for four vegetation classes.

Some research found that vegetation indices containing near infrared and mid-infrared bands were correlated with forest vegetation parameters [17]. Therefore, this paper selects vegetation indices consist of OLI band 5, 6 and 7, as far as possible. The used VIs are included in Table 2 [17]. These VIs were divided into four groups for convenient analysis [18].

Table 2. Vegetation indices used in the research

Type	No.	Vegetation indices	Equation
Simple ratio	1	OLI5/4	OLI 5/ OLI 4
	2	OLI6/4	OLI 6/ OLI 4
	3	OLI6/5	OLI 6/ OLI 5
	4	OLI6/7	OLI 6/ OLI 7
Normalized vegetation indices	5	NDVI	(OLI5 – OLI4)/(OLI5 + OLI4)
	6	NDWI	(OLI5 – OLI6)/(OLI5 + OLI6)
	7	ND6-4	(OLI6 – OLI4)/(OLI6 + OLI4)
	8	ND6-7	(OLI6 – OLI7)/(OLI6 + OLI7)
	9	ND4-3	(OLI4 – OLI3)/(OLI4 + OLI3)
	10	ND5-3	(OLI5 – OLI3)/(OLI5 + OLI3)
	11	ND5-36	(OLI5 – OLI3 – OLI6)/(OLI5+ OLI3+ OLI6)
	12	ND53-64	(OLI5+ OLI3 – OLI6 – OLI4)/(OLI5+ OLI3+ OLI6+ OLI4)
	13	ND53-67	(OLI5+ OLI3 – OLI6−OLI7)/(OLI5+ OLI3+ OLI6+ OLI7)
	14	ND5-46	(OLI5 – OLI4 – OLI6)/(OLI5+ OLI4+ OLI6)
	15	ND56-34	(OLI5+ OLI6 – OLI3 – OLI4)/(OLI5+ OLI6+ OLI3+ OLI4)
	16	ND5-67	(2 OLI5 – OLI6 – OLI7)/(OLI5+ OLI6+ OLI7)
Complex vegetation indices	17	ARVI	(NIR-2RED+BLUE)/ (NIR+2RED-BLUE)
	18	ASVI	$\left((2NIR+1)-\sqrt{(2NIR+1)^2-8(NIR-2RED+BLUE)}\right)/2$
	19	MSAVI	$\left((2NIR+1)-\sqrt{(2NIR+1)^2-8(NIR-2RED)}\right)/2$
	20	GEMI	$\xi(1-0.25\xi)-((RED-0.125)/(1-RED))$ $\xi=\left(2\left(NIR^2-RED^2\right)+1.5NIR+0.5RED\right)/\left(NIR+RED+0.5\right)$
Image transform	21	VIS234	OLI2+OLI3+OLI4
	22	MID67	OLI6+OLI7
	23	Albedo	OLI2+OLI3+OLI4+OLI5+OLI6+OLI7
	24	TC1	0.304OLI2+0.279 OLI3+0.474 OLI4+0.559 OLI5+0.508 OLI6+0.186 OLI7
	25	TC2	−0.285 OLI2−0.244 OLI3−0.544 OLI4+0.704 OLI5+0.084 OLI6−0.180 OLI7
	26	TC3	0.151 OLI2+0.197 OLI3+0.328OLI4+0.341 OLI5−0.711 OLI6−0.457 OLI7

3.4 Selection of texture measures

The texture measures consisting of mean, variance, homogeneity, contrast, dissimilarity, entropy, second moment, and correlation are used in this paper, which is obtained based on OLI band 5, band 6, band7 and panchromatic band 8. The moving windows include 3 × 3, 5 × 5, 7 × 7, 9 × 9, 11 × 11, 15 × 15, 19 × 19, 25 × 25, and 31 × 31.

4. Result

4.1. Separability analysis

The improved TD distances considering weights of vegetation sample were used to analyze the separability between different vegetation classes for each feature. The separability analysis values for all VI and textural measures were showed in Figure 2 and Figure 3, respectively. Not all features have high sparability values for vegetation types. The separability distances for VI (Figure 2) show that the simple ratios, OLI5/4 and OLI6/5, had the greatest potential to distinguish vegetation classes.

The separability distances for textural features varied greatly with bands and the size of moving window (Figure 3). In band 5, the textural feature based on entropy with 11×11 window size pixels had a better separability performance. In band 6, the entropy with 19×19 window size pixels had a higher TD_{bh} distance. In band 7, the texture based on mean with 19×19 window size pixels provide best separability for vegetation classes. In panchromatic band, the texture based on second moment with 31×31 window size had relatively high separability distances values. Different texture measures from different bands have various capabilities in separating vegetation classes. The entropy and second moment measures based on band 5 have high levels of separability and in band 6, the entropy, the second moment and mean measures have relaively high separability value. However, in the panchromatic band 8, the variance, second moment and correlation have a better performance in identifying the vegetation types. Therefore, there exists a best combination of texture measure and size of moving window in different spectral range suitable for the separation of vegetation.

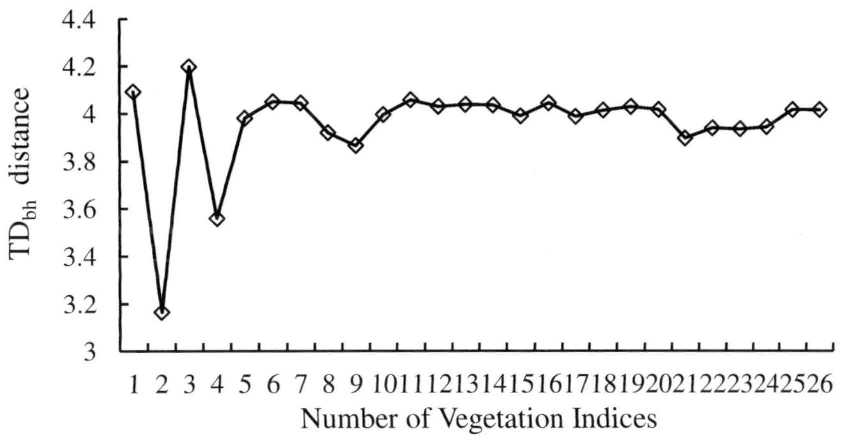

Figure 2. TD_{bh} distance values for each vegetation indices based on selected training sample plots of different vegetation classes. Notes: the number in the x-axis is the same as the number of the Table 2.

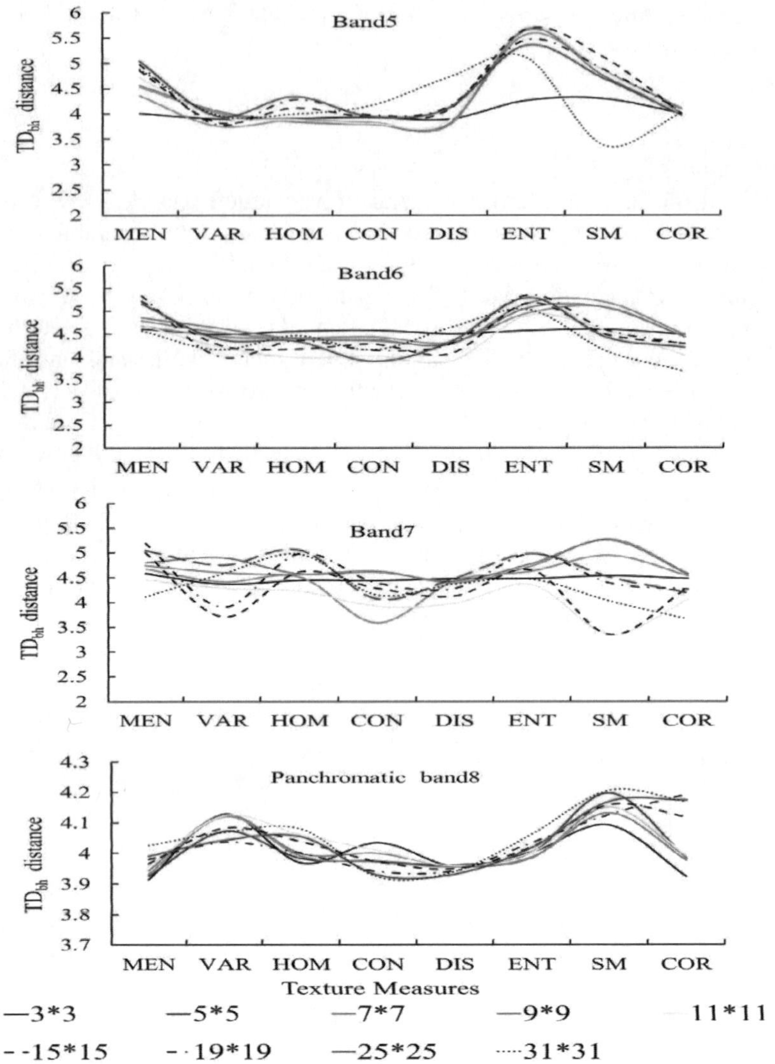

Figure 3. TD_{bh} distance values for each textural image from different bands based on selected training sample plots of different vegetation classes.

4.2. The optimal features combination

For determining the best number of feature combination, the single feature with the highest TD_{bh} distance was selected. Then, correlation coefficients analysis between vegetation indices and between the textural images identified least correction features. Finally, the optimal combination features were obtained by the feature selection method. The features with the best capability in vegetation separation but less correlation were identified using Best Feature Combination (BFC) analysis. Figure 4 shows the maximum TD_{bh} distances of different feature combination for VIs and textural measures from multi-spectral bands or panchromatic band. The identified four images of vegetation indices were from OLI6/5, ND6-4, ND53-67 and MSAVI. In the multi-spectral bands (OLI band5, 6 and 7), the selected two textural measures were from the entropy with 11×11 window size based on OLI band5 and second moment with 7×7 window based on OLI band7. And in the panchromatic band8, the selected two textural measures were from the variance with 5×5 window and correlation with 15×15 window. Four feature combinations for vegetation indices and two feature combinations for textural measures from multi-spectral bands or panchromatic band were selected as optimal combinations for vegetation classification.

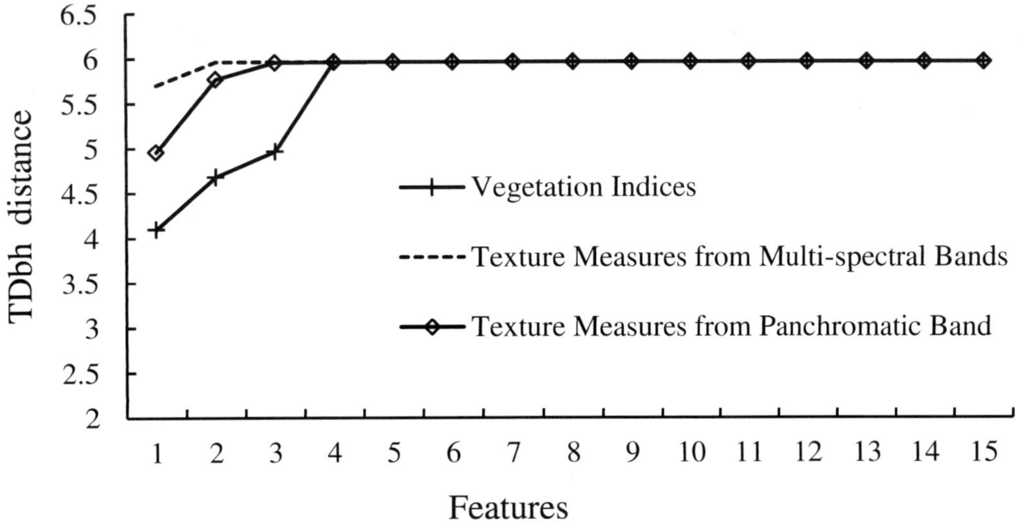

Figure 4. TD_{bh} distance values for VI and each textural image from different bands based on selected training sample plots of different vegetation classes.

5. Conclusion

This research proposes an approach to select optimal feature by TD analysis considering the weights of different classes and band correlation coefficient. For a large number of features, the approach provides an easy method to obtain the optimal feature combinations to improve the class separability. Our study found that four vegetation indices combination are suitable for improving vegetation classification performance, while two features combination from textural images based on OLI multi-spectral bands or OLI panchromatic band are suitable for improving vegetation separability.

Acknowledgments

This work was financially supported by The National Key Research and Development Program of China(2017YFC0803900). The author would like to express loyal thanks to reviewers and editors for their work.

References

[1] D. Lu, et al., *Linear mixture model applied to Amazonian vegetation classification.* Remote Sensing of Environment, 2003. 87 (4): p. 456-469.
[2] D. Chen, et al., *Examining the effect of spatial resolution and texture window size on classification accuracy: an urban environment case.* International Journal of Remote Sensing, 2004. 25 (11): p. 2177-2192.
[3] Y. Zhang, et al., *Coastal wetland vegetation classification with a Landsat Thematic Mapper image.* International Journal of Remote Sensing, 2011. 32 (2): p. 545-561.
[4] Z. Gong, et al., *Estimating wetland vegetation abundance based on spectral mixture analysis: a comparison between LSMA and FCM classification methods.* International Journal of Remote Sensing, 2013. 35 (1): p. 189-203.
[5] D. Lu, et al., *Fractional forest cover mapping in the Brazilian Amazon with a combination of MODIS and TM images.* International Journal of Remote Sensing, 2011. 32 (22): p. 7131-7149.
[6] J. Dash, et al., *Land cover classification using multi‐temporal MERIS vegetation indices.* International Journal of Remote Sensing, 2007. 28 (6): p. 1137-1159.

[7] J. Oldeland, et al., *Combining vegetation indices, constrained ordination and fuzzy classification for mapping semi-natural vegetation units from hyperspectral imagery.* Remote Sensing of Environment, 2010. 114 (6): p. 1155-1166.

[8] L. Zhong, et al., *Efficient corn and soybean mapping with temporal extendability: A multi-year experiment using Landsat imagery.* Remote Sensing of Environment, 2014. 140 (1): p. 1-13.

[9] R.S. Fletcher, *Using Vegetation Indices as Input into Random Forest for Soybean and Weed Classification.* 2016. 07 (15): p. 2186-2198.

[10] Z. Wentao, et al., Texture classification of vegetation cover in high altitude wetlands zone, in: IOP Conference Series: Earth and Environmental Science, 2014, pp. 682-691.

[11] V.K. Alilou, *Texture Classification.* International Journal of Systems Science, 2014. 28 (7): p. 669-682.

[12] G. Li, et al., *A comparative analysis of ALOS PALSAR L-band and RADARSAT-2 C-band data for land-cover classification in a tropical moist region.* ISPRS Journal of Photogrammetry and Remote Sensing, 2012. 70 (0): p. 26-38.

[13] D. Lu, et al., *A Comparative Study of Landsat TM and SPOT HRG Images for Vegetation Classification in the Brazilian Amazon.* Photogrammetric Engineering & Remote Sensing, 2008. 74 (3): p. 311-321.

[14] M. Pal, G.M. Foody, *Feature Selection for Classification of Hyperspectral Data by SVM.* Geoscience & Remote Sensing IEEE Transactions on, 2010. 48 (5): p. 2297-2307.

[15] Y. Zhang, et al., *Feature selection for classification with class-separability strategy and data envelopment analysis.* Neurocomputing, 2015. 166 (C): p. 172-184.

[16] P.W. Mausel, *Optimum band selection for supervised classification of multispectral data.* Photogramm Eng Remote Sens, 1990. 56 (1): p. 55-60.

[17] G. Li, et al., *Land-cover classification in a moist tropical region of Brazil with Landsat Thematic Mapper imagery.* International Journal of Remote Sensing, 2011. 32 (23): p. 8207-8230.

[18] D. Lu, et al., *Relationships between forest stand parameters and Landsat TM spectral responses in the Brazilian Amazon Basin.* 2004. 198 (1): p. 149-167.

IOP Conf. Series: Journal of Physics: Conf. Series **1284** (2019) 012002 doi:10.1088/1742-6596/1284/1/012002

Analysis of Vegetable Price Fluctuation Law and Causes based on Lasso Regression Model

Weige Yu[1,2,3,a], Chunjiang Zhao [2,3,4,b], Huarui Wu [2,3,4] and Cheng Peng [2,3,4]

[1] School of Information Engineering, Capital Normal University
[2] National Engineering Research Center for Information Technology in Agriculture
[3] Key Laboratory of Agri-informatics, Ministry of Agriculture
[4] Beijing Engineering Research Center of Agricultural Internet of Things

[a]yuweige888@126.com; [b]zhaocj@nercita.org.cn

Abstract. In recent years, the sharp price fluctuation of vegetables has attracted wide attention. In order to explore the causes of price fluctuation, the price fluctuation law of cucumber from 2010 to 2017 was deeply explored based on the empirical analysis of the price fluctuation law of vegetables. Most of the existing literatures use linear regression models to analyze the influence degree of various factors, but there are problems of multicollinearity among multiple variables and small influence degree of some variables, which affect the fitting accuracy. Lasso regression model is introduced to model and solve cucumber price data, and the influencing factors of price with smaller correlation are eliminated to obtain the main factors and their correlation. Compared with the least square method, the multicollinearity determination condition value is only 19.66 and the fitting coefficient reaches 0.8448, which proves that lasso regression model is suitable for vegetable price cause analysis and has better performance than traditional methods. It also provides basis for further vegetable price prediction.

1. Introduction

China is the country with the largest vegetable production and consumption in the world, and the vegetable industry is developing rapidly. In recent years, the vegetable market price fluctuates frequently, which has a great impact on people's lives, especially on low-income people. Farmers' untimely response to the market results in loss of benefits. Under this background, this paper deeply explores the law of vegetable price fluctuation and show the relationship between vegetable and influencing factors, which can be used to guide farmers to plan their planting rationally. It is of great significance to increase the economic benefits of vegetable farmers, reduce the living costs of urban residents and stabilize the prices of vegetables in the market.

At present, many scholars have done a lot of research on the law of vegetable price fluctuation and its causes. Literature[1] analyzes the characteristics and causes of the price fluctuation of vegetables in China since the new century and puts forward the idea of stabilizing prices. The article also draws the conclusion that the price change direction of vegetable consumption market is not completely consistent with the production price change direction. Literature[2] uses seasonal adjustment model and covariance analysis to analyze the degree of influence of factors affecting vegetable prices on price fluctuation. Literature[3] starts from the circulation system, deeply studies the characteristics and causes of changes in vegetable prices in China, further studies the industrial problems behind vegetable prices, and clarifies the influence of circulation factors. Literature[4] analyzes the law and influencing factors of price fluctuation of various vegetables, establishes a linear regression model, and obtains the correlation of various factors. Literature[5] explores the changes of vegetable daily price data and forecasts them through generalized regression neural network, which opens up new ideas for short-term price forecasting. Literature[6] combines time series decomposition and H-P filtering method to deeply analyze the law and influencing factors of vegetable price fluctuation in Beijing from 2003 to 2014, and establishes a linear regression model to analyze the influencing factors. However, neither the literature nor the literature[4] consider the problem of over-fitting and high correlation of the linear regression model, resulting in low interpretability of the results.

The above documents have all analyzed the composition of influencing factors of vegetable price from the change rule of vegetable price, but they have not quantified the influence degree of various factors on vegetable price. In addition, the existing researches mostly adopt multiple linear regression models to model vegetable prices, which often have multiple collinearities and over-fitting phenomena in the fitting process. Based on this, after deeply analyzing the advantages and disadvantages of lasso regression method, a lasso regression model was established to analyze the influencing factors of vegetable prices. The model has the characteristics of linear regression model describing the relationship between independent variables and dependent variables. At the same time, it effectively avoids the defects of over-fitting and high correlation and eliminates the problem of multicollinearity.

2. Empirical Analysis of Vegetable Price Fluctuation Law

According to the cucumber price data from eight markets, including Beijing shunxin agricultural products wholesale market and Xinfadi agricultural products wholesale market from 2010 to 2017, the average exportable price trend chart is shown in figure. 1.

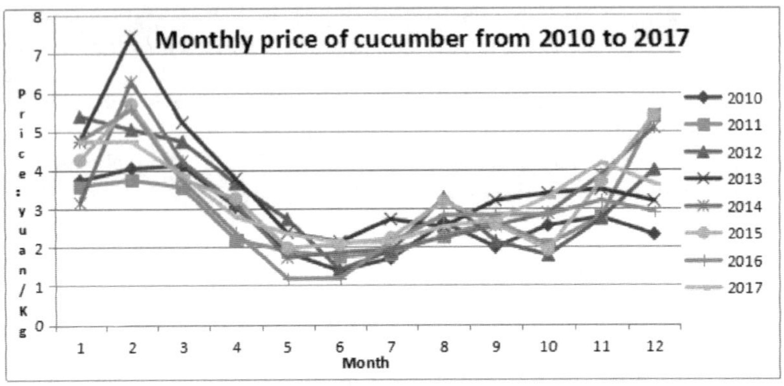

Figure 1 Monthly Average Data

From the price trend chart, it can be seen that cucumber price is highly correlated with time and fluctuates periodically and regularly every year. Vegetable prices are affected by the production period and consumers' holiday consumption habits and have obvious seasonal fluctuation characteristics. According to the theory of supply and demand in economics, when the supply decreases or the demand increases, the price will increase, while vice versa. Cucumber prefers warm environment but not cold. Summer is suitable for cucumber growth, and the yield increases greatly. In addition, there are many alternatives to choose from. The demand of the resident market decreases, and the price of cucumber falls to a low point. The data show that it conforms to the general law of supply and demand. On the other hand, the price increase during holidays is already a common phenomenon. It can be seen that the price of cucumbers will increase to varying degrees during major festivals such as Spring Festival. In addition to traditional festivals such as Spring Festival, Mid - Autumn Festival and National Day, holiday spending driven by the Internet such as " Double Eleven" is also occupying more and more important seats. With the combination of e-commerce and agricultural products becoming more and more close, the successive traditional festivals and online holidays in the second half of the year continue to promote residents' demand for vegetables and drive the prices of vegetables to rise[6].

The average annual cucumber price data can be plotted into a line chart as shown in Figure 2.

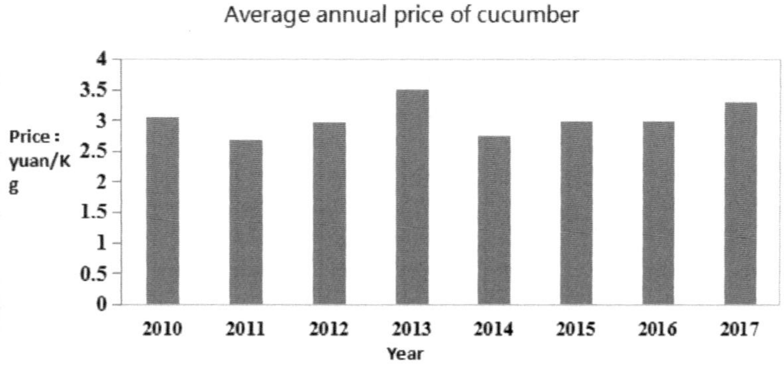

Figure 2 Annual Average Data

It can be seen from the figure that the overall cucumber price change trend from 2010 to 2017 is fluctuating and rising. The reasons for the rise and fall of specific years and months are multiple and complex. According to the previous research experience[1-6], combined with the existing data, the possible factors affecting the price fluctuation of vegetables are classified into four categories, namely, supply factors, demand factors, circulation factors and comprehensive factors. Many influencing factors have different degrees of influence on prices, some of which are extremely small or even negligible. Some factors are easy to be ignored, but in fact they will have a greater impact on price fluctuations. Therefore, in the next step, lasso regression method will be used to model and analyze the relationship between influencing factors and prices, select the influencing factors with high correlation and eliminate the influencing factors with low correlation.

3. Methods

When we have data (X_i, Y_i), where $X_i = (x_{i1}, x_{i2}, \ldots x_{im})^T$ and Y_i are the values of independent and dependent variables respectively obtained in the I times experiment, we can determine the general linear regression model as follows:

$$y = \beta_0 + \beta_1 x_1 + \ldots + \beta_m x_m + \varepsilon \tag{1}$$

Where m refers to the independent variable with m number, β_0 , β_1 ,..., β_m is the regression coefficient of the equation, and ε is the random error. The general solution of linear regression equation is the least square method, i.e. estimation of parameters using observed values.

$$Q=\sum \varepsilon_k^{\ 2} = \sum \left[y_k - (\hat{\beta}_0 + \hat{\beta}_1 x_{1k} + \ldots + \hat{\beta}_m x_{mk}) \right]^2 \tag{2}$$

The least square method is used to solve the problem. The method is usually to minimize the sum of squares of residuals and the Q value. Only the partial derivatives of the above formula $\hat{\beta}_0$ $\hat{\beta}_1$ ... $\hat{\beta}_m$ need to be calculated respectively, and the corresponding regression coefficients can be obtained.

Lasso regression model belongs to a kind of linear regression model[7]. The commonly used method of linear regression models is the least square method, but there are often problems when dealing with models with multicollinearity. There are mainly two defects[8]: first, the prediction accuracy is not high. When the least square method is used to solve the problem, if the linear correlation of independent variables is high, the variance is often large. The second is that the model does not have good interpretability. For the case with more independent variables, the estimated value has a large deviation from the actual situation. At the same time, the estimation results by the least square method will be affected by the multicollinearity between variables. This method cannot eliminate the collinearity and cannot reduce the dimension[9].

Lasso (*Least absolute shrinkage and selection operator, Tibshirani(1996)*) regression method[10] is essentially a compressed estimation. First make $z_{i1}, z_{i2}, ..., z_{im}$ the mean value 0, variance 1, dimensionless difference. Lasso estimation of the regression model is:

$$(\hat{\alpha}, \hat{\beta}) = \arg\min_{(\alpha,\beta)} \sum_{i=1}^{n} (y_i - \alpha - \sum_{j=1}^{p} x_{ij}\beta_j)^2$$
$$\text{subject to} \sum_{j=1}^{p} |\beta_j| \le s \qquad (3)$$

Among them, s≥0 is the harmonic parameter. For any S has the estimation of α , $\hat{\alpha} = \bar{y}$, only the harmonic parameter S needs to be adjusted in order to reduce the overall regression coefficient. When the $s^o = \sum_j |\beta_j|, s \le s^0$, the coefficients of some variables will be reduced, even close to 0 or equal to 0. These independent variables with little or no relation will be screened out, thus improving the accuracy and interpretability of the regression model. The Mallows Cp statistic[11] is generally used to solve the Lasso regression coefficient. This method can evaluate the advantages and disadvantages of the regression model. The method is defined as follows:

$$C_p = \frac{SSE_p}{\delta^2} - n + 2p,$$
$$SSE_p = \sum_{i=1}^{n} (Y_i - Y_{pi})^2 \qquad (4)$$

n represents the total number of samples, ***p*** represents the number of independent variables in the subset regression model, SSE_p represents the sum of squares of residuals after regression of dependent variable Y by p, and δ^2 represents the prediction of the mean value of variances when all independent variables regression of dependent variable Y. According to this, when the model[12] is obtained as C_p is the minimum value, the subset of variables that are best in the global range is obtained, and at the same time, the regression equation with the best effect is generated.

4. Analysis on the Causes of Vegetable Price Fluctuation Based on lasso Model

4.1 Establishment of lasso Regression Model

This paper selects the monthly price data of cucumbers in Beijing from 2010 to 2017 as the research object. Cucumber prices come from a total of 8962 price data on the website of AgSoSo (National Agricultural Science and Education Cloud Intelligent Search System) and the national agricultural products business information public service platform. The demand factor W (independent variable) and dependent variable Q that affect the price of cucumber are used as component residual plot[13].The supply factor X (independent variable), dependent variable Q, circulation factor Y (independent variable) and comprehensive factor Z (independent variable) which affect cucumber price are also made into component residual plot, which can analyze the linear correlation or nonlinear correlation between independent variable and dependent variable in regression model[13].It is observed that the demand factor, supply factor, circulation factor and comprehensive factor all present linear relationship. The cucumber price fluctuation and influencing factor model can be designed as the following linear regression model:

$$Q = \alpha_1 + \beta_1 W + \delta_1 X + \lambda_1 Y + \xi_1 Z + \varepsilon_1 \tag{5}$$

Among them, Q represents the price of cucumber, W represents the demand factor affecting the price of cucumber, X represents the supply factor, Y represents the circulation factor, and Z represents the comprehensive factor. β_1 represents the regression coefficient of demand factor, δ_1 represents the regression coefficient of supply factor, λ_1 represents the regression coefficient of circulation factor, ξ_1 represents the regression coefficient of comprehensive factor and ε_1 represents random error $\varepsilon_1 \sim N(0, \delta^2)$ of the established model.

Using the model (5) to model the acquired data, we can obtain the degree of influence of various factors on cucumber prices. The next section will show the difference between the general solution effect of using lasso regression method to solve the model and multiple linear regression model, proving that lasso regression model is effective in reducing the multicollinearity of data and removing the factors that have little correlation with cucumber price fluctuation.

4.2 lasso Regression Model Experiment and Analysis

In this paper, regression model (5) is used to model the experimental data. Lasso regression model was adopted to obtain the degree of influence of various factors on the fluctuation of cucumber price. Meanwhile, the experiment is compared with multiple linear regression model solved by the widely used least square method. The experimental program is programmed with R language.

Before the experiment, the data were processed into standardized data with a mean value of 0 and a variance of 1. The first two thirds of the data were taken as training sets to build the prediction model, and the last third was taken as test sets to evaluate the model performance. Firstly, a linear regression model is established to solve the problem.

The experimental results obtained from the solution are shown in Table 1. The coefficient in the table is greater than 0, indicating that this item has a positive effect on the price of cucumber. The larger the value, the greater the effect on the price of cucumber. If the coefficient in the table is less than 0, it indicates that this item has a negative effect on the price of cucumber. The smaller the value, the greater the negative effect on the price of cucumber. The smaller the standard deviation in the table, the better the effect.

Table 1 Estimated Regression Coefficient of Linear Regression Model

Variable	Regression coefficient	Standard deviation
Per capita disposable income	0.0862	0.2259
inhabitant	-0.0594	0.0985
Consumption level of residents	0.5265	0.0563
Inflation level	0.6859	0.3324
Sowing area of cucumber	-0.0085	0.0181
Cucumber yield	-0.0486	0.3098
Index of agricultural means of production	0.4582	0.0324
Price Index of Means of Production for Mechanized Farm Tools	0.2241	0.0094
Affected area	0.1256	0.2558
Diesel price	0.0495	0.0544
Driver service fee	0.0836	0.0674
seasonal variation	0.6229	0.0635
Level of economic development	0.0482	0.0649

The model has the following disadvantages: when the multicollinearity is checked by the conditional discriminant method, the multicollinearity determination condition k value of the multicollinearity model solved by the least square method is 118.4072.At present, it is generally believed that when the k value is more than 30, the k value can be considered too large, which indicates that there is a serious collinearity problem[12, 14, 15].When there is such a collinearity problem, it shows that there is a high correlation between the independent variables in the regression model, which will cause the confidence interval of the model parameters to be too large and the solution result of the model to be inaccurate. At the same time, it is difficult to explain a single regression coefficient, causing instability of the model[16].

In the experiment, the model (5) is further solved by lasso method, and redundant variables are eliminated by Mallows Cp method. The solution results of this model are shown in Table 2.The experimental results show that six variables are eliminated and the multicollinearity criterion value is 19.660. The multicollinearity problem is well optimized and the set of extracted variables is also better. The regression coefficient of a variable in Table 2 is greater than zero, which indicates that the variable plays a positive role in cucumber price, while it plays a negative role when it is less than zero. The coefficient of per capita disposable income, resident population, index of production data of mechanized farm tools, labor cost of drivers, and economic development level is 0, which is eliminated by lasso regression solution and Mallows Cp unified measurement method. It can be seen that lasso regression method can effectively solve the problem of multiple collinearities in the model, select the best variable set and obtain a regression model with better generalization ability.

Table 2 Regression Coefficient of Lasso Regression Model

Variable	Regression coefficient
inhabitant	-0.1552
Consumption level of residents	0.4679
Sowing area of cucumber	-0.0081
Cucumber yield	-0.0698
Index of agricultural means of production	0.4253
Affected area	0.0658
Diesel price	0.0365
seasonal variation	0.5965

We use the remaining one-third of the data set as the test set to test the multiple linear regression model and lasso regression model respectively, and observe the test error and model fitting performance. Table 3 shows the multicollinearity determination condition values, fitting errors and goodness of fit of the two methods. Obviously, the multicollinearity determination condition value of method 3 is only 19.660, and the fitting coefficient is 0.8448, which is better than that of method (2) on the whole, and the problem of multicollinearity is reasonably solved, and the fitting error is small, indicating that lasso regression method has better performance in the model of influencing cucumber price factors.

Table 3 Comparison of Fitting Results of Two Methods

Method	Multicollinearity determination condition value	regression error	regression coefficient
Least square method	118.4072	0.1169	0.7629
Lasso regression method	19.66	0.0903	0.8448

5 Conclusion

According to the empirical analysis of the price fluctuation rule, we can draw a conclusion that the cucumber market price in Beijing has obvious seasonal characteristics, with high price in winter and low price in summer. Besides seasonal factors, it is also affected by supply factors, demand factors, circulation factors and other comprehensive factors.In order to determine the degree of influence of each influencing factor, lasso regression method is used to model each influencing factor and price. Compared with conventional multivariate linear model, it is proved to have better effect, mainly reflected in higher precision of regression coefficient of influencing factors and smaller linear correlation.

The model regression results show that among the demand factors, the correlation degree between per capita disposable income and inflation level and cucumber price is smaller, and the consumer level and resident population have greater influence on cucumber price. Among the supply factors, the index of the means of production of mechanized farm tools had almost no effect on the price of cucumber, so it was eliminated, while the sowing area, the yield of cucumber, the index of the means of agricultural production and the affected area all had an effect on the price of cucumber to some extent. Among the circulation factors, the correlation between the driver's labor cost and the price is not much excluded, while the price of diesel oil will affect the price of cucumber. Among the comprehensive factors, the level of economic development has little to do with the price of cucumber.

Based on cucumber price data and influencing factor data, this paper applies lasso regression method to analyze the influencing factors of vegetable price and more accurate results have been achieved. It provides new ideas and analysis methods for the analysis of influencing factors of vegetable price and provides basis and reference for further research on vegetable price prediction.

Acknowledgments

This work was supported by Beijing Municipal Natural Science Foundation (4172025, 4172026) and the Natural Science Foundation of China (41501418) .

References

[1] Li Chongguang, Bao Yuze. Analysis on the Characteristics and Causes of Vegetable Price Fluctuation in China [J]. Chinese Vegetables, 2012 (9): 1 - 7.

[2] Song Changming, Xu Juan, Xiang Chaoyang. Analysis on the Causes of Vegetable Price Fluctuation Based on Time Series Decomposition [J]. statistics and decision, 2014 (3): 106 - 108.

[3] Zhou Zhenya, Li Jianping, Zhang Qing, et al. Analysis of Vegetable Price Problem and Its Causes in China [J]. Agricultural Economic Problems, 2012 (7): 91 - 95.

[4] Zhao Anping, Zhao Yousen, Wang Chuan. Influencing Factors and Evaluation of Vegetable Price Fluctuations in Beijing and Policy Recommendations [J]. Agricultural Modernization Research, 2012,33 (5): 000598 - 602.

[5] Chen Fang, Lou wengao. daily price forecast of vegetable market based on generalized regression neural network [J]. journal of Zhejiang agriculture, 2015,27 (7): 1253 - 1258.

[6] Xu Yiting, Mu Yueying, Zhao Yousen. Analysis of Vegetable Price Changes and Influencing Factors in Beijing [J]. Chinese Vegetables, 2017 (11): 62 - 69.

[7] Liu Yan. Mathematical Model of Multiple Linear Regression [J]. Journal of Shenyang Institute of Engineering (Natural Science Edition), 2005,1 (2): 128 - 129.

[8] Breiman L. Heuristics of instability and stabilization in model selection[J]. Ann Statist, 1996, 24(6):2350-2383.

[9] Li Haichao, Wang Kaijun. Mining Factors Affecting the Success of Online Loan Borrowing Based on LASSO Regression Model [J]. Computer System Application, 2017, 26 (7): 204 - 209.

[10] Tibshirani, Robert. Regression shrinkage and selection via the lasso: A retrospective[J]. Journal of the Royal Statistical Society: Series B Statistical Methodology, 2011, 73(3):273-282.

[11] Mallows C L. Some Comments on Cp[J]. Technometrics, 42(1):87-94.

[12] Wu Xizhi. Complex data statistical methods: R - based applications [M]. Renmin University of China Press, 2015.
[13] Fox J. Component+Residual (Partial Residual) Plots[J].
[14] Gao huixuan. several methods to deal with collinearity of independent variables in multiple linear REGression - use of reg and other process enhancement functions in --SAS/STAT software (6.12) [J]. mathematical statistics and management, 2000,19 (5): 49 - 460.
[15] He Xiaoqun, Liu Yiqing. Applied Regression Analysis [M]. Renmin University of China Press, 2001.
[16] RobertI.Kabacoff, Cabaco Fu, Wang Xiaoning, et al. R Language Practice [M]. People's Posts and Telecommunications Press, 2016.

Locality Kernel Canonical Variate Analysis for Fault Detection

Shujun Xiao[1]

[1] Beijing University of Chemical Technology, Beijing 100029, China

E-mail: 15117955800@126.com

Abstract. Most of the existing fault detection techniques merely consider the global structure information, but neglect the local structure details between different data points. However, local structure information plays an essential role in data mining and feature extraction. In this article, a locality kernel canonical variate analysis (LPP-KCVA) method for fault detection is proposed. This method inherits the merits of KCVA in coping with nonlinear as well as dynamic problems, and makes full use of the advantage of Locality Preserving Projections (LPP) in retaining local manifold structure. To demonstrate the accuracy of the LPP-KCVA approach, Tennessee Eastman process (TEP) is applied, and KCVA, KPCA methods serve as the reference. The outcome illustrates its power in fault detection.

1. Introduction

With the speedy development of the modern industry and the increasing complexity in control systems, the capability to ensure product quality as well as the safe operation has considerably enhanced. Once a fault occurs, it will exert a detrimental impact on the stability of process, which can result in significant economic losses even heavy casualties. Over the past decades of years, fault detection has not only received gradually increasing concern, but an ocean of achievements have emerged as well. Basically, rough classification of fault detection methods is knowledge-based, model-based and data-driven ones [1]. Among the massive data-driven methods, multivariate statistical process monitoring (MSPM) ones are celebrated by the means of setting up thresholds obtained from mining inner features of historical data to distinguish whether a fault has occurred. Well-known methods of MSPM are those known as principal component analysis (PCA) [2, 3], partial least squares (PLS) [4, 5] and canonical variate analysis (CVA) [6-8].

As tools of dimensionality reduction, PCA and PLS are effective under the assumption of time independence and Gaussian distribution in static process [9]. However, due to the influence of noise and disturbance, actual process contains serial correlations which severely weaken their ability of capturing abnormalities. To address the drawbacks, variants of PCA and PLS [10, 11] consider the serial correlations between variables, but it is still challenging to manage strong autocorrelation in some circumstances. Compared with PCA and PLS, CVA could handle dynamic issue more proficient thanks to its consideration of both input and output data. It can't be denied that CVA indeed possesses a wealth of excellence; however, it still bears some deficiency in most practical processes which exhibit nonlinear relationships between variables.

In an attempt to improve the insufficiency and uplift the performance of CVA, the kernel trick is integrated into CVA because it masters at disposing of nonlinear problems. The kernel method is popular for the fact that it does not call for settling down nonlinear optimization problem when

IOP Conf. Series: Journal of Physics: Conf. Series **1284** (2019) 012003 doi:10.1088/1742-6596/1284/1/012003

extracting nonlinear features [12]. As a novel method, kernel canonical variate analysis (KCVA) combines CVA with the kernel method to absorb their merits in accounting for dynamics and fully describing the nonlinear relationships between variables. On the basis of KCVA, some research productions have been conducted around the fault detection in recent years [13-16].

All the methods mentioned above merely focus on the information that the global structure holds, however, they ignore the local relationship between the data points. For the sake of inner data information and feature extraction, local structure methods have been presented to preferably interpret the deep links hidden behind the data. As an effective method for local information preservation, LPP can preserve the nonlinear structure of high-dimensional data after projection because of its ingenious combination of Laplacian Eigenmaps. Recently, combinations of MSPM methods and LPP have been extensively researched. Deng and Tian [17] extended LPP to KPCA for the sake of fault diagnosis and demonstrated its superiority over KPCA. Sun and Chen [18] proposed a LPCCA method which combines LPP with Canonical correlation analysis (CCA) for data visualization and pose estimation. Besides, Lu and Jiang [19] introduced a method for fault diagnosis based on the incorporation of CVA, FDA and LPP.

In this article, a method called locality kernel canonical variate analysis is suggested so as to monitor the process behaviour by taking both local and non-local structure information into consideration. The novel scheme is established on the basis of KCVA and LPP, which could handle nonlinearity, serial correlations issues and could retain local manifold structure simultaneously. The article can be separated into the following parts. A brief review of CVA, KCVA and LPP will be given in Section2. Section 3 will present a detailed introduction on LPP-KCVA. To illustrate the effectiveness of LPP-KCVA, Section 4 is used to focus on the simulation of the TEP. Finally, some conclusions are drawn in Section 5.

2. CVA, KCVA and LPP Method

2.1. CVA algorithm

As a subspace identification method, the linear state-space model of CVA is expressed as

$$x_{t+1} = \Phi x_t + G u_t + w_t \tag{1}$$

$$y_t = H x_t + A u_t + B w_t + v_t \tag{2}$$

where $u_t \in R^{m_u}$ is the input vector, $x_t \in R^d$ is the state vector, $y_t \in R^{m_y}$ is the output vector, w_t and v_t are independent stochastic white noise with zero mean, and Φ,G,H,A,B are state space matrices. Define p_t and f_t as the past and future vectors at time point t, then they can be stacked as:

$$p_t = [{y^T}_{t-1}, {y^T}_{t-2}, \cdots, {y^T}_{t-l}, {u^T}_{t-1}, {u^T}_{t-2}, \cdots, {u^T}_{t-l}]^T \tag{3}$$

$$f_t = [{y^T}_t, {y^T}_{t+1}, \cdots, {y^T}_{t+h}]^T \tag{4}$$

where l and h denote the lag number chosen to better address autocorrelation issues. Both p_t and $f_t$vectors are normalized to eliminate the dimension effect among different observation variables. If s observations are available, then the Hankel matrices for p_t and f_t are collected in the following form:

$$X_p = [\, p_t, p_{t+1}, \cdots p_{t+N-1}] \tag{5}$$

$$X_f = [f_t, f_{t+1}, \cdots f_{t+N-1}] \tag{6}$$

where $X_p \in R^{l(m_u+m_y)\times N}$, $X_f \in R^{h(m_y+1)\times N}$, and N is the column number in Hankel matrices, $N = s - l - h + 1$.

Then it is naturally to obtain the auto covariance matrices of X_p, X_f as well as their cross-covariance. Σ_{pp}, Σ_{ff}, Σ_{pf} denote the auto covariance matrices of X_p, X_f and their cross-covariance respectively. Singular value decomposition can provide a solution for the CVA method.

$$\Sigma_{pp}^{-1/2}\Sigma_{pf}\Sigma_{ff}^{-1/2} = U\Sigma V^T \tag{7}$$

For ease of calculation, only the first d (<r) canonical variables are selected which could account for most of the dynamic properties. Therefore, the state vector resembles the canonical variable can be calculated as:

$$x_d(t) = J_d p_t = U_d^T \Sigma_{pp}^{-1/2} p_t \tag{8}$$

where $J_d = U_d^T \Sigma_{pp}^{-1/2}$. U_d represents the column vectors corresponding to the first d maximum correlation coefficients in U.

2.2. KCVA algorithm

Generally, fewer studies have expanded the CVA method to the kernel field. The kernel method could transform the linear inseparable data in low dimension into the linear separable data in high dimension by suitable non-linear mapping functions. On combining CVA with the kernel method, KCVA provides relative ease to analyze large-scale and complex industrial process. Similarly, let $\varphi_1(X_p)$ and $\varphi_2(X_f)$ denote the matrices projected from the kernel matrices, and then the KCVA problem can be transformed into the following form:

$$\begin{bmatrix} 0 & \varphi_1(X_p)\varphi_2^T(X_f) \\ \varphi_2(X_f)\varphi_1^T(X_p) & 0 \end{bmatrix} \begin{bmatrix} \tilde{\alpha} \\ \tilde{\beta} \end{bmatrix} = \lambda \begin{bmatrix} \varphi_1(X_p)\varphi_1^T(X_p) & 0 \\ 0 & \varphi_2(X_f)\varphi_2^T(X_f) \end{bmatrix} \begin{bmatrix} \tilde{\alpha} \\ \tilde{\beta} \end{bmatrix} \tag{9}$$

$\tilde{\alpha}$ can be regarded as the vector spanned by the columns of $\varphi_1(X_p)$, $\tilde{\alpha} = \varphi_1(X_p)\alpha$. Similarly, $\tilde{\beta} = \varphi_2(X_f)\beta$.

Now, using the kernel trick like:

$$K_{pp} = \varphi_1^T(X_p)\varphi_1(X_p) \tag{10}$$

$$K_{ff} = \varphi_2^T(X_f)\varphi_2(X_f) \tag{11}$$

where

$$[K_{pp}]_{i,j} = k_p(X_{p,i}, X_{p,j}) = \langle \varphi_1(X_{p,i}) \cdot \varphi_1(X_{p,j}) \rangle \tag{12}$$

$$[K_{ff}]_{i,j} = k_f(X_{f,i}, X_{f,j}) = \langle \varphi_2(X_{f,i}) \cdot \varphi_2(X_{f,j}) \rangle \tag{13}$$

$X_{p,i}$ is the ith column in X_p and $X_{f,i}$ is the ith column in X_f, both K_{pp} and K_{ff} are Gram matrices which centered as $\widetilde{K}_{pp} = K_{pp} - K_{pp}E - EK_{pp} + EK_{pp}E$ and $\widetilde{K}_{ff} = K_{ff} - K_{ff}E - EK_{ff} + EK_{ff}$. E has the same dimension as K_{pp} or K_{ff} with each element equals to $1/N$. The kernel functions k_p, k_f usually choose radial basis function. By applying kernel function, the equation (9) can be rewritten as:

$$\begin{bmatrix} 0 & \widetilde{K}_{pp}\widetilde{K}_{ff} \\ \widetilde{K}_{ff}\widetilde{K}_{pp} & 0 \end{bmatrix} \begin{bmatrix} \alpha \\ \beta \end{bmatrix} = \lambda \begin{bmatrix} \widetilde{K}_{pp}\widetilde{K}_{pp} & 0 \\ 0 & \widetilde{K}_{ff}\widetilde{K}_{ff} \end{bmatrix} \begin{bmatrix} \alpha \\ \beta \end{bmatrix} \tag{14}$$

In order to tackle the above optimization problem, generalized eigenvalue decomposition can be applied. One thorny problem is that $\widetilde{K}_{pp}$ and $\widetilde{K}_{ff}$ are singular, thus $\widetilde{K}_{pp}\widetilde{K}_{pp} + \eta I_1$ and $\widetilde{K}_{ff}\widetilde{K}_{ff} + \eta I_2$ serve as the replacement of $\widetilde{K}_{pp}\widetilde{K}_{pp}$ and $\widetilde{K}_{ff}\widetilde{K}_{ff}$, where η is the regularization parameter with small value and I_1, I_2 are identity matrices which have the same dimension as $\widetilde{K}_{pp}\widetilde{K}_{pp}$ and $\widetilde{K}_{ff}\widetilde{K}_{ff}$.

The coefficient vectors α and β can be obtained by computing equation (14). For the past vector p_t, the canonical vector c can be calculated as:

$$c = \widetilde{A}^T\varphi_1(p_t) = A^T\varphi_1^T(X_p)\varphi_1(p_t) = A^T k_p(X_p, p_t) \tag{15}$$

where $\widetilde{A} = [\tilde{\alpha}_1, \tilde{\alpha}_2, \cdots, \tilde{\alpha}_N]$, $A = [\alpha_1, \alpha_2, \cdots, \alpha_N]$, and $k_p(X_p, p_t) = [k_p(X_{p,1}, p_t), k_p(X_{p,2}, p_t), \cdots, k_p(X_{p,N}, p_t)]^T$.

2.3. LPP algorithm

LPP is a method aiming to capture the local manifold structures among samples in the initial sample space and maintains their previous relationships after the projection. More specifically, if two data points are neighbors in the high-dimensional data space, they are close as well after projected to the low-dimensional data space. The purpose of LPP is to avoid the divergence of data points and maintain the original local structure.

Given the original data sample matrix $X = [x_1, x_2, \cdots x_n]$, where n is the number of samples. And define g as the projection vector that transforms X to Z, where $Z = [z_1, z_2, \cdots z_n]$, $z_i = g^T x_i$, $i = 1, \cdots, n$. The objective function of LPP algorithm is expressed in the following form:

$$J = \min \sum_{i=1}^{n} \sum_{j=1}^{n} (z_i - z_j)^2 w_{ij} = \min \sum_{i=1}^{n} \sum_{j=1}^{n} g^T (x_i - x_j) w_{ij} (x_i - x_j)^T g \quad (16)$$

where w_{ij} is a weighting parameter that represents the relationship between the two samples and it can be calculated by the heat kernel function:

$$w_{ij} = \begin{cases} \exp\left(-\frac{\|x_i - x_j\|^2}{\sigma}\right), \text{if } x_i \in \text{Neigh}(x_j) \text{or } x_j \in \text{Neigh}(x_i) \\ 0, \qquad \text{otherwise} \end{cases} \quad (17)$$

where $\text{Neigh}(x_j)$ denotes the neighbors of x_j and k-Nearest Neighbor method can be applied to judge the range of the neighbors. It can be seen that, for each data point, a non-zero weight is given to the data point next to it, and a zero value is given to the data point far away from it, then the goal of preserving the local structure of the data can be achieved. Formula (16) can be further deduced as follows:

$$J = \min\left(\sum_{i=1}^{n} g^T x_i D_{ii} x_i^T g - \sum_{i=1}^{n} \sum_{j=1}^{n} g^T x_i w_{ij} x_i^T g\right) = \min g^T X L_x X^T g \quad (18)$$

where $L_x = D_x - W$, W stands for the weighting matrix, L_x is the Laplace matrix and $D = \text{diag}\{d_1, d_2, \cdots d_n\}$ with $d_i = \sum_{j=1}^{n} w_{ij}$. In order to prevent over-fitting, some restrictions need to be added as follows:

$$z^T D z = 1 \quad (19)$$

then equation (18) can be written as:

$$g^T X D X^T g = 1 \quad (20)$$

That is to say, under the condition of $g^T X D X^T g = 1$, the value of $g^T X L_x X^T g$ should be minimized.

$$\begin{aligned} \min \quad & g^T X L_x X^T g \\ \text{s.t.} \quad & g^T X D X^T g = 1 \end{aligned} \quad (21)$$

In this paper, the main function of LPP is to retain the local structure and avoid losing the details.

3. Locality Kernel Canonical Variate Analysis (LPP-KCVA) method

After projected from the original low-dimensional space to high-dimensional space, the data may not be able to maintain the original local structure information due to the data divergence. In this section, a fault detection method called locality kernel canonical variate analysis (LPP-KCVA) is proposed to solve the above problem. With some constraints provided by LPP, the data could retain local manifold in low-dimensional space and avoid the loss of detailed information even if the dimension changes. In addition, LPP-KCVA absorbs the merits of LPP and KCVA, so it is proficient in addressing the mixed problems.

As defined in equation (5) and (6), X_p and X_f are the Hankel matrices respectively, $\varphi_1(X_p)$ and $\varphi_2(X_f)$ denote the matrices projected from X_p and X_f. Consider the LPP algorithm and the KCVA method introduced in section 2, if X in equation (21) is replaced by $\varphi_1(X_p)$, then we can obtain the following equation:

$$J_p = \min g^T \varphi_1(X_p) L_p \varphi_1^T(X_p) g \quad (22)$$

where L_p denotes the Laplace matrix for the past data. Similar to the KCVA method, g can be spanned by the column vectors of $\varphi_1(X_p)$.

$$g = \varphi_1(X_p)\alpha \quad (23)$$

By replacing g in equation (22), the objective function J_p can be written as:

$$J_p = \min\{\alpha^T \varphi_1^T(X_p) \varphi_1(X_p) L_p \varphi_1^T(X_p) \varphi_1(X_p) \alpha\} \quad (24)$$

After implementing standardization on K_{pp}, equation (24) can be rewritten as:

$$J_p = \min \alpha^T \widetilde{K}_{pp} L_p \widetilde{K}_{pp} \alpha \quad (25)$$

Similarly, the optimal function of Hankel matrix X_f can be obtained:

$$J_f = \min \beta^T \widetilde{K}_{ff} L_f \widetilde{K}_{ff} \beta \quad (26)$$

On combining with the conditions of KCVA, the following optimization problem can be obtained:

$$\max \alpha^{T}\widetilde{K}_{pp}\widetilde{K}_{ff}\beta \quad (27)$$
$$\text{s.t. } \alpha^{T}\widetilde{K}_{pp}L_{p}\widetilde{K}_{pp}\alpha = 1,\ \beta^{T}\widetilde{K}_{ff}L_{f}\widetilde{K}_{ff}\beta = 1$$

The above optimization problem can be transformed into the following equation:

$$\begin{bmatrix} 0 & \widetilde{K}_{pp}\widetilde{K}_{ff} \\ \widetilde{K}_{ff}\widetilde{K}_{pp} & 0 \end{bmatrix}\begin{bmatrix}\alpha \\ \beta\end{bmatrix} = \lambda \begin{bmatrix} \widetilde{K}_{pp}L_{p}\widetilde{K}_{pp} & 0 \\ 0 & \widetilde{K}_{ff}L_{f}\widetilde{K}_{ff} \end{bmatrix}\begin{bmatrix}\alpha \\ \beta\end{bmatrix} \quad (28)$$

The generalized eigenvalue decomposition can solve the above equation. Like the KCVA method, because $\widetilde{K}_{pp}$ and $\widetilde{K}_{ff}$ are singular matrices, $\widetilde{K}_{pp}\widetilde{K}_{pp} + \eta I_1$ and $\widetilde{K}_{ff}\widetilde{K}_{ff} + \eta I_2$ should be used rather than $\widetilde{K}_{pp}$ and $\widetilde{K}_{ff}$, then the canonical variable c of the LPP-KCVA method can be obtained:

$$c = \widetilde{A}^{T}\varphi_1(p_t) = A^{T}\varphi_1^{T}(X_p)\varphi_1(p_t) = A^{T}k_p(X_p, p_t) \quad (29)$$

where $\widetilde{A} = [\widetilde{\alpha}_1, \widetilde{\alpha}_2, \cdots, \widetilde{\alpha}_N]$, $A = [\alpha_1, \alpha_2, \cdots, \alpha_N], k_p(X_p, p_t) = [k_p(X_{p,1}, p_t), k_p(X_{p,2}, p_t), \cdots, k_p(X_{p,N}, p_t)]^{T}$.

Constructing statistics T_s^2, T_r^2 corresponding to the state space and the residual space

$$T_s^2 = c_s^{T}c_s \quad (30)$$
$$T_r^2 = c_r^{T}c_r \quad (31)$$

where c_s comprises the first s column of c while c_r comprises the rest of the column of c.

Accordingly, the control limits T_{sUCL}^2, T_{rUCL}^2 of T_s^2 and T_r^2 statistics can be calculated as below:

$$\int_{-\infty}^{T_{sUCL}^2} p(T_s^2)dT_s^2 = \gamma \quad (32)$$
$$\int_{-\infty}^{T_{rUCL}^2} p(T_r^2)dT_r^2 = \gamma \quad (33)$$

The control limits T_{sUCL}^2 and T_{rUCL}^2 can be estimated provided that $p(T_s^2)$, $p(T_r^2)$ and the significance level γ are determined beforehand. Through comparing the value of T_s^2, T_r^2 and their control limits, a fault can be captured if the value of T_s^2 is larger than T_{sUCL}^2 or the value of T_r^2 is larger than T_{rUCL}^2.

To sum up, the LPP-KCVA procedure can be briefly stated as below:

Off-line model establishment

1. Prepare the data under normal condition and normalize each variable.
2. Define the past and future vectors by using equation (3), (4) and put them into the past and future Hankel matrices X_p and X_f like equation (5), (6).
3. Select proper kernel functions k_p, k_f and calculate the Gram matrices K_{pp}, K_{ff} with equation (10), (11).
4. Implement the normalization on the matrices K_{pp}, K_{ff}.
5. Calculate the weighting matrices W_p, W_f, diagonal matrices D_p, D_f and Laplace matrices L_p and L_f.
6. Solve the generalized eigenvalue problem in equation (28) to obtain α and β.
7. Compute the T_s^2 and T_r^2 statistics with equation (30), (31) and calculate the control limits T_{sUCL}^2 and T_{rUCL}^2 using equation (32), (33).

Online monitoring

1. Scale the new test data with the mean and variance identical to the training data.
2. Form the online p_k and f_k vectors and calculate each k row of Gram matrices K_{pp}^{test} and K_{ff}^{test} with $[K_{pp}^{test}]_k = k_p^{T}(X_p, p_k)$ and $[K_{ff}^{test}]_k = k_f^{T}(X_f, f_k)$.
3. Center the Gram matrices K_{pp}^{test} and K_{ff}^{test} as follows: $[K_{pp}^{test}]_k = [K_{pp}^{test}]_k - [K_{pp}^{test}]_k E - 1^{test}K_{pp} + 1^{test}K_{pp}E$ and $[K_{ff}^{test}]_k = [K_{ff}^{test}]_k - [K_{ff}^{test}]_k E - 1^{test}K_{ff} + 1^{test}K_{ff}E$ where $1^{test} = 1/N\,[1, \cdots, 1] \in R^{1\times N}$.
4. Calculate $T_{s,k}^2$ and $T_{r,k}^2$ at each time point k with equation (30), (31).
5. Compare the obtained $T_{s,k}^2$ and $T_{r,k}^2$ with T_{sUCL}^2 and T_{rUCL}^2 to decide whether a fault has occurred. When a fault exists, $T_{s,k}^2 > T_{sUCL}^2$ or $T_{r,k}^2 > T_{rUCL}^2$.

4. Experiments on the Tennessee Eastman Process

In this section, the Tennessee Eastman process (TEP) is utilized to assess the effectiveness of the proposed LPP-KCVA fault detection technique. The TEP establishes a realistic nonlinear dynamic industrial platform for assessing newly developed fault detection methods. It has five major reactions and eight components. The process covers 21 faults, one normal condition and involves 52 variables in which 12 are manipulated variables and 41 are measurement variables. The training data include 500 samples under the fault-free condition and 480 samples under the faulty conditions. The testing data include 960 samples for each fault. Each fault begins at the 161th sample and each data is sampled every 3 minutes.

The performance of the developed LPP-KCVA method is assessed on the foundation of the missed detection rate which is based on the percentage of the data sample value smaller than the control limits in the last 800 faulty data sample in 21 faults. The missed detection rate reveals the sensitivity of each fault method, more precisely, if the missed detection rate of one method is lower than others, then the method can be considered better. Another index for monitoring the performance of the LPP-KCVA method is detection delay which measures the promptness. The detection delay is referred as the time interval which the fault introduced to the system until the fault to be detected. From the perspective of the promptness, if one fault detection method has shorter detection delay time, it can be considered better than others.

As the false alarms are inescapable, it might be considered faulty under the fault-free conditions. To reduce the false alarms, a fault can be identified if several continuous statistics have the value larger than the control limit. Both the missed detection rates and the detection delay are obtained under relatively low false alarm rates.

Table 1. Missed detection rates for the testing set

Fault	KCVA T_s^2	KCVA T_r^2	KPCA T^2	KPCA Q	LPP-KCVA T_s^2	LPP-KCVA T_r^2
1	0.0038	0.0013	0.0063	**0**	0.0025	**0**
2	0.0165	0.0114	0.0138	0.0063	0.0127	**0.0025**
3	0.6810	0.9899	0.9437	0.7612	0.8241	**0.3304**
4	0.8747	0.8430	0.3250	**0**	0.9304	**0**
5	0.5911	0.5557	0.7212	0.4163	0.6329	**0.3203**
6	0.0013	**0**	0.8962	**0**	0.0013	**0**
7	0.4278	0.4354	0.0213	**0**	0.1152	**0**
8	0.0089	0.0481	0.0262	0.0088	0.0114	**0.0013**
9	0.7203	0.9924	0.9425	0.7963	0.7456	**0.3456**
10	0.3430	0.3228	0.5225	**0.1737**	0.4228	0.2481
11	0.6962	0.7114	0.4150	0.1475	0.7215	**0.0139**
12	0.0177	0.1038	0.0250	0.0025	0.0152	**0**
13	0.0418	0.0418	0.0462	0.0413	0.0367	**0.0253**
14	0.9797	0.9886	**0**	**0**	0.0038	**0**
15	0.7456	0.6291	0.9263	0.7025	0.7899	**0.4671**
16	0.3937	0.8810	0.6987	**0.2000**	0.5127	0.2468
17	0.2873	0.2937	0.2338	0.0213	0.1570	**0.0127**
18	0.0835	0.0696	0.8375	0.0737	0.0924	**0.0215**
19	0.8949	0.9684	0.8712	0.3413	0.9316	**0.0975**
20	0.4291	0.4354	0.5300	0.2100	0.4342	**0.0595**
21	0.6177	0.7101	0.5550	0.3175	0.6165	**0.1367**

Table 1 displays the missed detection rates about the T_s^2, T_r^2 statistics in LPP-KCVA, the T_s^2, T_r^2 statistics in KCVA and the $T^2$and Q statistics in KPCA for 21 faults. The bold figures represent the best statistic in one fault. It can be seen from the table that, among all the 21 faults, the T_r^2 statistic in

LPP-KCVA acts superior to other statistics in terms of missed detection rates. The T_r^2 statistic in LPP-KCVA possesses the best performance in most fault conditions. However, the T_r^2 statistic still has some drawbacks when facing the Fault 3 and Fault 9. The superiority of the presented method tends to owe to the consideration of both local and global information.

Table 2. Detection delay (minutes) for the testing set

Fault	KCVA T_s^2	KCVA T_r^2	KPCA T^2	KPCA Q	LPP-KCVA T_s^2	LPP-KCVA T_r^2
1	9	3	12	**0**	6	**0**
2	39	27	33	**0**	30	6
3	**3**	1188	60	24	51	**3**
4	162	159	**0**	**0**	171	**0**
5	27	**0**	3	3	30	**0**
6	3	**0**	15	**0**	3	**0**
7	**0**	**0**	**0**	**0**	**0**	**0**
8	**0**	15	45	18	12	3
9	**0**	984	**0**	**0**	3	**0**
10	12	12	21	**0**	18	9
11	96	366	15	**3**	27	**3**
12	12	69	6	6	12	**0**
13	**0**	**0**	108	51	87	3
14	90	2112	**0**	**0**	**0**	**0**
15	261	**30**	273	168	1413	**30**
16	**0**	906	12	24	**0**	3
17	96	93	**3**	21	78	**3**
18	153	150	51	30	42	**0**
19	**0**	201	30	3	18	**0**
20	216	219	**0**	**0**	171	3
21	1236	1683	39	**15**	930	249

The details of the detection delay for the LPP-KVCA, KCVA and KPCA methods are tabulated in Table 2. As the Table 2 presents, the T_r^2 statistic in LPP-KCVA can capture the fault earlier than other statistics for most faults. In other words, the T_r^2 statistic in LPP-KCVA can contribute more time for operators to deal with the abnormalities.

Take Fault 7 as an example to implement fault detection. The real-time monitoring curves are displayed in Figure 1-6. For the KPCA method, the T^2 statistic appears missed detection for several times after the occurrence of Fault 7 while the Q statistic appears several false alarms in the first 160 samples. The T_s^2 and T_r^2 statistics of KCVA possess relatively high missed detection rates though no false alarms appear under normal conditions. The T_s^2 statistic in LPP-KCVA method can basically realize the function of fault detection for Fault 7. As for the T_r^2 statistic in LPP-KCVA method, it has the lowest missed detection rate among all the statistics and bears no false alarms. Thus, the T_r^2 statistic of the LPP-KCVA method can detect faults more accurately and timely.

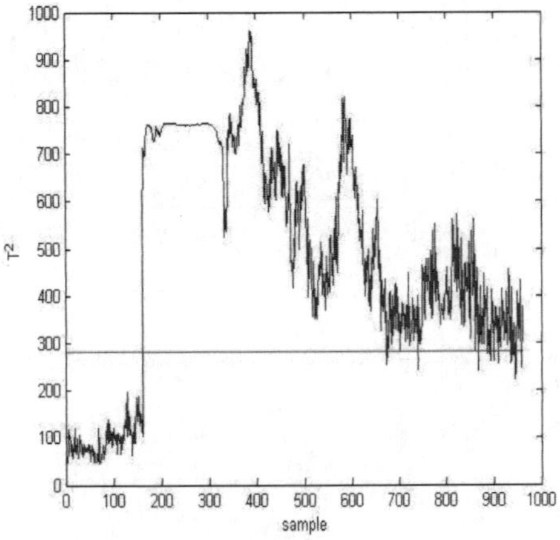

Figure 1. Fault 7 KPCA T^2 statistic

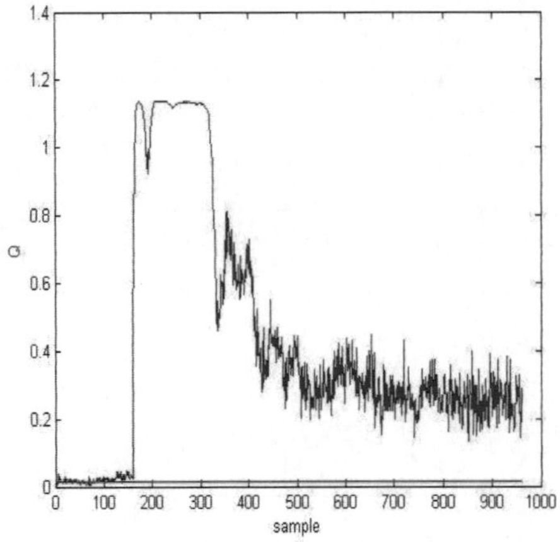

Figure 2. Fault 7 KPCA Q statistic

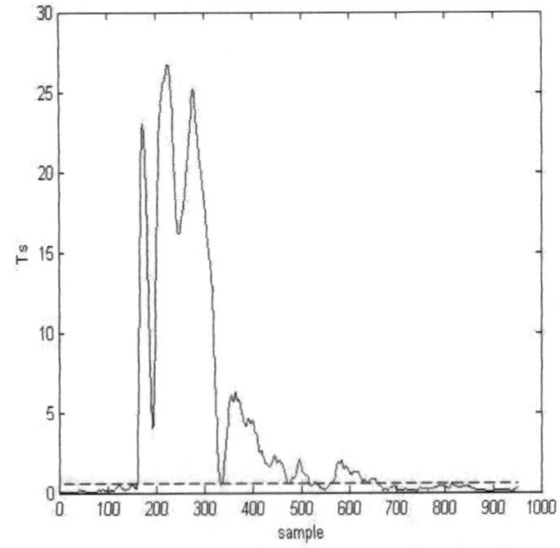

Figure 3. Fault 7 KCVA T_s^2 statistic

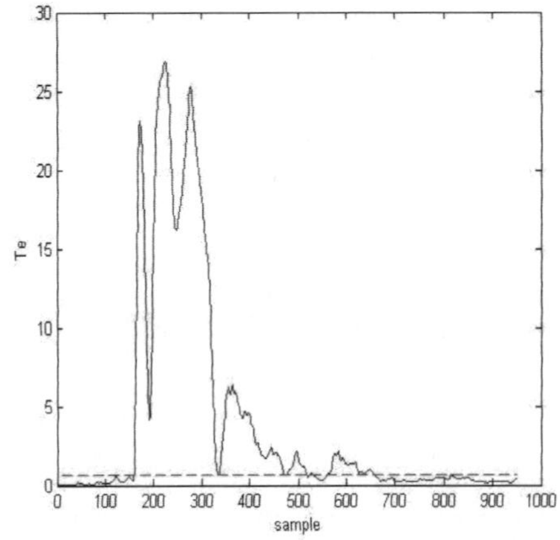

Figure 4. Fault 7 KCVA T_r^2 statistic

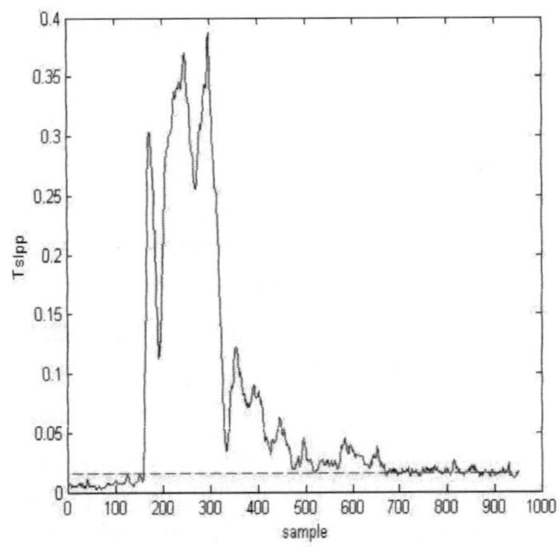

Figure 5. Fault 7 LPP-KCVA T_s^2 statistic

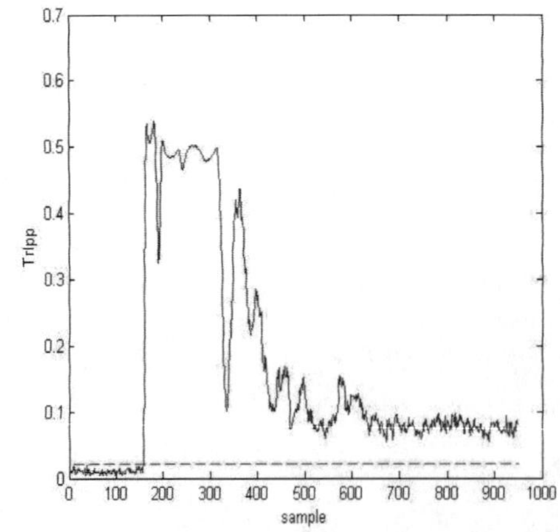

Figure 6. Fault 7 LPP-KCVA T_r^2 statistic

5. Conclusion

This paper proposed a novel fault detection method called the locality kernel canonical variate analysis (LPP-KCVA), which integrates the ability of KCVA in handling the nonlinearity and serial correlations with the capacity of LPP in preserving the local manifold structure. The LPP-KCVA method takes both local and non-local structure information into consideration to get more details. The efficiency of this method is verified by implementing experiments on Tennessee Eastman process in comparison with KCVA and KPCA. The outcome shows that the LPP-KCVA scheme presents a superior performance for fault detection than KCVA and KPCA.

References

[1] Chiang L H, Russell E L, Braatz R D. Fault detection and diagnosis in industrial systems[M]. Springer Science & Business Media, 2001.
[2] Nomikos P, MacGregor J F. Monitoring batch processes using multiway principal component analysis[J]. AIChE Journal, 1994, 40(8): 1361-1375.
[3] Bakshi B R. Multiscale PCA with application to multivariate statistical process monitoring[J]. AIChE journal, 1998, 44(7): 1596-1610.
[4] Ji H, He X, Shang J, et al. Incipient fault detection with smoothing techniques in statistical process monitoring[J]. Control Engineering Practice, 2017, 62: 11-21.
[5] Zhu Q, Liu Q, Qin S J. Concurrent canonical correlation analysis modeling for quality-relevant monitoring[J]. IFAC-PapersOnLine, 2016, 49(7): 1044-1049.
[6] Russell E L, Chiang L H, Braatz R D. Fault detection in industrial processes using canonical variate analysis and dynamic principal component analysis[J]. Chemometrics and intelligent laboratory systems, 2000, 51(1): 81-93.
[7] Juricek B C, Seborg D E, Larimore W E. Fault detection using canonical variate analysis[J]. Industrial & engineering chemistry research, 2004, 43(2): 458-474.
[8] Jiang B, Braatz R D. Fault detection of process correlation structure using canonical variate analysis-based correlation features[J]. Journal of Process Control, 2017, 58: 131-138.
[9] Joe Qin S. Statistical process monitoring: basics and beyond[J]. Journal of Chemometrics: A Journal of the Chemometrics Society, 2003, 17(8-9): 480-502.
[10] Ku W, Storer R H, Georgakis C. Disturbance detection and isolation by dynamic principal component analysis[J]. Chemometrics and intelligent laboratory systems, 1995, 30(1): 179-196.
[11] Lee G, Song S O, Yoon E S. Multiple-fault diagnosis based on system decomposition and dynamic PLS[J]. Industrial & engineering chemistry research, 2003, 42(24): 6145-6154
[12] Cho J H, Lee J M, Choi S W, et al. Fault identification for process monitoring using kernel principal component analysis[J]. Chemical engineering science, 2005, 60(1): 279-288.
[13] Odiowei P E P, Cao Y. Nonlinear dynamic process monitoring using canonical variate analysis and kernel density estimations[J]. IEEE Transactions on Industrial Informatics, 2010, 6(1): 36-45.
[14] Giantomassi A, Ferracuti F, Iarlori S, Longhi S, Fonti A, Comodi G. Kernel canonical variate analysis based management system for monitoring and diagnosing smart homes[C]. International Joint Conference on Neural Networks. IEEE, 2014: 1432-1439.
[15] Samuel R T, Cao Y. Improved kernel canonical variate analysis for process monitoring[C]. Automation and Computing. IEEE, 2015: 1-6.
[16] Linzhe H, Yuping C, Xuemin T, et al. A nonlinear quality-relevant process monitoring method with kernel input-output canonical variate analysis[J]. IFAC-PapersOnLine, 2015, 48(8): 611-616.
[17] Deng X, Tian X, Chen S. Modified kernel principal component analysis based on local structure analysis and its application to nonlinear process fault diagnosis[J]. Chemometrics and Intelligent Laboratory Systems, 2013, 127: 195-209.
[18] Sun T, Chen S. Locality preserving CCA with applications to data visualization and pose estimation[J]. Image and Vision Computing, 2007, 25(5): 531-543.
[19] Lu Q, Jiang B, Gopaluni R B, Loewen P D, Braatz R D. Locality preserving discriminative canonical variate analysis for fault diagnosis[J]. Computers & Chemical Engineering, 2018, 117: 309-319.

IOP Conf. Series: Journal of Physics: Conf. Series **1284** (2019) 012004 doi:10.1088/1742-6596/1284/1/012004

Artificial Neural Networks for Urban Water Demand Forecasting: A Case Study

Leandro L Lorente-Leyva[1,*], Jairo F Pavón-Valencia[1], Yakcleem Montero-Santos[1], Israel D Herrera-Granda[1], Erick P Herrera-Granda[1] and Diego H Peluffo-Ordóñez[2,3]

[1]Facultad de Ingeniería en Ciencias Aplicadas, Universidad Técnica del Norte, Ibarra-Ecuador.
[2]SDAS Research Group, Yachay Tech, Urcuquí, Ecuador.http://www.sdas-group.com/
[3]Corporación Universitaria Autónoma de Nariño - Colombia

[*]Email: lllorente@utn.edu.ec

Abstract. This paper presents an application of an artificial neural network model in forecasting urban water demand using MATLAB software. Considering that in any planning process, the demand forecast plays a fundamental role, being one of the premises to organize and control a set of activities or processes. The versatility of the short, medium and long-term prediction that is provided to the company that offers the water distribution service to determine the supply capacity, maintenance activities, and system improvements as a strategic planning tool. Shown to improve network performance by using time series water demand data, the model can provide excellent fit and forecast without relying on the explicit inclusion of climatic factors and number of consumers. The excellent accuracy of the model indicates the effectiveness of forecasting over different time horizons. Finally, the results obtained from the Artificial Neural Network are compared with traditional statistical models.

1. Introduction

At present, water forecasting is very important, being crucial for the supply in any distribution system. The consideration of forecasting time periods supports the planning of supply capacity, scheduling, tariff adjustment and optimization of the activities of a company that provides water distribution service [1-2].

One of the purposes of the forecast made by the companies is to match supply with demand for water and provide a quality service for consumers [3]. Responding efficiently to current needs, which have become very inconsistent, reason for which the flexibility and adaptability of companies is fundamental, especially in urban areas that trace high population growth as it will influence the future distribution of urban water service and should be included in a planning for activities have a logical and efficient development, in addition to sustainable management of natural resources [4-5].

The usefulness of making long-term forecasts points to obvious factors for evaluating the effectiveness of conservation measures that will influence the generation of policies and strategies such as the price of water [6]. In the development of projects of the companies that provide the distribution service, the projection of consumption is important, since areas are not always supplied from a single source; several sources can supply many consumers [7].

This work focuses on the forecast of water demand in the medium term, these allow optimal operations of extraction, collection and distribution of water resources. Helping to develop strategies for water demand management in the medium term. The stages of a system of this type include the collection of raw water, production, regulation and distribution of urban water. For this reason, various controls can be carried out within the process and the forecast supports decision-making in the medium term. As well as implementing restrictions on water use in times of drought or thermal stress [8-9].

Short-term demand forecasting has traditionally used linear regression models, such as multi-line regression and integrated autoregressive moving average model (ARIMA). These methods remain the most common according to Adamowski and Chan [7], but the invalidity of linear regression models to adequately account for nonlinear parameters. Non-stationary water demand data have generated an examination of other models. Jain and Ormsbee [2] applied artificial neural networks (ANNs) in the short-term forecast of urban water demand to address nonlinearity [4]. Research has generally shown that ANNs outperform linear regression techniques in forecasting urban water demand [10-12].

Medium-term (weekly and monthly) urban water demand forecasts help water managers make more informed water management decisions when balancing the needs of water supply, residential/industrial demands, and stream flows for fish and other habitats [13]. Accurate forecasts also aid with decision making [1-2, 6]. Urban water is generally supplied based on the experience of operators, although accurate and reliable forecasts of water demand helps operators provide water more sustainably [13]

2. Development of the urban water forecast model

The development of the ANNs model has the purpose of forecasting urban water demand taking into account the variables that are handled as the volume of urban water consumed, being fundamental to determine the time horizon of the prediction and also the level of reliability of the model that influences decision making in the company.

This model, neuron *j* receives a series of inputs (X_1, X_2, ..., X_n) the temporal variables of months and years of the volume of urban water consumption. Where each input signal must be multiplied by the weight associated with its connection (W_1, W_2, ..., W_n) depending on the weights found in relation to the learning process. Then, these weighted inputs are added, and the activation function is applied to them to return an output from the input values.

The inputs to the network generate a link with the next neuron that can pass through the connection of a hidden layer and that output is represented according to the weighting of the neuron connection and the previous layer, as shown below:

$$S_{2j} = F_{2j}(X_1 W_{1j}) \tag{1}$$

X_1 is the input vector of layer 1 and W_{1j} corresponds to the weights that relate to the first neuron, to neuron *j* of the second layer. The function F_{2j} is the one that corresponds to the activation of neuron *j* of the second layer. Likewise, with the neurons of the second layer. Each neuron *i* of the output layer must generate a value output as we can see in the following equation:

$$S_{si} = F_{si}(W_{2i} S_2) \tag{2}$$

Where W_{2i} is the vector of weights that are related to the connections that go from the neurons of the second layer to the neuron *i* of the output layer, S_2 is the vector of output of the neurons of layer two. This become inputs of the neurons of the output layer.

2.1. ANN forecast process

The neural network prediction model will be established according to the influencing factors, considering within the inputs the factors that affect demand, and the expected demand as an output. Figure 1 shows the process proposed for forecasting using neural networks. In it they indicate the factors to be considered as the identification and preprocessing of the information with which the time horizon of the prediction is analyzed.

Once the variables have been identified, the neural network technique is applied, taking into account the values that make up the training, validation and test sets, thus determining the architecture of the network and using these criteria, training is carried out on the artificial neural network model for urban water forecast model.

The database of urban water demand in the preparation of the forecast is very important for its decisive value in the prediction. The input variables are fundamental in the process, in this case, the amount of urban water demand, as a quantitative variable measured in m^3.

Customers, depending on the sector in which they are located, are coded by a plan and the route of the distribution system. The sectorization for the case study is the urban area. Based on the plan and the distribution route, the data are ordered and grouped based on a general matrix of customers for each year with the corresponding volumetric consumption.

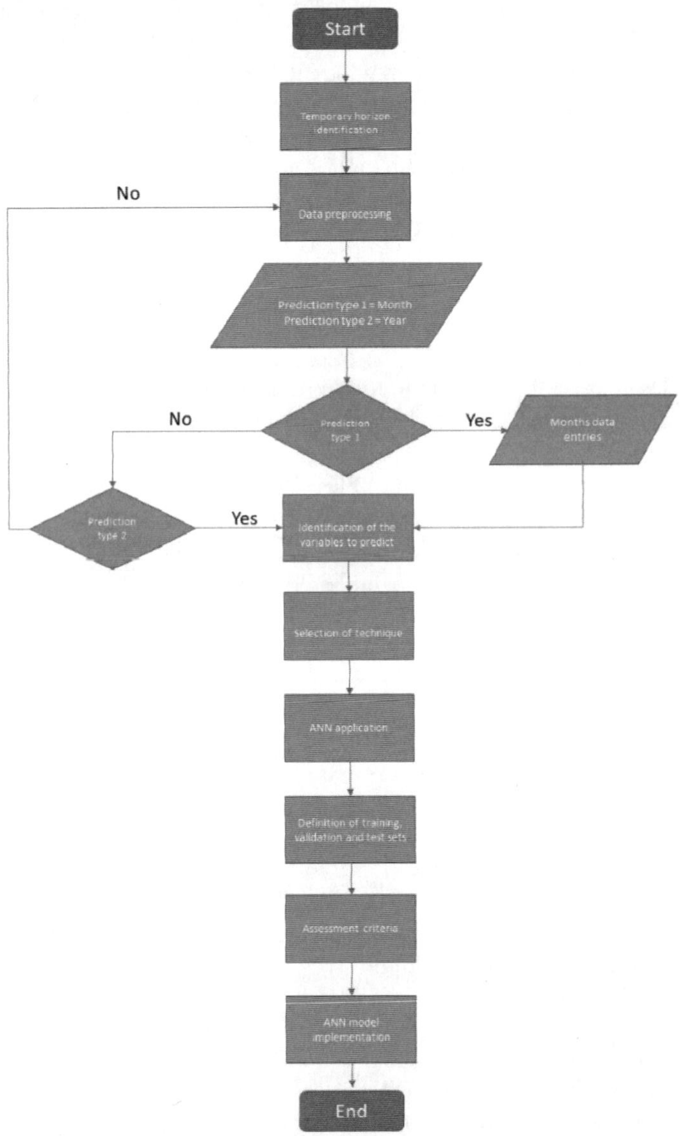

Figure 1. ANN forecast process

It is necessary to pre-process the data before entering them into the software so that they are in a noise-free format (without null or inconsistent values) in order to prepare them for the forecast development. Table 1 shown the volume consumed from 2013 to 2017 in a principal city of zone north of Ecuador.

Table 1. Summary of volume consumed (m^3)

Month	2017	2016	2015	2014	2013
January	214188	1089767	4396911	4131349	3928938
February	208742	923892	3691402	3710167	3719853
March	213828	984884	3973450	3701854	3402837
April	209964	969027	3842442	3655319	3681741
May	207155	974981	3988776	3840875	3757560
June	210661	988781	3996753	3739800	3549555
July	216438	1039832	4246655	3832171	3842532
August	215641	1036875	4214157	4091927	3834200
September	223899	947316	3708790	4051405	3824123
October	225073	1103062	4537799	3974116	4019610
November	215110	1028107	4124829	3934271	3865487
December	201456	938876	3739104	3540789	3524104

With the definition of the training sets, the network carries out the process of learning the trends and patterns of the data, in this case it is 70% of the data that represent the first 42 months, in this case it is 70% of the data that represents the first 42 months. The validation set is basically used for the conclusive verification stage of the network, where the data are the consequent values to those of the training which in this case represent 15% of the values that are the following 9 months. The test set is the data remaining after taking the values in the training process. The purpose of this set is to evaluate the accuracy of the network. It is composed of the last 9 months representing 15% of the data.

There are many ways to establish the architecture of a neural network, in most cases backpropagation training algorithms are used. In this case study, the Levenberg Marquardt retropropagation algorithm will be used.

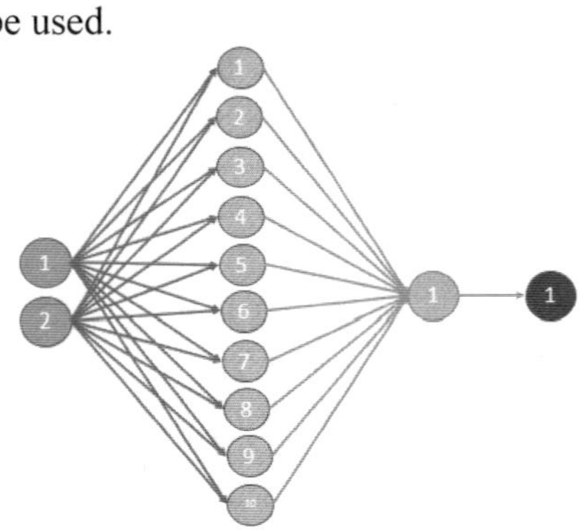

Figure 2. ANN Architecture

The input variables for the forecast of urban water that in this case will be two as shown in Figure 2, corresponding to the number of input neurons. A hidden layer is shown so that a better performance is generated since with the increase of hidden layers the processing time is increased and an overadjustment is obtained causing a deficient performance of the prediction. In this case there are 10 hidden neurons that represent the necessary condition so that the principle of entropy minimization is conserved in the presence of retropropagation. The number of output neurons in our network is one, because the value of the selected variable is predicted so that the network precision is not diminished.

2.2. Assessment criteria

The Levenberg Marquardt retropropagation algorithm is specifically designed to minimize the Mean Square Error (MSE) function with respect to the new weight of the vectors. The training performance

IOP Conf. Series: Journal of Physics: Conf. Series **1284** (2019) 012004 doi:10.1088/1742-6596/1284/1/012004

stops when the generalization halt improving due to an increase in the MSE of the validation samples. In this paper, the MSE and Coefficient of Correlation (R^2) were used to evaluate the prediction effect of the model.

$$\text{MSE} = \frac{1}{n}\sum_{t=1}^{n}(y_t - \hat{y_t})^2 \tag{3}$$

Where n is the number of samples and $\hat{y_t}$ is the estimate of y_t. From the above formula it follows that the measurement loss function is the quadratic or mean square error. Coefficient of Correlation (R^2), being:

$$R^2 = \frac{\frac{1}{n}\sum_{t=1}^{n}(y_t - \hat{y_t})2}{\frac{1}{n}\sum_{t=1}^{n}(y_t - \bar{y})2} \tag{4}$$

Where $\hat{y}$ is the forecasted demand, y_t is the measured demand, y_t is the mean of the demand forecasting, and n is the number of observations.

Low values are better while the closer to zero means there is no error. Regression analysis R is performed to measure the correlation between outputs and targets. The goal of training is to find the least error in the function of weights between neurons unless you have an over-adjusted model. The training uses a descending gradient technique of the retropropagation algorithm.

3. Result and discussion

Table 2 shows the results of the forecasting urban water demand model based on artificial neural networks. They are analyzed considering the two important criteria such as MSE and R^2 to compare them with those of the ARIMA time series model.

Table 2. Forecast urban water consumption volume (m^3)

Month	2018	2019
January	1.036.700	1.045.200
February	995.580	1.023.300
March	1.007.500	1.057.700
April	1.036.300	1.106.900
May	1.056.200	1.137.200
June	1.075.200	1.142.300
July	1.099.000	1.131.300
August	1.111.100	1.121.000
September	1.112.000	1.116.400
October	1.100.300	1.109.800
November	1.061.000	1.086.700
December	976.830	1.014.000

To conclude with the implementation of the model, the network performance graphs are presented in Figure 3, as training error decreases through iterations.

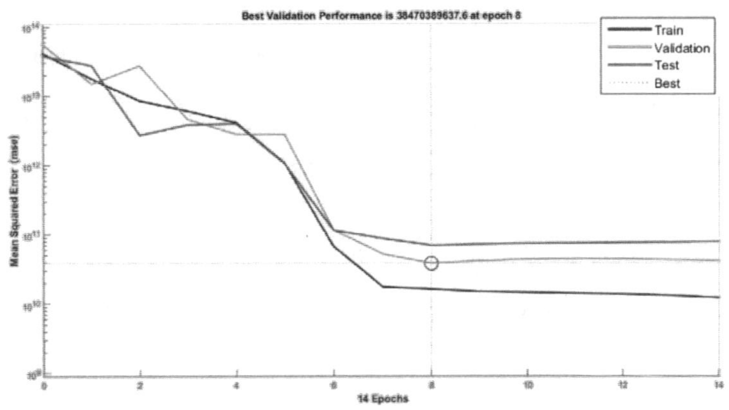

Figure 3. Training performance

In Figure 4 it is possible to see how the correlation coefficient is 0.99493 which is very close to 1, which indicates that there is a high level of correlation between real and predicted values.

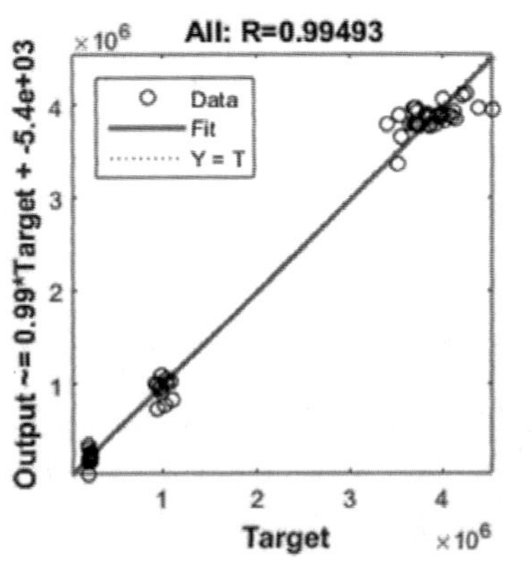

Figure 4. Correlation between actual values and forecast

The forecast of the artificial neural networks is presented with greater equivalence in the behavior of the real demand of urban water consumption of the company, since the behavior of the forecast in the ARIMA model tends to soften and overadjust the model, this by the moving averages with which the model is predicted and tendency to become a prediction of constant values. In the artificial neural networks model the data are self-trained and validate the patterns of behavior and at the end they are evaluated to progressively determine the best prediction option. As shown in the figure 5 below:

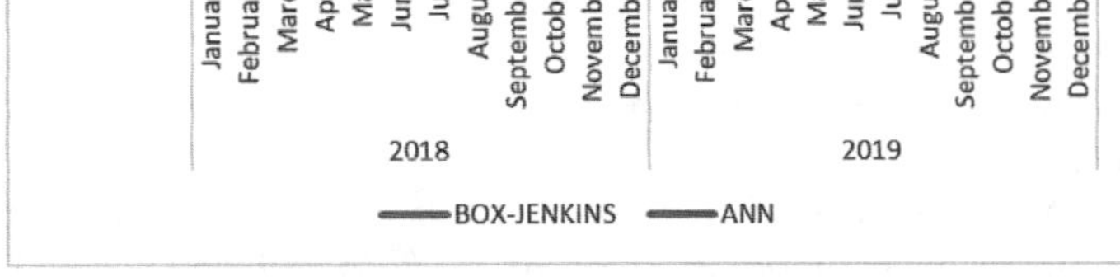

Figure 5. Urban water forecast behavior

The comparison between the ANN model and the ARIMA model, the fundamental criterion is that the MSE. It must be as close to zero as possible, and the predicted values are closer to the actual values. So, it is evident that the artificial neural networks model with a 0.471 is better than the one thrown by the ARIMA model of 0.769. In addition, it is appreciable that using the ANN model provides a much higher level of reliability than that exposed by the classical statistical model as shown in Table 3.

Table 3. Forecast comparison

Model	MSE	R
Artificial Neural Networks	0.471	0.98
ARIMA	0.769	0.55

As for the correlation between the predicted value and the real value determines the force and direction of the relationship between these values. As long as the correlation values are close to 1, it means that the predicted data are behaving with greater relation to the sense of the real values with the artificial neuronal networks model the correlation is 0.98, while the value of the correlation with the ARIMA model is 0.55, which indicates a better relation in the sense and force of the level of relation of the predicted values with the ANN model.

4. Conclusions

The application of Artificial Neural Networks as a prediction technique to develop the forecasting urban water demand, where dependent and independent variables were analyzed that were processed in the construction of the network through the processes of training, validation and testing of the network, obtaining results that behave with greater similarity to actual consumption.

With the analysis and comparison of the results obtained with the ANN model and the ARIMA the differences and similarities between them are determined and the interpretation of the results is facilitated. These results thrown by the ANN model presents greater similarity in the behavior of real consumption, while the results of the ARIMA model tend to soften and over fit, so they become constant predictions by the moving average with which they are calculated.

Acknowledgments

The authors acknowledge to the research project "Optimización de la Distribución Física y en Planta en la Cadena de Suministro aplicando Técnicas Heurísticas" supported by Agreement HCD Nro. UTN-FICA-2019-0149 by Facultad de Ingeniería en Ciencias Aplicadas from Universidad Técnica del Norte. As well, authors thank the valuable support given by the SDAS Research Group (www.sdas-group.com).

References

[1] Ghiassi M., Zimbra D K and Saidane H 2008 Urban Water Demand Forecasting with a Dynamic Artificial. *Journal of Water Resources Planning and Management* **134** pp 138-146

[2] Jain A and Ormsbee L E 2002 Short-term water demand forecast modeling techniques - Conventional methods versus AI *Journal/American Water Works Association* **94(7)** pp 64-72

[3] Firat M, Turan M E and Yurdusev M A 2009 Comparative analysis of fuzzy inference systems for water consumption time series prediction *Journal of Hydrology* **374** pp 235-241

[4] Caiado J 2010 Performance of combined double seasonal univariate time series models for forecasting water demand *Journal of Hydrologic Engineering* **15(3)** pp 215-222

[5] Babel M S, Gupta A and Pradhan P 2007 A multivariate econometric approach for domestic water demand modeling: An application to Kathmandu, Nepal *Water Resources Management* **21(3)** pp 573-589

[6] Herrera M, Torgo L, Izquierdo J and Pérez-García R 2010 Predictive models for forecasting hourlyurban water demand *Journal of Hydrology* **387** pp 141-150

IOP Conf. Series: Journal of Physics: Conf. Series **1284** (2019) 012004 doi:10.1088/1742-6596/1284/1/012004

[7] Adamowski J and Chan H F 2011 A wavelet neural network conjunction model for groundwater level forecasting *Journal of Hydrology* **407** pp 28-40

[8] Sahoo G B, Schladow S G and Reuter J E 2009 Forecasting stream water temperature using regression analysis, artificial neural network, and chaotic non-linear dynamic models *Journal of Hydrology* **378** pp 325-342

[9] Anele A O, Hamam Y, Abu-Mahfouz A M and Todini E 2017 Overview, comparative assessment and recommendations of forecasting models for short-term water demand prediction *Water (Switzerland)* **9(11)** 887

[10] Verma A, Karan A, Mathur A and Chethan S 2017 Analysis of Time-Series Method for Demand Forecasting *Journal of Engineering and Applied Sciences* **12** pp 3102-3107

[11] Piasecki A, Jurasz J and Kaźmierczak B 2018 Forecasting daily water consumption: A case study in Torun, Poland *Periodica Polytechnica Civil Engineering* **62(3)** pp 818-824

[12] Zhang X, Yue M, Yao Y and Li H 2018 Regional annual water consumption forecast model *Desalination and Water Treatment* **114** pp 51-60

[13] Tiwari M and Adamowski J 2015 Medium-Term Urban Water Demand Forecasting with Limited Data Using an Ensemble Wavelet–Bootstrap Machine-Learning Approach. *Journal of Water Resources Planning and Management* **141(2)** 04014053

IOP Conf. Series: Journal of Physics: Conf. Series **1284** (2019) 012005 doi:10.1088/1742-6596/1284/1/012005

Using Shopping and Time Attitudes to Cluster Food Shoppers: An Empirical Finding from Indonesia

A A Utami[1], A R Ginanjar[1], N Fadlia[1], I A Lubis[1] and M M Ulkhaq[1,*]

[1] Department of Industrial Engineering, Diponegoro University, Semarang 50275, Indonesia

[*]ulkhaq@live.undip.ac.id

Abstract. Retail market is claimed to be an important interface in channel management as it bridges the manufacturers and consumers. In Indonesia, retail industry has increased significantly. One of the main characteristics of this market is its heterogeneity. This heterogenous market could be segmented into different segments of consumers which have different characteristics, behaviours, and needs. This requires retail market to set proper marketing strategies by analyzing each segment. This research attempted to conduct market segmentation for food shoppers in Indonesia based on attitudes to shopping and time. Four clusters have been constructed, namely, apathetic but regular, convenience seekers, hedonist, and time-pressured convenience seekers. The ANOVA test confirmed that those four clusters were appropriate as it created more differentiated and consistent clusters. This research could give contributions to the theoretical perspective and empirical point of view on market segmentation where different marketing strategies could be generated for each segment.

1. Introduction

Retail industry in Indonesia is showing a substantial growth as the number of retailers has increased significantly by 11% for minimarket and 10.6% for super/hypermarket. This statistic is influenced by the growth in number of families. Middle income families are estimated to expand from 36% in 2015 to 58% in 2020. More than 60 million low income population is projected to join this middle class in the coming decade. Not surprisingly, with the huge number of populations, consumers spending in Indonesia are growing substantially for 11.8% per year in average for the period of 2012-2015 [1].

Among various products, food product is considered promising. In 2015, consumer spending on food is estimated at IDR 1.930 trillion, while the food product is amounted at IDR 4,369 trillion [1]. These numbers mean that the competition is fierce for the retailers, especially for food retailers. This led the service providers to pay more attention to the consumers' needs and wants. In order to do so, food retailers have to realize that market is heterogeneous. The market could be segmented into different distinct groups which have different characteristics, behaviors, and needs who might require separate services or marketing mixes to be served [2]. Thus, analyzing demand by segmenting the consumers would let marketing strategies more efficiently. Once the market is segmented, the retailer could target which segment that it can be served best and set the competitive positioning for the particular service rather than trying to compete in an entire market [2]. Moreover, noticing the rapid growth of retail markets, shoppers nowadays tend to be more selective [3]. Thus, in allocating the limited resources, it is important to focus to a certain segment so the impacts that the retail market gains will be multiplied [4].

Market segmentation refers to the process of dividing the whole market into a number of smaller, more homogeneous submarkets, i.e., segments [5]. It is widely used as it can help marketers, to develop better products and services that are targeted for a specific segment of customer [6]; see for example: [3], [7] used segmentation for identifying the particularities of shopping centers; [8], [9] utilized market segmentation in tourism sector; [10] for ecological products; while [11] in commercial fields.

One of the most important decisions when one segments a market is determining the most valuable variable that will be used to identify different groups of consumers. Some studies used the conventional socio-demographic characteristics [12], [13] and motivation [14], [15]. This research tried to utilize attitudes to shopping and time as they are noteworthy for their usefulness in the identification of consumer clusters or groups. Such attempt has been conducted to segment a market of food shoppers in the United Kingdom [16]. However, this paper sought to extend the scope of the previous study by conducting the research in Indonesia. Different attitudes of the consumers between two countries, could aid validating and generalizing those variables for segmenting food shoppers.

2. Research Design

The objective of this research is to perform market segmentation for food shoppers based on shopping and time attitudes. Such attempt has been conducted by [16] to segment food shoppers in the United Kingdom; thus, this research tried to make a generalization and validation of the selected variables, i.e., sopping and time attitudes, in the different sample. A total of forty-nine items from ten aspects related to attitudes to shopping and time were adopted from [16]. These aspects are: (i) future orientation, (ii) past orientation, (iii) time pressure, (iv) present orientation, and (v) succession from the time attitudes variable; and the rests are from the shopping attitudes variable, i.e. (vi) enjoyment, (vii) regular shopper, (viii) eating convenience shopper, (ix) shopping as an event, and (x) apathy.

A survey was conducted in Semarang, Indonesia. It is the capital city of Central Java Province. This city is considered as one of the cities that is believed to have a rapid development associated with the food business. The questionnaire that was used in this survey has two sections consisting of the respondent's personal information in the first section and then followed by the forty-nine questions. In the first part, the information conveyed are the respondent's name, age, gender, educational background, marital status, and income (in IDR) per month. The questions in the second section examine those ten aspects of food shoppers on a seven-point Likert scale. The participants were asked to show their approval or disapproval of the questions being asked. Score of 1 shows strong disapproval while score of 7 refers to a strong agreement. The list of those forty-nine items could be seen in the following section.

To do segmentation, *k-means* analysis [17] was performed to differentiate and specify the characteristics for each cluster by determining the relationship of variables, i.e., attitudes to shopping and time. This method partitions the information into the cluster so that the information that has the same characteristics are grouped into the same cluster and the information that has different characteristics are grouped into other clusters. The farther the distance between one to another cluster, the greater variant characteristics among groups.

3. Findings

To make generalization and to validate the selected variables that those could be used as a basis for segmenting food shoppers, a survey-based research was conducted in Semarang, Indonesia. We performed an online questionnaire to collect the data. The potential respondents are approached at first and requested if they were willing to partake in the survey. We have gathered 251 valid answers from the respondents. The profile of the respondents is shown in Table 1. As has been mentioned previously, the non-hierarchical *k-means* was used to explore the cluster solution. The number of clusters was found to be four as they have the greatest differences among the clusters. Validation procedure using ANOVA test shows that the four clusters was the best solution. The largest cluster member was cluster 3 (33.50%), followed by cluster 2 (26.30%), cluster 1 (21.50%), and the last is cluster 4 (18.70%). Table 2 shows the mean of the respondents' answers for the corresponding cluster. Note that the differences were statistically significant since all of *p-value* (sig.) are 0.000 or less than significance level of 0.05.

Table 1. Profile of the respondents.

Variables	Categories	Percentage
Gender	Male	91.63
	Female	8.37
Age	Under 15	0
	15 – 24	13.15
	25 – 34	34.66
	35 – 44	25.90
	45 – 54	22.31
	55 – 64	3.98
	Over 64	0
Marital status	Single	18.73
	Married	81.27
Educational background	Elementary school	1.20
	High school	37.45
	Diploma degree	19.82
	Undergraduate	37.85
	Master	3.19
	Ph.D.	0.40
Employment status	Student	9.16
	Civil servant	20.72
	Employee	24.70
	Self-employee	15.54
	Housewife	26.29
	Retired	0.40
	Others	3.19
Income per month	< IDR 1 million	19.92
	IDR 1 – 3 million	39.44
	IDR 3 – 5 million	27.89
	IDR 5 – 7 million	7.17
	IDR 7 – 9 million	2.39
	> IDR 9 million	3.19

The characteristics of the four constructed clusters are as follows. Cluster 1 is called apathetic but regular. It has 54 respondents. The members have characteristics such as they spend their time regularly for shopping; or in the other word, they consider shopping as a traditional activity as it is viewed as part of an everyday routine. However, they have low scores for convenience. It means that they tend to dislike shopping but not as much as the fourth cluster. Cluster 2 is called convenience seekers. It has 66 members which most of them are female (92.42%). The members of this clusters are concerned with convenience shopping as they got the highest scores in this aspect. Conversely, they do not see shopping as a regular activity as their scores in regular shopper aspect are the least one. Since they will not be influenced by traditional views about food shopping, they are therefore less past-oriented. Cluster 3 is hedonist. This cluster has the largest number of respondents. They have high scores in enjoyment and shopping as an event. The members of this cluster like to go to several stores to get the best value for money, spend time carefully comparing prices before buying items, and purchase different foods for special occasions. They do not see shopping as a traditional activity, and they concerned with what is happening now. Cluster 4 is called time-pressured convenience seekers. It has

47 members. This cluster usually tends to rush while shopping, delays spending for future needs, views time as a series of events and finds comfort while shopping. The members have lowest scores on regular shopping among other clusters since they do not have much time for shopping.

Table 2. Segmentation of food shoppers based on shopping and time attitudes.

	Variable and item questions	Sig.	Cluster 1 21.50%	Cluster 2 26.30%	Cluster 3 33.50%	Custer 4 18.70%	Total 100.0%
Future Orientation	The future is more important than the past to me	0.00	3.33	6.21	6.18	5.60	5.47
	The future is less interesting than the past	0.00	3.70	2.97	3.30	5.38	3.69
	I prefer to keep myself free and ready to go wherever the future may lead	0.00	3.74	4.91	5.73	5.91	5.12
	I spend less time thinking about last year	0.00	3.91	5.41	5.71	5.38	5.18
	I plan on a day-to-day basis	0.00	3.83	5.26	5.56	5.68	5.13
	I like thinking about what will happen in the future.	0.00	3.70	5.11	5.69	5.66	5.10
	I feel better of my life now than I did five years ago	0.00	3.89	5.71	5.96	5.77	5.41
Past Orientation	I like listening to old people talk about the "old days"	0.00	5.31	4.08	4.81	2.89	4.37
	Children should be taught the traditions of the past	0.00	5.81	4.61	5.51	3.06	4.88
	One's family history is important thing to know	0.00	5.76	5.35	5.86	3.38	5.24
	I like society the way it was	0.00	5.70	3.76	4.71	2.89	4.33
	I like to think about the past and to look back on what I have done	0.00	5.70	3.09	4.31	2.77	4.00
Time Pressure	I am always in a rush	0.00	2.19	3.23	4.56	5.26	3.83
	There is often not enough time in the day for me to do what I want to do	0.00	2.37	3.29	4.83	5.40	4.00
	I often feel I have no time	0.00	2.35	3.23	4.64	5.74	3.98
	Doing things fast is vital	0.00	2.46	4.98	5.86	5.74	4.88
	I always look for ways to save the time	0.00	2.52	4.77	5.64	6.00	4.81
	I regularly look at my watch	0.00	2.39	3.61	5.06	5.98	4.27
Present Orientation	I think mostly about what is happening here and now	0.00	2.81	3.82	4.35	4.60	3.92
	It is no use worrying about the future	0.00	3.33	4.00	4.42	4.62	4.11
	I live for today	0.00	3.44	4.38	4.98	4.77	4.45
	The best way to do new tasks well is to rely on what has been done in comparable occurrences	0.00	3.15	3.77	4.63	5.28	4.21

IOP Conf. Series: Journal of Physics: Conf. Series **1284** (2019) 012005 doi:10.1088/1742-6596/1284/1/012005

	in the past						
	I am mostly concerned about how I feel now	0.00	3.30	4.47	5.44	5.17	4.67

Table 2. Segmentation of food shoppers based on shopping and time attitudes (*Continued*)

	Variable and item questions	Sig.	Cluster 1 21.50%	Cluster 2 26.30%	Cluster 3 33.50%	Custer 4 18.70%	Total 100.0%
Succession	I use calendar to plan	0.00	2.50	4.24	4.27	5.23	4.06
	I use calendar to schedule well ahead of time	0.00	2.35	4.64	4.70	5.64	4.35
	I plan on a weekly basis	0.00	2.37	4.45	4.57	5.53	4.25
	Time is priceless	0.00	4.46	5.85	6.26	6.19	5.75
Enjoyment	I like going to many stores to get the best value	0.00	4.26	4.48	5.56	5.45	4.98
	I spend time cautiously comparing prices	0.00	4.63	3.94	5.19	5.68	4.83
	Food shopping is recreation	0.00	4.54	4.35	5.58	5.64	5.04
	I like to go to other stores for fresh products	0.00	4.63	4.65	5.13	5.85	5.03
	Food shopping is fun	0.00	4.46	4.42	5.50	5.47	4.99
	I spend time selecting the best quality food	0.00	4.59	4.55	5.42	5.57	5.04
Regular Shopper	I go shopping on a specific day of the week	0.00	5.48	3.09	4.44	2.51	3.95
	I go shopping for the same time of the day	0.00	5.78	3.08	4.04	2.45	3.86
	I take the same amount of time when I shop	0.00	5.81	3.82	4.93	2.94	4.45
	I never go shopping during busy hours	0.00	5.70	4.47	5.29	2.81	4.70
Eating Convenience	Buying take-away meals is saving the time	0.00	2.83	3.53	4.71	5.26	4.10
	I save time by buying convenience foods	0.00	2.67	3.15	4.01	5.68	3.81
	Eating out is saving time	0.00	2.69	3.23	4.01	5.68	3.83
	I shop quickly, buying the first product or brand I find	0.00	2.67	3.52	4.39	5.70	4.04
Shopping as An Event	I am not aware about time when shopping	0.00	5.15	3.94	4.43	3.38	4.26
	I have as much time as I need for shopping	0.00	5.37	3.95	4.44	3.19	4.28
	I spend much energy and time in shopping	0.00	5.22	3.35	3.90	3.53	3.97
	I am continuously buying different foods for special events	0.00	5.13	4.35	5.23	3.72	4.69

Apathy	Shopping is timewasting	0.00	3.07	3.26	3.85	4.68	3.68
	I get annoyed if I can't find what I need	0.00	3.22	3.71	5.07	5.11	4.32
	I prefer shopping where there are no lines	0.00	3.44	4.85	6.04	5.36	5.04
	I make a food shopping trip only when I need to	0.00	3.33	4.59	5.18	5.72	4.73

Table 3. Classification of the four clusters according to socio-demographic variables (in percentage).

Variables	Categories	Cluster 1 21.50%	Cluster 2 26.30%	Cluster 3 33.50%	Custer 4 18.70%	Total 100.0%
Gender	Female	98.15	92.42	84.52	95.74	91.63
	Male	1.85	7.58	15.48	4.26	8.37
Age	Under 15	-	-	-	-	-
	15 – 24	-	16.67	21.43	8.51	13.15
	25 – 34	11.11	28.79	30.95	76.60	34.66
	35 – 44	61.11	16.67	17.86	12.77	25.90
	45 – 54	27.78	25.76	27.38	2.13	22.31
	55 – 64	-	12.12	2.38	-	3.98
	Over 64	-	-	-	-	-
Marital status	Single	-	16.67	33.33	17.02	18.73
	Married	100.00	83.33	66.67	82.98	81.27
Educational level	Elementary school	-	4.55	-	-	1.20
	High school	64.81	24.24	44.05	19.15	37.45
	Diploma degree	12.96	9.10	26.19	31.91	19.82
	Undergraduate	22.22	50.00	33.33	46.81	37.85
	Master	-	10.61	1.19	-	3.19
	Doctorate degree	-	1.52	-	-	0.40
Employment status	Student	-	9.09	19.05	2.13	9.16
	Civil servant	16.67	21.21	20.24	25.53	20.72
	Employee	24.07	24.24	19.05	36.17	24.70
	Self-employee	18.52	15.15	8.33	25.53	15.54
	Housewife	40.74	25.76	27.38	8.51	26.29
	Retired	-	-	1.19	-	0.40
	Others	-	4.55	4.76	2.13	3.19
Income per month	< IDR 1 million	16.67	18.18	28.57	10.64	19.92
	IDR 1 – 3 million	51.85	42.42	32.14	34.04	39.44
	IDR 3 – 5 million	27.78	18.18	23.81	48.94	27.89
	IDR 5 – 7 million	3.70	6.06	10.71	6.38	7.17
	IDR 7 – 9 million	-	6.06	2.38	-	2.39
	> IDR 9 million	-	9.09	2.38	-	3.19

After segmenting food shoppers based on their shopping and time attitudes, additional information could be added based on socio-demographic variable of individuals, i.e., gender, age, marital status, employment status, educational background, and income (see Table 3 for the detail). It can be seen that the majority of respondents in this study are women (91.63%). Cluster 1 has the most female respondents (98.15%). In terms of age, the respondents of this survey are most widely between 25-34 years (34.66%) and followed by 35-44 years old (25.90%). Cluster 4 presents the greatest number between 25-34 years old (76.60%) and cluster 1 presents the greatest number between 35-44 years old (61.11%). In cluster 1, all respondents have married (100%). The cluster that has the highest percentage of single respondents is cluster 3 with 33.33%. In terms of educational background, most of the respondents in clusters 1 and 3 are high school graduated, while in clusters 2 and 4 have the largest percentage in bachelor's degree. In terms of employment status, cluster 1 is the only cluster that does not have members of student, while the others share variety of employment status. The majority of the respondents based on income ranged from IDR 1 million to 3 million (39.44%), followed by IDR 3 to 5 million (27.89%). There is no respondent with income of IDR 7 to 9 million and more than IDR 9 million in cluster 1 and 4.

4. Conclusion, Limitations, and Future Research Directions

The objective of this study was to determine the segment of food shoppers based on time attitudes (past, present, and future orientation, as well as succession and time pressure) and shopping attitudes (convenience, enjoyment, apathy, shopping as a routine activity, and as an event). Based on practical perspective, this study could give such insight to the retail store managers since establishing strategical plans based on customers segmentation is considered as essential [18], [19]. Based on the theoretical point of view, this study could give such contribution toward the related literature of segmentation study especially for segmenting food shoppers in Indonesia.

The result of this research stated that there are four clusters of Indonesia food shoppers, i.e., apathetic but regular, convenience seekers, hedonist, and time-pressure convenience seekers. The members of the first cluster consider shopping as a traditional activity since they spend their time regularly for shopping. The second cluster enjoys shopping as a convenience activity as there is no time pressure for doing it. The third one has the highest score of enjoyment as the members of this cluster consider shopping as an event. The last cluster tends to rush while shopping since the members seem to do not have much time for shopping.

Limitations of this study are twofold. First is related to the selected variables, i.e., time and shopping attitudes. For the upcoming research, more variables have to be considered, such as demographics, shopping motivation, and distance as [16] stated that there was no evidence that those two selected variables are better than the later variables to segment food shoppers. The second limitation is related to the area (location) of this research, i.e., the study was conducted in Semarang, Indonesia. As suggested for the future research is a research with a larger sample size so that a more detailed analysis would be obtained. In addition, a more nationally representative sample, not only geographically but also demographically, could provide higher generalizability. Therefore, to make a generalization, the upcoming research should broaden and expand the research area.

References

[1] Duniaindustri.com Online: http://duniaindustri.com/downloads/data-industri-minimarket-supermarket-hypermarket-di-indonesia/
[2] Kotler P, Armstrong G and Wong V 1996 *Principles of Marketing* (London: Prentice Hall)
[3] El - Adly M I 2007 *Int. J. Retail Distrib. Manag.* **35** 936–950
[4] Dibb S 1998 *Mark. Intell. Plan.* **16** 394–406
[5] Danneels E 1996 *Eur. J. Mark.* **30** 36-51
[6] González-Hernández E M and Orozco-Gómez M 2012 *Int J Retail Distrib Manag* **40** 759–777
[7] Gilboa S 2009 *J Retailing Consum Serv* **16** 135–44
[8] Cho M, Bonn M A and Brymer R A 2017 *J Hosp Tour Res* **41** 415–444
[9] Iversen N M, Hem L E and Mehmetoglu M 2016 *J Travel Tour Mark* **33** 38–66
[10] Barr S, Shaw G, Coles T and Prillwitz J 2010 *J Transp Geogr* **18** 474–481
[11] Sarki I, Bhutto N and Arshad I 2012 *Interdiscip J Contemp Res Bus* **3** 643–654
[12] Fraj E and Martínez E 2006 *J Consum Mark* **23** 133–144
[13] Jain S K and Kaur G 2006 *J Int Cons Mark* **18** 107-146
[14] Fung C K and Jim C Y 2015 *Int J Sus Dev World* **22** 76-88
[15] Guttentag D, Smith S, Potwarka L and Havitz M 2017 *J Travel Res* **57** 342-359
[16] Chetthamrongchai P and Davies G 2000 *Brit Food J* **102** 81–101
[17] Capó M, Pérez A and Lozano J A 2017 *Knowl-Based Syst* **117** 56–69
[18] Griva A, Bardaki C., Pramatari K and Papakiriakopoulos D 2018 *Expert Syst Appl* **100** 1–16
[19] Kabadayi S and Paksoy B 2016 *Int Rev Retail Distrib Consum Res* **26** 456–476

Construction of the Monitoring and Early Warning System of Beijing Pork Market

Jinying Yu[1], Xiaodong Wang[*2] and lixin Sun[1]

[1.]Institute of Agricultural Information and Economics, Beijing Academy of Agriculture and Forestry Sciences, Beijing, 100097;
[2.]Information Center of Beijing Agriculture Bereau, Beijing, 100029;

[*]630696858@qq.com

Abstract. In view of the sharp fluctuation of pork prices in the market in recent years, this paper integrates the large data from pig production to pork consumption. Based on economic theory, analysis of time series and modern information technology, the monitoring and early warning system of Beijing pork market is designed by using B/S model and SSH framework. It realizes the main functions of pork price trend analysis, cycle identification, forecast and early warning, regulation and reference, and provides information support for better grasping the market pork price change law and ironing pork price fluctuation.

1. Introduction

In recent years, the sharp fluctuation of pork prices on the market has brought a great impact on the supply of "vegetable basket" and pig-breeding industry. Based on The Data Monitoring Pilot Project For Beijing Pig Industry Chain Of Consumption Trend, the monitoring and early warning system of Beijing pork market was designed. The purpose was to build a comprehensive, timely and accurate monitoring and early warning system. It can give full play to the prediction function of large data of pig industry, accurately identify abnormal fluctuations and carry out risk early warning. It has important practical significance to iron the excessive fluctuation of pig supply and demand, and push forward the structural reform of supply side of pig industry.

Since 2007, many domestic scholars have realized the importance of monitoring and early warning of pork market, and have conducted in-depth research on pork market price analysis, prediction and early warning model, information system construction. Jie Lu has analyzed the causes of periodic fluctuation of pork prices [1]. Xuefeng Mao [2], Shaodong Yu [3] and others scholars have used H-P filtering and other time series analysis models to quantitatively identify the fluctuation period of pork prices. Chuangguo Luo has adopted the ARIMA method to build a prediction model of hog prices[4]. Rong Yang has used the structural equation to discuss the construction of hog price early warning model [5]. Tianzhong Wang [6], Quan Liu [7], Liang Qian [8], Anping Zhao [9] and others scholars have explored and studied the framework design and functional design of other agricultural products and pig market monitoring and early warning system. However, the combination of information technology and pig price analysis, market price forecasting and early warning model is still lack of comprehensive and in-depth elaboration. On the basis of the foregoing work, based on the economic theory, time series analysis and modern information technology, this research adopts B/S model and SSH framework to design and complete the monitoring and early warning system of Beijing pork market. This system is an organic combination of information technology and pig price analysis, market prediction and early warning model.

2. System overall design

The laws of pig price fluctuation, the forecasting and early warning model are the theoretical and analytical basis of the monitoring and early warning system of Beijing pork market. The main functions of the system are based on electronic database and information technology to realize the price fluctuation of pig market, the operation of forecasting and early warning model and the visualization of the results.

2.1. System structure

The monitoring and early warning system of Beijing pork market draws on the design ideas of Spring, Struts and other open source projects. The system adopts B/S mode and SSH framework.It is developed based on Java language and has integrated the accumulation of business functions for many years.The system is embedded the part functional modules of Eviews 6.0 and Easyfit software through the middleware. Its technical framework is mainly composed of data base layer, system support layer, application service layer and system user layer(Figure 1.). The system design level is clear and easy to maintain and expand.

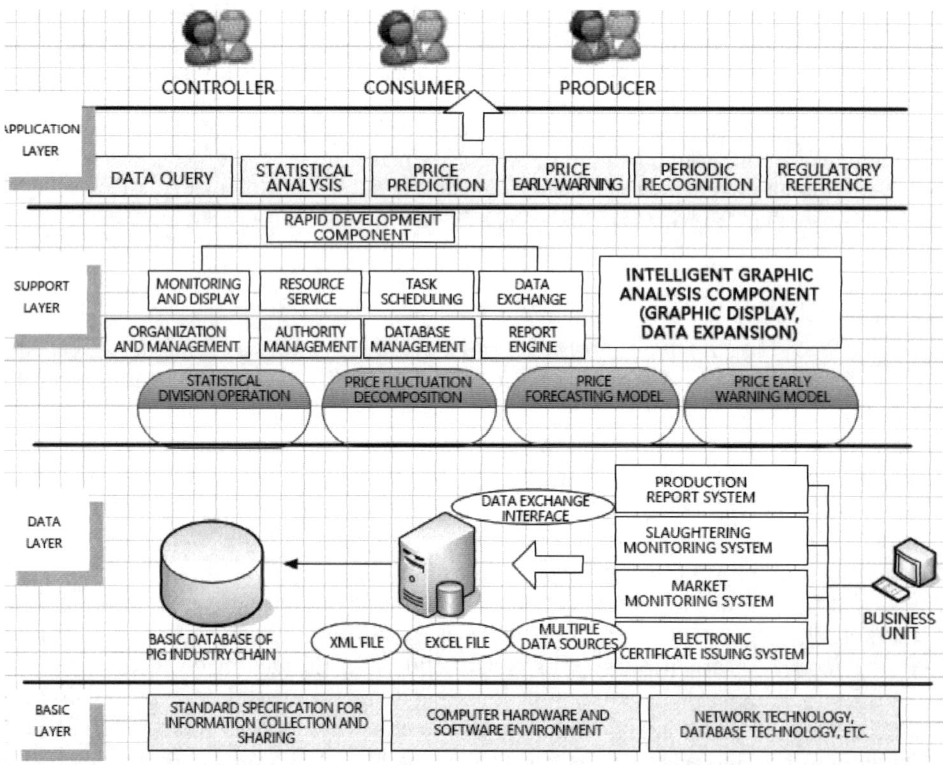

Figure 1. Overall framework of the market monitoring and early warning system.

The data base layer includes data management center, self-service platform and virtual application component. It mainly completes data exchange, extraction and management. It can transform and connect with EXCEl, XML, TXT and other data sources. It also develops a standardized open interface for data. Through the data exchange and sharing platform of Beijing Agricultural Bureau, data exchange and sharing can be realized with the “Business Report System of Agricultural Bureau”, “Monitoring System of Agricultural Products Origin”, “Agricultural Products Market System”, “Electronic Certification System of Livestock and Poultry Products”.

The system support layer which is the core of the system includes two parts: technical support and analysis model. The technology support part provides rapid development components with good expansibility and maintainability. It builds and implements the technology support of system business application through various service functions it provides. In the analysis model part, the basic algorithms such as formula, model and probability distribution function are provided for business function modules such as forecasting, early warning and market analysis.

The application service layer mainly realizes the management of the business data and the system. The User level of the system is to provide unified data display, query, analysis and reporting services such as government management departments, pig breeding, slaughtering, sales and other enterprises, as well as departments or individuals concerned about pig industry data.

2.2. Database construction
The business datas of the system include key indicators from the whole chain of pig breeding to terminal consumption. Datas from different links come from different business systems. Through the information resources catalogue and exchange system of Beijing Agricultural Bureau, the data exchange and sharing with various business systems are completed. At the same time, the transformation, docking and import of EXCEl, XML, TXT and other data sources are supported.

2.3. Man-machine interface
The human-machine interaction interface consists of 4 parts: main window, sub window, dialog box and form control. The main window is responsible for guiding users to select function menu. Sub windows are used to receive user information and display simulated running results. Dialog box mainly provides users with error information in system operation. Drop-down menus, toolbars, tables and graphs provided by form controls can interact with users.

3. Main functions and technical principles of the system

3.1. Statistical analysis module
This module mainly realizes the basic statistical functions such as classification, aggregation, comparison, average value, standard deviation, coefficient of variation, index generation, and fast calculation of technical indicators. JSP is used to develop the front page and open source Eclipse BIRT technology is used to improve the expansibility and renewal of the system.

3.2. Cycle identification module
The data used in the price cycle identification module are the monthly price fixed base index series. In the process of implementation, the time span of pork price from one peak (trough) to one adjacent trough (peak) is automatically arranged by system program positioning. The period (monthly number) is calculated.

3.3. Price prediction module
The main techniques used in price forecasting are exponential smoothing model and differential autoregressive moving average (ARIMA) model in time series to realize pork price forecasting. In the process of implementation, Eviews is embedded in middleware to determine the time dimension, and the system will automatically display data processing mode and model equation.

3.4. Price warning module
This module realizes early warning of pork price exceeding threshold. The purpose of pork price alarm is to determine the threshold of price fluctuation, which is regarded as normal price fluctuation within the threshold range and abnormal fluctuation beyond the threshold. To determine the threshold, we need to simulate the statistical distribution characteristics of time series data (such as normal distribution) and delimit the threshold interval at a certain confidence level. For pork, two kinds of

time series data are usually used: the sequence of pork price itself, and the sequence of pig grain ration (pork price to corn price ratio).

3.5. Control reference module

This module uses data visualization technology to visualize the period of abnormal fluctuation of pig price in history (since 2000), major natural and social events affecting the supply and demand of pig market, key data indicators of pig production and market in special period, and regulatory measures of pig market introduced by the government.

4. Development and implementation of the system

4.1. System operation and development environment

The system runs in the network environment and consists of database server, application server and corresponding network equipment. The system is using Oracle 10g as database platform, Tomcat 5.0 as Web server and JSP engine, Windows 2003 Sever as server operating system. Java5.0 is the development platform, J2EE is the development framework, and Ellipse4.0 is the foreground development tool. Call Eviews6.0's time series analysis component to realize friendly and easy operation of human-machine interface.

4.2. System function example

4.2.1. Market analysis interface. This functional module completes the calculation of key statistical indicators of market data for different segments of live pigs. It mainly includes discrete analysis (average value, standard deviation, coefficient of variation, etc.), contribution rate analysis (feed price, piglet price, number of piglets and so on) and decomposition analysis (decomposition of time series data of price by H-P filter).

4.2.2. Price prediction interface. This module can complete time series prediction, multi factor prediction and other prediction. Taking the operation of combined forecasting model in time series forecasting as an example, the system automatically displays the monthly forecasting values in the form of data tables (Figure 2.).

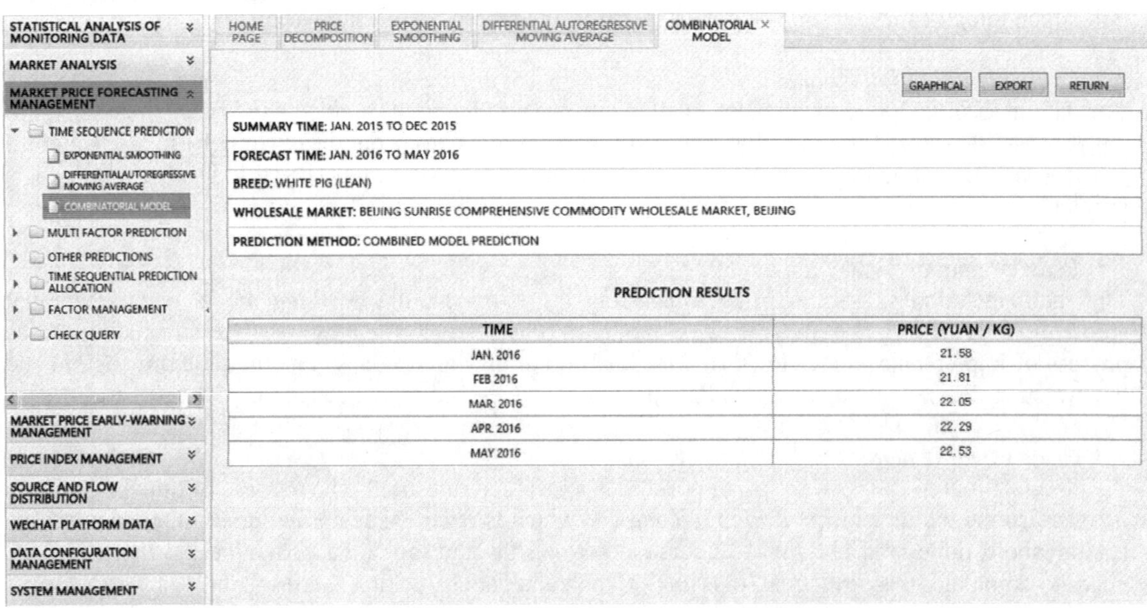

Figure 2. Time series prediction price function display interface.

4.2.3. Price warning function module. This module completes the alarm analysis of abnormal fluctuation of hog prices. Taking the operation of Value-at-Risk method as an example, click the "Price Warning-VAR" button on the left toolbar and select the time period in the right dialog box. The system shows the results in the form of data tables and data probability distribution maps (Figure 3.).

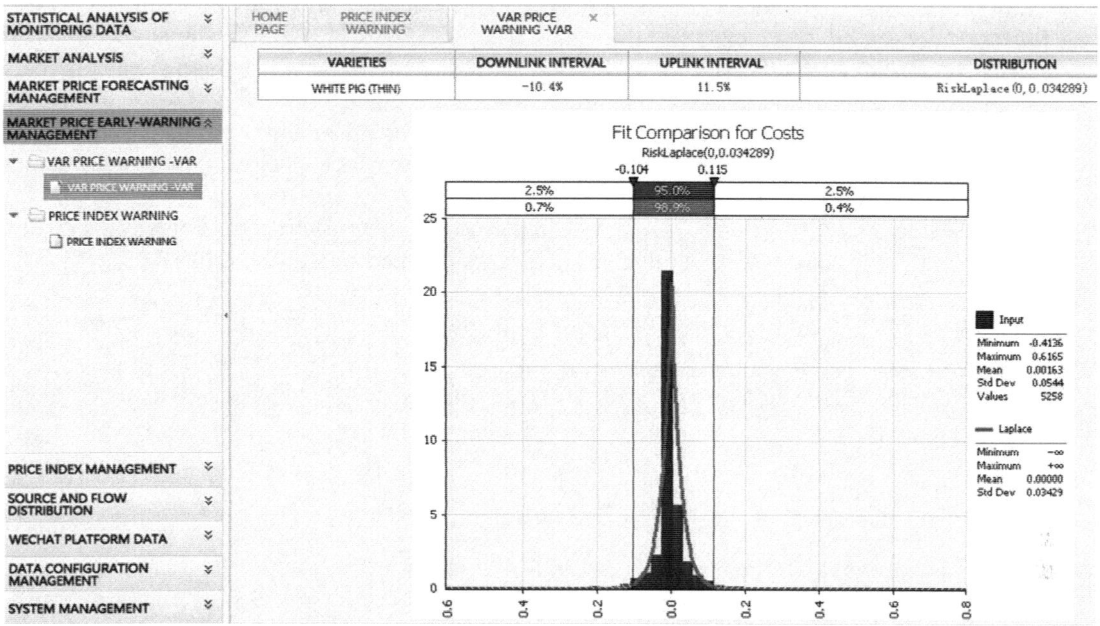

Figure 3. Price warning function display interface.

5. Conclusion and discussion

The monitoring and early warning system of Beijing pig market uses modern information technology to realize the integration and integration of multiple and decentralized data, forming a basic database including pig breeding, slaughtering, market circulation, terminal consumption, etc. The system grasps the core of market price fluctuation analysis and embeds economic analysis model to realize the decomposition, prediction and early warning of pork price fluctuation. The research and development of the system and its application are useful explorations for the analysis and research of pig industry based on big data. It has positive significance for improving the scientific, pertinence and effectiveness of macro-control of pig market and promoting the sustained, stable and healthy development and transformation and upgrading of pig industry.

There are still some problems to be solved in the process of system development and application, for example, data collection standards and standards for pig industry chain need to be improved. Data is the core of system operation. It is necessary to gather important data indicators which have great impact on production and consumption, strong representativeness, high update frequency and high accuracy around the key links of pig industry chain. At the same time, the data acquisition standards and norms of various indicators are formulated, and the data verification mechanism is established to improve the accuracy of data.

References

[1] Lv,J.,Qi,Y. Economic Analysis of Periodic Fluctuation of Pig Market Price. Agricultural Economic Issues 7, 89-92. (2000)
[2] Mao,X., Zeng,Y. Identification of pig price cycle based on time series decomposition. Rural Economy of China12, 4-13. (2008)
[3] Yu,S.Analysis of the fluctuation period of pork prices in Beijing. agricultural economic problems2, 75-78. (2012)

[4] Luo,C., Zhang,M., Xue,J. Short-term prediction of pig prices in China based on ARIMA model. World Agriculture10, 45-48. (2010)
[5] Yang,R. Early warning of pig price based on structural equation modeling. livestock economics18, 6-9,14. (2011)
[6] Wang,T. Design and implementation of price monitoring and early-warning information system. System practice9,40-42,37. (2010)
[7] Liu,Q., Cheng,B., Zheng,G., et al. The periodic analysis and prediction and early warning model of pig production in Liaoning. Livestock Economics10,22-26. (2011)
[8] Qian,L.,Wu,S., Leng,Y., et al. Research and development and practice of pig production and price monitoring and early warning system in Sichuan Province. Sichuan Animal Husbandry and Veterinary Medicine10,6-7, 9. (2012)
[9] Zhao,A.,Wang,X., Xiao,J., etc. Design and Implementation of Short-term Forecasting System for Agricultural Products Price. Agricultural Network Information4,5-10. (2014)

Centrifugal Pump Bearing Fault Diagnose Based on Time-Frequency Domain Analysis and BP-Neural Network

Wang Chaowei[1, a], Zhou Guojing[1], Wu Qian[2, b, *], Ding Lvhui[1] and Wei Yi[1]

[1]China Ship Develop and Design Center, Wuhan 430064, China
[2]China National Institute of Standardization, Beijing 100191, China

E-mail: [a]wangcwbu@163.com; Corresponding Author: [*, b]ustbwuqian@163.com

Abstract. Fault refers to the loss of function of equipment during operation. When a fault occurs, the equipment would usually show some surface features, which can be also called fault signals. Thus, the faults of the equipment can be diagnosed by deep analysis. Among the mechanical and electrical products, the centrifugal pump is one of the most typical ones, which has the majority characteristics of equipment in various kinds of ordnance. Through the study of centrifugal pump fault diagnose, the route of diagnose of mechanical and electrical products and the establishment of diagnose system can be clarified. In the study, the composition and function of centrifugal pump is firstly introduced. So is the fault diagnose system. Then the concept of the fault diagnose experiment is given. The data gathered by the diagnose system are processed in time and frequency domain. The BP- neural network is also applied to analyse the pump's fault which is injected during the experiment. The analysis result reveals that the BP- neural network can effectively estimate the condition of centrifugal pump and distinguish the different fault modes.

1. Introduction

Fault or failure refers to the event or inoperable state, in which any item or part of an item does not, or would not, perform as previously specified. It usually causes the loss of functions of the equipment or system, which can lead to the failure of scheduled missions or the damage on person and environment [1]. The capability of rapid recovery much depend on whether the faults can be detected or isolated, or can be diagnosed in other word. So to improve the fault diagnose level is one of the key way to increase the operational availability of equipment [2].

In the paper, centrifugal pump, a typical mechanical and electrical product, is chosen as an example to study the feasible way of equipment fault diagnose. The composition and function of centrifugal pump and the fault diagnose system is introduced firstly. Then the concept of the fault diagnose experiment is given, which includes experiment time, section, fault component, fault inject method and the rules of data gathering. The bearing fault of centrifugal pump is the focus. The data of bearing fault is processed in time domain in advance. In addition, the feature parameter of bearing fault signals is also analyzed in frequency domain. Then the BP- neural network is applied to analyse the bearing 's different fault, including inner ring, outer ring and rolling elements. The analysis result reveals that the BP- neural network can effectively estimate the condition of centrifugal pump and distinguish the different fault modes. The consequence shows that the method of combination of time- frequency domain analysis and BP-neural network is efficacious and meet the engineering practical need.

2. Centrifugal pump and fault diagnose system

2.1. Centrifugal pump

2.1.1. Function. The centrifugal pump is used to transfer the given liquid with the centrifugal force which caused by the movement of impeller. The liquid in the channel would be driven into the pump shell by the centrifugal force, which lead to the decrease of the stress around the centre of impeller. The gap of stress made the liquid get into the shell again.

2.1.2. Composition. The common centrifugal pump usually consists of six parts: impeller, shell, bearing, main axis, seal ring and electrical motor. The alternating current drives the electrical motor runs. The electrical motor is linked to the main axis. So is the impeller. Finally, the impeller is moved and controlled by electrical motor, as shown in Figure 1.

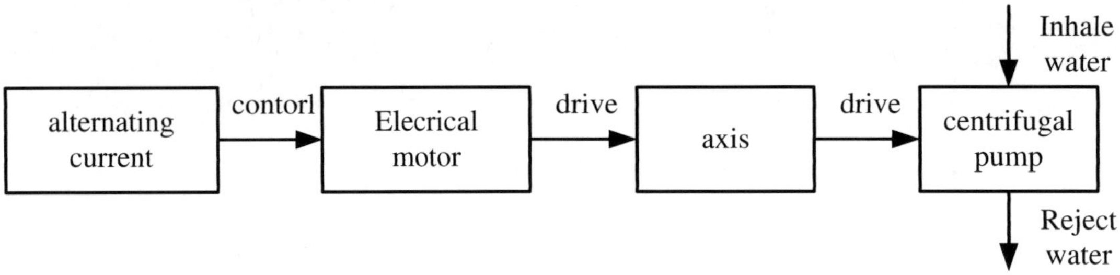

Figure 1. Centrifugal pump composition

2.1.3. Performance parameter. The performance parameters of centrifugal pump in the experiment are as followings: 1)Specified flow rate: $18m^3/h$, 2)Specified lift:24.5m, 3)NPSH (net positive suction head):< 3m H_2O, 4) specified rotate speed: 2950r/min, 5) specified power rate:3kW, 6)weight:94kg

2.1.4. Requirements for functions and performance checking before experiment Failure signal. The centrifugal pump have to be checked after being fixed onto the rack, which includes two parts: performance parameter, especially flow rate, and operation condition. The flow rate must reach the degree of $18m^3/h$ without leakage of oil and water. In the meantime, there is no abnormal noise in specified rotate speed.

2.2. Fault diagnose system

2.2.1. Composition. The fault diagnose system consists of three parts. The first one is data perception module that contains electric current sensor, flow rate sensor, vibration sensor and displacement sensor. The second one is data gather and primary-processing module, which contain the four outcomings and inputs of four kinds of sensors. It can be divided into two parts: data recuperating module and data gathering module. The inputs are linked by plugs. The third one is main fault diagnosis system, which contains original data module, time domain and frequency domain analysis module, feature value extract module and fault diagnosis module. The configuration of the fault diagnose system is shown in Figure 2.

2.2.2. Operation requirements of fault diagnose system. The part number and operation requirements of different module in the system is given in table 1. The sensors is chosen according to the monitoring requirements. And the data gather and primary-processing module and main fault diagnosis system is developed according to the fault diagnose experiment requirements. The electric current sensor, vibration sensor and displacement sensor is fixed in the pump, and the flow rate sensor is fix on the pipe.

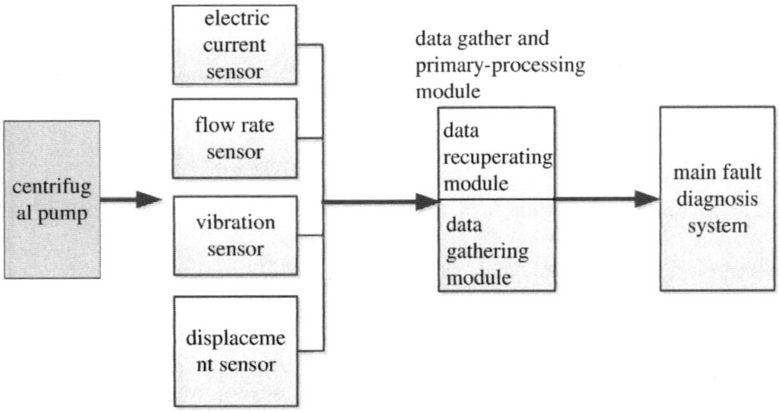

Figure 2. Fault diagnose system configuration

Table 1. Part number and operation requirements of fault diagnose system

No.	Name	Part No.	Operation requirement
1	vibration sensor	YD	frequency: 10~1kHz。
2	electric current sensor	YY-I	range of measurement:±30A。
3	flow rate sensor	LWGY	1)diameter of pipe:50mm; 2)transmitter: water。 3) flow rate: 18 m^3/h
4	displacement sensor	PMP	range of measurement:±10mV
5	data recuperating module	AD620/TLC2262(YY-16)	maximum gain:2000
6	data gathering module	STM32F103ZE(YY-STM32)	1) sampling rate:10kHz 2) number of data channel:10

3. Fault diagnosis experiment concept

3.1. Establish of experiment system. The centrifugal pump must operation in a certain experiment system in order to accomplish the imitate of normal and abnormal condition. The centrifugal pump and the electrical motor is fixed to the iron framework with bolts. The side face of framework is also linked with substrate with bolts. The substrate's function is to support pump. Then the fault diagnose system should be connected with pump. Then four kinds of sensors should be fixed to the corresponding place so as to monitoring the fault signals effectively.

3.2. Experiment section. The centrifugal pump fault diagnosis experiment contains five kinds of different sections. The first one is centrifugal pump under normal operation, which amounts to 100h. All the four kinds data should be recorded every 2.5h.Hence 40 group data would be gathered. The second one is electrical motor fault injection, which amounts to 7.5h. The electric current signal should be recorded every 2.5h.Hence 3 group data would be gathered. The third one is main axis fault injection, which amounts to 30h. The displacement signal should be recorded every 2.5h. Hence 12 group data would be gathered. The fourth one is impeller fault injection, which amounts to 10h. The flow signal should be recorded every 2.5h. Hence 4 group data would be gathered. The last one is bearing fault injection. It can be divided into three part, inner ring fault injection, outer ring injection and rolling element injection. Each is 17.5h, which amounts to 52.5h. The vibration signal should be recorded every 2.5h. Hence 21 group data would be gathered, in which there are three kinds according to the experiment of bearding. In sum, the total time of experiment is 200h.

3.3. Requirement for experiment data gathering, recording and processing. 1)method of data storing: with the software(labVIEW) in main fault diagnosis system, 2)type of data storing: xlsx format, 3) sampling interval time:2.5h, 4) sampling frequency:20000Hz, 5)the data would be recorded with the fuction of report generation. It will help generate the report in word or excel format automatically, including date, environment, operator and fault diagnoses result.

4. Pump bearing fault diagnosis based on time- frequency domain analysis and BP-neural network

In the study, the impeller, main axis, and electrical motor's fault can be detected according to the monitoring signals directly. The problem of above three elements' fault diagnose is relatively simple. Therefore, the paper focuses on the diagnosis of bearing. The signal of vibration is applied to the diagnosis. The vibration would be preprocessed and time- frequency domain analyzed. The BP-neural network conduct the diagnosis finally.

4.1. Noise reduction and time domain of original analysis data. The three-dimensional vibration signal is gathering by the vibration sensor in the experiment. The original data's noise of three kinds of bearing fault has been reduced by the method of wavelet denoising . It shows that the noise level has been reduced. The same process progress is applied to another two group of vibration data. Then the relatively pure data can be used to extract and calculated the fault feature valve. The Fault feature valve of three-dimensional vibration is shown in table 2. This step is the foundation of fault diagnosis. Whereas it is difficult to determine the fault only from the time domain [4].

Table 2. Fault feature valve of three-dimensional vibration

Fault \ **direction**	**X**	**Y**	**Z**
Inner ring	**-0.6600**	-0.3381	-0.0907
Outer ring	0.7007	**1.1241**	0.47057
Rolling element	0.4554	2.2888	**1.0012**

4.2. Frequency domain analysis. The frequency domain analysis of vibration signal is to calculate the bearing's main frequency. The analysis result of three kinds of monitoring data is shown in Figure 3, which is processed by Fourier transform (FFT). It can be noted that the main frequency of inner ring is 236.6Hz in the experiment, which basically coincide with the theoretical value 237.3Hz [5]. Through there is some distraction around 90Hz and 270 Hz, they wouldn't influence the main frequency. The same process progress has been conduct also in other two group of data, which are outer ring fault data and rolling element fault data.

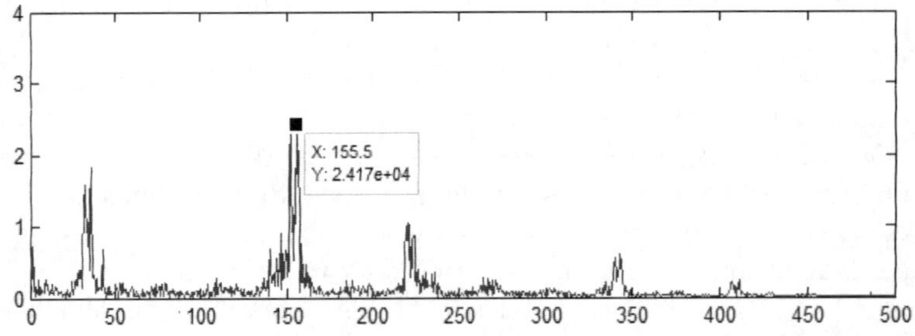

Figure 3. Frequency domain analysis of inner ring fault vibration signal

4.3 BP-neural network analysis.

4.3.1. Training of sample data. It must be figured out that the assemblage should be full of the sample space when choosing the sample data for training. Aiming at the bearing fault diagnose, six elements of vibration signal is chose as feature valves to train the BP-neural network based on the different kinds of fault and data gathered. These feature valves are marked as V1-V6 for short. The first feature valve is range maximum [6]. The second one is range minimum. The third one is range mean value. The fourth one is core frequency. The fifth one is the number of feature frequency. The sixth one is mean value of variation. The expected outcome of the BP--neural network is expressed by the binary system "0"&"1". The four condition of bearing is expressed as (1,0,0,0),(0,1,0,0),(0,0,1,0)and(0,0,0,1). The corresponding condition is normal, inner ring fault, outer ring fault and rolling elements fault. The target outcome of BP-neural network analysis is marked with T1-T4. The matlab (LM training function) is used to create the BP-neural network to meet the requirements above. The number of training time is set as 1000.

4.3.2. Test of BP-neural network analysis. After the training of BP-neural network analysis, the fault vibration signal data should be normalized before diagnosing. According to the experiment concept, seven group data has been recorder under the state of inner ring fault. All the seven group data are processed by the well-trained BP-neural network. As shown in table 3, the diagnose error can be summarized in the outcome of BP-neural network. The fault detecting error is nearly 4%, as shown in Figure 4. In other word, the accuracy rate of inner ring fault detect is 96%.

Table 3 Analysis data of inner ring fault

No.	V1	V 2	V 3	V 4	V 5	V 6	Outcome of BP-neural network
1	7.10	-10.03	0.77	241.96	25	8.56	(0.0092,0.9584,0.0051,0.0062)
2	7.48	-10.04	0.74	241.96	15	8.76	(0.0087,0.9632,0.0068,0.0113)
3	7.22	-10.12	0.79	241.94	26	8.67	(0.0096,0.9620,0.0057,0.0074)
4	7.64	-10.45	0.77	241.99	16	9.04	(0.0088,0.9752,0.0062,0.0056)
5	7.46	-9.92	0.75	241.94	24	8.64	(0.0079,0.9654,0.0071,0.0106)
6	6.78	-10.02	0.71	242.13	25	8.40	(0.0069,0.9589,0.0072,0.0089)
7	6.80	-10.02	0.78	241.86	21	8.41	(0.0073,0.9687,0.0069,0.0073)

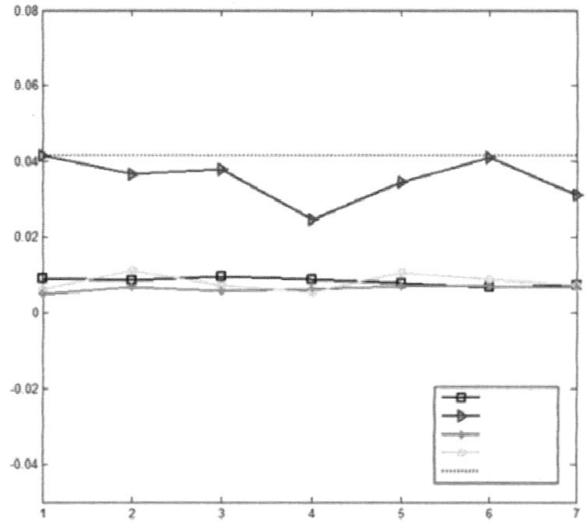

Figure 4. Error curve of inner ring fault diagnose

IOP Conf. Series: Journal of Physics: Conf. Series **1284** (2019) 012007 doi:10.1088/1742-6596/1284/1/012007

5. Conclusion

This study presented a method of mechanical and electrical products analysis based on time-frequency domain analysis and BP-neural network. Based on the analysis of centrifugal pump composition, function and performance parameters, the corresponding fault diagnosis system is built. So is the experiment system. The concept of experiment is also given in the paper. The validity and valid of the method of time-frequency domain analysis and BP-neural network is verified through experiment. It reveals that BP-neural network has a certain practical significance for the water pump fault diagnosis research. The study way can guide the design of fault diagnosis system and other condition monitoring system.

References

[1] Chaowei Wang, Peng Li , Boren Chen, Xingpan Zhou, Design of Pump Fault Diagnosis System Based on T-FMEA[C]. 2018 2nd International Conference on Data Mining, Communications and Information Technology,2018

[2] Ante Bukša, IvicaŠegulja. Ship system reliability[J]. Scientific Journal of Maritime Research 2008.8 .

[3] JIN Qiao,FANG Shuai,YAN Shi,LI Hong Nan. Methods to Improve BP Network[J]. Journal of Sheny ang Ar ch. and Civ. Eng. Univ.China. 2001.7 : 197-200.

[4] WU Y Y,XING W S,ZHU X M,et al.Study on vibration characteristic of centrifugal pump [J],Chinese Journal of Ship Research,2008, 3(1): 51-54

[5] Zhou Zheng. Survey of Current Progress in BP Neural Network [J]. ,Shanxi Electronic Technology. China.2008.2:90-92.

[6] LI Jun qing, CHEN He nian, YAN Lili, JI Wentian Evaluation Index for the Management Level of Computer Lab Based on BP Neural Network [J]. RESEARCH AND EXPLORATION IN LABORATORY.2011.04: 71-73.

IOP Conf. Series: Journal of Physics: Conf. Series **1284** (2019) 012008 doi:10.1088/1742-6596/1284/1/012008

Technology and Application for Substation Automatic Inspection Based on "Robot +" Mode

Zhao Bincai[1,*], **Wang Feifei**[1] **and Han Fucheng**[2]

[1]State Grid Weifang Power Supply Company, Weifang, China
[2]Shandong Luneng Software Technology Co., Ltd., Jinan, China

[*]3342038921@qq.com

Abstract. Daily substation equipment inspection is important for safety production of the power grid. To improve the safety and reliability of the traditional substation instrument inspection mode, this paper aims on the technology and application for substation automatic inspection based on "robot +" mode. Upon the foundation of substation inspect robot, we expand the cover range and types of monitoring devices. We first introduce the system architecture and inspect data type of the proposed "robot+" mode, and then show the data access architecture of all the monitoring devices related to the system. The processing strategies for two typical kinds of structured and unstructured data which are not from the robot are explained. For structured data, we adopt the ARIMA model to predict the time series data. Meanwhile for unstructured data, we take advantage the concept of sparse representation to recover the deformation of the monitoring object. With the newly automatic inspection mode, we can achieve an all-round three-dimensional inspection of substation equipment and environment, which can improve the coverage, the quality and efficiency of substation operation and maintenance. Through the work of this paper, we can reduce the working pressure of team members and enhance the condition control and operation inspection management penetration of substation equipment.

1. Introduction

It is an important daily work for electric operation and maintenance personals to inspect substation equipment, which directly affects the production safety and reflects the level of production management. It is a huge work for them to carry out regular inspections on the substation equipment to meet the demands of current management requirements [1-3]. With the development of power systems and the growth of power networks, the shortage of operation and maintenance personals has become increasingly prominent, and the mode of traditional substation inspection is obviously backward. The intelligent inspection robot is an effective technical way for substation to operate in an unattended environment [4]. However, the automatic detection of the equipment status is still incomplete and the inspection points cannot be fully covered. It is still a question to optimize the inspection mode and content under the premise of improving the quality and ensuring the safety of electric power production.

This paper develops a new substation automatic inspection management system based on the further study from the substation robot named "Robot +" automatic inspection mode. The system accesses various inspection data and images through various interface modes to realize remote inspection instead of on-site inspection. It alleviates the work pressure of operation and maintenance personnel and improves the management level of substation operation and maintenance.

2. The proposed mode

2.1. System structure

The whole system architecture is shown in figure 1. The application logic can be divided into four layers: the communication processing layer, the data processing layer, the business application layer and the presentation layer.

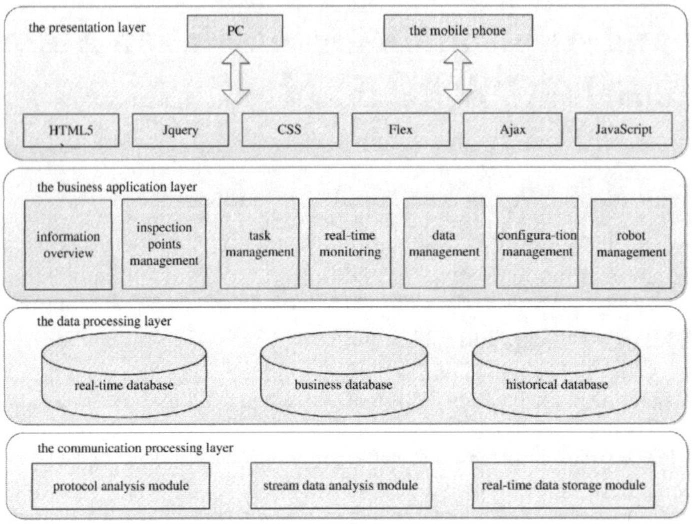

Figure 1. System architecture.

The communication processing layer is responsible for protocol analysis, stream data analysis, and real-time data storage. The data processing layer classifies the data according to its generation time and priority. The business application layer is the core module, which consists of information overview, inspection points management, task management, real-time monitoring, data management, configuration management, and robot management. The presentation layer provides two interfaces, one is a browser-oriented WEB interface, and the other is a service interface for smartphone platforms. The mobile APP accesses the service server through interfaces such as Flex, Ajax, and JavaScript.

2.2. "Robot +" inspection data

The "Robot+" inspection mode is based on robot inspection technology, supplemented by other data such as main equipment monitoring, auxiliary equipment monitoring, and video surveillance to achieve full coverage inspections indoors and outdoors of the substation. The inspection data collected is as follows:

(a) Robot inspection data: visible light images such as the equipment appearance, opening and closing indication, meter reading, and secondary equipment status, thermal images of equipment bodies and connectors, and operating equipment sounds such as transformers and reactors;

(b) Main equipment monitoring: mainly focus on operating data which cannot be covered by the robot, such as meter reading value of GIS and arresters, voltage and internal resistance of batteries and so on;

(c) Auxiliary equipment monitoring: mainly collects temperature and humidity, indoor SF6/O2 values, fire and flooding warning signals, etc.;

(d) Video surveillance: mainly collects visible light images of equipment appearance, opening and closing indication and meter reading for areas that cannot be covered by the robot.

2.3. Data access

There is an intelligent gateway device in the substation, which integrates the monitoring data of all stations and transmits it to the main server of the city company. The server has an interface service to transmit data to the database. The data storage module stores the system data, the scheduling data of the massive platform accessed through the interface and the power status monitoring data in the database. The data access architecture is shown in figure 2.

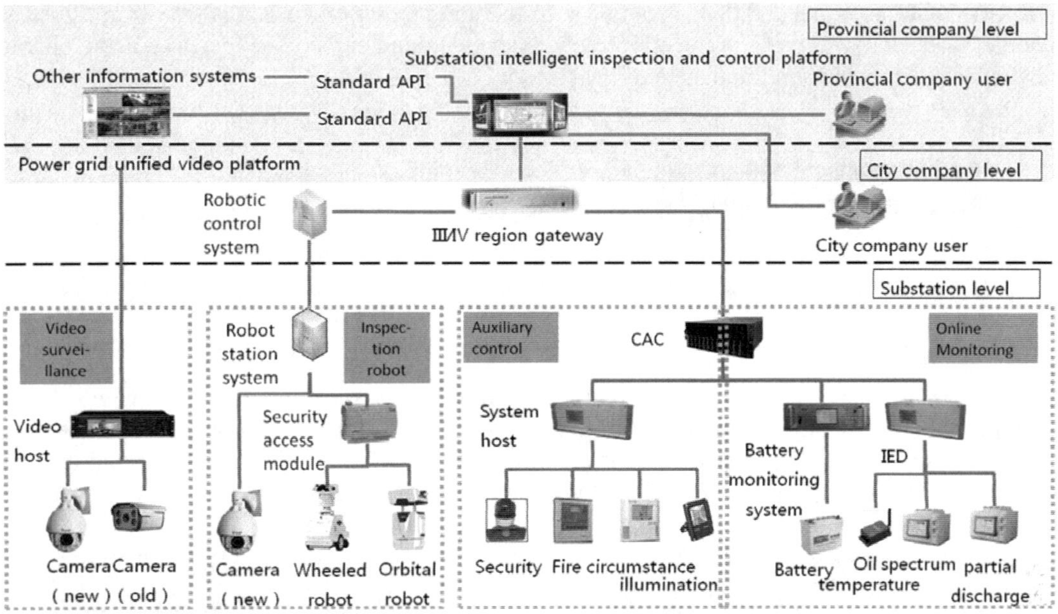

Figure 2. Data access architecture.

3. Real-time data analysis

It should be noted that the data processing methods for different kind of data from the robot is relatively mature. On the other hand, the indoor environment information detected by the auxiliary equipment also has a corresponding alarm mechanism. Therefore, in this section, we mainly discuss the analysis of data which is not from the robot.

3.1. Data categories to be analyzed

The data should be analyzed can be divided into two categories as described in section 2.2:

(a)Structured data: mainly contains numerical data of time series such as battery voltage, internal resistance, as well as meter values of GIS and arresters. The alarm rules of meter values have corresponding specifications, but for the time series data such as the battery voltage and internal resistance, it needs to be analyzed according to the characteristics of the data itself.

(b)Unstructured data: mainly refers to visible light images which from the robot un-covered area as shown in section 2.2(d).

3.2. Data Processing Strategy

3.2.1. Structured data. Take the monitoring data of a battery voltage as an example. The battery voltage is around 2~ 2.2V and the actual monitored value is definitely not guaranteed to be at a constant value (as shown in figure 3(a)). Here, we use the ARIMA [5] prediction model to analyze the historical monitoring data of the voltage and predict the reasonable data interval that the voltage data should be in.1000 monitoring points were used to verify the effectiveness of the method, within which 800 monitoring points were used to fit the model. In the ARMA (p,q) model, the search range of p and q orders is [0,3], and 200 points are predicted forward. See figure 3 (b) for a comparison of predict and real data. We calculate the accuracy of the prediction method by using the concept of root mean square error.

IOP Conf. Series: Journal of Physics: Conf. Series **1284** (2019) 012008 doi:10.1088/1742-6596/1284/1/012008

$$\text{RMSE}\ (X)=\sqrt{\frac{1}{M}\sum_{i=1}^{M}(x_i-\hat{x}_i)} \qquad (1)$$

The RMSE of the model is 0.004. It can be seen that the prediction curve (as shown in figure 3(b)) is smoother than the real curve, and can effectively reflect the trend of the voltage in a controllable range, thereby rapidly detect the abnormal state of the battery.

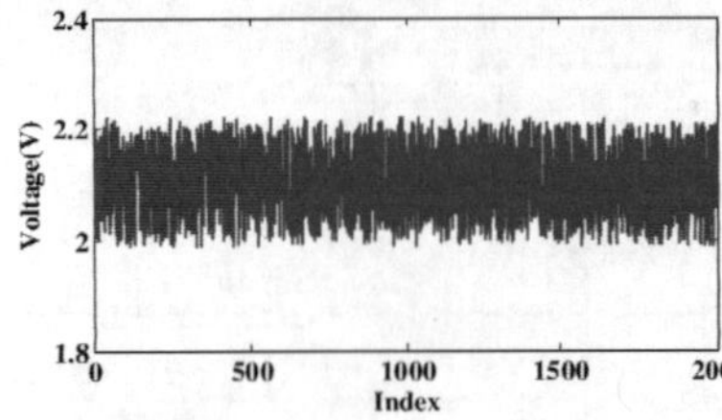

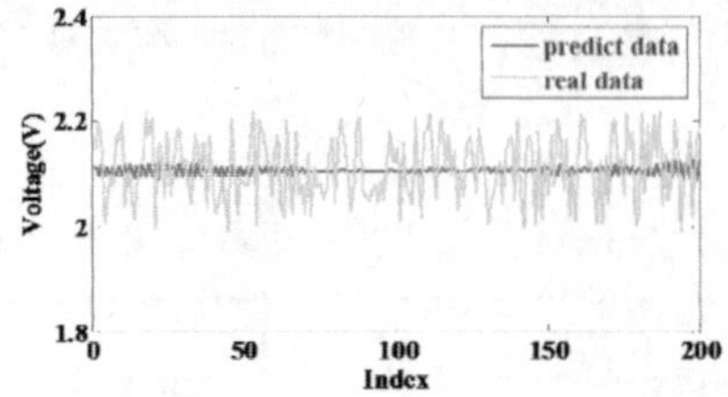

Figure 3. Data Prediction Result. (a) is the original curve, (b) is the prediction curve.

3.2.2. Unstructured data. As shown in figure 4, there are several typical unstructured images that need to be processed. The specific information of the equipment is covered by yellow blocks. It can be seen that we can judge whether the appearance of the device changes by historical data template matching for figure 4(a). Meanwhile, and the traditional edge detection and morphological processing can handle the meter reading of figure 4(b) (c).

(a)

(b)

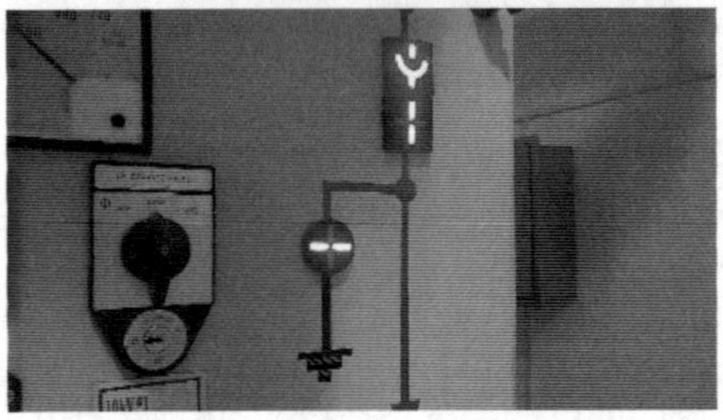
(c)

(d)

Figure 4. Typical unstructured images. (a) is the equipment appearance monitoring, (b) is the meter reading monitoring and (c) (d) are two types of opening and closing indication monitoring.

Different from other images, there is a rotation within the indicator region in figure 4(d), which makes the traditional ways useless. So a certain process is required to correct the rotation. The proposed process is as follows:

1) Get the edge image with canny operator, as shown in figure 5(a);

2) Use the Hough transform to detect lines $L=\{l_i\}, i=1,2,\cdots k$ as shown in figure 5(b). Four parameters to represent a line as $l=\{p^s, p^e, \theta, r\}$, where p^s, p^e are the starting and ending points of the line in the image. θ is the angle in degrees of l, and r is the distance from the origin to the closest point on l;

3) According to the characteristics that the left and right edges of the indicator region. We can reconstruct the border edge of the region.

3.1) Select the relatively straight lines among the detected lines, where *L* is all the straight lines detected, θ_L are angles in degrees of *L*, and *n* is the number of straight lines of L_1.

$$L_1 = \{l_{1,1}, l_{1,2}, \cdots l_{1,n}\} \ where \ \theta(L_1) < 5^o \tag{2}$$

3.2) Go through L_1 to find straight line pairs in which two lines within the pair are parallel. Furthermore, two lines determined by their starting and ending points are also parallel.

$$\begin{aligned} P_1 &= \{\{l_{i,1}, l_{i,2}\}, i = 1,2,\cdots m\} \ where \ |\phi(l_i^s) - \phi(l_i^e)| < 5^o \\ P_2 &= \{\{l_i^s, l_i^e\}, i = 1,2,\cdots m\} \end{aligned} \tag{3}$$

Here P_1 represents the set of line pairs obtained from L_1. Each line pair $\{l_{i,1}, l_{i,2}\}$ contains two straight lines. $\phi(l_i^s)$ is the angle of line determined by $p_{i,1}^s, p_{i,2}^s$ and $\phi(l_i^e)$ is the angle of line determined by $p_{i,1}^e, p_{i,2}^e$. P_2 represents the set of line pairs decided by $p_{i,1}^s, p_{i,2}^s$. *m* is the number of lines pairs within in P_1 and P_2.

3.3) Go through the lines in *L* and P_2 to calculate *b* in (4). Find the line pair in P_1 with the largest *b*, which are the final obtained lines indicating the left and right edges of the indicator region.

$$\begin{aligned} a_{i,j} &= \begin{cases} 1 & if \ (|\theta_i - \phi_j^s| < 5 \ or \ |\theta_i - \phi_j^s| < 5) \ and \ (dist(l_i, l_j^s) < 10 \ or \ dist(l_i - l_j^s) < 10) \\ 0 & otherwise \end{cases} \\ b_j &= \sum_{i=1}^{n} a_{i,j} \end{aligned} \tag{4}$$

The two lines named as H_1, H_2 finally obtained are marked with red lines in figure 5(c). The green lines in figure 5(c) are straight lines determined the red lines. The four lines reconstruct the border edge of the indicator region.

4) According to the concept of sparse representation, the rank of the indicator region is lowest when its rotation is removed. We here use the TILT method in [6] meet the demand.

We take two points as the initial input points of TILT as marked in figure 5(c) with blue cross. After iterative optimization, the optimized image with rotation removed can be obtained. As shown in figure 5(d), the red windows denote the original input, green windows denote the deformed texture. Figure 5(e) is the rectified region relative to the original input region. Figure 5(f) is the final image with rotation removed. With the rectified image, the position of the opening and closing indicator can be easily recognized.

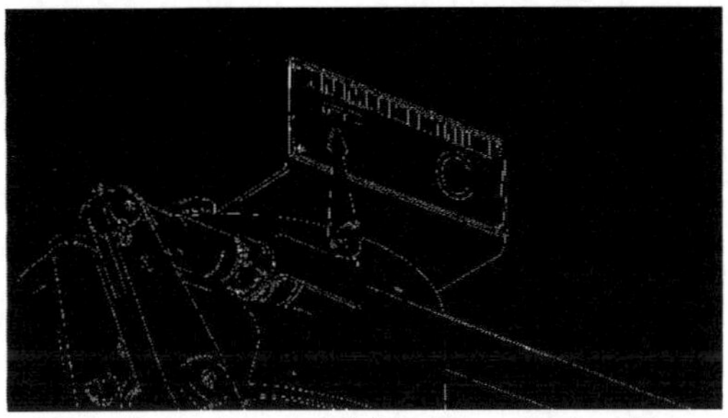

(a)

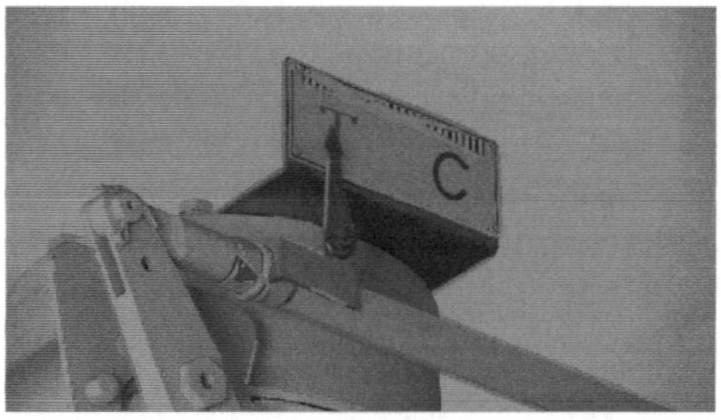

(b)

IOP Conf. Series: Journal of Physics: Conf. Series **1284** (2019) 012008 doi:10.1088/1742-6596/1284/1/012008

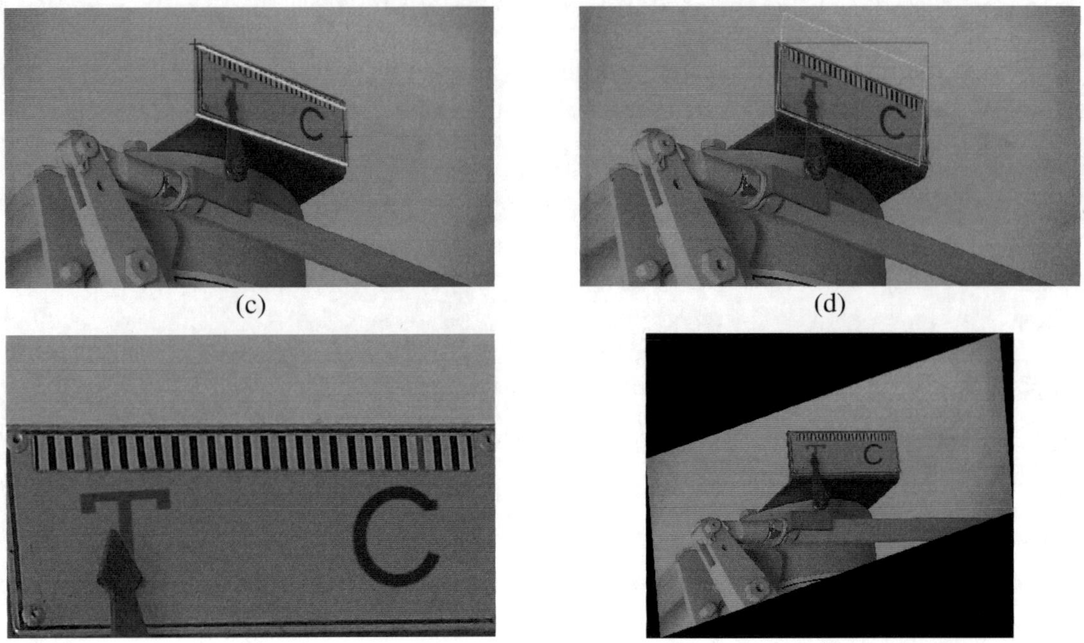

(c) (d) (e) (f)

Figure 5. Treating process of figure 4(d). (a) is the edge image, (b) is the vertical edge lines detected, (c) is the reconstructed border edge. (d) (e) (f) are the results of TILT.

4. Conclusion

In this paper, we proposed a new substation inspection mode named "Robot+" mode. It focuses on expanding the cover range and types of monitoring devices on the basis of the existing substation robot. For the complex kinds of data, we introduce the system and data architecture of the substation automatic inspection management system. Furthermore, ARIMA model and TILT method are adopted to process data which are different from the robot detected. At present, the technical solution has been deployed at several 220kV substations of Shandong province, China. The average inspection time is only about 30% relative to the traditional inspection mode. Therefore, the work of this paper contributes high practical value.

References

[1] Wang Binhai, Guo Rui, Li Bingqiang, et al. SmartGuard: an autonomous robotic system for inspecting substation equipment[J]. Journal of Field Robotics, 2012, 29 (1): 123-137.

[2] Nhan H. Pham, Hung M. La. Design and implementation of an autonomous robot for steel bridge inspection, Communication Control and Computing (Allerton) 2016 54th Annual Allerton Conference on, 2016: 556-562.

[3] Nakashima M, Yakabe H, Maruyama Y. Application of semi-automatic robot technology on hot-line maintenance work[C]. Proceedings of the IEEE International Conference on Robotics and Automation. Nagoya, Japan: 1995: 843-850.

[4] Shouyin Lu, Yanping Li, Tao Zhang. Design and implement of control system for power substation equipment inspection robot[P]. Intelligent Robots and Systems, 2009. IROS 2009. IEEE/RSJ International Conference on,2009.

[5] G.E.P. Box, G.M. Jenkins, G.C. ReinselTime series analysis: forecasting and control (revised edition)J. Market. Res., 14 (2) (1994), pp. 199-201.

[6] Zhengdong Zhang, Xiao Liang, Arvind Ganesh, and Yi Ma. Proc. of ACCV, 2010.

Study on Reliability Calculation of Repairable System Based on Monte-Carlo Simulation

Wang Chaowei[1], Hao Xun[1], Wu Qian[2,*], Qie Yutao[1] and Fang Yuanyuan[1]

[1]China Ship Develop and Design Center, Wuhan 430064, China
[2]China National Institute of Standardization, Beijing 100191, China

Corresponding author: *ustbwuqian@163.com

Abstract. Nowadays, the complexity of engineering system grows with the higher requirement of quality. Reliability, one of the most important quality characteristics, has been drawn a lot of attention. The level of reliability decides whether the system can accomplish mission successfully or not, as well as the safety of system and operator. The reliability calculation (prediction) is the most important item in design phrase, as an approach to measure the reliability. There is much study which focuses on the unrepairable system, such as aircraft and satellite. The study presents a reliability calculation method for repairable system based on monte-carlo simulation. The inputs of reliability calculation, which are different from unrepairable system, are analysed firstly. Then, the simulation principles are introduced aiming at various reliability models. The effectiveness is verified through a case of vessel in the end, which shows the simulation calculation of repairable system is valid. It can make a contribution to the improvement of system reliability in design phrase and aid the developer to locate the weakness parts.

1. Introduction

With the development of science and technology, the engineering system is becoming lager and more complex, which brings risks and challenge with more investment and longer research period. It is much harder to keep this kind of system in the same reliability level as well as previous ones. Meanwhile, the environment which the systems applied is unceasingly expanded and gets more critical. It puts forward more requirement of high reliability level. Reliability calculation (prediction) is one of the most important works in design to verify the existing state and is an effective approach to improve reliability level. The repairable systems' reliability calculations are much different from the unrepairable ones. Aiming at the characteristics of repairable systems, reliability calculation method must be upgraded [1-2].

The paper firstly analyses the inputs of reliability calculation of repairable system which differs from the unrepairable one, such as equipment attribute, mission profile, RMS indexes and so on. Maintenance margin, an important concept, is introduced in the calculation. The different reliability model is also introduced, as well as the advantages and disadvantages. The fault tree analysis (FTA) is chosen as the best way to establish the repairable system reliability model in doer to conduct the monte-carlo simulation. The simulation strategies aiming at various models are also expatiated. The reliability of a certain vessel in the typical mission profile, as an example, is conducted, compared to the unrepairable algorithm. The consequence shows that the system's can be evaluated more accurately and is more approximate to the real system's reliability level, which can aid the designers to make alternatives and optimization.

2. Input of Reliability Calculation of Repairable System

2.1. Equipment Attribute

The repairable attribute of system can be divided into two kinds: one is repairable without the system halting, the other is system can stop work in specified time, which can be also called maintenance margin (explained in 2.2). The operation mode and environment will have an effect on whether the system can be repaired or not during the mission. The aircraft is one of the unrepairable typical systems. Other aerospace systems, such as satellite, missile, are unrepairable, too. Whereas the vessel and vehicle is usually can be repaired. It is related to mission profiles. The vessel will be unrepairable during combat operation period. The support structure would also be influential. The support resources in the operational level (O-level) are decisive, including personnel, support equipments, tools, spares and technical manual. They are must be supplied immediately during the operation. If there is no maintenance support resources, the system's state (repairable or not) would switch. So the ISS (international space station) can be repairable if the support resources are stored in the station. In the end, the equipments (or called parts) must be designed to be easy to remove or maintenance by the operator or elementary maintenance person, which is usually in the O-level. The general arrangement of all parts should be reasonable. In other words, the maintenance accessibility of the system must be favourable, as shown in Figure 1.

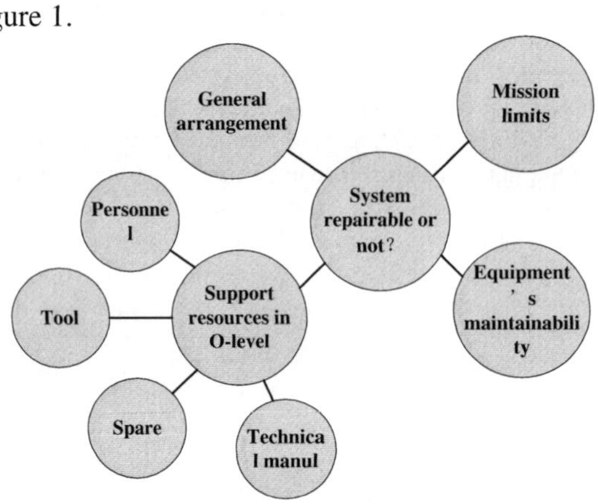

Figure 1. Influence factors for repairable system.

2.2. Mission Profile

Mission profile is the description of all events and environment's in order which system would go through during specified tasks. It is the basic input of reliability calculation. The reliability of the same system varies with the change of different mission profile. The mission profile consists of system operation time, operation mode of constituent parts and duty cycle in common. Whereas, all above factors in mission profile is inadequate for repairable system reliability calculation. Maintenance factors should be added, such as maintenance margin. The maintenance margin refers to the maximum time which the system can halt and be repaired well. The maintenance margin usually divided into several parts according to the mission profile. Because the whole mission profile consists of several parts, and different sub-mission profile (parts or phases of the whole profile) has the corresponding limits. When taking vessel as an example, maintenance is not allowed during the operation phase. On the contrary, sailors can repair the breakdown parts with the equipments and spares on board when sailing.

2.3. RMS Index

The indexes of each equipments in unrepairable system reliability calculation consists of mean time between failures (MTBF), mean time between critical failures (MTBCF) or probability of mission success (P). Whereas, the other RMS (reliability, maintainability and supportability) indexes are not taken into consideration. The meantime to repair (MTTR) needs to add at least. If the calculation need to be more accurate, other RMS indexes, such as the number of spares (in operational level), the MTTR of each failure mode of equipments (from FMEA), should be also attached to the equipment's attribute [3].

2.4. Common System Reliability Model

2.4.1. Reliability Block Diagram. The reliability block diagram is one of the most popular modelling methods applied in practical project, which is also called RBD. The advantages of RBD method concludes that easy to read or learn and can reflect the work relationship among every single constituent part in the system, which is more or less like a kind of circuit diagram. As shown in Figure 2, the series, paralleling and sequentially model is built with the method of RBD, which is the most common relationship in the system. It is need to analyse carefully to confirm the system's state once one or more units are to be failure 3.

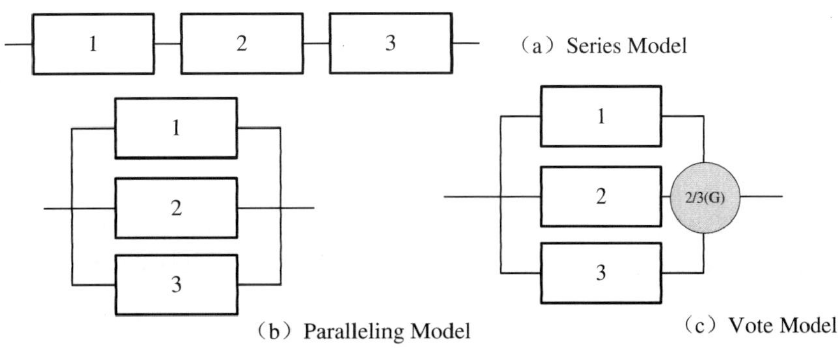

Figure 2. Typical reliability model in RBD.

2.4.2. Fault Tree Analysis. The fault tree analysis (FTA) is a kind of logic diagram which refers the parts or parts assembly that lead to the certain fault. Both qualitative and quantitative analysis can be conduct through FTA. The event and logic gate is the basic element of FTA. Event represents parts or failure modes and logic gate represents relationship. This kind of modelling is clear and easy to figure out which parts' failure would lead to the whole system's failure. And the relationship among the parts which causes fault is distinct. It is very convenient to analyse the complicated relationship among much kinds of events. The followings are the common modes of FTA (according to the RBDs in Figure 3). It refers that the two kinds of modes can be converted to each other. The block No.0 is called top event, which represents the most unbearable consequence of the system. In reliability calculation, it usually represents failure.

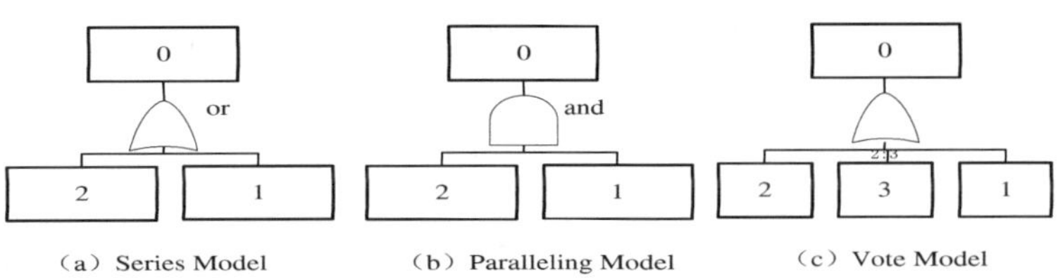

Figure 3. Typical reliability model in FTA.

IOP Conf. Series: Journal of Physics: Conf. Series **1284** (2019) 012009 doi:10.1088/1742-6596/1284/1/012009

3. Simulation calculation on Monte-Carlo

3.1. Monte-Carlo method
The Monte-Carlo method is also called computer random simulation method, which is based on "random number". It is oriented from the project of atomic bomb in World War II. The moderator of the project named it after the city of Monte Carlo, which is famous for gambling. The basic theory of the method is to determine whether an event would occur or not by the random number. And the occurrence probability obeys a certain kind of distribution, which is much like the failure. The occurrence of failure is also random and the distribution of failure is called failure rate.

3.2. Minimal cut sets of FTA
The minimal cut sets are several sets of events, and the top event would occur when minimal cut sets all occur. If any events are removed from a cut set, the group of events cannot be a cut set. From the definition, it can be inferred that once the top event occurs, at least one minimal cut set should occur. So the simulation mode of system is based on the parts of system. The system's operational state is simulated by the parts simulation with a certain relationship, which is reflected by FTA. On the other hand, the relationship of FTA can be summarized by its structure function as followings. It is not hard to draw a conclusion that the function is the union set of all minimal cut sets [4].

$$\psi(X) = \prod_{i}^{K} M_i(X) \tag{1}$$

3.3. Simulation procedure
Base on the analysis above, it comes to a conclusion that the earliest minimal cut sets occurred is the cut sets which lead to the system failure. So it is same to make a random sample of top events and the cut sets. And the certain cut set can be easily expressed with array. Thus, the system operation state can be simulated by the digital logic of cut sets based on the FTA. The basic simulate calculation procedures are shown in Figure 4. Often, the whole simulate calculation procedure are conducted in special software to aid designer get the final result.

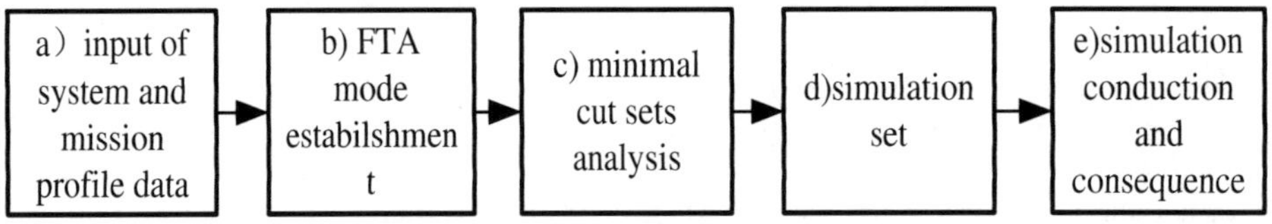

Figure 4. Simulate calculation procedures based on FTA.

From the analysis, in can be easily inferred that the simulate calculate method based on Monte-Carlo is not only suitable to the repairable system (with or without maintenance margin) but also suitable to the unrepairable system. N

4. Case of repairable system reliability calculation

4.1. Basic hypothesis
In the study, a typical repairable system, vessel's ventilating system is taken as a example. Before calculating, some basic hypothesises are made as followings: a) only parts which would influence the function or performance of system are taken into consideration; b) all the parts have only two state: normal or failure; c) all the parts' failures are independent; d) all the parts' failure rates obey exponential distribution or are approximate to exponential distribution; e) operators are completely reliable and the interactive between system and person is no problem; f) the maintenance support resource on board is

adequate; g) all the failures are repairable on board unless special statement; h)all the parts' different operation modes which may influence duty cycles are not taken into consideration [5-6].

4.2. Input parameters of system

The total time of mission profile is set in 10h. And operation time is set in 8h, and maintenance margin is set in 2h. The ventilating system consists of six parts, with the preconditions of ignore of secondary equipments. The RBD model of the system is shown in Figure 5 and the specific RMS indexes are shown in table 1.

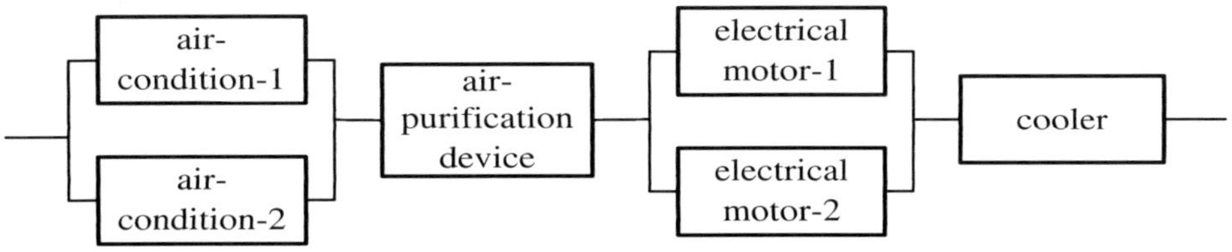

Figure 5. RBD of ventilating system.

Table 1. RMS indexes of ventilating system.

RMS index / parts	failure rate (1/h)	MTTR (h)	repairable or not
air-condition-1 (A1)	0.001	0.5	yes
air-condition-2 (A2)	0.001	0.5	yes
electrical motor-1 (B1)	0.004	1.5	yes
electrical motor-2 (B2)	0.004	1.5	yes
air-purification device (E1)	0.002	1	yes
cooler (F1)	0.002	/	no

4.3. FTA

The FTA model of the ventilating system is shown in Figure 6 according to the RBD. In the model, the top event represents system failure and the bottom blocks (A1-F1) represent the parts. The block A/B is the middle event, which represent the two-paralleling model in the system. The FTA can be more logical and distinct by this way. The RMS attribute of parts are attached to the model when the FTA is established.

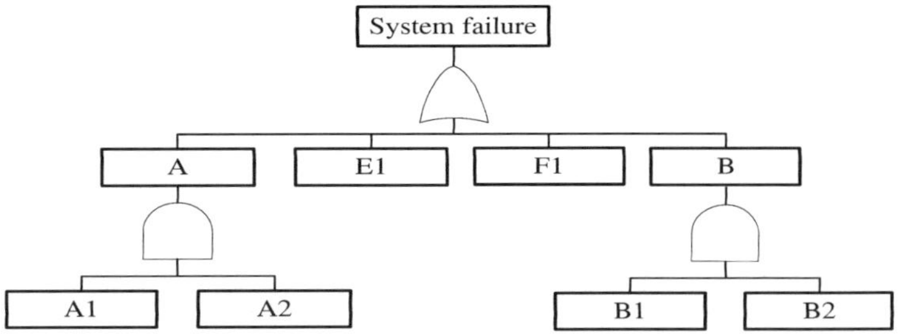

Figure 6. FTA model of the ventilating system.

4.4. *Simulate Calculation*

In order to compare the repairable and unrepairable calculation, the study presents the three calculation consequence: unrepairable, repairable without maintenance margin and repairable with maintenance margin. According to the analysis above, the reliability of repairable with maintenance margin is the highest, the repairable without maintenance margin is second and the unrepairable is third. The unrepairable and repairable without maintenance margin calculation can be obtained by mature commercial software. In the study, "Relex" of PTC is selected to calculate and the epairable with maintenance margin calculation is conduct by the self-made programme. The calculation consequeces are various, such as mission reliability (Rm), MTBF and failure rate. However, the latter two parameters are system's instantaneous value, which cannot reflect the whole mission. Rm is the common parameter. The calculation consequences with 3 times simulation (10000, 100000 and 1000000 times) are shown in table 2. It needs to point out that the consequences of every calculation would change a bit due to the simulation method. Its influence can be eliminated by the way of improving simulate times, repeated simulations and averaging values. The system reliability of repairable calculation is higher than the unrepairable [7]. And the reliability can be improved effectively once the maintenance margin considered.

Table 2. system reliability calculation consequences.

No.	Repairable mode	Rm 1	Rm 2	Rm 3
1	unrepairable	0.591	0.589	0.592
2	repairable without maintenance margin	0.670	0.671	0.669
3	repairable with maintenance margin	0.752	0.749	0.751

5. Conclusion

This study presented a method of repairable system reliability calculation with maintenance margin analysis based on monte-carlo simulation. The differences of calculation inputs between repairable systems and unrepairable ones are analysed. And the simulation theory and procedures are introduced based on the FTA cut sets. Finally, the validity of the method is verified by an case of ventilating system on the vessel. The method can aid the developers to locate the weakness of system and take serviceable measures to improve the reliability. And it can also help developers to consider the maintainability of equipment as well as the operation mode, which can put forward operation advice to the final user.

References

[1] D. N. P. Murthy. New research in reliability, warranty and maintenance [A]. In: The4th Asia-Pacific International Symposium on Advanced Reliability and Maintenance Modeling [C]. Wellington, New Zealand: 2010, 504-515.

[2] D. Kececioglu. Reliability Engineering Handbook [M]. New Jersey: PTR Prentice Hall, 1991, 293-303.

[3] E. A. Elsayed. Reliability Engineering [M]. New Jersey: Prentice Hall, 1996.

[4] R. Jiang. A General Proportional Model and Modelling Procedure [J]. Quality and Reliability Engineering International, 2012, 28 (6): 634-647.

[5] R. D. Gupta, D. Kundu. Generalized exponential distribution: different method of estimations [J]. Journal of Statistical Computation and Simulation, 2001, 69(4):315-317..

[6] J. K. Chan, L. Shaw. Modeling repairable systems with failure rates that depend on age and maintenance [J]. IEEE Transactions on Reliability, 1993, 42(4): 566-571.

[7] A. Biswas, J. Sarkar. Availability of a system maintained through several imperfect repairs before a replacement or a perfect repair [J]. Statistics & Probability Letters, 2000, 50(2): 105-114.

A Loose Coupling Method on the Twist Angle Optimization of Jig Shape Wing

Yang Zhao[1, a] **and Pan Cheng**[1, b]

[1] Shanghai Airplane Design and Research Institute, COMAC. 5188 Jinke Rd, 201210, Shanghai, PRC

Corresponding author: [a]zhaoyang2@comac.cc; [b]chengpan@comac.cc

Abstract. When we design the 3D wing of an airplane, a cruise-shape wing will be designed first, which has the most ideal aerodynamic pressure distribution. In the real flight the wing will be deformed by the aeroelasitcity, which changes the pressure distribution. The twist angle will be optimized to redistribute the pressure on the deformed wing along the spanwise, and thus the pressure distribution can be similar to the ideal cruise shape. A loose coupling method is developed to optimize the twist angle of the jig shape wing, which can be divided into two phase and is able to lower the time and computation resource compared to the tight coupling effectively.

1. Introduction

Aeroelasiticity problem is quite complicated to solve in aerospace engineering as CFD (computational fluid dynamics) is coupled with FEM (finite element method) and thus a robust computation cycle need to be designed to transfer the information between the structure and the aerodynamic computation, but coupled numerical method has the potential to be the powerful tool in the jig-shape optimization algorithm [1].

Jig-shape optimization is one popular application of coupled aerodynamic and structural optimization. First, the cruise shape configuration of an aeroplane is designed for the ideal aerodynamic performance. Aerodynamic loads computed using jig shape configuration, however, can cause structural deformation due to the flexibility of the structure. These structural deformations then change the aircraft configuration for the aerodynamic load computation, affecting the structural deformation once again, and finally lowering the aerodynamic performance [2]. The target of the jig shape optimization is to compensate the aeroelastic effect and to approach the designed cruise performance around the cruise condition [3]. The optimization procedure is coupled in a tightly manner, which means the coupled aerodynamic and structural problems are solved simultaneously [4] [5]. For the aerodynamic solver in the coupling procedure, the RANS model is necessary as the Euler solver cannot predict the flow separation, which is time and resource consuming for the convergence of the aerodynamic solution and the coupling method [1]. Timothée Achard, Christophe Blondeau, Roger Ohayon put forward an uncoupled idea for the optimization [6], which reduce the time and the computation resource cost in the optimization.

An alternative approach except the method above is to develop a loose coupling optimization method. The method splits the optimization into two phases, and in the first phases the iteration is based on the FEM solution and in Phase II the iteration is based on the CFD solution. The coupled method connect the two phases. The iteration history and the Cp distribution on the wing prove it is an effective and efficient method.

2. Model Description

For the aeroplane design, we have two different models with a same configuration: cruise-shape model and original jig-shape model. The cruise-shape is the ideal shape when the airplane flies in the cruise condition, the jig-shape is the airplane manufactured from the assembling line. The original jig-shape is same with the cruise-shape, and the twist angle alone the spanwise will be adjusted for a better aerodynamic performance to offset the aeroleastic effect.

2.1. Configuration for the cruise-shape and the original jig-shape

An aeroplane configuration has been designed, which is a left-half model and includes body, wing, horizontal tail, vertical tail, pylon and nacelle, and this is the cruise-shape configuration and original jig-shape configuration, which is shown in Figure 1.

Figure 1. Configuration for the Cruise-Shape and the Original Jig-Shape

2.2. Aerodynamic Model

Discretizing the half space with the half model of the cruise shape/original jig-shape aeroplane, a structured mesh is set up for Navier-Stocks equations with 22 million mesh cell. In the CFD computation, k-omega SST is selected as the turbulence model. The mesh is shown in Figure 2.

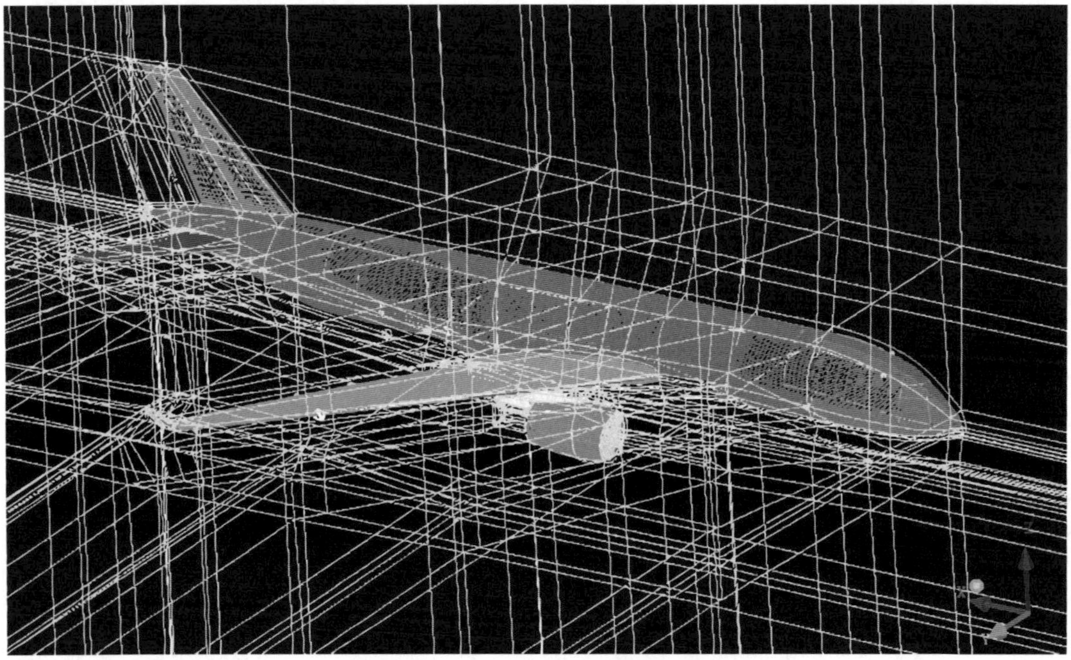

Figure 2. Topology of the CFD Mesh

2.3. Structure Model

The FEM mesh is set up independently, and the grid point is not one on one corresponded with the aerodynamic surface mesh, so the load transfer between two different meshes is necessary.

The four edges of the wing box are beams which definite the deflection of the wing. The upside front beam is the axis of the twist angle of the wing. In this optimization, only twist angle will finally be updated but not the deflection of upside front beam because the factory will manufacture the same beam as the cruise shape.

The surface mesh of the aerodynamic surface will be partitioned into a lot of subsurface which involves only one FEM surface point. The load on the FEM point will be integrated from the distributed pressure on the aerodynamic subsurface. Considering the thrust of the engine, a point force is added. The FEM model is shown in Figure 3.

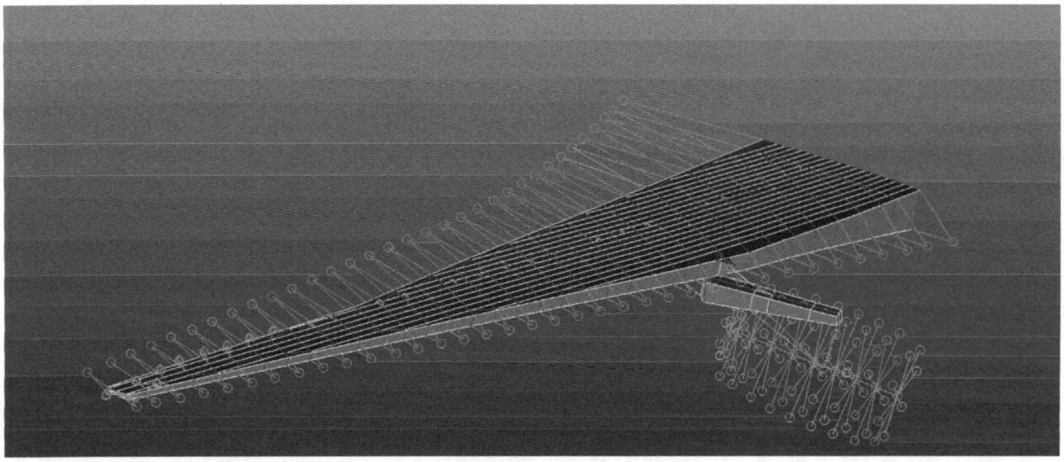

Figure 3. FEM Mesh

3. Methodology: Two Phases of the Inverse Optimization Procedure

The optimization procedure can be divided into two phases and four more configurations are generated except the cruise shape and the original jig shape. The new configurations are the preliminary jig shape, the elastic cruise shape and the optimized elastic shape and the optimized jig shape.

3.1. Phase I: Inverse Loading

The target of Phase I will generate a jig shape which is optimized for the cruise shape. The preliminary jig phase is generated and it is the starting point of the optimization in phase II.

The procedure of Phase I is straight forward and is as follows:

1. Calculate the aerodynamic fluid field of the cruise shape in the cruise condition in a certain Mach number and altitude.
2. Extract the surface pressure on the aeroplane of the cruise shape.
3. Integrate the aerodynamic load of the surface pressure of the jig shape and transfer to the surface node of the FEM mesh.
4. Record the geometry information of the upside front beam from the FEM mesh.
5. Along the spanwise of the wing, cut the wing with some planes vertical to the front beam and obtain the airfoil slices of the wing in different spanwise stands.
6. Calculate the twist angles of the different slices.
7. Iterative process on generating the preliminary jig shape:
   a) Add the aerodynamic load on the FEM nodes.
   b) Calculate the displacement of the FEM nodes compared with the initial coordinates of the node.
   c) Reverse the vector of the displacement.
   d) Add the reversed displacement on the coordinates of all the FEM nodes and update the coordinates of the node.
   e) Go to the step a) or stop the loop if the change of displacement is small enough.
8. Calculate the displacement on each FEM node between the final status and the initial coordinates.
9. Update the upside front beam information temporarily.
10. Cut the wing with the same ratio stand along the wing, vertical to the temporary upside front beam of the deformed cruise shape and calculate the twist angles in the new slices.
11. Calculate the difference of the twist angle in the corresponding spanwise stands and add the difference to the cruise shape configuration.

After the iteration process, the new configuration is the preliminary jig shape. Notice the deflection of the preliminary jig shape is same with the cruise jig shape because in Phase I only the twist angle but not the deflection of the wing is updated.

3.2. Phase II: Direct Optimization of Twist Angle

This phase starts from the preliminary jig shape and, finally achieves the optimized jig shape. The optimization process is time and resource consuming in this phase because of the convergence of the CFD computation. The target of the optimization is minimizing the sum of the variance of the Cp distribution between the cruise shape and the shape we are optimizing. The variables are the twist angle on the different stands of the wing. The only limit of the variables is the twist angles are monotone decreasing from the root to the tip of the wing.

The following steps are operated in Phase II:

1. Cut the cruise shape wing with planes vertical to the upside front beam and obtain the airfoil slices of the wing in different spanwise stands, and record the Cp distribution on each airfoil .
2. Generate the CFD mesh of the preliminary jig shape configuration.
3. Start a fluid/structure coupling computation of the preliminary jig shape and obtain the elastic cruise shape. The elastic cruise shape is the deformed preliminary jig shape as the result of the aerodynamic load, which is the shape in the flight.
4. Update the upside front beam information of the elastic cruise shape.

5. Cut the elastic cruise shape wing with planes vertical to the upside front beam and obtain the airfoil slices of the wing in different spanwise stands, and record the Cp distribution on each airfoil .
6. Iterative process on optimizing the elastic cruise shape
   a) alculate the variance of the Cp distribution of these two shapes.
   b) Adjust the twist angle of the elastic cruise shape to minimize the variance.
   c) Use moving-mesh method to generate new CFD mesh
   d) Calculate the CFD result of the temporarily elastic cruise shape.
   e) Obtain Cp distribution on each airfoil.
   f) Go to step a) or stop the loop if the variance can be lowed any more.
7. Calculate the twist angle difference between final elastic cruise shape and the initial elastic cruise shape and add it to the preliminary jig shape.

The new shape is the optimized jig shape.

4. Implementation

In the case following, we apply the model in the methodology above. In the cruise condition of flight, Mach number equals to 0.85 and the altitude equals to 35000ft. The Cp of the aeroplane is fixed at 0.48 in the entire optimization procedure, and the fluid solver will look for the angle of attack of the aeroplane in the aerodynamic computation. Gradient Decent is selected in the optimization procedure.

4.1. Phase I

In the Figure 4, the plot shows the difference of the twist angle between the cruise shape (red) and the preliminary jig shape (green).

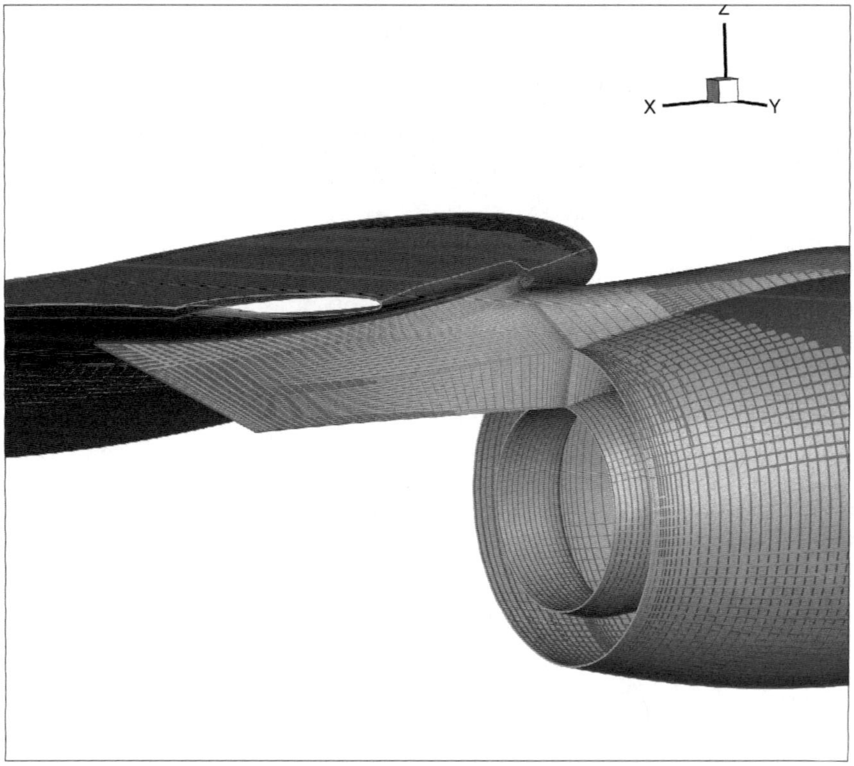

Figure 4. The Cruise Shape and the Preliminary Jig Shape

4.2. Phase II

Phase II starts from the preliminary jig shape with the fluid structure coupled computation in the cruise condition. The we select 27 different slices to calculate the variance of the Cp distribution between the cruise shape and the optimizing elastic cruise shape. To reduce the number of the variables, we define two more limits: the twist angle of the inner wing will not change, which is also corresponding to the physics of the aeroelasticity of the aeroplane; we do not change the twist angle directly, but we select three stands on the wing and set their increments, use the nurbs spline to interpolate all the increment to all the twist angle of the outer wing to generate a new twist angle distribution. To balance the efficiency in this problem, we chose three twist angle stands to optimize the entire Cp distribution along the wing span, so the Cp of the outer wing of the elastic cruise shape will not be totally same with the cruise shape.

Figure 5 shows that shape (green) comparison to the preliminary jig shape (blue).

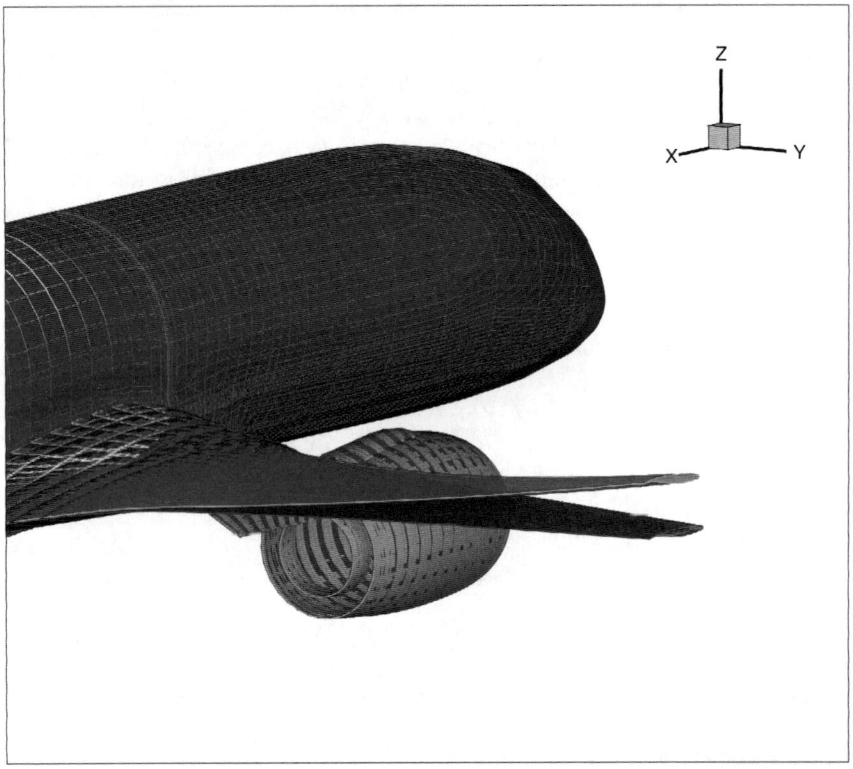

Figure 5. The Elastic Cruise Shape and the Preliminary Jig Shape

Figure 6 is the variance history in optimization iteration of the twist angle.

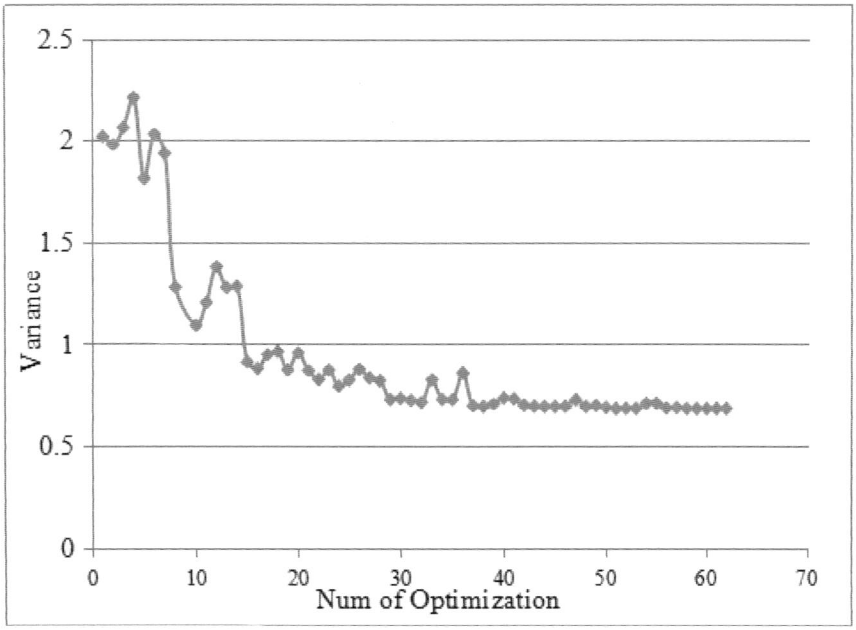

Figure 6. The Variance History

Some typical slices are chosen and number series of the stand position increases from the root to the tip of the wing. The blue line “std_cp” is the Cp of the cruise shape, the red line “test_cp” is the Cp of the optimized elastic cruise shape and the green line “start_cp” is the Cp of the elastic cruise shape.From the comparison, it is obviously that the optimization is effective, which is shown in Figure 7.

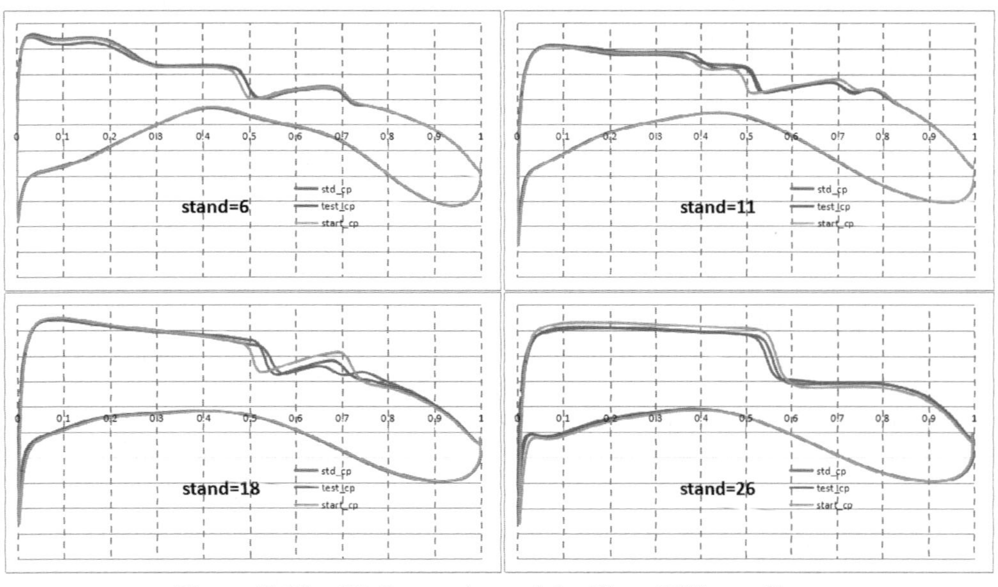

Figure 7. The CP Comparison of the Three Different Shape

Figure 8 shows the change of the twist angle in the optimization.

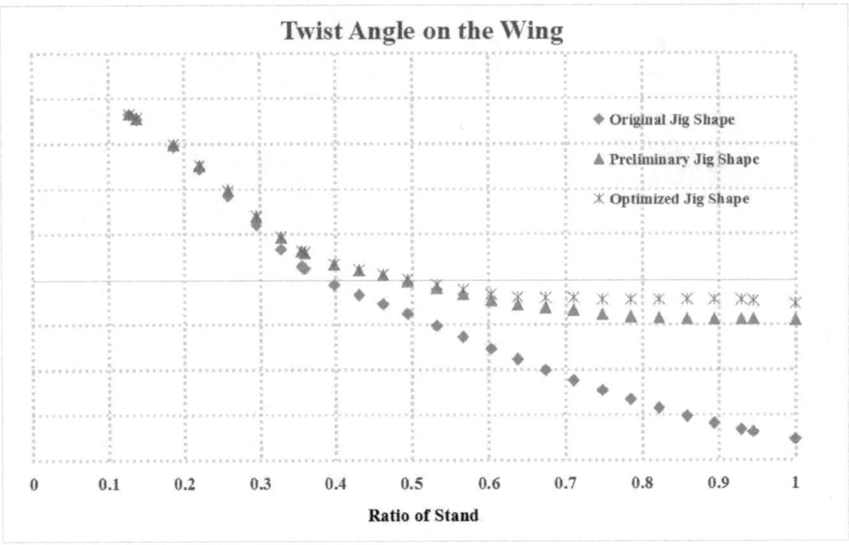

Figure 8. Twist Angle in the Optimization from Phase I to Phase II

5. Conclusion

To recover the performance of the wing from the jig shape which is manufactured in the factory and is also deformed because of the aerodynamic load in the flight, to the cruise shape which is the ideal shape when the aeroplane is designed, the optimization on the twist angle of the wing is applied with two phases: the displacement iteration and the optimization algorithm. In this paper, we design an aeroplane CFD model and FEM model, develop a loose coupling method and test the methodology in the cruise condition, which is effective and valuable in industry.

In most of the case, the aeroplane flies in the cruise condition, but for the robust design, in the future research, other flight condition will be considered for this optimization, for example, non-cruise condition as Mach number equals 0.7 and the altitude equals 15000 feet, which is robust in multi flight condition, or on the boundary of the flight envelop as Mach number equals 0.95 and the altitude equals 30000 feet, which is the designed for the safety. For these cases, multi objective optimization method is necessary as NSGAII and we need evaluate the weight of different flight condition in the optimization. Thus the method developed above is a good step stone for the future research in the aeroelasticity design of the aeroplane.

References

[1] Kenway GKW, Kennedy GJ, and Martins JRRA 2014 Aerostructural Optimization of the Common Research Model Configuration *15th AIAA/ISSMO Multidisciplinary Analysis and Optimization Conference, AIAA AVIATION Forum* **2014** 3274

[2] Pak C 2018 Jig-Shape Optimization of a Low-Boom Supersonic Aircraft *AIAA/ASCE/AHS/ASC Structures, Structural Dynamics, and Materials Conference, AIAA SciTech Forum, AIAA* **2018** 2154

[3] Vázquez M, Dervieux A, and Koobus B 2006 A Methodology for the Shape Optimization of Flexive Wings *Engineering Computations* **23** 4

[4] Shubin GR 1995 Application of Alternative Multidisciplinary Optimization Formulation to a Model Problem for Static Aeroelasticity *Journal of Computational Physics* **118** 1

[5] ARIAN E 1997 On the Coupling of Aerodynamic and Structural Design On the Coupling of Aerodynamic and Structural Design *Journal of Computational Physics* **135** 1

[6] Achard T., Blondeau C. and Ohayon R. 2016 An Uncoupled Approach to Compute Aero-Structure Gradients Using High-Fidelity CFD-CSM *17th AIAA/ISSMO Multidisciplinary Analysis and Optimization Conference, AIAA AVIATION Forum AIAA* **2016** 4121

Research on the Dynamic Characteristics of the High Air Speed Nozzle in Isentropic Flow

Dong Xiang, Yingting Wang and Chenjing Li

School of Mechatronic Engineering, Harbin Institute of Technology, Harbin, Heilongjiang, 230103, China

Email: xiangdong@hit.edu.cn;

Abstract: The application of microsatellites has been development with the continuous expansion of space missions in resents years. The microsatellites have several characteristics such as light weight, small size and low cost. The platform flotation table is one of the simulation ways on the ground. With the continuous development of space technology, satellite ground simulation platform flotation table as a key technology which attracts lots of researchers. In this paper, the highspeed nozzle has been analyzed based on the research of Laval nozzle. The state of shock waves can be researched by the Laval nozzle and the air speed from the nozzle can be calculated. The flow field of the high speed nozzle has been simulation by the FLUENT software. The nozzle mode has been built by the GAMBIT and the model is meshed in structured grid. The simulation result shows that the air in the nozzle has many speeds in different place. The stable speeds are 390 m/s and 150m/s in the larynx and trailing end position.The maximum are reached into the nozzle and the speed is 578m/s, which is deviated from the theoretical analysis. At the constant pressure of 4 bar, the mass of nozzle flow is 0.000318758 kg/s. The mass of outlet flow is 0.000318798 kg/s, and the calculation error is 10^{-8}. The thrust of the nozzle is also stable at 3ms, and its stability value is 50.1 mN. The maximum force of the outlet is 176 mN before the nozzle reaches stability. The simulation result can guided for the nozzle design and itis more significance for the high precision control and measurement of nozzles.

1. Introduction

The application of microsatellites has been development with the continuous expansion of space missions in resents years[1-4]. The microsatellites have several characteristics such as light weight, small size and low cost[5-6]. The platform flotation table is one of the simulation ways on the ground. With the continuous development of space technology, satellite ground simulation platform flotation table as a key technology which attracts lots of researchers[7-8]. The nozzle is one of the most parts of the platform. At present, the design and research of supersonic nozzle mainly depends on simulation optimization and experiment. The internal flow field of the nozzle should be simulated by the finite element analysis software. The study of supersonic nozzles focuses on the static characteristics. The pressure and velocity can be influenced by the optimizing structure. Some results shows that the flow field changes slightly in the tapering section and the change is much smaller[9]. The most drastic changing part is concentrated near the throat. The expanding section changing is relatively gentle, which generally satisfies the process of decompression, acceleration and air cooling. The wall friction has a great influence on the velocity and temperature. The outlet velocity of the nozzle has a great relationship of the outlet diameter. With the increase of the outlet diameter, the outlet velocity increases obviously. This part of results can guild to the design the nozzle in this work.

It is very important in the microsatellite that research on low thrust measurement system of Laval nozzle. In 2016 year, researcher Lu used FLUENT calculation software to study the low thrust measurement system of Laval nozzle. It is analyzed on the influence of the baffle on the nozzle[10]. Three kinds of Laval nozzles with outlet diameters of 5.1 mm, 3.6 mm and 2.3 mm were numerically calculated, and their thrust values were 10N. The results show that changing pressure ratios of nozzle can influence the outlet pressure of nozzle when the baffle fixed. When the total inlet pressure is constant, each type of nozzle has a suitable baffle location area. There are many researchers has study on the nozzle. However, the research on supersonic nozzle is mostly focused on the static state of the nozzle. The finite element analysis and simulation software such as Fluent is used to analyze the changes of pressure, specific impulse and velocity of supersonic nozzle under different conditions, so as to get the different sizes of Laval nozzle under specific conditions. However, little research has been focus on the initial period of the nozzle impulse. The main problem is that the force produced by the micro nozzle is too small and the response time is too short to be measured and studied.

Nowadays, the most widely measurement technology of micro force is the micro-electronic systems (MEMS). It has been already development the micro sized nozzles with μN thrust force[11-14]. Due to the force of the micro high speed nozzle is too small, it can be stabilized from initial period in a very short time. At present, there are mainly two kinds of test methods for micro high speed nozzle which can be call the static thrust and impulse measurement. Impulse measurement is an indirect method to measure the impulse size. This may not be accurate enough for small satellites and air flotation platforms with stricter force requirements. When the time scale is large, the dynamic characteristics of the nozzle can be neglected. If the time scale reaches ms level, it is difficult to evaluate the accuracy of the experimental results because there has less theoretical calculation. Therefore, it is very important to study the dynamic characteristics of the nozzle, especially the change rules of the nozzle force before the flow field reaches stability.

In this paper, the high speed nozzle has been analyzed based on the research of Laval nozzle. The state of shock waves can be researched by the Laval nozzle and the air speed from the nozzle can be calculated. The flow field of the high speed nozzle has been simulation by the FLUENT software. The nozzle mode has been built by the GAMBIT and the model is meshed in structured grid. The simulation result shows that the air in the nozzle has many speeds in different place. The stable speeds are 390 m/s and 150m/s in the larynx and trailing end position. The maximum are reached into the nozzle and the speed is 578m/s, which is deviated from the theoretical analysis. At the constant pressure of 4 bar, the mass of nozzle flow is 0.000318758 kg/s. The mass of outlet flow is 0.000318798 kg/s, and the calculation error is 10^{-8}. The thrust of the nozzle is also stable at 3ms, and its stability value is 50.1 mN. The maximum force of the outlet is 176 mN before the nozzle reaches stability. The simulation result can guided for the nozzle design and it is more significance for the high precision control and measurement of nozzles.

2. The Engineering Theoretical Calculation of Laval Nozzle

The steady state theoretical calculation of the nozzle is based on the pressure, density, temperature and velocity at one interface. The parameters of different sections can be calculated. The basic equations are as follows:

$$\begin{cases} \frac{d\rho}{\rho}+\frac{dA}{A}+\frac{dV}{V}=0 \\ \frac{dp}{p}+\gamma Ma^2\frac{dV}{V}=0 \\ \frac{dT}{T}+(\gamma-1)Ma^2\frac{dV}{V}=0 \\ \frac{dp}{p}-\frac{d\rho}{\rho}-\frac{dT}{T}=0 \end{cases} \tag{1}$$

The basic equations of pipe flow with variable cross section can be obtained by using the flow characteristics.

$$\left.\begin{aligned} \frac{dV}{V}&=-\frac{1}{1-M_a^2}\frac{dA}{A} \\ \frac{dM_a}{M_a}&=-\frac{2+(\gamma-1)M_a^2}{2(1-M_a^2)}\frac{dA}{A} \\ \frac{dp}{p}&=\frac{\gamma M_a^2}{1-M_a^2}\frac{dA}{A} \\ \frac{d\rho}{\rho}&=\frac{M_a^2}{1-M_a^2}\frac{dA}{A} \\ \frac{dT}{T}&=\frac{(\gamma-1)M_a^2}{1-M_a^2}\frac{dA}{A} \end{aligned}\right\} \tag{2}$$

It can be concluded the some parameters relationship on the cross section area such as velocity, Mach number, density and temperature. For supersonic flow, the speed increases when the area increase, and the pressure density and temperature are all getting lower. For subsonic flow, the speed decreases when the area increase, and the pressure density and temperature are all getting much higher. The energy equation is

$$h+\frac{1}{2}V^2=\frac{a_0^2}{\gamma-1} \tag{3}$$

The isentropic equation is

$$\frac{p}{p_0}=\frac{\rho^\gamma}{\rho_0^\gamma}=(\frac{T}{T_0})^{\frac{\gamma}{\gamma-1}}=(1+\frac{\gamma-1}{2}M_a^2)^{-\frac{\gamma}{\gamma-1}} \tag{4}$$

The outlet velocity can be calculated through the equation (3) and (4).

$$V_e=\sqrt{2\frac{\gamma R}{\gamma-1}T_0\left[1-(\frac{p_e}{p_0})^{\frac{\gamma-1}{\gamma}}\right]} \tag{5}$$

$$M_e=\frac{V_e}{a_e}=\sqrt{\frac{2}{\gamma-1}\left[(\frac{p_0}{p_e})^{\frac{\gamma-1}{\gamma}}-1\right]} \tag{6}$$

The mass flow in the isentropic condition is

$$\dot{m}=\rho_e V_e A_e=A_e\rho_0 a_0\sqrt{\frac{2}{\gamma-1}}\left[(\frac{p_e}{p_0})^{\frac{2}{\gamma}}-(\frac{p_e}{p_0})^{\frac{\gamma+1}{\gamma}}\right]^{\frac{1}{2}} \tag{7}$$

Critical section is defined as A_*. If this section appeared, the mass flow rate of the nozzle reaches the maximum value, and the equation as follows:

$$\dot{m}_{\max}=\rho_* A_* V_*=(\frac{\gamma}{R})^{\frac{1}{2}}(\frac{2}{\gamma+1})^{\frac{\gamma+1}{2(\gamma-1)}}\frac{p_0}{\sqrt{T_0}}A_* \tag{8}$$

At this time, the maximum throat flow is

$$\dot{m}_{max} = \rho_* A_* V_* = (\frac{\gamma}{R})^{\frac{1}{2}} (\frac{2}{\gamma+1})^{\frac{\gamma+1}{2(\gamma-1)}} \frac{p_0}{\sqrt{T_0}} A_* = 0.0003598 kg/s \quad (9)$$

It can be calculated the speed is

$$a_e = \sqrt{\gamma RT} = \sqrt{1.4 * 286 * 107.1} = 207.08 m/s \quad (10)$$

$$V_e = M_e a_e = 3 * 207.08 = 621.24 m/s \quad (11)$$

3. The Simulation of the High Air Speed Nozzle

The Gambit software is used to build the nozzle model. Then the actual flow field of the simulated nozzle is selected in FLUENT software. The cylindrical injection area was added to the tail of the nozzle in order to simulate the actual working conditions. The monitoring surfaces were set up to detect the internal flow field. The model is meshed. Figure 1 is the simulation model of the high air speed nozzle.

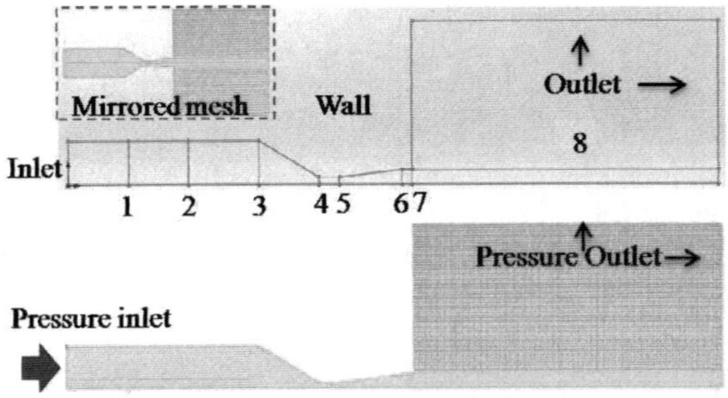

Figure 1 the simulation model of the high air speed nozzle.

There are 7 monitoring surfaces on the model. The surfaces 3, 5 and 7 are retract, larynx and trailing end position. The flow type of the inlet and outlet are the pressure. The mesh has been checked and there is no negative volume and area. The pressure basedmode has been select. The viscous model is k-epsilon. The inlet condition has been set up as 4 bar. The outlet is 1bar and the air has been defined as ideal gas PISO arithmetic has been selected.

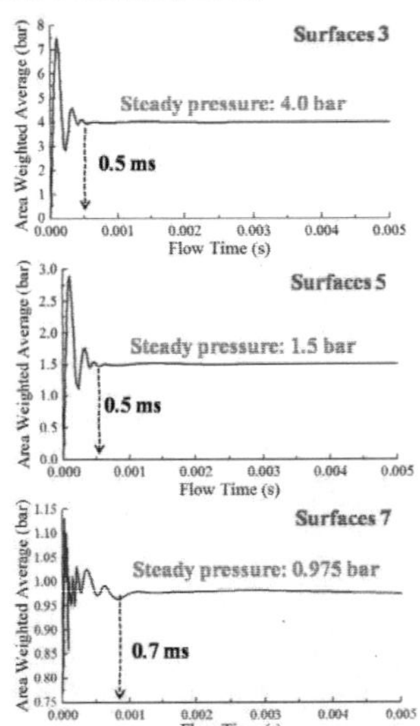

Figure 2 pressure-time curve of the Laval nozzle

The surfaces 3, 5 and 7 have been selected to present the different flow characteristic. Figure 2 shows the pressure-time curve of the Laval nozzle. The retract position has much higher pressure. The steady pressure is 4.0 bar from 0.5ms. When the air flow across the larynx position the pressure are getting down quickly as shown in surfaces 5. The steady pressure is 1.5 bar. After this time the air flow in the trailing end, the pressure steady at 0.975 bar. With the pressure changing, the velocity changes obviously. Figure 3 shows the velocity-time curve of the Laval nozzle. In surfaces 3, the velocity is stable at 0.5 ms and the speed is 5 m/s. The velocity is fluctuated considerably before reaching stability. In surfaces 5, the velocity is stable at 2.5 ms and the steady speed is 390 m/s. In surfaces 7, the velocity reaches stability at 3ms and the steady speed is 150 m/s. Pressure and velocity is fluctuated considerably before reaching stability.

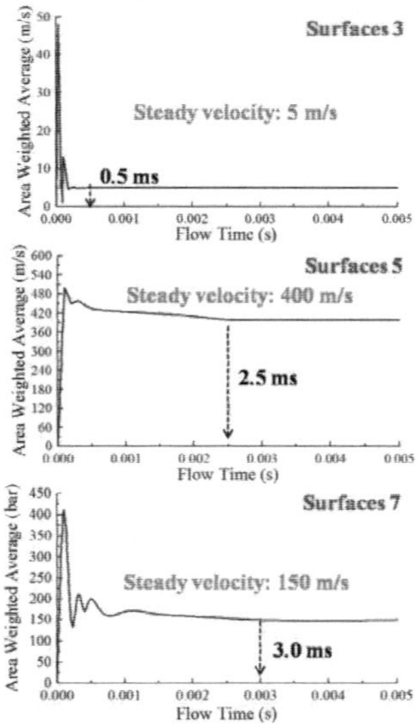

Figure 3 the velocity-time curve of the Laval nozzle

Through the FLUENT software, the dynamic velocity and pressure nephograms are also made in Figure 4 in different times. The changing pressure and velocity of nozzle can be seen intuitively from whole stages. The time 1 ms and 3 ms are selected in the Figure. Velocity does not maximize at the outlet. The maximumare reached into the nozzle and the speed is 578m/s, which is deviated from the theoretical analysis.

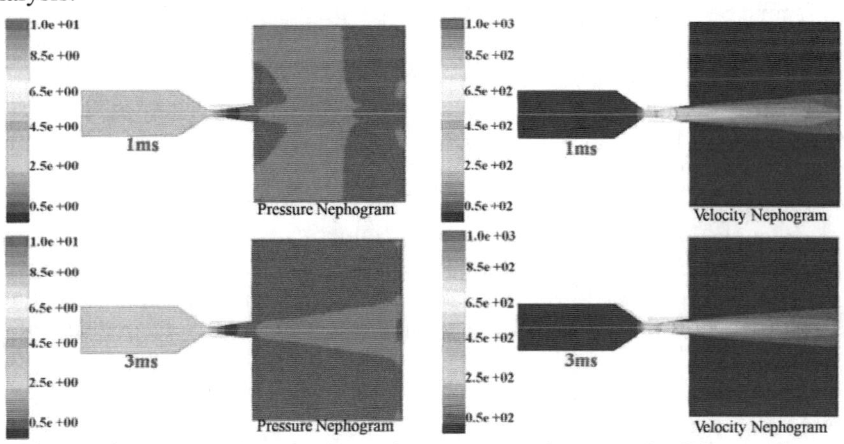

Figure 4 the dynamic velocity and pressure nephograms in different time.

The reason for the error may be that the engineering calculation considers the ideal isentropic flow in the nozzle, and there is deviation between the setting and the actual flow in the simulation. The mass of nozzle flow is calculated. At the constant pressure of 4 bar, the mass of nozzle flow is 0.000318758 kg/s. The mass of outlet flow is 0.000318798 kg/s, and the calculation error is 10^{-8}. It is basically in agreement with the theoretical calculation. The thrust of the nozzle is also stable at 3ms, and its stability value is 50.1 mN. The maximum force of the outlet is 176 mN before the nozzle reaches stability.

4. Conclusion

In this paper, the high speed nozzle has been analyzed based on the research of Laval nozzle. The state of shock waves can be researched by the Laval nozzle and the air speed from the nozzle can be calculated. The simulation result can guide for the design. After the acceleration, the air speed can reach the isentropic flow. The flow field of the high speed nozzle has been simulation by the FLUENT software. The nozzle mode has been built by the GAMBIT and the model is meshed in structured grid. The surfaces 3, 5 and 7 are retract, larynx and trailing end position. The flow type of the inlet and outlet are the pressure. The mesh has been checked and there is no negative volume and area. The simulation result shows that the air in the nozzle has many speeds in different place. The stable speeds are 390 m/s and 150m/s in the larynx and trailing end position. The maximum are reached into the nozzle and the speed is 578m/s, which is deviated from the theoretical analysis. At the constant pressure of 4 bar, the mass of nozzle flow is 0.000318758 kg/s. The mass of outlet flow is 0.000318798 kg/s, and the calculation error is 10^{-8}. The thrust of the nozzle is also stable at 3ms, and its stability value is 50.1 mN. The maximum force of the outlet is 176 mN before the nozzle reaches stability. The simulation result can guided for the nozzle design and it is more significance for the high precision control and measurement of nozzles.

References

[1] Lee J K and Kwon S J.Mixing efficiency of amultil amination micro mixer with consecutive recirculation zones[C] Chemical Engineering Science 64 (2009) 1223-1231.

[2] Lee J K, Kim K and Kwon S J.Design, fabrication, and testing of MEMS solid propellant thruster array chip on glass wafer[C]. Sensors and Actuators A 157 (2010) 126-134.

[3] Janson S W and Helvajian H, MEMS Microengineering and Aerospace Systems[C] Aiaa Paper, 1999

[4] Larangot B, Conédéra V, Dubreuil P, Conto T D and RossiC,Solid. Propellant Micro Thruster : an alternative propulsion device for nanosatellite[J]. Journal of Propulsion & Power, 2002, 22(1):56-63.

[5] NagaiR H, NaraokaK, Sawadaand KAsai. Pressure-Sensitive Paint Measurement of Pressure Distributionin a Supersonic Micronozzle[J]. AIAAJourna. 2008, 46: 1.

[6] TanakaaS,Kondoa K and Habub H. Test of B/T multilayer reactive igniters for a micro solid rocket array thruster[J]. Sensors and Actuators A,2008, 144: 361–366.

[7] Daniel W. Son Y. Lu T. MEMS Mega-pixel Micro-thruster Arrays for Small Satellite Stationkeeping[J]. MEMS Mega-Pixel Micro-Thruster Array, 2000.

[8] Gamero M and Torsional A. Balance that Resolves Sub-micro-Newton Forces[J]. Electric Rocket Propulsion Society,2001(01).

[9] Gao Q J, Tang H J, Wang Zhao H, Yong H E. Numerical simulation and structure optimization of supersonic nozzle based on Fluent[J]. Manufacturing Automation,2015,(04):88-90.

[10] Lu Y H, Wei W Q, Shao F X and Sun Y X. Numerical calculation for the small thrust measurement system of Laval nozzle[J]. Journal of Mechanical & Electrical Engineering,2016, (10):1188-1192.

[11] Eccardt P, lofink A, Garssen H. Analysis of crosstalk between fluid coupled CMUT membranes[J]. Ultrasonics Symposium, 2005: 593-596.

[12] Na S, Chen A I H and Wong L P P. Capacitive micromachined ultrasonic transducers based on annular cell geometry for aircoupled applications[J]. Journal of Ultrasonics, 2016, 71: 152–160.

[13] Merkowitz S M, Maghami P G and Sharma A. A Newton thrust-stand for LISA[J]. Classical & Quantum Gravity, 2002, 19(7):1745-1750.

[14] Pierre Pennarun, Carole Rossi, Daniel Estève, VéroniqueConédéra,Development of MEMS based safe electro-thermal pyrotechnic igniter for a new generation of microfuze[J] Microtechnologies for the New Millennium, 2005, 5836

Coupled Inductance Three-Level Cuk Converter

Xin-ping DING [1], Lei YU[1], Dai-ling YU and Hong-xing Zhang[1]

[1]College of Automation Engineering, Qingdao University of Technology, Qingdao, 266520, China

792999234@qq.com

Abstract. A three-level Cuk converter with coupled inductance is proposed in this paper. Based on the research and design of existing three-level Cuk converter and Cuk converter with coupling inductance, this paper combines three-level technology with coupling inductance, which is called the "three-level coupled inductance Cuk converter" in this paper[1-3]. Compared with the existing converter, it has the advantages of boost capability and energy conversion efficiency. This kind of converter integrates the low voltage stress of the switch with three-level DC-DC converter and the boost function of the DC-DC converter with coupling inductance[4-6]. The three-level structure reduces the voltage stress when the switch is turned off, while the coupling inductance structure greatly improves the boost capability of the circuit. The working principle and periodic mode of three-level coupled inductor Cuk circuit are analysed, and the mathematical formula of voltage gain is deduced.

1. Introduction

At present, DC-DC converter is widely used in industrial production. For example, photovoltaic power generation, multi motor HVDC distribution system, UPS system. Apart from high voltage gain, it also requires high efficiency, low electromagnetic interference noise, high switching frequency, fewer components and so on.

The basic topology of Cuk converter can satisfy the requirement of fewer components, but the duty must be infinitely close to one hundred when we need high voltage gain, which makes the efficiency of the circuit dramatically reduced and the safety factor reduced, correspondingly. In the titled as "Research on high gain three-level zeta converter"[1] is proposed to increase voltage gain by introducing coupling winding. This can avoid the emergence of extreme duty ratio, while circuit output high-gain voltage. However, the voltage stress on the switch will rise sharply due to the release of leakage inductance energy, which will affect the the components's service life and topological efficiency. The three level Buck-Boost converter proposed in paper titled as "Research on a new quasi-Z source three-level Boost DC converter"[2], reduced the stress of the switch and capacitor by half by adding a switch and cap capacitor correspondingly, but it can not meet the requirement of high voltage gain.

On the basis of summarizing all the characteristics of the above topology, a three-level Cuk DC converter with coupling inductor is proposed for meeting the industrialization requirements of this kind of topology. For the converter in this paper, the high voltage gain is achieved by adding coupling windings, and a switch and capacitor is introduced symmetrically on the basis of the original circuit, which can reduce the stress of the switch and capacitor. Under the same conditions, the limit duty ratio can be avoided in this way, the switching frequency of the circuit can be increased, and the security of the topology can be guaranteed. By adding a diode, the leakage inductance energy of the topology is

IOP Conf. Series: Journal of Physics: Conf. Series **1284** (2019) 012012 doi:10.1088/1742-6596/1284/1/012012

transmitted to the output capacitor through the diode, thus further improving the efficiency of the circuit. On the basis of this new topology, a series of theoretical analyses are carried out through the detailed description of its work process, the analysis and derivation of voltage gain formula, the establishment simulation of the main circuit model. Finally, the new topology proposed in this paper is verified by the test prototype.

2. Working principle and steady state analysis

Figure 1 shows the circuit diagram of the leakage inductance and the parasitic capacitance of the switch. Compared with the basic circuit, the topology first adds a switching and capacitor by three-level transformation to the basic circuit and keeps symmetry, so that the voltage stress of the switch and capacitor can be reduced by half accordingly; then by introducing coupling windings, the limiting duty ratio can be avoided when the requirements of increasing voltage is needed. Finally, a diode is connected in parallel at the end of the coupling winding to reduce the voltage stress on the switch and the energy loss during charging when the leakage inductance discharge occurs, so as to improve the efficiency of the circuit.

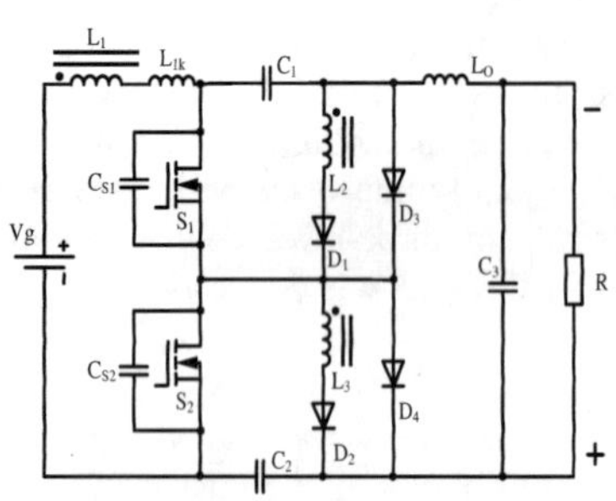

Figure 1. Circuit principle

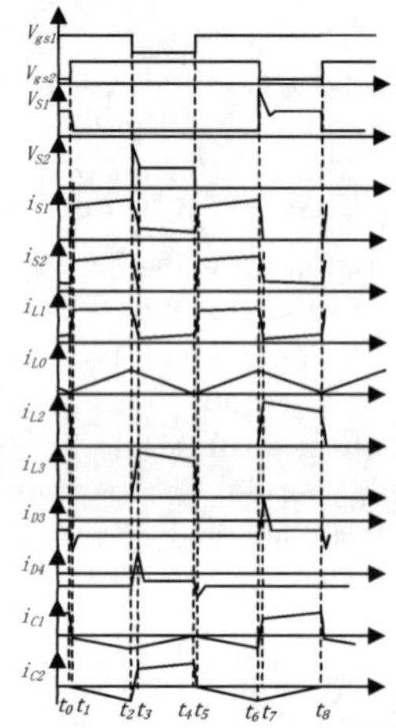

Figure 2. Key device voltage current waveform.

When considering the leakage inductance and parasitic capacitance of the switch, the work cycle can be divided into the following eight stages, in which stage 3 is divided into two small stages: A and B. Figure 2 is the voltage and current waveform of the key components when considering the leakage inductance of the circuit.

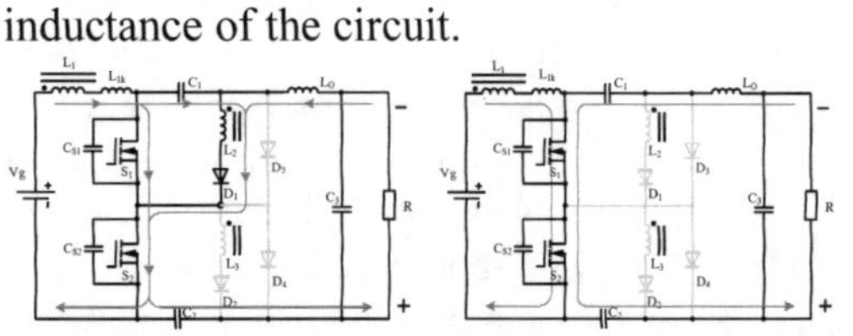

Figure 3. Work status of Phase 1 and Phase2

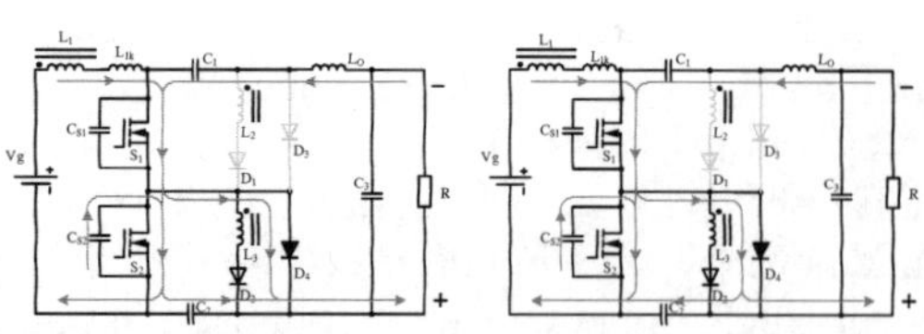

(a) (b)

Figure 4. Phase 3 working state

Stage 1[t_0~t_1]: as shown in Figure 3. Switch conducts. The excitation inductor L_1 is charged through the loop V_g - L_1 - L_{1k} - S_1 -S_2, while inductor L_2 is discharged and inductor L_O is charged. Capacitor C_1, C_2 is discharged slowly, parasitic capacitance C_{S1} is charged. The discharge of leakage inductance L_{2k} increases the reverse voltage of diode D_3 at this stage, which leads to the increase of leakage current.

Stage 2[t_1~t_2]: as shown in Figure 3. Switch S_1 and S_2 continue to turn on. The leakage inductance ends the discharge, and the excitation inductance L_1 continues to charge through the original current

IOP Conf. Series: Journal of Physics: Conf. Series **1284** (2019) 012012 doi:10.1088/1742-6596/1284/1/012012

loop, and the inductance L_0 continues to charge. This stage is the main energy storage stage of the whole cycle. C_1 and C_2 end the process of slow discharge in the Stage1. The discharge speed increases. The released energy is used for the inductor L_0 to storage energy, which is prepared for the discharge of voltage rise later.

Stage 3[t_2~t_3]: as shown in Figure 4. Switch tube and all conduction. In phase 3, inductors L_0 and L_1 discharge, inductance L_3 charges. The C_{S2} which is the parasitic capacitance of switch S_2 is discharged. As a result of the discharge of leakage inductance L_{1k}, the peak voltage of the switch S_2 is generated, which reduces the peak voltage of S_2 greatly. Thus diode D_4 is turned on. In stage 3, the capacitor C_2 is charged, and the current direction is reversed once. The states of the other devices remain unchanged. The state before the reverse of the i_{C2} is called state a, and after the reverse is called state B.

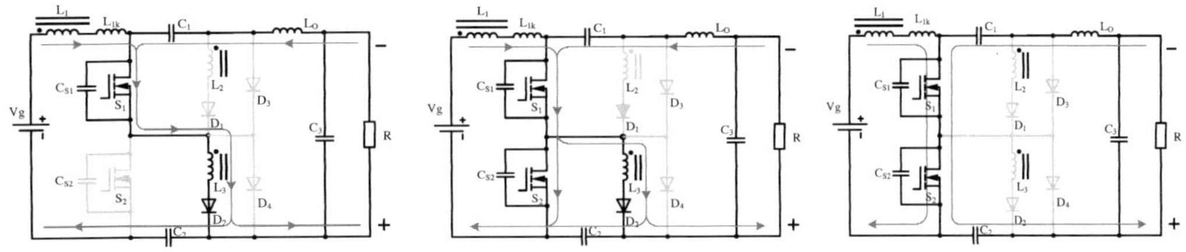

(phase 4) (phase 5) (phase 6)

Figure 5. Working state of phase 4、 phase 5、 phase 6

Stage 4[t_3~t_4]: as shown in Figure 5. The switch S_1 is turned on and the S_2 is turned off. The inductor L_1 is charged by input voltage V_g and inductor L_3, inductor L_0, L_3, capacitor C_1 are discharged. The load side will be charged by the original energy storage of the previous stage.

Stage 5[t_4~t_5]: as shown in Figure 5. Switch conducts. The excitation inductor L_1 is charged through the current loop V_g - L_1 - L_{1k} - S_1 -S_2, inductor L_3 is discharged, inductor L_0 is charged. Capacitor C_1, C_2 is discharged slowly, parasitic capacitance C_{S1} is charged. As the result of the leakage inductance L_{3k} discharge, the reverse voltage of diode D_4 increases at this stage, which leads to the increase of leakage current.

Stage 6[t_5~t_6]: as shown in Figure 5. Switch tube S_1 and S_2 continue to turn on. The leakage inductance ends the discharge, and the excitation inductance L_1 continues to be charged through the original current loop, and the inductance L_0 continues to be charged. This stage is the main energy storage stage of the whole circuit. Capacitors C_1 and C_2 end the process of slow discharge in the last stage. The discharge speed is increased. The released energy is used for the energy storage of inductor L_0, which is ready for the discharge of voltage rise later.

Stage 7[t_6~t_7]: as shown in Figure 6. Switch conducts. In stage 7, inductors L_0 and L_1 are discharged, inductance L_2 is charged. The C_{S1} whitch is the parasitic capacitance of switch S_1 is discharged. The leakage L_{1k} discharge causes the peak voltage of the switch S_1. The peak voltage generated by the discharge of leakage inductance L_{0k} is added to the diode D_3 to turn it on, which enables the capacitor C_2 to be charged faster and reduces the peak voltage borne by the switch S_1. In stage 7, the capacitor C_1 is charged and the current direction is reversed once. The states of the other devices remain unchanged. The state before the reverse of the i_{C1} is called state a, and after the reverse is called state B.

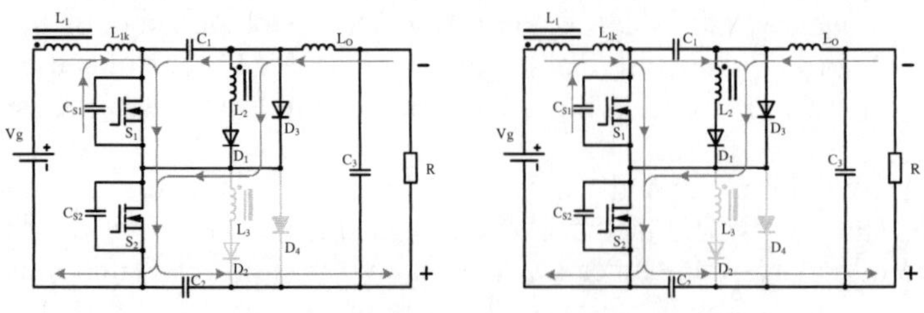

(a) **(b)**

Figure 6. Phase 7 working state

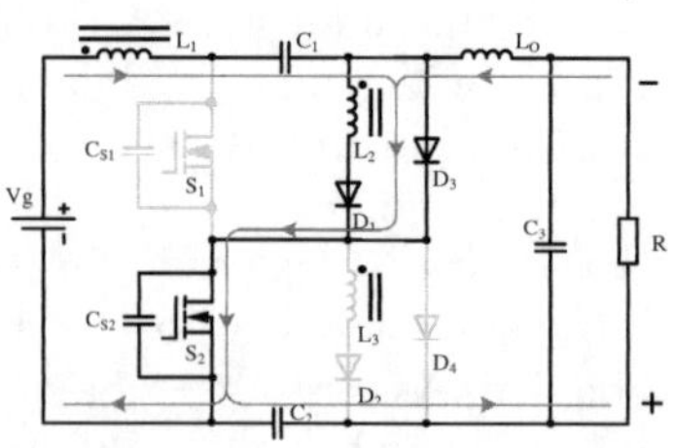

Figure 7. Phase 8 working state

Stage 8[t_7~t_8]: as shown in Figure 7. The switch S_2 is turned on and the S_1 is turned off. The inductor L_1 are charged by input voltage V_g and inductor L_2 are charged. Inductor L_0, L_2, capacitor C_1 are all discharged. At the previous stage, the original energy storage will discharge to the load side.

Without considering leakage inductance and some parasitic parameters, the static operating point of the system can be expressed as:

$$\begin{bmatrix} di_{L0}/dt \\ di_{L1}/dt \\ dv_{C1}/dt \\ dv_{C2}/dt \\ dv_{C3}/dt \end{bmatrix} = A \begin{bmatrix} i_{L0} \\ i_{L1} \\ v_{C1} \\ v_{C2} \\ v_{C3} \end{bmatrix} + B \begin{bmatrix} v_{D3} \\ v_{D4} \\ v_g \end{bmatrix} \tag{1}$$

In this paper, steady state analysis of four states is carried out: the two switches of S_1 and S_2 are conducted at the same time[$t_1 \sim t_2$], the switch S_1 is turned on, S_2 is turned off[$t_3 \sim t_4$], the two switches of S_1 and S_2 are conducted at the same time[$t_5 \sim t_6$], the switch S_1 turns off and the S_2 turns on[$t_7 \sim t_8$]. A_1, B_1, A_2, B_2, A_3, B_3, A_4 and B_4 can be obtained.

according to: $A = A_1\left(D-\frac{1}{2}\right)*2 + A_2(1-D) + A_4(1-D)$ $\quad B = B_1\left(D-\frac{1}{2}\right)*2 + B_2(1-D) + B_4(1-D)$

we can get A、B:

$$A = \begin{bmatrix} A_{11} & A_{12} & A_{13} & A_{14} & A_{15} \\ A_{21} & A_{22} & A_{23} & A_{24} & 0 \\ A_{31} & A_{32} & 0 & 0 & 0 \\ A_{41} & A_{42} & 0 & 0 & 0 \\ 0 & 0 & 0 & 0 & 0 \end{bmatrix} \tag{2}$$

$$B = \begin{bmatrix} \frac{1-D}{L_0(n+1)} & \frac{1-D}{L_0(n+1)} & \frac{2n(1-D)}{L_0(n+1)} \\ -\frac{1-D}{L_1(n+1)} & -\frac{1-D}{L_1(n+1)} & -\frac{2nD-n+1}{L_1(n+1)} \\ 0 & 0 & 0 \\ 0 & 0 & 0 \\ 0 & 0 & 0 \end{bmatrix} \tag{3}$$

After substitution, without considering leakage inductance, the voltage gain and circuit efficiency of the topologies obtained by several simultaneous formulas are as follows:

$$\begin{cases} B = \frac{V_o}{V_g} = \frac{n(1-2D) - D + (1-D)(V_{D3}+V_{D4})/V_g}{ar_{L0} - br_{L1} - c(r_{L2}+r_{L3}) - d(r_{S1}+r_{S2}) - (1-D)} \\ \eta = \frac{V_o i_o}{V_g i_{in}} = \frac{n(1-2D) - D + (1-D)(V_{D3}+V_{D4})/V_g}{ar_{L0} - br_{L1} - c(r_{L2}+r_{L3}) - d(r_{S1}+r_{S2}) - (1-D)} \cdot \frac{1-D}{D[(1+n)(2D-1)+2(1-D)]} \end{cases} \tag{4}$$

$$\begin{cases} a = \dfrac{1-D}{R} \\ b = \dfrac{D(2D-1)(D+n)(n+1)+2D^2(1-D)}{(1-D)R} \\ c = \dfrac{2D-1}{R} \\ d = \dfrac{(2D-1)[D(n+1)^2 - n(1-D)]}{(1-D)R} \end{cases} \tag{5}$$

By formula (4), the relationship between duty and voltage gain, duty and topological efficiency in different turn-to-turn ratios can be plotted, when taking parasitic parameters such as on-resistance of switch, parasitic resistance of coupling inductance and voltage drop of diode into account. Figure 8 shows the relationship between duty and voltage gain under different turns ratio. As can be seen from the figure, with the increase of duty, the voltage gain basically shows an upward trend, and the rising rate is faster and faster, but when exceeding a certain limit value, the voltage gain begins to decline rapidly; and under the same duty, the larger the turn ratio of the coupling winding is, the greater the voltage gain can be achieved. This shows that the addition of coupling windings can improve the voltage gain of the topology and ensure the avoidance of the occurrence of limit duty.

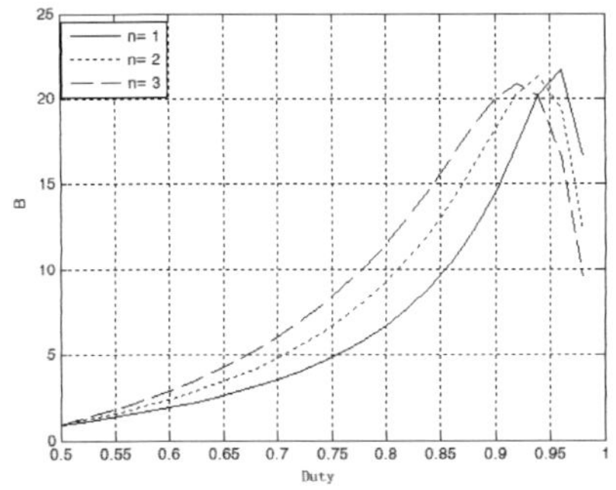

Figure 8. Duty cycle curve

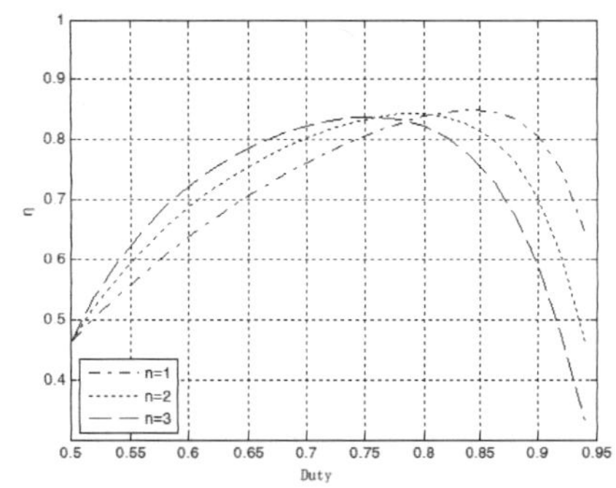

Figure 9. Efficiency graph

Figure 9 shows the relationship between duty and topology efficiency under different turns ratio. It can be seen from the graph that the efficiency of the topology increases with the increase of duty when the turn-to-turn ratio is fixed, but it decreases rapidly when the duty exceeds its limit value. When the duty is fixed, the efficiency of the topology increases with the increase of the turn-to-turn ratio, and the greater the turn-to-turn ratio, the topology efficiency reaches its maximum value. The duty will decrease correspondingly according to Saber simulation

Figure 10 is a picture of Saber simulation.

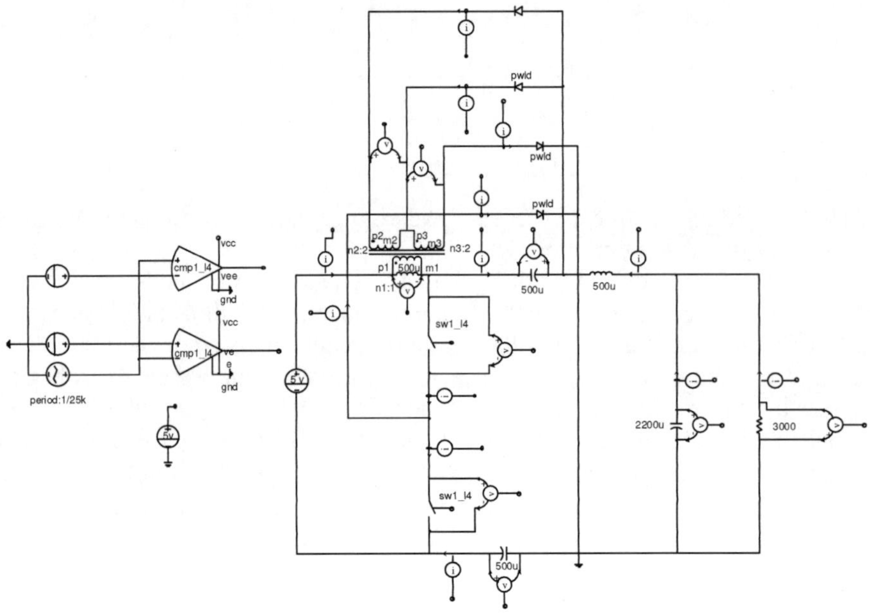

Figure 10. Saber simulation

3. Conclusion

It can be seen that under the same conditions, the efficiency and voltage gain increase with the increase of the topological turn ratio. Moreover, regardless of the turn ratio of the coupling winding, the corresponding voltage gain curve and efficiency curve will decrease sharply after a certain maximum value appears. This shows that the optimal duty should be selected according to the curve of voltage gain and topological efficiency to ensure that the circuit can work in the optimal state without limiting duty.

References

[1]Guan Enhui. Research on high gain three-level zeta converter [D]. Harbin Institute of Technology, 2015.

[2]Zhang Qiliang, Liu Qianying. Research on a new quasi-Z source three-level Boost DC converter[J].Shandong Electric Power Technology,2018, 45 (02):40-43.

[3]SONG Qi-zhen, LIU Wen-an, CHEN Li, YANG Wei. A Novel Switch-Coupled Inductive Quasi-Z Source Inverter[J].Electrified Railway, 2017,28 (05):26-30.

[4]Yuan Hong. Research on high gain boost converter based on switch-coupled inductor [D]. Qingdao Technological University, 2016

[5]Ji Yuxiong. Research on high-boost ratio switch-coupled inductor quasi-Z source inverter [D]. Harbin Institute of Technology, 2015.

[6]WANG Ting, TANG Yu, HE Yaohua, FU Dongjin. Multi-unit switching inductor/switching capacitor active network converter[J]. Proceedings of the CSEE, 2014, 34(06): 832-838.

IOP Conf. Series: Journal of Physics: Conf. Series **1284** (2019) 012013 doi:10.1088/1742-6596/1284/1/012013

Scheduling Complexity of an Automated Cell Manufacturing Process

Jinjuta Kaowtain[1,*], Hong-ming Ku[1] and Kwanchanok Pasuwat[1]

[1]Department of Chemical Engineering Faculty of Engineering, King Mongkut's University of Technology Thonburi, Bangkok, Thailand

[*]E-mail: z.jinjuta@gmail.com

Abstract. Osteoarthritis is currently the most prevalent chronic joint disease and is expected to increase continually due to aging population and obesity epidemic. Many methods exist to treat this disease but the most promising one is cell-based therapy which uses chondrocyte cells. Chondrocyte cells are cultivated in an automated cell manufacturing system (ACMS). However, ACMS suffers from long idle time caused by incubation times required to cultivate cells in different samples in different incubators. But before optimization can be used to find an optimal schedule, the complexity of the ACMS scheduling problem is established. Parallel operation of the ACMS was able to process a maximum of 26 samples in a year, while sequential operation yielded only 14. Moreover, if we produce more sample in a year the number of different schedules will increase. Then, the problem has more complicated.

1. Introduction

Cell-based therapy or regenerative medicine is capable of restoring the function of damaged cells, tissues, and organs using patients' own cells [1]. The outcome of this treatment relies heavily on the number and quality of the cells used in the therapy. For future, the applications of cells are used in a tissue construction to transplant tissue to patient A. Jain and R. Bansal reported [2], or a cell banking is an alternative David T. Harris reported [3]. Other treatment methods based on disease-modifying drugs such as non-pharmacological, pharmacological, and surgical approaches are not effective when applied too late [4]. Conventional manufacturing of cells involves manual operation which requires highly skilled technicians and is very common at the laboratory scale [5]. However, manual operation is labour-intensive, time-consuming, prone to human errors, and unable to meet commercial needs of patients. In addition, cell processing and culturing in laboratories must be carried out under aseptic condition to avoid bacterial contamination, which is difficult for technicians to control. Therefore, an automated cell manufacturing system (ACMS), as shown in figure 1, is employed to overcome problems associated with manual operation. The ACMS consists of two robot arms housed in an isolator to carry out tasks normally done by humans. The isolator is connected to two incubators to handle the processing of two different samples simultaneously. The aseptic environment inside the system is assured by performing periodic decontamination to prevent bacterial contamination and cross-contamination between samples.

With the above-mentioned benefits, our research group has used the ACMS (CellPROi, Shibuya Cooperation, Japan) for large-scale production of human autologous chondrocytes to treat cartilage defects. The manufacturing process of chondrocyte cells involves several steps whose sequence must follow the standard operating protocol (SOP) as shown in figure 2. The procedure is divided into three main parts, namely cell expansion (cell seeding and subculture), medium replacement, and cell collection (for subsequent applications). In “Cell Seeding”, cells initially isolated from human articular cartilage tissue (passage 0) are plated onto two 100 mm tissue culture petri dishes before transferring them into the connected incubator. The culture medium is replaced every 3 days using “Medium Replacement 1” mode. Once the cells reach 80% confluency or 80% of the culture area is covered by a monolayer of cells, “Subculture 1” is activated to expand the number of cells by transferring all the cells to two new 150 mm culture petri dishes, resulting in the cells at passage 1 with a split ratio of 1:1. “Medium Replacement 2” mode is used until the cells reach 80% confluency before starting “Subculture 2” to yield the cells at passage 2 at a split ratio of 1:3. These steps are repeated for “Medium Replacement 3” and “Subculture 3”. At the end of the SOP, the chondrocytes at passage 4 in eighteen 150 mm culture dishes are collected for further use [6].

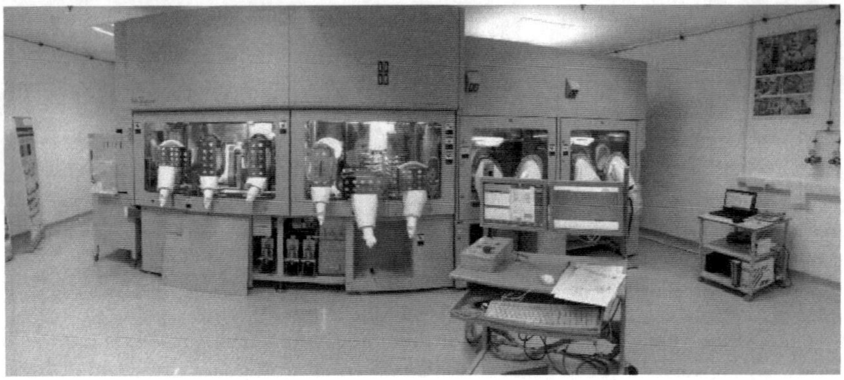

Figure 1. The automated CellPROi machine

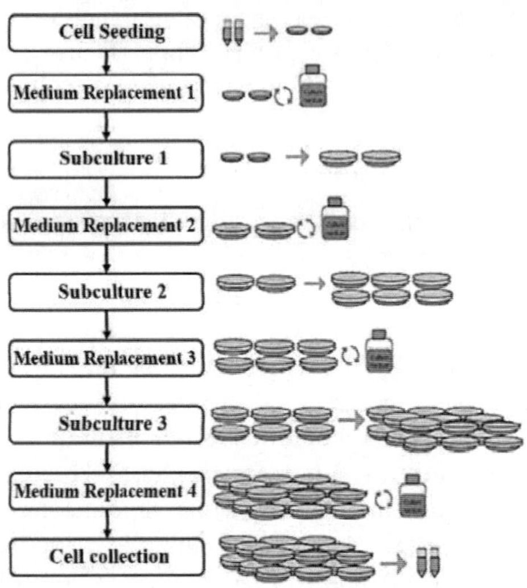

Figure 2. Flow chart of automated cell culture

Although the automated cell culture has several advantages, as previously mentioned, it suffers from long waiting time between each processing step, leading to inefficient utilization of the system and opportunity loss. Consequently, an important problem that arises is to find an optimal cell manufacturing schedule to maximize the number of samples while minimizing the operating cost of the two incubators in a given year, given that there are clearly many feasible schedules. At this point, waste treatment, product quality testing, and transportation of cell products are not taken into account. Optimization to find an optimal schedule is beyond the scope of this study. Instead, this work explores and addresses the complexity of the ACMS scheduling problem which shows that subsequent optimization is needed to maximize the efficiency of the system.

In general, the complexity of an optimization problem is defined by the number of decision variables and the number of constraints (collectively known as the size of the problem) and how its feasible solution space increases with the problem size. For example, the solution space can increase linearly or exponentially as the problem size increases [7]. Obviously, the latter is much more difficult to solve. In certain optimization problems, the problem solution space or the number of feasible solutions can be established precisely as a function of the number of decision variables. Unfortunately, this often is not the case, and the ACMS scheduling problem is no exception. As a result, before optimization can be carried out, this study seeks to establish the complexity of the scheduling problem.

2. Experimental

2.1. Production of human primary chondrocytes

The ACMS (CellPROi) manufactured by Shibuya Cooperation Co., Ltd. (Japan) was installed in a laboratory at the Bangkhuntien Campus of KMUTT. The machine consists of two robotic arms and two incubators, which allow to process two cell samples simultaneously. The process procedure following the SOP is shown in figure 2. The cells must be subcultured when their confluency reaches 80% to avoid contact inhibition. If the cell confluency does not reach 80%, the medium will be replaced again. A flow chart in figure 3 shows the steps involved in achieving a confluency of 80% in a cell culture.

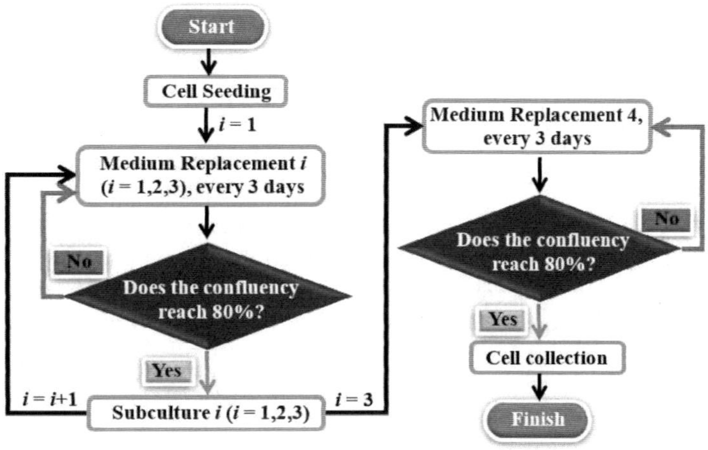

Figure 3. Flow chart of the cell culture procedure

2.2. Simulation

All data used in the calculations involving schedules come from actual experiments at the Bangkhuntien laboratory of KMUTT. First, it should be noted that the ACMS scheduling problem is multi-level. The first issue is to maximize the number of samples that can be produced in one year, given the number of incubators. After that, two sub-objective functions will be considered, which are to minimize the total production time and to minimize the operating cost. For simplification, the incubation time is assumed to be the same for all sample although cells in different samples behave differently depending on the type and the quality of cell. All samples must follow same processing sequence as shown in figure 3.

In the ACMS, a feasible schedule can be found by simple hand calculations for both sequential and parallel production. A feasible schedule is one that does not violate any constraints in the system such as sequences of process follow the standard operating protocol and all product can be processed in one year. These calculations can be programmed and automated in software such as MATLAB. Random feasible schedules or solutions are then generated for both sequential and parallel production. For each feasible schedule, the value of its objective function, e.g. the operating cost, is calculated. To establish the complexity of the scheduling problem, the total number of random schedules, N_R, is varied, each of which is treated as one case. For each case, the number of schedules N_F that give different values of the objective function is collected. The trend for N_F is then observed. If N_F starts to plateau quickly as N_R increases, it indicates that the scheduling problem does not contain many feasible solutions.

First, the maximum number of samples that can be processed in one year via sequential and parallel production was determined by trial-and-error. This was achieved by randomly generating 10,000 feasible schedules for each case of N_S, the number of samples to be processed, with $N_S = 2, 3, 4, \ldots N_{max}$. When N_{max} was reached in which all 10,000 schedules generated required more than one year to process, $(N_{max} - 1)$ was taken the maximum number of samples that the ACMS can process in one year.

3. Results and discussion

The scheduling of the ACMS is based on two existing incubators for the production of human primary chondrocyte cells. Because the cell cultivation must follow the SOP, the production is by nature sequential. The incubation time of each sample is the same, and we assume that the specific growth rate of each sample is the same.

3.1. Preliminary calculation

The chondrocyte cell production can be carried out either sequentially or in parallel. In sequential production, the ACMS handles one samples at a time, which allows the use of only one incubator. Decontamination is performed only once at the beginning of the sample. When the cells of sample 1 are collected, sample 2 will be processed in a sequential manner. The processing time of each sample is 37.78 hours. The maximum sample in one year was found to be 14 samples which resulted in an idle time of 8,032 hours and a production time of 8,560.97 hours. Then the operating cost, which consists of material, chemical and electricity cost, was evaluated. The operating cost is 68,550 Baht/sample. On the other hand, in parallel production, the ACMS handles two samples at a time by using two incubators. However, the SOP must still be followed which means that cross-contamination between samples must be avoided. The decontamination step must be performed every time when there is a switch between samples. This leads to a total of 18 decontaminations during one run of two samples. The processing time of the two samples is 459.57 hours. The maximum number of samples for year is found to be 26 which resulted in an idle time of 2,316.98 hours and a production time of 8,291.35 hours. The operating cost is 87,450 Baht/sample. The results of the two types of production are summarized in Table 1.

Table 1. Comparative results between two types of production

Parameters	Sequential production	Parallel production
Number of samples	14	26
Production time (hour)	8,560.97	8,291.35
Idle time (hour)	8,032	2,316.98
Expense (Baht)		
- Materials	18,000	36,400
- Chemicals	43,700	46,500
- Utilities	6,850	4,550
Operating cost (Baht/sample)	68,550	87,450

Comparative calculations between sequential and parallel production show that the latter can reduce the idle time by 5,715 hours due to its higher efficiency and the ability to produce 12 more samples in a year. Even though parallel production incurs a higher operating cost than sequential production because of the need for more decontaminations, the average utility cost per sample in parallel production is cheaper because more samples are produced.

3.2. Randomization method

With two incubators, the ACMS can process up to a maximum of 26 samples per year. The simulation shows that if 27 samples are to be processed, total production time of 370.88 days is needed. Clearly, more samples to be processed means there are more possible ways or schedules to process them. To establish this relationship and the complexity of the AMCS scheduling problem, random feasible schedules are generated for each N_S, starting with $N_S = 2$ up to $N_S = 26$. For each N_S, a total number of 10,000 random feasible schedules are generated based on coding in MATLAB. The total number of different feasible schedules N_F is then recorded for each N_S, which are then plotted in figure 4. Note that $N_F \leq 10{,}000$ for all cases as it should be figure 4 shows that as N_S increases, N_F also increases and goes up rapidly when $N_S > 10$, suggesting that when N_S is large and approaches 26, the number of possible feasible schedules also increases. At $N_S = 24$, only about 20% of the random schedules generated are repeats or the same. This randomization of feasible solutions indicates that simply picking the best schedule from the 10,000 random schedules does not guarantee an optimal schedule or even a good schedule. Hence, optimization is needed which will lead to an optimal or a near-optimal schedule using a lot less computation time than the randomization method.

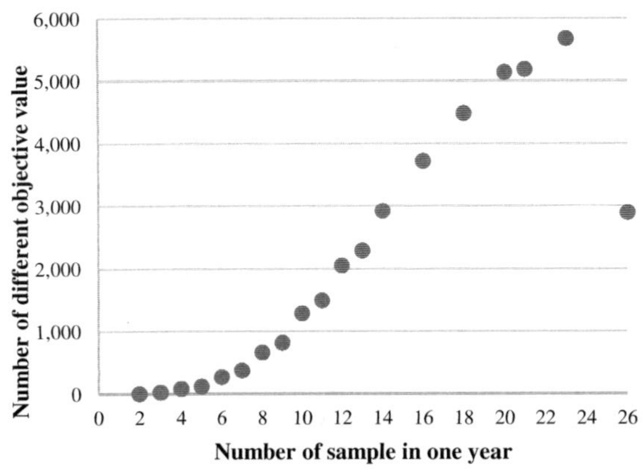

Figure 4. N_F in each N_S

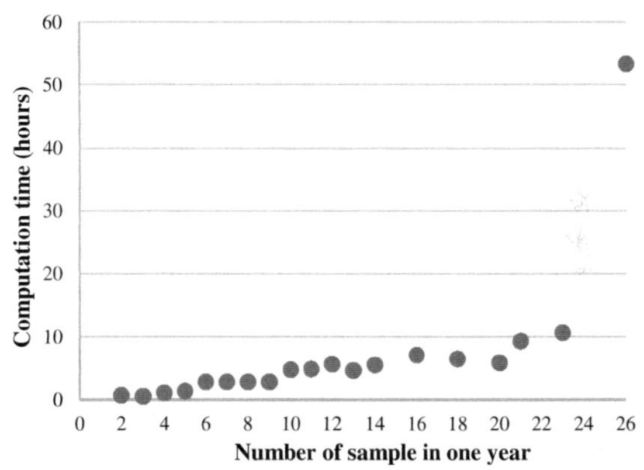

Figure 5. The computation time of each N_S

To further illustrate the need for optimization, the computation times required to generate 10,000 solutions in each case of N_S are recorded and shown in figure 5. It is clear that when $N_S = 26$, it took a laptop computer (Intel$^{(R)}$ Core$^{(TM)}$ i7-7700HQ CPU @ 2.80GHz 2.81 GHz, 64-bit operating system, x64-bases processor) about 10 hours to complete the solution generation. With optimization, the computation time is expected to be just a fraction of that of the randomization method with almost a guarantee that the optimized solution would be superior to the best solution from randomization.

It is possible that different feasible schedules result in the same objective value. In the context of optimization, these two solutions with the same objective values are equivalent. Moreover, it is also important to note the variance in the values of the objective function for the schedules generated. figure 6 shows the distribution of feasible schedules for the case of $N_S = 26$ in terms of the frequency of feasible schedules with the same objective values and the variance of the objective values. The range of the production time is between 8,291.5 - 8,760 hours (345.47 – 365 days), while the range of the operating cost is between 86,260 - 87,450 Baht/sample. As the operating cost depends on the utility cost and the decontamination cost, different production schedules will yield different production duration and operating cost. At the maximum number of samples per year ($N_S = 26$), the relationship between the production time and operation cost from each randomly generated feasible schedule was investigated, as shown in figure 7. As expected, the operating cost is inversely proportional to the production time. A shorter production time requires the decontamination step to be performed more frequently, leading to a higher cost. This result shows that the overall production cost of each sample can be more affected by the frequency of the decontamination process. Therefore, optimization is necessary to determine a processing schedule that can provide a lower operating cost within the desirable production duration.

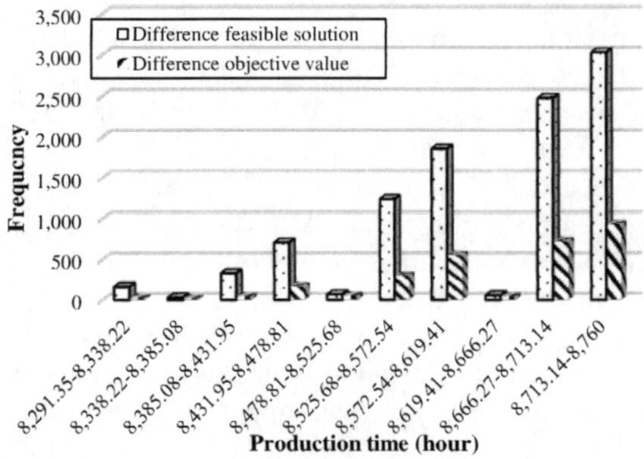

Figure 6. Frequency of feasible solution at $N_S = 26$

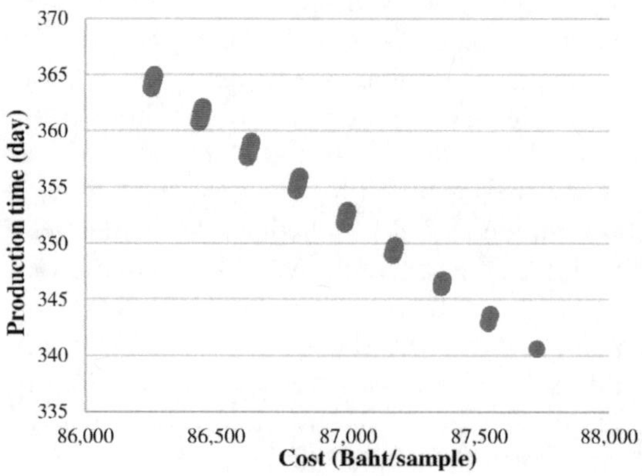

Figure 7. The relation of sub-objective function between production time and operating cost at $N_S = 26$

4. Conclusions

In this work, the scheduling problem of an ACMS with two incubators to produce chondrocyte cells was studied. Because many feasible schedules exist, ACMS scheduling is an optimization problem, whose optimal solution depends upon the objective function chosen to be minimized or maximized. In this study, the complexity of the scheduling problem was explored and quantified. First, two types of schedules are possible, namely sequential production and parallel production. By generating a large number of feasible manufacturing schedules, it was found that the maximum number of samples that could be produced in a year was 14 for sequential production and 26 for parallel production with the operating costs for each sample of 68,550 Baht and 87,450 Baht, respectively. Parallel production was more desirable, as it can produce more samples per year despite a higher production cost, Therefore, an application of optimization would be particularly beneficial to the ACMS as it can help determine an optimal production schedule that yields a large number of samples while minimizing the operating cost.

Acknowledgments

This project has been partially funded by Thailand Center of Excellence in Life Sciences (TCELS) and the funding provided by Faculty of Engineering, King Mongkut's University of Technology Thonburi.

References

[1] Mao A.S. and Mooney D.J. 2015 "Regenerative Medicine: Current Therapies and Future Directions", Proceedings of the National Academy of Sciences, vol. 112, No. 47, pp. 14452-14459

[2] Jain, A. and Bansal, R., 2015, "Applications of Regenerative Medicine in Organ Transplantation", Journal of pharmacy & bioallied sciences, Vol. 7, No. 3, p. 188

[3] Harris, D.T., 2014, "Stem Cell Banking for Regenerative and Personalized Medicine", Biomedicines, Vol. 2, No. 1, pp. 50-79

[4] Bijlsma J.W., Berenbaum F., and Lafeber F.P. 2011 "Osteoarthritis: An Update with Relevance for Clinical Practice", The Lancet, vol. 377, No. 9783, pp. 2115-2126

[5] Ozturk S.S. 2006 Cell Culture Technology for Pharmaceutical and Cell-Based Thrapies, Taylor & Francis Group

[6] K. Pasuwat, T. Chanthasopeephan and A. Khantachawana 2016 Automated Tissue Kulture

[7] Eduardo D. Glandt, Michael T. Klein and Thomas F. Eddgar 1988 Optimization of Chemical Processes, McGraw-Hill

Research on Global Path Planning Method of Mobile Robot Based on BAS

Xu-Mei LIN, Yu-Lu WANG*, Yun-Fei LIU and Jing-Xue LU

Information and Control Engineering, Qingdao University of Technology, Qingdao, 266520, China

*17864226753@163.com

Abstract. In this paper, a path planning algorithm based on the beetle antennae search (BAS) is presented, which can effectively improve the speed, obstacle avoidance and environmental adaptability of the robot path planning. The distance between the beetle and the target point is the fitness function, and the changing step size is to avoid the local optimization. According to compare random coordinates of the beetle's left and right antennae, the direction of walking is determined to plan the better path in the static environment. The obstacle avoidance is judged by whether the beetle's position and walking trail are in the effective area. Multiple beetles are set at each path point, and the optimal direction which approaches the theoretical optimal path, is chosen after comparing with every beetle's trail. A large number of results of the experiments based on RRT and BAS algorithm indicate that BAS algorithm is more capable of searching its path and avoiding obstacles. Even in a complex environment, compared with other algorithms, BAS is more likely to rapidly plan an optimal path which can successfully avoid obstruction.

1. Introduction

Path planning is significant in the research of Mobile Robot Control and is fundamental to task completion [1]. Path planning means that the robot which is compatible with certain restriction conditions and evaluation conditions can find out an approximate optimal or suboptimal path from the starting point to the terminating point and avoid the obstacles. According to environmental information known by the robots, path planning has been divided into global path planning based on the all-known environment information and local path planning based on unknown or partly known environment information [2]. At present, there are many common global path planning methods for mobile robots, such as traditional grid method and artificial potential field method [3]. Although grid method is suitable for global path planning, when the environmental space increases, the required storage space soars, the computational efficiency and the decision speed decreases. On the other hand, the artificial potential field method is likely to cause local optimum and deadlock phenomenon. With the development of intelligent control technology, algorithms such as A* algorithm, genetic algorithm (GA), ant colony algorithm (ACA), particle swarm algorithm (PSO) and rapidly-exploring random tree algorithm (RRT) have appeared [4-8]. When solving the problems of robots planning the paths, the artificial potential field method, the neural network method and the genetic algorithm need to model the obstacle in definite space. However, these methods are always inefficient in practice. RRT has the characteristics of large randomness, strong obstacle avoidance ability, but it is weak in real time and inefficient in search. Compared with other algorithms, BAS [9] only requires one individual

IOP Conf. Series: Journal of Physics: Conf. Series **1284** (2019) 012014 doi:10.1088/1742-6596/1284/1/012014

and can replan the path by observing every step whether bumping into the obstruction or not. Therefore, there's no need modeling and analyzing on the current environment in advance, which greatly decrease the computational complexity while largely increase the efficacy.

BAS used in this paper is a heuristic search algorithm. The algorithm has the features of strong robustness, rapidity, randomness and no need of preprocessing. At the same time, it can effectively solve the problem of path planning in high dimensional space and complex environment with simple algorithm. This intelligent optimization algorithm can realize optimization without knowing the specific form and gradient information. This paper uses the BAS algorithm to do research on the path-planning of the mobile robots, with the emphasis on the simulation analysis.

2. The description of BAS algorithm

2.1. Basic principle of the algorithm

Bionics: The beetle does not know where the food is when it is preying, but it can adjust the movement by feeling the smell of food. Beetle has two long antennae, when the left antennae detects a higher concentration of odour, the beetle would turn to the left, otherwise, it would turn to the right until the food is found [10]. The simplified model diagram of foraging for a beetle is shown in figure 1.

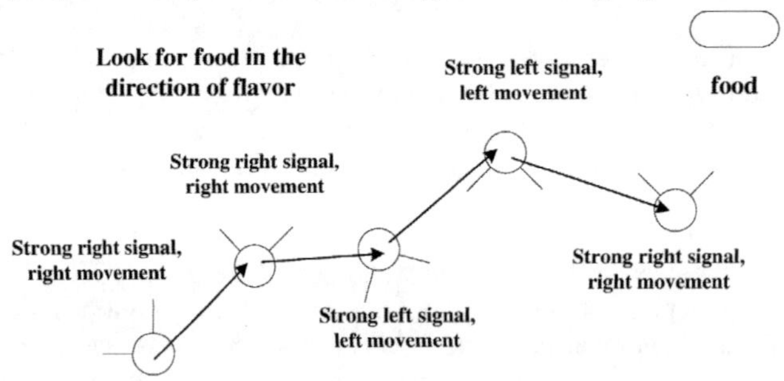

Figure 1. Simplified model diagram of foraging for a beetle.

The hypothesis of a simplified model of a beetle:

- The head orientation of the beetle is arbitrary and can be represented by a random vector.

$$b = rands(k,1) \tag{1}$$

The normalization follows as:

$$\vec{b} = \frac{rands(k,1)}{\|rands(k,1)\|} \tag{2}$$

where rands( ) represents a random function and k represents the spatial dimension.

- Position of the right and left antennae of the beetle.

$$\begin{cases} x_r = x^t + d_0 * \vec{b} / 2 \\ x_l = x^t - d_0 * \vec{b} / 2 \end{cases} (t = 0,1,2,\cdots,n) \tag{3}$$

where x_l and x_r represent the antennae coordinate; x^t represents the centroid coordinate of the t iteration and d represents the distance between two antennae.

- The fitness function of the path planning which needs optimizing is used to calculate the fitness value of both left and right antennae.

$$\begin{cases} f_r = f(x_r) \\ f_l = f(x_l) \end{cases} \tag{4}$$

- Update the position of the beetle.

Making comparison between the left and right values, if f_l is less than f_r, the beetle moves step to the direction of left antennae and vice versa.

$$x^{t+1} = x^t - \delta^t * \vec{b} * sign(f(x_r) - f(x_l)) \quad (5)$$

where δ^t is the step size of t iteration and sign ( ) is symbolic function.

2.2. Environmental modeling

The working environment of the robot is two - dimensional space. During the operation of the robot, the obstacles are stationary and the size remains unchanged. The images are drawn according to the actual environment, in which shapes in black are the obstacles and the white area is the passable place. The shapes in black are larger than the actual obstacles, trying to make the environmental model consistent with the real situation, so that the robot can move smoothly in the planned path. After the image is transferred into a binary image, the starting and ending points of the robot can be set at the feasible interval.

2.3. Algorithm execution steps

The specific flow chart of BAS path planning is demonstrated as shown in figure 2:

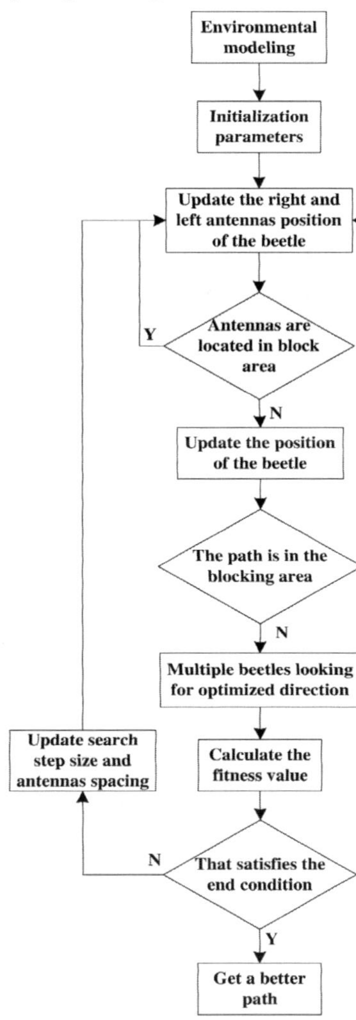

Figure 2. BAS path planning process.

The steps of BAS path planning follow as:

- Modeling is on the basis of the working environment of the robot.
- Detailed parameters of BAS algorithm are to set: step (δ), the distance between two antennae (d), number of iterations (n), initial position (x_0) and spatial dimension (k).
- Set the initial position of the beetle as the starting point and determine the coordinates of the left and right antennae to ensure that it is within the effective area of the environmental model.
- Calculate the value of the two antennae coordinates with fitness function and then compare the value. If the fitness value of the left antennae is smaller than that of the right antennae, then the position of the direction of left antennae is the next position of the beetle and vice versa.

Fitness function:

$$fitness(x^t, y^t) = \sum \sqrt{(x-x^t)^2 + (y-y^t)^2} \tag{6}$$

where (x, y) is the coordinates of the end point of the path and (x^t, y^t) is the position of beetle at t iteration.

- During the update the position information of beetle, if the search position exceeds the set space range, it will select the set space range as the position information.
- The position of the right and left antennae is checked whether it is in the feasible region. If not, the coordinates of the left and right antennae need to be obtained again. The path is judged whether it is valid according to the constraint conditions. If not, it needs to be reinitialized until the path is guaranteed to be valid.
- This paper has set a number of beetles at each planned position and each of which will randomly search for one direction and select the optimal direction by comparison.
- When the beetle reaches a certain distance from the target point or the maximum number of iterations, the beetle exits the loop and a better path will be displayed in the figure.

3. Simulation analysis of path planning based on BAS algorithm

Under the two circumstances of constant and variable step size, the path planed by the BAS algorithm and the number of iteration are demonstrated in figure 3 and figure 4:

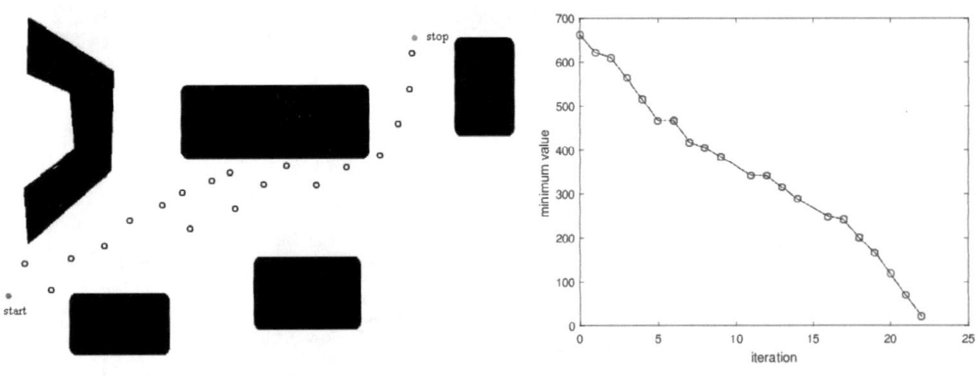

Figure 3. BAS path planning based on constant step size.

It can be seen from figure 3 and figure 4 that the path planning can be completed after about 25 generations. If the step size is unchanged, the robot will move in the same length as the step size, which will largely deviate from the shortest path. While in the case of a variable step size, with the gradually reduced step size the robot will be carefully searched near the target point, so that the obtained path is closer to the optimal value.

This paper puts 30 beetles at each planned location. Multiple beetles are looking for their next landing spot at the same time. The beetle that ran into obstacles is abandoned. It is more efficient to re-search the path than one single beetle does. The path planning based on 30 beetles and the number of iteration are shown in figure 5.

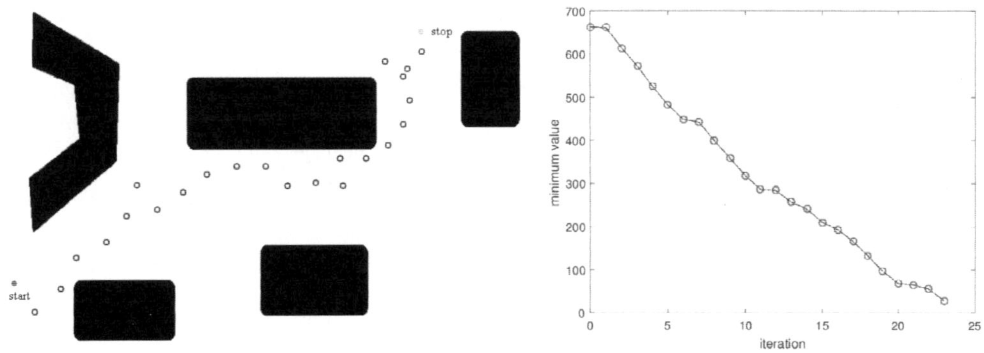

Figure 4. BAS path planning based on variable step size.

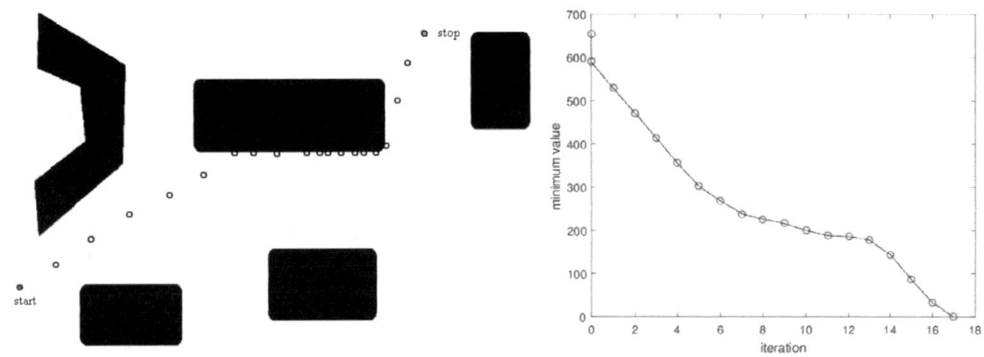

Figure 5. BAS path planning based on multiple beetles.

In figure 5, compared with one single beetle, the planned path of multiple beetles is more stable, so that a better path can be selected.

In order to test whether BAS algorithm is superior to other intelligent optimization algorithms in robot path planning, comparisons of robots' path-planning based on RRT and BAS algorithms are made, in which the starting point and the environment are guaranteed to be the same. See figure 6.

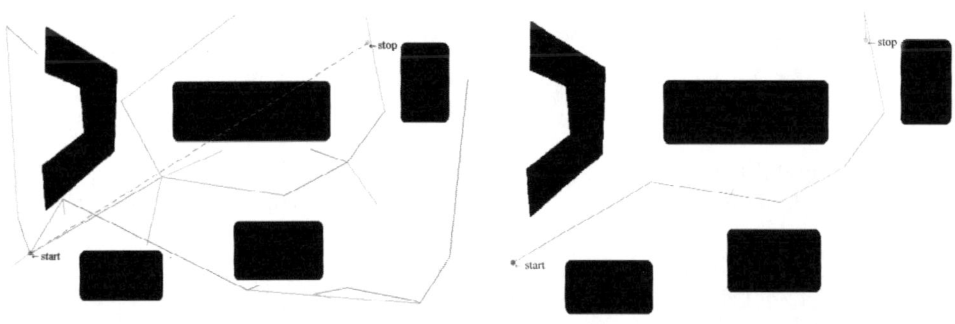

Figure 6. Path planning based on RRT algorithm.

According to the figure 5 and figure 6, both algorithms can successfully avoid the obstacles and plan a path. The path planning based on RRT algorithm will randomly extend the tree node until the new node covers the target point or enters the region containing the target point. It will form a path consisting of random tree nodes between the starting point and the target point. However, many redundant nodes will be generated. The path planning based on BAS algorithm can correct the direction step by step according to the fitness value, so that the time of the algorithm is reduced and the planned path is closer to the optimal one. The time taken by RRT algorithm and BAS algorithm to complete the path planning is demonstrated in table 1:

Table 1. Program running time comparison table based on different algorithms.

	Total time (s)	Self-use time (s)
RRT algorithm	4.174	1.883
BAS algorithm with constant step size	2.618	0.479
BAS algorithm with variable step size	2.516	0.425
BAS algorithm based on multiple beetles	2.636	0.616

By comparing the runtime of the programs based on different algorithms, not only the total time of the program but also the algorithm itself, the BAS algorithm runs 1.5s faster than the RRT algorithm. Also, the difference in the total time of the BAS algorithm with variable and constant step size is of little difference. Multiple beetles take a little more time than one single beetle.

In conclusion, the BAS algorithm based on multiple beetles can plan a better path in a short time, reflecting that the BAS algorithm has good applicability in path planning.

4. Conclusion

In this paper, a new algorithm of path planning for the robots based on the beetle antennae search is proposed. The complexity of this algorithm is not much affected by environmental obstacles. Simulation results demonstrate that the proposed algorithm has strong environmental adaptability and can rapidly plan a path and even though in a complex case of obstacles. On this basis, the planned path is not prone to fall into local optimum, and the iteration speed is fast. However, there are also some deficiencies. For example, the optimal solution of the path needs to be further improved.

References

[1] Bing Fu,Lin Chen,Yuntao Zhou,Dong Zheng,Zhiqi Wei,Jun Dai,Haihong Pan. An improved A* algorithm for the industrial robot path planning with high success rate and short length[J]. Robotics and Autonomous Systems,2018,106.

[2] LI L, YE T, TAN M. The present situation and the future of mobile robot technology research [J]. Robot, 2002, 24(5) : 475–480.

[3] Yuan Xin, Zhu Qidan, Yan Yongjie.Collision avoidance planning in multi-robot system based on improved artificial potential field and rules[J].Journal of Harbin Institute of Technology,2009,16(03):413-418.

[4] Zhang Hui,Rong Xuewen,Li Yibin,Li Bin,Zhang Junwen,Zhang Qin.Path planning based on sliding window and variant A* algorithm for quadruped robot[J].High Technology Letters,2016,22(03):334-342.

[5] Chaymaa Lamini,Said Benhlima,Ali Elbekri. Genetic Algorithm Based Approach for Autonomous Mobile Robot Path Planning[J]. Procedia Computer Science,2018,127.

[6] Tan Guanzheng, He Huan,Aaron Sloman.Global optimal path planning for mobile robot based on improved Dijkstra algorithm and ant system algorithm[J].Journal of Central South University of Technology(English Edition),2006(01):80-86.

[7] YONG T, QING L, LI J W. An improved PSO for path planning of mobile robots and its parameters discussion [C]. International Conference on Intelligent Control and Information Processing, Dalian, China, 2010: 34–38.

[8] SHANG Song-tian. Vehicle's RRT path planning algorithm under dynamics constraint[A]. Chinese Society of Inertial Technology. Proceedings of 2016 China International Conference on Inertial Technology and Navigation[C]. Chinese Society of Inertial Technology:,2016:5.

[9] JIANG X Y, LI S. Beetle antennae search without parameter tuning (BAS-WPT) for multi-objective optimization[J]. arXiv preprint arXiv:1711. 02395,2017.

[10] Wang T T, Liu Qiang. Prediction of storm surge disaster loss based on BAS-BP model [J]. Marine environmental science, 2018,37(03):457–463.

Research on Simulation Decision of Logistics in Flexible Factory

Luo Wentao[1], Ren Yan[2], Zhang Jianfu[1,*] and Du Linjiang[2]

[1]Department of Mechanical Engineering, Tsinghua University, Beijing 100084;
[2]Beijing North Vehicle Group, Beijing 100072

[*]Corresponding Email: zhjf@tsinghua.edu.cn

Abstract. Aiming at problems that the flexible scheduling capability of the production line in the existing factory is poor due to islanding effect and existing scheduling methods ignore non-cutting time and dynamic change of time when solve the minimum completion time, this paper studies the optimization of multi-variety and small batch workflow scheduling, and establishes an optimal job flow allocation model based on dynamic time parameters. Taking a flexible flow shop as an example, the model is applied to the implementation, and the EM-PLANT simulation software is used to make digital decision to realize the scientific scheduling of workpiece flow in cyber environment of production line. The simulation results show that the optimal allocation model can reduce the average completion time of workflow and improve production line efficiency.

1. Introduction

The modeling and simulation of machine and logistics behavior in the factory has always been a hot topic among scholars[1]. The digital analysis capability of the factory is also an important part of intelligent manufacturing. As factory modeling and simulation technology matures, modeling and simulation technology begins to play a role in providing decision support. Research based on system behavior for optimal decision making[2] or based on historical and real-time data to predict future behavioral patterns[3] began to sprout. Since the accuracy of decision-making depends on the amount of training data, data analysis based on industrial software has begun to develop in the industry[4].

At present, the decision analysis research on shop scheduling is more and more in-depth and detailed. Markus P. Roessler pointed out the challenges of factory decision making and provided solutions on joint decision-making in production and logistics from a management perspective[5]. Nour El Houda Tellache et al. considered some conflicting operations that could not be processed simultaneously in different robots, and established a mathematical model to provide a polynomial solution to this problem[6]. M. A. Dulebenets et al. considered the difference in the size of cargo carried by different ships and established a mixed integer non-linear model to minimize the total ship run time[7]. XinNianWang establishes the estimation distribution algorithm for solving the minimal completion time of benchmark problems[1]. Although the research on scheduling algorithms is getting more and more in-depth, most scheduling studies only take processing time into consideration, and non-processing auxiliary time is often ignored. Many scheduling studies are based on the benchmark problems[8]. These problems lack the consideration of logistics time and assume that the processing time is constant. It is an idealized mathematical problem and still has a long distance from engineering applications[9].

Based on the above review, it is known that constructing the reasonable mathematical model of details of factory, and solving the model based on simulation software can achieve the purpose of comprehensively simulating the on-site production status and providing basis for factory behavior decision[10]. The independence between the same type of production line of the factory is still obvious, and the simulation granularity is relatively low, which becomes a bottleneck restricting the further improvement of the production efficiency and logistics efficiency of the factory. Therefore, this paper studies the optimization of the logistics flow based on dynamic time parameters based on EM-Plant software platform.

2. Problem Description

The production lines with the same function in the factory are independent of each other, resulting in poor modularity and flexibility of production lines. The problem is manifested in the following aspects:

- Islanding Effect. The islanding effect means that after the workpiece enters a production line, all production tasks must be completed before it can be shipped out of this production line. Due to the different busy time of each production line equipment, it will encounter the situation the workpiece still stay at waiting area but it cannot be dispatched to the adjacent production line with idle equipment due to islanding effect.
- Non-cutting time of flexible workpiece flow. Flexible workpiece flow refers to the flow of various types of workpieces. Different types of workpieces need to be reloaded before they can be processed by specific equipment. Due to the order of workpiece flow is irreversible, the dynamic reloading time needs to be considered into the distribution process of workpiece flow.

In order to solve the above problems, the rules of the workpiece inflow line are redefined, the workpiece can be exchanged between different production lines of the same function, breaking the islanding effect of the factory. At the same time, the dynamic time parameters such as the reloading time, the cross-line logistics time, the processing time and the buffer waiting time are introduced into the research to establish optimization allocation model.

3. Logistics Optimization Allocation Model

Suppose there are M production lines with the same function in the factory. Each production line has N pieces of processing equipment. L is the number of workpieces in logistics flow. The total time T^{o} of each workpiece entering and exiting the flexible mixing shop according to one of strategies can be expressed by the following formula:

$$T^{o}=\sum_{i=1}^{M}\sum_{j=1}^{N}\left(P_{Hij}^{o}+P_{Gij}^{o}+P_{Kij}^{o}\right)X_{ij}^{o}+S_{X}^{o},(o=1,2,\cdots,L) \tag{1}$$

Where, P_{Hij}^{o} indicates the reloading time of the o-th workpiece in the j-th equipment of the i-th production line. P_{Gij}^{o} indicates the processing time of the o-th workpiece in the j-th equipment of the i-th production line. P_{Kij}^{o} indicates the waiting time of the o-th workpiece in the j-th equipment of the i-th production line. X_{ij}^{o} indicates whether the o-th workpiece has selected the j-th device of i-th production line for processing. X_{ij}^{o} can form an N × M size reachability matrix X. When $X_{ij}^{o}=0$, it means that the o-th workpiece will not be processed in the j-th equipment of the i-th production line. When $X_{ij}^{o}=1$, it means that the oth workpiece will be processed in the jth equipment of the i-th production line. S_{X}^{o} represents the total length of the path traveled by the o-th workpiece between devices with element equals 1 in reachability matrix.

Taking a workpiece z as an example, the workpiece needs to be processed from the corresponding equipment of j=1, 2, ..., N before it can go out of the production line, but z has M options for each device (because there are M identical production lines), the feasible solution for the selection of the workpiece z pass through the production line is shown in Figure 1.

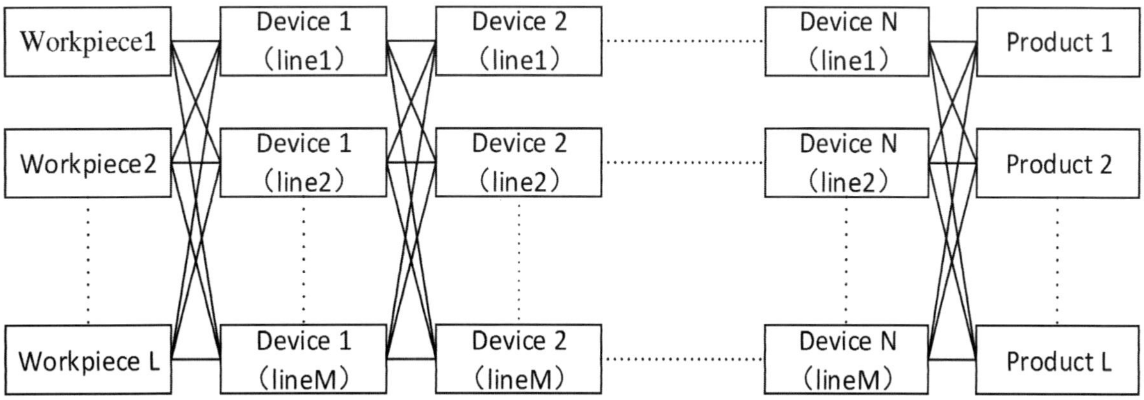

Figure 1. Logistics plan.

According to selection principle of production line, the constraint equation is established as follows:

$$\sum_{i=1}^{M} X_{ij}^{o} = 1, \mathrm{j} = 1, 2, \cdots, \mathrm{N} \tag{2}$$

$$X_{ij}^{o} = \{0,1\} \tag{3}$$

The goal of optimal allocation model is to acquire the reachability matrix solution X while it also satisfies the condition of minimizing the total consuming time $\min T^{o}$ and the corresponding constraint equation.

For a workpiece, there are a total of M^{N} feasible solutions under the flexible production rules; for the workpiece flow, there are $O \times M^{N}$ feasible solutions. The time complexity of the algorithm are all the order of magnitude level $o\left(M^{N}\right)$. Not only that, because the types of workpieces are versatile and the complex environment interferes, the time parameters of each workpiece in the same device appear to change dynamically. For the above reasons, it is impossible to traverse the feasible solution with the traditional heuristic algorithm. Therefore, it is necessary to combine the simulation technology with the heuristic algorithm to solve the dynamic NP hard problem integrally.

4. Case Study

The hybrid flow shop J specializes in the production of various parts of the automotive transmission part. The factory is widely used in unit layout. Each unit uses U-shaped lines to form an independent production line. The production lines with similar processes contents and little difference in processing equipment are arranged together to form several modules, each of which is complete and isolated with each other. The production lines of the original hybrid flow shop are independent of each other, that is, once the workpiece enters a production line, all processing tasks must be completed before the workpiece leaves the production line, therefore the flexibility between the lines is insufficient.

For example, in the besides-star wheel production line, the J factory has five complete besides-star wheel production lines with the same equipment configuration, which are line6, line7, line8, line9 and line10 respectively. When the workpiece enters one of the besides-star wheel production lines, it can only be shipped out after completing all the processes in the production line. Even if some work stations on this production line are occupied, it cannot leave the production line to another idle production line .The factory layout and process flow are shown in Figure 2.

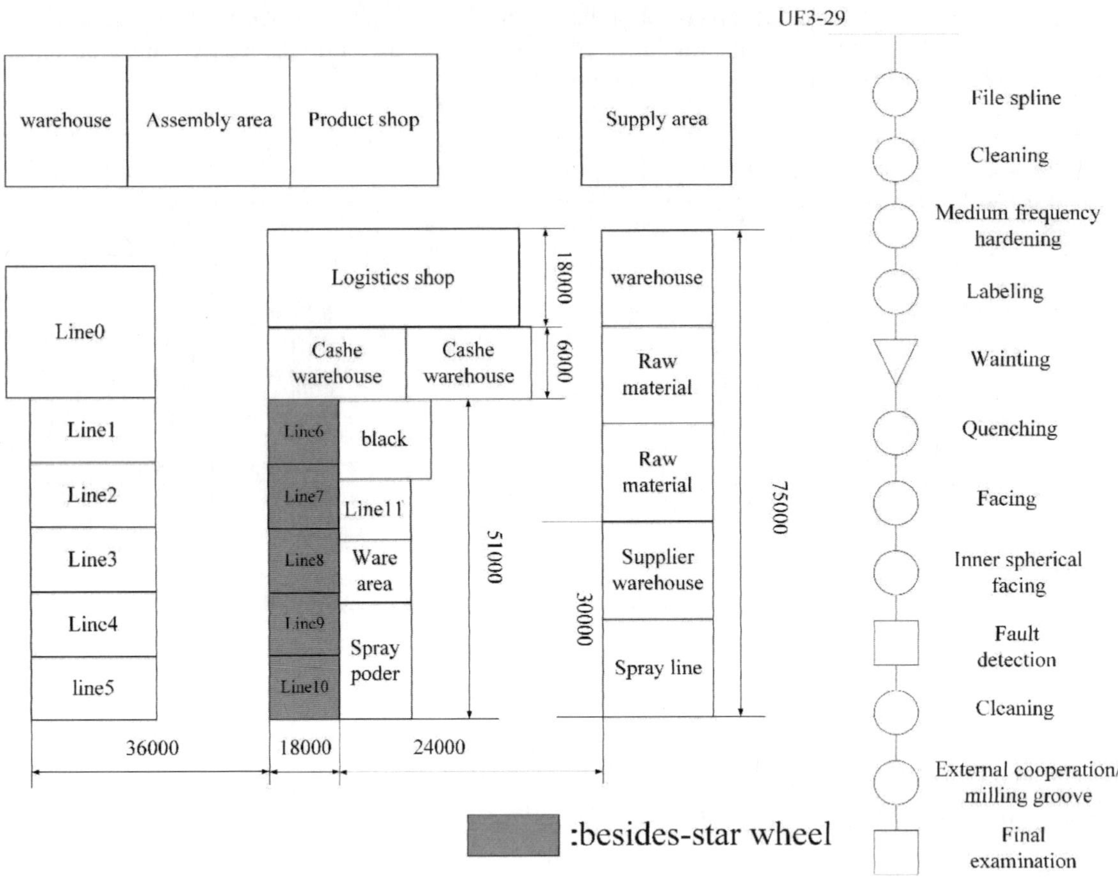

Figure 2. Besides-star wheel production line process.

It is known that there are nine pieces of equipment for each besides-star wheel production line. According to the principle of optimized distribution, the independence between the lines is now removed, and the semi-finished products processed by each equipment can be freely entered into the next equipment of other production lines. Since the flow between the lines is turned on, the time for moving the workpiece between the lines becomes a time factor that cannot be neglected. Since the inter-line logistics cost is significantly higher than the logistics between the devices in the line, it is assumed that the in-line logistics cost is neglected. The line-to-line equipment logistics statistics is shown in the table 1.

Table 1. Logistics statistics table (unit:mm)

Line No	Line6	Line7	Line8	Line9	Line10
Line6	0	10000	20000	30000	40000
Line7	10000	0	10000	20000	30000
Line8	20000	10000	0	10000	20000
Line9	30000	20000	10000	0	10000
Line10	40000	30000	20000	10000	0

This production line produces two typical models of besides-star wheel: UF3-29 and AC2000i. There are only minor differences between the two besides-star wheel process, such as size and spraying technology. However, it is necessary to change the fixture tools of the besides-star wheel for different equipment. This process is manually completed. The average time for manual reloading is $\overline{t}$=1min. Whether reloading the fixture tools of the besides-star wheel depends on whether the present workpiece type on the processing equipment is consistent with the previous one. Therefore, the reloading time can be calculated by algorithm logic:

$$T^{o}(n)=\begin{cases}\frac{1}{\sqrt{2\pi}\sigma}\exp\left(-\frac{(x-\overline{t})^2}{2\sigma^2}\right), s(n)\neq s(n-1)\\ 0 \qquad\qquad , s(n)=s(n-1)\end{cases} \tag{4}$$

Where, $s(n)$: the type of workpiece processed at the n-th time step ; $T^{o}(n)$: reloading time at n-th time step ; $\overline{t}$ is the mean time of reloading manually; σ is the time deviation of reloading manually.

After the above resources are defined, the simulation software can be used to solve the optimal reachable set. The simulation is based on the Tecnomatix Plant Simulation 9 version of the EM-PLANT software platform. The simulation time is set from Wednesday, March 1, 2017 to Friday, March 31, 2017. And the daily production plan is 2,700 pieces. The shift change system uses two shifts to ensure that they are running 24 hours a day. Rest time is three times per shift, each break for 15 minutes.

In addition, to simulate device failures, the availability ratio and average repair time are set to simulate the failure. The simulation fault parameters of each device are shown in Table 2.

Table 2. Machine equipment simulation fault parameters

Device	Availability ratio	MTTR
File spline	90%	60min
Quench	90%	50min
Tempering	70%	120min
Facing	90%	50min
Spherical Face	90%	48min
Flaw detection	95%	48min

After the above data is entered into the model, the simulation model of the besides-star wheel production line is established by using EM-PLANT software, as shown in Figure 3.

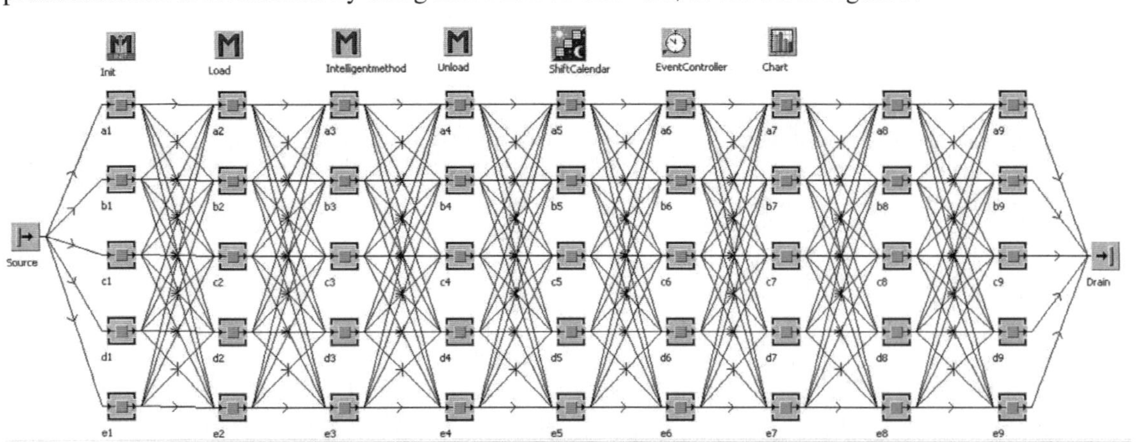

Figure 3. Besides-star wheel production line simulation model

The simulation model ‘source’ is a workpiece flow generator, which can generate a workpiece flow according to a predetermined flow according to a workpiece flow scheme. The model has a total of 45 machines, and each row representing a simplified model represents all the equipment on a production line (actually it is a U-shaped line). 'Drain' is the output of the workpiece stream. Each arrow line indicates the reachable path of the workpiece flow. 'Method' is the core part of the software, which can control the production line to operate according to the rules in the Method module input code program. The optimized distribution management model is input into the 'Intelligent Method' to simulate an optimal reachable path. Due to the large number of scenarios, the traversal search of the scheme uses genetic algorithms. . After the model is stable, the 'Chart' can be used to observe the operation status of each process.

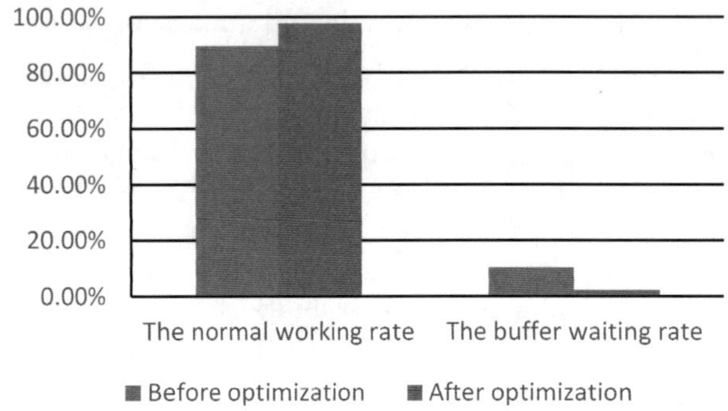

Figure 4. Comparison of workpiece flow optimization indicators before and after improvement

5. Conclusion

In this paper, the EM-PLANT simulation platform is used to combine the simulation technology with the heuristic algorithm to solve the dynamic NP hard problem in an integrated way. The normal working rate and the buffer waiting rate of the workpiece are compared under the condition of line-to-line isolation and inter-line fusion. The simulation comparison results show that the reachability scheme obtained by the optimized allocation algorithm can ensure the overall waiting rate of the workpiece is 2.17%, which is 8.14% lower than that before, as shown in Figure 4. The results verify that feasibility of the model.

Acknowledgement

We gratefully acknowledge the financial support from the National Natural Science Foundation of China (Grant No. 51575301) and Shenzhen Foundational Research Project (Subject Layout) (Grant No.JCYJ20160428181916222).

References

[1] Wang, X.N., et al., *An estimation of distribution algorithm for scheduling problem of flexible manufacturing systems using Petri nets.* Applied Mathematical Modelling, 2018. **55**: p. 776-788.
[2] Gabor, T., et al. *A Simulation-Based Architecture for Smart Cyber-Physical Systems.* in *IEEE International Conference on Autonomic Computing*. 2016.
[3] Verner, I., et al., *Robot Online Learning Through Digital Twin Experiments: A Weightlifting Project.* 2018.
[4] Akan, B., et al. *Intuitive industrial robot programming through incremental multimodal language and augmented reality.* in *IEEE International Conference on Robotics & Automation*. 2011.
[5] Roessler, M.P., et al., *Simulation-Based Assessment of Lean Production Methods: Approaches to Increase Volume and Variant Flexibility*, in *Dynamic and Seamless Integration of Production, Logistics and Traffic: Fundamentals of Interdisciplinary Decision Support*, E. Abele, M. Boltze, and H.-C. Pfohl, Editors. 2017, Springer International Publishing: Cham. p. 83-104.

[6] Tellache, N.E.H. and M. Boudhar, *Flow shop scheduling problem with conflict graphs.* Annals of Operations Research, 2018. **261**(1): p. 1-25.

[7] Dulebenets, M.A., *The Vessel Scheduling Problem in a Liner Shipping Route with Heterogeneous Fleet.* International Journal of Civil Engineering, 2018. **16**(1): p. 1-14.

[8] Wu, G., L. Yao, and S. Yu, *Simulation and Optimization of Production Line Based on FlexSim.* 2018: p. 3358-3363.

[9] Alam, K.M. and A.E. Saddik, *C2PS: A Digital Twin Architecture Reference Model for the Cloud-based Cyber-Physical Systems.* IEEE Access, 2017. **5**(99): p. 2050-2062.

[10] Feldmann, K. and O. Meedt, *Determination and evaluation of the optimal end of life strategy for products based on simulation of disassembly and recycling*. 1997.

Research on the Method that Improving the Space Headway Simulation Accuracy of Gipps Model

Dongyang Sui[1], Tongfei Shang[1,2] and Bo Li[1,*]

[1] The School of Electronic and Information Engineering, Xi`an Jiaotong University, Xi`an, Shaanxi, 710049, P.R China
[2] College of Information and Communication, National University of Defense Technology, Xi`an 710106, P.R China

[*]E-mail: boblee@xjtu.edu.cn

Abstract. Gipps model is a typical car-following model based on safe distance and is widely used in computer simulation. However, in order to avoid collision with the preceding vehicle, the model has strict restrictions on the following spacing, that makes the model has a certain error at the simulation of vehicle space headway in complex traffic environment. Due to that, based on the mathematical expression of the following spacing in Gipps model, this paper proposes a method of adding influence coefficient to improve the simulation accuracy of space headway. Furthermore, the effectiveness of the method is proved by simulation experiments, and the experimental results illustrate that the simulation results of the improved model are closer to the field data in complex traffic environment than the original model.

1. Introduction

As one of the basic car-following models in traffic simulation, Gipps model has many advantages and is widely used. But the model has strict restrictions on the following spacing ensure that continuous vehicles will not collide, which is inconsistent with the reality, and the consequence is that in complex traffic environment, the simulation accuracy of model about the space headway is not ideal. Irene Soria et al. [1] evaluated Gipps model using field data from different traffic (congested vs. uncongested) and weather conditions (rain vs. sun) and various driver types (aggressive, average and conservative). The results showed that under different traffic and weather conditions, Gipps model is relatively accurate in simulating the velocity of subsequent vehicle, but the error in vehicle space headway is larger; In experiment with different driver types, Gipps model can well simulates the average type driver's vehicle trajectory, however, for aggressive and conservative drivers, the simulation result of space headway is deviate greatly. Therefore, by studying the mathematical expression of following spacing in Gipps model, then proposes a method that adding influence coefficient to the function expression of the following spacing, and this method solves the problem that the original model is not accurate for the space headway simulation in the complex traffic environment to a certain extent.

2. Gipps Model

Gipps model [2] is based on the assumption that the required acceleration and braking acceleration are limited for each following vehicle. In order to prevent collision in the longitudinal distance of vehicle running, the velocity of next time step will be calculated by the following spacing relationship between the following vehicle and the preceding vehicle. That is, when two continuous vehicles stop, the tail of

IOP Conf. Series: Journal of Physics: Conf. Series **1284** (2019) 012016 doi:10.1088/1742-6596/1284/1/012016

front vehicle n-1 should exceed the head of rear vehicle n. The formula of this process is as follows.

$$x_{n-1}(t) + \frac{v_{n-1}{}^2(t)}{2b_{n-1}} - S_{n-1} \geq x_n(t) + \frac{v_n(t)+v_n(t+\tau)}{2}\tau + v_n(t+\tau)\theta + \frac{v_n{}^2(t+\tau)}{2b_n} \tag{1}$$

Where $x_{n-1}(t)$ is the head position of preceding vehicle n-1 at time t, and $v_{n-1}(t)$ is the velocity of vehicle n-1 at time t, b_{n-1} is the braking acceleration of vehicle n-1, s_{n-1} is the length of vehicle n-1 and the spacing naturally maintained when the vehicle stops. $x_n(t)$ is the head position of the following vehicle n at time t, $v_n(t)$ is the velocity of vehicle n at time t, τ is the driver reaction time of vehicle n, $v_n(t+\tau)$ is the velocity that the next simulation time step $t+\tau$ of vehicle n, and θ is an additional safety reaction time generally taken as $\tau/2$, and b_n is the maximum braking rate of vehicle n.

Equation (1) is transformed to obtain the expression of vehicle n's velocity at $t+\tau$:

$$v_n(t+\tau) \geq -b_n\tau + \left\{b_n^2\tau^2 + b_n\left[2[x_{n-1}(t) - S_{n-1} - x_n(t)] - v_n(t)\tau + \frac{v_{n-1}(t)^2}{b_{n-1}}\right]\right\}^{\frac{1}{2}} \tag{2}$$

Taking the equality part in Eq. (2) as the velocity expression of the Gipps model's congestion-following state, combined with the equation representing the free-running state, composes the complete model, and selects the minimum value in the two states as the velocity of the next simulation time step of the following vehicle n. The complete formula is expressed as follows:

$$v_n(t+\tau) = min\left\{\begin{array}{c} v_n(t) + 2.5a_n\tau\left(1 - \frac{v_n(t)}{V_n}\right)\left(0.025 + \frac{v_n(t)}{V_n}\right)^{\frac{1}{2}}, \\ -b_n\tau + \left\{b_n^2\tau^2 + b_n\left[2[x_{n-1}(t) - S_{n-1} - x_n(t)] - v_n(t)\tau + \frac{v_{n-1}(t)^2}{\hat{b}}\right]\right\}^{\frac{1}{2}} \end{array}\right\} \tag{3}$$

Among them, a_n is the maximum acceleration of vehicle n, V_n is the expect velocity that vehicle n want to reach, and $\hat{b}$ is the maximum braking acceleration of the preceding vehicle n-1 which estimated by vehicle n.

3. Influence Coefficient

The focus of this paper is to find a way to improve the setting of following spacing in the model, indirectly affects the space headway in the simulation, so that the model can cope with complex traffic environment. Through the analysis of different conditions, it can be concluded that good weather and road conditions have a positive impact on the driver's keeping space headway. For example, the drivers usually keep a small space headway with good weather conditions, while bad weather is the opposite. According to the type of drivers, aggressive driver will keep a smaller space headway than average driver, and conservative driver always leave a larger space headway than average driver and so on.

Therefore, an influence coefficient acting on the expression of the following spacing function is needed. Its function is to change the space headway without affecting the accuracy of the model according to different traffic conditions, so that the simulation results of space headway can meet the forecasting requirements under complex traffic conditions.

According to the above discussion and analysis, transform the Eq. (1) at first:

$$x_{n-1}(t) - x_n(t) - S_{n-1} = \frac{v_n(t)+v_n(t+\tau)}{2}\tau + v_n(t+\tau)\theta + \frac{v_n{}^2(t+\tau)}{2b_n} - \frac{v_{n-1}{}^2(t)}{2b_{n-1}} \tag{4}$$

Equation (4) corresponds to the inter-vehicle following spacing at time t, takes the equal sign on both sides is because that in development of the model, only the minimum following spacing condition, that is, the equality part in Eq. (2) is considered. The velocity $v_n(t+\tau)$ which needs to be calculated is used to maintain the safe following spacing in the model. Equation (4) can also be expressed as:

$$Ds_n(t) = x_{n-1}(t) - x_n(t) - S_{n-1} = \frac{v_n(t)+v_n(t+\tau)}{2}\tau + v_n(t+\tau)\theta + \frac{{v_n}^2(t+\tau)}{2b_n} - \frac{{v_{n-1}}^2(t)}{2b_{n-1}} \quad (5)$$

$Ds_n(t)$ is the safe following spacing between vehicle n and n-1 at time t. It is now assumed that there exists a certain size relationship γ (influence coefficient) between the actual space headway $Dr_n(t)$ and $Ds_n(t)$ at time t. This makes it possible that:

$$Dr_n(t) = \gamma * Ds_n(t) \quad (6)$$

Move γ in Eq. (6) to the left of the equal sign, and use the method of zooming in and out to get the $Ds_n(t)'$:

$$\frac{Dr_n(t)}{\gamma} = \frac{Ds_n(t)'}{\gamma} = Ds_n(t) \quad (7)$$

The $Ds_n(t)'$ in Eq. (7) is the safe following spacing improved by the influence coefficient γ, which guarantees invariance of the element equilibrium relationship in the original model while changing the following spacing. Combining Eq. (3), (5), (7), a new model is obtained, which is called IC Gipps in this paper.

$$v_n(t+\tau) = min\left\{\begin{matrix} v_n(t) + 2.5a_n\tau\left(1-\frac{v_n(t)}{V_n}\right)\left(0.025+\frac{v_n(t)}{V_n}\right)^{\frac{1}{2}}, \\ -b_n\tau + \left\{b_n^2\tau^2 + b_n\left[\frac{2[x_{n-1}(t)-S_{n-1}-x_n(t)]}{\gamma} - v_n(t)\tau + \frac{v_{n-1}(t)^2}{\hat{b}}\right]\right\}^{\frac{1}{2}} \end{matrix}\right\} \quad (8)$$

As an extension, and increasing the persuasiveness of simulation experiment, the enhanced safety distance model [3] was improved using the influence coefficient method. This paper refers to it as IC ESD, and the mathematical expression is as follows:

$$v_n(t+\tau) = min$$

$$\left\{\begin{matrix} v_n(t) + 2.5a_n\tau\left(1-\frac{v_n(t)}{V_n}\right)\left(0.025+\frac{v_n(t)}{V_n}\right)^{\frac{1}{2}}, \\ -b_n\tau + \left\{b_n^2\tau^2 + b_n\left[\frac{2[x_{n-1}(t)-s_{n-1}-x_n(t)]}{\gamma(e^{-\alpha_1\Delta v_n(t)}H(\Delta v_n(t))+e^{\alpha_2\Delta v_n(t)}H(-\Delta v_n(t)))} - v_n(t)\tau + \frac{v_{n-1}(t)^2}{\hat{b}}\right]\right\}^{\frac{1}{2}} \end{matrix}\right\} \quad (9)$$

Where, α_1 and α_2 are two parameters that need to be calibrated, $\Delta v_n(t)$ is the velocity difference between vehicle n-1 and n at time t, and $H(x)$ is the Heaviside function which indicates a jump state of 0 to 1. The coefficient γ before the velocity difference function expression is the assumed influence coefficient, and the rest of the model remains unchanged.

4. Simulation Analysis

In this section, IC Gipps, IC ESD, Enhanced Safety Distance Model and Gipps Model will be used to simulate realistic vehicle trajectories, then compared with actual vehicle trajectory data, to discuss the simulation effect of influence coefficient method on space headway in complex traffic environment.

The selection of field data set is the US Highway 101 Dataset collected by NGSIM [4], on the one hand, this data set records real vehicle trajectory data for different traffic conditions and driver types under the same weather conditions, and on the other hand because this data set has been used in the research of YANG et al. [3] to calibrate the parameters of Gipps model and enhanced safe distance model. On this basis, the value of influence coefficient γ is calibrated. Meanwhile, the effective range of γ is studied, and the range is determined to be between 0.9 and 3.5. When γ is less than 0.9, the equilibrium conditions [5] of the original equation will be destroyed and the calculation results will be wrong. When γ is greater than 3.5, the model will always in a free-running state, which makes the model meaningless. The calibration results of the four models are shown in Table 1.

Table 1. The Parameter calibration result

Model	a_n $(m \cdot s^{-2})$	b_n $(m \cdot s^{-2})$	b_{n-1} $(m \cdot s^{-2})$	α_1	α_2	γ
Gipps	3.59	-3.3	-3.0	-	-	-
Enhance Safe Distance	3.37	-5.3	-4.2	0.073	0.069	-
IC Gipps	3.59	-3.3	-3.0	-	-	1.211
IC ESD	3.37	-5.3	-4.2	0.073	0.069	1.298

For the simulation experiment, using the four parameter calibrated models simulate 100 vehicles on a straight road respectively, the velocity of these vehicles oscillated through 300 simulation time steps and eventually stabilized at a congestion state about 5 m/s, a slightly congested state of about 10 m/s and a free-following state about 16 m/s. The average value of the simulated space headway in different states is compared with the average of the corresponding state in the field data set. The result is shown in Figure 1.

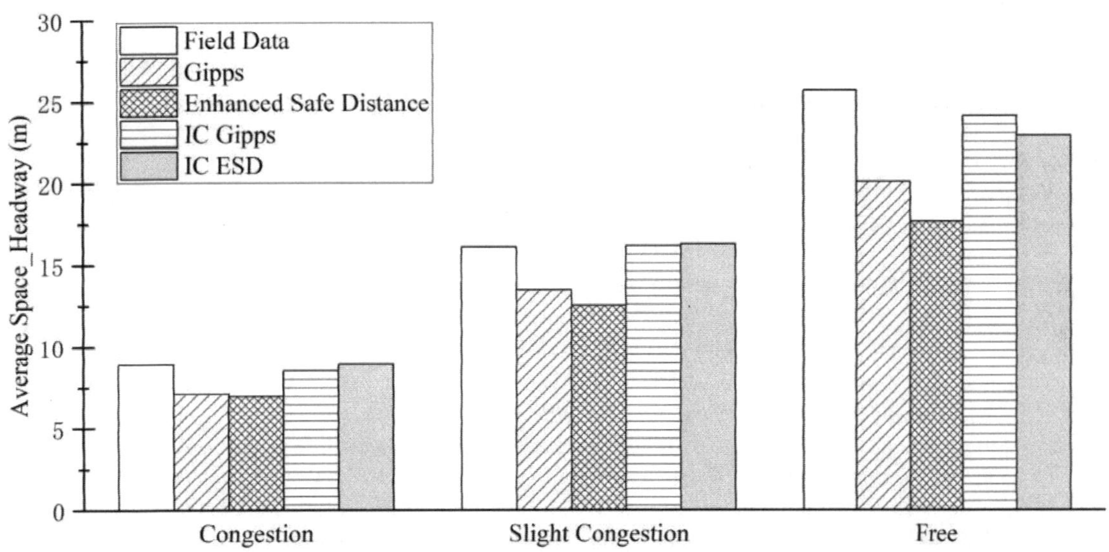

Figure 1. Comparison the space headway between different model and field data

From the experimental results of Figure 1, compared with the original models, the improved two models using the influence coefficient method can effectively improve the simulation accuracy of vehicle space headway in congestion state and slightly congested state, and compared with the field data, the error is smaller. The simulation accuracy of space headway in free-following state is also improved, but the lifting effect is not as good as the first two states, the reason is that the acceleration equation of the free-running state of the model is involved in this state, and the specific mean error is shown in Table 2. From the overall perspective, the simulation experiment proves that the addition of influence coefficient has a positive effect on improving the simulation accuracy of the model.

Table 2. The mean error of space headway in different states

Model	Mean Error		
	Congestion	Slight Congestion	Free
Gipps	1.80	2.63	5.59
Enhanced Safe Distance	1.94	3.58	8.04
IC Gipps	0.38	-0.07	1.57
IC ESD	-0.02	-0.16	2.75

5. Conclusions

In this paper, the main problem of Gipps model in the simulation of vehicle space headway is analyzed, that is, the simulation accuracy of the space headway in complex traffic environment cannot be well satisfied. Although Biagio et al. [6] proposed that calibrate the model for different specific situations, the accuracy of the model can be improved. However, there are so many factors affecting the simulation accuracy in the real traffic environment, and it is impossible to calibrate each factor in turn. In view of the above reasons, the method of adding influence coefficient to the safe following spacing expression of the model is proposed. By changing the influence coefficient, the model can meet the prediction requirements of the space headway very well under different conditions and improve the simulation precision.

Although the method of adding influence coefficient to the model can achieve better results, the improvement of simulation accuracy is limited only by simply adding influence coefficients. A better way is to allocate the proportion of vehicles with different influence coefficients in the simulation reasonably, and combine with multi-agent simulation system to make it more flexible and accurate to reflect the real traffic situation, which is also the future research direction.

Reference:

[1] Irene Soria, Lily Elefteriadou, Alexandra Kondyli. Assessment of car-following models by driver type and under different traffic, weather conditions using data from an instrumented vehicle. *Simulation Modelling Practice and Theory* **40** (2014) 208–220.

[2] Gipps P G. A behavioural car-following model for computer simulation. *Transportation Research. Part B: Methodological*, 1981, **15(2):** 105-111.

[3] YANG Da, ZHU Li-Ling, YU Dan. An Enhanced Safe Distance Car-Following Model. *J. Shanghai Jiaotong Univ. (Sci.)*, 2014, **19(1):** 115-122

[4] NGSIM Homepage. FHWA. http://ngsim.fhwa.dot.gov.

[5] Wilson R E, Ward J A. Car-following models: Fifty years of linear stability analysis — A mathematical perspective *[J]. Transportation Planning and Technology,* 2011, **34(1):** 3-18.

[6] Biagio Ciuffo, Vincenzo Punzo, Marcello Montanino. Thirty Years of Gipps' Car-Following Model Applications, Developments, and New Features. *Transportation Research Record: Journal of the Transportation Research Board*, **2315:**89–99, December 2012.

Study on Penetration Resistance of Ceramic Composite Armor under Different Contact Stiffness

Li Feng[1], Shi Quan[2] and Chen Cai[3]

[1]Department of equipment command and management, Shijiazhuang campus, Army Engineering University, Shijiazhuang, 050003
[2]Department of equipment command and management, Shijiazhuang campus, Army Engineering University, Shijiazhuang, 050003
[3]Department of equipment command and management, Shijiazhuang campus, Army Engineering University, Shijiazhuang, 050003

E-mail address: 1121373995@qq.com

Abstract. The finite element software ANSYS/LS-DYNA is used to simulate the penetration of tungsten alloy long rod projectile into ceramic composite armor. The penetration effect of tungsten alloy long rod projectile penetrating ceramic composite armor was obtained under different contact stiffness. At the same time, the theoretical model is used to calculate the residual velocity of tungsten alloy long rod projectile. It is found that the simulation results and theoretical calculation results are better.According to the fitting of simulation data, the relationship between penetration depth and loss velocity and incident velocity is obtained according to the different contact stiffness. It will provide some reference and reference value for studying the penetration effect of tungsten alloy long rod projectile on ceramic composite armor at a certain speed.

1. Introduction

In the future war, with the rapid development of science and technology, the speed of military attack is faster and faster, the probability of hitting weapon system is higher and higher, and the ability of damage is stronger and stronger. The protective performance of single material armor can no longer meet the requirements of future war. Therefore, the research of composite material armor is getting more and more popular.Ceramic composite armor has the characteristics of high strength, high hardness, corrosion resistance, high wear resistance and light weight. It is widely used in the protection of tanks, aircraft, ships, vehicles and other fields[1]. Experiments have proved that the protective performance of many advanced tank armors in the world has been greatly improved by using high-performance ceramics. Ceramics have become one of the indispensable materials for composite armor[2]. Wu Hai Ling[3] obtained the best match between ceramic and panel by changing the thickness of the panel. Sun Sujie[4], Cao Bing[5] ,Wang Jinhu[6]and others reasonably analyzed the anti-elastic performance of ceramic composite target by changing the different materials of target plate by numerical simulation and experiment. Xiong Fei[7] established homogeneous steel equivalent target of ceramic composite armor by changing penetration speed. However, these studies do not analyze the influence of contact stiffness on the penetration resistance of the projectile.

2. Establishment of finite element model

2.1. Finite element model

The finite element model is composed of tungsten alloy long rod projectile and three layers of target plate. Each part is composed of eight-node hexahedron element. The length of tungsten alloy long rod projectile is 93mm, the head of the missile body is hemispherical, the diameter is 6mm, and the aspect ratio is 15. The ceramic composite target plate is sandwich structure. The first layer and the third layer are 4043 armored steel of 100mm x 100mm x 5mm and 100mm x 100mm x 15mm respectively. The second layer is 100mm x 100mm x 30mm ceramic target plate. In order to ensure the calculation accuracy of the model, the mesh of the contact area between the projectile and the target plate is 1 mm, and the other meshes are 2 mm. Lagrange algorithm is used for all elements, which attaches the mesh to the material. The deformation of the mesh can well reflect the deformation of the material, thus accurately describing the structural changes of the material. Based on the symmetry of the model, the calculation time is saved, and the whole 1/4 of the model is used to model and calculate, and the symmetrical boundary is applied on the symmetrical surface. In order to ensure that the contact can still be carried out in the remaining units after the failure of the projectile unit, ERODING _SURFACE _TO _SURFACE is used to contact the projectile and the target plate. The finite element model is shown in Figure 1.

Figure.1 finite element calculation model.

2.2. Selection of material parameters

The constitutive model of tungsten alloy long rod elastic material is Plastic_Kinematic. The mechanical properties of 4043 armor steel are described by Johnson_Cook material model and Gruneisen equation of state. The ceramic material is described by Johnson_Holmquist_Ceramics. The determination of material parameters is shown in table 1~ table 3.

Table 1. Material parameters of 4340 steel[7]

Parameter	Numerical value	Parameter	Numerical value	Parameter	Numerical value
$\rho/(g\cdot cm^{-3})$	7.83	B/MPa	510	TM/K	1793
G/GPa	77	n	0.26	TR/K	294
A/MPa	792	c	0.014	C	477

Table 2. Material parameters of tungsten alloy projectile[8]

Parameter	Numerical value	Parameter	Numerical value	Parameter	Numerical value

ρ/(g·cm^{-3})	17.5	σ/GPa	0.23	C	40
E/GPa	310	Et/GPa	0	P	5
v	0.3	β	1	Fs	0.3

Table 3. Material parameters of alumina ceramic[9]

Parameter	Numerical value	Parameter	Numerical value	Parameter	Numerical value
ρ/(g·cm^{-3})	3.25	n	0.6	K1/GPa	220
G/GPa	90.16	HEL/GPa	10	K2/GPa	0
A/GPa	0.93	PHEL/GPa	5	K3/GPa	0
B/GPa	0.31	T/GPa	0.3	Fs	1.6
c	0	D1	0.005	ε	0
m	0.6	D2	1		

3. Theoretical model

3.1. General form of de Mar formula

In experimental study on the equivalent relationship between anti-fragment penetration of target plates of different materials, the experiment of tungsten ball penetrating armored steel is carried out, and the conclusion that the main penetrating factor of tungsten ball penetrating armored steel is shear punching plug is obtained, and the composite coefficient of armored steel is 52935. Therefore, the De Marr formula which can well reflect this phenomenon is used to calculate. The general formula of the de Mar formula for the vertical penetration of tungsten alloy long rod projectile is formula (1):

$$V_j = K\frac{d^{0.75} \times b^{0.7}}{m^{0.5}} \tag{1}$$

In the formula, Vj is the limit penetration velocity (m/s), d is the projectile diameter (m), b is the thickness of the first target plate (m), and K is the composite coefficient of penetration.

De Marr formula can calculate the ultimate penetration velocity more accurately when tungsten alloy long rod projectile penetrates armor steel of the same material. In experimental study on penetration failure modes of targets with different materials, the residual velocity of projectile is calculated by using the formula. Through comparative analysis, it is concluded that the residual velocity calculated by this algorithm is basically the same as that obtained by test. But when tungsten alloy long rod projectile penetrates ceramic composite target, the ultimate penetration velocity of long rod projectile can not be calculated accurately only by using the general De Mar formula because of the different material of target.

3.2. Revised De Mar formula

The general form of de Mar formula reflects a comprehensive coefficient of the physical properties of armored materials. For ceramic composite armor, the armor is composed of many kinds of materials. Because of the different materials, the shear strength of each layer of target plate is different, and the value varies with the material. At the same time, when the tungsten alloy long rod projectile penetrates the ceramic composite armor at different speeds, the ceramic composite armor resists the tungsten alloy long rod projectile. The elastic energy is also different, so it is necessary to correct the general form of De Mar formula.

The revised DeMar formula is shown in equation (2):

$$V_j = \left(\sum_i^n K_i \frac{d^{0.75} \times b_i^{0.7}}{m^{0.5}} \right) \times (A+B), i = 1,2,3\Lambda\ n \tag{2}$$

$$A = \frac{G_i}{G_{\text{armored steel}}} \tag{3}$$

$$B = f_s \frac{V_i - \mu}{\sigma} \tag{4}$$

In the formula, Vj is the ultimate penetration velocity (m/s), d is the diameter of projectile (m), bi is the thickness of the first target plate (m), Ki is the composite coefficient of armor piercing, which comprehensively reflects the factors affecting armor piercing, such as the properties of target plate and projectile materia, A is the ratio of shear strength of the first target plate to that of armor steel, B is the penetration of projectile by velocity. When the target is obstructing velocity, fs is the coefficient of dynamic friction.

4. Analysis of simulation results

4.1. Comparison between Theory and Simulation Results

Table 4. Comparison between theoretical calculation and simulation test results

Incident velocity (m/s)	Theoretical results (m/s)	Simulation results (m/s)	Error(%)
1000	402.86	369.42	8.30%
1100	420.35	391.18	6.94%
1200	443.59	440.18	0.77%
1300	451.65	453.15	-0.33%
1400	465.71	454.87	2.33%

Table 4 is a comparison between the theoretical calculation residual velocity and the simulation test value of the penetration of ceramic composite target plate by tungsten alloy long rod projectile with different velocity under the condition of contact stiffness 8. From the results in Table 4, it can be seen that the theoretical calculation model is in good agreement with the simulation test results, and the errors under the above conditions are less than 10%. The rationality of the simulation model, parameters, algorithm and the reliability of the results are explained.

4.2. Contact Stiffness Analysis

According to the finite element model (Figure. 1) established above, the numerical simulation is carried out by changing the incident velocity so that the projectile penetrates vertically into the target plate with different contact stiffness at different incident velocities. A total of 50 cases are calculated, and the ceramic composite assembly of the projectile body with different contact stiffness at different incident velocities is obtained. A penetration depth and loss velocity. The results are shown in Table 5 and Table 6.

Table 5. Penetration depth of projectiles penetrating different contact stiffness targets at different incident velocities

Contact stiffness	Incident velocity (m/s)				
	1000	1100	1200	1300	1400
1	36.25	37.00	37.95	38.75	40.32

2	43.70	48.50	49.40	50.00	50.00
3	48.60	49.30	50.00	50.00	50.00
4	48.57	50.00	50.00	50.00	50.00
5	48.40	49.00	50.00	50.00	50.00
6	49.65	50.00	50.00	50.00	50.00
7	49.73	50.00	50.00	50.00	50.00
8	50.00	50.00	50.00	50.00	50.00
9	50.00	50.00	50.00	50.00	50.00
10	50.00	50.00	50.00	50.00	50.00

As shown in Table 5, with the increase of contact stiffness, the anti-penetration ability of the target decreases, and the target penetrates the target eventually. Laterally, the penetration depth increases with the increase of incident velocity. When the contact stiffness is 1, the penetration depth is less than 50 mm and does not penetrate the target plate. This shows that the target plate can achieve a good protection effect under this contact stiffness.

Table 6. LOSS VELOCITY OF PROJECTILE PENETRATING DIFFERENT CONTACT STIFFNESS TARGETS AT DIFFERENT INCIDENT VELOCITIES

Contact stiffness	Incident velocity (m/s)				
	1000	1100	1200	1300	1400
1	1000.00	1100.00	1200.00	1300.00	1400.00
2	1000.00	1100.00	1200.00	845.94	906.45
3	1000.00	1100.00	766.78	853.02	934.73
4	1000.00	713.63	777.86	847.89	923.27
5	1000.00	718.49	784.97	849.27	940.92
6	1000.00	689.27	786.09	870.93	936.39
7	1000.00	706.02	772.79	862.90	935.95
8	630.58	708.82	759.82	846.85	945.13
9	666.06	690.12	775.77	835.02	924.66
10	645.90	711.90	781.35	854.37	937.95

From Table 6, it can be seen that, in the case of penetrating the target plate, the velocity of projectile loss varies slightly, but slightly, due to the different contact stiffness. From the transverse point of view, in the case of penetrating the target plate, the greater the incident velocity of the projectile, the greater the loss of the projectile velocity caused by the target plate. This shows that the elastic performance of the composite target plate is not the only determinate one. The elastic performance of the composite target plate varies greatly under different working conditions. The greater the incident velocity of the projectile in this range of velocity, the velocity of the target plate against the projectile will be increased. The ability to weaken is stronger.

Generally speaking, the penetration effect of tungsten alloy long rod projectile with different contact stiffness is different. When the contact stiffness is set to 1, even if the incident velocity of the projectile is changed, the projectile does not penetrate the target plate, which achieves good protection effect. When the contact stiffness is set to 8, 9 and 10, the protective effect of the target plate is the worst. Even at the speed of 1000m/s, the projectile can penetrate the target plate.

In the case of non-penetrating target (contact stiffness is 1), the penetration depth of projectiles at different velocities is fitted by polynomials, and the normalized incident velocities are obtained as shown in Figure 2:

$$h = 0.1178x^3 + 0.2661x^2 + 1.308x + 37.84$$

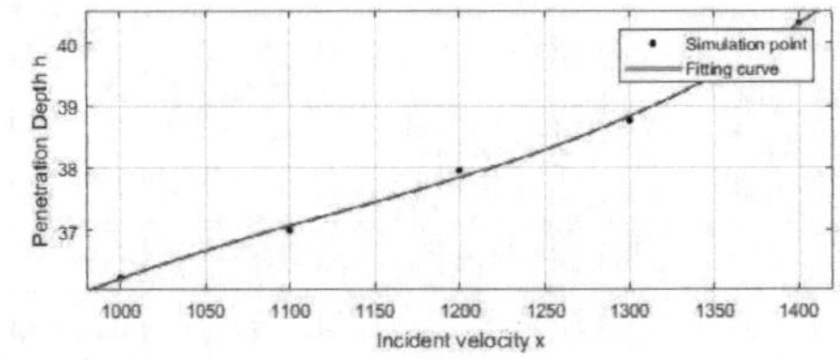

Figure. 2 Fitting curve of penetration depth and incident velocity.

In the case of penetrating the target plate (i.e. contact stiffness is 8, 9 and 10 respectively), the projectile loss velocities at different velocities are polynomially fitted, and the normalized incident velocities are obtained as shown in Tables 7, Figure 3, Figure 4, and Figure 5.

Table 7. Relationship between projectile loss velocity and incident velocity under different contact stiffness

Contact stiffneSs	Projectile loss velocity (m/s)	Goodness of fit
8	$V_{s8} = 12.68x^3 + 13.59x^2 + 104.1x + 767.4$	0.998
9	$V_{s9} = -10.28x^3 + 18.71x^2 + 118.7x + 763.4$	0.993
10	$V_{s10} = 2.342x^3 + 6.916x^2 + 111.7x + 780.8$	0.999

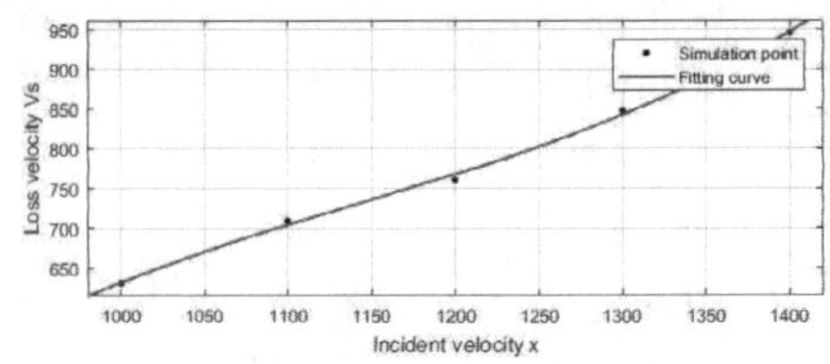

Figure. 3 Fitting curve of relation between projectile loss velocity and incident velocity when contact stiffness is 8.

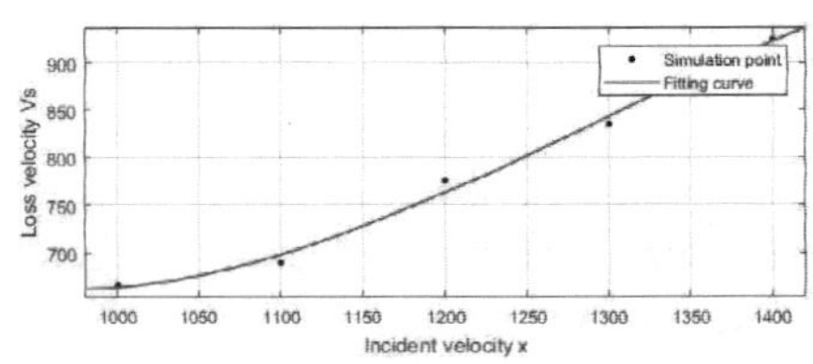

Figure.4 Fitting curve of relation between projectile loss velocity and incident velocity when contact stiffness is 9.

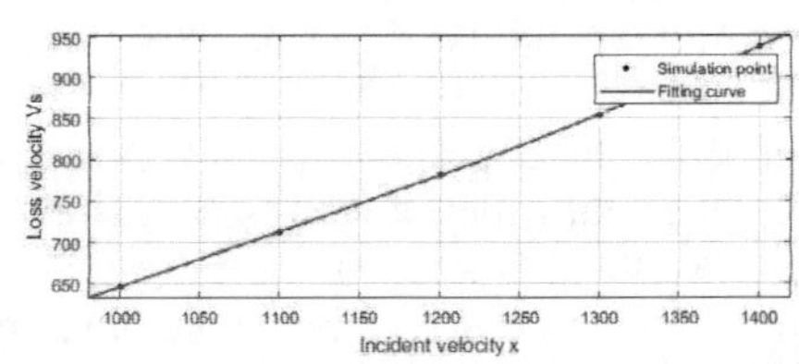

Figure.5 Fitting curve of relation between projectile loss velocity and incident velocity when contact stiffness is 10.

When the contact stiffness is set to 1, the penetration depth of tungsten alloy long rod projectile can be estimated according to the fitting formula. When the contact stiffness is set to 8, 9 and 10, the loss velocity of tungsten alloy long rod projectile can be estimated according to the fitting formula. The calculated formula can save a lot of time and get more accurate results.

5. Conclusion

The penetration of tungsten alloy long rod projectile into ceramic composite armor was simulated by finite element software, and the residual velocity of tungsten alloy long rod projectile was obtained. Based on the theoretical model, the limit penetration velocity of tungsten alloy long rod projectile is calculated, and then the residual velocity of target plate is calculated. The rationality of the model is verified by comparing the two results. The penetration depth and loss velocity of tungsten alloy long rod projectile penetrating ceramic composite armor at different incident velocities under different contact stiffness were calculated by simulation, and the results were analyzed. The results show that when the contact stiffness is set to 1, the anti-penetration ability of ceramic composite armor is the best. With the increase of contact stiffness, the anti-penetration ability of ceramic composite armor decreases. When the contact stiffness is greater than 8, the anti-penetration ability of ceramic composite armor is the worst, and it can not protect tungsten alloy long rod projectile whose velocity is more than 1000m/s.

References

[1] Kang Y and Chai X 2011 Research and development of ceramic composite armor materials. *Foshan ceramics,21* (1), 44-45.
[2] Shen F 1992 Research progress of vehicle and human protective materials. *Ordnance material science and engineering,22*(6): 53-57.
[3] Wu H L, Wu X, Wang J H, Wang Q, Zhong T, Miao C, Li S T and Bai L H 2012 Experimental study on the effect of panel thickness on the elasticity of ceramic composite targets. *Ordnance material science and engineering, 35* (1), 62-65.
[4] Sun S J, Zhao B R, Wang J and Wang Z Q 2006 Study on the influence of different backplanes on the elastic resistance of ceramic composite armor. *Ordnance material science and engineering, 29* (2),70-72.
[5] Cao B 2006 Experimental study on the equivalent relationship between anti-fragment penetration of target plates of different materials. *Journal of Missile and Guidance, 26* (4), 113-114.
[6] Wang J H, Jiao J H and Li Q 2013 Experimental study on penetration failure modes of targets with different materials. *Journal of Zhongbei University (Natural Science Edition), 34* (6), 623-627.
[7] Xiong F, Shi Q, Wang G Y and Chen Cai 2015 Establishment of equivalent homogeneous steel target plates for ceramic composite armor at different penetration velocities. *Fire control and command control*(4), 72-75.
[8] Wu Y H 2006 ceramic composite target. *Beijing. Beijing Institute of Technology.*
[9] Ma Z Y, Zeng S Y and Jiang Z G 2007 Numerical simulation and analysis of the resistance of ceramic composite target to long-rod projectiles. *Journal of China University of Science and Technology, 37* (7), 727-731.

Armature Optimization Design Based on Orthogonal Test

Qing Rong Chen[1, a], Tao Shu[2, b], Ming Liu[1, c] and Meng Yang Cui[1, d]

[1] Graduate College, Air Force Engineering University, Xi'an 710051,China;
[2] Air Defense and Missile Defense College, Air Force Engineering University, Xi'an 710051, China

[a]1034021926@qq.com; [b]st9711@sina.com; [c]929237749@qq.com; [d]1104086310@qq.com,

Abstract. Armature is the core component of electromagnetic launcher. It not only plays the role of constituting current loop, but also provides launching power for projectile. Therefore, armature optimization is a key step to improve the comprehensive performance of six-pole orbital electromagnetic launcher. In this paper, the finite element simulation software Ansoft Maxwell is used to simulate and analyze the armature, and the influence of the main size of armature on the maximum current density and electromagnetic thrust of armature are obtained. At the same time, the optimal design of armature size is carried out by using matrix analysis method in orthogonal experiment. Finally, the optimized armature is verified by simulation. The results show that the optimized armature has the smallest maximum current density and the largest electromagnetic thrust in all size combinations.

1. Introduction

Electromagnetic railgun is a new concept weapon that uses electromagnetic force to launch projectiles. It breaks through the speed limit of traditional guns and can easily accelerate projectiles to several kilometers per second, thus destroying targets with great kinetic energy[1]-[3]. In recent years, in order to further enhance the killing efficiency and precision strike ability of electromagnetic railgun, scholars at home and abroad have begun to study the use of railgun to launch intelligent ammunition[4]-[6]. In the traditional bipolar orbital electromagnetic launcher, the electromagnetic field environment of the projectile location is bad, which can not meet the launching requirements of intelligent ammunition having high precision instruments. six-pole orbital electromagnetic launcher can make use of its own structural advantages to generate weak magnetic field in the armature center area to realize electromagnetic shielding protection for intelligent ammunition. At the same time, it can provide more electromagnetic thrust for armature at lower current, which has strong research significance.

Whether the traditional bipolar railgun or the six-pole orbital electromagnetic launcher, there are problems such as excessive concentration of armature current, ablation, melting and so on, which can reduce the service life and launch efficiency of the launcher[7]. Reasonable design and optimization of armature structure of six-pole orbital electromagnetic launcher will help to improve the uniform distribution of armature current, reduce the maximum current density, inhibit the occurrence of ablation, increase the electromagnetic force on armature, and improve the emission efficiency. Hughes[8] et al revealed the current distribution characteristics of armature with simple shape under steady state conditions. Satapathy[9] et al studied the effect of armature geometry change on magnetic field and force. Ferrero R[10] et al studied the inductance gradient of the launcher and the distribution of current density in the armature.

This paper focuses on the influence of armature structure size change of six-pole orbital electromagnetic launcher on the maximum current density and armature force. orthogonal test is used to optimize armature structure to determine the optimal size of current distribution and armature force.

2. Basic principles of six-pole orbital launcher

The model of six-pole orbital electromagnetic launcher is shown in figure 1 and the cross section of orbital is shown in figure 2. During the launching process, the current flows into the 1, 3 and 5 orbits with symmetrical distribution of 120 degrees and equal spacing, and flows back to the 2, 4 and 6 orbits through the plum blossom shaped armature to form a closed loop. The circular magnetic field generated by orbital current interacts orthogonally with the current in armature, which generates Z-direction ampere force and promotes the armature to emit outward at high speed.

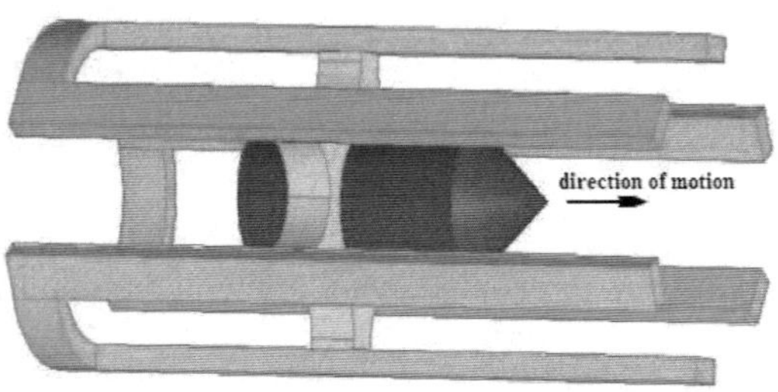

Figure 1. Six-pole orbiter launcher model. The armature is driven by electromagnetic force, and push the missile out.

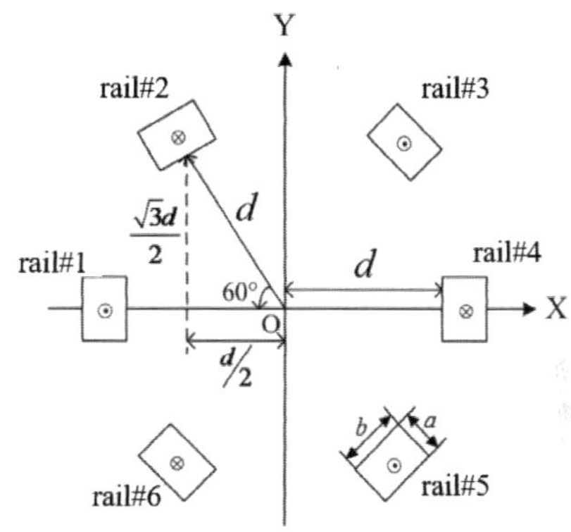

Figure 2. Rail cross section. It is looking from the z axis.

Assuming that the orbital current source point is $S(x, y, z)$, the field point on the armature is $P(x_1, y_1, z_1)$ [11]. The current element $Id\boldsymbol{l}$ located in the source point S, so the magnetic flux density generated by the current element $Id\boldsymbol{l}$ at the field point P is

$$d\boldsymbol{B} = \frac{\mu_0 I}{4\pi R^3} d\boldsymbol{l} \times \boldsymbol{R} \tag{1}$$

Among them, μ_0 is the vacuum permeability, I is the intensity of the current flowing through the guide rails, $\boldsymbol{R}$ stand for S to P distance vector, $\boldsymbol{R} = (x_1 - x)\boldsymbol{i} + (y_1 - y)\boldsymbol{j} + (z_1 - z)\boldsymbol{k}$.

Considering the skin effect, it is assumed that the current is concentrated on the inner surface of the orbital. According to the definition of surface current element, the current density on the orbital surface is known as $J_s = I/b$ [12]. Assuming the current source point of rail 1 is $S(-d, y, z)$, the field point on the armature is $P(0,0,z_1)$, which is on the center point of the armature. The thickness of armature is ignored in calculation. Rail 1 current element $Id\boldsymbol{l} = J_s dydz\boldsymbol{k}$, $\boldsymbol{R} = d\boldsymbol{i} - y\boldsymbol{j} + (z_1 - z)\boldsymbol{k}$, $R = \sqrt{d^2 + y^2 + (z_1 - z)^2}$.

$$\begin{aligned} d\boldsymbol{B} &= \frac{\mu_0 I}{4\pi R^3} d\boldsymbol{l} \times \boldsymbol{R} = \frac{\mu_0 J_s}{4\pi R^3} [ddydz\boldsymbol{j} + ydydz\boldsymbol{i}] \\ &= dB_x \boldsymbol{i} + dB_y \boldsymbol{j} \end{aligned} \tag{2}$$

IOP Conf. Series: Journal of Physics: Conf. Series **1284** (2019) 012018 doi:10.1088/1742-6596/1284/1/012018

The scope of z is $(0, z_1)$, the scope of y is $(-b/2, b/2)$. The magnitude of the magnetic induction intensity generated by rail 1 along the X axis B_{1x} at the armature and along the Y axis B_{1y} are as follows

$$B_{1x} = \frac{\mu_0 J_s}{4\pi}\int_0^{z_1}\int_{-b/2}^{b/2}\frac{y}{R^3}dydz = \frac{\mu_0 J_s}{4\pi}\int_0^{z_1}\int_{-b/2}^{b/2}\frac{y}{[d^2+y^2+(z_1-z)^2]^{3/2}}dydz \tag{3}$$

$$B_{1y} = \frac{\mu_0 J_s}{4\pi}\int_0^{z_1}\int_{-b/2}^{b/2}\frac{d}{R^3}dydz = \frac{\mu_0 J_s}{4\pi}\int_0^{z_1}\int_{-b/2}^{b/2}\frac{d}{[d^2+y^2+(z_1-z)^2]^{3/2}}dydz \tag{4}$$

It can be seen from figure 2 that rail 1 and rail 4 are symmetrical about the YOZ plane, so the magnetic field generated by the currents in rail 1 and rail 4 at point $P(0,0,z_1)$ in the x direction is zero, and the size in the Y direction is equal, the direction is the same. So we can solve the magnetic field of one rail in the Y direction, then multiply by 2. Supposing B_{14} is the magnetic field at $P(0,0,z_1)$ generated by rail 1 and rail 4, then

$$\boldsymbol{B}_{14} = (B_{1y} + B_{4y})\boldsymbol{j} = 2B_{1y}\boldsymbol{j} \tag{5}$$

Similarly, the magnetic field $\boldsymbol{B}_{25}$ produced by rail 2 and rail 5 in $P(0,0,z_1)$, and the magnetic field $\boldsymbol{B}_{36}$ produced by rail 3 and rail 6 in $P(0,0,z_1)$ are obtained. Because the current in each rail is the same size, the magnetic fields generated by each rail are equal in magnitude, as shown in figure 3.

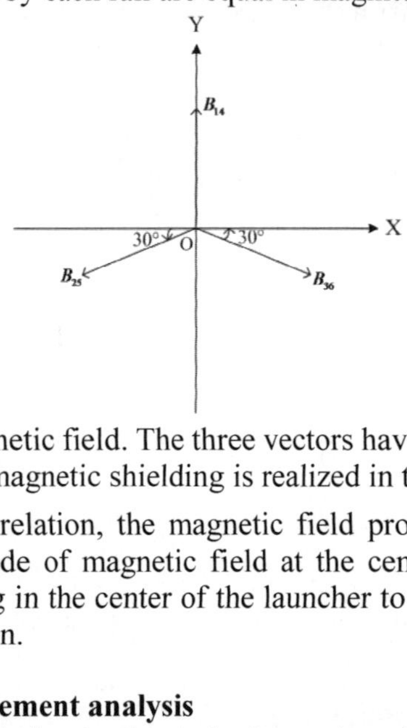

Figure 3. Sketch map of magnetic field. The three vectors have the same size and in different direction. Electromagnetic shielding is realized in the armature center.

According to the geometrical relation, the magnetic field produced by the six rails is mutually counteracted, that is, the magnitude of magnetic field at the center of the armature is zero, which makes the magnetic field shielding in the center of the launcher to be realized, and is beneficial to the launching of intelligent ammunition.

3. Simulation model and finite element analysis

The armature model used in simulation is shown in figure 4, and the one-sixth model is selected as shown in figure 5. In order to simulate the launch process of a large volume object, $h = 0.2$m is the thickness of armature head, length of tail is $d_s = 0.2$m and thickness of tail is $d_0 = 0.1$m, the fillet radius of the connection between the armature tail and the head is $r = 0.1$m .The total length of the rail is 4 m , and in cross section $a = 0.16$m, $b = 0.24$m. The armature material is aluminum and the material of the rail is copper.

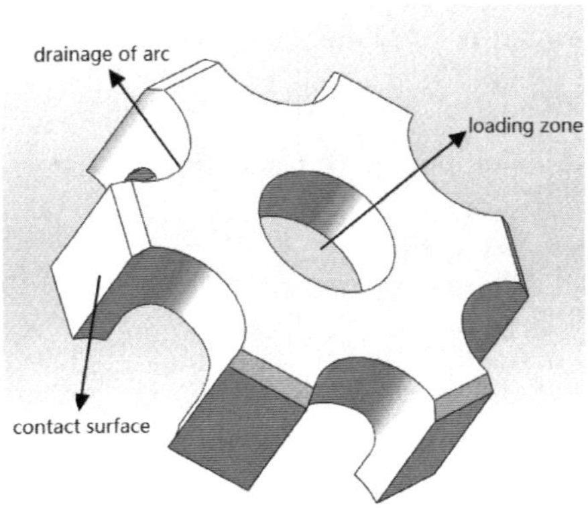

Figure 4. Armature model. Loading zone is used to place smart ammunition like missile. Armature and rail meet through contact surface. The current is guided by the arc.

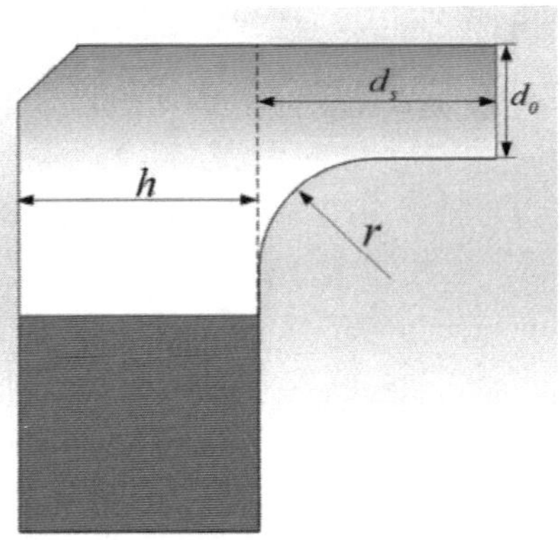

Figure 5. One-sixth of Armature model. It shows four main size of armature, including thickness of armature head, length of tail, thickness of tail and the fillet radius.

The peak value of pulse current applied in orbit is 500kA, and the maximum length of the grid of armature is 0.05m. The solution domain of vacuum boundary condition and 5 times model size are selected. Maxwell 3D module in ANSYS software was used to analyze the model. In this paper, the armature is mainly analyzed, and the current distribution cloud diagram of the armature surface is shown in figure 6, and the electromagnetic force of the armature is shown in table 1.

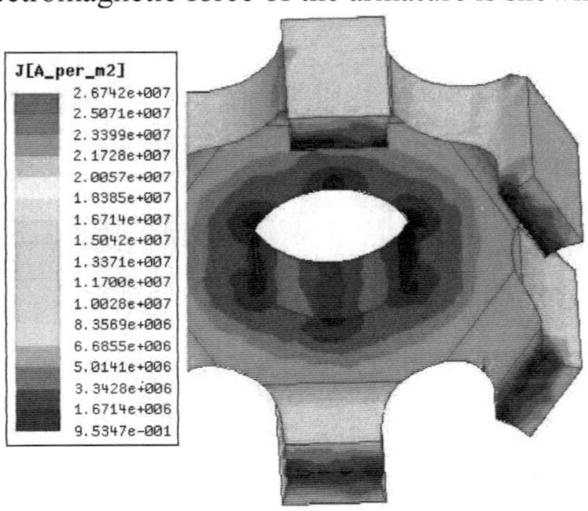

Figure 6. Current distribution cloud map. It shows that the current concentrate on the corner and the maximum current density is $2.6742\times10^7\,A/m^2$.

Table 1. Electromagnetic force of armature.

Electromagnetic force	Armature
F_x	870N
F_y	4720N
F_z	2.13×10^5 N
Mag F	2.14×10^5 N

It is easy to see from figure 6 that the current distribution on the armature is relatively uniform, and the current distribution is symmetrical about the center of the circle. The current is less distributed in the middle of the armature, and the darker area appearing at the connection between the tail fin and the armature main body is very small. The maximum current density appears in the corner area, and the maximum current density is $2.6742\times10^7\,A/m^2$. The armature is mainly subjected to the z-direction thrust, which propelling armature outward emission, and the interference force in the x and y directions is small.

Keep the other conditions constant. The maximum current density of armature and the variation of electromagnetic force can be obtained by changing the size of the armature, such as the head thickness, tail length, tail thickness and fillet radius, as shown in figure 7 and figure 8.

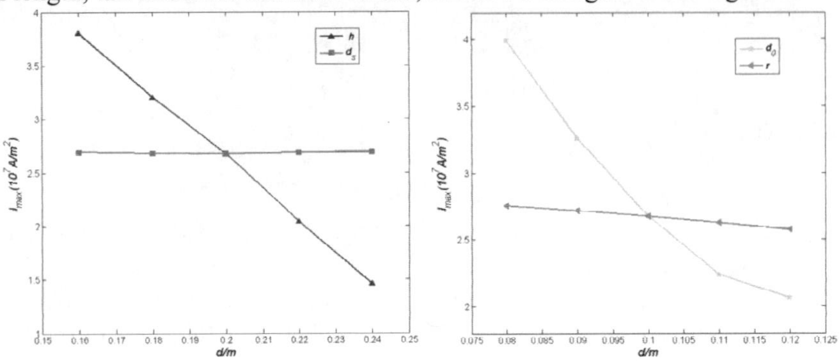

(a) (b)

Figure 7. Change law of maximum current density. The above two figures show that the maximum current density change with four main structure of the armature.

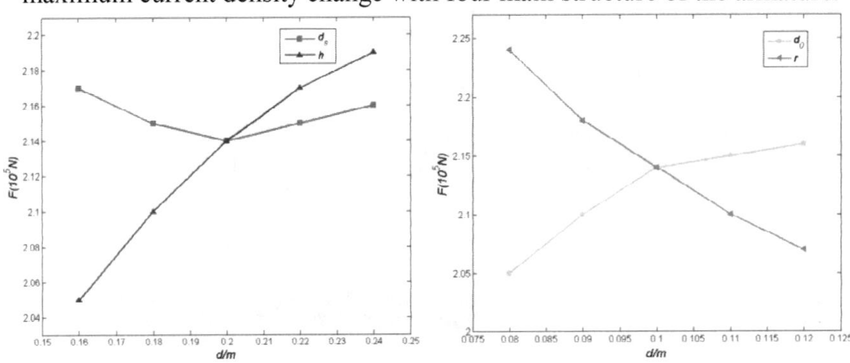

(a) (b)

Figure 8. Change law of electromagnetic force. The above two figures show that the electromagnetic force change with four main structure of the armature.

It is easy to get from the figure that the maximum current density decreases with the increase of the thickness of the head, the thickness of the tail, and the radius of the fillet. By comparison, with the increase of radius, the maximum current density decrease amplitude is lower. The change in tail length has little effect on the maximum current density. The force of the armature increases with the increase of the thickness of the head and the thickness of the tail, and decreases with the increase of the fillet radius. The change of the length of the tail makes the electromagnetic force decrease first and then increase, but the change is small.

4. Orthogonal test to optimize armature

According to the experimental procedure of the orthogonal test, first determine the test index, factors and levels, the test indicators for the maximum current density and the electromagnetic force, there are four factors that affect the test index, there are three levels, the thickness of the head (A): 0.18m, 0.20m, 0.22m. Length of tail (B): 0.18m, 0.20m, 0.22m. Tail thickness (C): 0.09m, 0.10m, 0.11m. Corner radius (D): 0.09m, 0.10m, 0.11m. Secondly, select the appropriate orthogonal test table according to the factors and the number of levels. $L_9(3^4)$ orthogonal test table selected. Redesign the test plan and conduct the test. Fill the test results into table 2. According to the test results, the range analysis table 3 and table 4 of the two indexes are obtained. Finally, the test results are analyzed by matrix analysis method, and the armature size is obtained, which makes the maximum current density as small as possible and the electromagnetic force as large as possible.

Table 2. Orthogonal test scheme.

Test Serial Number	A	B	C	D	Maximum Current Density ($\times10^7$A/m^2)	Electromagnetic Force ($\times10^5$N)
1	1	1	1	1	3.82	1.94
2	1	2	2	2	3.19	1.93
3	1	3	3	3	2.71	1.93
4	2	1	2	3	2.63	2.19
5	2	2	3	1	2.28	2.21
6	2	3	1	2	3.26	2.11
7	3	1	3	2	1.61	2.40
8	3	2	1	3	2.56	2.28
9	3	3	2	1	2.08	2.41

Table 3. Range analysis table of maximum current density.

Test Serial Number	A	B	C	D
$\overline{K}_1$	3.24	2.69	3.21	2.73
$\overline{K}_2$	2.72	2.68	2.63	2.69
$\overline{K}_3$	2.08	2.68	2.20	2.63
R	1.16	0.01	1.01	0.10

Table 4. Range analysis table of electromagnetic force.

Test Serial Number	A	B	C	D
$\overline{K}_1$	1.93	2.18	2.11	2.19
$\overline{K}_2$	2.17	2.14	2.18	2.15
$\overline{K}_3$	2.36	2.15	2.18	2.13
R	0.43	0.04	0.07	0.06

By using the method of orthogonal test matrix analysis, the comprehensive weight matrix of the two performance indicators is obtained as follows

$$\begin{aligned} \boldsymbol{k} &= \begin{bmatrix} 0.1782 & 0.2048 & 0.2405 & 0.0117 & 0.0115 & 0.0115 & 0.0785 & 0.0923 & 0.1067 & 0.0236 & 0.0233 & 0.0233 \end{bmatrix}^{\mathrm{T}} \\ &= \begin{bmatrix} A_1 & A_2 & A_3 & B_1 & B_2 & B_3 & C_1 & C_2 & C_3 & D_1 & D_2 & D_3 \end{bmatrix}^{\mathrm{T}} \end{aligned} \quad (6)$$

The greater the weight, the greater the impact on the results, so the optimal collocation scheme for the two performance indexes is $A_3B_1C_3D_1$, the armature head thickness h=0.22m, tail length d_s=0.18m, tail thickness d_0=0.11m, corner radius r=0.09m. After the simulation of the optimized armature, the armature maximum current density is $1.6123\times10^7\,\mathrm{A/m^2}$, the electromagnetic force exerted on the armature is $2.43\times10^5\,\mathrm{N}$. Compared with other sizes of armature, the maximum current density and the electromagnetic force are all optimal.

5. Conclusion

In this paper, the model of Hexapole orbital electromagnetic launcher is established by using finite element analysis software Ansoft Maxwell, and the superiority of the model in launching intelligent ammunition is proved by theoretical analysis and simulation verification. Secondly, by changing the armature structure size of Hexapole track electromagnetic launcher, the effects of armature head thickness, tail length, tail thickness and fillet radius on armature current density maximum and armature force are obtained. The maximum current density decreases with the increase of head thickness, tail thickness and fillet radius, and the change of tail length has little effect on the maximum current density. The armature force increases with the increase of the thickness of the head and tail, and decreases with the increase of the radius of the fillet. The armature force decreases first and then increases with the change of the length of the tail, but the change is not significant.Finally, the armature size is optimized by using the orthogonal test which takes the maximum current density and the armature force as the performance index. The optimum size of armature is the head thickness h=0.22m , tail length d_s=0.18m , tail thickness d_0=0.11m , corner radius r=0.09m .Simulation verification shows that the optimized armature has the smallest current density maximum and the largest electromagnetic force in all size armatures.

References

[1] R. A. Marshall and Y. Wang, "Railguns: their science and technology," China Machine Press, Beijing, China, 2004.
[2] J. Li, P. Yan, W. Q. Yuan, "Electromagnetic gun technology and its development," High Voltage Engineering, vol. 40, pp. 1052-1064, April .2014.
[3] I. R. McNab, "Development of a naval rail gun," IEEE Trans. Magn, vol. 45, pp. 206-210, Jan. 2015.
[4] P. Lehmann, B. Reck, "Acceleration of a suborbital payload using an electromagnetic railgun," IEEE Trans. Magn, vol. 41, pp. 480-485, Jan. 2007.
[5] R. Ciolini, M. Schneider, B. Tellini, "The use of electronic components in railgun projectiles," IEEE Trans. Magn, vol. 45, pp. 578-583, Jan. 2009.
[6] Z. Alex, D. L. Calvin, A. B. John, "In-bore electric and magnetic field environment," IEEE Trans. Magn, vol. 35, pp. 457-462, Jan. 1999.
[7] Y. Guo, X. Liu, W. P. Zhou, "The revolution of The US navy's power railguns," Modern Ships, vol. 31, pp.30-32, May. 2004.
[8] W. F. Hughes, F. J. Young, "Rail and armature current distribution in electromagnetic launchers," IEEE Trans. Magn, vol. 18, pp. 33-41, Jan. 1982.
[9] S. Satapathy, T. Watt, "Effect of geometry change on armature behavior," IEEE Trans. Magn, vol. 43, pp. 408-412, Jan. 2007.
[10] R. Ferrero, M. Marracci, B. Tellini, "Characterization of inductance gradient and current distribution in electromagnetic launchers," IEEE Trans Instrumentation and Measurement, vol. 60, pp. 1795-1801, May. 2011.
[11] Shen Max Chung, Y. C. Chuang.Characteristics of electromagnetic radiation of a railgun at the final firing stage[J]. IEEE Transactions on Plasma Science,2016,44(1):49-58.
[12] Z. Y. Yang, G. Feng, X. P. Xue. An Electromagnetic Rail Launcher by Quadrupole Magnetic Field for Heavy Intelligent Projectiles[J].IEEE Transactions on Plasma Science,2017,45(1):1095-4099.

IOP Conf. Series: Journal of Physics: Conf. Series **1284** (2019) 012019 doi:10.1088/1742-6596/1284/1/012019

Research on Decision-making Model for Transport Means of Equipment Based on Multi-factor Comprehensive

Wu Qian[1], Wang Chaowei[2], Xu Yingcheng[3,*] and Lu Xiaowei[4]

[1]China National Institute of Standardization, Beijing 100191, China, ustbwuqian@163.com;
[2]China Ship Develop and Design Center, Wuhan 430064, China
[3*]Correspondence author, China National Institute of Standardization, Beijing 100191, yingcheng_xu@126.com;
[4] China National Institute of Standardization, Beijing 100191, China

Email: yingcheng_xu@126.com

Abstract. This paper systematically analyzes the influencing factors of the transport means of equipment. Based on the principles of completeness, maneuverability, rationality and comparability, this paper puts forward a multi-factor comprehensive evaluation index system covering safety index, timeliness index and economy index. Combining the multi-factor comprehensive evaluation model, the decision-making model of the transport means of equipment and the decision mode of combined transportation mode are constructed, and the construction and solution process of the combined transportation mode decision model is analyzed by assuming the case. The multi-factor comprehensive evaluation model of the transport means of equipment proposed in this paper provides a quantitative analysis method for the transport means of equipment decision-making, which lays a foundation for further evaluation of equipment transportation efficiency.

1. Introduction

The transport means of equipment used during routine training and in transit in wartime are generally classified into roads, railways, waterways, and air. Each of them has applicable conditions and advantages, and has been compared with others and analyzed by many experts and scholars. Yu Danya [1] put forward an evaluation system for advantage index of waterway transport and analyzed and commented each index. Zheng Yi [2], Zhao Ruijia [3] and Xu Junhua [4] analyzed the advantages and applicability of waterway transport and road transport from different angles. Zhao Wenjuan [5] and Zhang Xueqin [6] considered that each of them has its most favorable application scope through analyzing the technical and economic characteristics of four transport means. However, how to select one or a combination of transport means from four transport means for a specific transport task. It is not enough to decide the most favorable transport means by considering the advantages and disadvantages of one of them, it is also necessary to analyze evaluation factors of transport means comprehensively and quantitatively analyze transport means by constructing a multi-factor comprehensive evaluation model.

IOP Conf. Series: Journal of Physics: Conf. Series **1284** (2019) 012019 doi:10.1088/1742-6596/1284/1/012019

2. Multi-factor comprehensive evaluation index system

Due to different design characteristics of equipment for different transport means and the safety, time and cost required for different transport means used for a specific transport task are different; therefore, it is necessary to construct an evaluation index system for transport means based on various factors to determine the best transport means for the task.

2.1 Principle of index system construction

Identify and establish a systematic and reasonable evaluation index system based on certain construction principles is the premise of constructing a decision-making model of transport means of equipment. The principles [1] adhered to construct a multi-factor comprehensive evaluation index system of transport means of equipment are as follows:

2.1.1 Completeness. The ev*aluation* index system should reflect all characteristics of a transport means as much as possible.

2.1.2 Maneuverability. The evaluation index should be quantitative if possible, to perform an intuitive comparison based on values.

2.1.3 Rationality. The evaluation index must have clear boundary and reasonable content, to clearly express the contents to be compared.

2.1.4 Comparability. The evaluation index must be comparable among different transport means.

2.2 Frame of index system

In this paper, the frame of a multi-factor efficiency evaluation index system of transport means is constructed based on the construction principles from aspects of safety, timeliness and economy. Safety includes reliability of transport means; timeliness includes transport time, on-time delivery rate, intact delivery rate; economy includes generality of equipment and transport cost [7, 8, 9].

2.2.1 Safety index. Safety index refers to the probability of a transport means to complete a task under a specific environment and transportation condition. It is represented by the reliability of transport means.

2.2.2 Timeliness index. Timeliness of a transport means are indicated by transport time, on-time delivery rate and intact delivery rate. Transport time is represented by time required for each link of transport [7] and waiting time, on-time delivery rate is represented by the ratio of on-time delivery times to total delivery times of equipment, and intact delivery rate is represented by the ratio of intact delivery times to total delivery times of equipment to the destination. Under the same conditions, when the shortest transport time is, the higher the on-time delivery rate is, and the higher the intact delivery rate is, the higher the timeliness index is.

2.2.3 Economy index. Economy index can be divided into generality of equipment and transport cost. Generality index refers to the general condition of transported resource during transport of equipment, and the transport cost refers to the expenses arising from personnel, equipment, maintenance and energy consumption in the transport of equipment. Under the same conditions, when the fewer the type and quantity of special and newly-researched equipment used is and the lower the transport cost is, the higher the economy of the transport means is.

3. Decision-making Model for Single Transport Means

Single transport means refers that only one of road transport, railway transport, waterway transport and air transport is used to complete a specific task.

Definition 1: u_i refers to the valuation index of transport means; u_1 refers to the reliability of transport means; u_2 refers to transport time; u_3 refers to on-time delivery rate; u_4 refers to intact delivery rate; u_5 refers to generality of equipment; u_6 refers to transport cost.

Definition 2: t_j refers to the transport means of equipment; and t_1 refers to the road transport, t_2 refers to the railway transport, t_3 refers to the waterway transport, and t_4 refers to the air transport.

Definition 3: $u_i(t_j)$ refers to i evaluation index of j transport means, and f_{ij} refers to the single factor evaluation value of this index.

The specific formula of f_{ij} is as follows:

$$f_{ij} = u_i(t_j)/\sum_{j=1}^{4} u_i(t_j) \quad i=1,2,3,4,5,6;j=1,2,3,4 \tag{1}$$

Definition 4: w_i refers to the weighted value of importance of i transport means evaluation index, and $\sum_{i=1}^{6} w_i = 1$; F_j refers to the multi-factor integrated evaluation value of each transport means.

The specific formula of F_j is as follows:

$$F_j = \sum_{i=1}^{6} w_i f_{ij} \quad i=1,2,3,4,5,6;j=1,2,3,4 \tag{2}$$

Based on the formula (1) and (2), the multi-factor comprehensive evaluation values of road transport (t_1), railway transport (t_2), waterway transport (t_3) and air transport (t_4) are calculated and expressed as followsF_1、F_2、F_3、F_4:

$$F_1 = w_1f_{11} + w_2f_{12} + w_3f_{13} + w_4f_{14} + w_5f_{15} + w_6f_{16} \tag{3}$$
$$F_2 = w_1f_{21} + w_2f_{22} + w_3f_{23} + w_4f_{24} + w_5f_{25} + w_6f_{26} \tag{4}$$
$$F_3 = w_1f_{31} + w_2f_{32} + w_3f_{33} + w_4f_{34} + w_5f_{35} + w_6f_{36} \tag{5}$$
$$F_4 = w_1f_{41} + w_2f_{42} + w_3f_{43} + w_4f_{44} + w_5f_{45} + w_6f_{46} \tag{6}$$

The higher the multi-factor integrated evaluation value is, the better the comprehensive evaluation result of this type of transport means is, which is represented by F_j^*. The specific calculation method of F_j^* is as follows:

$$F_j^* = \max\{F_j | j = 1,2,3,4\} \tag{7}$$

Then, t_j refers the best transport means selected after comprehensive comparison and evaluation for above-mentioned factors, and F_j^* refers the integrated evaluation value for the best transport means.

4. Decision-making Model for a Combination of Transport Means

Instead of just one transport means, two or more transport means are often required for a task due to complex and changeable natural environment and battlefield situation. We select the best combination of transport means by constructing a selection model for combined transport means and considering various factors.

4.1 Steps of model construction

1) Draw a network diagram for multi-path transport of equipment with the origin and destination clearly specified;

2) Determine the integrated evaluation value of the best transport means from the node of the previous layer to the destination in turn by deducting from the destination reversely and referring to the selection model of single transport means;

3) Determine the best combination of transport means between the nodes of each layer by deducting from front to back, and then determine the corresponding transport route.

4.2 Example of model construction

The decision-making model of combination of transport means is complex, which is illustrated briefly through giving examples.

4.2.1 Problem description. In order to meet the reserve demand of materials, it is necessary to transport a batch of equipment system from A to E. Due to long distance, different levels of transit warehouses are required during the transport, and the specific route is shown in Figure 1 below. It is assumed the distance for various transport means used are calculated with d_i.

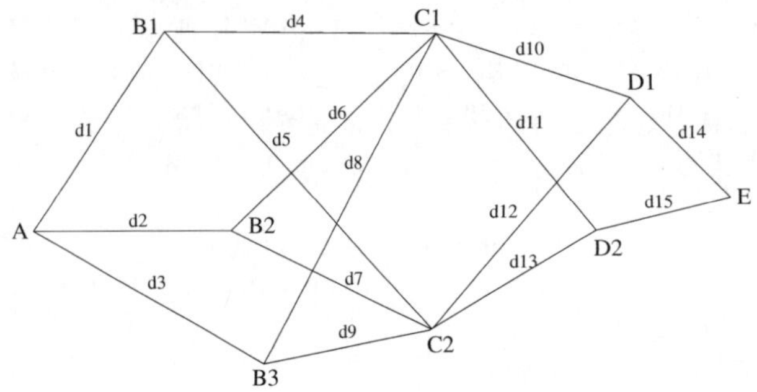

Figure.1 Network Diagram for Transport of Equipment

4.2.2 Model construction.

4） The model is solved by the single transport means of Equation 1 - Equation 7 to obtain the integrated evaluation values $F(D_1)$ and $F(D_2)$ of the most reasonable transport means for D_1-E and D_2-E, so as to determine the most reasonable transport means for each of them respectively.

Then:

$$F(D_1) = N_1, S(D_1) = E; F(D_2) = N_2, S(D_2) = E.$$

Among them, N_i indicates the integrated evaluation value of the best transport means from a transfer station to a terminal at a certain level of transport, $S(D_1) = E$ means that the transport route corresponding to the most reasonable transport means from the transfer station D1 to E is $(D_1 - E)$.

5） The optimal transport means and corresponding transport route algorithm of the two transfer stations on the layer C are as follows:

$$F(C_1) = \max\begin{Bmatrix} \dfrac{d_{10}F(C_1 - D_1)}{d_{10} + d_{14}} + \dfrac{d_{14}F(D_1)}{d_{10} + d_{14}} \\ \dfrac{d_{11}F(C_1 - D_2)}{d_{11} + d_{15}} + \dfrac{d_{15}(D_2)}{d_{11} + d_{15}} \end{Bmatrix}$$

$$F(C_2) = \max\begin{Bmatrix} \dfrac{d_{12}F(C_2 - D_1)}{d_{12} + d_{14}} + \dfrac{d_{14}F(D_1)}{d_{12} + d_{14}} \\ \dfrac{d_{13}F(C_2 - D_2)}{d_{13} + d_{15}} + \dfrac{d_{15}F(D_2)}{d_{13} + d_{15}} \end{Bmatrix}$$

Make:

$$\frac{d_{10}F(C_1-D_1)}{d_{10}+d_{14}} + \frac{d_{14}F(D_1)}{d_{10}+d_{14}} = N_3, \frac{d_{11}F(C_1-D_2)}{d_{11}+d_{15}} + \frac{d_{15}(D_2)}{d_{11}+d_{15}} = N_4, \text{ and } N_3 > N_4;$$

$$\frac{d_{12}F(C_2-D_1)}{d_{12}+d_{14}} + \frac{d_{14}F(D_1)}{d_{12}+d_{14}} = N_5, \frac{d_{13}F(C_2-D_2)}{d_{13}+d_{15}} + \frac{d_{15}F(D_2)}{d_{13}+d_{15}} = N_6, \text{ and } N_5 > N_6.$$

Then you can get,

$S(C_1) = D_1$, that is, the transport route corresponding to the optimal transport means of the transfer station C_1 to D is: C_1-D_1 ;

$S(C_2)= D_1$, that is, the transport route corresponding to the optimal transport means of the transfer station C_1 to D is: C_1-D_1.

6） The optimal transport means and corresponding transport route algorithm of the three transfer stations on the layer B are as follows:

$$F(B_1) = \max\begin{cases}\dfrac{d_4F(B_1-C_1)}{d_4+d_{10}+d_{14}}+\dfrac{(d_{10}+d_{14})F(C_1)}{d_4+d_{10}+d_{14}}\\ \dfrac{d_5F(B_1-C_2)}{d_5+d_{12}+d_{14}}+\dfrac{(d_{12}+d_{14})F(C_2)}{d_5+d_{12}+d_{14}}\end{cases}$$

$$F(B_2) = \max\begin{cases}\dfrac{d_6F(B_2-C_1)}{d_6+d_{10}+d_{14}}+\dfrac{(d_{10}+d_{14})F(C_1)}{d_6+d_{10}+d_{14}}\\ \dfrac{d_7F(B_2-C_2)}{d_7+d_{12}+d_{14}}+\dfrac{(d_{12}+d_{14})F(C_2)}{d_7+d_{12}+d_{14}}\end{cases}$$

$$F(B_3) = \max\begin{cases}\dfrac{d_8F(B_1-C_1)}{d_8+d_{10}+d_{14}}+\dfrac{(d_{10}+d_{14})F(C_1)}{d_8+d_{10}+d_{14}}\\ \dfrac{d_9F(B_1-C_2)}{d_9+d_{12}+d_{14}}+\dfrac{(d_{12}+d_{14})F(C_2)}{d_9+d_{12}+d_{14}}\end{cases}$$

Make:

$$\frac{d_4F(B_1-C_1)}{d_4+d_{10}+d_{14}}+\frac{(d_{10}+d_{14})F(C_1)}{d_4+d_{10}+d_{14}}=N_7, \frac{d_5F(B_1-C_2)}{d_5+d_{12}+d_{14}}+\frac{(d_{12}+d_{14})F(C_2)}{d_5+d_{12}+d_{14}}=N_8, \text{ and } N_7>N_8;$$

$$\frac{d_6F(B_2-C_1)}{d_6+d_{10}+d_{14}}+\frac{(d_{10}+d_{14})F(C_1)}{d_6+d_{10}+d_{14}}=N_9, \frac{d_7F(B_2-C_2)}{d_7+d_{12}+d_{14}}+\frac{(d_{12}+d_{14})F(C_2)}{d_7+d_{12}+d_{14}}=N_{10}, \text{ and } N_9>N_{10};$$

$$\frac{d_8F(B_1-C_1)}{d_8+d_{10}+d_{14}}+\frac{(d_{10}+d_{14})F(C_1)}{d_8+d_{10}+d_{14}}=N_{11} \frac{d_9F(B_1-C_2)}{d_9+d_{12}+d_{14}}+\frac{(d_{12}+d_{14})F(C_2)}{d_9+d_{12}+d_{14}}=N_{12}, \text{and } N_{11}>N_{12}.$$

Then you can get,

$S(B_1)= C_1$, that is, the transport route corresponding to the optimal transport means of the transfer station B1 to C is: B1-C1;

$S(B_2)= C_1$, that is, the transport route corresponding to the optimal transport means of the transfer station B2 to C is: B2-C1;

$S(B_3)= C_1$, that is, the transport route corresponding to the optimal transport means of the transfer station B3 to C is:B3-C1.

7） Same as above, the optimal transport means of the transfer station A and the corresponding transport route algorithm are as follows:

$$F(A) = \max\begin{cases}\dfrac{d_1F(A-B_1)}{d_1+d_4+d_{10}+d_{14}}+\dfrac{(d_4+d_{10}+d_{14})F(B_1)}{d_1+d_4+d_{10}+d_{14}}\\ \dfrac{d_2F(A-B_2)}{d_2+d_6+d_{10}+d_{14}}+\dfrac{(d_6+d_{10}+d_{14})F(B_2)}{d_2+d_6+d_{10}+d_{14}}\\ \dfrac{d_3F(A-B_3)}{d_3+d_8+d_{10}+d_{14}}+\dfrac{(d_8+d_{10}+d_{14})F(B_3)}{d_3+d_8+d_{10}+d_{14}}\end{cases}$$

Make:

$$\frac{d_1F(A-B_1)}{d_1+d_4+d_{10}+d_{14}}+\frac{(d_4+d_{10}+d_{14})F(B_1)}{d_1+d_4+d_{10}+d_{14}}=N_{13},$$

$$\frac{d_2F(A-B_2)}{d_2+d_6+d_{10}+d_{14}}+\frac{(d_6+d_{10}+d_{14})F(B_2)}{d_2+d_6+d_{10}+d_{14}}=N_{14},$$

$$\frac{d_3F(A-B_3)}{d_3+d_8+d_{10}+d_{14}}+\frac{(d_8+d_{10}+d_{14})F(B_3)}{d_3+d_8+d_{10}+d_{14}}=N_{15}, \text{ and } N_{13}>N_{14}>N_{115}$$

Then you can get,

$S(A) = B_1$, that is, the transport route corresponding to the best transport means of transfer stations A to B is: A-B_1;

8） Determine the combination of transport means.

Because of $S(A) = B_1$, the transport route from A to B is: A-B_1;
Because of $S(B_1) = C_1$, the transport route from B to C is: B_1-C_1;
Because of $S(C_1) = D_1$, the transport route from C to D is: C_1-D_1;
Because of $S(D_1) = E$, the transport route from D to E is: D_1-E.

Therefore, the transport route under the best combination of transport means is: $A - B_1 - C_1 - D_1 - E$.

In summary, the most reasonable combination of transport means is:

The transport means corresponding to $F(A - B_1)$ is adopted for A-B_1 road section;

The transport means corresponding to $F(B_1 - C_1)$ is adopted for B_1-C_1 road section;

The transport means corresponding to $F(C_1 - D_1)$ is adopted for C_1-D_1 road section;

The transport means corresponding to $F(D_1)$ is adopted for D_1-E road section.

5. Conclusion

The decision making of equipment transport means is a difficult point in equipment transport research. In recent years, China has recognized the importance of equipment transport, carried out research on equipment transport and obtained some research results, but it is still in the development stage. There is still a certain gap compared with foreign countries. From the aspects of safety, timeliness and economy, this paper comprehensively analyzes the influencing factors of equipment transport means, and builds a decision-making model of equipment transport means based on multi-factor integrated evaluation, which solves the decision making of transport means of equipment transport problem to some extent. The paper is hoped to provide reference for relevant scholars to carry out research on equipment transport related content.

Acknowledgment

We would like to acknowledge that this research are supported and funded by the National Science Foundation of China under Grant No. 71571191, 91646122 and 91746202, the Basic Scientific Research Business Projects 552017Y-5266-2017, the National Key R&D Program of China under Grant No.2016YFF0202600, No.2016YFF0202604, 2017YFF0209603, 2017YFF0209604.

References

[1] Yu Danya. Research on the Evaluation Index System of Comparative Advantages of Waterway Transport [J]. Journal of Transport Research, 2018, 4(2): 1-6.
[2] Zheng Yi. Advantages of Economic Development of Waterway Transport under Low Carbon Economy [J]. Economic and Trade Practice, 2017(17): 57.
[3] Zhao Ruijia, Xie Xinlian, Wei Zhaokun. Quantitative Analysis and Comparison of Energy Saving and Emission Reduction for Highway and Waterway Transport [J]. Journal of Energy Conservation and Environmental Protection, 2017, 13(3): 46-50.
[4] Xu Junhua. Analysis of Economic Development Advantages of Waterway Transport under Low Carbon Economy [J]. Modern Business, 2018(8): 11-12.
[5] Zhao Wenjuan. Comparison of Technical and Economic Characteristics of Various Transport Means [J]. Cooperative Economy and Technology, 2017(3): 38-39.
[6] Zhang Xueqin, Cao Lixin. Comparative Analysis of Technical and Economic Characteristics of Various Transport Means [J]. Journal of Transport and Transport (Academic Edition), 2013(1): 170-172.
[7] Qian Wu, Xiaowei Lu, Shuang Wang and Dong Zheng. Analysis of Influencing Factors on the Transportation of Equipment [C] . ICSSD 2018,2018: 364-374.
[8] Qian Wu, Ma Lin, Chaowei Wang. Study on Evaluation Index System of Equipment System Transportability [C], WCEAM2013, 2015: 1539-1547.
[9] Wu Qian. Research on the Transport Analysis Method of UAV System [D]. Beijing: Beihang University, 2014.

An Improved RANSAC Surface Reconstruction Study

Ma Li, Zhang Jinming*, Ma Dongchao and Fu Yingxun

School of Information, North China University of Technology, Beijing China

*E-mail: jimzhangx@qq.com

Abstract. Surface reconstruction technology has always been an important research area in graphic image. However, efficiently processing massive amounts of high-precision point cloud data is still a problem worth studying. Based on the characteristics of point cloud data and the existing research results, this paper proposes a RANSAC algorithm based on voxel segmentation and applies it to 3D surface reconstruction. By analyzing the characteristics of the original point cloud data, the method improves the reconstruction speed of the 3D model while retaining a large amount of effective feature information and ensuring high precision. The experimental results show that compared with the reconstruction algorithm based on the original RANSAC algorithm, the proposed algorithm can effectively reduce the surface reconstruction time of 47.51%, and the model error after processing is only one thousandth compared with the original model.

1. Introduction

Surface reconstruction technology has been widely used in industrial high-precision parts error detection, digital cultural relics engineering, urban and rural construction and planning, precision medical analysis, virtual reality and augmented reality technology, military simulation training and other fields. With the rapid development of 3D laser scanning instruments, it is easy to collect a large amount of three-dimensional data on the surface of the object, that is, point cloud data, and the point cloud data set contains the spatial geometric features and surface characteristics of the measured object. Use a large amount of point cloud data to reconstruct a surface map to build a high-precision 3D model. However, due to various irresistible factors such as the error of actual operation, there are bound to be many redundant points and outliers in the original point cloud data set. Therefore, it has important practical significance to optimize the processing of point cloud data sets and improve the efficiency of surface reconstruction.

Based on the improved random sampling consistency algorithm of voxel segmentation, this paper analyzes the feature properties of point cloud data and retains a large number of effective feature information to ensure the high precision of the model and reduce the denoising time of point cloud. The method proposed in this paper is compared with the application of the original RANSAC algorithm in surface reconstruction. The experimental results show that the proposed method can reconstruct the high-precision 3D model more quickly and efficiently.

2. Related Works

Shi and other researchers proposed a simplified method based on clustering, which is to classify the point cloud data by 3D raster and iteratively calculate the normal vector deviation to ensure the point cloud data characteristics, but the calculation is very large[1]. Chu et al. proposed to use two-dimensional SIFT to perform multi-view cloud data registration and splicing, extracting and matching feature points in effective texture images, and removing the interference of holes and invalid data in point cloud data. The calculation process is simple and improved accuracy and effect[2]. It is proposed in Simon Kriegel to iteratively determine the 3D surface model by boundary detection and surface trend estimation of the acquired model. When performing a new laser scan, point cloud data is integrated into the triangle mesh and the probabilistic voxel space in time. This method can quickly collect information on the surface of the object, but the algorithm takes a long time[3]. Anette Eltner summarizes the state of the art in SfM workflows in photogrammetry and geosciences, as well as the development of high-performance digital sensors and important software innovations developed in the field of computer-based vision and visual perception research. Processing extends to 3-D to generate point cloud data from a series of non-calibrated images[4]. For underwater scanning technology, Ricard Campos proposed a full 3D surface reconstruction method for underwater environment to denoise unstructured point cloud data[5]. Wu Mengqi et al. proposed the use of Point Feature Histograms (FPFH) and SIFT to achieve adaptive stitching of point clouds, while considering geometric features and image features, the algorithm has good stability[6]. JUAN LI et al. used random sampling consistency to eliminate erroneous point cloud data, and finally reconstructed the surface of point cloud data by singular value decomposition[7]. EDS and Raindrop companies in the United States developed Imageware software[8] and Geomagic Studio software[9] respectively. Imageware can quickly fit point cloud data directly into high-quality surfaces, while Geomagic Studio can efficiently use point cloud data to create polygonal grids and finally generate NURBS surfaces.

The surface reconstruction algorithm mentioned above and the existing common reconstruction methods (poisson reconstruction, etc.) can realize the prototype reconstruction of point cloud data from different degrees, but these methods do not consider improving the running speed of reconstruction.

3. Improvement of RANSAC Algorithm Based on Voxel Segmentation Idea

3.1 Algorithm Background and Problem

Fischler and Bolles first proposed the RANSAC (random sampling consistency) algorithm in 1981 [10]. They use the RANSAC algorithm to solve the LDP problem in the 3D model reconstruction. At present, RANSAC has been applied to geometric primitive detection, wide baseline stereo matching, motion segmentation, stitching, robust feature image matching and other fields. It estimates the parameters of a mathematical model from a set of observation data sets by iterative calculations. However, it is an indeterminate algorithm as it only has a certain probability to get a reasonable result, the number of iterations must be increased for better results.

The random sampling consistency algorithm is widely used in point cloud data preprocessing.It has a good effect on some isolated points and redundant points generated when processing the scan.It is a process of constructing a basic subset consisting of only intra-point data by randomly sampling out-of-office points. The basic idea is: in the parameter estimation, instead of treating all possible input point cloud data indiscriminately, firstly, a judgment criterion model is designed for the specific problem, and the judgment criterion is used to iteratively eliminate those and the estimated the point cloud data with inconsistent parameters is then estimated by the correct point cloud data.This algorithm is a well-established robust estimation method.In computer vision and many other research fields, robust estimation of model parameters is a core issue. Guo Jidong et al. mentioned that the random sampling consistency algorithm can still obtain ideal processing results for data with error rate exceeding 50%, which is one of the most effective robust estimation algorithms[11].

3.2 Basic Idea of Improved Algorithm

Based on the existing RANSAC random sampling consistency algorithm, this paper introduces the idea of voxel grid partition space, which realizes the effective speed improvement of point cloud denoising. Firstly, according to the existing original point cloud data N, spatial voxel segmentation is performed in the three-dimensional virtual space, that is, the point cloud data set is wrapped by many unit voxels, so that each point cloud data is included in the unit body. Within the prime, there may be multiple point cloud data within each unit voxel. Figure 1 is a schematic diagram of voxel segmentation of a point cloud data set; Figure 2 is a spatial state of point cloud data in a unit voxel.

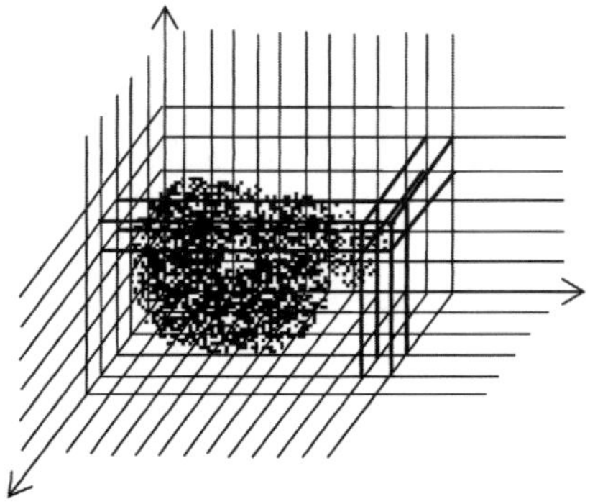

Figure 1. Schematic diagram of voxel segmentation

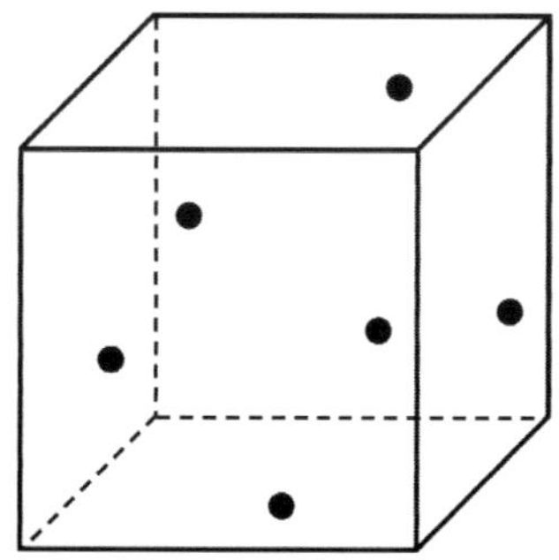

Figure 2. Point cloud data in unit voxels

After the voxel partitioning is completed, the original point cloud data set is divided into several point cloud data subsets. It is necessary to compress the point cloud data in each unit voxel according to certain criteria, but at the same time as the point cloud compression, it is necessary to retain some of the regional features described in the subset of point cloud data within the unit voxel. The algorithm selects the data point $\mathrm{G}(x, y, z)$ closest to the unit voxel center of gravity $\mathrm{Q}(x, y, z)$ as the feature information retention point of data points in the subset of point cloud data.

Set the vertex closest to the origin of the coordinate system in each voxel as $\mathrm{D}_{ijh}(x_i, y_j, z_h)$, The voxel is the i-th voxel in the x-axis direction, the j-th voxel in the y-axis direction, the z-th voxel in the z-axis direction, the position vector is $d = \begin{pmatrix} i & 0 & 0 \\ 0 & j & 0 \\ 0 & 0 & h \end{pmatrix}$, and the coordinate vector is $p = (x_i, y_j, z_h)$. The voxel side length is $l = (l_x, l_y, l_z)$, which represents the voxel side length in the x, y, and z axis directions, respectively. Therefore, the vertex $\mathrm{D}_{ijh}(x_i, y_j, z_h)$ obtained by $p = dl$ is obtained. The center of gravity $\mathrm{G}(x, y, z)$ and the coordinate vector $g = (x, y, z)$ in the unit voxel can be calculated by $g = dl + 0.5l$.

The side length $l = (l_x, l_y, l_z)$ of a unit voxel directly affects the amount of point cloud data within a unit voxel, and thus requires a reasonable estimate. It can be seen from the conclusion of the literature[12] that when the side length l is large, the efficiency of the point cloud compression is affected. When the side length l is too small, an empty body element is generated. The unit voxel is a cube, that is $l_x = l_y = l_z$, and the calculation formulas (1) to (4) are as follows. Where α is the scale factor used to adjust the side length of the unit voxel; θ is the scale factor; n' is the number of point cloud data per unit volume in the ideal state; L_x, L_y, L_z are the maximum coordinate values of the entire point cloud dataset along the x-axis, y-axis, and z-axis, respectively, and V is the minimum voxel required to cover all point cloud data.

$$l' = \left(\frac{\theta\alpha^3}{n'}\right)^{1/3} \tag{1}$$

$$n' = \frac{N}{V} \tag{2}$$

$$V = L_x \times L_y \times L_z \tag{3}$$

The final formula of unit voxel side length is expressed as:

$$l' = \left(\frac{\theta L_x L_y L_z \alpha^3}{N} \right)^{1/3} \tag{4}$$

The center of gravity $G(x, y, z)$ coordinate vector $g = (x, y, z)$ is:

$$g = ((i+0.5)l', (j+0.5)l', (h+0.5)l') \tag{5}$$

Traversing each point cloud data $Q_k(x_k, y_k, z_k)$, $(k = 1, 2, \ldots, n)$ within the voxel. Calculate the spatial Euclidean distance s_k, $(k = 1, 2, \ldots, n)$ of each data point to the voxel center of gravity $G(x, y, z)$.

$$s_k = \left((x - x_k)^2 + (y - y_k)^2 + (z - z_k)^2\right)^{1/2} \tag{6}$$

Get the smallest s_k, $(k = 1, 2, \ldots, n)$, retain the point cloud data, and discard the remaining $n-1$ point cloud data in the subset of point cloud data. Figure 3 shows the variation of voxel partitioning point cloud compression before and after a unit voxel.

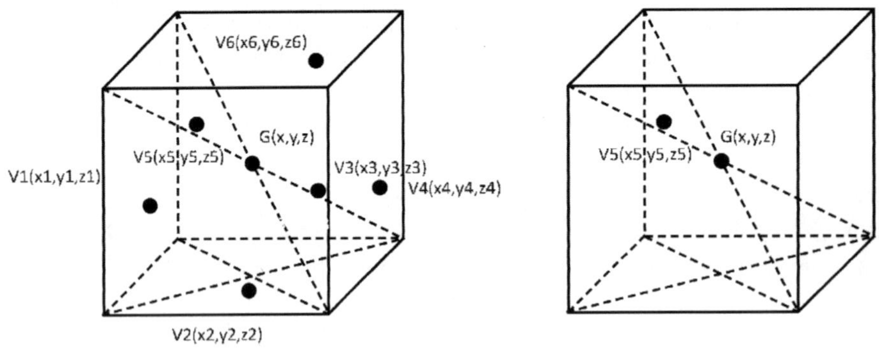

Figure 3. Changes in data points within voxels

In the classic RANSAC formula, the key to solving the model is to find all valid interior points from a set of data points. The so-called inner points are the data points that match the "best" model. The number of valid interior points is usually unknown. The RANSAC algorithm finds all the required interior points and corresponding models with a certain probability by repeatedly drawing random samples from the input data point set. m is the number standard used to determine the suitability of the model in the data set.

Suppose t is the number of times the current iteration is calculated, and the initial value is 0.

① When t is smaller than the target iteration number r, num data points are randomly selected from the data set, and then a model mb_model suitable for num is constructed.

② The remaining $(N - num)$ points are traversed, from which data points that satisfy the model mb_model are found and counted in the number of data points applicable to the model.

③ When the number of data points applicable to the model is greater than the set standard number m, it is considered that the optimal model b_model in this iterative calculation is found.

④ Then through a certain strategy to determine the applicability of b_model for all points in the data set.

⑤ If the applicability of b_model is higher than the optimal model $model$ in the previous $(t-1)$ iterations, then the $model$ and current best applicability are updated.

⑥ Repeat ① to continuously find the optimal *model* until the end of the iterative calculation.

In the original RANSAC algorithm, it is assumed that w is the probability that each time a point is selected from the point cloud data set N is exactly an inner point. The value of w is usually unknown, but can be roughly estimated by equation (7).

$$w = \frac{m}{N} \tag{7}$$

P is the ideal probability that the original RANSAC algorithm provides an effective model after running. According to the theoretical results of[13], the number of iterations r is a function of P. $\boldsymbol{Min}$ is the minimum value for the number of data points in the model. Satisfy formula (8)

$$r = \frac{\ln(1-\mathrm{P})}{\ln(1-w^{\min})} \tag{8}$$

In the improved RANSAC algorithm integrated with the voxel partitioning idea, the number of unit voxel N' in V can be calculated according to formula (1), (2) and (7), that is

$$N' = V \div l'^3 = V \div \frac{\theta\alpha^3 V}{N} = \frac{N}{\theta\alpha^3} \tag{9}$$

Since the improved algorithm performs point cloud compression within the unit voxel, $N' < N$ must hold. Under the premise that the model fitness degree m of the original RANSAC algorithm is constant, the probability w' of selecting the inner point from the point cloud data set is higher than w. This means that after voxel partitioning, the applicability of the improved RANSAC algorithm generation model to the same point cloud dataset is effectively improved.

$$w' = \frac{m}{N'} > w = \frac{m}{N} \tag{10}$$

Similarly, voxel segmentation optimizes the selection of finite data points in a unit voxel. From the formula deduction, we can draw another conclusion: the number of iterations of the RANSAC algorithm after r' improvement, the ideal probability P of the effective model and the applicable Under the premise that the minimum number min of the model data points is constant, the number of iterative calculations is reduced, and the efficiency of the algorithm is improved.

$$r' = \frac{\ln(1-P)}{\ln(1-w'^{\min})} < r = \frac{\ln(1-P)}{\ln(1-w^{\min})} \tag{11}$$

The improved RANSAC algorithm based on the fusion of voxel segmentation can solve the effective de-noising model with higher probability and faster speed, it also provides a new point cloud data set that is more regular and easier to process for the subsequent point cloud reconstruction.

4. Experimental results and analysis

The experiment was implemented using the VS2010. In order to verify the excellent characteristics of the above algorithm, the optimization effect of the RANSAC algorithm based on voxel segmentation on the de-noising and surface reconstruction of point cloud datasets was investigated.

Figure 4 is a comparison of the effects of a set of point cloud data sets in the first set of experiments after the improved RANSAC algorithm. It can be seen from Figure 4(a) that the point cloud density of the set of point cloud data at the edge is higher than the internal density, and there are some separation points that are outside the overall point cloud data set. Figure 4(b) shows the processing effect of the improved RANSAC algorithm. The density of the edge point cloud data is greatly reduced, but it is basically consistent with the internal density, and the geometric features of the edge regions are preserved.

The second set of experiments uses the commonly used surface reconstruction algorithms, namely the Greedy Projection Triangulation reconstruction algorithm, the Poisson reconstruction algorithm and the Moving Cube reconstruction algorithm to compare and improve the optimization effect of the improved RANSAC algorithm on the surface reconstruction algorithm. The point cloud dataset used in the experiment was acquired by industrial part scanning, and Figure 5 is the visualization effect of opening the point cloud data directly in Geomagic Studio software. and Figure 5 is the visualization effect of opening the point cloud data directly in Geomagic Studio software. Many redundant points and outliers can be seen very intuitively.

(a) (b)

Figure 4. Improved RANSAC algorithm processing renderings

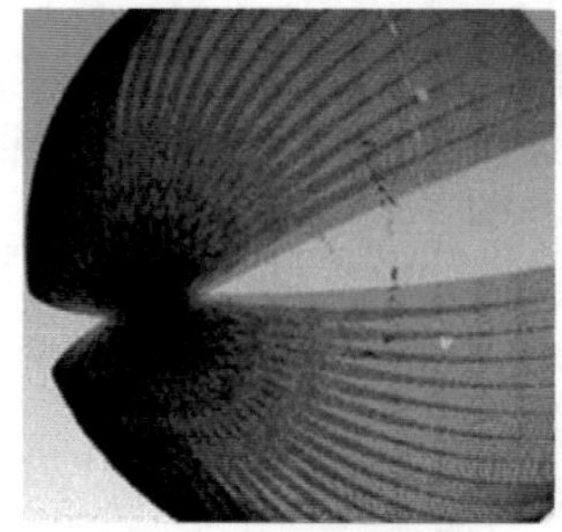

Figure 5. Original point cloud data

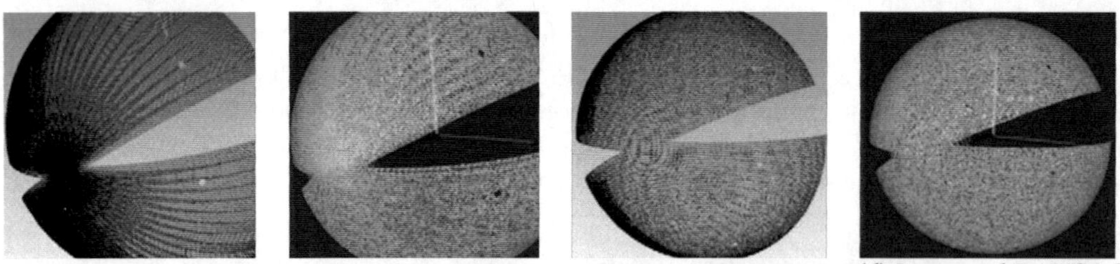

(a) Original RANSAC algorithm point cloud
(b) Original RANSAC algorithm grid diagram
(c) Improved RANSAC algorithm point cloud
(d) Improved RANSAC algorithm grid diagram

Figure 6. Greedy Projection Triangulation reconstruction algorithm point cloud and grid diagram

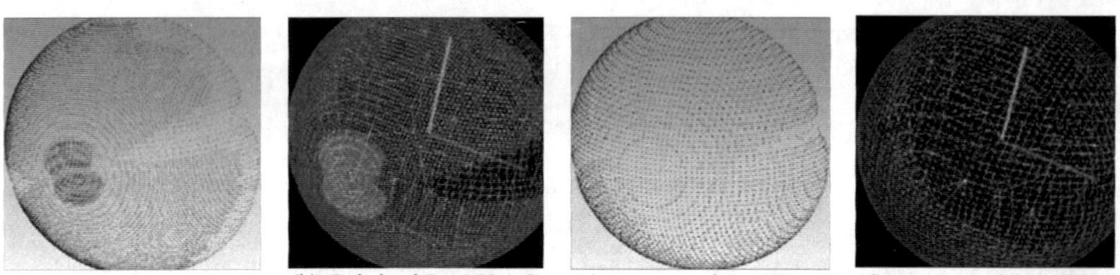

(a) Original RANSAC algorithm point cloud
(b) Original RANSAC algorithm grid diagram
(c) Improved RANSAC algorithm point cloud
(d) Improved RANSAC algorithm grid diagram

Figure 7. Poisson reconstruction algorithm point cloud and grid diagram

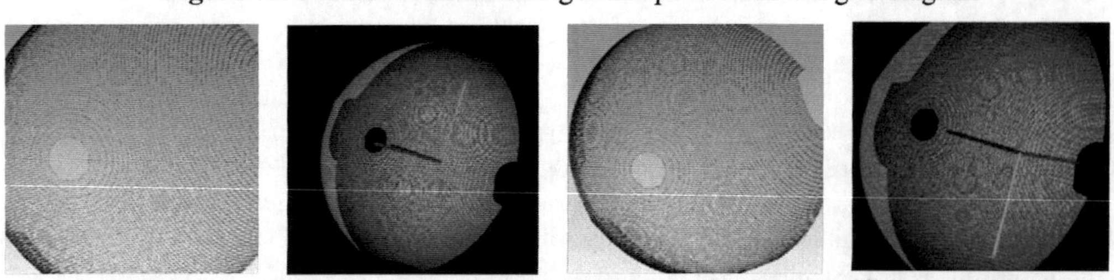

(a) Original RANSAC algorithm point cloud
(b) Original RANSAC algorithm grid diagram
(c) Improved RANSAC algorithm point cloud
(d) Improved RANSAC algorithm grid diagram

Figure 8. Moblie Cube reconstruction algorithm point cloud and grid diagram

IOP Conf. Series: Journal of Physics: Conf. Series **1284** (2019) 012020 doi:10.1088/1742-6596/1284/1/012020

Figure 6 (a)(b), 7 (a)(b) and 8 (a)(b) respectively show the Greedy Projective Triangular surface reconstruction algorithm, Poisson reconstruction algorithm and Mobile Cube reconstruction algorithm based on the original RANSAC point cloud data denoising processing point cloud renderings and grid renderings.

Figure 6(c)(d) 7(c)(d) and 8(c)(d) respectively shows the point cloud renderings and grid renderings of the Greedy Projection Triangulated surface reconstruction, Poisson reconstruction algorithm and Mobile Cube reconstruction algorithm based on the fusion voxel partitioning idea's RANSAC algorithm.

In the comparison of the three sets of pictures, the optimization effect of Figure 6 and Figure 7 is the most intuitive, It can be clearly seen that the redundant area on the circumference surface is reduced, and the outliers in the center of the sphere are completely removed. Moreover, in Figure 9, it can be intuitively seen that the time taken by the algorithm is reduced. The visual effect in Figure 8 is not obvious, but the improved RANSAC algorithm is more efficient from the algorithm execution time comparison in Figure 9.

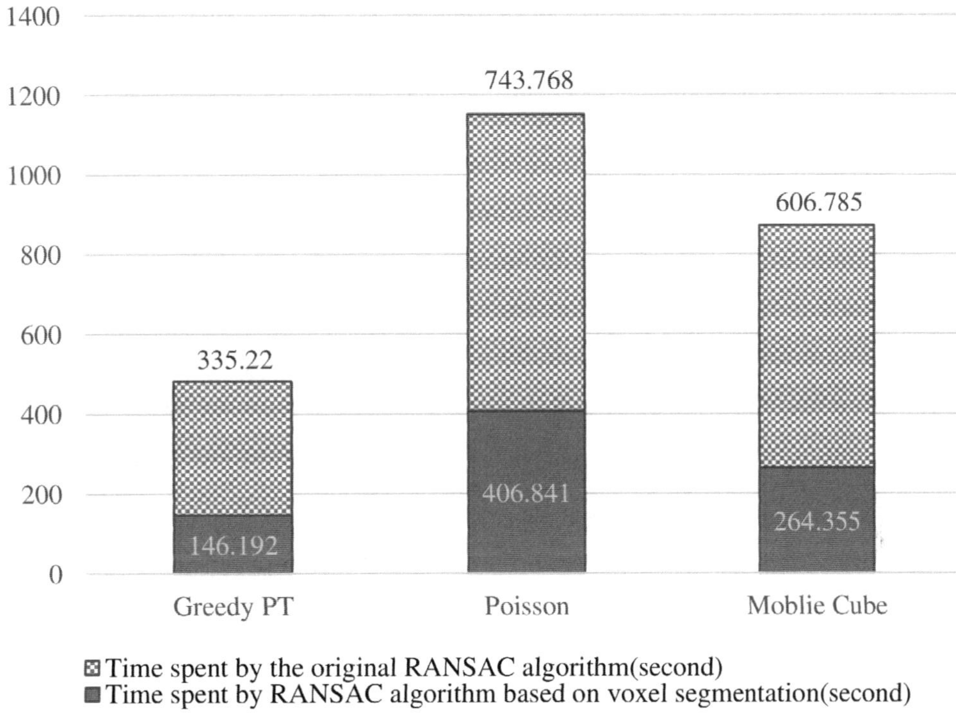

Figure 9. Algorithm running time comparison

Table 1. Error comparison between model based on improved RANSAC algorithm and original model

	coordinates of the spheric center	radius of the spheric
original model	(-0.038,0.085,1.173)	110.002
model based on improved RANSAC algorithm	(-0.032,0.086,1.177)	110.008

Through the time-consuming data obtained from the experiment, the average optimization time-consuming rate of the improved RANSAC algorithm is 47.51%. The relevant experimental data shows that the point cloud data set obtained by the improved RANSAC algorithm is used to calculate the radius of the spheric and the coordinates of the spheric center. Compared with the result calculated by the original RANSAC algorithm, the error is only in the thousandth position, maintaining the accuracy of 96.48% of the model. The detailed data is shown in Table 1.

5. Conclusion

In the process of surface reconstruction, RANSAC (Random Sample Consensus) is frequently applied to denoise processing, but the traditional RANSAC algorithm takes a lot of time in this process. In this paper, the improved RANSAC algorithm based on voxel segmentation is proposed. The improved algorithm effectively reduces the number of iterative calculations and achieves the purpose of reducing time consumption. The experimental result shows the improved algorithm makes the model error after processing only in the thousands of decimal places and reduces the algorithm execution time by 47.51%. However, the algorithm proposed in this paper still needs to be further studied in point cloud data hole repair.

Foundation items

National Natural Science Foundation of China(6172023,61371143)

Key B of Natural Science Foundation of Beijing(KZ201810009011)

Natural Science Foundation of Beijing(4172019)

Science and technology innovation project of North China University of Technology(18XN053)

References

[1] Shi Baoquan, Liang Jin, and Zhang Xiaoqiang. Research on Point Cloud Reduction Technology for Feature Preservation[J]. *Journal of Xi'an Jiaotong University*, 2010.

[2] Chu Wei, Nie Chunmei, and Wang wei. Multi-viewpoint Cloud Data Registration and Splicing Algorithm Based on SIFT Features[J]. *Semiconductor Optoelectronics*, 2011, **32(3)**: 442-447.

[3] Kriegel S, Rink C, and Tim Bodenmüller. Efficient next-best-scan planning for autonomous 3D surface reconstruction of unknown objects[J]. *Journal of Real-Time Image Processing*, 2015, **10(4)**:611-631.

[4] Anette E, Andreas K, Carlos C, et al. Image-based surface reconstruction in geomorphometry – merits, limits and, developments[J]. *Earth Surface Dynamics*, 2016, **4(2)**:359-389.

[5] Campos R, Garcia R, Alliez P, et al. A surface reconstruction method for in-detail underwater 3D optical mapping[J]. *The International Journal of Robotics Research*, 2015, **34(1)**:64-89.

[6] Wu Mengqi, Li Zhongwei, Zhong Kai, et al. Point cloud adaptive stitching method based on geometric features and image features[J]. *Acta Optica Sinica*, 2015, **35(02)**.

[7] Li J, and Jiang G. Point clouds registration with surfaces of low curvatures[C] *2015 IEEE International Conference on Signal Processing, Communications and Computing (ICSPCC)*. IEEE, 2015.

[8] Niu Jianwei. Imageware reverse modeling technology and 3D printing [M]. *Publishing House of Electronics Industry*, 2014.

[9] Guo Peizhen,and Du Liming. Surface reconstruction and error analysis of point cloud data using Geomagic Studio[J]. *GeoInfo World*, 2015(1): 57-60.

[10] Fischler M A, and Bolles R C. Random sample consensus: a paradigm for model fitting with applications to image analysis and automated cartography[J]. *Communications of the ACM*, 1981, **24(6)**:381-395.

[11] Guo Jidong, and Xiang Hui. A Robust Estimation Algorithm for a Basic Matrix [C] *National Conference on Image Graphics*. 2005.

[12] Lengyel E S. Voxel-Based Terrain for Real-Time Virtual Simulations[J]. *Dissertations & Theses - Gradworks*, 2010.

[13] Chen Fuxing, Wang Runsheng. Fast Random Sampling Consistency Algorithm Based on Pre-Test[J]. *Journal of Software*, 2005, **16(8)**: 1431-1437.

IOP Conf. Series: Journal of Physics: Conf. Series **1284** (2019) 012021 doi:10.1088/1742-6596/1284/1/012021

Research on Overall Evaluation Model for Traffic Transportation Efficiency of Equipment based on Rough Set Theory

Wu Qian[1],Wang Chaowei[2],Lu Xiaowei[3*] and Xu Yingcheng[4]

[1]China National Institute of Standardization, Beijing 100191, China, ustbwuqian@163.com, 13811184659
[2]China Ship Develop and Design Center, Wuhan 430064, China
[3*]Correspondence author, China National Institute of Standardization, Beijing 100191, luxw@cnis.gov.cn, 18510511168
[4]China National Institute of Standardization, Beijing 100191, China

Email: luxw@cnis.gov.cn

Abstract. On the basis of overall consideration on influence factors of traffic transportation efficiency of equipment, based on Completeness, Maneuverability, Rationality and Comparability, this paper proposes a safe, cost-effective, universal and timely evaluation index system for traffic transportation efficiency of equipment, puts forward a quantitative calculating approach for the weight of evaluation index system for traffic transportation efficiency of equipment and creates an overall evaluation model for traffic transportation efficiency of equipment based on rough set theory. The evaluation index system and evaluation model for traffic transportation efficiency of equipment proposed in this paper are quantitative calculation approaches to improve and evaluate traffic transportation efficiency of equipment, which paves the way for further research on traffic transportation efficiency of equipment.

1. Introduction

The R&D and usage of equipment is a complex system engineering, which are highly exploratory and integrated, with a large number of evaluation and decision-making issues involved [1]. Equipment efficiency analysis is an important theory and method for judging, selecting and optimizing the equipment's combat readiness, transportation, reliability and other related issues [2]. Traffic transportation efficiency of equipment directly affects maneuverability of equipment. No matter how sophisticated the equipment is, if it cannot be assembled quickly and effectively, its combat effectiveness cannot be fully exerted [3]. Therefore, it is particularly important to scientifically evaluate the traffic transportation efficiency of equipment.

In recent years, Chinese scholars have realized that it's important to evaluate the traffic transportation efficiency of equipment, and experts and scholars have also carried out extensive exploration and thorough research in this field. Zhang Lei et al. [4] have created an evaluation model of wartime military airlift efficiency based on the entropy approach, which solves the problem of incomplete transport efficiency evaluation data to a certain extent. Sheng Youwen et al. [5] have made a quantitative analysis on the factors affecting the traffic transportation efficiency of military container in respect of the features of military container transportation, and have created a model for selecting optimal traffic transportation efficiency for transporting military containers in combination with the fuzzy neural network approach. Zhang Jianing [2] has made a comparative analysis on traditional efficiency evaluation approaches and created an overall evaluation model for traffic transportation efficiency of warehoused goods with analytical hierarchy process and ABC approaches. Based on the existing research results, this paper takes into account the influence factors of traffic transportation efficiency of equipment, creates the evaluation index system for traffic transportation efficiency of equipment, applies the rough set theory to the weight calculation of the evaluation index system for traffic transportation efficiency of equipment, and also uses an integrated evaluation method to create the overall evaluation model for traffic transportation efficiency of equipment based on the rough set theory.

2. Construction of an Evaluation Index System for Traffic Transportation Efficiency of Equipment

The evaluation index system for traffic transportation efficiency of equipment is the outcome of evaluation objectives, and should cover safety, cost-effectiveness, universality, timeliness and other indexes to evaluate the traffic transportation efficiency of equipment. It improves the strategic maneuverability of equipment, reduces transport maintenance costs and shortens the transportation time.

2.1 Principles for creating the evaluation index system

Identifying and constructing a systematic and reasonable evaluation index system based on certain construction principles is the premise of creating an evaluation model for traffic transportation efficiency of equipment. The principles [6, 7] adhered to construct an overall evaluation index system for traffic transportation efficiency of equipment are as follows:

2.1.1 Completeness. The traffic transportation efficiency of equipment should be fully reflected in the evaluation index system.

2.1.2 Maneuverability. The evaluation index must be quantitative if possible, to perform an intuitive comparison based on values.

2.1.3 Rationality. The evaluation index must have a clear boundary and reasonable contents, to clearly express the contents to be compared.

2.1.4 Comparability. The same evaluation index must be comparable among different traffic transportation efficiency.

2.2 Evaluation index system for traffic transportation efficiency of equipment

In this paper, the frames [7, 8 and 9] of the evaluation index system for traffic transportation efficiency of equipment are constructed based on the construction principles from aspects of safety, cost-effectiveness, universality and timeliness, as detailed in the following figure1.

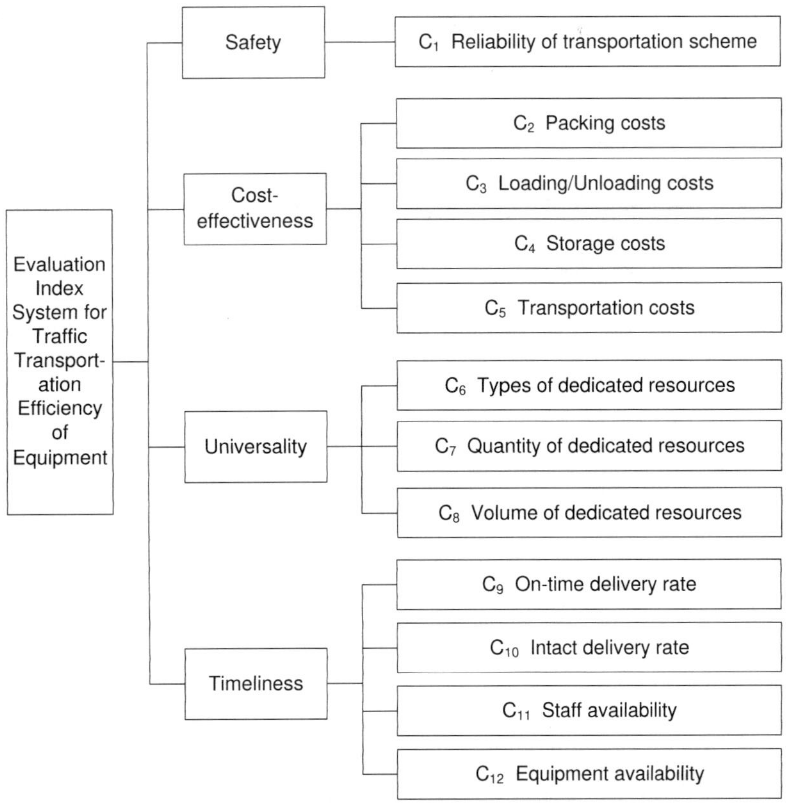

Figure 1. Evaluation Index System for Traffic Transportation Efficiency of Equipment

2.2.1 Safety index. Safety index refers to the probability that the transportation task of the equipment is completed under conditions specified in a scheme of traffic transportation efficiency, which is expressed by the reliability of traffic transportation efficiency. Given the same conditions, the higher the reliability value, the better the traffic transportation efficiency.

2.2.2 Cost-effectiveness index. The cost-effectiveness includes costs of packaging, loading/unloading, storage and transportation, specifically referring to the sum of staff, procurement, maintenance, energy consumption and leasing costs incurred in transportation. Given the same conditions, the less the sum of costs, the better the traffic transportation efficiency.

2.2.3 Universality index. The universality index is related to the number, type and volume of the special transportation resources involved in the traffic transportation efficiency, and is expressed by the ratio of the number, type and volume of the general transport resources to that of the total transportation resources. Given the same conditions, the larger the universality value is, the better the traffic transportation efficiency.

*2.2.4 Timeliness index.*Timeliness of traffic transportation efficiency is expressed by transportation time, on-time delivery rate, intact delivery rate, staff and equipment availability. On-time delivery rate is expressed by the ratio of on-time delivery times to total delivery times of military equipment; intact delivery rate is expressed by the ratio of intact delivery times to total delivery times of military equipment to the destination; staff and equipment availability is expressed by the ratio of the actual number of staff and equipment to the total number of staff and equipment. Given the same conditions, the higher the timeliness value, the better the traffic transportation efficiency.

3. Creation of an Evaluation Model for Traffic Transportation Efficiency of Equipment based on Rough Set Theory

3.1 Weight calculation method [8, 9, 10 and 11]

In this paper, the rough set theory combining qualitative and quantitative methods is used to analyze the evaluation decision table of traffic transportation efficiency of equipment, in which the evaluation indexes of traffic transportation efficiency are discretized through formulating the discrete data processing rules. Then the weights of evaluation indexes of traffic transportation efficiency are calculated according to rough set theory.

3.1.1 Information system. In rough set theory, the collection of study objects is called a domain of discourse that is usually expressed by U. In the domain of discourse, a two-dimensional information table consisting of the attribute values of the objects is called an information system on U.

Definition 1 $S = (U, C, V, f)$ is an information system; wherein, $U = \{x_1, x_2, \dots\dots, x_n\}$, C, V and $f: U \times C \to V$ refer to domain of discourse, conditional attribute set, collection of attribute values and information function that assigns an information value for each attribute of each study object respectively.

3.1.2 Indiscernibility relation. For a set of objects in the domain of discourse $U = \{x_1, x_2, \dots\dots, x_n\}$, if they are assigned with the same values on an attribute (set), they cannot be discerned according to the attribute (set), which is called indiscernibility relation in the rough set.

Definition 2 In the information system $S = (U, C, V, f)$, in terms of $P \subseteq C$, if each of a set of objects has the same attribute $p \in P$, such objects are viewed to be in an equivalence relation based on attribute subset P. In rough sets, this equivalence relation is defined as an indiscernibility relation, which is expressed as follows:

$$IND(P) = \{(x_i, x_j) \in U \times U | \forall a \subseteq P, f(x_i, x_j) = f(x_j, x_i)\} \quad (1)$$

Relation IND(P) constitutes a division of U and is expressed by $\frac{U}{IND(P)} = \{p_1, p_2, \dots\dots, p_n\}$, abbreviated as $U/_P$.

3.1.3 Attribute importance. In the information system, the importance of attributes is characterized by the change of information content in the information table before and after the attributes are removed.

Definition 3 In $S = (U, C, V, f)$, $\forall c \in C$. Supposing $U/c = \{C_1, C_2, \dots, C_n\}$, the importance of attribute c is defined as follows:

$$sig(\{c\}) = \sum_{j=1}^{n} \frac{|C_j||U-C_j|}{|U|(|U|-1)} \quad (2)$$

3.1.4 Attribute weight. In the HU-based discernibility matrix, $|C_i||U - C_i|$ refers to the total discernibility elements generated by attribute $\{c\}$ in the discernibility matrix, which is usually regarded as the heuristic information, that is, if the total discernibility attribute produced by an attribute $\{c\}$ is larger, the attribute $\{c\}$ is more important.

Definition 4 In the information system $S = (U, C, V, f)$, $\forall c_j \in C$, and the weight of attribute c_j in the information system can be defined as per the following formula:

$$W(\{c_j\}) = \frac{sig(\{c_j\})}{\sum_{i=1}^{n} sig(\{c_j\})} \quad (3)$$

3.2 Evaluation model

In this paper, the comprehensive evaluation method is used to sort design schemes for traffic transportation efficiency of equipment. Based on the determination of attribute weight above, it is assumed that the weight vector of evaluation indexes of traffic transportation efficiency of equipment is denoted as $W = (w_1, w_2, \dots\dots, w_m)^\tau, w_s \geq 0$ and $\sum_{s=1}^{m} w_s = 1$ after normalization. Vector $R_{m \times n}$

refers to a normalized index of decision vector, and the comprehensive evaluation model for traffic transportation efficiency in design schemes for traffic transportation efficiency of equipment can be expressed as follows:

$$Z_i = \sum_{i=1}^{m} w_s r_{ij} \quad (4)$$

4. Examples

4.1 Case scenario

In order to meet the strategic requirements of a certain army, a new research equipment needs to be developed, and eight design schemes for traffic transportation efficiency of equipment are given, as shown in Table 1. It is expected to evaluate the transportation efficiency of the eight design schemes for traffic transportation efficiency of equipment from safety, cost-effectiveness, universality and timeliness, so as to find out the optimal scheme.

Table1 Information of Design Schemes for Traffic Transportation Efficiency of Equipment

Scheme	Safety (%)	Cost-effectiveness (RMB 10,000)	Universality (%)	Timeliness (%)
A_1	97	10	70	80
A_2	95	15	80	80
A_3	96	11	60	90
A_4	95	12	70	80
A_5	94	13	80	70
A_6	97	12	80	80
A_7	98	13	70	80
A_8	97	9	70	80

4.2 Evaluation for traffic transportation efficiency

4.2.1 Calculation of importance. Based on the above-mentioned evaluation index system of traffic transportation efficiency and the information of the design scheme for traffic transportation efficiency of equipment, 10 experts are arranged to score the evaluation indexes of traffic transportation efficiency of each alternative scheme, with a maximum of 5 scores and a minimum of 0 score. The scores of the 10 experts are averaged to get the scores of the indexes of each alternative scheme, as shown in Table 2:

Table2 Initial Scoring Decision

Scheme	C_1	C_2	C_3	C_4	C_5	C_6	C_7	C_8	C_9	C_{10}	C_{11}	C_{12}
A_1	4.56	4.53	4.45	3.45	3.56	4.25	3.24	4.03	4.23	3.98	3.96	4.04
A_2	4.85	4.23	4.38	2.87	3.12	4.32	3.87	4.14	4.86	3.56	3.87	4.04
A_3	4.12	3.93	4.65	4.28	4.29	4.52	4.56	4.56	4.13	4.56	4.78	4.08
A_4	3.53	4.65	4.12	3.56	4.45	3.87	4.21	3.87	3.87	4.02	3.98	4.21
A_5	4.82	4.12	3.86	2.96	3.26	4.51	4.02	4.06	4.81	4.03	4.23	3.96
A_6	2.78	4.86	4.66	4.65	4.89	3.03	3.35	3.02	3.02	3.04	3.67	3.87
A_7	3.28	3.69	4.68	3.63	4.86	2.98	4.03	3.86	3.86	3.89	3.67	4.02
A_8	3.67	4.46	4.07	3.97	4.52	4.56	3.98	4.02	4.02	4.76	4.23	4.56

The weight of each efficiency evaluation index is determined on the basis of rough set theory. First, it is necessary to discretize each efficiency evaluation index shown in Table 2. In this paper, by rounding up to an integer, the indexes with scores of (4.5, 5], (4.0, 4.5], (3.5, 4.0], (3.0, 3.5] and below 3.0 are respectively judged to be excellent and recorded as 5, good and recorded as 4, medium and recorded as 3, poor and recorded as 2 and very poor and recorded as 1. Thus, the set of all conditional attribute values after discretization can be obtained, as shown in Table 3:

Table 3 Decision after Discretization

Scheme	C_1	C_2	C_3	C_4	C_5	C_6	C_7	C_8	C_9	C_{10}	C_{11}	C_{12}
A_1	5	5	4	3	3	4	2	4	4	3	3	4
A_2	5	4	4	1	2	4	3	4	5	3	3	4
A_3	4	3	5	4	4	5	5	5	4	5	5	4
A_4	3	4	4	3	4	3	4	3	3	4	3	4
A_5	5	4	3	1	2	5	4	4	5	4	4	3
A_6	1	5	5	5	5	2	2	2	2	2	3	3
A_7	2	3	5	3	5	1	4	3	3	3	3	4
A_8	3	4	4	3	5	5	3	4	4	5	4	5

According to rough set theory, the equivalence relation of 12 condition attributes namely C_1, C_2, C_3, C_4, C_5, C_6, C_7, C_8, C_9、C_{10}, C_{11} and C_{12} in Table 3 can be obtained:

$$U/\{C_1\} = \{\{A_1,A_2,A_5\},\{A_3\},\{A_4,A_8\},\{A_7\},\{A_6\}\};\ U/\{C_2\} = \{\{A_1,A_6\},\{A_2,A_4,A_5,A_8\},\{A_3,A_7\}\};\ U/\{C_3\} = \{\{A_1,A_2,A_4,A_8\},\{A_3,A_6,A_7\},\{A_5\}\};\ U/\{C_4\} = \{\{A_1,A_2,A_5\},\{A_3\},\{A_4,A_8\},\{A_7\},\{A_6\}\};\ U/\{C_5\} = \{\{A_6\},\{A_3\},\{A_1,A_4,A_7,A_8\},\{A_2,A_5\}\};\ U/\{C_6\} = \{\{A_3,A_5,A_8\},\{A_1,A_2\},\{A_4,\},\{A_6\},\{A_7\}\};\ U/\{C_7\} = \{\{A_3\},\{A_4,A_5,A_7\},\{A_2,A_8\},\{A_1,A_6\}\};\ U/\{C_8\} = \{\{A_3\},\{A_1,A_2,A_5,A_8\},\{A_4,A_7\},\{A_6\}\};\ U/\{C_9\} = \{\{A_2,A_5\},\{A_1,A_3,A_8\},\{A_4,A_7\},\{A_6\}\};\ U/\{C_{10}\} = \{\{A_3,A_8\},\{A_4,A_5\},\{A_1,A_2,A_7\},\{A_6\}\};\ U/\{C_{11}\} = \{\{A_3\},\{A_5,A_8\},\{A_1,A_2,A_4,A_6,A_7\}\};\ U/\{C_{12}\} = \{\{A_8\},\{A_1,A_2,A_3,A_4,A_7\},\{A_5,A_6\}\}。$$

According to the formula (2) for calculating importance based on rough set theory, the importance of the above condition attributes can be calculated. Taking attributes C_1 and C_2 as examples, the specific algorithm of importance of its attributes is as follows:

$$\begin{aligned}\mathrm{sig}(\{C_1\}) &= \frac{|\{A_1,A_6\}||U-\{A_1,A_6\}|}{|U|(|U|-1)} + \frac{|\{A_3\}||U-\{A_3\}|}{|U|(|U|-1)} + \frac{|\{A_4,A_8\}||U-\{A_4,A_8\}|}{|U|(|U|-1)} \\ &+ \frac{|\{A_7\}||U-\{A_7\}|}{|U|(|U|-1)} + \frac{|\{A_6\}||U-\{A_6\}|}{|U|(|U|-1)} \\ &= \frac{3\times 5}{8\times 7} + \frac{1\times 7}{8\times 7} + \frac{2\times 6}{8\times 7} + \frac{1\times 7}{8\times 7} + \frac{1\times 7}{8\times 7} = \frac{48}{56}\end{aligned}$$

$$\begin{aligned}\mathrm{sig}(\{C_2\}) &= \frac{|\{A_1,A_2,A_5\}||U-\{A_1,A_2,A_5\}|}{|U|(|U|-1)} + \frac{|\{A_3,A_7\}||U-\{A_3,A_7\}|}{|U|(|U|-1)} \\ &+ \frac{|\{A_2,A_4,A_5,A_8\}||U-\{A_2,A_4,A_5,A_8\}|}{|U|(|U|-1)} \\ &= \frac{2\times 6}{8\times 7} + \frac{4\times 4}{8\times 7} + \frac{2\times 6}{8\times 7} = \frac{40}{56}\end{aligned}$$

In a similar way, the calculation results of importance of other attributes are:

$$\mathrm{sig}(\{C_3\}) = \frac{38}{56}, \mathrm{sig}(\{C_4\}) = \frac{42}{56}, \mathrm{sig}(\{C_5\}) = \frac{46}{56}, \mathrm{sig}(\{C_6\}) = \frac{46}{56}, \mathrm{sig}(\{C_7\}) = \frac{46}{56}, \mathrm{sig}(\{C_8\}) = \frac{42}{56},$$

$$\mathrm{sig}(\{C_9\}) = \frac{46}{56}, \mathrm{sig}(\{C_{10}\}) = \frac{46}{56}, \mathrm{sig}(\{C_{11}\}) = \frac{34}{56}, \mathrm{sig}(\{C_{12}\}) = \frac{34}{56}$$

4.2.2 Weight calculation. According to rough set theory, we can get:

$$\sum_{i=1}^{m} \mathrm{sig}(\{c_j\}) = \frac{508}{56}$$

Then according to the weight calculation formula (3) of rough set theory, the weight values of 12 condition attributes namely C_1, C_2, C_3, C_4, C_5, C_6, C_7, C_8, C_9、C_{10}, C_{11} and C_{12} can be obtained, as shown in Table 4:

Table 4 Weight Value of Traffic Transportation Efficiency Evaluation Index for Traffic Transportation Efficiency of Equipment in Alternative Design Schemes

Attribute	Weight	Attribute	Weight	Attribute	Weight	Attribute	Weight
C_1	0.0945	C_4	0.0827	C_7	0.0906	C_{10}	0.0906
C_2	0.0787	C_5	0.0906	C_8	0.0827	C_{11}	0.0669
C_3	0.0748	C_6	0.0906	C_9	0.0906	C_{12}	0.0669

4.2.3 Analysis of evaluation results. Based on the weight results above and Table 3, the comprehensive evaluation results of traffic transportation efficiency in alternative design schemes for traffic transportation efficiency of equipment can be obtained by the comprehensive evaluation model calculation formula (4), as shown in Table 5:

Table 5 Comprehensive Evaluation Results of Alternative Design Schemes for Traffic Transportation Efficiency of Equipment

Alternative schemes	Comprehensive evaluation values of traffic transportation efficiency	Ranking of evaluation results
Alternative design scheme A_1 for traffic transportation efficiency	3.6620	4
Alternative design scheme A_2 for traffic transportation efficiency	3.5085	5
Alternative design scheme A_3 for traffic transportation efficiency	4.4183	1
Alternative design scheme A_4 for traffic transportation efficiency	3.4928	6
Alternative design scheme A_5 for traffic transportation efficiency	3.7055	3
Alternative design scheme A_6 for traffic transportation efficiency	3.0201	8
Alternative design scheme A_7 for traffic transportation efficiency	3.2132	7
Alternative design scheme A_8 for traffic transportation efficiency	4.0717	2

According to the comprehensive evaluation results of the design scheme for traffic transportation efficiency of equipment obtained from the above table, the advantages and disadvantages of the alternative design scheme for traffic transportation efficiency of equipment can be ranked as: A_3、A_8、A_5、A_1、A_2、A_4、A_7、A_6, in which A_3 is the optimal design scheme, A_8 is second only toA_3, and A_6 is the worst. Therefore, it can be concluded that the alternative A_3 shall be selected as the design scheme for traffic transportation efficiency of the equipment.

5. Conclusion

The evaluation of traffic transportation efficiency of equipment is a difficult point in the research of equipment transportation. In recent years, China has recognized the importance of research on equipment transportation, carried out a series of related researches on equipment transportation and obtained some research results, but it is still in the stage of development research, with certain gap compared with foreign countries. From the aspects of security, timeliness, cost-effectiveness and universality, this paper synthetically analyzes the influence factors of traffic transportation efficiency of equipment, introduces rough set theory, gives the quantitative calculation method of evaluation index of traffic transportation efficiency of equipment and creates a comprehensive evaluation model of traffic transportation efficiency of equipment based on rough set theory, which solves the problem of efficiency evaluation for traffic transportation efficiency of equipment to a certain extent, provides a reference method for selecting the optimal design scheme of traffic transportation efficiency of equipment, and has a certain reference significance.

Acknowledgment

We would like to acknowledge that this research are supported and funded by the National Science Foundation of China under Grant No. 91646122, No.91746202 and No.71571191, the Basic Scientific Research Business Projects 552018Y-5927, the National Key R&D Program of China under Grant No.2016YFF0202604, 2017YFF0209603, 2017YFF0209604, the quality inspection technology project 2017QK014.

References

[1] Zilong Cheng, Li Fan, and Yulin Zhang. A Framework for Equipment Systems-of-systems Effectiveness Evaluation Using Parallel Experiments Approach [J]. Journal of Systems. Engineering and Electronics, 2015, 12(2): 292-300.
[2] Jianing Zhang. Research on Evaluation Method of Warehouse Cargo Transportation Efficiency and Its Application [D]. Beijing: Harbin Engineering University, 2017.
[3] Qian Wu. Research on Transportation Analysis Method of Unmanned Aerial Vehicle System [D]. Beijing: Beihang University, 2014.
[4] Lei Zhang, Yunpeng Chen, Minglu Guan. Evaluation Model of Wartime Military Airlift Efficiency Based on Grey-fuzzy Theory [J]. Fire Control & Command Control, 2013, 38 (3): 69-75.
[5] Youwen Sheng, Yanbin Li, Tongbin Zhang, Zhenguo Long. Selection of Optimal Transportation Scheme for Military Cargo Containers in Complex Environment Based on Fuzzy Neural Network [J]. Port & Waterway Engineering, 2012, 6: 63-68.
[6] Qian Wu, Xiaowei Lu, Shuang Wang and Dong Zheng. Analysis of Influencing Factors on the Transportation of Military Equipment [C]. ICSSD 2018,2018: 364-374.
[7] Qian Wu, Ma Lin, Chaowei Wang. Study on Evaluation Index System of Equipment System Transportability [C], WCEAM2013, 2015: 1539-1547.
[8] Guofang Fu, Xinshui Bao. Research on Safety Awareness Enhancement Strategy of Construction Workers Based on Rough Set [J]. Industrial Engineering and Management, 2019: 1-11.
[9] Lei Zhang, Chenxing Zheng, Yanjiu Zhong, Xu Qin. Mechanical Product Green Design Knowledge Update Based on Rough Set [J]. China Mechanical Engineering, 2019: 1-9.
[10] Sen Wu, Ping Wang, Mengguo Xu, Aiping Cheng, Xiangdong Li. Weight Analysis of Ground Pressure Developing Factors in Stope Based on Rough Set Theory [J]. Mine Engineering, 2019, 2: 1-4.
[11] Wu Chen, Guohua Zhang, Hao Wang, Libiao Chen. Risk Assessment of Mountain Tunnel Collapse Based on Rough Set Conditional Information Entropy [J]. Rock and Soil Mechanics, 2019: 1-12.

Virtual Simulation of Automatic Quality Measurement System for Tapered Roller Bearing Based on 3D Automate Software

Alhadi Khlil[1, a], B Ma[1, b] and Z Shi [1, c, *]

[1] School of Mechanical Engineering, Hebei University of Technology, Tianjin, 300130, China

[a] 201740000010@stu.hebut.edu.cn; [b] lymabotong@126.com; [c, *] z_shi@hebut.edu.cn

Abstract. In present years, the simulation technique has been extensively used in manufacturing processes, solving complex problems with least time, effort and cost. This paper presented a virtual simulation of automatic quality measurement system for tapered roller bearing to aid the decision-making processes, observe system operation processes and provide an initial design for constructing processes using 3D modeling process and Python programming language, the proposed system was intended to check the tapered roller bearing dimensions and to achieve the prescribed goals, the 3D model was built, the virtual environment has been created and the motion control process was applied in 3DAutomate software.

1. Introduction

Nowadays, small and medium industrial companies have the vision to improve their production capability through the automation of various manufacturing facilities and computerization of numerous functions. Since the 1990s, the increasing demand for enhancing the quality of production and efficiency has progressively led industrials to seek solutions of computer software to master the various processes of production [1]. In this context, virtual simulation technology has a huge contribution, it allows simulating the modelling of the production systems, machine installation, and testing the different setup configurations of the virtual system before the actual installation, another major function of the virtual simulation is to provide the real-3D world for decision-maker to observe the running of system [2].

Virtual simulation is "a technology that combines software tools and other equipment to set up an immersive interactive experience that simulates a real-world scenario in virtual mode", it used in various fields such as mechanical engineering, mining, civil engineering, etc. this technology can empower the industry decision-making process to be faster and more powerful, and it changed the ways scientists and engineers look at computers for performing data visualization and mathematical simulations [3]. Generally, virtual simulation intended to reduce production cost, time and effort, which can aid small and medium enterprises to adopt global competition, improve their production capability without a huge investment. Therefore, this paper introduces an automatic quality measurement system for bearing. The proposed system was predicted to measure the tapered roller bearing dimensions that shown in figure 1b such as outer diameter, inner diameter, etc. then evaluate and save the process data to build up a production database for the product life cycle. The 3D Automate simulation tool will be used to creating the virtual simulation processes.

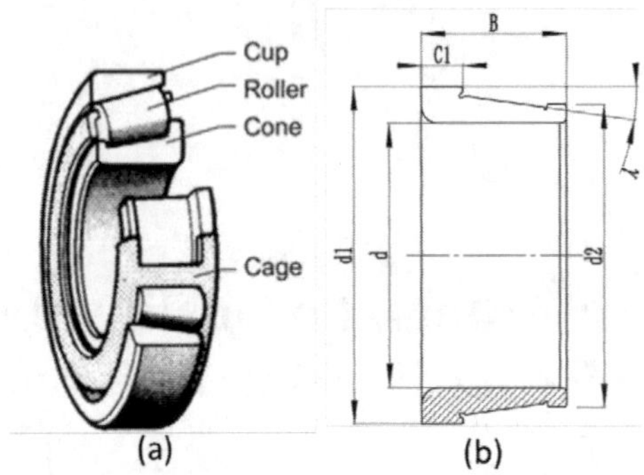

Figure 1. Tapered roller bearing component and parameters

2. Related works

Virtual simulation technique involves building a model that represents a real or an existing system [4]. People from various backgrounds and expertise are called upon to uses Virtual simulation to solve issues that arise during the design process and it facilitates Communication across different disciplines in the project, getting a diverse group to fully understand a particular issue and contribute input from each perspective to achieve good design. A survey conducted in industry sector use of Virtual simulation in product design and manufacturing, the Results indicate that Virtual simulation has extremely flourished in the last twenty years [5]. Another study by [6] used modeling simulation technique with ProModel software to measure the capacity of an existed mattress production line and simulate the ability of the improvement for the next five years, The result was clearly shown the suggested requirement of the improvement scenario and its percentage within the exact period. Also [7] are used the virtual simulation for iCIM3000 flexible manufacturing system, to discover the efficiency improvement possibilities of the whole system and the effectivity of its individual devices, CATIA CAD software had been used to create a static CAD model and HTC Vive glasses headset with Simulation software, the result illustrate the optimum simulation scenarios.

3. The structural design of the virtual simulation

This study considered the tapered roller bearing during the manufacturing processes as experimentally sample shown in figure 1a, which consist of a cup, roller, cone, and cage. Through the recommendations made by the manufacturer, the Cone is the main part of the tapered roller bearing. Therefore, this study will adopt the Cone dimensions illustrated in figure 1b as the references for design and create the virtual simulation of the system, figure 2 shows the Structural design of this virtual simulation.

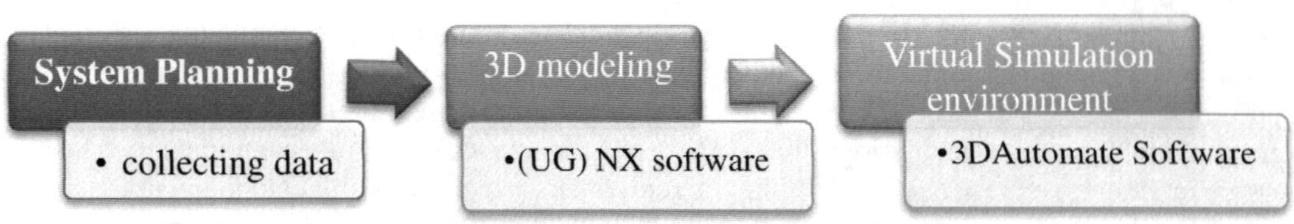

Figure 2. The structural design of the simulation system.

3.1. System planning

According to the experiment sample (Cone) dimensions illustrated in figure 1b, the Initial design of the system mainly consists of three Platforms, the first Platform uses two electro-pneumatic sensors to measure the outer diameter of the cone d1and d2, then detect another two points to check the tapered slop angle γ. The second Platform uses other sensors to measure the height of the Cone (B) and the height of the back flange (C1). The third Platform uses an electronic plug gauge to measure the inner diameter of the Cone (d). Then all measurement data were collected will be sent to the control unit to analyse, evaluate and save. Table 1 shows the simulation components, and figure 3 illustrates the processes sequences between the system components.

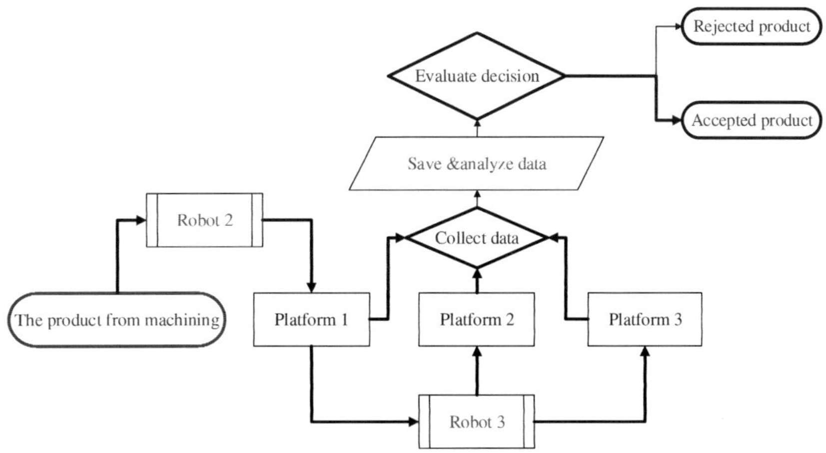

Figure 3. The measurement process system sequences

Table1. The system components,

Component Name	Numbers	source
Lathe machine	2	3D Automate
Quality Measurement Platform	3	(UG) NX software
Conveyer Belt	4	3D Automate
Robot Arm	3	3D Automate
Work piece	1	(UG) NX software

3.2. 3D Modeling

The 3D Modeling is "the process of manually creating geometric object models which meet desired design criteria" [8] and Computer Aided Design (CAD) software are used. The increasing power of CAD software has a significant impact on the product development process, allowing improved quality, reduced cost, and aids products to get to market faster [9]. Nowadays, there are many different CAD software packages for different purposes are available such as CATIA, SolidWorks, and AutoCAD. In this study UNIGRAPHICS (UG) NX software package from Siemens has been chosen. It is one of the world's most advanced and tightly integrated CAD/CAM/CAE product development solution [10]. The software has been used for building and visualizing the experimental prototype components, figure 4 shows the assembled model of Platform 1 in order to export it to 3D Automate software.

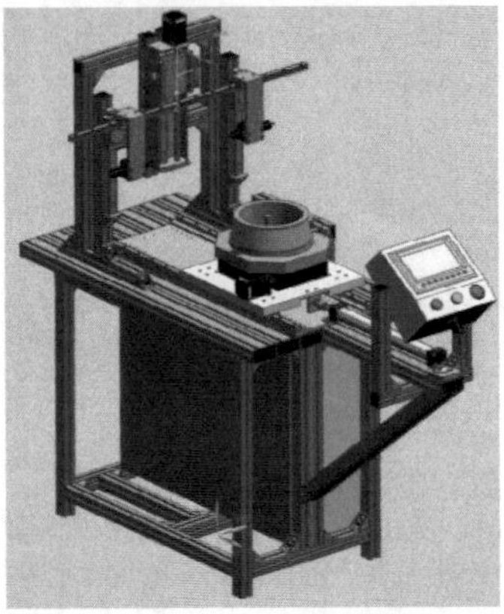

Figure 4. The assembled model of Platform 1

3.3. Virtual Simulation Environment

There are several issues associated with the proper development of virtual simulation environments including the translation of data generated by (CAD) systems, implementing the physical behaviour and constraints of objects, collision detection [11]. The 3D Automate software will be used, it is the powerful, flexible, and scalable platform for 3D manufacturing simulation solutions to support critical planning and decision-making processes. To create the intended Virtual Simulation the steps shown in figure 5 will be followed:

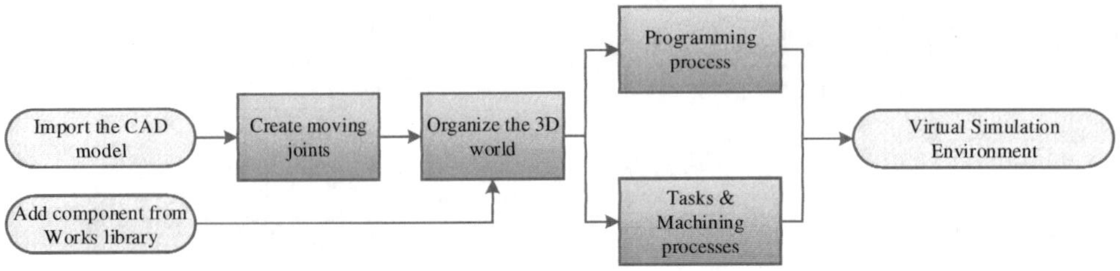

Figure 5. The steps of creating the Virtual Simulation

- Import the CAD model into 3D Automate software. Rendering and organizing the models set in one parent geometry set.
- Create moving joints, which means split the joint geometry set form the parent geometry and transfer it to a new link to become an independent interactive joint.
- Add components from works library, 3D Automate software provides hundreds of industrial components and facilities which can be used to create a virtual simulation where we used several components such as lathe machines and conveyors.
- Organized the 3D world, rearrange the imported and added components into the simulation interface.
- Tasks and machining processes, these processes will be done for the components which added from works library, by creating a virtual network to implement the intended tasks and exchanging signals.
- Programming processes, it will apply on the imported models to implement the desired functions using an object-oriented programming language is called Python.
- Virtual Simulation Environment, testing and modifying the system to confirm the required motion control and proper operation sequences

4. Simulation Result

Based on the stated objective, the final virtual simulation environment was built as shown in figure 6, including the tow CNC lathe machines feed it by Conveyor 1 and Robot1, conveyor 2 received the product from the production line and send a Boolean Signal to Robot 2 to pick the product to feed the Platform 1 and send a signal to execute the measuring processes, Robot 3 received a signal from Platform 1 to remove the product and feed Platform 2 and Platform 3 respectively, it also controlled the signals to execute the measuring processes, and received the signals to remove the product. The system ended with tow conveyors, used for separating the accepted or rejected products as a result of measuring and evaluation process. ATA _Report Output storage are attached to the conveyers to calculate the production rate.

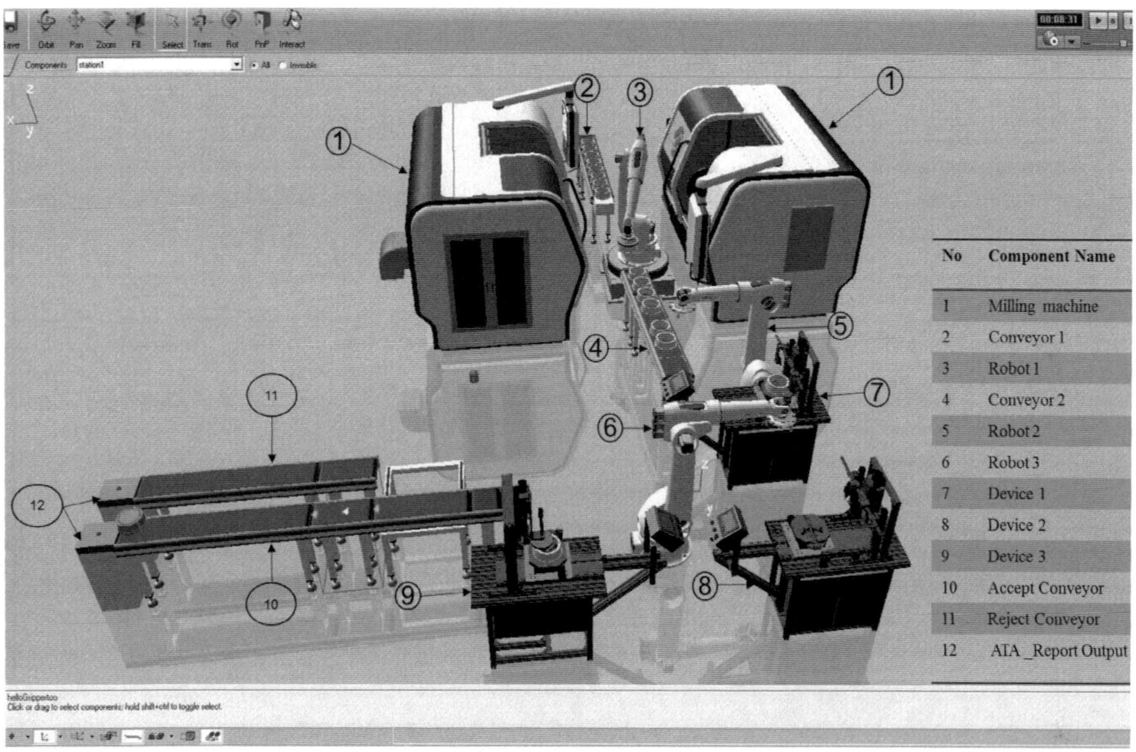

Figure 6. The virtual simulation environment

Finally, it can be said that the simulation provides the product flow process for the entire system in the real world which allow the intended user can observe and check the operation process sequences and design of the system structural elements to support decision-making processes. In addition, many different scenarios have been implemented to get the optimum design and simulation result which will strongly aid the construction process of the experimental prototype.

5. Conclusions

Virtual simulation one of the most recent useful tool in the industry sector, it used to develop a simulation of an Automatic Quality Measurement System for tapered roller Bearing attached with an existed production line, the construction of 3D modeling has been done in (UG) NX, 3D Automate is considered for the virtual environment. The current simulation providing the real world to aid the decision-making process and optimum scenario for the experimental prototype construction process. In general, the developed system was intended to reduce the production cost, time and effort, improve the production capacity without a huge investment.

References

[1] Mohamed A, Barbara L, Pierre C, Rosario T, and Patrick B 2015 *Robotics, Automation, and Control in Industrial and Service Settings* (Lyon University, France) chapter 5 pp 142–172

[2] Yang X, Feng Y, and Li T 2002 Solving sequential decision-making problems under virtual reality *Proceeding of the 2001 Winter Simulation Conf* (Arlington, USA) Cat. No.01CH37304

[3] Abhishek S, Judy M V, and James H O 2011 Virtual Reality for Assembly Methods PrototypingA Review, *Virtual Reality* Vol15(London: Springer-Verlag) pp 5–20

[4] Surbhi R available at http://www.businessmanagementideas.com/decision-making/top-10-techniques-of-decision-making/337

[5] Leif P B, and Judy M V 2017 Industry use of virtual reality in product design and manufacturing: a survey *Virtual Reality* Vol 21(London: Springer) pp 1–17

[6] Mohammad H, and Farhad Z 2013 Modeling and Simulation of a Mattress Production Line Using ProModel *Proceedings of the 2013 Winter Simulation Conference* (Washington, DC, USA)

[7] Radovan H, Daynier R D S, Peter K, Roman R and Karol V 2017 Using Virtual Reality tools to support simulations of manufacturing instances in Process Simulate: The case of an iCIM 3000 system *MATEC Web Conf.* Vol 137(2017)

[8] Norman I B, and Andrew S G 2017 *3D Object Modeling Microsoft Research* WA 98052 (Philadelphia, USA)

[9] Adekunle A A and Adejuyigbe S B 2012 Computer Aided Design Software development for Welding hollow cylinder *J Am Sci 2012*vol8(7)pp82-86

[10] Ming L, Amir G, and Krishna NX 10 for Engineering Design *Missouri University of Science and Technology*

[11] Abdulrahman M A, Mustufa H A, Ali A and Saber D 2016 Development of a virtual manufacturing assembly simulation system *J advances in Mechanical Engineering* Vol. 8(3) 1–13

Research on Modelling and Simulation Method of Low Pressure and Temperature Environment Chamber

Ju shichao

Shanghai Aircraft Design and Research Institute, COMAC, No.5188 Jinke Road, Shanghai, China

symseer@163.com

Abstract. This paper provides modelling and simulation method for low pressure and temperature environment chamber. The environment chamber internal air-flow organization with air deflector and anisotropic damping plate are calculated and optimized. The results show that under the condition of low pressure, adopting the above airflow organization control method can ensure that the temperature difference of the core area is less than 1 degree. At the same time, a preliminary analysis of the environmental chamber loaded with test parts is carried out. The flow field and temperature field in the environmental chamber are greatly affected by large size test pieces. In order to keep the temperature uniform, the wind speed in the test section should be adjusted according to the test pieces. The results show that the method of the temperature uniform optimization is effective. The results in this paper have a certain reference value for the design of large environmental chambers.

1. Introduction

At present, the cruise height of large passenger aircraft can reach more than 40000ft, and it will last for more than ten hours. During the flight, the atmospheric environment condition will change greatly. The temperature will change from 65 degree below zero to more than 50 degree, and the pressure can less than 10 kPa[1-3]. From the point of view of design and verification, functional design is to separate the various professions, systems, and physical quantities to achieve their specific functions. For example, pneumatics provide lift, rudder surfaces are stable, and electronic control is achieved separately. On the other hand, system verification is bottom-up, and ultimately it is to be verified as a whole, in addition to system integration, environmental integration, and installation integration, for example, cabin temperature control, achieved by low-pressure pipeline air supply, system integration refers to The integration of the upstream mixing chamber air conditioning package, the environment integration refers to the hot and cold environment outside the machine, the installation integration refers to the low-pressure pipeline installed in the passenger cabin, the "installation" in FAR 25.1309, including the above three aspects, to achieve the equivalent These test environments are needed for verification as a whole.

In order to verify the reliability of the design, one of the important ways is to carry out the high altitude environment simulation test on the ground. The high altitude environment simulation system is the main test equipment used to build the external temperature environment of the aircraft, and, enhance the performance test of the aircraft environmental control system and the thermal comfort test of the aircraft cabin. For the environment chamber, the temperature uniformity is one of the most important technical indicators, which directly determines the performance of the environment chamber[4]. For using air convection heat transfer, the temperature distribution is directly related to the air flow organization in the environmental chamber.

2. High altitude environment simulation

The high altitude environment is a low pressure environment, and low air density environment, therefor, the heat transfer coefficient is extremely small. And generally the internal volume of environment chamber, can reach thousands of cubic meters, to meet temperature uniformity in such a large space under low pressure environment is very difficult[5]. At home and abroad usually use orifice plate air supply or cold plate radiation cooling mode, or a combination of both, but when there is a big test of heating power, it is difficult to control the temperature in the chamber, because the small average wind speed and cabin air flow resistance[6,7]. In this paper, the structure of air flow with anisotropic damping laminates is studied, in Oder to optimizes the temperature uniformity in a large space environmental chamber.

3. Technology realization

The physical model of the large environmental chamber is shown in Figure 1. The main characteristics are: (a) the air distribution in the test area of the environmental chamber is a horizontal air supply mode, and the internal circulation is carried out through the top air supply cavity and the bottom air supply cavity, as shown in Figure 2. (b) the top air supply cavity and the bottom air supply cavity contain a dynamic section. The power section includes the air cooling heat exchanger component and the air re-circulation fan component. (c) there are air deflectors and anisotropic damping laminates on the front and rear ends of the test area of the environmental chamber. The damping characteristics of the damping laminates in horizontal and vertical directions can be designed according to needs.

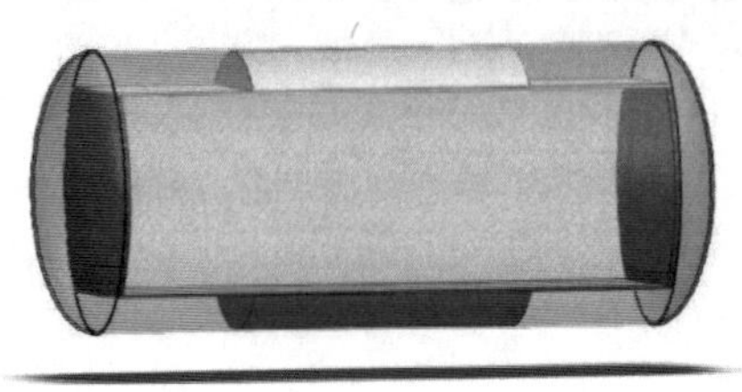

Figure 1. Environmental chamber

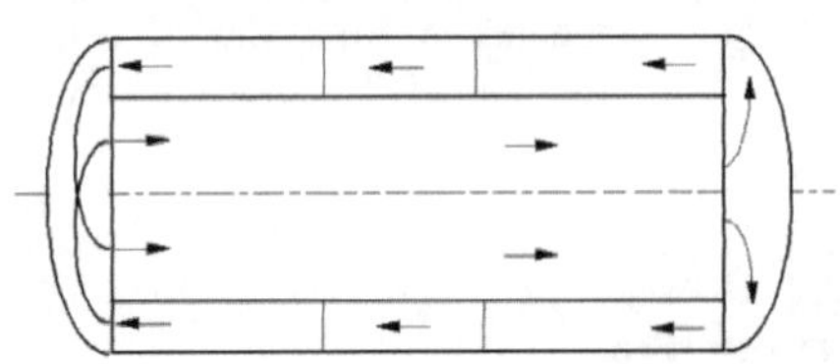

Figure 2. Air flow organization

4. Sections, subsections and subsubsections

Using CFD technology, the flow field and temperature uniformity are simulated and analysed. The analysis shows that by controlling the air flow organization of the damping layer, the flow field can be realized under the condition of small pressure loss, which ensures the high / low temperature uniformity of the low pressure environment in the high-altitude cabin.

4.1. Mathematical model and boundary condition

The grid is divided into three parts: the dynamic section of the environmental chamber, the damping layer, the test section and the section of the head. and the grid with the test piece contains one test area more. The whole grid size is set to 0.2m, the head of rotary section of the fluid near the wall region, etc. for partial encryption, as shown in Figure 3.

(a) without test piece

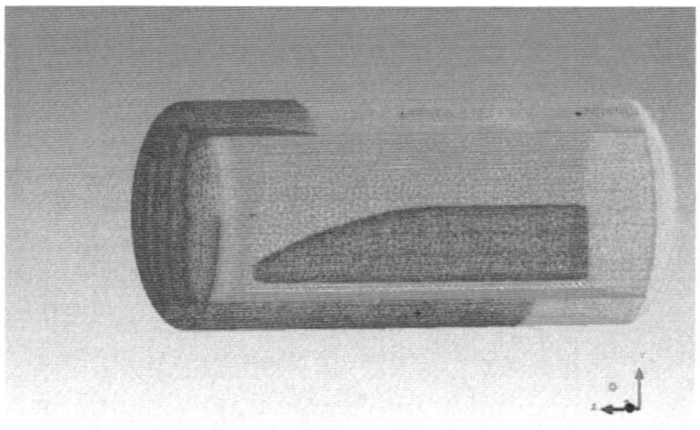

(b) with test piece

Figure 3. Grid of environmental chamber

In the calculation, only the steady state operation is considered. The K-Epsilon turbulence model is adopted. The damping layer area is porous media model, and the corresponding viscous resistance coefficient and inertia drag coefficient are set up. The power cabin heat exchanger model, dynamic inflow and outflow end section adopts a fan wall condition, heat flow setting cabin surface boundary conditions. The definition of each working condition is shown in Table 1.

Table 1. Conditions and damping layer features

	Fan pressure Pa	Attitude ft	horizontal inertial resistance coefficient 1/m	vertical inertia resistance coefficient 1/m
Case1	800	0	40	20
Case2	800	0	50	20
Case3	800	0	100	20
Case4	1000	0	60	20
Case5	500	35000	50	20

4.2. Characteristics of flow field and temperature field

The flow field calculation result is shown in Figure 4. Taking CASE1 and CASE2 as example, the CASE1 horizontal inertia resistance coefficient is small, in the center a high speed jet format, the core area of the jet can be extended to the regional test center, test flow field uniformity is poor; in CASE2 maintain the vertical inertia resistance coefficient, improve the level of horizontal inertia resistance coefficient, flow field uniformity is obviously improved, and the average air speed of test area is 4.5m/s ,which is 4.6m/s in CASE1 , the changing is little. In CASE 3 continued to increase horizontal inertial resistance coefficient to 100/m. The central location of test area was cavitating and the flow field became uniform. The mean velocity 3.9m/s of test area was significantly lower than that of CASE1 average flow velocity 4.6m/s.

In order to get higher average velocity of test area, CASE 4 increased fan static pressure and adjusted horizontal inertia resistance coefficient, and the average velocity of test area increased to 5.6m/s. In conclusion, the flow field control can be realized by adjusting the resistance characteristics of the damping plate. Proper horizontal inertial resistance and vertical inertia resistance can significantly improve the uniformity of the flow field and have no obvious effect on the average speed.

IOP Conf. Series: Journal of Physics: Conf. Series **1284** (2019) 012023 doi:10.1088/1742-6596/1284/1/012023

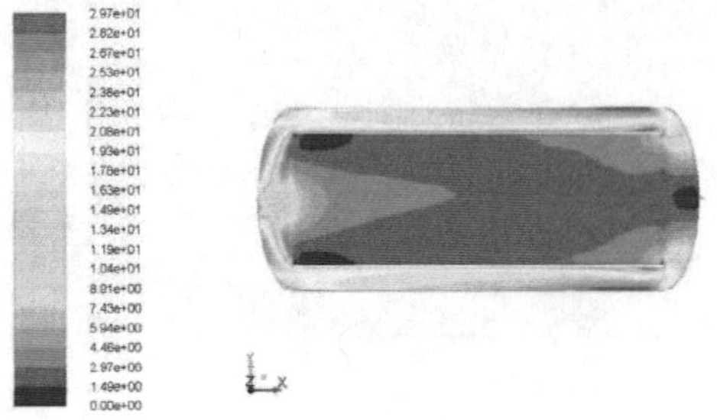

(a) Case 1

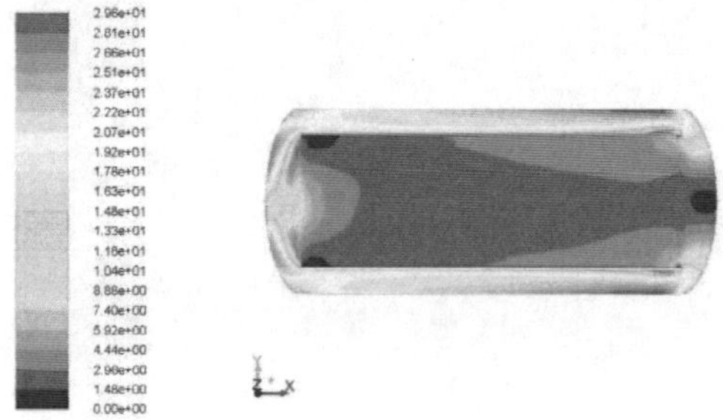

(b) Case 2

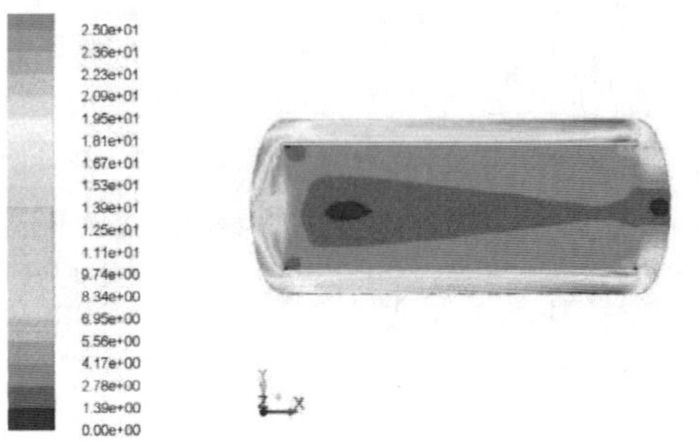

(c) Case 3

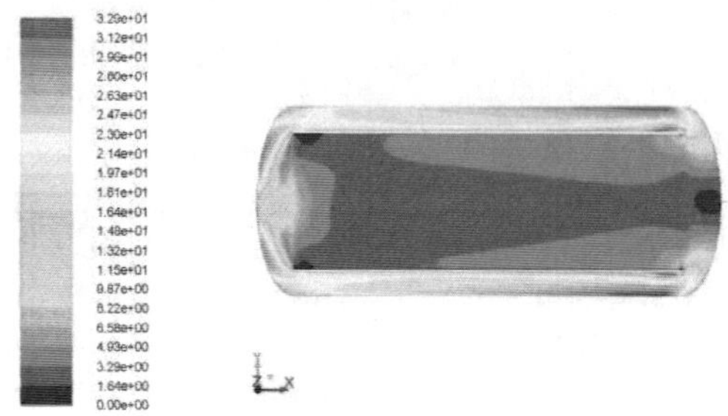

(d) Case 4

Figure 4. Z=0 Cross section flow field

4.3. Temperature field of low pressure flow

For the low pressure environment, because the air density is low, the air density at 35000 feet is only about 1/4 of the sea level. The static pressure provided by the recirculating fan will decrease, and the flow resistance will also decrease. Taking CASE5 as an example, the average velocity of the test area is 5m/s, and the uniformity is better. Near the cabin wall temperature is high, but the test area was not significantly affected by wall temperature, as shown in Figure 5.

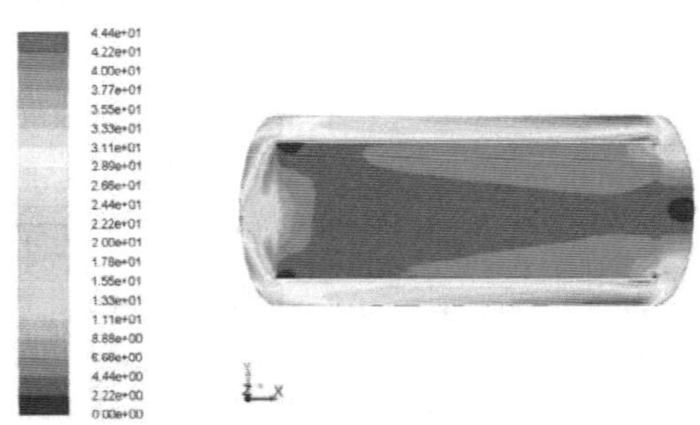

(a) Case 5 flow field

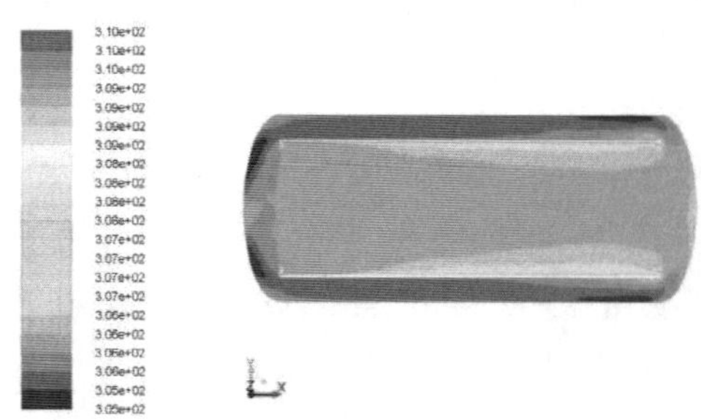

(b) Case 5 Temperature field

Figure 5. Case 5 Z=0 Cross section flow field

4.4. Temperature uniformity of low pressure flow

In order to evaluate the environment chamber temperature uniformity, a certain area which is constant distance to cabin wall, roof and floor is a distance is define as the core region. The center of the core area, the center of the surface and the center of the body are taken as the object of investigation. Calculation of extreme cold (XC) -55℃, cold temperature(C) -10 ℃, hot (H) 40 ℃, extreme hot (XH) 70 ℃ situation, examine the temperature contrast, the result is shown in Figure 6, the temperature difference ΔT of each point in the core area is were less than 1℃.

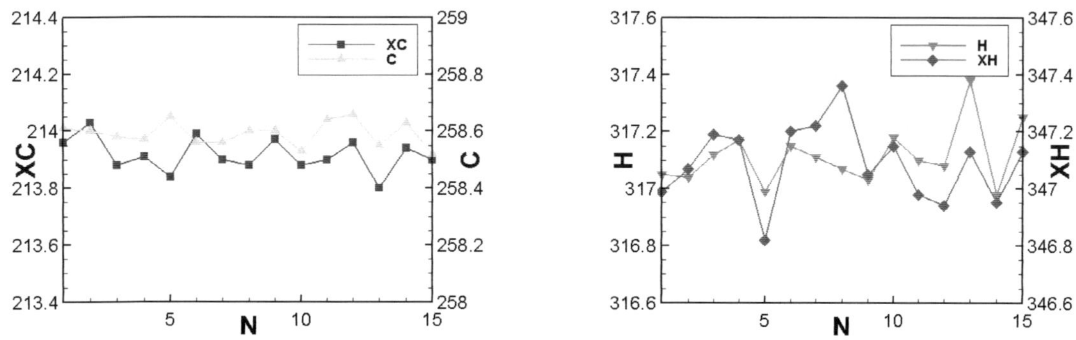

(a) extreme cold and cold

(b) extreme hot and hot

Figure 6. Temperature contrast in the core area

5. Conditions with test piece

In order to verify the environmental chamber design, working conditions with the test parts were studied. The ambient temperature set to 276K, the surface temperature of the test parts was set to 320K, and the surface area of the test parts was 138m^2. The calculation results show that the flow field in the test area is well uniform and presents a laminar flow state as shown in Figure 7. The temperature uniformity around the test parts is better. The average air velocity is 3.2m/s, and the total heat transfer is 65kW. Surface streamline and heat flux density of test parts is shown in Figure 8.

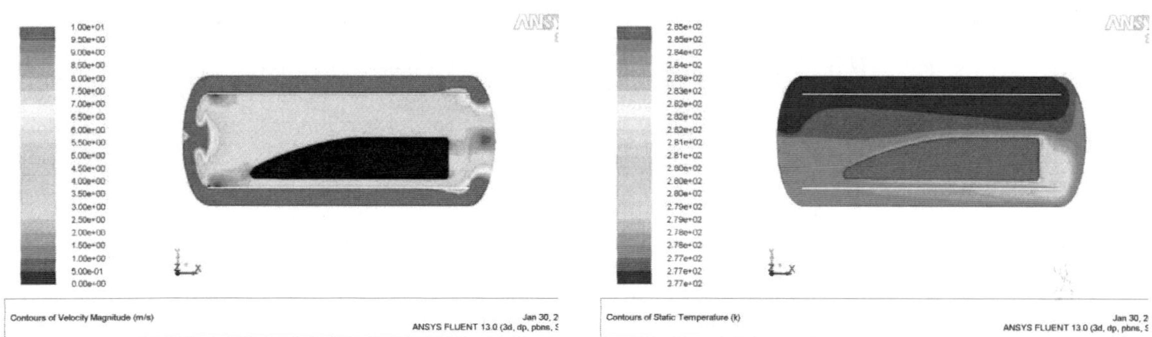

(a) Case 5 flow field

(b) Case 5 Temperature field

Figure 7. Cross section flow field with test piece

When the test parts are placed, especially the large size test parts, the obstruction in the environment cabin is greater, which will significantly change the flow field in the environment chamber. When the heat source exists in the test piece, the heat load generated by the heat source in the test chamber is usually larger than that in the environmental chamber itself. After placing the internal heat source, the temperature field in the environmental chamber will be significantly changed. Therefore, after placing test pieces in the environmental chamber, the temperature uniformity in the cabin will be greatly affected. In order to keep the temperature uniform, it is necessary to adjust the wind speed in the test section; and the degree of the specific influence needs to be analysed according to the calorific conditions of the test piece.

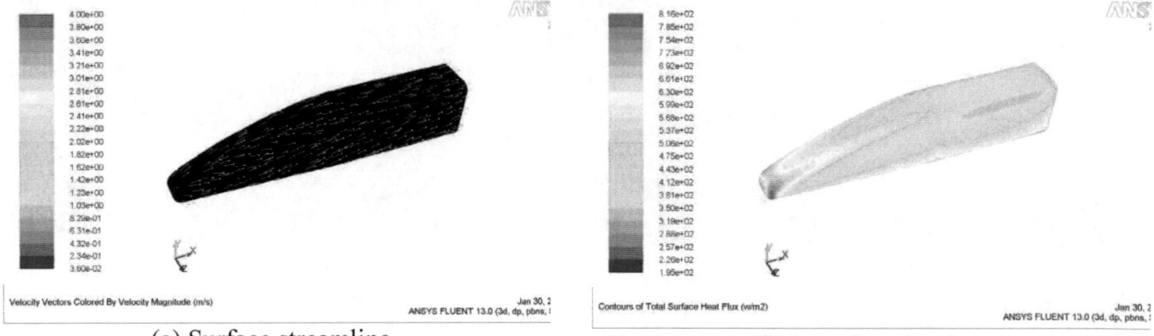

(a) Surface streamline (b) heat flux density

Figure 8. Surface streamline and heat flux density of test parts

6. Conclusions

As the degree of integration of complex systems increases, the performance range of each system that can be verified is declining, and the boundaries of each system cannot be verified. This is an inherent contradiction. The solution is to think about the transfer of some flight tests to the ground laboratory. In order to achieve uniform design and requirements of the internal temperature of the environmental chamber, and to meet the real test scenario, In this paper, a horizontal wind blows and air distribution scheme using anisotropic damping laminates is designed for the environment chamber. By adjusting the drag characteristics of the damping laminates, the flow field can be controlled, and the temperature homogeneity can be simulated in the large space high altitude environmental chamber. The results show that in low pressure environment the chamber core area temperature different can less than 1℃.

When placing large size test pieces in the environmental chamber, the flow field and temperature field in the environmental chamber will be greatly affected. In order to keep the temperature uniform, the wind speed in the test section should be adjusted according to the calorific value of the test piece.

References

[1] Liu Meng, Li Li, Wang Jun, Pang Liping, Research on temperature uniformity simulation in a big low atmosphere pressure environment, CRYOGENICS,2010, (06), p33-36.

[2] Wang Shanshan, Study on a High-Low Temperature Environment Simulization System with Helium as Working Gas, Shanghai Jiao Tong University, 2010.

[3] John J. Bertin, Russel M. Cummings. Critical Hypersonic Aerothermodynamic Phenomena[J]. Annu Rev Fluid Mech, 2006.38:129-157.

[4] He Chao, Zhou Ying, Li Gao, The environment temperature control method in the low pressure chamber. CSAA/IET International Conference on Aircraft Utility Systems, AUS 2018, CP743, p205-209, 2018.

[5] S. N. Ashwindran, A. A. Azizuddin, Design of small thermal vacuum chamber for 125-U CubeSat satellite, IPCME 2018, IOP Conf. Series: Materials Science and Engineering 469 (2019) 012108.

[6] Stoecker W F, Design of Thermal Systems, McGraw-Hill, 1980.

[7] Goulding.P, Verification and Expansion of the Hensel Blockage Correction in the National Full Scale Aerodynamics Complex 80- by 120-foot Wind Tunnel, 52nd AIAA Aerospace Sciences Meeting, National Harbor, MD, January 13-16, 2014.

IOP Conf. Series: Journal of Physics: Conf. Series **1284** (2019) 012024 doi:10.1088/1742-6596/1284/1/012024

Modeling Investigation of Concurrent-flow Chemical Extraction Process

Miaomiao Nie[a], Jing Tan [a,*], Wen-Sheng Deng [a] and Yue-Feng Su [b]

[a] School of Chemistry and Chemical Engineering, Beijing Institute of Technology, Beijing, 100081, China
[b] School of Material Science and Engineering, Beijing Institute of Technology, Beijing, 100081, China

[*]tanjing@bit.edu.cn

Abstract. In this study, modeling investigation of mass transfer in concurrent-flow chemical extraction process is carried out. Continuum Model is used to describe mass transfer process. Concentration curves in organic droplet and aqueous liquid film are firstly calculated and rate-limiting step is determined to be mass transfer of solute in aqueous phase. Mass transfer performance in aqueous phase is further calculated. Volumetric mass transfer coefficient and volumetric mass transfer coefficient during mass transfer time before solute concentration decreases to 5%, are calculated for evaluating mass transfer characteristics. The effect of organic droplet size is studied. The result demonstrates that increase in aqueous/organic volume ratio significantly decreases mass transfer performance, decrease in droplet size enhances mass transfer process through both reducing mass transfer resistance and increasing contacting area.

1. Introduction

Solvent extraction (liquid-liquid extraction) is known as a process for separating compounds based on the solubility difference between two immiscible liquids (usually a water phase and an organic solvent phase).[1] Solvent extraction can be divided into physical extraction and chemical extraction[2]. Physical extraction only contains physical process, such as polarity difference between two solvents while chemical extraction contains chemical reactions, which might provide more significant distribution of solute. Due to high separation efficiency, extraction process containing reversible chemical reactions is catching more and more attention from researchers and is applied in wastewater treatment[3], pharmaceutical industry[4] and nuclear fuel treatment[5]. In extraction process with chemical reactions, there is usually an extractant (E) contained in organic solvent, which coordinates with the solute (S) in aqueous phase. When the organic solvent and aqueous solution contacts, a hydrophobic compounds generates (Equation 1).

$$[S]_{aq.} + [E]_{org.} \leftrightarrow [A.E]_{org.} \quad (1)$$

In extraction process with chemical reactions, the rate-limiting step is usually the mass transfer, because fast or even instant reactions are usually selected. On the other hand, in extraction process with chemical reactions, distribution coefficient is usually higher than 10 or even higher than 50 due to the presence of chemical reactions. Therefore, the water/organic phase ratio is also usually higher 10:1

or even 50:1. High water/organic phase ratio leads to slow mass transfer of solutes in aqueous phase, which is caused by decreased interface area and increased mass transfer resistance in aqueous phase. Understanding of mass transfer mechanism is helpful for process intensification of extraction process with chemical reactions.

Since the last decades, microsized liquid-liquid systems were applied in solvent extraction processes and have caught much attention from researchers. In microsized liquid-liquid systems, microsized droplets were generated by dispersion medium, such as microchannel devices [6] and microfiltration membrane [7]. Experimental results demonstrate that solvent extraction process is significantly intensified when microsized liquid-liquid systems were generated. [7, 8] The mechanism that how the decrease in droplets affects liquid-liquid mass transfer process is still need further understanding. Better understanding of mass transfer mechanism as well as prediction of mass transfer characteristics of liquid-liquid systems with different droplet size are important for designing extraction processes with chemical reactions.

In this study, a physical model is presented for understanding mass transfer mechanism in extraction process with chemical reactions. The effect of droplet size on mass transfer coefficients is then calculated.

2. A general model

Figure 1(a) shows a schematic diagram of organic-droplet-in-water system. After the organic phase was dispersed into droplets, the liquid-liquid system flows along the flowing channel concurrently. The relative velocity of organic droplets to continuous water phase is low and therefore, the organic-droplet-in-water system can be divided into a large number of mass transfer units. As shown in Figure 1(b), in each mass transfer unit, an organic droplet and a liquid film (aqueous phase) are contained. In each mass transfer unit, mass transfer of solute (S) in aqueous phase and extractant (E) in organic phase take place independently, which could be calculated according to Continuum Model, based on mass balance.

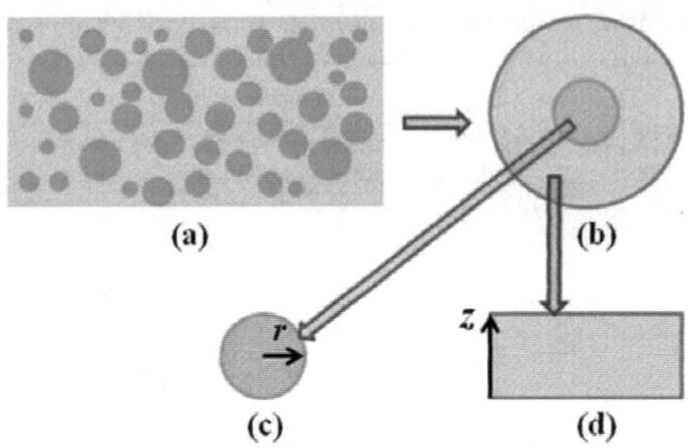

Figure 1. Schematic diagram of (a) flowing system in concurrent extraction process; (b) mass transfer unit; (c) mass transfer in organic droplets; (d) mass transfer in aqueous liquid film.

2.1. Assumptions

Before calculation, the following assumptions are made:

a) The dispersed organic droplets are in the middle of mass transfer unit and uniform. Averaged-droplet-diameter, d_{org}, is used in modeling.

b) The thickness of aqueous liquid films are uniform and averaged-liquid-film-thickness, L_{aq}, is used in modeling, which is calculated according to d_{org} and organic/aqueous flow rate ratio, F_{org}/F_{aq}.

c) d_{org} and L_{aq} are both constant during contacting period, where average values before and after contacting period are applied and therefore coalescence of organic droplets can be ignored.

d) During contacting period, both organic droplets and aqueous liquid films are stagnant, and therefore, only effects of diffusion of solute in aqueous phase and extractant in organic phase are considered, respectively.

e) Reaction as Equation (1) takes place between solute in aqueous phase and extractant in organic phase, intrinsic dynamics of which is fast enough for consuming solute (S) as soon as it reaches droplet-liquid-film interface.

f) In each mass transfer unit, only process of mass transfer though interface between organic droplet and aqueous liquid film around it is considered, while process of mass transfer though outer surface of aqueous liquid film to adjacent mass transfer units is not taken into account, by considering uniformity of mass transfer units.

Based on Assumption a)-f), in each isolated mass transfer unit, mass transfer of solute in aqueous phase and extractant in organic phase can be calculated, respectively.

2.2. Mass transfer in the organic droplets

Mass transfer of extractant in the organic droplet in mass transfer unit can be calculated according to the Continuum Model in spherical coordinate, as Equation (2) (Figure 1(c)).

$$\frac{\partial C_{\text{E-org}}}{\partial t} = D_{\text{E-org}}[\frac{1}{r^2}\frac{\partial}{\partial r_{\text{org}}}(r^2\frac{\partial C_{\text{E-org}}}{\partial r}) + \frac{1}{r^2 \sin\theta}\frac{\partial}{\partial\theta}(\sin\theta\frac{\partial C_{\text{E-org}}}{\partial\theta}) + \frac{1}{r^2\sin^2\theta}\frac{\partial^2 C_{\text{E-org}}}{\partial\phi^2}] \quad (2)$$

in which $C_{\text{E-org}}$ is extractant concentration in the organic droplet in mol/m^3; $D_{\text{E-org}}$ is diffusion coefficient of extractant in organic phase in m^2/s; t is contacting time in s.

Equation (2) can be simplified as Equation (3) due to the symmetry of assumed mass transfer unit.

$$\begin{aligned}
&\frac{\partial C_{\text{E-org}}}{\partial t} = D_{\text{E-org}}(\frac{\partial C_{\text{E-org}}}{\partial r^2} + \frac{2}{r}\frac{\partial C_{\text{E-org}}}{\partial r}) \\
&I.C.\ \ t = 0, 0 \le r \le R_{\text{org}}, C = C_{\text{E-org},0} \\
&B.C.\ \ r = 0, \forall t, \frac{\partial C_{\text{E-org}}}{\partial r} = 0 \\
&\qquad\ r = R_{\text{org}}, \forall t, C_{\text{E-org}} = 0
\end{aligned} \quad (3)$$

in which $C_{\text{E-org},0}$ is initial extractant concentration in organic droplet in mol/m^3.

Equation (3) is solved as Equation (4), in which instantaneous concentration profiles is calculated.

$$C_{\text{E-org}} = \sum_{k=1}^{\infty}\frac{2C_{\text{E-org},0}(-1)^{k-1}\sin(\frac{k\pi}{R_{\text{org}}}r)}{\frac{k\pi}{R_{\text{org}}}r}\exp\left\{-\left(\frac{k\pi}{R_{\text{org}}}\right)^2 D_{\text{E-org}}t\right\} \quad (4)$$

According to Fick's First Law, mass transfer rate through the outer surface of organic droplet can be calculated as Equation (5).

$$N_{\text{E-org}}(t) = 8\pi R_{\text{org}} D_{\text{E-org}} C_{\text{E-org}} \sum_{k=1}^{\infty}\exp\left[-\left(\frac{k\pi}{R_{\text{org}}}\right)^2 D_{\text{E-org}}t\right] \quad (5)$$

IOP Conf. Series: Journal of Physics: Conf. Series **1284** (2019) 012024 doi:10.1088/1742-6596/1284/1/012024

2.3. Mass transfer in aqueous liquid film

For simplifying calculation, aqueous liquid film surrounding is approximately spread to be an infinite plate. In this case, mass transfer of solute in aqueous liquid film in mass transfer unit can be calculated according to the Continuum Model, as Equation (6) (Figure 1(d)).

$$\frac{\partial C_{\text{S-aq}}}{\partial t} = D_{\text{S-aq}}\left(\frac{\partial^2 C_{\text{S-aq}}}{\partial x^2} + \frac{\partial^2 C_{\text{S-aq}}}{\partial y^2} + \frac{\partial^2 C_{\text{S-aq}}}{\partial z^2}\right) \tag{6}$$

in which $C_{\text{S-aq}}$ is solute concentration in aqueous liquid film in mol/m^3; $D_{\text{S-aq}}$ is diffusion coefficient of solute in aqueous phase in m^2/s.

Equation (6) can be simplified as Equation (7) due to symmetry of assumed mass transfer unit.

$$\begin{aligned}
&\frac{\partial C_{\text{S-aq}}}{\partial t} = D_{\text{S-aq}}\frac{\partial^2 C_{\text{S-aq}}}{\partial z^2} \\
&I.C.\ \ t = 0, 0 \le z \le L_{\text{aq}}, C_{\text{S-aq}} = C_{\text{S-aq,0}} \\
&B.C.\ \ z = 0, \forall t, \frac{\partial C_{\text{S-aq}}}{\partial z} = 0 \\
&\qquad\quad z = L_{\text{aq}}, \forall t, C_{\text{S-aq}} = 0
\end{aligned} \tag{7}$$

in which $C_{\text{S-aq,0}}$ is initial solute concentration in aqueous liquid film in mol/m^3.

Equation (7) is solved as Equation (8), in which instantaneous concentration profiles is calculated.

$$C_{\text{S-aq}} = \sum_{k=0}^{\infty} \frac{2C_{\text{S-aq,0}}(-1)^k}{\mu_k L_{\text{aq}}} \cos(\mu_k z)\exp(-\mu_k{}^2 D_{\text{S-aq}} t) \tag{8}$$

in which

$$\mu_k = \frac{2k+1}{2L_{\text{aq}}}\pi \tag{9}$$

According to Fick's First Law, mass transfer flux through interface between organic droplet and aqueous liquid film can be calculated as Equation (10).

$$N_{\text{S-aq}}(t) = \frac{8\pi R_{\text{org}}^2 D_{\text{S-aq}} C_{\text{S-aq,0}}}{L_{\text{aq}}} \sum_{k=0}^{\infty} \exp(-\mu_k{}^2 D_{\text{S-aq}} t) \tag{10}$$

in which $N_{\text{S-aq}}$ is instantaneous mass transfer rate of solute in aqueous liquid film in mol/s.

3. Modeling results and discussion

3.1. Comparison of mass transfer performance in organic droplet and aqueous liquid film

Firstly, mass transfer performance in organic droplet and aqueous liquid film are calculated, respectively, for determining the rate-limiting step in concurrent-flow chemical extraction. The decrease in concentration of extractant in organic droplet and solute in aqueous liquid film are determined respectively by assuming all substrate is consumed instantaneously as it reaches liquid-liquid interface. When chemical extraction is carried out, $F_{\text{org}}/F_{\text{aq}}$ is usually higher than 10 because of high distribution coefficient. Thus, results under operating conditions of $F_{\text{org}}/F_{\text{aq}}$ =10, 20, 40, 80 and 160 are calculated. Also, droplets size affects mass transfer characteristics significantly, thus, situation with droplet radius of 100 μm and 1 mm are calculated, respectively, as shown in Figure 2(a) and (b). During the calculation in this section, $D_{\text{E-org}}$ and $D_{\text{S-aq}}$ are both set as $1\times10^{-9}\text{m}^2/\text{s}$, considering the similarity of solute and extractant in liquids.

As shown in Figure 2 and 3, at the same droplet size, the curve of $C_{\text{E-org}}/C_{\text{E-org,0}}$ is significantly steeper than curves of $C_{\text{S-aq}}/C_{\text{S-aq,0}}$. For instant, contacting time for $C_{\text{S-aq}}/C_{\text{S-aq,0}}$ to reach 5% in aqueous phase when F_{org}/F_{aq}=10:1 is about 50 times longer than contacting time for $C_{\text{E-org}}/C_{\text{E-org,0}}$ to reach 5% in organic phase and even longer at higher F_{org}/F_{aq} value. This result indicates concurrent-flow chemical extraction process is limited by mass transfer of solute in aqueous phase. Comparison among curves with different F_{org}/F_{aq} values at the same droplet size demonstrate that $C_{\text{S-aq}}/C_{\text{S-aq,0}}$ curves become flatter as the increase in F_{org}/F_{aq} value, indicating weakening in mass transfer characteristics. Comparison among curves with different droplet size at the same F_{org}/F_{aq} value indicates significant intensification in mass transfer performance caused by the decrease in droplet size, which is in coincidence with experimental results in literatures [7, 8].

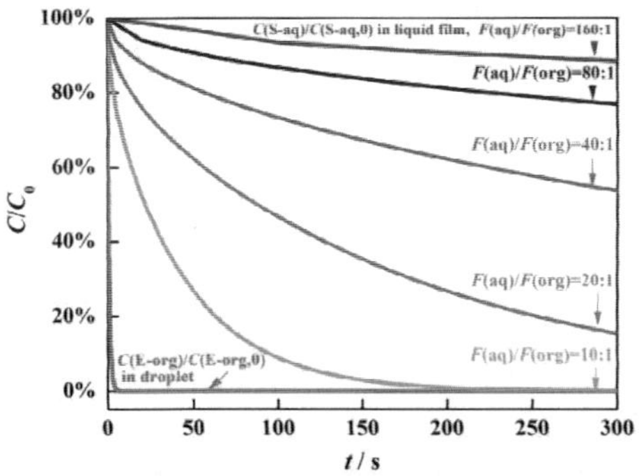

Figure 2. Comparison of concentration curve in organic droplet and aqueous liquid film with droplet radium 100 μm

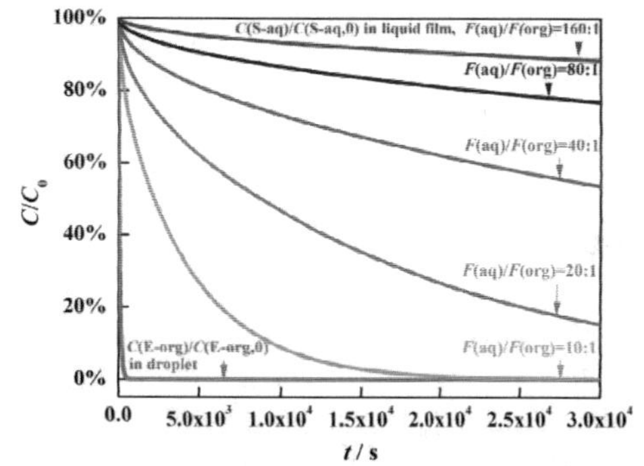

Figure 3. Comparison of concentration curve in organic droplet and aqueous liquid film with droplet radium 100 μm

3.2. Influence of droplet dispersion size on mass transfer characteristics in aqueous liquid film

As shown in *Section 3.1*, current-flow chemical extraction process is limited by mass transfer of solute in aqueous phase. Thus, modeling of mass transfer performance in aqueous liquid film is conducted for better understanding of process intensification caused by decreasing dispersion size. Mass transfer coefficients as well as specific surface area are calculated for evaluating mass transfer characteristics in different systems. The volumetric mass transfer coefficient during mass transfer process before $C_{\text{S-aq}}/C_{\text{S-aq,0}}$ reaching 5%, $k_{\text{W}}a$, is determined as Equation (11).

$$k_{\text{W}}a = \frac{(C_{\text{S-aq,0}} - C_{\text{S-aq,1}})}{t_1 \cdot \overline{\Delta C_{\text{S-aq}}}} \tag{11}$$

in which $C_{\text{S-aq,1}}$ is solute concentration in aqueous phase after mass transfer in mol/m^3; t_1 is mass transfer time for solute concentration to reach $C_{\text{S-aq,1}}$ in s; $\overline{\Delta C_{\text{S-aq}}}$ is the logarithmic mean concentration difference during mass transfer process in mol/m^3.

Mass transfer coefficient during mass transfer process before $C_{\text{S-aq}}/C_{\text{S-aq,0}}$ reaching 5%, k_{W}, is determined as $k_{\text{W}} = k_{\text{W}}a/a$, in which a is specific surface area of dispersed system in m^{-1}.

As shown in Figure 4 and 5, at different F_{org}/F_{aq} value, both k_W and $k_W a$ increase significantly as decrease in organic droplet size, in which the $k_W a \sim R_{org}$ curve is significantly steeper than the $k_W a \sim R_{org}$ curve. This result indicates that realization of microdispersion intensifies extraction process by both decreasing mass transfer resistance as well as increasing mass transfer area. The modeling result also demonstrates significant increase in mass transfer resistance as the increase in phase ratio.

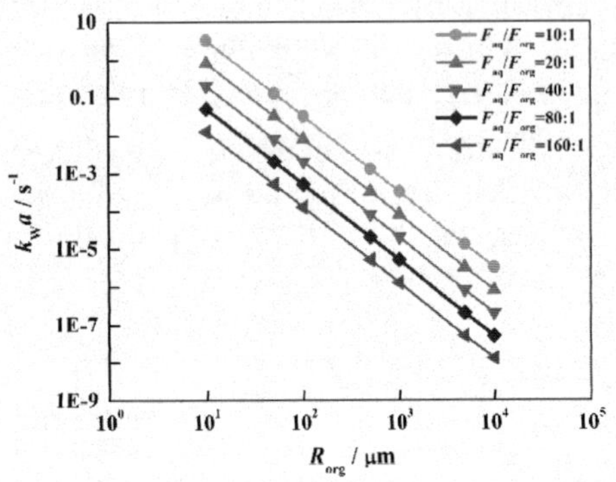

Figure 4. Influence of droplet radiu on $k_W a$ in aqueous phase.

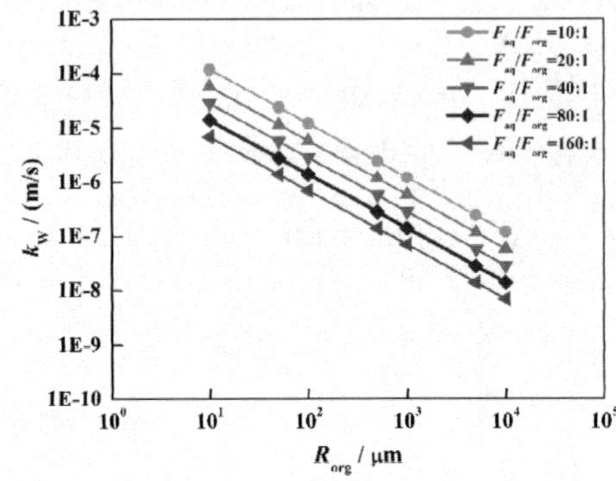

Figure 5. Influence of droplet radiu on k_W in aqueous phase.

4. Conclusion

In this study, modeling investigation of mass transfer in concurrent-flow chemical extraction process is conducted using the Continuum Model. The rate-limiting step is firstly determined by calculating concentration curves in organic droplet and aqueous liquid film, respectively. Mass transfer characteristics in aqueous liquid film is calculated further, which is evaluated by volumetric mass transfer coefficient and mass transfer coefficient during mass transfer time before solute concentration decreases to 5% of initial concentration. Influence of organic droplet size is studied. The result demonstrates the increase in aqueous/organic volume ratio significantly decreases mass transfer performance, the decrease in droplet size enhances mass transfer process through both reducing mass transfer resistance and increasing contacting area.

Acknowledgments

The authors gratefully acknowledge the support of National Natural Science Foundation of China (21878019, 51472032, 21573017, U1664255) and Beijing Natural Science Foundation (2182063).

References

[1] S. X. Lu, Z. T. Mi, L. Wang, Y. Q. Wang, Z. H. Zhu and S. B 2005 Fu Experimental investigation and simulation of gas-liquid-liquid reactive extraction process for the production of hydrogen peroxide, Chem. Eng. Sci. **60** 6298-306

[2] King, C.J 1987 Handbook of Separation Process Technology *John Wiley & Sons* (New York)

[3] Urasaki N and Wong CP 1999 Separation of low molecular siloxanes for electronic application by liquid-liquid extraction IEEE Trans Electron Packag Manufact. **22** 295−8

[4] Goklen KE and Hatton TA 1987 Liquid-liquid extraction of low molecular-weight proteins by selective solubilization in reversed micelles *Sep Sci Technol.* **22** 831−41

[5] Pai SA, Lohithakshan KV, Mithapara PD and Aggarwal SK 2000 Solvent extraction of uranium(VI), plutonium(VI) and americium-(III) with HTTA/HPMBP using mono- and bi-functional neutral donors: synergism and thermodynamics J Radioanal Nucl Ch. 245 623−8

[6] H. Hotokezaka, M. Tokeshi, M. Harada, T. Kitamori and Y. Ikeda 2005 System for high-level radioactive waste using microchannel chip - Extraction behavior of metal ions from aqueous phase to organic phase in microchannel *Prog. Nucl. Energ.* **47** 439-47
[7] J. H. Xu, G.S. Luo, G.G. Chen and B. Tan 2005 Mass transfer performance and two-phase flow characteristic in membrane dispersion mini-extractor J. Membrane. Sci. **249** 75-81
[8] A. Muto, M. Maruta, T. Bhaskar and Y. Sakata 2004 Extraction properties of a Y-pattern microchannel reactor with a settler Kagaku Kogaku Ronbun **30** 159-63

Analysis of the Click through Rate of Chinese Netizens through Social Software based on DeepFM Algorithm

Li Bo[1,2], Zhang Xiao[1,2], Li Kewei[1,2]and Yan Jingyi [3]

[1.]School of Computer Science, Northwestern Polytechnical University, Xi'an Shaanxi 710129, China;
[2.]Ministry of communications Key Laboratory of big data storage and management, Xi'an Shaanxi 710129, China;
[3] School of Management, Northwestern Polytechnical University, Xi'an Shaanxi 710129, China

Corresponding author: e-mail:zhangxiao@nwpu.edu.cn

Abstract. Advertising as an important means of marketing, plays an important role in the promotion of commercial services, but its profit efficiency has not been optimistic. How to reduce the push cost and improve the accuracy rate of advertising is the algorithm problem of various enterprises. Chinese netizens have a large base and a large proportion use social software. It is a novel idea to analyze the click-through rate of Chinese netizens'advertisements through the data provided by social software. Based on the idea of lookalike crowd expansion, this paper uses DeepFM algorithm and designs the corresponding experimental process to analyze and predict the clicks of Chinese netizens through social software.

1. Introduction

With the development of life, we are surrounded by advertising, but people are indifferent to the overwhelming advertising. It has become an important research direction that how to put the right advertisement to the right crowd or find the right potential customers through advertising data mining. When Facebook began using the Facebook Custom Audiences feature in the advertising space in 2012, audiences found that the concept was really being used on a large scale. Audience discovery can find potential users according to existing users, carry out appropriate advertising promotion, reduce the cost of promotion[1]. A recommendation algorithm like Facebook that discovers and extends other users through a group of existing users is called Lookalike. Each company has its Lookalike algorithm designed and kept secret. For example, Tencent hosted the Tencent Advertising Algorithms Competition in 2017 and 2018 to discover talents and more accurate algorithms in this area. Aiming at the problem of advertising click-through rate in Lookalike algorithm, this paper analyzes the complicated social scene, various advertising forms, and huge crowd data, and explores the current click-through situation of Chinese netizens through social software[2].

2. Research status

Lookalike, also known as similar crowd extension, seeks similar audiences based on seed user portraits and social links by opening up rich data capabilities.This algorithm requires a relatively large order of magnitude, suitable for the current era of data blowout development. In other words, it's a technology that seeks out similar groups of people through seed users, which can improve the accuracy of crowd orientation, as shown in the Figure 1.

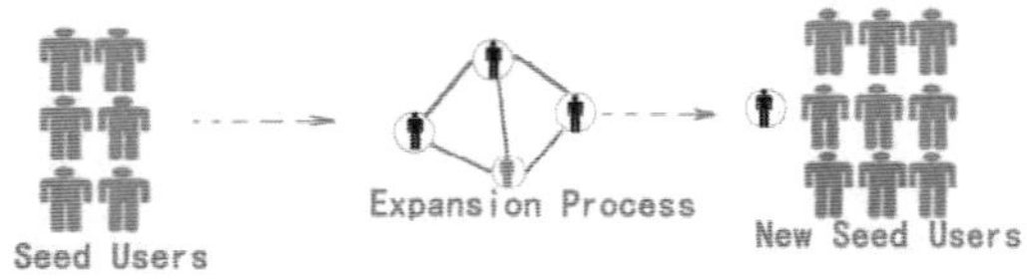

Figure 1. The picture of population expansion

Ads lookalike crowd expansion algorithm, in Weixin, Weibo, Facebook, Google has a very wide range of applications.Advertising, as a very important part of all kinds of services, can be analyzed as a separate part, because even close friends may be interested in different advertisements. Delivering corresponding advertisements to specific users can reduce the operation and maintenance costs of the company. What's more, clustering analysis of clicks on certain advertisements can lead to people with similar interests and then recommend similar advertisements[3-4].

3. Algorithm

The analysis of advertisement click rate algorithm in lookalike algorithm can be regarded as CTR problem. CTR (Click-Through-Rate) refers to the click-through rate of an online advertisement divided by the actual click-through rate of the advertisement by the amount of display of the advertisement[5]. CTR prediction is to predict whether users click on the ads. In fact, it can be seen as a two-class problem, that is, point and no point. Aiming at this problem, the following algorithms are mainly studied at present.

3.1. Existing algorithms

3.1.1. FM algorithm. FM (Factorization Machine) is mainly to solve the problem of how to combine features when data is sparse. In the problem of advertising classification, according to some characteristics of users and advertising space, to predict whether users will click on the advertisement[6]. The objective function of the two element cross FM (2-way FM) algorithm is as follows:

$$\hat{y} := w_0 + \sum_{i=1}^{n} w_i x_i + \sum_{i=1}^{n} \sum_{j=i+1}^{n} \langle V_i, V_j \rangle x_i x_j \tag{1}$$

Among them, $w_0 \in R$, $w \in R^n$, $V \in R^{n \times k}$, W is the parameter of input feature, $\langle v_i, v_j \rangle$ is the cross-parameter between input features I and j, and V is the k dimensional vector $\langle V_i, V_j \rangle := \sum_{f=1}^{k} v_{i,f} \cdot v_{j,f}$.

Under sparse conditions, the independence of features is broken through the learning method of vector v, which can better mine the correlation between features. FM takes into account the relationship between features by learning a hidden variable for each feature. Its advantages are: considering the relationship between features, enhance the generalization ability of the model; through clever decomposition and merging of the objective function, can be completed under $o(n)$ time complexity; suitable for processing sparse data.

3.1.2. FFM algorithm. FFM (Field-aware Factorization Machine) improves the FM model by classifying the same features into one field. FFM. In the FM model, each feature corresponds to a hidden variable. But in FFM model, the feature should be divided into multiple fields, each feature corresponds to a hidden variable for each field. The model of FFM is:

$$f(x) = w_0 + \sum_{i=1}^{n} w_i x_i + \sum_{j_1=1}^{n-1} \sum_{j_2=i+1}^{n} (v_{j_1,f_2}^T v_{j_2,f_1}) x_{j_1} x_{j_2} \quad (2)$$

j_1, j_2 represent the index of features.

The FFM algorithm is based on the FM model, and the performance of the model is further improved. It is also suitable for dealing with sparse data. But FFM can not decompose and merge the objective function like FM and complete the calculation of the function in linear time. Time complexity is $o(k * n^2)$ which more complex than FM. For large data, the time complexity of FFM will increase significantly, and the performance of computer CPU will be tested very much.

3.2. Algorithm analysis

Most CTR problems are consistent with the Long Tail Effect. "Head" and "tail" are two statistical nouns. The middle part of the normal curve is called "head", and the relatively gentle part on the two sides is called "tail". From the point of view of people's needs, most of the demand will be concentrated in the head, which we can call popular, and distributed in the tail of the demand is personalized, scattered small amount of demand. The relationship between users and advertisements can be abstracted as the relationship between point and point. A pink circle with a green triangle and a black line indicate that the user clicked on the ad, resulting in Figure 2. But the fact is that everyone doesn't have to click on just one ad, so it turns out that Figure 2 down Figure 3.

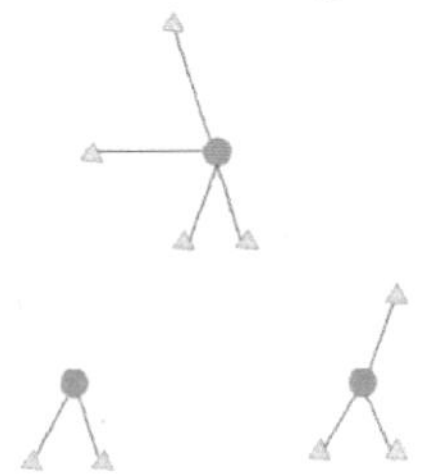

Figure 2. The chart of the basic relationship between users and advertisements.

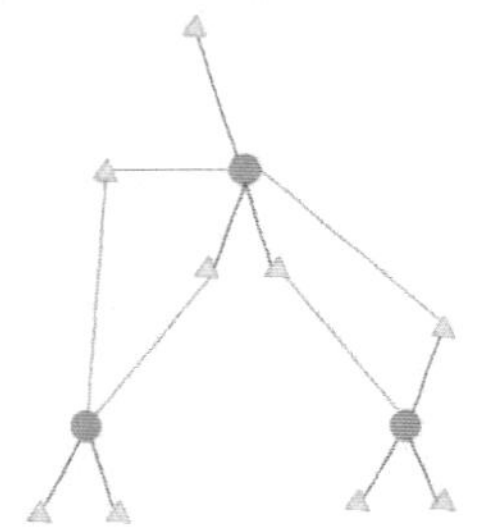

Figure 3. The chart of the complex relationship between users and advertisements.

In the current situation, one person can correspond to more than one advertisement, one advertisement can correspond to more than one person, but there is no relationship between advertising and advertising, between people. The ultimate goal is to explore the relationship between people and then recommend the right ads according to the group or the person's preferences. We can add the weight of the degree of association between advertisements, such as the new yellow line in the Figure 4, and the thickness of the line represents the degree of association between advertisements. The thicker the yellow line represents the greater the degree of association between the two advertisements.

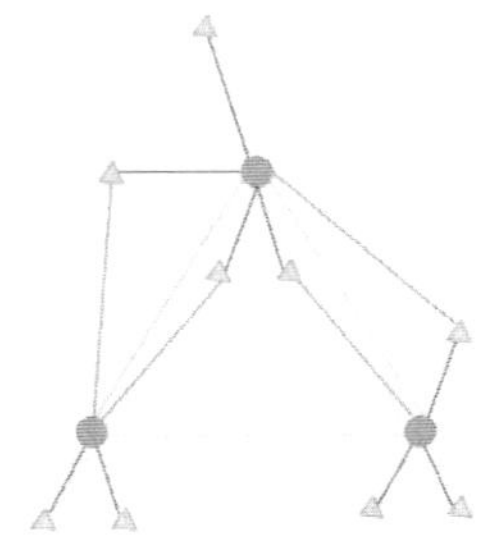

Figure 4. The chart of the complex relationship between users and advertisements, which considers the relevance of advertisements

The previous three scenarios assumed that the number of clicks per person on an advertisement was fixed, but the actual number of clicks per person was significantly different, and according to the long tail effect, the number of clicks per person on an advertisement would be significantly different, more than 90% of the advertisement clicks would not exceed five, and the remaining 10% of the advertisement clicks would be 90%. Advertising hits several times. In Figure 4, it is not for the first time to consider the difference between people clicking on advertisements and the relevance of advertisements. At this point, to recommend potential ads to the same group of people, you only need to recommend ads that the user does not pay attention to, but are most relevant to the ads that user has already paid attention to.

Similarly, by introducing the difference in click-to-click advertising rates, the thicker the black line, the greater the user's attention to the advertisement, and this leads to Figure 5. According to the analysis of this situation, only need to recommend to the user that he did not click on the ads, and he clicked on the ads associated with the greatest degree of click, and the same crowd clicked the greatest degree of advertising.

Figure 5. The chart of the complex relationship between users and advertisements, which considers the relevance and customer differences of advertisements.

In Figure 4 and Figure 5, because of a lot of restrictions, the problem will be simplified. At this point, through data pre-processing, we can classify advertising, and introduce the related functions of image recognition and character recognition. But there are actually two methods:

- Advertisers in advertising when the tag, so that advertising can be directly classified so that it can be completed in accordance with the above analysis(Figure 4).
- According to the relevance between users and advertisements, the relationship between users and advertisements is analyzed by data mining.

In this paper, we analyze the correlation between users and advertisements. In view of this problem, such a huge amount of data for general graphics, are very inappropriate, so this paper chose the DeepFM algorithm to Chinese netizens through social software advertising clicks.

3.3. DeepFM algorithm

DeepFM is an algorithm proposed by Google to solve the CTR problem. This paper applies this algorithm to analyze the click-through rate of Chinese netizens'social media ads and uses the data set provided by Tencent Advertising Algorithms Competition to verify the algorithm. The core idea of the DeepFM algorithm is:

- Using FM Component + Deep Component. FM extracts low order combination features, and Deep extracts higher order combination features. DeepFM is end-to-end training without artificial feature engineering.
- Analysing feature embedding. FM and Deep share input and feature embedding to make training faster and more accurate.

As can be seen, the whole model is divided into two parts: FM and DNN, which are responsible for extracting low-order features and high-order features respectively. With the idea of FNN, FM is used to embedding, and then the wide and deep models share the results after embedding. The input of DNN is exactly the same as that of FNN. When combined in a certain way, the model simulates the effect of FM completely on the wide. Finally, the results of DNN and FM are combined to activate the output. DeepFM's prediction results are:

$$\hat{y} = sigmoid(yFM + yDNN) \tag{3}$$

Compared with other algorithms, DeepFM does not pre-train the hidden vector V with FM and initializes the neural network with V. The FM module is not independent, it is trained and learned with the whole model. No feature engineering is needed, and the training efficiency is high. No pre-training shares Feature Embedding, no feature engineering, and captures both low-high-order interaction features. The DeepFM algorithm can effectively reduce the singularity and uncertainties in the data set, and get more accurate prediction results.

4. Experiment

In order to verify the click-through rate of Chinese netizens'social software advertisements discussed in this paper, the data set is provided by Tencent Advertising Algorithms Competition in 2018. The dataset has no time information, so it is difficult to construct time-dependent features. This dataset mainly includes information such as users and advertising categories, so we explore these data.

4.1. Data set analysis

4.1.1. Evaluation indicators. For the extended similar users, if there is relevant effect behavior (click or transform) in the advertising, it is considered as a positive case; if there is no effect behavior, it is considered as a negative case. Each seed pack to be evaluated will provide the following information: the ad aid corresponding to the seed pack and its characteristics, and the corresponding candidate user set (userid and its characteristics). So the accuracy of the algorithm is tested.

4.1.2. Data problem. According to the five features of the data set, click, ratio, conversion rate, special conversion rate, multi-value length, tagging tags are needed for each class feature. It is noteworthy that the conversion rate uses the pre-processed block tags to separate a block verification set without adding statistics, the remaining blocks do drop out cross-statistics, and the test set uses all the training set data for statistics.

4.2. Characteristic structure

According to the five features of the data set, click, ratio, conversion rate, special conversion rate , multi-value length, tagging tags are needed for each class feature. It is noteworthy that the conversion rate uses the pre-processed block tags to separate a block verification set without adding statistics. We found that the importance of some multi-value fields is very high and extract the top 20 coding features for storage and analysis, as shown in the Figure 6.

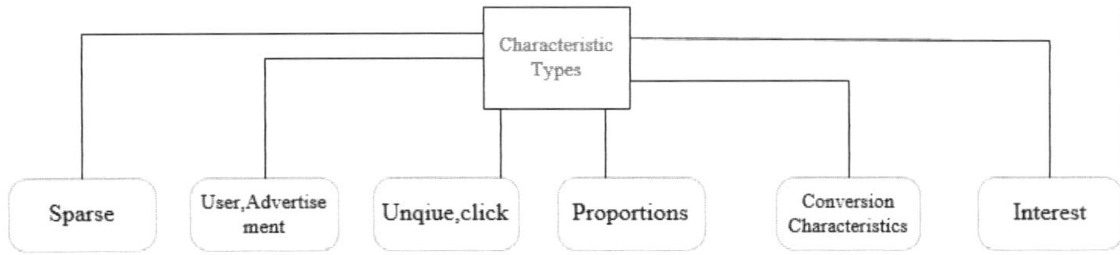

Figure 6. Characteristics selected in experiments

As the structure of the data set, this paper selects the following features: the original features of the data set (user characteristics, advertising features), sparse features, unqiue and click features, as well as proportional features. The proportional feature is used to characterize the preferences of a class of users, and better distinguish users according to preferences.Because no time feature is given in this data set, it is easy to generate data traversal and over-fitting when constructing the conversion rate feature. In order to solve data traversing, we can use the block conversion rate or Bias smoothing method.

4.3. Feature selection

After the above process of constructing features, the data set can achieve hundreds of features, but it is impossible to train all features, because it may contain many redundant features, and we need to achieve the effect of multiple features in the case of few features . The most commonly used method is the correlation coefficient method and the method of modeling the importance of output characteristics. This paper divides the data set into 10 groups, screening each group of features, sorting the importance of the characteristics within the group.

4.4. Model training and assessment

DeepFM model method is used in this paper. After AUC sorting according to the multi-group prediction values of the verification set, the weights are traversed in turn (list (range (0,101)*0.01) to obtain the best weights, and then the same weights are applied to the prediction results of the test set, so that each of them is more. By weighting a sub model, the AUC of the validation set will only be greater than or equal to the AUC before the weighted sub model. In fact, the whole weighting process is similar to a linear fitting. It can also use the verification set of each sub-model and the prediction results of the test set as the characteristics, use the label of the verification set as the real label, and use the xgboost and other models to train, so the effect is similar to the previous ergodic weighting. According to aid, a set of 20% training sets is used as validation set to verify the accuracy of the algorithm.

4.5. Experimental results

Through the algorithm described above, and the corresponding experimental steps, rent the Ali cloud server, configure the Python environment under the server Linux, and then do the experimental analysis. Although the research in this paper is still inadequate, but through analysis, it can greatly improve the accuracy of the GTR problem. Comparing the top ten results of the 2018 Tencent Advertising Algorithms Competition, all above 0.7728, with the highest accuracy of 0.781422. In this paper, the DeepFM algorithm is used to do three experiments, the prediction accuracy is 0.74456, 0.75375, 0.75221. The algorithm presented in this paper is less accurate than that of the Tencent Advertising Algorithms Competition in 2018, but it provides a way to analyze the click-through rate of Chinese netizens on social software. Moreover, in the competition, if the excessive collection of features, may also make a higher accuracy, but this is not desirable, because it is not conducive to generalization of research, so the results of the 2018 Tencent advertising algorithm contest can only be a reference result, can not be fully judged according to this result.

5. Conclusion

The current era belongs to the era of data, mainstream social software companies, such as Tencent, Facebook have a strong interest in lookalike algorithm, and in the search for a better algorithm. GTR problem is a problem that appears in recent years. Its precision is generally low, and it has great research prospects. In this paper, DeepFM algorithm is used to analyze and predict the potential interest of Chinese netizens in social software advertisements. Through the algorithm analysis, the relationship between users and advertising is still in line with the long tail distribution, Chinese netizens are still mainly young people, so for games, entertainment software advertising click rate is higher than other advertising, if further excavation, there will be great potential. In most experimental scenarios of recommended categories, if only to achieve good prediction rate, there is a relatively simple feature mining idea: the more fine-grained features are important, the closer the features are, the more important the features are, and the feature weights about the future are the largest. This is also a drawback of data mining in recent years.

Acknowledgments

This work is supported by the MIIT Key Laboratory of Big Data Storage and Management, the NFS of China under Grant No.61502392 and No.61272323 and the Ministry of Science and Technology of China, National Key Research and Development Program (No.2018YFB1004401).

References

[1] Jiesi Guo,Herbert W. Marsh,Philip D. Parker,Theresa Dicke. Cross-cultural generalizability of social and dimensional comparison effects on reading, math, and science self-concepts for primary school students using the combined PIRLS and TIMSS data[J]. Learning and Instruction,2018,58.

[2] Daisuke Ichikawa,Toki Saito,Waka Ujita,Hiroshi Oyama. How can machine-learning methods assist in virtual screening for hyperuricemia A healthcare machine-learning approach[J]. Journal of Biomedical Informatics,2016,64.

[3] Colin E. Vize,Katherine L. Collison,Joshua D. Miller,Donald R. Lynam. Using Bayesian methods to update and expand the meta-analytic evidence of the five-factor model's relation to antisocial behavior[J]. Clinical Psychology Review,2018.

[4] M. Vulfin,K. F. Tagirova. Enhancement of accuracy of deep-pumping equipment based on data minning[J]. Optical Memory and Neural Networks,2015,24(1).

[5] Dzyuba, Vladimir, and M. V. Leeuwen. "Learning What Matters – Sampling Interesting Patterns." Pacific-Asia Conference on Knowledge Discovery and Data Mining Springer, Cham, 2017:534-546.

[6] Dzyuba, Vladimir, and M. V. Leeuwen. "Learning What Matters – Sampling Interesting Patterns." (2017):534-546.

IOP Conf. Series: Journal of Physics: Conf. Series **1284** (2019) 012026 doi:10.1088/1742-6596/1284/1/012026

A Comparison of Feature Selection Methodology for Solving Classification Problems in Finance

Xia Xiaomao[1,*], Zhang Xudong[1] and Wang Yuanfang[1]

[1]JD Digits

[*]baiyouxxm@163.com

Abstract. Classification is one of the most important tasks in real world with the intention of finding the underlying patterns of the data and making use of the found patterns. Other than considering lots of parameters that may influence the classification model accuracy, the influence of the feature selections has been rarely discussed. SHAP (SHapley Additive exPlanation) value is a novel ensemble learning measurement that is the unique consistent and locally accurate attribution value. Its applications to feature selection are not mentioned in the literature. In this paper, we compare the prediction accuracy of several feature selection techniques on a set of Polish companies. We show that new feature selection method based on the SHAP values is superior (not worse) to other widely used prediction methods.

1. Introduction

There are two famous special methods of dimensionality reduction. The first one is feature extraction where the input data is transformed into a reduced representation set of features, so new attributes are generated from the initial ones. The second category is feature selection. In this category a subset of existing features is selected without a transformation. Generally, feature selection is preferred over feature extraction since it keeps all information about the importance of each single feature while in feature extraction obtained variables are usually not interpretable [1].

Conserving the information of each feature provides much simplicity and interpretability to financial data processing. Hence, feature selection is more appropriate to our study. Feature selection is an important framework in knowledge discovery and specially in financial applications, not only for the insight gained from determining predictive modelling features but also for the improved performance, understandability and accuracy of the financial models.

Feature selection is a process of finding an optimal subset which is given by a feature subset selection

algorithm that provides the highest possible accuracy. The following definition is presented by Koh avi and John (1997): "Given an inducer I and a dataset Δ with features $X_1, X_2, \ldots, X_n$ from a distribution D over the labeled instance space, an optimal feature subset, X_{opt} , is a subset of the features such that the accuracy of the induced classifier X = I(Δ) is maximal (p. 276)" [2]. An optimal feature subset is not necessarily unique because when one feature can be replaced by another feature since they are perfectly correlated to each other, the accuracy derived by different combination of features is the same.

Until recently the dominating methods for predicting corporate bankruptcies have been based on statistical modelling, however, lately models based on machine learning have been proposed [3].

Recently, machine learning models have successfully been used for many classification and regression problems and these models have often outperform traditional classification methods [4]. The purpose of bankruptcy prediction is to assess the financial health status and future perspectives of a company. For a given period of time this problem can be modelled as a two-class classification problem. Companies either survive the given time period, or go bankrupt during it. The problem is to predict which of these two possible outcomes that is the more probable one.

Financial reporting laws and public expectations on transparency means that there is a lot of data available concerning companies' financial status [5]. The large amount of data makes the area well suited for sophisticated data intensive computation methods [6].

Estimating the risk of corporate bankruptcies is of large importance to creditors and investors. There are large indirect and direct costs associated with financial distress and bankruptcies [7] and for this reason bankruptcy prediction has for a long period of time constituted an extensive area of research [3]. Corporate bankruptcies can have serious effects both locally and globally [8], employees, investors, customers, suppliers and their financiers are all affected when a company disappears [9]. In some cases, a corporate bankruptcy can cause an entire industry to suffer [10].

According to Rodriguez et al. (2010) a successful feature selection: a) reduces the feature space's dimensionality, b) speeds up and reduces the cost of a learning algorithm and c) obtains the feature subset which is the most relevant to classification [11]. Feature selection algorithms are typically composed of the following four components: search direction, search strategy, evaluation function and the stopping point.

Random forest is an ensemble learning method. The algorithm is proposed by Ho [12]. The training algorithm is based on bagging, which is short for bootstrap aggregating. Each decision tree is trained on a data set randomly sampled from the training data with replacement. Such an approach means that each tree learner is shown a different subset of the training data and that same observation can be chosen more than once in a sample [13].

Extreme Gradient Boosting (XGBoost) is similar to gradient boosting framework but more efficient. It has both linear model solver and tree learning algorithms. So, what makes it fast is its capacity to do parallel computation on a single machine. This makes XGBoost at least 10 times faster than existing gradient boosting implementations. It supports various objective functions, including regression, classification and ranking. It also has additional features for doing cross validation and finding important variables [14].

SHapley Additive exPlanation (SHAP) value, a unified measure of feature importance was first proposed by Lundberg and Lee [15]. They are based on ideas from game theory and local explanations. It can be proved that SHAP values are the only consistent and locally accurate individualized feature attributions.

The paper is organized as below. In section 1, we describe the concepts of SHAP values, the importance of feature selections when we solve classification problems and issues on bankruptcy predictions. In section 2, we describe the dataset and samples of our research. Then, in section 3, we discuss our outcomes of different feature selection methods. In section 4 of conclusion and discussion part, we make a conclusion of the study and discuss other potential applications.

2. Dataset

The dataset used in this paper consisted of financial condition about polish companies in the manufacturing sector. Many Polish manufacturing companies went bankrupt in this sector since 2004 [16]. The dataset contained information for both bankruptcy companies and still operating ones. The financial information describing the health of the 700 bankrupt companies (almost 2,400 financial statements) was gathered during the period 2007-2013 and more than 10,000 (more than 65,000 financial statements) operating companies was collected during the period 2000-2012.

The data is divided into five different subsets, which are described in Table 1. The subsets are created to allow for different lengths of forecasting period. The task in each of them is to predict whether or not a company goes bankrupt within five, four, three, two and one year respectively based on information that could be retrieved from 64 different financial features (regressor variables or indicators). All five data sets are, as expected also in real life, heavily imbalanced. There are much fewer bankrupt companies compared to still operating ones. 64 different financial indicators are used to describe the financial status of the companies. Almost all of the features are financial ratios constructed from information found in the companies' economic statements. The use of ratios should mean that the predictors are not too heavily correlated with the size of the companies. A complete list of the features' descriptions can be found in UCI Machine Learning Repository.

Table 1 Description of the data pertaining to the five different classification tasks

Dataset	Features from	Bankruptcy after	No. of bankruptcy	No. of not bankruptcy	Total
1st year	1st year	5 years	271	6,756	7,027
2nd year	2nd year	4 years	400	9,773	10,173
3rd year	3rd year	3 years	495	10,008	10,503
4th year	4th year	2 years	515	9,277	9,792
5th year	5th year	1 year	410	5,500	5,910

3. Methods, applications and results

3.1. Methods

This paper considers only supervised learning. Corporate bankruptcy prediction is treated as a two-category classification problem. A company either belongs to the class of companies that faces bankruptcy or to the class that still operates after a given time period.

Here we consider several methods to do feature selection. The focus is on the importance ranking of different features. (1) Information Value ranking (**IV**) (2) Random Forest with the default mean decrease in impurity (or Gini importance) ranking (**RF**) (3) Random Forest with SHAP values ranking (**RFS**) (4) Extreme Gradient Boosting with split count importance ranking (**XGBSC**) (5) Extreme Gradient Boosting with gain importance ranking (**XGBG**) (6) Extreme Gradient Boosting with coverage importance ranking (**XGBC**) (7) Extreme Gradient Boosting with SHAP values ranking (**XGBS**).

To compare results from different methods, cross validation is implemented in this study. The dataset is divided into n equally large subsets, $n-1$ of those are applied to the training set and the last subset is applied to testing. We repeat the procedure n times, with every of the n sets acting as test subset once each. In this way we get n measurements of the scoring metric that we are concerned with, and we can calculate both the mean and standard deviation of this scoring metric.

Most machine learning algorithms does not return a binary prediction (e.g. still operating or bankrupt), but rather a probability. To calculate true/false positive rate from the returned probabilities we have to choose a threshold, which we use to decide which class we consider a sample to belong to. For example, all companies that the machine identifies will face bankruptcy by more than 70% probability we will consider as bankrupt. As we increase this threshold both the true/false positive rate will decrease. If we set the threshold to 100% we will have TPR (true positive rate) = FPR (false positive rate) = 0 and if it is set the threshold to 0% we will get TPR = FPR = 1. In between these two endpoints we have a space of opportunity upon which we can decide how we want our classifier to perform. The curve which connects these two values are called a Receiver operating characteristic (ROC) curve. Values closer to the upper left corner are better since they mean a higher true positive rate as well as a lower false positive rate [17]. From just looking at a graph it is hard to measure the performance of two or more types of algorithms. To overcome this a metric called Area Under Curve (AUC) has been derived. It is, as the name suggests, defined as the area under the Receiver operating characteristic (ROC) curve. An algorithm that guesses randomly will obtain a value of 0.5, since the area under a triangle with legs of length one is precisely 0.5. AUC can be interpreted as how much opportunities our algorithms give us for different scenarios, where potentially different measures are of importance.

3.2. Applications and results

The first 30 most important features are chosen for its own method except for IV rank. Particularly for IV rank, the first 30 chosen features are used for training random forest model. All the seven models are fit and computed in Python.

Hyperparameters values typically have large influence on the performance and the behaviour of a machine learning algorithms. For each method, the hyperparameters are explored and tuned as below. For information value (IV), random forest (RF) and random forest with SHAP values (RFS) method, the following numbers are assigned. The maximum depth of tree is 6. The number of trees in the forest is 300. The minimum number of samples required to be at a leaf node is 100. The "balanced" mode uses the values of y to automatically adjust weights inversely proportional to class frequencies in the input data. For four Extreme Gradient Boosting (XGboost), we set the numbers as follows. Maximum tree depth for base learners is 3. Minimum sum of instance weight(hessian) needed in a child is 6. Number of boosted trees to fit is 300. L1 regularization term on weights is 0. L2 regularization term on weights is 0.05.

There is a large discrepancy in the distribution of missing values between the two different classes. In other words, information about the missing values should be very helpful in the classification. We try to (1) impute missing values as the average (2) impute missing values as very large or small values (3) add the total number of missing values as a synthetic feature (4) add a dummy variable for each possible missing value (1-hot). There is no significant difference between these four methods in AUC and KS values. In the following tables of the study, we choose to report the results from the first method.

Table 2 Experimental results for considered classification models. The numbers are the area under curve (AUC) of the ROC curve. STD is short for the standard deviation.

	1st Year		2nd Year		3rd Year		4th Year		5th Year	
	MEAN	STD	MEAN	STD	MEAN	STD	MEAN	STD	MEAN	STD
XGBSC	0.953	0.023	0.918	0.037	**0.930**	0.020	0.934	0.024	0.952	0.015
XGBG	**0.955**	0.024	**0.931**	0.028	0.923	0.022	0.934	0.020	**0.954**	0.015
XGBC	0.830	0.044	0.724	0.027	0.746	0.037	0.823	0.042	0.934	0.015
XGBS	0.951	0.016	0.930	0.025	0.924	0.026	**0.936**	0.021	0.953	0.016
IV	0.872	0.044	0.833	0.058	0.842	0.043	0.832	0.025	0.893	0.018
RF	0.899	0.045	0.865	0.050	0.861	0.039	0.847	0.025	0.887	0.018
RFS	0.896	0.041	0.856	0.053	0.857	0.041	0.843	0.027	0.891	0.018

Table 3 Experimental results for considered classification models. The numbers are the Kolmogorov-Smirnov (KS) statistic. STD is short for the standard deviation.

	1st Year		2nd Year		3rd Year		4th Year		5th Year	
	MEAN	STD	MEAN	STD	MEAN	STD	MEAN	STD	MEAN	STD
XGBSC	0.792	0.076	0.703	0.095	**0.729**	0.067	**0.745**	0.063	0.793	0.043
XGBG	**0.798**	0.086	**0.733**	0.074	0.721	0.062	0.735	0.048	**0.812**	0.029
XGBC	0.553	0.080	0.366	0.049	0.416	0.068	0.526	0.084	0.743	0.042
XGBS	0.794	0.060	0.728	0.073	0.723	0.061	0.741	0.051	0.811	0.053
IV	0.633	0.097	0.564	0.091	0.566	0.058	0.547	0.047	0.666	0.036
RF	0.686	0.106	0.606	0.086	0.620	0.062	0.566	0.044	0.675	0.039
RFS	0.693	0.103	0.593	0.089	0.602	0.062	0.558	0.047	0.678	0.040

Table 4 The p-values of signed rank Wilcoxson test of the KS/AUC of XGBS vs the best model

	1st Year		2nd Year		3rd Year		4th Year		5th Year	
XGBS vs Best	AUC	KS	AUC	KS	AUC	KS	AUC	KS	AUC	KS
P-value	0.545	0.762	0.880	0.821	0.496	0.940	NA	0.880	0.880	0.880

From Table 2 and Table 3, the mean of the best model and XGBS are very similar. To further explore the difference significance between results obtained by XGBS vs Best model, signed rank Wilcoxson test (also known as 'Mann-Whitney' test) is applied to AUC and KS on the test dataset from 10-fold cross validation. The p-values of signed rank Wilcoxson test of the KS/AUC of XGBS vs the best model are in Table 4. Assuming the significance level equal to 0.05, for the Wilcoxson test, the null distribution difference hypotheses cannot be rejected. In conclusion, XGBS performs as good as the best models in terms of AUC or KS values. It is summarized that the performance of XGBS is very close to the best model which is in bold if XGBS is not the best feature selection model. XGBS is proven to be a very effective feature selection method.

4. Conclusions and discussions

The predictive performance achieved on the dataset is very good. One possible reason for the superb performance is the information contained in the missing values. Future interesting research topics in the development of algorithms for classification problems in finance could be to incorporate new types of data. For example, unstructured data, such as texts in annual reports, could be of interest. Also, apply described algorithms to another dataset would also be of interest to better understand the validity of the results.

In conclusion, we introduced a new feature selection method based on SHAP values for identification of informative variables. The SHAP framework identifies the class of additive feature importance methods and shows there is a unique solution in this class that adheres to desirable properties. SHAP values consistently attribute feature importance, better align with human intuition, and better recover influential features. When applied to the Polish bankruptcy data, the method not only discovered all the important features but also produced the correct classifier. With better interpretability, the SHAP framework performs not worse than any other common feature selection methods in our empirical financial classification study.

References

[1] Giudici, P. (2003). *Applied Data Mining: Statistical Methods for Business and Industry*. West Sussex PO19 8SQ, England: John Wiley & Sons Ltd, The Atrium, Southern Gate, Chichester.

[2] Kohavi, R., & John, G. H. (1997). Wrappers for feature subset selection. *Artificial Intelligence*, 97(1–2), 273-324.

[3] Heikki Linden et al. Synthesis of research studies examining prediction of bankruptcy. 2015.

[4] Alex Krizhevsky, Ilya Sutskever, and Geoffrey E Hinton. Imagenet classification with deep convolutional neural networks. In *Advances in neural information processing systems*, pages 1097–1105, 2012.

[5] Xavier Bredart. Bankruptcy prediction model using neural networks. *Accounting and Finance Research*, 3(2):124, 2014.

[6] Junfei Qiu, Qihui Wu, Guoru Ding, Yuhua Xu, and Shuo Feng. A survey of machine learning for big data processing. *EURASIP Journal on Advances in Signal Processing*, 2016(1):1–16, 2016.

[7] Edward I Altman. A further empirical investigation of the bankruptcy cost question. *The Journal of Finance*, 39(4):1067–1089, 1984.

[8] Ralph De Haas and Neeltje Van Horen. International shock transmission after the lehman brothers collapse: Evidence from syndicated lending. *The American Economic Review*, 102(3):231–237, 2012.

[9] Stefan Engström. Kan nyckeltal påvisa framtida betalningsoförmåga? *Balans*, 3:36–45, 2002.

IOP Conf. Series: Journal of Physics: Conf. Series **1284** (2019) 012026 doi:10.1088/1742-6596/1284/1/012026

[10] S Alex Yang, John R Birge, and Rodney P Parker. The supply chain effects of bankruptcy. *Management Science*, 61(10):2320–2338, 2015.
[11] Rodriguez , I. , R. Huerta , C. Elkan , and C. S. Cruz . (2010). Quadratic programming feature selection. *Journal of Machine Learning Research* 11 (April) : 1491 – 1516
[12] Tin Kam Ho. Random decision forests. *In Document Analysis and Recognition, 1995., Proceedings of the Third International Conference on*, volume 1, pages 278–282. IEEE, 1995.
[13] Leo Breiman. Bagging predictors. *Machine learning*, 24(2):123–140, 1996.
[14] Chen, T., and Guestrin, C. (2016). "XGBoost: A scalable tree boosting system," in *Proceedings of the 22nd ACM SIGKDD International Conference on Knowledge Discovery and Data Mining - KDD '16* (New York, NY: ACM Press),785–794.
[15] Scott M Lundberg and Su-In Lee. 2017. A Unified Approach to Interpreting Model Predictions. In Advances in *Neural Information Processing Systems 30*. Curran Associates, Inc., 4768–4777.
[16] Maciej Zieba, Sebastian K Tomczak, and JakubMTomczak. Ensemble boosted trees with synthetic features generation in application to bankruptcy prediction. *Expert Systems with Applications*, 58:93–101, 2016.
[17] Christopher M Bishop. Pattern recognition and machine learning. springer, 2006.

Research on the Private Customized Information Retrieval based on Hadoop Cluster

Li Bo[1,2], Zhang Xiao[1,2] and Yan Jingyi [3]

[1] School of Computer Science, Northwestern Polytechnical University, Xi'an Shaanxi 710129, China;
[2] Ministry of communications Key Laboratory of big data storage and management, Xi'an Shaanxi 710129, China;
[3] School of Management, Northwestern Polytechnical University, Xi'an Shaanxi 710129, China

Corresponding author: e-mail:zhangxiao@nwpu.edu.cn

Abstract. With the development of science and technology, it is no longer a technical difficulty for users to find the answers they need from massive data of information retrieval. At present, with information retrieval methods, the most urgent need is how to find personalized search results that satisfy users and meet their needs. Therefore, private customized information retrieval technology has great research value. This paper builds a Hadoop distributed cluster search engine, improves the PageRank algorithm, adds classified attributes such as user tone, and further classifies and ranks the web pages. At the same time, the C4.5 algorithm is used to classify users, thus it provides private customized search service. Through comparative experimental analysis, it is proved that the improved method proposed in this paper is effective and has profound research value.

1. Introduction

With the development of science and technology, people are in the era of a big data explosion. According to the "37th Statistical Report on Internet Development in China" released by China Internet Network Information Center (CNNIC) in June 2016, the number of Chinese Internet users has reached 710 million at that time. At present, the digital content of the world is about 487 billion GB, which has created new research topics. How to search for satisfactory answers? Searching for answers is not difficult. The key is to get a "satisfactory" answer, which is suitable for the searcher's "private" order. When dealing with the current massive amount of data, even a single high-performance supercomputer will have a big burden. In order to improve the experimental accuracy and load balance, this paper chooses to build a Hadoop computer cluster. Hadoop is a very classic distributed computering platform, including the distributed file system HDFS (Hadoop Distributed File System) and the distributed computing framework MapReduce. By building a Hadoop cluster and optimizing search algorithms, semantic recognition, speech (tonal) recognition, generalized search, and providing customers with privately-tailored search services to optimize search results[1]-[2]-[3]-[4].

2. Research status

In the field of information retrieval, researchers began research in the mid-20th century. Many articles describes the information retrieval as follows: Information retrieval is a process in which users use some methods to search according to their own information needs. The current definition is to find information that meets user needs from a collection of large-scale unstructured data. In order to deal with massive amounts of data, search engines hecome an indispensable tool in people's lives. Currently, the search resources are no longer single text or words, and the users' needs are more complex. The data format of information retrieval is more extensive. Voice search, image recognition, and multimedia resource-based search technology are also developed[5]. For example, Photo of Baidu is a typical image search engine.

Distributed computing is a commonly used method when dealing with massive data. Hadoop has the advantages of high scalability, high reliability, and low cost of use[6].The research in this paper is based on Hadoop's distributed personalized retrieval. According to the user's behavior habits and the sentiment attitude when sending the search information, search engine will combine the classification algorithm C4.5 algorithm and provide the private customized search service, which best meet the user's desire.

3. Hadoop distributed file system

The retrieval structure of this paper adopts the architecture of distributed file system Hadoop to build a distributed cloud computing platform. Distributed computing is an effective solution to deal with big data problems, which important significance is each node handles huge data to achieve load balancing, resource sharing, and good scalability[7]. It is suitable for the retrieval requirements under the current big data era and can provide information retrieval.

3.1. Hadoop framework

Hadoop is currently the most widely used and most recognized parallel computing framework. It has two core components, HDFS and MapReduce. Due to the characteristics of Hadoop, users can concentrate on distributed upper-layer applications, do not care about how the distributed underlying layer is implemented, and make full use of the high performance of the cluster to achieve high-speed distributed computing.

3.1.1. Simple. Hadoop abstracts, modularize, and implements all of the underlying design details, which greatly increases development efficiency and makes development easier.

3.1.2. Fault tolerance. Hadoop's task allocation strategy can balance load for each node, so processing speed is fast and execution efficiency is high.

3.1.3. Efficient. Hadoop abstracts, modularize, and implements all of the underlying design details, which greatly increases development efficiency and makes development easier.

3.2. MapReduce concept

The current mainstream algorithm for information retrieval uses the PageRank algorithm. The PageRank algorithm mainly uses the MapReduce technology. MapReduce applies the ancient Chinese divide-and-conquer ideas to computer programming, which is inspired by traditional programming. For MapReduce, the common explanation is to compose a bunch of unstructured data into key-value pairs. Map parses these messy data extraction keys and values. The process of parsing is also called extracting the characteristics of these data. After the key-value pair is established, it will There is a Shuffle process. The essence of this Shuffle process is to clean up the memory, make the system more efficient, and internally sort these key-value pairs. The Reduce stage filters and organizes these key-value pairs according to the rules to get the required Data fragment. The working principle of MapReduce is shown in Figure 1.

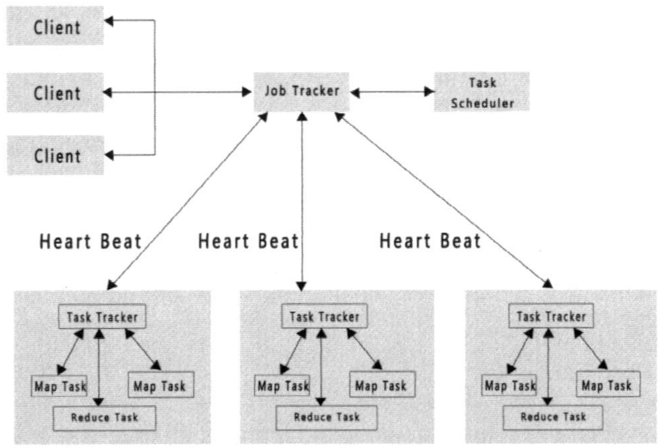

Figure 1. Architecture diagram of Hadoop MapReduce

Hadoop MapReduce uses a master-slave architecture consisting of four core components: Client, JobTracker, TaskTracker, and Task. In MapReduce, the Client submits the job to the JobTracker and the program begins execution. JobTracker acts as the core controller in the execution of the job and acts as the commander. During the execution of the job, it monitors the status of each node. Once an exception occurs, it will take immediate action and assign it to other nodes for execution. The job was successfully completed. The TaskTracker performs the JobTracker assigned tasks and periodically reports the execution status of the job to the JobTracker in the form of a heartbeat. The task is divided into MapTask and ReduceTask. MapTask parses the input information into key-value pairs to generate intermediate data. After the Shuffle and Merge processes, the sorted intermediate data is passed to the ReduceTask. ReduceTask organizes the intermediate results on demand and saves the results to HDFS for further processing.

3.3. HDFS Distributed File System

HDFS is an open source file system developed by researchers. The advantage of HDFS is that it is highly fault-tolerant, especially suitable for massive data sets, with low cost and reliable service. Figure 2 shows the architecture of HDFS.

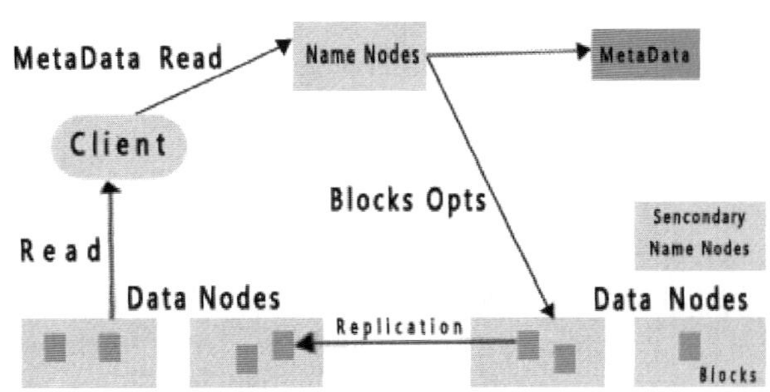

Figure 2. Architecture diagram of HDFS

HDFS adopts the master-slave architecture mode, which has structural components such as NameNode, SecondaryNameNode, DataNode, and Block. The main task is to manage the metadata of HDFS and master the file directory tree of the entire HDFS and its directories and files. The NameNode is a control center that manages the namespace and controls the client's access to files to initiate monitoring. SecondaryNameNode

IOP Conf. Series: Journal of Physics: Conf. Series **1284** (2019) 012027 doi:10.1088/1742-6596/1284/1/012027

The metadata of the NameNode is backed up periodically, sometimes called checkpoint. The main purpose is to set up a special node to check and back up the metadata. The DataNode is responsible for saving data. All DataNodes perform file operations under the command of NameNode. HDFS is most suitable for processing large data sets. The data basically follows the rules of one-time write and multiple-read. The data on HDFS is saved in the form of a block on the DataNode.

4. Improvement of the search algorithm

In order to realize the "private" search service, the search algorithm is combined the search algorithm with the data mining algorithm.

4.1. Information Retrieval Algorithm

For search engines, the core search algorithm has a great impact on the performance of the platform. Voice input recognition and search is a technology that is now emerging as compared to traditional pure text search. At present, artificial intelligence is not good enough to integrate human emotions into data analysis. This paper introduces the measure of "tone" in the algorithm to carry out more accurate and humanized analysis. For the same sentence, the different tone may have different meanings or ideas. In addition to adding "tone" and user attitudes, this article also considers the starting point of the current search. Such as a user wants to search for related content, he maybe will exist, but if search engines still return he common responses and do not consider his tone, the user also is disappointed. According to the form of data mining, the search engine set the corresponding search according to the user's behavior and preferences, and applies the data mining method to the information retrieval, thereby providing the user with a private search service.

Information retrieval algorithm is the core component of the search engine. This paper uses the PageRank algorithm as search sorting algorithm and adds some new key points proposed in this paper. PageRank usually uses two hypotheses of quantity hypothesis and quality hypothesis to calculate. The simple PageRank algorithm flow can be expressed as a process in which a web page constitutes a directed graph. It is assumed that A, B, C, and D are the nodes of four web pages. If web page A has a link relationship with web page C, there is a directed edge A-> C, as Figure 3 show.

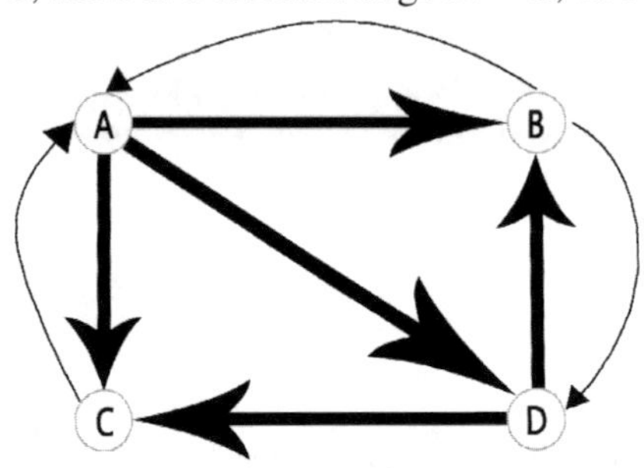

Figure 3. Graph of RagePank algorithm

If the user currently stays on the B page, the probability of the Internet user jumping to A, C, and D is 1/3, and 3 means that B has 3 outgoing links. To abstract this problem, it can be represented by a convex combination of matrix P and matrix E, which is defined as follows:

$$G = \alpha P + (1-\alpha)E \tag{1}$$

Here a < (0, 1) is the damping factor, the matrix P is defined by the hyperlink structure of the network. N is the dimension of the matrix P, and the PageRank problem is the eigenvector corresponding to the first eigenvalue 1 of the matrix G. The iterative tensor of the higher-order PageRank problem can be expressed as:

$$\overline{M}(i, j, ..., l, k) = \alpha \overline{P}(i, j, ..., l, k) + (1-\alpha)v_i \tag{2}$$

Through the underlying PageRank algorithm, combined with the data mining and classification algorithms, the accurate information retrieval process of data analysis can be realized, what's more, these also are the experiment of the "private" search information retrieval service.

4.2. Customer Classification Algorithm

In the data mining part of the overall algorithm, it is first necessary to analyze whether it is a "classification" or a "clustering" problem.In theory, each user is a separate cluster in search, they may have the same starting point in some specific searches, but in fact, each user's needs may not be exactly the same, so the classification algorithm is more appropriate. In this paper, the C4.5 algorithm is used as the algorithm for classifying users. C4.5 constructs a classifier in the form of the decision tree processes a large number of data that needs to be classified and tries to predict the category to which the new data belongs.

4.2.1. Information Gain. The information gain is actually used in the ID3 algorithm to perform attribute selection metrics, and the attribute with the highest information gain is selected as the split attribute of the node N. The selected attributes minimize the amount of information required for tuple classification in the result partition. The expected information required to classify tuples in D is as follows:

$$Info(D) = -\sum_{i=1}^{m} p_i \log_2(p_i) \tag{3}$$

Info (D) is also known as entropy. It is assumed that the tuple in D is divided according to attribute A, and attribute A divides D into v different classes. After this division, the information needed to obtain an accurate classification is measured by the following equation:

$$Info_A(D) = \sum_{j=1}^{v} \frac{|D_j|}{|D|} \times Info(D_j) \tag{4}$$

The information gain is defined as the difference between the original information requirement (ie, based on the class ratio only) and the new requirement (that is after the A is divided):

$$Gain(A) = Info(D) - Info_A(D) \tag{5}$$

4.2.2. Information gain rate. In order to avoid the disadvantage of information gain, C4.5 behind ID3 adopts the concept of information gain rate. The information gain rate normalizes the information gain using the "split information" value. The classification information is similar to Info(D) and is defined as follows:

$$SplitInfo_A(D) = -\sum_{j=1}^{v} \frac{|D_j|}{|D|} \times \log_2(\frac{|D_j|}{|D|}) \tag{6}$$

This value represents information generated by dividing the training data set D into v partitions corresponding to the v outputs of the attribute A test. Information gain rate definition:

$$GainRatio(A) = \frac{Gain(A)}{SplitInfo(A)} \tag{7}$$

Through this improvement, the classification accuracy of C4.5 is improved and can be better used in user classification. Compared with other algorithms, C4.5 algorithm are easy to understand and have high accuracy. In this article, we use C4.5 as a classification algorithm to classify users and provide them with more humanized search results. The algorithm performs iterative update after each search by the user. If the user searches for data very frequently during this time period, it will perform iterative recalculation after the search peak is reduced. Through the combination of these two algorithms, it is possible to return accurate search results for a specific user.

5. Experiment

The design of this paper is verified by the Hadoop environment analysis in Section 3 of this paper, the algorithmic logic analysis in Section 4. The experiment runs on a cluster of 4 nodes, 1 master node, 3 slave nodes, Hadoop version is Hadoop-0.21.0, and JDK version is jdk1.6.0-21. Table 1 is the parameter configurations for each node.

Table 1. *Experimental configuration*

	Master node	Slave node
cpu	Inter Core Processor	Inter Pentium Processor
cpu GHz	2.20GHz	2.66GHz
amount of cpu	2	1
memory	16G	8G
hard disk	2T	1.5T

The search engine based on distributed clustering and unoptimized PageRank algorithm, the search engine based on a distributed cluster and PageRank algorithm optimized by this paper are tested and analyzed according to network data. The experimental process is divided into three stages: web crawling, web page analysis, and web page query. For the PageRank algorithm optimized by this paper, the process of classifying customers is also included.

5.1. Web crawling

To give users the most comprehensive information, search engines first need to obtain a large number of web pages, based on the download page, analysis, download, and analysis. By starting from the search engine's seed site, and constantly crawling the link pages, search engines will achieve web crawling.

5.2. Web page analysis

Search engines call the calculation module to calculate the topic of the web page, the degree of mirroring, the degree of entry, the degree of ranking. The core keywords of the web page will be analyzed For the search engine, the originality of the website is very important. You can calculate which pages of each page of a website are linked to, and which pages link to the website so that each page has a weight value. In this process, the PageRank algorithm proposed in this paper adds classification factors such as "tone". The web crawling classification will be more refined and will be based on the searcher's preference.

5.3. Web page query

In this process, the role of the PageRank algorithm is reflected, and the improvements studied in this paper are mainly reflected in this stage. According to the search cache mechanism, the webpage has been downloaded and analyzed in the early stage, and then each webpage is indexed. In this process, the unimproved and improved PageRank algorithm will provide customers with different search results. We will analyze the pros and cons of the algorithm according to the recommended order of users clicking on the webpage. Remove the ad page. If the user clicks on the lower ranked page and clicks on different searches to get more pages or even research, the algorithm has lower recommendation accuracy. In this paper, the first result weight of the search is set to 1, and the order is incremented. If the search is researched, 10 is added, and the clock frequency of the user each time the search is completed is obtained, thereby obtaining the experimental result.

When the sample space is small, the traditional PageRank algorithm does not have too much difference with the improved PageRank algorithm. However, as the number of data increases, the improved PageRank algorithm will be greatly improved. The improved PageRank algorithm is more traditional. The PageRank algorithm has an obvious improvement. Overall, the algorithm proposed in this paper is effective and has great development value.

6. Conclusion

By constructing the experimental environment, simulating the actual experimental platform, conducting scientific verification and analysis, and drawing the experimental conclusions, through the addition of some new indicators, the improved PageRank algorithm has a great effect. By this method, the user's search accuracy, and satisfaction are higher, which proves the paper's improvement is effective, the idea is correct. Following works will be done in the future:

- The essential purpose of information retrieval is to give users a quick presentation of the data they want. No algorithm can adapt to the current environment 100%. Combining the attitudes of speech, tone, and speaker, this is very useful for human beings, but it still takes time to develop for computers.
- Most of the current information retrieval is the search for short or one-sided data, and there is no continuous data retrieval for continuous data. For example, the ability of a computer to read an article's paragraphs, or to make judgments on continuous searches, and the ability to logically analyze, this article will continue this module for analysis.
- The current SSD is a very common storage structure and is suitable for the case where the number of writes is small and the number of reads is large. For this type of storage device, this article will try to integrate SSD technology into the Hadoop file system to speed up the reading and give the user a better experience. However, the author believes that the prospect of SSD development is not particularly optimistic. The traditional disk system as the storage structure may be more stable and will ensure the stability of the data.

Acknowledgments

This work is supported by the MIIT Key Laboratory of Big Data Storage and Management, the NFS of China under Grant No.61502392 and No.61272323 and the Ministry of Science and Technology of China, National Key Research and Development Program (No.2018YFB1004401).

References

[1] Azzedin, Farag and Muthucumaru Maheswaran. "Towards Trust-Aware Resource Management in Grid Computing Systems." 2nd IEEE/ACM International Symposium on Cluster Computing and the Grid (CCGRID'02) (2002): 452-452.
[2] Sun Yongchao. Research on information retrieval system based on Hadoop [J]. intelligence exploration, 2016, 1 (8): 125-130.
[3] Lee, Kyong-Ha et al. "Parallel data processing with MapReduce: a survey." SIGMOD Record 40 (2011): 11-20.
[4] Li Bo, Zhu Chenqi, Liu Jiamin, He Jing, Ma Tiao. Wireless Charging Device Based on Magnetic Resonance Technology [J]. Electronic Design Engineering, 2016, 24 (10): 111-113.
[5] Beloglazov, Anton and Rajkumar Buyya. "Energy Efficient Resource Management in Virtualized Cloud Data Centers." 2010 10th IEEE/ACM International Conference on Cluster, Cloud and Grid Computing (2010): 826-831.
[6] Wang Peng. Research and application based on Hadoop full-text retrieval [D]. Chongqing University of Posts and Telecommunications, 2017.
[7] Schwarzkopf, Malte et al. "Omega: flexible, scalable schedulers for large compute clusters." EuroSys (2013).

An Entity Linking Method Based on Entity Category and Word Embedding

HaiBo Ma, Yue Li and Yang Li

Beijing University of Technology, 100124, Beijing

email:bjut@foxmail.com

Abstract. Named entity linking is a process of linking a given reference in a document to a knowledge base. In natural language processing, entity linking can enhance the computer's understanding of unstructured text data. Applying traditional entity linking methods, especially entity linking methods for person names and organization names, has its limitations. Similar vocabulary as an entity to be linked is difficult to make full use of its contextual semantic information for ambiguity elimination. This paper makes full use of the entity's category attribute and the semantic information contained in the context to design an entity linking method based on entity category and semantic word embedding. First, Training text classification model based on corpus to obtain entity attributes. Then the semantic feature is extracted by the word vector template to perform entity disambiguation through the semantic classification model. Finally, the results of the entity linking are predicted by means of model ensemble. Experiments show that the accuracy of the method after fusion on the entity linking dataset has improved.

1. Introduction

In the era of rapid development of the Internet, the rapid growth of text data has caused the problem of information explosion. Strengthening the analysis and understanding of unstructured text data by computer through the relevant methods of natural language processing is of great value to the use of this data. In natural language, nouns often have a polysemy. In this case, entity linking can eliminate ambiguity and improve computer text comprehension. An entity linking task refers to extracting named entities in a document, including names of people, organizations, etc., and linking them to unambiguous entities in the existing knowledge base. The main task of the entity linking is to take two steps, firstly identify the reference item to be linked in the text, obtain the candidate list in the corresponding knowledge base, and then link the reference to the unambiguous entity in the knowledge base based on the entity disambiguation algorithm.

At present, the methods of entity linking mainly include a machine learning classifier based method, a topic model based method, a graph model based method, and a vector space model based method. As stated in the reference [1], the classifier-based method is to use the multiple entities corresponding to the reference item as the result of classification, use the classifier to train and predict the result, but this method generally uses noun co-occurrence as a feature, lacking contextual semantic information dataset noise impact big. As stated in the reference [2], the topic model performs entity disambiguation based on the document theme, but it is difficult to extract deeper contextual semantic information of higher dimensions by linking only through the topic. As stated in the reference [3], the graph model-based method constructs the semantic network to extract the relevant features to link the entities. However, if the amount of data is large, the computational cost will be doubled. As stated in the reference [4], the vector space model converts the context of the referential terms into vectors, and the entity disambiguation is performed by the similarity calculation. This method does not fully apply the important information in the external corpus and knowledge base.

The main contribution of this paper is to propose a named entity linking method based on entity attributes and word embedding, mapping context nouns in the text and corresponding entities to be linked to the same semantic topic space. In this way, the classification model is used to semantically disambiguate the entity. The innovation is that from the perspective of deep learning, the entity disambiguation method is proposed in the way of classifier classification prediction. Not only can the context semantic information be fully utilized, but also the semantic classification information of the entity can be used for disambiguation, and the accuracy rate can be improved. The structure of this paper is as follows: Section 1 is introduction. Section 2 introduces the method proposed in this paper; Section 3 is the entity linking part; Section 4 is the experimental part; the last part is the conclusion and outlook.

2. Ensemble classification model

2.1. Semantic word embedding

Semantic classification model is in Figure 1. The neural network part of the semantic classification model adopts the CBOW model [5]. CBOW is a three-layer neural network model, from left to right, the input layer, the hidden layer and the output layer. The basic idea is to train each words are mapped into a K-dimensional real number vector containing semantic and grammatical information, K is an optional parameter, and the distance between the vectors such as Euclidean distance and cosine similarity is used to judge the semantic similarity between them. The model is to model the language model and obtain the representation of words in the distributed vector space while modelling. By solving the objective function by stochastic gradient descent, the V words in the corpus can be represented as distributed vectors with deep semantic features. Its corresponding objective function is shown below.

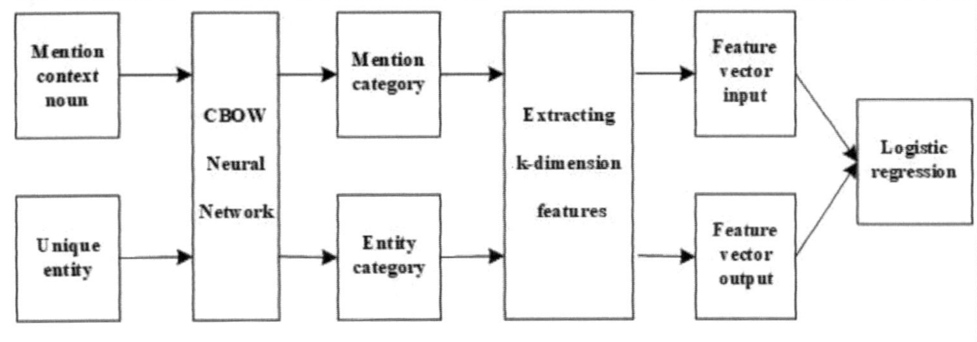

Figure 1. Semantic classification model

$$l(\theta) = \text{argmax}\frac{1}{v}\sum_{j=1}^{S}\left(\sum_{ij=1}^{D_j}\log p\left(w_{ij}|context_{ij}\right)\right) \tag{1}$$

2.2. Extract features

Feature selection for a given training data set, D= (d_1, d_2, ⋯, d_n) is used to represent each piece of text in the training data set, and S=(S_1, S_2, ⋯, S_n) indicates that the text is Correspondingly, a collection of unambiguous entities that have been linked to the knowledge base. Based on the assumption that all nouns in a text, including entity references, are located in similar semantic spaces, extract the nouns in the training set, and obtain the word vector template by the method in Section 2.1, and represent the extracted nouns as distributed vector, get the noun vector set N. The distributed vector contains deep semantic information, and the set N is clustered by k-means [6], and k center points $P=(P_1, P_2, \cdots, P_k)$ are obtained as k features where k is k -means clusters the number of cores. At the same time, the category label of each word in the set N is obtained by calculating the distance of each word to k center points.

The training data characterizes the label of each noun obtained in the previous section. The text d_i in the set D can be represented as a k-dimensional vector. The frequency at which each noun category in d_i is present is used as the weight on the dimension feature. Select k=10 to select the 10-dimensional feature. "Apple, Google and Microsoft are the world's largest technology companies, often compared by people." In this sentence, the clustering label corresponding to Apple and Google is 5, and the corresponding clustering label of the technology company is 1, people correspond the clustering label is 9. So this text can be represented as (0,1,0,0,0,1,0,0,0,1) corresponding to this text, the unambiguous entity that has been linked to the knowledge base is "Apple" Find the corresponding vector from the vector template, calculate the closest category to Apple by cosine similarity, and find that Apple belongs to category 5, so Apple can be represented as a vector (0,0,0,0,0,1,0,0,0,0). Through the above process, we can represent the text in the training set and the corresponding entity to be linked as a k-dimensional vector pair, and k is the number of selected features.

2.3. Semantic word embedding classification model

Related work shows that in practical applications, logistic regression classifiers are similar to SVM and random forest classification models [7]. The logistic regression classifier algorithm has the lowest complexity. Therefore, in this part we use a logistic regression classifier to build a classification model. Use $D(d_1^*, d_2^*, \cdots, d_n^*)$ to represent the characterization of the text set, and use $S(S_1^*, S_2^*, \cdots, S_n^*)$ to represent the characterization that a collection of unambiguous entities that correspond to the text and are linked to the knowledge base. Where d_i^* and S_i^* are both k-dimensional vectors. After characterizing the training data and presenting it in a vector, the problem can be transformed into a multi-class problem. The significance of this is that it not only utilizes the relationship between the text to be linked entity and the nouns in the text, but also utilizes the semantic features of the entity's word vector.

Let $x= d_i^*$ and $y= S_i^*$, Then (x, y) is an observation sample, and the feature selected in Section 2.2 can be obtained, which ranges from $[0,1,2,\cdots, k-1]$. The conditional probability can be expressed as follows.

$$P(y = c_i|x) = \frac{e^{g_{ei}(x)}}{1+\sum_{j=1}^{k-1} e^{g_{ei}(x)}} \tag{2}$$

Solving the model by constructing a likelihood function, and the sample above, yields the likelihood function as follows. The likelihood function is solved by the gradient descent method, and the trained semantic model of word vector is obtained.

$$L(\beta) = \sum_{j=0}^{k-1}\sum_{i=1}^{n} y_{ij}\log(\pi_j(X_i)) \tag{3}$$

2.4. Entity attribute model

Based on the assumption that entities often recur in the same domain, this paper uses the classification properties of entities to disambiguate entities. The text categorization task is mainly to let the computer automatically divide the text into pre-defined categories. Because the Text CNN [8] model has high classification accuracy and high training efficiency, the text classification model adopts the Text CNN model. The CNN model mainly includes a convolutional layer, a pooling layer, an activation function, and a fully connected layer. The external corpus required for training comes from THUCNews. Training through backpropagation and annotated data classification to obtain an entity context classification model. The entity is linked based on the results of the entity context classification and the category attributes of the entities in the knowledge base.

3. Entity linking

3.1. Task details

An entity linking is the process of linking a given entity's reference in the text to an unambiguous entity in the knowledge base. In this paper, three features are selected for entity disambiguation, word embedding classification model (WECM), entity category feature and feature of entity popularity [9]. The entity linking process is expressed as follows. Where E represents the final score of the candidate entity, e_{ij} represents the candidate entity corresponding to the m_i, f_{tc} represents entity classification model, $f_{ef}(e_{ij})$ represents the candidate entity popularity score, and A B represents the weight of the semantic classification feature of the word vector.

$$E = \arg\max\left[A\cos\left(WECM_{mi}, e_{ij}\right) + Bf_{tc}(m_i, e_{ij}) + (1 - A - B)f_{ef}(e_{ij})\right] \quad (4)$$

The whole process can be divided into three parts: entity referral standardization, candidate entity expansion and entity disambiguation. Many entities in the text have a number of different names, some are aliases such as magicians, nicknames such as Dayao, some are part of the full name or abbreviations such as PKU. Therefore, it is first necessary to map the representations that appear in the text to a standard expression. Specifically, a synonym vocabulary [10] is constructed to solve this problem. Among them, the Key value represents the irregular reference of the entity, and the Value represents the standard entity. After standardizing the entity references, a list of candidate entities needs to be built for the named entities to be disambiguated. This paper constructs an ambiguous words table that stores the standard form of the entity and its corresponding list of unambiguous entities. The entity popularity feature is measured by the number of times the entity appears, and the weight of the entity popularity is set according to experience. An example are shown in Table 1.

Table 1. An example of entity popularity.

Unambiguous entity	Entity frequency	Weights
Apple(enterprise)	168	1.0
Apple(fruit)	15	0.8
Apple(figure)	2	0.6

3.2. Entity linking

A text entity linking method based on word vector semantic classification, the input time text and its corresponding entity to be linked, corresponding to the unambiguous entity linked to the knowledge base. In the first step, according to the synonym table in the knowledge base, the description of the reference in M is standardized, and the standardized reference set is obtained. In the second step, E = $(e_1, e_2, \cdots, e_n)$, and the candidate entities in e_i are sorted by popularity. When $|e_i|=0$, it means that there is no corresponding entity in the knowledge base, that is, the reference mi is unlinkable, and the label NIL is returned; when $|e_i|=1$, the unique candidate entity is directly returned as the final link entity; when $|e_i|>1$, perform steps 3 and 4. In the third step, the semantic classification features of each candidate entity are obtained by calculating the cosine distance of each candidate entity to the tag. In the fourth step, the weighted sum of the two features of each candidate entity is calculated according to the equation, and the candidate entity with the highest result is output as the final link entity. If the weighted sum is less than the threshold, the flag NIL is returned.

4. Experiment

4.1. The data set

The data used in this paper to establish synonym tables, ambiguous vocabularies, entity popularity tables, and trained word vector templates are all Wikipedia [11], and the data version used is Chinese

Wikipedia in 2015. The data is extracted from the knowledge base through rules and statistics are obtained, and the data size is as shown in Table 2.

Table 2. Experimental data

Unambiguous entity	Data amount
Synonym table key number	531526
Synonym table value number	183727
Ambiguous words table key number	19284
Ambiguous words table value number	241467

4.2. Comparison experiment

The comparison experiment is carried out to verify whether the algorithm is effective. This part selects the entity link method based only on the entity classification attribute, and the entity link method based only on the semantic vector classification model, and compares the current method after the model fusion. During the experiment, the same resource template and preprocessing were used for all methods. Firstly, the multi-class regression model is trained by training set data, and then the model set is used to obtain the optimal weight α=1.58, A=0.52, B=0.27. Finally, the method is verified in the test set. Select Precision, Recall, and F1 as the evaluation indicators. Among them, the in-KB part indicates the accuracy rate of the included entities in the knowledge base, and NIL indicates the entity link accuracy rate not included in the knowledge base. The results of the comparative experiments are shown in Tables 3.

Table 3. Experiment result

	in KB			NIL		
Model	Precision	Recall	F1	Precision	Recall	F1
WECM model	0.7927	0.8488	0.8198	0.8871	0.8648	0.8758
Entity attribute model	0.7341	0.7942	0.7884	0.8143	0.8194	0.8361
Ensemble model	0.8137	0.8593	0.8358	0.9134	0.8763	0.8949

In the experimental results, the bold portion indicates the method herein. Experiments show that the fusion method is superior to the other two algorithms in terms of accuracy and recall rate, especially the accuracy rate is significantly improved. The method of this paper is mainly a method of combining classification attributes and semantic word vectors to fully apply context semantic information. The relationship between the semantic classification feature of word vector and the number k of clustering features. Only the WECM model is used for entity linking, and different k values are selected to observe the relationship between the model and k. When k=10, the model obtains the highest F1 value. Between 5~15, the change of k value has little effect on the prediction efficiency of the model. However, when k=20, the F1 value of WECM drops significantly. By counting the number of nouns contained in each text in the evaluation data, it is found that there are an average of 8 nouns in each text. The analysis considers that when k=20, the decrease of F1 value is due to excessive selection of features and sparse training data.

5. Conclusion and outlook

Based on the assumptions that entities appear in the same field, and the nouns and contexts in the text have similar semantics, this paper proposes the idea of semantic disambiguation of text entities by using word vector semantic classification, and designs a complete entity linking method and performs on the data. verification. The experimental results show that using the entity link method based on classification attribute and word vector proposed in this paper, the link effect uses the effect of word vector and entity attribute alone, and the link accuracy rate is significantly improved. The follow-up work is mainly focused on two points. One is to combine the word vector and the graph model for physical linking.

Reference lists

[1] Gattani A, Lamba D S, Garera N, et al. Entity extraction, linking, classification, and tagging for social media: a wikipedia-based approach[J]. Proceedings of the VLDB Endowment, 2013, 6(11): 1126-1137.
[2] Han X, Sun L. An entity-topic model for entity linking[C]//Proceedings of the 2012 Joint Conference on Empirical Methods in Natural Language Processing and Computational Natural Language Learning. Association for Computational Linguistics, 2012: 105-115.
[3] Han X, Sun L, Zhao J. Collective entity linking in web text: a graph-based method[C]//Proceedings of the 34th international ACM SIGIR conference on Research and development in Information Retrieval. ACM, 2011: 765-774.
[4] Monteiro L B, Li W, Saleh A A. An Approach of Vector Space Model to Link Concrete Concepts with Wiki Entities[C]// IEEE International Conference on Computer & Information Technology; Ubiquitous Computing & Communications; Dependable. 2015.
[5] Mikolov T, Chen K, Corrado G, Dean J. Efficient estimation of word representations in vector space. arXiv: 1301.3781,2013.
[6] Hartigan J A, Wong M A. Algorithm AS 136: a k-means clustering algorithm. Journal of the Royal Statistical Soci-ety — Series C (Applied Statistics), 1979, 28(1): 100–108
[7] Fernandez-Delgado M, Cernadas E, Barro S, Amorim D. Dowe need hundreds of classifiers to solve real world classifi-cation problems Journal of Machine Learning Research, 2014,15: 3133–3181
[8] Krizhevsky A, Sutskever I, Hinton G E. ImageNet classification with deep convolutional neural networks[C]// International Conference on Neural Information Processing Systems. 2012.
[9] Shen W, Wang J Y, Han J W. Entity linking with a knowl-edge base: issues, techniques, and solutions.IEEE Trans-actions on Knowledge and Data Engineering, 2015,27(2):443–460
[10] Guo Y H, Qin B, Li Y Q, Liu T, Lin S. Improving can-didate generation for entity linking. In: Proceedings of the18th International Conference on Applications of Natural Language to Information Systems. Salford, UK: Springer, 2013. 225–236
[11] Wikipedia[Online], available:http://download.wikipedia.comzhwikilate-stzhwiki-latest-pages articles. xml. bz2. Octo-ber 31, 2015

IOP Conf. Series: Journal of Physics: Conf. Series **1284** (2019) 012029 doi:10.1088/1742-6596/1284/1/012029

Experimental study on long-distance SPAD-based VLC systems

Jingshu Xue[a], Yijun Zhu, Chao Wang[b], Dongfang Zhang and Yu Mu
Information Engineering University, China
E-mail: `xjslovesstudy@163.com`[a], `xxgcwangchao@163.com`[b]

Abstract. Long-distance visible light communication (VLC) based on single photon avalanche diode (SPAD) can supplement the existing wireless communication mechanism in many senarios. Although there are many researches on VLC in recent years, the experiments on long-distance VLC are relatively scarce. In this paper, we use the experimental method to study the long-distance VLC, so as to apply it to two new application scenarios — VLC between buildings and in mountainous areas. A system model is established for both of the scenarios. Based on the model, a low energy consumption device for long-distance VLC is completed to realize the long-distance transmission of voice and data. At the end of the paper, we simulate the performance of the proposed system. The results indicate that the system is feasible to realize the duplexing voice and data transmission within the range of 2 km when the data rate is below 200 kbps.

1. Introduction

Visible light communication (VLC) is based on the high-frequency flash of light-emitting diode (LED), which makes it has the advantages of sufficient frequency resources, energy conservation and environmental protection compared with traditional wireless communication, thus, it is easier to achieve the goal of higher speed and lower energy consumption using VLC. However, as visible light is located in the terahertz band, its energy decays sharply in space, so the long-distance VLC has always been a perplexing problem. Since single photon avalanche diode (SPAD) can detect optical signals on photon, it greatly improves the sensitivity of the VLC receiver [1–4]. Therefore, we use SPAD to realize the application of VLC in two typical long-distance scenes.

Outdoor VLC is mainly affected by atmospheric absorption and turbulence, etc. Under weak atmospheric turbulence, the channel of long-distance VLC can be characterized by Lognormal model [5]. Up to now, the research on VLC focuses on indoor transmission, and there are few experiments on outdoor long-distance VLC. Most of the engineering implementation processes of long-distance VLC focus on underwater communication [6, 7] and lighthouse communication. China has achieved real-time underwater communications with a distance of 1,244m and a data rate of 1Mbps. Japan has implemented one-way lighthouse communication with a data rate of 1,022 bps and a transmission distance of 2 km.

In this paper, firstly we introduce two kinds of application scenarios of long-distance SPAD-based VLC which are not well performed by using traditional wireless communication methods. We establish a SPAD-based VLC system model to meet the specific requirements of communication. Based on the model, we devise a new kind of long-distance duplexing VLC device. Finally, in order to verify the feasibility of the scheme, we simulate the proposed system under different communication distances and data rates.

2. Application scenarios

We apply the long-distance VLC based on SPAD to the scene as shown in figure 1 and figure 2. In the first scenario, VLC is used to realize high-speed voice and data transmission between buildings, so as to overcome the poor quality of communication service caused by electromagnetic shielding and reflection of tall buildings. In figure 2, SPAD-based VLC is applied to mountainous areas where the base station layout is not very practical. The using can be of certain significance to field research and rescue work ect in practice.

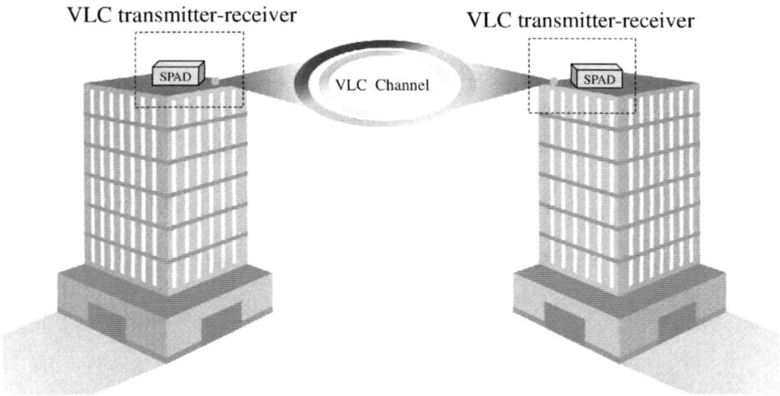

Figure 1. Diagram of inter-building VLC.

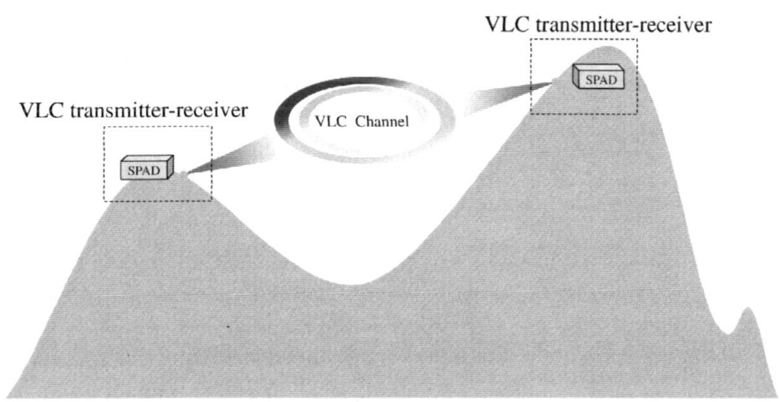

Figure 2. Diagram of VLC in mountainous areas.

To some extent, long-distance SPAD-based VLC can complement the traditional wireless communication. It can solve the wiring difficulties, electromagnetic interference, inconvenient construction of base stations and other problems. Meanwhile, it has the advantages of higher quality, lower cost, easier installation and more safety, etc. Furthermore, on the basis of point-to-point communication, multiple VLC transceivers can be used to form a network to achieve communication coverage in more areas.

3. System model of long-distance VLC

The model of the long-distance VLC system we develop is shown in figure 3 [8, 9]. The input voice or digital signals become bitstreams to be sent after the pre-processing process such as analog-to-digital conversion, encoding and framing etc. After on-off keying (OOK) modulation, the LED transmitter sends out optical signals. At the receiving terminal, the optical signals are detected by SPAD detector after the convergence of optical lens, after which the converted electrical signals are judged. Finally, the signals are output through decoding, digital-to-analog conversion and other processes.

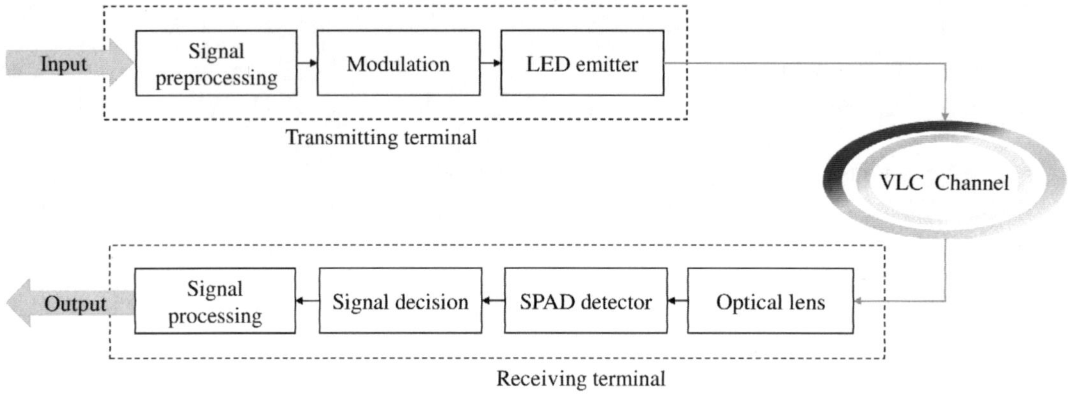

Figure 3. The system model of long-distance VLC.

4. Experiment and system setup

4.1. The structure of the long-distance VLC device

In order to meet the requirements of the proposed scenarios, we completed a long-distance SPAD-based VLC device, which can be used to realize voice and data transmission within the range of 0.4∼2 km. The average power consumption of the device is 0.5 watts.

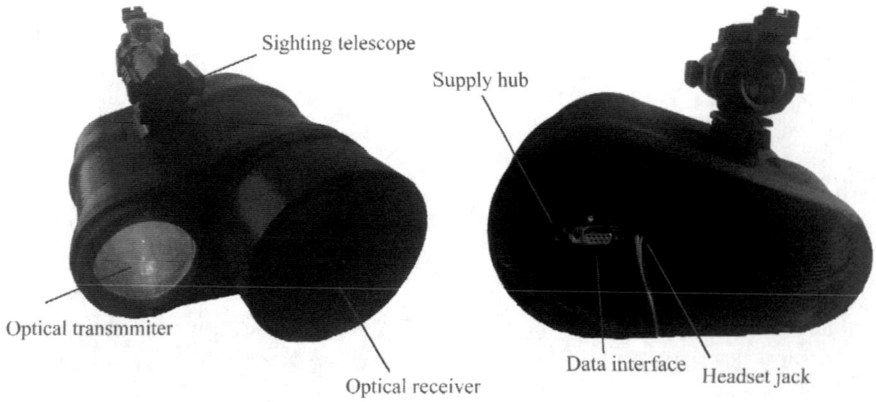

Figure 4. Long-distance wireless VLC device.

Figure 5. Schematic diagram of device setting in communication link.

As is shown in figure 4, in addition to the optical signal transmitter and receiver, the device also comprises a sighting telescope for the optical alignment between the sending and receiving ends. There is a supply hub, a data interface, a headset jack and a power indicator on the back of the device. They are set up to connect to some external devices or display the status of the system. The corresponding receiver and transmitter parts are aligned using the telescope to realize real-time duplexing communication as in figure 5, but the distance between them will be much greater than the figure shows in practice.

4.2. The working process of long-distance VLC device

Figure 6 is the block diagram of the device, which is mainly composed of ARM+AEM2000 platform, FPGA platform, optical front-end circuit board and some peripherals and interfaces. It adopts OOK modulation and Reed-Solomon (RS) codes. Users can transmit voice signals

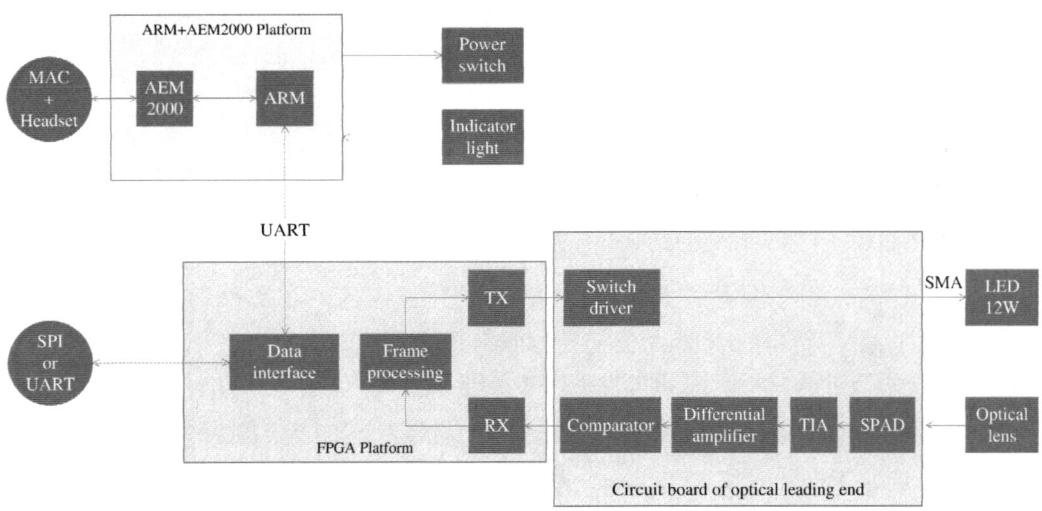

Figure 6. The block diagram of long-distance VLC device.

through microphone (MIC). After the processing of AEM2000 and advanced RISC machine (ARM), the signals are sent to data interface via universal asynchronous receiver/transmitter (UART). Then for transmitting purpose, the frame handle (FH) is available for framing. Finally, the switch driver circuit controls LEDs to flash to transmit signals based on the modulated electrical signals. Data transfer can also be achieved through Serial Peripheral Interface (SPI) or UART which is directly connected to the data interface, and the rest of the process is basically the same as voice transmission.

When receiving data, the optical signals enter the device after the convergence of the lens. The light is perceived by SPAD and then stirs up electrical signals. Subsequently, the signals are amplified by trans-impedance amplifier (TIA) and differential amplifier, and then determined by comparator. After that, the frame structure of signals is removed by FH and bitstream data can be transmitted to the data interface, which are then received through SPI or UART for data or converted into voice by ARM+AEM2000 platform and played through headset.

What is more, further improvements will be made on ARM to increase the functions of the device and enhance its adaptability in different scenarios.

5. Simulation results and discussions

In order to verify the feasibility of the proposed scheme, we made a simulation of the bit error rate (BER) under different data rates and communication distances, as is shown in figure 7. The simulation uses Lognormal model to depict the long-distance propagation of light in free space.

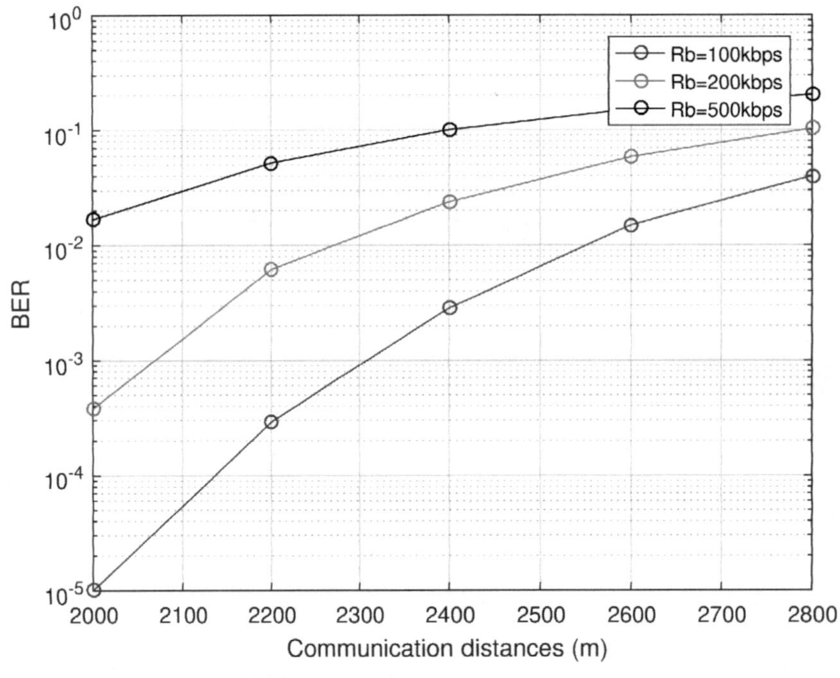

Figure 7. BER under different propagation distances.

Obviously, when the communication distance is about 2 km, a relatively reliable data transmission under the rate of 200kbps can be achieved. It is noticeable that the simulation is implemented without channel coding, and the signal decision mode has been greatly simplified [10], which means that the performance of the actual system still has room for improvement. However, the channel and electronic device is relatively ideal, thus the actual performance may be of some difference from the simulation result.

6. Conclusion

In this paper, we applied the long-distance SPAD-based VLC in two new scenarios: inter-building communication and mountainous belt communication. We built a system model in these two scenarios, based on which a long-distance VLC device has been designed. The results of the simulation indicate that the system is feasible within the communication distance of 2 km and under the date rate of 200 kbps.

Funding

Natural National Science Foundation (NSFC) of China (61671477); Major Scientific and Technological Project of Guangdong Province, China (2015B010112001); Major Scientific and Technological Project of Henan Province, China (161100210200).

References

[1] Jovicic A, Li J and Richardson T 2013 *IEEE Communications Magazine* **51** 26-32
[2] Li Y, Safari M, Henderson R and Haas H 2015 *IEEE Photonics Technol. Lett* **27** 943–6
[3] Zhang J, Si-Ma L H, Wang B Q, Zhang J K and Zhang Y Y 2016 *IEEE Photonics Technol. Lett.* **28** 1799–802
[4] Chitnis D and Collins S 2014 A *Journal of Lightwave Technology* **32** 2028–34
[5] Karp S, Gagliardi R M and Moran S E 1988 *Optical Channels* (Boston: Springer)
[6] Wang C, Yu H-Y, Zhu Y-J, Ji Y-W and Wang T 2017 *Optics Communications* **410** 889-95
[7] Wang C, Yu H-Y, Zhu Y-J and Wang T 2017 *IEEE Communications Lett.* **21** 2602–5
[8] Wang C, Yu H-Y, Zhu Y-J, Wang T and Ji Y-W 2017 *Optics Express* **25** 28783–93
[9] Zhang J, Si-Ma L H, Wang B Q, Zhang J K and Zhang Y Y 2016 *IEEE Photonics Technol. Lett* . **28** 1799–802
[10] Chatzidiamantis N D, Karagiannidis G K and Uysal M 2010 *IEEE Transactions on Communications* **58** 3381-5

IOP Conf. Series: Journal of Physics: Conf. Series **1284** (2019) 012030 doi:10.1088/1742-6596/1284/1/012030

An Improved Distance Metric Clustering Algorithm for Association Rules

LIANG Futai[1, 2,*], LI Hongquan [1], ZHANG Weifeng[2] and ZHANG Chenhao [1]

[1] Air Force Early Warning Academy, Wuhan, China
[2] 31121 PLA Troops, Nanjing, China

[*]Corresponding author:417516911@qq.com

Abstract. By mining association rules in large data, we can reveal useful information contained in the data and find out the relationship between things or the law of motion. However, because of the huge transaction data, the association rules obtained by mining are complex and massive. It is difficult to find useful association relations, especially when the re-demand is uncertain. To solve this problem, this paper first uses Apriori algorithm to mine association rules from a data set, then defines similarity measure between association rules, and applies DBSCAN clustering algorithm to association rules analysis. The analysis results show that this method is effective in association rules analysis.

1. Introduction

Association rule mining, also known as association analysis, is to find frequent patterns, associations, or causal structures that exist between project collections or object collections in relational data, transaction data, or other information carriers [1]. Association analysis is one of the methods of big data mining [2].

In the process of mining and analyzing Association rules, the analysis and interpretation of association rules is a key link. Because the association rule set extracted from a large amount of data is very large, it is very difficult to analyze and interpret. In order to solve this problem, one method is to cluster transaction data sets, such as such as shown in literature [3, 4]. This method clusters the items with high correlation according to a given threshold, divides transaction data sets into several parts, and then mining association rules in each transaction data set. This method reduces the size of transaction data sets, but from in the perspective of large data, it affects the integrity of the data, which is not conducive to the fullness of useful information. Another method is to analysis association rule sets by clustering algorithm, such as shown in literature [5, 6], literature [6] defined the distance between rules, and cluster association rule sets through K-means method. Literature [5] defined the similarity measure between different association rules. Density-based DBSCAN clustering algorithm is used to cluster Association rules, which greatly reduces the number of output rules. However, the support and confidence factors are not taken into account in the definition of distance and similarity measure, which affects the clustering effect.

In view of the above problems, an improved distance metric clustering algorithm for association rules is proposed in this paper to improve the clustering effect.

2. Basic Concept

Association analysis can mine and analyze association rules such as "the occurrence of some events causes other events to occur" from a large number of historical data [7]. For example, by investigating what customers buy in the mall, it is found that 30% of the customers will buy sheets and pillowcases at the same time, and 80% of the people who buy the sheets buy pillowcases, which hides a connection: {sheets} $\Rightarrow$ {pillowcases}, that is to say, a large number of customers will buy pillowcases while purchasing sheets, so for the mall, sheets and pillowcases can be placed In the same shopping area of the same mall, which will convenient customers to shop.

Based on the experimental data listed in Table 1, the related concepts of association analysis are explained as follows:

Table 1. Experimental data of association analysis.

Tid	Sheets	pillowcase	broom	bread	meat	noodles	...	cabbage
1	1	1	1	0	0	0		0
2	0	0	0	1	0	1		0
3	1	1	0	0	1	0		1
4	0	0	1	1	0	1		0
...	...	...	...	...	...	...		...
m	0	0	0	1	1	0		1

(1) Item: A commodity in a transaction in the experimental data is one item. For example, sheets, pillowcases, hamburgers, etc.

(2) Transaction: A transaction represents the sum of all commodities involved in an exchange. For example, the first transaction record includes sheets, pillowcases and brooms. A transaction T is an itemset, and each transaction has a unique identifier Tid. All transactions together constitute the transaction data set of association rule mining.

(3) Itemsets: A set containing one or more items is called an itemset, such as {sheet, pillowcase, broom}.

(4) K-Itemsets: An itemset containing K items is called a K-Itemset. For example, {pillowcase, bread} is called a 2-itemset.

(5) Support count: also known as the frequency and count of itemset, which is the number of transactions containing the itemset. For example, {spaghetti, bread} appears in transactions with Tid 2 and Tid 4, so its support count is 2.

(6) Support: The support of association rules $A \Rightarrow B$ refers to the probability that events A and B occur simultaneously, which is equal to the ratio of support count to total transactions [8]. For example, the total number of transactions listed in Table 1 is 5, and the support count of itemset {noodles, bread} is 2, so its support is 40%, indicating that noodles and bread appear in 40% of transactions.

(7) Frequent itemsets: The itemset whose support is greater than or equal to a threshold are called frequent item set. For example, when the threshold is set to 30%, the support of the itemset {noodles, bread} is 40%, so it is a frequent itemset.

(8) Right (RHS) and Left (LHS): For rules {bread} $\Rightarrow$ {noodles}, {bread} is called left (also called the former), and {noodles} is called the right part (also called the later).

(9) Confidence: Confidence of association rules $A \Rightarrow B$ is the probability of occurrence of event B on the basis of occurrence of event A. The formula is as follows:

$$\text{confidence} = P(B \mid A) = P(AB)/P(A) \tag{1}$$

For the rule {noodles} $\Rightarrow$ {bread}, the support count of itemset {noodles, bread} divided by the support count of {noodles} is the confidence of the rule, which is 100%, indicating that the person who bought the noodles 100 % will buy bread..

IOP Conf. Series: Journal of Physics: Conf. Series **1284** (2019) 012030 doi:10.1088/1742-6596/1284/1/012030

(10) Strong Association rules: is an association rules that is greater than or equal to the minimum support and the minimum confidence threshold.

The goal of association analysis is to find strong association rules.

3. Association Rule Mining

Apriori algorithm and FP-growth algorithm are commonly used in association rules mining. Apriori algorithm is a classical data mining algorithm for frequent item set and association rules [9]. The purpose of association rules is to find out the relationship between items in a data set, also known as shopping basket analysis, because "shopping basket analysis" expresses a scenario suitable for the algorithm. The core of the algorithm is the iteration based on the idea of two-stage frequent itemsets:

The first step is to find all frequent itemsets.

The steps of the process are as follows:

Scan all transactions to generate candidate 1-itemset C1. Then, according to the minimum support threshold, the item satisfying is selected from C1 (as shown in Table 2), that is, the frequent itemset L1.

Table 2. First iteration result.

1-itemset C1	Support count	Support(%)	frequent itemset L1	Support(%)
{bread}	3	60	{bread}	60
{sheets}	2	40	{sheets}	40
{broom}	2	40	{broom}	40
{meat}	2	40	{meat}	40
{noodles}	2	40	{noodles}	40
{pillowcase}	2	40	{pillowcase}	40
{cabbage}	2	40	{cabbage}	40

By pruning the set generated by L1's own connection, candidate 2-itemset C2 are generated. By scanning all transactions, and satisfying items are selected from C2 according to the minimum support threshold(as shown in Table 3), that is, frequent sets L2.

Table 3. Second iteration result.

2-itemset C2	Support count	Support(%)	frequent itemset L2	Support(%)
{bread, broom}	1	20	{bread, broom}	20
{bread, cabbage}	1	20	{bread, cabbage}	20
{bread, meat}	1	20	{bread, meat}	20
{bread, noodles}	2	40	{bread, noodles}	40
{bread, pillowcase}	0	0		
{broom, cabbage}	0	0		
{broom, meat}	0	0		
{cabbage, meat}	2	40	{cabbage, meat}	40
{cabbage, noodles}	0	0		
{cabbage, pillowcase}	1	20	{cabbage, pillowcase}	20
{meat, noodles}	0	0		
{meat, pillowcase}	1	20	{meat, pillowcase}	20
{noodles, pillowcase}	0	0		

IOP Conf. Series: Journal of Physics: Conf. Series **1284** (2019) 012030 doi:10.1088/1742-6596/1284/1/012030

By pruning the set generated by L2's own connection, candidate 3-itemset C3 are generated. By scanning all transactions, and the satisfied items are selected from C3 according to the minimum support threshold (as shown in Table 4), which means frequent 3-itemset L3.

Table 4. Third iteration result.

3-itemset C3	Support count	Support(%)	frequent itemset L3	Support (%)
{sheets, broom, cabbage}	0	0		
{sheets, broom, meat}	0	0		
{sheets, broom, pillowcase}	1	20	{sheets, broom, pillowcase}	20
{sheets, cabbage, meat}	1	20	{sheets, cabbage, meat}	20
{sheets, cabbage, pillowcase}	1	20	{sheets, cabbage, pillowcase}	20
{sheets, meat, pillowcase}	1	20	{sheets, meat, pillowcase}	20
{bread, broom, cabbage}	0	0		
{bread, broom, meat}	0	0		
{bread, broom, noodles}	1	20	{bread, broom, noodles}	20
{bread, cabbage, meat}	1	20	{bread, cabbage, meat}	20
{bread, cabbage, noodles}	0	0		
{bread, meat, noodles}	0	0		
{broom, noodles, pillowcase}	0	0		
{cabbage, meat, pillowcase}	1	20	{cabbage, meat, pillowcase}	20

Through an iterative operation, frequent k-itemsets Lk are obtained from L(k-1).

Now find frequent itemsets from the transactions listed in Table 1. Set the minimum support threshold to 0.2, and finally get the frequent 4-itemset is:

{cabbage, meat, pillowcase, Sheets}, and its support is 0.2.

The second step is to generate strong association rules from frequent itemsets.

When frequent itemsets are found, strong association rules are generated directly from them:

All non-empty subsets (frequent itemsets) are obtained from each frequent itemset Ln. For each non-empty subset P of Ln, if

$$\frac{\text{support(Ln)}}{\text{support(P)}} \geq \text{min\_conf} \tag{2}$$

Then strong association rules are obtained, where min_conf is the minimum confidence threshold.

For the frequent itemsets obtained by the first iteration {cabbage, meat, pillowcase, Sheets}, made the minimum confidence threshold is 0.8, and strong association rules whose RHS is limited to a single item can be obtained, as shown in Table 5:

Table 5. Strong association rules.

strong association rules	Support (%)	Confidence (%)
{cabbage, meat, pillowcase}⇒{Sheets}	20	100
{cabbage, meat, Sheets}⇒{pillowcase}	20	100
{cabbage, pillowcase, Sheets}⇒{meat}	20	100
{meat, pillowcase, Sheets}⇒{cabbage}	20	100

4. Association Rule Analysis

The association rules which satisfy the minimum support and confidence are mined to form the association rules set [10].

For Association rules $\mathrm{LHS} \Rightarrow \mathrm{RHS}$, LHS and RHS can be itemsets containing multiple items. Generally speaking, association rule mining is demand-driven. Once the user' needs is determined, RHS is usually limited to a single item, while the number of items contained in LHS is uncertain. For example, in order to determine the storage location of meat cabinet, it is necessary to find out the rules and related factors of customers buying meat. At this time, RHS is {meat}, and items of LHS can be one or many. Considering that the association rules with high confidence are often concerned by users, the rules with high confidence are chosen to facilitate users to interpret association rules and analyse the rules of activities.

In most cases, mining and analysis needs are not clear or general. In this case, clustering algorithm can be used to mine and analyse association rules set [11]. Clustering similar rules in the same cluster helps to reduce the size of rules and facilitate users to analyse and understand the meaning of association rules [6].

In this paper, the similarity measure of association rules is defined according to the relationship between the former and the latter. Combining with density-based DBSCAN clustering algorithm, the association rules are clustered, and the similarity rules in clustering cluster are selected to output. Based on the density distribution of the sample, the DBSCAN clustering algorithm continuously checks the connectivity between samples, and finally obtains the clustering result. It can cluster clusters of arbitrary shapes, but the user must select the appropriate distance metric [12].

For a given rule R_1: $X_1 \Rightarrow Y_1$ and $\mathrm{R}_2: X_2 \Rightarrow Y_2$, the distance measure between the two can be defined as:

$$\mathrm{dist}_{\mathrm{R}}(\mathrm{R}_1,\mathrm{R}_2)=\alpha \mathrm{dist}_s(X_1 \cup Y_1, X_2 \cup Y_2)+\beta[\mathrm{dist}_s(X_1,X_2)+\mathrm{dist}_s(Y_1,Y_2)] \tag{3}$$

$$\text{Where, } \alpha=\left|\mathrm{support}_{\mathrm{R}_1}-\mathrm{support}_{\mathrm{R}_2}\right|+0.1$$

$$\beta=\left|\mathrm{confidence}_{\mathrm{R}_1}-\mathrm{confidence}_{\mathrm{R}_2}\right|+0.1$$

$\mathrm{dist}_s(X_1 \cup Y_1, X_2 \cup Y_2)$ represents the Jacquard distance of frequent itemsets corresponding to two rules. The Jaccard Distance measures the distinction between two sets by the proportion of different elements in the two sets to all elements:

$$\mathrm{J}_\delta(A,B)=\frac{|A \cup B|-|A \cap B|}{|A \cup B|} \tag{4}$$

$\mathrm{dist}_s(X_1,X_2)$ represents the Jacquard distance of the former of two association rules, $\mathrm{dist}_s(Y_1,Y_2)$ represents the Jaccard Distance of the latter of two association rules, $\mathrm{support}_{\mathrm{R}_1}$ and $\mathrm{support}_{\mathrm{R}_2}$ are the supports of the two rules. $\mathrm{confidence}_{\mathrm{R}_1}$ and $\mathrm{confidence}_{\mathrm{R}_2}$ are the confidences of the two rules. In order to reflect the difference degree of support between rules, the distance of frequent itemsets corresponding to rules is multiplied by coefficient α. In order to reflect the difference degree of confidence between rules, the distance between the former and the latter of rules is multiplied by coefficient β. Adding 0.1 to α and β is to avoid their values becoming zero. Thus, the distance between the two rules is compared at the same level of support and confidence. By clustering experimental data, the central rules of three kinds of clusters in association rules are obtained (as shown in Table 6).

Table 6. Central rules of association rules clusters.

Category	The former	The later	support	Confidences
1	{sheets, broom, pillowcase}	{meat}	0.3	0.8
2	{cabbage, meat}	{sheets, pillowcase}	0.2	0.9
3	{bread, cabbage, meat}	{noodles}	0.3	1.0

By observing the three association rules in cluster category 2 (as shown in Table 7), we can find that there are strong similarities among them.

Table 7. Association rules with high confidence.

Tid	The former	The later	support	confidences
R1	{sheets, broom, pillowcase}	{meat}	0.3	0.8
R2	{sheets, cabbage, pillowcase}	{meat}	0.2	0.9
R3	{sheets, bread, pillowcase}	{meat}	0.2	0.7

5. Conclusion

In this paper, Apriori algorithm is used to mine association rules from a shopping mall transaction data set. Aiming at the problem that the association rules sets are so big that it's difficult to analyse and interpret, a definition system of rule distance is established. A large number of association rules are clustered by density-based DBSCAN clustering algorithm, and different similar rule clusters are formed to make the classification of association rules. Analytical comprehension becomes easy and effective. The results show the effectiveness of this method.

References:

[1] Cui Y, Bao Z Q 2016 Survey of Association Rule Mining *Application of Research Computers* **33** pp 330-334.

[2] Xu Y, Hu W M and Yang Z F 2016 A Multivariate Partial Least Squares Approach to Joint Association Analysis for Multiple Correlated Traits *The Crop Journal* **4** pp 21-29.

[3] Mao Y X, Chen T B and Shi Bole 2011 Efficient Method for Mining Multiple-Level and Generalized Association Rules *Journal of Software* **22** pp 2965-2980.

[4] He Q 2017 Research and Improvement of Clustering-based Multi-level Association Rules Mining Algorithms *Shanghai Normal University Master of Science*.

[5] Zhang L L, Zhou Q L and Tang G G 2015 Research on Algorithm of Post-processing Association Rules Based on Clustering and Domain Knowledge *Chinese Journal of Management Science* **23** pp 154-161.

[6] Wang Zh, Xun Y L, Zhang J F 2016 An Association Rule Clustering Algorithm Based on K-means *Journal of Taiyuan University of Science and Technology* **37** pp 429-437.

[7] Song Ch X 2016 Research of Association Rule Algorithm based on Data Mining[A]. IEEE Beijing Section.Proceedings of 2016 IEEE International Conference on Big Data Analysis (ICBDA) IEEE Beijing Section pp 4.

[8] Zhao Ch Y 2016 Efficient Association Rule Mining Algorithm Based on User Behavior for Cloud Security Auditing IEEE Computer Society.Proceedings of 2016 IEEE International Conference of Online Analysis and Computing Science(ICOACS 2016) IEEE Computer Society pp 5.

[9] Cheng M C 2016 Research on Audit Log Association Rule Mining Based on Improved Apriori Algorithm IEEE Beijing Section.Proceedings of 2016 IEEE International Conference on Big Data Analysis (ICBDA)[C]. IEEE Beijing Section pp 7.

[10] Liu J Sh, Li J W and Liu L 2019 Ships' Association Rule Minging Based on Improved Interestingness *Ship Electronic Engineering* **39** pp 78-82.

[11] Yan M Q,Guo Zh M and Ren Zh H 2017 Spatio-temporal analysis of bus pickpocketing using association rules based on clustering *Journal of East China Normal University(Natural Science)* pp 145-152.

[12] Yang Y L 2017 The PGST-DBSCAN Algorithm for Mining Clusters from Massive Spatial-temporal Data Wuhan Zhicheng Times Cultural Development Co., Ltd.Proceedings of the 2nd International Conference on Artificial Intelligence and Engineering Applications(AIEA 2017)[C].Wuhan Zhicheng Times Cultural Development Co., Ltd pp 12.

IOP Conf. Series: Journal of Physics: Conf. Series **1284** (2019) 012031 doi:10.1088/1742-6596/1284/1/012031

Exploiting Meta-Path with Attention Mechanism for Fine-Grained User Location Prediction in LBSNs

Zhixiao Wang[1,2,3,4,a], Wenyao Yan[5,b], Wendong Wang[6,c], Ang Gao[7], Lei Yu[1], Shaobo Yang[1] and GaoYang Nie[1]

[1] School of Computer Science and Engineering, Xi'an University of Technology, 710048 Xi'an, China

[2] Telematics Group, the University of Goettingen, 37077 Goettingen, Germany

[3] Shaanxi Key Laboratory of Network Computing and Security, 710048 Xi'an, China

[4] School of Electronic and Information Engineering, Xi'an Jiaotong University, 710048 Xi'an, China

[5] College of Xi'an Innovation, Yan'an University, 710100 Xi'an, China

[6] Yan'an University, Yan'an 716000, China

[7] National Satellite Meteorological Center, China Meteorological Administration, Beijing 100081, China

E-mail: [a]wangzhx@xaut.edu.cn, [b]yadxywy@126.com, [c]yadxwwd@163.com

Abstract. Based on the huge volumes of user check-in data in LBSNs, users' intrinsic mobility patterns can be well explored, which is fundamental for predicting where a user will visit next given his/her historical check-in records. As there are various types of nodes and interactions in LBSNs, they can be treated as Heterogeneous Information Network (HIN) where multiple semantic meta-paths can be extracted. Inspired by the recent success of meta-path context based embedding techniques in HIN, in this paper, we design a deep neural network framework leveraging various meta-path contexts for fine-grained user location prediction. Experimental results based on two real-world LBSN datasets demonstrate the best effectiveness of the proposed approach using various evaluation metrics than others.

1. Introduction

The rapid advancement of smart mobile terminals and communication technologies has witnessed the prosperity of Location-based Social Networks (LBSNs) in recent years. On their heels, the typical LBSN platforms such as Foursquare and Yelp are also attracting thousands of users joining in every day. Through the location-aware services provided by these platforms, a large amount of user data (e.g. comments on certain restaurants and check-in records at some scenic spots.) which is associated with geo-tagged identifiers has been generated. Such data is rich in temporal-spatial and semantic information, which can be exploited for mining user mobility patterns and thus predicting where a user will go for his/her next visit. The application of user location prediction has proved beneficial for intelligent travel route recommendation [1, 2], personalized product advertising [3, 4], to name a few.

Although predicting users' next visit location has been one of hot issues in recent years and lots of embedding frameworks [5–8] have been proposed to learn users' distributed representation and locations in the latent space, these literatures mainly treat LBSNs as the homogeneous networks, which neglects the hidden semantics in interactions between different types of nodes.

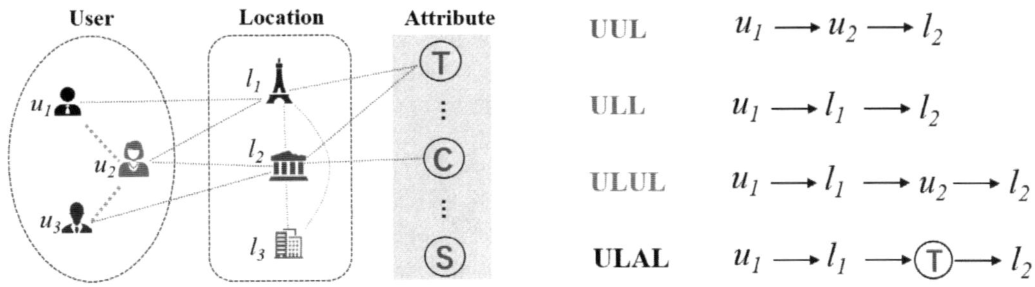

(a) The graphical illustration for LBSNs. (b) An example of the meta-path contexts for $< u_1, l_2 >$.

Figure 1. The schematic illustration of LBSNs and meta-path contexts in a HIN manner. Specifically, in Figure 1(a), there are three types of nodes, i.e. user node, location node and attribute node, and there exist four types of relations, namely, User-User, User-Location, Location-Location and Location-Attribute. In Figure 1(b), we extract four types of meta-paths for the $< u_i, l_j >$ pair, and an example for $< u_1, l_2 >$ pair is given.

Indeed, as noted in [9, 10], LBSNs are heterogeneous networks where different types of nodes and relationships can be observed. Therefore, the heterogeneous information network (HIN) embedding techniques, which are newly emerging learning frameworks with flexibility in characterizing diverse heterogeneous data [11], should be applicable for modeling interactions between user and location nodes in LBSNs.

To illustrate the motivation, we present a visual example in Figure 1. As we can see from Figure 1(a), there are three types of nodes in LBSNs, i.e. user nodes, location nodes and attribute nodes. Specifically, we use the venue category information associated with a venue in LBSNs as the corresponding attributes. LBSNs platforms such as Foursquare divide all Point-of-Interests (POI) into 10 categories. Following this convention, we give each category an abbreviation with one capital letter, where 'A' stands for 'Arts & Entertainment', 'C' stands for 'College & University', 'F' stands for 'Food', etc. Besides, there are four types of relations in LBSNs, i.e. user-user relation which represents social friendships, user-location relation which indicates check-in behavior, location-location relation which reflects successive transition pattern or geographical influence, and location-attribute relation which indicates what specific type a location belongs to. With regard to meta-path contexts, as we focus on the challenge of user location prediction, we only explore the interactions between users and locations, i.e. meta-paths for $< u_i, l_j >$ pairs. Considering that long meta-paths are likely to introduce noise [12], we only select short meta-paths with sequence length up to four, and for a specific user-location pair $< u_i, l_j >$, four types of meta-paths are extracted, i.e. 'UUL', 'ULL', 'ULUL' and 'ULAL'. The detailed example of the meta-path contexts for $< u_1, l_2 >$ pair can be seen in Figure 1(b). As can be seen, we merge four meta-path contexts in an interaction between a user and a location, in other words, we use the triple-tuple $< u_i, meta\text{-}paths, l_j >$ to characterize the contexts why user u_i visits venue l_j. To address user location prediction problem, we conceive a deep neutral network framework incorporating meta-path contexts as well as an attention mechanism.

The remainder of this paper is scheduled as follows. Section 2 briefly reviews related works. Section 3 introduces the deep neutral network framework. In section 4, we evaluate the prediction performance, and finally section 5 concludes our work.

2. Related Works

The location prediction of users has been a tough and excited problem during these years. Quite a lot of previous works deal with this problem based on collaborative filtering (CF) strategies such as matrix factorization [13–15]. The basic objective of matrix factorization methods is to factorize a user-location matrix into two low rank matrices, with each matrix representing the latent representation of users and locations, respectively. Then a user's visit probability at a candidate location is fitted by the inner product of two vectors. However, CF based methods usually suffer from cold-start problem, which means they are generally infeasible for new users or new locations in the training data.

In recent years, embedding techniques, especially Heterogeneous Information Network (HIN) based embedding techniques [11, 12, 16] have drawn much attention from researchers. As it is able to depict semantic relations between different kinds of nodes using meta-paths [17], HIN embedding is becoming increasingly popular in venue recommendation, and thus can be effective for location prediction with minor modification. Hu *et al.* [12] propose to characterize a three-way interaction $< u_i, \textit{meta-paths}, l_j >$ for semantic item recommendation. Particularly, they use a co-attention mechanism to capture the weights for each meta-path context. Inspired by [12], we conceive a deep neutral network framework incorporating meta-path contexts as well as an attention mechanism for user location prediction. Unlike [12], we adopt a quite different random walk method to sample meta-path instances based on predefined rules, and simplify the attention mechanism for meta-path contexts learning. Besides, as we only have positive samples due to implicit feedback in user check-in data, we use a pairwise learning algorithm to maximize the margin between visited and unvisited locations, after which the involved parameters can be optimized.

3. Methodologies

3.1. Preliminaries

In this paper, we strive to predict the next visit location for each user relied on his/her historical check-in records in LBSNs. For a given LBSN dataset, suppose we have M users, i.e. $\mathfrak{U} = \{u_1, u_2, .., u_i, ...u_M\}$, and we have N locations, i.e. $\mathfrak{L} = \{l_1, l_2, .., l_i, ...l_M\}$. As a social network dataset, it also contains social relations (e.g. friendship relation) among some users. A venue in LBSN is defined as a uniquely identified site such as a cinema or a coffee shop, and it is associated with geographical coordinates as well as the category information indicating which type of location it belongs to. Depended on the data, we can construct a user-venue check-in matrix $X \in \mathbb{R}^{M \times N}$, where each entry $x_{u,i} \in \{0, 1\}$ represents whether user u has visited location i. Given the above preliminaries, we can formulate the location prediction problem to be solved in this work.

Definition 1. ***User Next Location Prediction.*** *Given the implicit feedback matrix* $X \in \mathbb{R}^{M \times N}$ *with check-in records of each user* u *being* $C_u = \{< u, l_1, t_1 >, < u, l_2, t_2 >, < u, l_3, t_3 >, ... < u, l_n, t_n >\}$*, where each* $< u, l_i, t_i >$ *denotes user* u *check-in at location* l_i *and time* t_i*. We strive to predict the next venue where user* u *will visit after time* t_n*. To do so, we sort all the* N *feasible locations, thus the precise location that* u *will be visited next is sorted at the supreme position of the ranking list.*

3.2. Our Model

Following the main idea of [12], we also leverage meta-path contexts to model the semantics in a user-location pair $< u_i, l_j >$. Specifically, we extract four meta-paths (see Figure 1(b)) and fuse the meta-path contexts with user and location embeddings through an attention mechanism. Due to the incorporation of such semantic information, the model is expected to deliver more interpretability than previous works. The overall framework of the deep neural network can be seen in Figure 2.

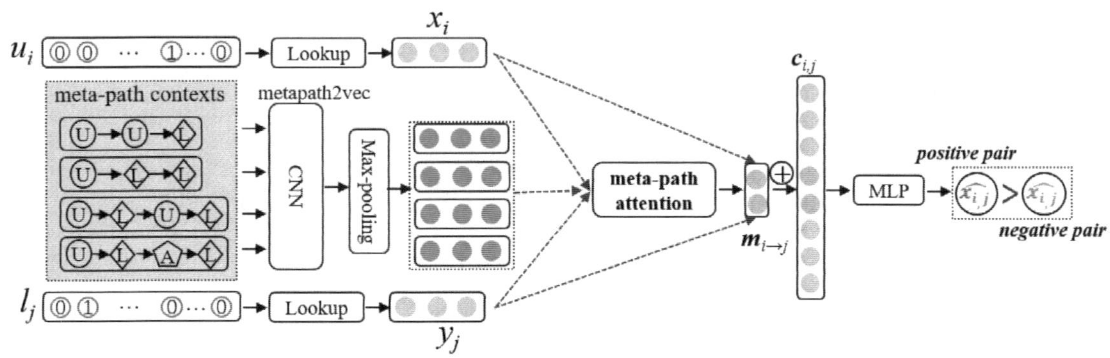

Figure 2. Framework of the deep neural network for user location prediction, where '$\oplus$' means the vector concatenation operation, and $\hat{x}_{i,j}$ represents the approximated real value of user u_i's preference for location l_j. Besides, $m_{i \to j}$ indicates the aggregate meta-path contexts embedding for the input $< u_i, l_j >$ pair.

4. Experiments

4.1. Datasets

We utilize two real world LBSN datasets for experimental evaluation, both of which are crawled from Foursquare. The first Foursquare dataset is provided by Xu *et al.* [18] and it includes 10,901 users who live in New York (NY) with more than 764,328 check-ins in total. The second Foursquare dataset comes from Yin *et al.* [19] and it contains the check-in history of 4,163 users who live in California (CA), USA. Both of the datasets contain explicit undirected friendship relations. To filter the noisy data in both datasets, we eliminate users who have less than 10 check-ins, as well as venues with fewer than 10 visitors. The basic statistics of both datasets and are shown in Table 1. Note that for each user in the datasets, we use his/her first 80% check-ins as the training data, the following 10% check-ins for validation, and the last 10% check-ins for testing. The overall framework is implemented using Keras with TensorFlow as the backend, where all hyper-parameters are set according to [12].

Table 1. Basic statistics of two Foursquare datasets.

Dataset	# Users	# Venus	# Check-ins	# 'U → U'	# 'U → L'	# 'L → L'	# 'L → U'	# 'A → L'
NY	7,678	13,045	983,984	48,917	595,743	170,172,025	595,743	13,045
CA	2,259	3,244	111,817	5,266	68,439	10,523,536	68,439	3,244

4.2. Evaluation Metrics

In this work, we adopt two commonly used evaluation metrics in user location prediction [9], one of which is *Acc@N* and the other one is *APR*. With regard to *Acc@N*, it's often employed to evaluate the recommendation or predication accuracy. Once sorting that location in the top-N list, we will be able to predict the next visit location for each user successfully. Herein, *Acc@N* is calculated by

$$Acc@N = \frac{\#hit@N}{|L|}, \tag{1}$$

where $\#hit@N$ indicates the hits number over the all test set. We take the mean value in the whole test cases as the final result.

With regard to APR, it's usually used to estimate the extent of how one coming check-in location can be precisely sorted. Herein, PR is calculated by

$$PR = \frac{|L| - rank(k)}{|L|}, \tag{2}$$

where $|L|$ is the total number of candidate locations and $rank(k)$ is the rank of the ground-truth location in the list. We again take the mean value in all the prediction instances as the final one, i.e. *Average PR* (APR).

4.3. Comparison Methods
We select two models as comparison, i.e. (1) **PRME** [20], which embeds user and location into the same latent space to capture the user transition patterns. The geographical influence is incorporated through a simple coefficient. (2) **GE** [6], which jointly learns the embeddings of locations, regions as well as venue categories.

4.4. Result Analysis
First of all, we compare the location prediction performance over all methods. To distinguish our model from other comparison models, we denote it as **MP_Loc_Pre**, which is short for *Meta-Path* based *Location Prediction*. Table 2 summarizes the numerical results in terms of *Acc*@10, *Acc*@50 as well as APR. As we can see, for one thing, location prediction models considering meta-path contexts achieve the highest prediction performance at all facets.

Table 2. Location prediction performance comparison among various models, where the bold fonts reveal the best performance for each metric.

	NY			CA		
	Acc@10	*Acc*@50	*APR*	*Acc*@10	*Acc*@50	*APR*
MP_Loc_Pre	**0.283**	**0.411**	**0.773**	**0.248**	**0.347**	**0.830**
PRME	0.226	0.336	0.762	0.174	0.242	0.755
GE	0.257	0.373	0.768	0.220	0.305	0.824

5. Conclusion
In this paper, we investigate the issue of user next location prediction depended on user check-in data in LBSNs. Inspired by the recent success of HIN embedding techniques, we conceive a deep neural network framework incorporating meta-path contexts and a simple attention mechanism. To deal with the implicit feedback, we optimize the model with negative sampling and use a pair-wise training method to maximize the margin between visited and unseen locations. Experimental results verify the effectiveness of the proposed approach as it can achieve comparable and even better performance than PRME and GE.

Acknowledgments
The work is supported in part by the grants which include: Youth Fund of the Ministry of Education of China (No.16YJCZH109), Scientific Research Program Funded by Shaanxi Provincial Education Department (Program No.16JS080), College Innovation Research and Development (No. 2016JXKY-20), YANAN Science and Technology Project (No. 2018KG-02), Xian Science technology (No.2017080CG/RC043(XALG027)), the CSC201608615074, NSFC (No.61772407, No.61771387), Shaanxi Social Science (No.2014M18), Shaanxi Department of Education (No.18JK1216), CERNET Innovation Project (NGII20171202, NGII20170303), Ph.D Innovation (No.112-451115006), College Innovation Research and Development (No.310/252051835).

References

[1] Jiang Shuhui, Qian Xueming, Mei Tao and Fu Yun 2016 Personalized travel sequence recommendation on multi-source big social media *IEEE Transactions on Big Data* **2** 43-56.

[2] Arain Qasim, Memon Hina, Memon Imran, *et al.* 2017 Intelligent travel information platform based on location base services to predict user travel behavior from user-generated GPS traces. *International Journal of Computers and Applications*, **39** 155-68.

[3] Zhu Yu and Li Hao and Liao Yikang, *et al.* 2017 What to do next: modeling user behaviors by time-lstm. *In Proc. of IJCAI* 3602-08.

[4] Meng Xiangwu, Li Ruichang, Zhang Yujie and Ji Weiyu 2018 Survey on mobile recommender systems based on user trajectory data. *Ruan Jian Xue Bao/Journal of Software(in Chinese)*, Chinese Academy of Sciences, **29** 3111-33.

[5] Jarana Manotumruksa, Craig MacDonald and Iadh Ounis 2016 Modelling user preferences using word embeddings for context-aware venue recommendation. *CoRR* http://arxiv.org/abs/1606.07828.

[6] Xie Min, Yin Hongzhi, Xu Fanjiang, *et al.* 2016 Graph-based metric embedding for next poi recommendation. *In Proc. of the International Conference on Web Information Systems Engineering* 207-22.

[7] Xie Min, Yin Hongzhi, Wang Hao, *et al.* 2016 Learning graph-based poi embedding for location-based recommendation. *In Proc. of the 25th ACM International on Conference on Information and Knowledge Management*, 15-24.

[8] Long Yan, Zhao Pengpeng, Sheng Victor S, *et al.* 2017 Social personalized ranking embedding fornext poi recommendation. *In Proc. of the International Conference on Web Information Systems Engineering*, 91-105.

[9] Cao Jiuxin, Xu Shuai, Zhu Xuelin, *et al.* 2018 Effective fine-grained location prediction based on user check-in pattern in lbsns. *Journal of Network and Computer Applications*, Elsevier, **108** 64-75.

[10] Cao Jiuxin, Xu Shuai, Zhu Xuelin, *et al.* 2017 Efficient fine-grained location prediction based on user mobility pattern in lbsns. *In Proc. of the 5th International Conference on Advanced Cloud and Big Data*, 238-43.

[11] Shi Chuan, Binbin Hu, Wayne Xin Zhao and Philip S. Yu 2019 Heterogeneous information network embedding for recommendation. *IEEE Transactions on Knowledge and Data Engineering*, **31** 64-75.

[12] Hu Binbin, Shi Chuan, Wayne Xin Zhao and Philip S. Yu 2018 Leveraging meta-path based context for top-n recommendation with a neural co-attention model. *In Proc. of the 24th ACM SIGKDD international conference on knowledge discovery and data mining*, 1531-40.

[13] Lian Defu, Zhao Cong, Xie Xing, *et al.* 2014 GeoMF: joint geographical modeling and matrix factorization for point-of-interest recommendation. *In Proc. of the 20th ACM SIGKDD international conference on Knowledge discovery and data mining*, 831-40.

[14] Manotumruksa Jarana, Macdonald Craig and Ounis Iadh 2016 Regularising factorised models for venue recommendation using friends and their comments. *In Proc. of the 25th ACM International on Conference on Information and Knowledge Management*, 1981-84.

[15] Zhao Shenglin, Zhao Tong, Yang Haiqin, *et al.* 2016 STELLAR: spatial-temporal latent ranking for successive point-of-interest recommendation. *In Proc. of the 13th AAAI conference on artificial intelligence*, 315-21.

[16] Fu Taoyang, Lee Wang-Chien and Lei Zhen 2017 HIN2Vec: explore meta-paths in heterogeneous information networks for representation learning. *In Proc. of the 26th ACM International on Conference on Information and Knowledge Management*, 1797-806.

[17] Dong Yuxiao, Nitesh V. Chawla and Ananthram Swami 2017 Metapath2Vec: scalable representation learning for heterogeneous networks. *In Proc. of the 23rd ACM SIGKDD international conference on knowledge discovery and data mining*, 135-44.

[18] Xu Shuai, Cao Jiuxin, Zhu Xuelin, *et al.* 2018 Community discovery based on social relations and temporal-spatial topics in lbsns. *In Proc. of the Pacific-Asia Conference on Knowledge Discovery and Data Mining*, 206-17.

[19] Yin Hongzhi, Hu Zhiting, Zhou Xiaofang, *et al.* 2016 Discovering interpretable geo-social communities for user behavior prediction. *In Proc. of the 32nd International Conference on Data Engineering (ICDE)*, 942-53.

[20] Feng Shanshan, Li Xutao, Zeng Yifeng, *et al.* 2015 Personalized ranking metric embedding for next new poi recommendation. *In Proc. of the 24th International Joint Conference on Artificial Intelligence (IJCAI 2015)*, 2069-75.

IOP Conf. Series: Journal of Physics: Conf. Series **1284** (2019) 012032 doi:10.1088/1742-6596/1284/1/012032

SSUP-Growth: A Novel Mining High Utility Algorithm Itemset with Single-Scan of Database

Naji Alhusaini, Li Jing, Ammar Hawbani and Abdullah Alhusaini

University of Science and Technology of China Hefei, China
School of Management, Hefei University of Technology Hefei, China

E-mail: naji@mail.ustc.edu.cn

Abstract. High Utility Itemset Mining (HUIM) alludes to the identification of itemsets of high utility in the value-based database UP-Growth algorithm is a standout amongst the best algorithms for overcome the challenge of candidate generation and scan database reputedly of previous algorithms. However, it needs scan database twice to actualize the UP tree. Regarding of the updating existing data with new information, UP-growth needs for twofold scanning of new information and existing information. The fundamental motivation behind this work is to build up another algorithm, Single-Scan Utility Pattern Tree (SSUP-tree), for mining high utility itemsets from transaction database through only single-scan of database. In our algorithm, the details of high-utility itemsets is preserved in a particular data structure of the SSUP-Tree after a single-scan of database. Consequently, it can retrieve the identical UP-tree with a fixed minimum utility threshold. The proposed algorithm required to scan the new data only to update SSUP-tree. In this regard, in order to estimate the execution of the proposed algorithm, the SSUP-tree algorithm has been implemented on synthetic and real datasets. The results of this study revealed that SSUP-tree shows a significant enhancement in the execution in terms of runtime since it keeps the huge databases details in a compact format and it avoids repetition of database scanning.

1. Introduction

A framework of HUIM has emerged to address the limitation of Frequent Itemset Mining (FIM), which only considers the frequency of the items without regard to the utility of items such as profit and quantity. HUIM has many scientific and commercial applications and is one of the topics of recent active research [1] [2] [3] [4] [5]. The importance of HUIM framework is to provide the decision makers with greater flexibility to reflect the utility of items, such as profit and quantity. This allows more interesting and feasible patterns to be generated In HUIM framework, each item has two values to determine its utility, external utility which represents the significance of the items that given separately by the users and the internal utility which represents the significance of the item in the transaction. Several methods have been applied in previous research [2, 4, 5, 6, 8] to enhance the performance of utility mining. Arguably, Regarding the method used, the HUIM framework approaches can be further classified as follows: level-wise approaches [1, 6, 7], Tree-based (pattern-growth) approaches [8, 9, 10, 11, 2, 12], Projection-based approaches [4, 13], and utility list based methods [3, 14, 5, 15].

Tree-based methods such as Utility Pattern Growth (UP-Growth) [11] seeks to overcome dilemma of the candidates generation. Nonetheless, this algorithm still needs to scan the database double time. Additionally, while the database has been updated, UP-Growth [11] needs to re-implement the whole process for new and old database together once more to update the rules. In other words, it ignores the past double scan of database and start the whole process again. In this paper, we propose a new method, SSUP-Growth, to generate High Utility Itemset with only single-scan of database. In case of updating the database, our proposed method needs a single-scan of the new data only.

Based on the literature aforementioned and the best knowledge of the authors, UP-growth approach is one of the best approaches that avoids the costly candidates generation. However, it is still needs to scan database double time. Furthermore, it handles the update of the database by rescan the old and the new data double time again. This motives the current work to create a novel algorithm that needs single-scan of database and handles the update of database by single-scan of the new data only.

The rest of this paper is structured in such a way: in Section 2, we present utility itemset mining related work. In Section 3, we explain the relative definitions and the problem statement. The proposed algorithm and the data structure are comprehensively introduced in Section 4. Finally, section 5 concludes this work.

2. Preliminaries and Problem Statement

2.1. Problem Definition

Let I= $\{y_1, y_2, ..., y_m\}$ represent a limited set of distinct items, and D= $\{T_1, T_2, ..., T_n\}$ depict a set of transactions, where each transaction $T_i(1 \leq i \leq n$ portrays a subset of I and has a distinctive identifier i known as Tid. Any subset of I includes K items $\{y_1, y_2, ..., y_k\}$ where $y_j \in I, 1 \leq j \leq k$ is named as the k-itemset. Table 1 provides an example transaction database D, and Table 2 lists the items of external profit details.

Table 1. An example database

TID	Transaction	TU
1	(A,1) (C,1) (D,1)	8
2	(A,2)(C,6)(E,2)(G,5)	27
3	(A,1)(B,2)(C,1)(D,6)(E,1)(F,5)	30
4	(B,4) (C,3) (D,3)(E,1)	20
5	(B,2) (C,2) (E,1) (G,2)	11

Table 2. Items Profit table

Item	A	B	C	D	E	F	G
Profit	5	2	1	2	3	1	1

Definition 1. Item internal utility refers to the significance of item y_j in the transaction T_i such as quantity provided in the transaction and denoted as $Q(y_j, T_i)$. For example in Table 2, $Q(A, T_2) = 2$.

Definition 2. Item external utility presents the prominence of item y_i provided in the utility table, such as profit. For example, in Table 2, $P(A) = 5$.

Definition 3. The utility of an item $y_j \in T_i$ remains distinct as$U(y_j, T_i$ is calculated as the product of the internal and external profit of items in the transaction, $U(y_j, T_i) = P(y_j) * Q(y_j, T_i)$

For example, in Table 1, $U(D, T_1) = 2 * 1 = 2$.

Definition 4. The utility of an itemset Y in T_i is denoted by $U(Y, T_d)$ and defined by, $\sum_{Y_i \in Y \wedge Y \subseteq T_i} U(Y_i, T_i)$

For example, in Table 1, $U(AC, T_1) = U(A, T_1) + U(C, T_2) = 5 + 1 = 6$.

Definition 5. The utility of the itemset Y in D is signified by $U(Y)$ and defined by $\sum_{T_i \in D} \sum_{y_i \in Y \wedge Y \subseteq T_i} U(y_i, T_i)$

For example, in Table 1, $U(AC) = U(AC, T_1) + U(AC, T_2) + U(AC, T_3) = 6 + 16 + 7 = 29$.

Definition 6. The user-specified minimum utility percentage value is denoted by β while the absolute minimum utility value, signified by (min_util), is calculated as $min\_util = \beta * \sum_{T_i \in D} TU(T_i)$

For the illustrative example in Table 1, when $\beta = 20\%$, the absolute minimum utility value is $min\_util = 0.20 * 71 = 42$.

Definition 7. The HUI depicts the itemset with the utility value greater than or equal to the user-defined minimum utility threshold (min_util). Otherwise, the itemset reveals an LUI.

Definition 8. The transaction utility of a transaction T_i is portrayed by $TU(T_i)$ and specifies the total utility of all items it contains. It is defined by, $TU(T_i) = \sum_{Y \subseteq T_i \wedge y_i \in Y} U(y_i, T_i)$

For example,in Table 1, $TU(T_5) = U(B, T_5) + U(C, T_5) + U(E, T_5) = 4 + 2 + 3 = 9$.

Definition 9. The transaction-weighted utilization of the itemset Y represents the sum of the transaction utilities of all the transactions including Y, indicated as $TWU(Y)$ and displays as,$TWU(Y) = \sum_{Y \subseteq T_i \wedge T_i \in D} TU(T_i)$

For example, $TWU(AC) = TU(T_1) + TU(T_2) + TU(T_3) = 8 + 27 + 30 = 65$.

If $TWU(Y)$ is not smaller than the minimum utility threshold, Y is considered as a high transaction-weighted utilization itemset (HTWUI).

Definition 10. The transaction-weighted downward closure, ($TWDC$ for short), is declared as follows: For any itemset Y, if Y does not equal an HTWUI, any superset of Y is not HTWUI, too. By this definition, we preserve the downward closure by employing transaction-weighted utilization. For example, in Table 1, since $TWU(AC) > min\_util$, any superset of AC depicts a high utility itemset.

3. Proposed Approach

3.1. The Components of SSUP-Tree

In SSUP-Tree, the root node is labelled as Null. Each other node N contains five components, N.parent, N.name, N.nu, N.count, and a set of children nodes, where N.parent designates the parent of N, N.name denotes the original item in the transaction database, N.nu refers to the utility value, and N.count represents the support count. In SSUP-Tree, each path represents one or more transaction(s), and its corresponding utility(s) is considered as the utility of its least high utility item(s). The utility of a given node is greater than the utility of its children or descendants.

A prefix in SSUP-tree represents the common identical patterns appeared in several transactions.

Instead of the Header Table in UP-Growth, in our proposed method there is an Item Utility List (IUL) to record the actual utility value for each item. IUL contains the names of items and their estimated utility values. Moreover, there is a High Utility Item List (HUIL), which includes the items with high utility only.

3.2. Building the SSUP-Tree

Our proposed approach constructs a compressed tree structure by taking advantage of the best usage of the appearance of the common prefixes in the transactions and the similarity of the transactions. As mention above, the primary SSUP-tree contains all of the items occur in the database. The SSUP-tree is built in two steps (see Algorithm 1):

- In the initial (and only) scan, the transactions will be inserted based on predefined item order (such as lexicographical order) one by one into the empty SSUP-Tree. Meanwhile, recording the actual utility value TWU of each item into the IUL.
- IUL is ascendingly sorted according to the items utility. Next, the nodes in each path are descendingly resorted according to their utility values recorded in IUL. Afterward, the sorted paths are inserted into a new tree (NSSUP-tree).

3.3. Achieving the UP-tree from the SSUP-tree
Based on the characteristics of the SSUP-tree, it is worth to mention that the SSUP-tree is included the high and low utility items. Thus, the UP-tree is a sub-tree of the SSUP-tree with a fixed (min_util) threshold, which involves the high utility items that meet the (min_util) threshold. In this regard, the mechanism of the UP-tree generation from SSUP-tree is proposed and analyzed in this section.

After constructing the primary SSUP-tree, the High Utility Items list (HUIL) is simply obtained according to a given specific (min_util). All we need to do is discard the unpromised (low utility) items from IUL. Afterward, we examine the utility of each node in the paths, from the root until the leaves, in order to trim the primary SSUP-tree by eliminating the unpromised nodes. Since the children or its descendant have less utilities than their parents, the unpromised nodes will be eliminated with its subtree.

3.4. Update the SSUP tree with new data
Since SSUP-tree contains all the items in the database. Therefore, SSUP tree can be updated by scanning new data without additionally scanning old data. The update process runs in two steps. First, the new transactions are inserted into the primary SSUP-tree based on IUL, and meanwhile the IUL is updated. Second, the updated SSUP-tree is rebuilt based on the updated IUL. In the case of a new transaction containing new items, which do not appear in the existing database, it can be handled as a new branch of the main SSUP tree with its actual utility values.

4. Experimental Evaluation
4.1. Settings
This section shows our experiments results after examining the execution of our algorithm and comparing it to the UP-Growth algorithm [11]. The experiments have been executed on a 2.66 GHz Intel Core 4 Quad Processor with 4G memory, on operation system Windows 7. The code of the algorithms are developed in Java. Furthermore, the algorithms performance is estimated by utilizing synthetic and real datasets. Moreover, T10I6D100K and chess datasets published in [16] have been considered in this work. Both of the datasets used in this study already contained unit profits and purchase quantities. Table 3 shows the properties of the datasets.

Table 3. The properties of the datasets.

Dataset	Items	Avg. Len	Transaction
T10I6D100K	1,000	10	100,000
Chess	76	37	3,196

4.2. Evaluation on Synthetic and Real Datasets
First, we use the synthetic dataset T10I6D100K to show the evaluation results. We employed five minimum utility thresholds, 40%, 50%, 60%, 70%, and 80%. The evaluation results of the phase I (i.e., tree construction) on the dataset T10I6D100K are plotted in Figure 1 the results showed that our algorithm takes less execution time as it needs single-scan of dataset. On the other hand, the execution time of phase II (i.e., get the corresponding UP-tree) is shown in Figure 2. At the beginning, both algorithms generate the corresponding UP-tree for the first minimum utility threshold (40%). After that, when the minimum utility threshold increases from 40% to 50% or 70%, the corresponding UP-tree can easily obtained through trimming the UP-tree with the minimum utility threshold of 40% with no need to scan the dataset again. However, when the minimum utility threshold has diminished, UP-Growth algorithm ignored the past double

scan of dataset and re-scanned dataset double time again that requires more time. As shown in Figure 2, when the (*min_util*) threshold decreased from 80% to 70%, UP-Growth spends more time to get the corresponding UP-tree. On the other hand, our algorithm gets the corresponding UP-tree by trimming the primary SSUP-tree, regardless how the minimum utility thresholds are modified, without having to re-scan the dataset. Second, we evaluated the execution time of UP-growth and SSUP-Tree for Phase I and Phase II on the chess dataset. Figure 3 and Figure 4 show the runtime of both algorithms for Phase I and Phase II on the dataset chess using different minimum utility thresholds. It is clear that our proposed algorithm spends less time in each phase for the same reason explain in the first experiment.

4.3. Scalability

In this subsection, we provide experiments of evaluate scalability of our proposed algorithm. The dataset size of the T10I6 dataset has been varied to evaluate the scalability of UP-Tree and SSUP-tree algorithms. Figure 5 and 6 show the performance of both methods on different size of dataset and with (*min_util*) is 85%. As shown in Figure 5 and 6, the execution time of the SSUP-Tree is less than the UP-tree. In case of the database is size increased, the execution time for distinguish high-utility itemsets also increases. Consequently, the UP-Growth algorithm requires more processing time than the SSUP-tree.

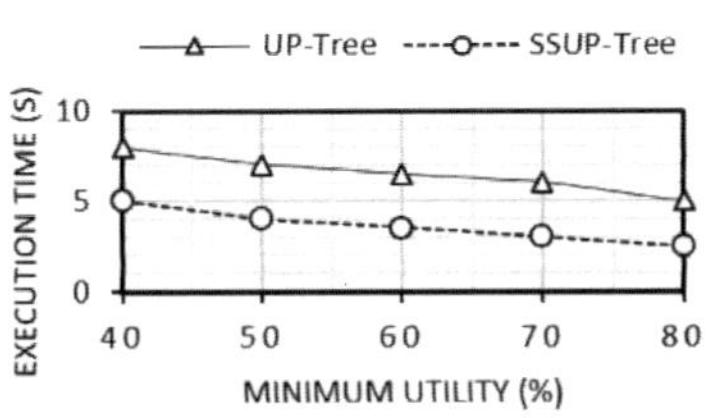

Figure 1: Execution time for Phase I on T10I6D100K.

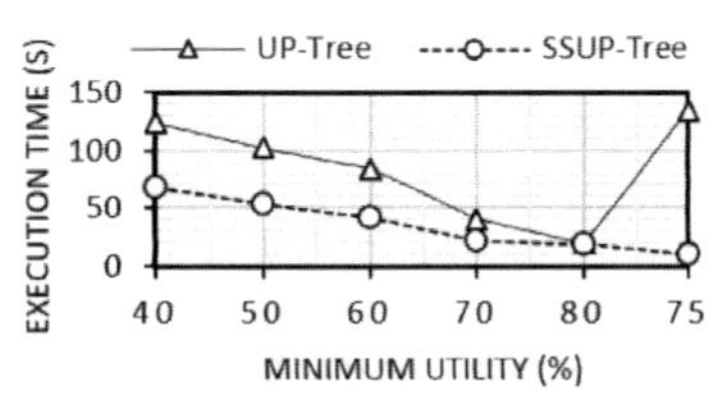

Figure 2: Execution time for Phase II on T10I6D100K.

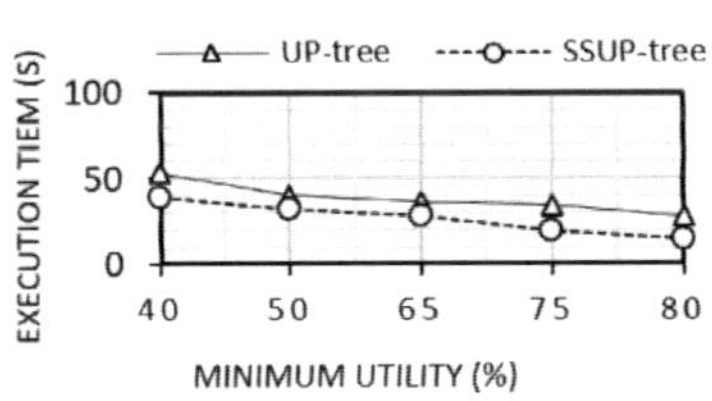

Figure 3: Execution time for Phase I on Chess.

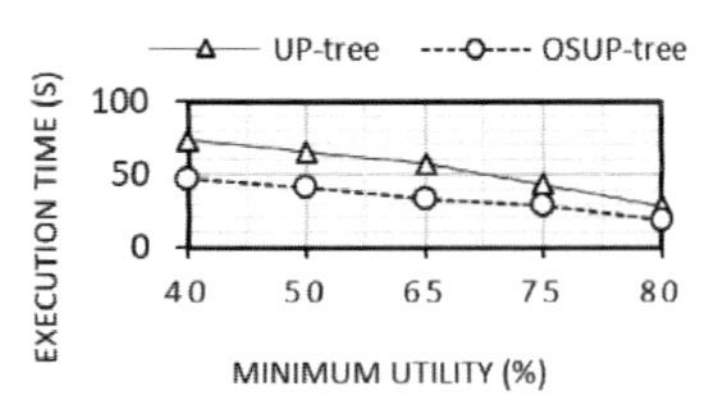

Figure 4: Execution time for Phase II on Chess.

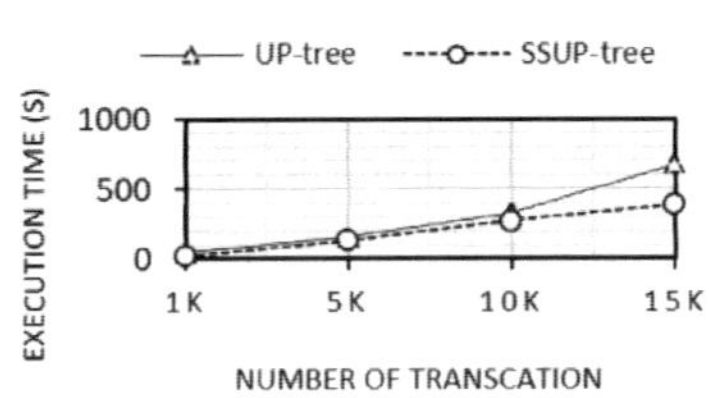

Figure 5: Scalabilityś runtime for Phase II.

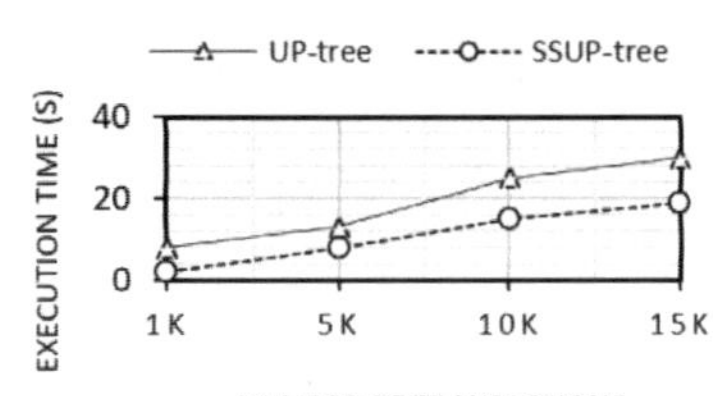

Figure 6: Scalabilityś runtime for Phase I.

5. Conclusions

In this paper, we present an algorithm called SSUP-tree for mining efficient itemsets from transactional databases. In addition, we discussed how the SSUP-tree is obtained by a single scan of the database and how it is updated by a single scan of the new data. In order to evaluate the runtime of the proposed algorithm, SSUP-tree algorithm and UP-growth implemented on real data sets and synthetic datasets. The experimental results show that our proposed algorithm, SSUP-tree, continuously exceeds the UP-tree algorithm in the term of runtime because SSUP-tree preserves the information of items in a compact structure and avoids scanning the database again, especially when the database contains a large number of transactions with shared prefixes. However, the limitation of the proposed method is that it based on a single minimum threshold, which has an adverse effect on the mining results. Therefore, the future work is to improve this approach to mine high utility itemsets based on multiple minimum thresholds.

Acknowledgement

This paper is supported by the "Strategic Priority Research Program of Chinese Academy of Sciences (A Class) NO. XDA19020102 ".

References

[1] Ying Liu, Wei-keng Liao, and Alok Choudhary. A two-phase algorithm for fast discovery of high utility itemsets. In *Pacific-Asia Conference on Knowledge Discovery and Data Mining*, pages 689–695. Springer, 2005.

[2] Vincent S Tseng, Bai-En Shie, Cheng-Wei Wu, and S Yu Philip. Efficient algorithms for mining high utility itemsets from transactional databases. *IEEE transactions on knowledge and data engineering*, 25(8):1772–1786, 2013.

[3] Philippe Fournier-Viger, Cheng-Wei Wu, Souleymane Zida, and Vincent S Tseng. Fhm: Faster high-utility itemset mining using estimated utility co-occurrence pruning. In *International symposium on methodologies for intelligent systems*, pages 83–92. Springer, 2014.

[4] Guo-Cheng Lan, Tzung-Pei Hong, and Vincent S Tseng. An efficient projection-based indexing approach for mining high utility itemsets. *Knowledge and information systems*, 38(1):85–107, 2014.

[5] Souleymane Zida, Philippe Fournier-Viger, Jerry Chun-Wei Lin, Cheng-Wei Wu, and Vincent S Tseng. Efim: a fast and memory efficient algorithm for high-utility itemset mining. *Knowledge and Information Systems*, 51(2):595–625, 2017.

[6] Hong Yao, Howard J Hamilton, and Cory J Butz. A foundational approach to mining itemset utilities from databases. In *Proceedings of the 2004 SIAM International Conference on Data Mining*, pages 482–486. SIAM, 2004.

[7] Yu-Chiang Li, Jieh-Shan Yeh, and Chin-Chen Chang. Isolated items discarding strategy for discovering high utility itemsets. *Data & Knowledge Engineering*, 64(1):198–217, 2008.

[8] Chowdhury Farhan Ahmed, Syed Khairuzzaman Tanbeer, Byeong-Soo Jeong, and Young-Koo Lee. Efficient tree structures for high utility pattern mining in incremental databases. *IEEE Transactions on Knowledge and Data Engineering*, 21(12):1708–1721, 2009.

[9] Jianying Hu and Aleksandra Mojsilovic. High-utility pattern mining: A method for discovery of high-utility item sets. *Pattern Recognition*, 40(11):3317–3324, 2007.

[10] Chowdhury Farhan Ahmed, Syed Khairuzzaman Tanbeer, Byeong-Soo Jeong, and Young-Koo Lee. Huc-prune: an efficient candidate pruning technique to mine high utility patterns. *Applied Intelligence*, 34(2):181–198, 2011.

[11] Vincent S Tseng, Cheng-Wei Wu, Bai-En Shie, and Philip S Yu. Up-growth: an efficient algorithm for high utility itemset mining. In *Proceedings of the 16th ACM SIGKDD international conference on Knowledge discovery and data mining*, pages 253–262. ACM, 2010.

[12] Wei Song, Yu Liu, and Jinhong Li. Mining high utility itemsets by dynamically pruning the tree structure. *Applied intelligence*, 40(1):29–43, 2014.

[13] Alva Erwin, Raj P Gopalan, and NR Achuthan. Efficient mining of high utility itemsets from large datasets. In *Pacific-Asia Conference on Knowledge Discovery and Data Mining*, pages 554–561. Springer, 2008.

[14] Junqiang Liu, Ke Wang, and Benjamin CM Fung. Direct discovery of high utility itemsets without candidate generation. In *2012 IEEE 12th international conference on data mining*, pages 984–989. IEEE, 2012.

[15] Heungmo Ryang and Unil Yun. Indexed list-based high utility pattern mining with utility upper-bound reduction and pattern combination techniques. *Knowledge and Information Systems*, 51(2):627–659, 2017.

[16] P Fournier-Viger, A Gomariz, A Soltani, and T Gueniche. Spmf: open-source data mining platform, 2013.

Appendix: Algorithms Pseudocode

```
ALGORITHM 1: SSUP-Tree Generation
Input: (1) A transaction database DB (2) A minimum utility threshold min_util
Output: (1) A Single-Scan Utility Pattern tree (SSUP-tree (2) Item utility list (IUL)
Procedure 1: Generate SSUP-tree and IUL
    1. Create SSUP- tree with "null" root
    2. Create empty Item Utility List IUL
    3. FOREACH transaction T_i in the transaction database DB
        1.Insert ([t |T_i], SSUP) t is refers to the first item in T_i
        2.Update IUL with items in [t|T_i]
    ENDFOR
Insert ([t |T_i], SSUP) function executes as follows.
  FOREACH SSUP's child node N
     IF T_i has child node N such that t.Name = N.Name
       THEN
         N. utility -N.utility +t.utility
         IF T_i is non empty
           THEN Insert (T_i , N)
         ENDIF
         RETURN
     ELSE
      Generate a new Node N^2
      N^2.Name← t.Name
      N^2.utility ← t.utility
      SSUP.childList← SSUP.childlist + N^2
      IF T_i is not empty
          THEN Insert (T_i, N^2)
      ENDIF
     ENDIF
  ENDFOR
  RETURN
Procedure 2: Restructure the initial SSUP-tree
Create new SSUP with "null" root not of tree
FOREACH path p_i from the root until a leaf in the primary SSUP-tree SSUP
   IF p_i is not empty do:
      1. Sort p_i according to IUL
      2. Call function (p_i^n, NSSUP) to insert the sorted p_i^n into the new SSUP-tree.
   ENDIF
 ENDF
```

```
ALGORITHM 2: Achieve UP-tree from the SSUP-tree
Input: AN SSUP-tree, the item utility list, min_util
Output: An UP-tree
1. Create High Item Utility List (HUIL) set as empty
2. FOREACH item x in IUL
    IF x.utility in≥ ε
        Insert x to HUIL
    ENDIF
   ENDFOR
3. Sort HUIL in utility descending order
4. Call Test (P). The description of function test as follows.
Function Test (N)
        FOREACH child node c_n of the node N
            IF c_n ∈ HUIL
            THEN
                Test (c_n)
            ELSE
                Delete c_n (and the potential subtree beginning from c_n)
            ENDIF
        Endfor
        Return
```

```
ALGORITHM 3: Update SSUP-Tree
Input: (1) A new database DB (2) Primary SSUP-Tree (3) Item Utility List IUL
Output: (1) An Updated One Scan Utility Pattern tree (NSSUP-Tree) (2) An Updated Item utility list IUL
Procedure 1: Update the primary SSUP-tree with new data as well as update IUL
    FOREACH transaction T_i in DB
        1. Insert ([t |T_i], SSUP)
        2. Update IUL with items in [t |T_i]
    ENDFOR
    Sort IUL in descending order according to utility values.
Procedure 1: reconstruct the updated SSUP-tree depend on updated IUL
Set: New SSUP-tree
  FOREACH path P_i in primary SSUP-tree
    IF P_i is not empty
        1. Sort P_i according to IUL
        2. Insert (P_i, New SSUP-tree )
    ENDIF
  ENDFOR
  RETURN
```

IOP Conf. Series: Journal of Physics: Conf. Series **1284** (2019) 012033 doi:10.1088/1742-6596/1284/1/012033

Sensor Based Affective Learning Information Process System for Intelligent Tutoring System

Kwang Sik Chung[1]

[1] Dept. Of Computer Science, Korea National Open University, Jongno-gu, Dongsung-Dong, Seoul, Republic of Korea

E-mail: kchung0825@knou.ac.kr

Abstract. As smartphones like televisions or cars are used every day and everybody, smartphones are used to access to digital content and various ICT services. Especially, smartphone is equipped with many types of sensors so that various multimedia content can use smartphone sensors. If users' information through sensors would be analysed, users' movement, emotion, and states can be inferred. Especially, on e-learning environment, students' learning states can be decided by students' information that are collected by smartphone sensors, since more than 90% of students of Korea National Open University have access to learning contents through their smartphones in mobile learning environment. However, previous researches focused on interactions between students and learning contents. And there are few methods, technology, and decision models for students to track their learning activity, learning interests, concentration and emotions. In this paper, we propose a learning reaction analysis state model and Student Activity Analysis System. In this research, sensed learning activity information is collected by sensors of smartphone. Students' learning activity information can be classified into mandatory and optional information. In Student Activity Analysis System, students' learning emotional state is decided by learning reaction analysis state model. The students' learning emotional state could be used for intelligent tutoring system (ITS) that construct personal learning strategy. Since students' learning activity information is collected and analyzed from the information technology viewpoint, it is possible to measure the concentration and interest of learning contents of students regardless of pedagogical model.

1. Introduction

In the offline classroom, the instructor can evaluate the student's affective emotional state with the higher reliability based on facial expression, gesture language, and posture. Skilled instructors can assess the student's affective emotional state by observing the characteristics of a student, including the student's knowledge level as well as the student's emotional state. IOT(internet of things) technology has been developed and various ICT services based on it have been provided. In particular, various information and communication services using sensors have provided digital contents that are more user friendly and meet user requirements than existing information and communication services. These technological advances have been focused on the development of the intelligent tutoring system (ITS), and the pursuit of learning contents of student in the field of e-learning and application of learning contents in accordance with the learning condition [1]. ITS has been studied to evaluate students' emotional state in terms of educational psychology and to develop learning strategies that can improve learning outcomes and learning outcomes using this information. However, the previous method of analyzing the students' emotional state is mainly following the educational psychological approach.

These learning emotional state analysis methods have not been studied on the basis of sensors, or have limited use of various sensors.

In this paper, sensor based student's emotional state for learning activity is proposed. The student's emotional state analysis system is built based on smartphone and his sensors. The proposed student's emotional state analysis system would decide the value of student's learning concentrativeness, learning interest, effectiveness. For the value of student's learning concentrativeness, learning interest, and effectiveness, all of student's interaction with smartphone are collected and analyzed by the proposed Sensor based Affective Learning Information Process System.

This paper consists of 5 chapters. We reviewed previous student's emotion state analysis research for e-learning learning and we show that proposed Sensor based Affective Learning Information Process System has several differences with them. The proposed Sensor based Affective Learning Information Process System and affective learning activity information process model are showed in chapter 3. And the Affective Learning Information Process Engine architecture, system messages flows and function modules of Affective Learning Information Process Engine of Sensor based Affective Learning Information Process System are described in chapter 4. We concluded the research paper in chapter 5.

2. Related Works

Previous ITS(intelligent tutoring system) was built based on affective pedagogical model. Thus some of them did not use sensors. And some of them used only 1~3 sensors of external equipment. Cognitive Tutor Algebra [2], CRYSTAL ISLAND [3], Inq-ITS [4], INES [5], MathSpring [6], WaLLis [7] do not use sensors and analyze log files and features of students that are recorded during learning activity and are related to the student's learning behavior, and learning activity history is also analyzed in the learning process. They analyze student's affective state basically based on pedagogical model. Prime Climb [8], Easy with Eve[9], FERMAT [10] and PAT2Math [11] used external equipped video camera so that facial feature, posture and eye pattern, eye tracking and gaze pattern are collected and analyzed. Prime Climb [8] used physical sensors so that skin conductivity, heart rate, muscle activity are collected and learning activity log files are analyzed. Most of the previous works are related with affective pedagogical model for learning activity. Thus student's emotion model is simply used in order to analyze and infer student's emotion state with only one or only two sensors. This research proposes various sensors with smartphone and collects many kinds of student's learning activity information through various sensors of smartphones and propose student's emotional state decision model. Various sensed student's activity information would be analyzed. With these sensed learning activity information, more detailed analyzed student's affective state could be decided by ITS.

3. Sensor based Affective Learning Activity Information Process Model

Proposed Sensor based Affective Learning Activity Information Process Model(SALAI model) collects various information about a student who interacts with learning contents trough smartphone. Some student's personal information are closely related with student's learning situations on smartphone as learning device. But some student's personal information is closely related with student's situations or environment or smartphone situations or environment. With those kind of information, SALAI model decides student's affective learning emotion state. The proposed SALAI model decision strategy comes from Automated Tutoring Engine that was proposed in [12]. Student's learning activity patterns, Student's personal environment state information, student's learning emotional state information and students' state decision conditions are defined in [13] and redefined.

Basically, SALAI model considers students' personal environment state, student's learning emotional state and student's learning activity patterns within smartphone that was proposed in [12,13]. Student's learning emotional state is defined as student's emotional state against learning contents and environment that student could feel as like boring, fun, difficult, etc.. Students' personal environment state is defined as students' environment information as like noisy environment, walking on the street, alone in a room, etc.. Student's learning activity patterns are defined as students' learning activity levels as like active, unconcerned, interested, etc.. According to the SALAI model, student's affective learning emotion state is decided by Affective Learning Information Process Engine. Affective Learning Information Process Engine analyzes students' personal environment state information and student's environment information and decides student's affective learning emotion state. Lastly, Affective Learning Information Process Engine sends student's affective learning emotion state to ITS. According to student's affective learning emotion state, ITS construct personal learning strategy.

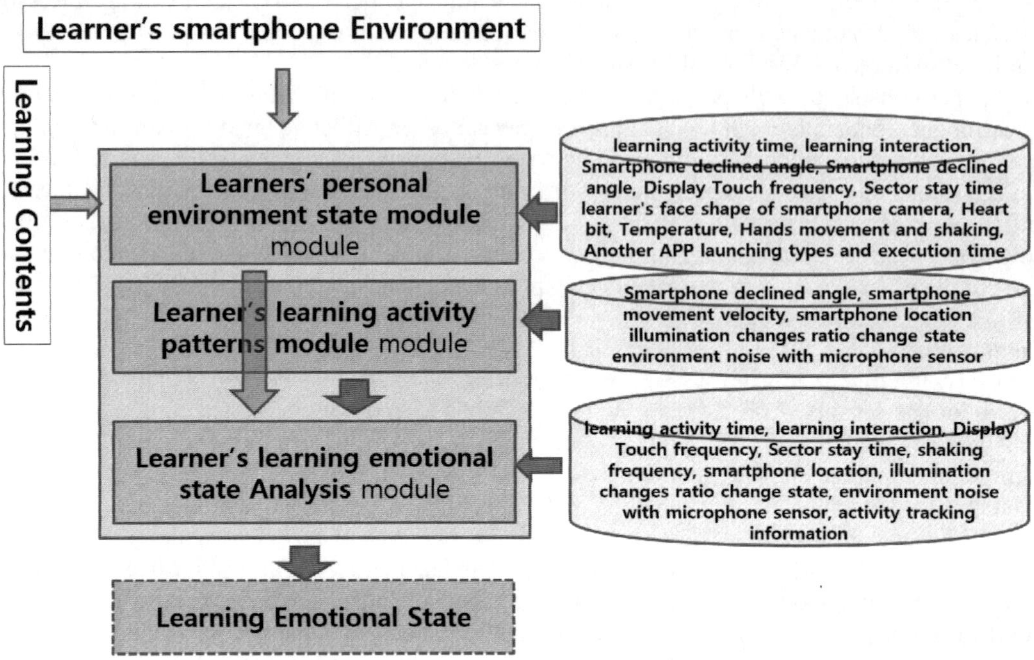

Figure 1. Architecture of Affective Learning Information Process Engine

Proposed Affective Learning Reaction Analysis Engine that is shown in figure 1, consists of Learning Reaction Analysis Engine, Students' Emotional State Model, Students' state database, Bio-Signals DB and Learning Activity database [12, 13].

- student and educational devices : source of students' learning activities and emotional information with learning agents
- student's affective learning emotion state : collection of students' learning emotional state information, and infer and decision of students' learning contents
- Learning activity : activity level of learning time
- Students' Learning Emotional State Model: model of learning interests, learning contents reaction and students' environment
- Bio-signals : student's bio-signals for learning contents

Learning emotional state means learning interests or learning contents intimacy that mean closeness of learning contents and one student. If learning emotional state level is active, it means that a student is well-matched with learning contents, regardless of difficulty, formats and type of learning contents.

4. System Message Flow of Affective Learning Information Process Engine

We assume that students' smartphone device has various sensors and seamless communication network that can collect and detect bio-signals. We also assume that there is no middleware to convert and manage noise of bio-signals.

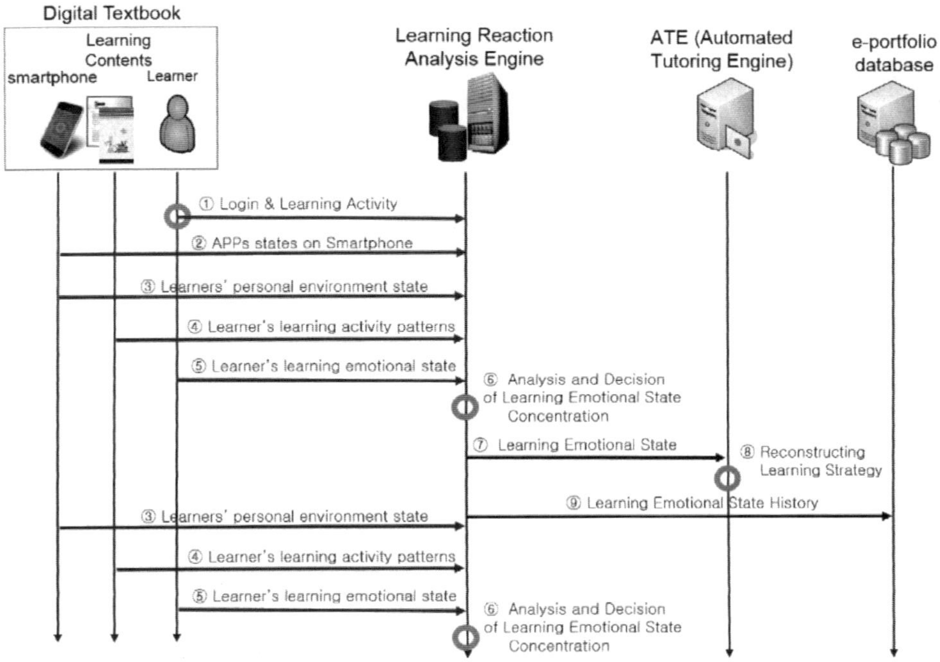

Figure 2. Affective Learning Information Process Engine Message Flow

① A student logs on a smart learning portal server (web server) by his/her learning devices as like as a smartphone. At the same time, a student logs on Learning Management Server (LMS) with single-sign-on protocol. In figure 2, a smart learning portal server and LMS are omitted for simple service message flow. And LMS delivers students' previous learning emotional state history, previous Students' personal environment state history and Student's learning activity patterns history.

② Digital textbook Application on smartphone sends App state and student's user information to ALIPE. ALIPE analyzes student's user information (smartphone specification information, etc.) and previous learning emotional state history and decides initiative learning emotional state of the student.

③ After delivering learning subject contents, learning personal environment state is sent to ALIPE that infer current emotional state of the student. The emotional state information is periodically sent to ALIPE.

④ Learning activity pattern information is sent to ALIPE that infer current emotional state of the student. Learning activity pattern information is periodically sent to ALIPE.

⑤ Student's learning emotional state information is sent to ALIPE that infer current emotional state of the student. Student's learning emotional state information is periodically sent to ALIPE.

⑥ ALIPE analyzes Student's learning personal environment state information and Learning activity pattern information and saves the results in the support learning emotional state information(e-portfolio). ALIPE analyzes learning emotional state information with the support learning emotional state information, and decides student's affective learning emotion state.

⑦ and ⑧ ALIPE sends student's affective learning emotion state to Intelligent Tutoring Engine(System) [13] (ITE). ITE decides and reconstructs personalized learning strategy with student's affective learning emotion state.

⑨ ALIPE sends student's affective learning emotion state to e-portfolio database.

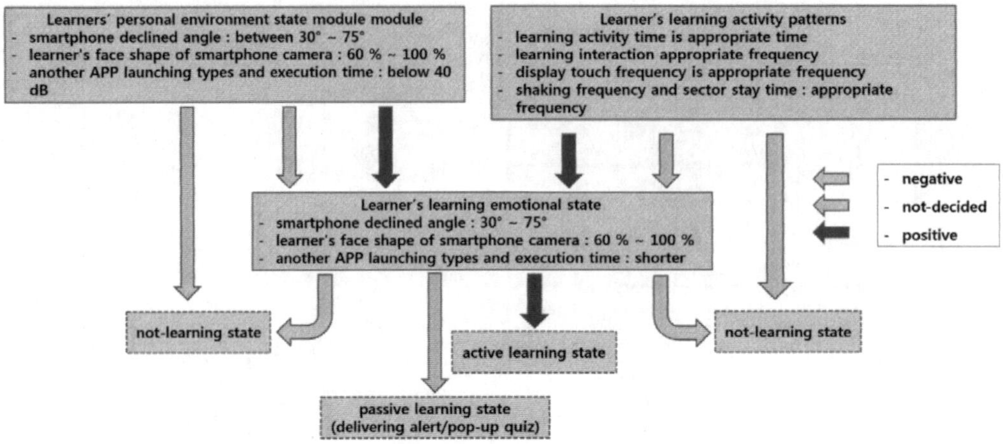

Figure 3. Students' affective learning emotion state Decision Algorithm

Learning Reaction Analysis Engine (ALIPE) deals with three kinds of information (students' personal environment state information, students' learning activity pattern information, students' learning emotional information) in figure 3.

At the first, ALIPE receives students' personal environment state information and students' learning activity pattern information. Each of students' personal environment state information and students' learning activity pattern information has mandatory information so that ALIPE checks the mandatory information of students' personal environment state information and students' learning activity pattern information. If all mandatory information satisfies conditions of students' personal environment state information and students' learning activity pattern information, a student does active learning activity. In this case, ALIPE need not receive optional information of students' personal environment state information and students' learning activity pattern information. And if all mandatory information of students' personal environment state information and students' learning activity pattern information does not satisfy conditions of each information, a student could have two kinds of learning state, one is not-learning state and the other is passive learning state. If a student's learning state is passive learning state, intelligent tutoring engine (ATE) sends alert or pop-up quiz to the student. If a student's learning state is in not-learning state, intelligent tutoring engine (ATE) sends this situation to a learning instructor or lecturer. Lastly, ALIPE receives students' learning emotional state information. ALIPE checks the mandatory information of students' learning emotional state information. If all mandatory information of students' learning emotional state information satisfies conditions of students' learning emotional state information, a student does active learning activity and ALIPE need not receive optional information of students' learning emotional state information. But, if all mandatory information of students' learning emotional state information does not satisfy conditions of students' learning emotional state information, ALIPE receives optional information of students' learning emotional state information and combines mandatory information of students' learning emotional state information, optional information of students' learning emotional state information and mandatory information of students' personal environment state information and students' learning activity pattern information. If the combined information analysis results show negative, it means the student is in 'not-learning' state. If the combined information analysis results show positive, it means the student actively do learning activity and is in 'active-learning' state. If the combined information analysis results show 'not-decided' state, it means the student's learning emotional state is determinant state. If a student's affective learning emotion state is 'not-learning' state or 'passive' state, ITS sends alert or pop-up quiz to the student and send a student's affective learning emotion state to a learning instructor or lecturer. The learning instructor personally contacts with the student and warns the student.

5. Conclusion

As smartphone get popular to students, most students become to use smartphone and smartphone becomes the most important learning devices. In this research, students' emotion state information is collected and analysed from sensors equipped with smartphone. We defined Sensor based Affective Learning Activity Information Process Model and Affective Learning Information Process Engine. And we used the decision conditions of three kinds of information, as like student's learning emotional state information, Students' personal environment state information student's learning activity patterns information that were proposed in [13]. In Sensor based Affective Learning Activity Information Process Model, various student related information comes from smartphone sensors. The student related information is classified intothe three kinds of student's information and Affective Learning Information Process Engine decides student's affective learning emotion state. Proposed Affective Learning Information Process Engine closely relates with ITE(Intelligent Tutoring Engine) that was proposed in [13]. Lastly we propose student's affective learning emotion state decision algorithm that would be used to decide student learning state. Student's affective learning emotion state could be used to construct personal learning environment by ITE. The future research have plan to combine pedagogical model with Sensor based Affective Learning Activity Information Process Model and Affective Learning Information Process Engine so that we could have more detail and more accurate learns' learning state.

Acknowledgement
This work was supported by 2018 Korea National Open University Research Fund.

References

[1] Sintija Petrovica, Alla Anohina-Naumeca, Hazım Kemal Ekenel, Recognition in Affective Tutoring Systems: Collection of Ground-truth Data, Procedia Computer Science, Volume 104 Issue C, March 2017, Pages 437-444
[2] Baker RSJ et al. Towards Sensor-free Affect Detection in Cognitive Tutor Algebra. In: Proc of the EDM'2012. p. 126-133.
[3] Sabourin JL et al. Considering Alternate Futures to Classify Off-Task Behavior as Emotion Self-Regulation: A Supervised Learning Approach. J Educ Data Mining. 5 (9); 2013. p. 9-38.
[4] Paquette L et al. Sensor-Free Affect Detection for a Simulation-Based Science Inquiry Learning Environment. In: Trausan-Matu S, Boyer KE, Crosby M, Panourgia K. (eds.). Proc of the ITS'2014. Switzerland: Springer International Publishing; 2014. p. 1-10.
[5] Heylen D, Nijholt A, Akker HJ. Affect in Tutoring Dialogues. J Appl AI. 19 (3-4); 2005. p. 287-310.
[6] Wixon M et al. The Opportunities and Limitations of Scaling Up Sensor-Free Affect Detection. In: Proc of the EDM'2014. p.145-152.
[7] Porayska-Pomsta K et al. Diagnosing and Acting on Student Affect. User Model User-Adap. 18 (1-2); 2008. p. 125-173.
[8] Amershi S et al. Using Feature Selection and Unsupervised Clustering to Identify Affective Expressions in Educational Games. In: Ikeda M, Ashlay KD, Chan TW. (eds.). Proceedings of the ITS'2006. Springer-Verlag; 2006. p. 21-28.
[9] Sarrafzadeh A et al. How Do You Know that I Don't Understand? Comput Hum Behav. 24; 2008. p. 1342-1363.
[10] Zataraín-Cabada R et al. Affective Tutoring System for Android Mobiles. In: Proceedings of the ICIC'2014; 2014. p. 1-10.
[11] Jaque PA et al. Rule-Based Expert Systems to Support Step-by-Step Guidance in Algebraic Problem Solving: The Case of the Tutor PAT2Math. Expert Syst with Appl. 40 (14); 2013. p. 5456-5465.
[12] Kwang Sik Chung, Learning Reaction Analysis Engine for Interactive Digital Textbook Platform FutureTech 2017: Advanced Multimedia and Ubiquitous Engineering pp 459-465, 2017
[13] Kwang Sik Chung, Sungho Chin, Affective Learning Reaction Model based on Smartphone and Sensors, Journal of Physics: Conference Series, Volume 1060

Construction and Application of an Online Child Portfolio Assessment System

Jingrong Liu[1,*] and Miao Liu[2]

[1]Guangdong Academy of Education, Guangzhou, 510035, China
[2]School of Computer Science and Educational Software, Guangzhou University, Guangzhou, 510006, China

[*]E-mail: 1728550606@qq.com

Abstract. Portfolio assessment is a comprehensive evaluation method for children's development, and effective evaluation is based on professional children's growth portfolios. The children's growth portfolio mainly includes paper, electronic, and online forms, each with its own pros and cons. The online portfolios can avoid defects of paper portfolios, with large capacity, diversified evaluation body, flexible and customized design, professional leading significance, more integration and commemorative features. We have designed and implemented the "Ning Xiner Growth Records" online portfolio, which has been tested in several kindergartens, and achieved remarkable results. With continuous innovation of information technologies, the online portfolio will have broader development and application prospects. It is suggested that governments at all levels allocate special funds to support kindergartens to carry out online portfolio evaluation, develop targeted information technology training for teachers, and increase professional training and guidance for teachers in order to use Portfolio assessment effectively.

1. Introduction

The earliest use of Portfolio assessment was the "Project Zero" conducted by the Harvard Graduate School of Education[1].

In this project, children evaluate their learning effects and abilities through portfolios. The emphasis is not on the quality of their final works, but on the development and understanding of learning, which can help children understand what they should pursue. The Portfolio, as a way of evaluating students' learning process and results in the project, has been attached great importance by educators all over the world. In the mid-20th century, it emerged from the evaluation reform and gradually spread from primary and secondary schools to kindergartens. Portfolio assessment, as a practical form of developmental evaluation concept, has been widely used in the world. Renée Forgette-Giroux and Marielle Simon defined Portfolio as " a cumulative and ongoing collection of entries that are selected and commented on by the student, the teacher and/or peers, to assess the student's progress in the development of a competency"[2]. Black, Harrison and et.al. pointed out that Portfolio assessment, on the one hand, helps students to develop their subjectivity, students become more active and initiative in learning by communicating their ideas with teachers, on the other hand, helps teachers to change their views on students and learn to listen to students[3].

In 2001, the Ministry of Education of China issued the "Outlines for the Guidance of Kindergarten Education" (hereinafter referred to as Outlines)[4]. The Outlines pointed out in the "Education

Evaluation" that "education evaluation is an important part of preschool education, which is also one of necessary means to understand the suitability and effectiveness of the education, carry out the adjustment and improvement, promote the development of each children, and improve the quality of education". Besides, "all administrative personnel, teachers, children and their parents are participants of the kindergarten educational evaluation. In addition, the evaluation process shall be such a type of the process which involves the participation, mutual support and cooperation of each party, and it shall go with the accompany of the whole education process naturally". Furthermore, it adopts "observation, conversation, analysis of works and other methods" comprehensively. Nothing is better than Portfolio assessment to achieve all these requirements which integrates a variety of forms. Therefore, the establishments of children's growth portfolios have become routines of many kindergartens in China since the 21st century.

Portfolio assessment refers to the evaluation method which collects and selects the children's works, and records the children's typical words, deeds and emotional expressions by the planned and purposeful way, and displays growths and development processes of children in a period of time so as to promote emotions, attitudes, abilities, knowledge, skills developments of children effectively based on educational objectives by adults. Undoubtedly, such a type of evaluation method needs to use portfolios. Generally speaking, the portfolio used for evaluation is an item collection system that reveals processes of children's developments over time and is also the "living container" with continuity. However, the capacity of this kind of "container" is limited, the information inside the "container" may not complete and other possible situations are bound to affect the objectivity and authenticity of evaluation results, which may also result in loss of scientific nature in the evaluation and leads to invalid evaluation results ultimately. Therefore, construction of portfolios has played a key role in the process of implementation of Portfolio assessment.

2. Existence forms of Portfolios

Since the birth of the Portfolio, its original form is paper. With the advancement of information technologies, portfolios exist in the form of not only electronic portfolios, but also online portfolios. The following is a detailed analysis of pros and cons of various forms of portfolios.

2.1. Paper portfolios

Paper portfolios are the initial forms of portfolios, which exist extensively within kindergartens. More specifically, the form of paper portfolios may be a manual (an upgraded version of Contact Manual Between Parents and Kindergarten), an accordion-style folder, a box or an album of paintings. In addition, the collected materials are mainly photographs, written records, charts, paintings and so on. Moreover, This form of portfolios is basically handmade by teachers, usually stored in the kindergarten classroom activity room. They can be shared and read by parents or children occasionally and will be sent back to each family at the end of a semester or a school year.

The advantage of paper portfolios is that all materials can maintain their own "original" states, which means that portfolios are completely authentic and can be accessed at any time. However, the shortcomings are also obvious: 1. The paper portfolios are basically manmade by teachers. Making portfolios occupies a lot of teachers' time and the workload is huge. Under the premise of ensuring quantity, teachers have no time to take account of the quality of portfolios. 2. The columns and frameworks of manuals limit the items, content, and quantity of materials collection, and lack of flexibility, so that materials with personalized characteristics cannot enter portfolios, and the collection amount is not sufficient. 3. The handwritten texts are difficult to modify and some teachers' handwritings are not beautiful, which may not only affect the overall aesthetic degrees of paper portfolios, but also affect professional figures of kindergarten teachers. 4. Paper portfolios are basically static photos and texts rather than dynamic audio-visual material, which cannot reflect the development path of children truly and fully; 5. Paper portfolios need a large storage space. However, kindergartens usually have no sufficient spaces for storing them forever, which cannot form a perfect management system of children's growth portfolios. 6. As paper portfolios are usually stored in kindergartens during making, parents' participation in production and evaluation is not high enough. 7. Both the parents and children are not interested in the paper portfolios for theirs poor overall qualities both in the forms and the contents. Regardless of the form and content, the overall quality of paper portfolios is often not good.

2.2. Electronic portfolios
Considering the disadvantages of the children's paper growth portfolios, some kindergartens began to attempt to apply information technologies and software to produce electronic portfolios, such as Powerpoint, Pocomaker, iebook, etc. Generally speaking, electronic portfolios collect and organize materials in multimedia form, including audio, video, picture, text and so on. Moreover, such a form of portfolios can be stored in the computer, can be burned into CDs, and can also be copied to mobile hard disks or U disks.

Compared with paper portfolios, electronic portfolios have their own irreplaceable advantages: 1. The minimum storage space. 2. The convenience to backup. 3. The portability. 4.They can be stored for a long time. 5. Low cost. 6. Modifiable. 7. They can store audio and visual materials. 8. Perfect vision effect.

Electronic portfolios are generally in the stand-alone version, which means that they do not connected to the Internet. Therefore, in the process of production, teachers are still the main body, parents and children's participation is limited, and so electronic portfolios are not satisfactory in terms of timeliness and accessibility. At the same time, although the workload of teachers has been reduced to some extent, but teachers still need to make an electronic portfolio for each individual child, which can only be operated on computers. The amount of work for making electronic portfolios is still very large for teachers.

2.3. Online portfolios
With the popularity of Internet and hardware devices such as computers and smart phones, online portfolios have emerged. This kind of portfolio is usually developed for kindergartens by professional IT service companies, which cannot be produced by teachers via some software themselves. Besides, its operation is also carried out on the Internet instead of stand-alone computer software. Moreover, online portfolios can be described as an upgraded version of electronic portfolios, which can be used at any time and at any place in the Internet environment. Obviously, the well-designed and properly used online portfolios have the basic advantages and more powerful functions over electronic portfolios, which can completely avoid drawbacks of traditional portfolios and truly realize the significance of Portfolio assessment.

Chi-Cheng Chang, Chaoyun Liang, and Yi-Hui Chen, Chi-Cheng[5] explored the reliability and validity of online portfolio self-assessment, and their research samples were from 72 senior high school students enrolled in a computer application course. Their research results showed that online portfolio self-assessment can faithfully reflect learning achievements, i.e. it is a reliable and valid assessment method.

At present, online growth portfolios of children are still in the initial stage. Moreover, some related products appearing in the market, because of the lack of professionalism, are mainly based on the commemorative meaning, and do not have professional values of Portfolio assessment. In order to solve the most difficult problem of children's development evaluations in the curriculum evaluations, we cooperated with a professional information technology team and developed the children growth portfolios named “Ning Xiner Growth Records” in our project research. The portfolios have also been on trial at several experimental kindergartens, which was welcomed by children, parents, teachers and kindergarten administrators.

3. Design and application of the online portfolios
In order to implement the new Education Act of China [6] , the Outlines and Guidelines for Learning and Development of Children Aged 3 to 6 Years Old [7] (hereinafter referred to as Guidelines), the “Ning Xiner Growth Records” online portfolio is designed according to the basic concept of “love and

trust" of the "Learning Story" which is one form of New Zealand national curriculum evaluation. The "learning story" was put forward by Margaret Carr, and is outlined in a number of publications [8-11]. Learning stories is both a method of evaluating children and a method of research. It is a structural observation and record done in a real situation. It provides a continuous pictures of the development of children and can be used to record and communicate the complexity of children's learning.

The "Ning Xiner Growth Records" highlights the atmosphere of "love and trust" in children's growth, and focuses on "the growth of such a cute child" ("Ning Xiner" means "such a cute child" in Chinese). The basic principles of the column design, naming and operation of the portfolios are positioned as concern and fully affirm the shining point of children, and stimulate their great potential of self-growth. Collectors (usually parents and teachers) are required to be more concerned with what children "can do", and things that they are interested in, instead of what children "cannot do" and their shortcomings. The focus of evaluation is on merits and interests of children's learning and development rather than weaknesses; the second person pronouns or epistolary style are used in the text records to show the care of adults and the equal and harmonious relationship between adults and children.

Besides, we studied dozens of different types of children growth portfolios, family-kindergarten contact manuals and kindergarten graduation memento books during constructing columns and framework of the "Ning Xiner Growth Records" portfolio, and carried out the overall design in accordance with the professional, cooperative, incentive, interactive, and monumental principles. In general, the ultimate objective of the "Ning Xiner Growth Records" construction is to use the Internet as a platform for parents and teachers to jointly record and evaluate development processes of children, and enhance the family-kindergarten interactions and equal dialogues, so as to achieve the in-depth cooperation and establishments of good environments for the healthy growths of children.

3.1. Basic framework and contents

The "Ning Xiner Growth Records" portfolio consists of four parts: a front cover, a catalog, inner pages and a back cover. In particular, the inner pages are designed with 16-20 columns according to the age characteristics of the toddler class, junior class, middle class and senior class. Besides, the same column has different names in the portfolio of different classes, and the content and operation requirements are adjusted accordingly. The design makes the entire portfolio both systematic and complete, and highlights the different requirements and characteristics of different age groups.

The content design of the portfolio can be summarized into 8 modules: Basic information, Kindergarten activities, Family activities, Children's works, Children's words, Memorial columns, Growth reports and Reviews, which combined with graphics and text, interspersed with audio and video. More specifically, the "Basic information" includes children, families, kindergartens and classes information. The "Kindergarten activities" includes the typical parts of the one-day activities, function rooms and regional activities, large-scale activities and theme activities. The "Family activities" includes the daily activities, none-routine activities, outdoor activities and other different types of activities which deserve to be recorded in the family. The "Children's works" includes various representative pictures, videos and sound recordings created by children in the kindergartens, families and communities, which include paintings, handmade works, construction works, self-compiled songs, dances, stories, games and so on. The "children's words" are interesting and memorable speeches and deeds that reflect children's thinking and cognitive characteristics. They can be written records or recorded audios or video clips by adults. The "Memorial columns" includes the birthday activities, graduation activities, friends souvenirs and so on. The "Growth reports" is a 11-page fuzzy rating evaluation form designed according to targets of the Guidelines, and combined with educational recommendations from the Guidelines. The "Growth reports" is also equipped with about 69,000-word of online assessment guidance, and is proposed for teachers and parents for each specific goal so as to make daily observations and evaluations of both teachers and parents in the kindergartens and homes to become more operational. The "Reviews" not only includes the overall evaluation and message of parents, teachers, friends and relatives writing or recording on the birthday of the child or at the end of the school year, but also includes the anecdotal records of teachers and parents in the form of learning stories and educational stories. Figure 1 shows the catalog page of the senior class in the "Ning Xiner Growth Records" portfolio.

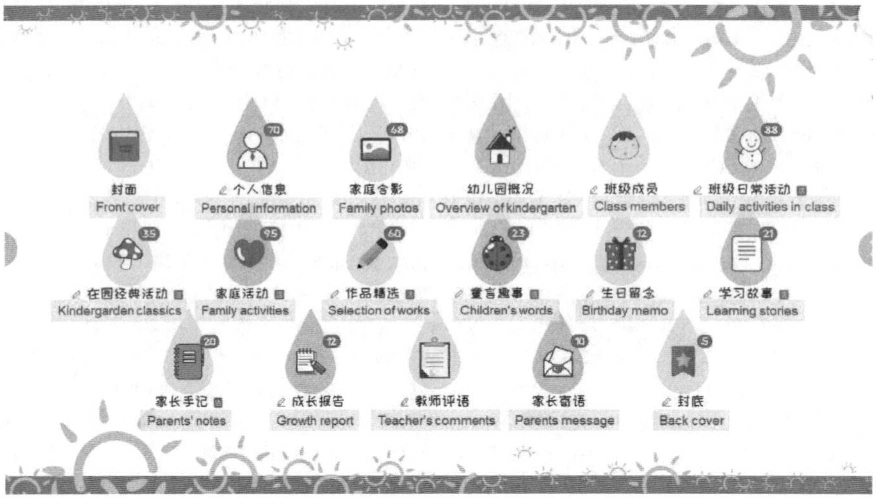

Figure 1. The catalog page of the senior class

The "Ning Xiner Growth Records" portfolio interface pages are written in Chinese, and the English words in Figure 1 is the translation of Chinese words nearby.

3.2. The portfolio creation and use

The "Ning Xiner Growth Records" portfolios run on the "NingXin Preschool Education Net" [12]. Parents or teachers can carry out the routine operations after logging in the website according to pre-assigned roles. Guangzhou Ningxin Information Technology Co., Ltd. is responsible for leasing servers and network spaces, maintaining software systems, and guiding parents and teachers to use the system in a correct manner. Moreover, the growth portfolios provide a detailed column operating guidelines for parents and teachers at the appropriate location of each page and explain the collection and editing methods of each column information. Figure 2 is the parent's message column page of the senior class. The 5 boxes in the middle are used to store parents' input of text, audio, or uploaded pictures and videos, and the upper right corner is the popped operating guidelines.

Figure 2. The parent's message column page of the senior class

The rich contents of the growth portfolios come from the activity records of children both in kindergartens and at homes. Therefore, the archival materials are provided by parents and teachers at any time throughout a whole school year. In the meantime, the system has also developed smart phone APP (Android version and iOS version) at the same time, which enables both teachers and parents to use them to take pictures or videos, upload , access and input comments anytime, anywhere. Figure 3 is the three screenshots on the APP.

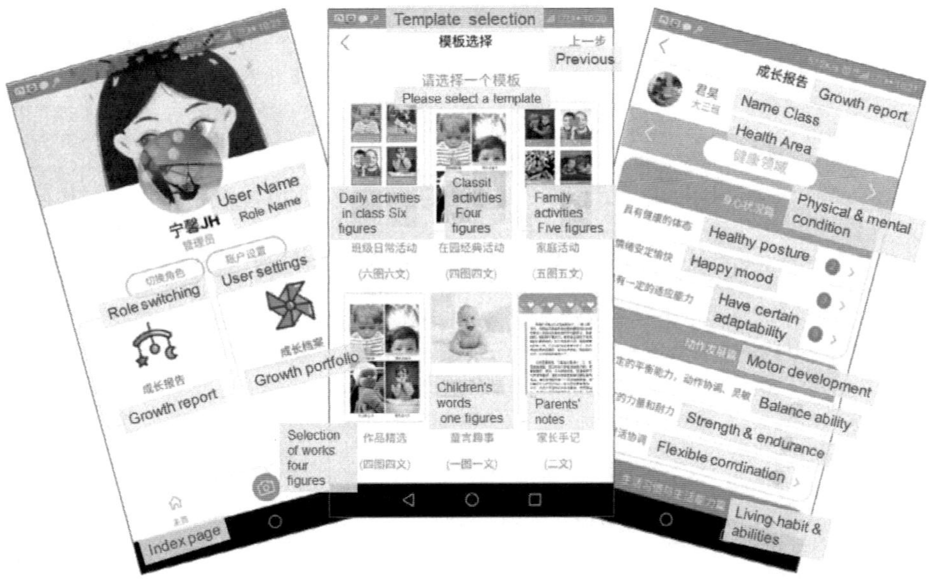

Figure 3. The three screenshots on the APP

Teachers are responsible for editing a few columns. The main editorial work of the whole portfolio is undertaken by parents. It will take a small amount of time each week for teachers to concentrate on the children's portfolios. At a fixed time every weekend, parents may discuss with their children, choose which materials to add to the portfolio, and edit relevant pages at home. This can also help to develop good parent-child relationships. Furthermore, children's growth reports and birthday souvenirs require parents and teachers to carry out observations, evaluations and reports in the week of children's birthday.

At the end of each school year, the whole growth portfolios can be downloaded and saved by parents and kindergartens. Besides, the growth portfolios are downloaded in PPT files form, which can be more delight to watch accompanying with background music chosen by parents and children if necessary. In addition, families can decide whether to print and bind the portfolio into a beautiful album according to actual situation. kindergartens can store the electronic portfolios of all children according to the grades and classes easily.

Teachers should read every portfolio at regular interval to get a thorough understanding of each child and his family's understanding, expectations and evaluation of him, compare the performance of children in kindergartens, make a professional analysis of each child, and conduct a comprehensive self-evaluation of the effectiveness of this class on the basis of portfolios, so as to formulate the more targeted teaching and caring activities work plans of the next stage and promote the case education. Furthermore, teachers should also have professional sensitivity, catch the information passed on by parents in time, communicate with parents at any time and on their own initiative when necessary, and provide professional support for the healthy and harmonious development of children.

Parents should check the materials uploaded by teachers at any time, keep abreast of the activities of children in the kindergarten and class, evaluate the development status of children objectively, and cooperate with teachers actively to complete the work plan for education. Once a week, parents should insist on editing portfolios, review and exchange with children. These can promote parent-child relationship and cultivate children's multiple abilities.

4. Characteristics and values of the portfolios

The "Ning Xiner Growth Records" portfolio has been tested in several kindergartens in Guangzhou, China for two school years. It has been well welcomed by teachers and parents. Its characteristics and values are mainly reflected in the following four aspects.

(1) Large capacity and comprehensive information: The design of the unlimited number of pages in the online portfolio enables the portfolio to display unlimited capacity, to store all kinds of documents adequately, and to meet needs of large capacity archives for both qualitative and quantitative evaluation. The degree of parental involvement has been significantly increased, and the number of archives of children's family activities and kindergarten activities in the complete set of portfolios is basically the same, which can objectively and comprehensively reflect the true state of children, and provide sufficient information for accurately evaluating the development of children.

(2) Diversification of evaluation subjects: The wide use of the Internet and smart phones enables the construction of online portfolios to be carried out anytime and anywhere, the immediacy and accessibility of portfolios can be realized, and parents become main collaborators and finishers of portfolio production. At the same time, the family participates in the process of selection, description, evaluation and editing of archival materials in the form of parent-child activities. The children are truly transformed from the object of evaluation to one of the evaluation subjects.

(3) Flexible and customized portfolio design: The application of mature information technology makes the portfolio is no longer the same for all kindergartens, but can be customized according to requirements of kindergartens with distinctive layout and content, to meet customization of different kindergartens and brand building requirements.

(4) Professional leading significance: In the process of guiding parents to fill the archives and correctly use the portfolio, the kindergarten promotes parents to study in depth and implement the Guidelines to help parents establish correct parenting concepts and master scientific family education methods. In turn, it will achieve mutual education and achieve deep cooperation between families and kindergartens. Through the construction of online portfolios, kindergartens can also form research projects to improve professional competence of teachers and modern management level of kindergartens, such as observation and evaluation of the status of early childhood development, parents and kindergarten cooperation based on network platform, etc.

5. Conclusion

In the future, we will continue to improve the "Ning Xiner Growth Records" portfolio, such as introducing AI technologies to automatically generate evaluation reports, recording the evaluation history of children on various evaluation indicators, predicting the development trend of children, and reminding parents and teachers to take targeted measures early.

In short, the advantages of the online portfolio are obvious, and has broader development and application prospects. We believe that it is necessary to give full play to the role of the online portfolio in kindergartens, the following issues need to be further concerned. For example, kindergartens with special funds for informatization construction should be included in their allocation category to support their informatization construction independently. Professional teams should be encouraged, supported and subsidized to develop information products needed by kindergartens, reduce their product prices and improve their service quality. The education departments shall carry out targeted professional training and guidance on information technology training and Portfolio assessment, improve the information literacy of teachers and the practical operation ability of the use of Portfolio assessment.

Acknowledgements

The paper's work is supported by Guangzhou Education Science 13th Five-Year plan Key project 1201720422("Internet +" kindergarten informatization construction research).

References

[1] Harvard Graduate School of Education (2018). Exploring the concept of student project

assessment to develop fair and effective grading methods. http://www.pz.harvard.edu/projects/apple/.
[2] R Forgettegiroux, and M. Simon. "Organizational Issues Related to Portfolio Assessment Implementation in the Classroom. "Practical Assessment Research & Evaluation7.8 (2000):3183-90.
[3] Black, P. , Harrison, C. , Lee, C. , Marshall, B. and Wiliam, D. . (2004). Working inside the black box: assessment for learning in the classroom. Phi Delta Kappan, 86(1), 8-21.
[4] Ministry of Education of China (2001). Outlines for the Guidance of Kindergarten Education. http://www.moe.gov.cn/publicfiles/business/htmlfiles/moe.
[5] Chang, Chi Cheng, C. Liang, and Y. H. Chen. "Is learner self-assessment reliable and valid in a Web-based portfolio environment for high school students?" Computers & Education 60.1(2013):325-334.
[6] Ministry of Education of China (2015). Education Act of China. http://www.moe.gov.cn/s78/A02/zfs__left/s5911/moe_619/201512/.
[7] Ministry of Education of China (2012). Guidelines for Learning and Development of Children Aged 3 to 6 Years Old. http://www.moe.gov.cn/srcsite/A06/s3327/201210/t20121009_143254.html.
[8] Carr, M., May, H., Podmore, V. N., Cubey, P., Hatherly, A., & Macartney, B. (2002). Learning and teaching stories: action research on evaluation in early childhood in new Zealand. European Early Childhood Education Research Journal, 10(2), 115-125.
[9] Bradbury, Alice. "Early childhood assessment: observation, teacher 'knowledge' and the production of attainment data in early years settings." Comparative Education 50.3(2014):322-339.
[10] Khetani, M. A., Mcmanus, B. M., Arestad, K., Richardson, Z., Charlifuesmith, R., & Rosenberg, C., et al. (2018). Technology-based functional assessment in early childhood intervention: a pilot study. Pilot & Feasibility Studies, 4(1), 65.
[11] Logan, Helen. "Retaining meanings of quality in Australian early childhood education and care policy history: perspectives from policy makers." Early Years 38.2(2018):1-17.
[12] Guangzhou Ningxin Information Technology Co., Ltd. (2019). https://nxyjw.net/.

An Ensemble Learning Imbalanced Data Classification Method Based on Sample Combination Optimization

Yuxin Wang

School of Software, Xi'an Jiaotong University, China

E-mail: 1010120806@qq.com

Abstract. Imbalanced data classification is one of the hot topics in data mining and machine learning in recent years. In practice, Imbalanced data classification is very common, such as cancer detection, spam discrimination, credit card fraud detection, etc. Because of the large difference in the number of categories and Imbalanced distribution, traditional classification algorithms have poor classification effect on minority classes, and correct identification of minority classes often brings greater value. Therefore, how to effectively identify minority classes in Imbalanced data is of great importance. Practical significance. Aiming at the problems that the Bagging-based Imbalanced data classification method cannot guarantee the validity and existence of classification boundaries by adding redundant noise information and sampling, an ensemble learning GABagging method based on sample combination optimization is proposed. Firstly, the sample combination optimization algorithm uses genetic algorithm to select a subset from most classes and construct a new data set with a few classes. Subsequently, several sample combinatorial optimization algorithms are used to train and integrate several classifiers. The experimental results show that GABagging can improve the correct recognition ability of minority classes on 19 Imbalanced datasets compared with other similar methods such as TPR and AUC, without excessive loss of recognition ability of majority classes. It is proved that GABagging can compensate for the shortcomings of related Bagging-based methods such as easy loss, increasing samples and not guaranteeing the validity and existence of classification boundaries after sampling.

1. Introduction

Imbalanced data appears in every field of real life, which is very common. However, traditional classification algorithms are based on the assumption that data is balanced. When encountering Imbalanced data, they have poor ability to recognize a few classes correctly. The most intuitive performance is that the predicted results are all classes of most samples [1]. Such results are obviously meaningless, and at the same time, compared with most sample classes, a few sample classes carry more important information [2]. The value of identifying a correct minority sample is much higher than that of most sample classes. When facing the problem of Imbalanced data classification, ensemble learning often combines with data-level methods to construct multiple data sets through multiple sampling processes, which can alleviate the adverse effects of data imbalance on classification [3]. Common ensemble learning methods include Bagging and Boosting. In the related Bagging method, Barandela et al. proposed the UnderBagging method [4], which combines the undersampling technique with the Bagging method, retains all the minority samples at a time, and takes the playback and non-playback sampling from the majority samples, extracts the majority samples equal to the minority samples, and forms a data subset, the number of basic learners, that is, the number of data subsets. It is determined by the ratio of the number of samples in most classes to that in a few classes. Molinara et al. proposed the EBBag method [5]. EBBag combines a small number of classes with a random subset of most classes, so that there is little difference in the number of classes in the training data set for each model in the integration. Wang et al. proposed Over Bagging and SOMTE Bagging methods [6]. Both of them combine the over-sampling method with Bagging method. Over Bagging is different from the previous Bagging methods in retaining all the minority samples, but adopts the same sampling strategy for most and minority categories. Random over-sampling method is adopted to extract a certain proportion of samples from various categories, and its proportion is set to majority samples. The ratio of the number of samples to the number of the current category is the product of an artificial proportional parameter a; SOMTEBagging retains all the majority of samples and generates a small number of samples by random oversampling and SOMTE method, with the same sampling proportion as the former. Zhou et al. proposed the Easy Ensemble method [7], which combines random undersampling, Adaboost and Bagging. The random undersampling method is used to generate a number of data subsets. It retains all minority samples and generates a majority of samples with the same number of minority samples. Adaboost is used as the base learner. Yang et al. proposed HSBagging method [8]. Firstly, bagging method was used to generate in-pocket data and out-of-pocket number. Then, under-sampling and SOMT method were used to sample in-pocket data at different sampling ratios for several times to obtain several classifiers. Then, the out-of-pocket data was used to evaluate the classifier and select the best classifier until several base classifiers were generated. Rong et al. proposed LBUS method [9], which uses ITQ method [10] to generate several subsets of data and train classifiers on this basis.

The above method improves Bagging's random sampling process from the data level, which basically uses random undersampling or oversampling to generate data subset. However, the shortcomings of undersampling and oversampling bring great problems to these Bagging Imbalanced data classification methods based on sampling. Because of the need to delete part of the majority class samples, under-sampling directly results in the loss of part of the majority class sample information, which contains important information that plays a decisive role in the classification boundary division, thus reducing the classification boundary pair [11]. Oversampling will inevitably increase redundancy and noise samples due to the addition of a few classes of samples, resulting in over-fitting phenomenon [12], and the classification boundary will be over-matched to the current number. At the same time, under-sampling and over-sampling methods only make the number of samples of different categories no longer vary greatly, the validity and existence of classification boundaries of data sets cannot be guaranteed.

2. Sample combination optimization algorithm

Following are the specific steps of the sample combination optimization algorithm.

Algorithm: Sample combination optimization algorithm
Input: Data $S=\{(x_1,y_1),(x_2,y_2),(x_3,y_3),\ldots\ldots,(x_n,y_n)\}$ Output: Data S^* 1: The maj of most class samples and min of a few class samples in S were extracted, and the number of samples was N and M, respectively. 2: Initialize T to the current algebra, Initialize n as the number of individuals per generation. Initialize max_gen as the largest evolutionary algebra. Initialize last_population to record the population of the previous generation of the current generation. Initialize last_average = 0 to record the average fitness of the previous generation. Initialization stop_time represents the number of stops. Initialization stop_time represents the number of stops. Initialize final_population as the final output group 3: Initialization divides maj into M parts and combines with min to form the first generation population. 4: While stop_time != 10 or T != max_gen

IOP Conf. Series: Journal of Physics: Conf. Series **1284** (2019) 012035 doi:10.1088/1742-6596/1284/1/012035

4.1: Calculate the fitness of each individual in the T generation and descend to get queue A
4.2: Calculate the difference between each individual in queue A and last average, and exclude the negative individual.
4.3: 30% of the individuals excluded from queue A
4.4: if len(A)>=2, then jump to Step 4.5
    else jump to Step 4.6
4.5: Calculate the average fitness s of generation T
    if s == last average,then stop_time←stop_time+1
    else stop_time←0
        last populaition←A
        last average←s
        T←T+1
4.6: Roulette betting chooses two individual points from last population to intersect and produces m/3 individuals
4.7: Roulette betting chooses two bodies from last population to intersect at two points and produces m/3 individuals.
4.8: Roulette betting varies from last population to m/3 individuals
To Step 4
5: Output the data set of the best individual S* in last population for the final sampling

When the sample combination optimization algorithm performs cross-over operation, it randomly selects the same location point from the majority of the sample data of two individuals, divides the majority of the sample data into two parts, and completes single-point cross-over by exchanging data sets of different parts of two individuals. When crossing two points, similar to single point crossing, we select two points at the same position for minority and majority samples of two individuals to complete the exchange. The mutation is aimed at an individual in a population. 10% of the samples are randomly removed from most and a few samples, and the remaining samples are retained to form a new individual. To evaluate an individual's fitness, a logistic regression model is trained by using the current individual, i.e. the current data set, to classify and predict the validation set, and to obtain the AUC value. The AUC value is used as the current individual's fitness. The higher the fitness value is, the higher the AUC value is, the clearer and more effective the classification boundary of the current individual is. The convergence conditions of the whole algorithm are two. First, if the evolution times of the population reach the maximum iteration threshold, the algorithm terminates. Second, when the average fitness of the continuous N-generation population remains unchanged, the algorithm is considered to be convergent. When either of the two convergence conditions is satisfied, the algorithm terminates and the individuals with the highest fitness value in the current population are output.

To overcome the shortcomings of sampling technology that cannot effectively remove overlapping areas and guarantee the existence and validity of classification boundaries, sample combination optimization algorithm is used to remedy the shortcomings. Random under-sampling and over-sampling as well as SOMTE algorithm are used in Bagging-based Imbalanced data classification method. For random under-sampling, the number of samples to be cut is selected according to random probability, with one. For random oversampling, only a few samples have been copied, which is no different from increasing the overlapping region. At the same time, SOMTE algorithm has a certain randomness in selecting synthetic samples, and the overlapping region still exists. Sample combination optimization algorithm regards the problem as a sample combination problem. An optimal sample combination has clear and effective classification boundary and the overlapping area is eliminated. The genetic algorithm is used to find the optimal combination. The current AUC is used as fitness evaluation to guide the genetic algorithm to find the sample combination with the highest AUC value, that is, to find an effective and clear data set with classification boundary. Compared with the sampling technology, the data set obtained by this method has removed the overlapping area, which ensures the existence and validity of the classification boundary.

3. Ensemble learning classification method based on sample combination optimization

In Figure 1, we introduce how GABagging overcomes the shortcomings of Bagging Imbalanced data classification methods in combination with Figure 1. In Figure 1, in the last Imbalanced data DN, the minority classes correspond to most classes of D1-valued DN-1 data set, and most of them correspond to a few classes of the previous data set. This is because the sample combination optimization algorithm is essentially under-sampled, and it will reduce part of the majority of class samples. As the number of iterations increases, the number of most class samples in the data set decreases, while a few. The number of classes remains unchanged, so there will always be a time when the number of most class samples in the data set will be less than that of a few classes. At this time, the last Imbalanced data set DN to be processed will be the last one. Taking the classes with fewer samples as a few classes and the classes with more samples as the majority classes, the sample combination optimization algorithm will be carried out to obtain the last balanced data set DN and join the set S, GA. The Bagging sampling process is over. In the data set S obtained by GABagging, each sample corresponds to a sample in the original Imbalanced data set without adding new samples, and the set S includes all samples in the original Imbalanced data set without reducing the original samples. At the same time, the classification boundary of each data set in the set S is clear and effective, and the number of samples in the set S is the same as that in the original Imbalanced data set. The specific samples are the same, and the information content is identical with that of the original Imbalanced data. It does not increase or decrease any sample information, which makes up for the shortcomings of the above-mentioned Bagging Imbalanced data classification method that is liable to lose important sample information and increase redundant noise sample information.

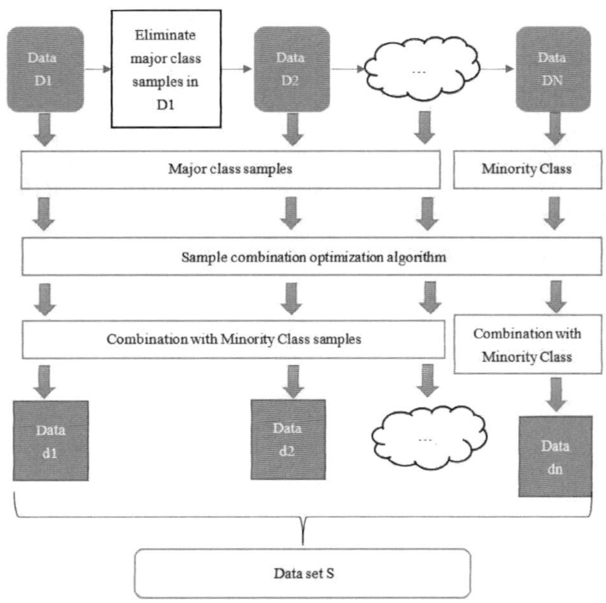

Figure 1. Sampling process of GABagging.

4. Experimental results and analysis

The experimental data of GABagging simulation are derived from KEEL Imbalanced data set in Table 1. The training set, validation set and test set of each experiment were obtained from source data D, which accounted for 50%, 30% and 20% of source data D respectively. Then the classification model was trained with these data and the classification effect was tested. In order to eliminate the influence of random factors and prove the effectiveness and promotion of GABagging method, 500 experiments were carried out, and the mean value was taken as the final experimental result. Six methods are selected to compare OverBagging, UnderBagging, ABBag, HSBagging and RBSBagging. Similarly, these five methods are experimented with the same methods mentioned above. The experimental indexes were TNR and AUC, and the final results were the average of 500 experiments for each index.

Table 1. Imbalanced Data.

Data	Size	Min/Maj	Min/Size
b1	683	239/444	0.35
breast	789	220/569	0.28
yeast6	1485	35/1450	0.02
yeast4	1485	51/1434	0.03
yeast3	1484	163/1321	0.11
yeast1	1485	429/1054	0.29
wisconsin	684	239/445	0.35
vehicle3	847	212/635	0.25
vehicle2	846	218/638	0.26
vehicle1	847	217/630	0.26
vehicle0	846	199/647	0.24
segment0	2309	329/1980	0.14
pima	769	268/501	0.35
haberman	306	81/225	0.26
glass6	214	29/185	0.14
glass4	215	13/202	0.06
glass2	214	17/197	0.08
glass0	214	70/144	0.33
spect	80	26/54	0.33

Table 2. Comparison of TPR values of different methods.

Data	OverBagging	UnderBagging	ABBag	HSBagging	RBSBagging	GABagging
b1	0.55	0.76	0.692	0.6911	0.45	**0.8**
breast	0.75	0.7	0.755	0.5424	0.6523	**0.95**
yeast6	0.59	0.83	0.59	0.7986	0.59	**0.88**
yeast4	0.7	0.85	0.7011	0.7585	0.76	**0.9**
yeast3	0.59	0.79	0.743	0.6993	0.81	**0.91**
yeast1	**1.0**	0.9	0.9915	0.8991	0.8	**1.0**
wisconsin	0.5667	**0.9333**	0.8	0.84	0.9	**0.9333**
vehicle3	0.8267	0.9	0.8633	0.7887	0.89	**0.9293**
vehicle2	0.3308	0.585	0.5551	0.5686	0.5212	**0.6842**
vehicle1	0.8571	0.9714	0.8771	0.9818	0.8	**1.0**
vehicle0	0.8694	0.9196	0.8259	0.8979	0.9	**0.995**
segment0	0.5391	0.7792	0.4682	0.7535	0.7952	**0.8526**
pima	0.9131	0.9586	0.9487	0.9662	0.9127	**0.9909**
haberman	0.2666	0.7	0.3788	0.7897	0.45	**0.9222**
glass6	0.6855	0.96	0.7988	0.8647	0.81	**0.98**
glass4	0.4612	0.7012	0.4883	0.7977	0.6537	**0.8482**
glass2	0.7339	0.9131	0.7555	0.8953	0.6874	**0.9443**
glass0	0.1945	0.8236	0.4112	0.8786	0.79	**0.9**
spect	0.4	0.8	0.8275	**0.87**	0.7539	0.8571

In Table 2, the bold font of each line is the maximum value of the line, and the accuracy of TPR value for minority categories represents the correct recognition ability of the model for minority categories. The larger the TPR value, the stronger the correct recognition ability of the model for minority categories. It can be seen from the table that the TPR value of GABagging method is the highest on other data sets except SPECT data set, which is slightly lower than the maximum value of 0.0129. Through the above data, it can be proved that GABagging can be effectively and correctly recognized. Compared with other Bagging-based Imbalanced data classification methods, a few classes have improved significantly. Next, in order to prove that GABagging improves the correct recognition ability of a few classes without excessive loss of the overall classification ability, relevant experiments are carried out.

Table 3. Comparison of AUC values of different methods.

Data	OverBagging	UnderBagging	ABBag	HSBagging	RBSBagging	GABagging
b1	0.1696	0.1566	0.3385	0.4578	**0.9165**	0.9041
breast	0.1138	0.1285	0.3543	0.4977	0.9592	**0.9834**
yeast6	0.1826	0.0928	0.2425	0.4224	0.9158	**0.9286**
yeast4	0.1233	0.1207	0.3587	0.2958	0.9666	**0.9793**
yeast3	0.2066	0.1696	0.4225	0.2156	0.9126	**0.9085**
yeast1	0.0062	0.0206	0.3854	0.2664	**0.9802**	0.9725
wisconsin	0.1812	0.1329	0.2454	0.4345	**0.9275**	0.8458
vehicle3	0.0615	0.0635	0.3578	0.4556	**0.981**	0.9623
vehicle2	0.3688	0.4107	0.2657	0.3587	0.564	**0.6853**
vehicle1	0.0488	0.0107	0.2479	0.3556	0.9687	**0.9984**
vehicle0	0.0256	0.0276	0.3457	0.5355	0.9855	**0.9931**
segment0	0.1934	0.1926	0.2147	0.3335	0.8142	**0.8803**
pima	0.0103	0.0168	0.2523	0.4125	0.9972	**0.9978**
haberman	0.3142	0.2271	0.3541	0.2578	0.7568	**0.779**
glass6	0.0806	0.0416	0.2147	0.3688	0.9723	**0.9842**
glass4	0.2902	0.2716	0.4525	0.2456	0.7059	**0.7836**
glass2	0.081	0.0568	0.2146	0.3357	0.9491	**0.975**
glass0	0.3178	0.1253	0.2567	0.3558	0.8848	**0.9348**
spect	0.2119	0.1331	0.2227	0.1488	0.8897	**0.931**

In Table 3, the bold font of each line is the maximum value of the line. AUC represents the overall classification ability of the model, including the correct recognition ability of a few and most categories. The larger the AUC value, the stronger the overall classification ability of the model. It can be seen from the table that except for b1, yeast1, Wisconsin and vehicle3 data sets, the AUC value of GABagging is the highest in other data sets, and the TPR value of GABagging is also the highest in the corresponding data of Joint Table 2, which indicates that GABagging has a high recognition rate of a few classes, and the overall classification ability of GABagging is higher than other methods.

5. Conclusion

Aiming at the problem that the Bagging-based Imbalanced data classification method cannot guarantee the validity and existence of classification boundary by adding redundant noise information and sampling, an integrated learning Imbalanced data classification method GABagging based on sample combination optimization is proposed. In GABagging, sample combination optimization algorithm is used to sample source data, which can remove overlapping regions of different categories of data subsets after sampling and ensure an effective and clear classification boundary. Secondly, by using the sample combination optimization algorithm many times and eliminating most of the samples that have been sampled before each use, until all samples are sampled, to preserve all data sample information. However, GABagging method has good prediction effect, but its time complexity is too high and

IOP Conf. Series: Journal of Physics: Conf. Series **1284** (2019) 012035 doi:10.1088/1742-6596/1284/1/012035

running time is too long. Considering introducing the idea of parallelization, the parallelization of genetic algorithm is studied to further improve the performance of GABagging method and shorten the running time.

References

[1] S. Barua, M. M. Islam, X. Yao and K. Murase, "MWMOTE--Majority Weighted Minority Oversampling Technique for Imbalanced Data Set Learning," in IEEE Transactions on Knowledge and Data Engineering, vol. 26, no. 2, pp. 405-425, Feb. 2014.

[2] B. Wang and J. Pineau, "Online Bagging and Boosting for Imbalanced Data Streams," in IEEE Transactions on Knowledge and Data Engineering, vol. 28, no. 12, pp. 3353-3366, 1 Dec. 2016.

[3] M. R. Hasan, H. Gholamhosseini and N. I. Sarkar, "A new ensemble classifier for multivariate medical data," 2017 27th International Telecommunication Networks and Applications Conference (ITNAC), Melbourne, VIC, 2017, pp. 1-6.

[4] H. R. Sanabila and W. Jatmiko, "Ensemble Learning on Large Scale Financial Imbalanced Data," 2018 International Workshop on Big Data and Information Security (IWBIS), Jakarta, 2018, pp. 93-98.

[5] M. Molinara, M. T. Ricamato and F. Tortorella, "Facing Imbalanced Classes through Aggregation of Classifiers," 14th International Conference on Image Analysis and Processing (ICIAP 2007), Modena, 2007, pp. 43-48.

[6] R. Nicole, "Title of paper with only first word capitalized," J. Name Stand. Abbrev., in press.

[7] S. Wang and X. Yao, "Diversity analysis on imbalanced data sets by using ensemble models," 2009 IEEE Symposium on Computational Intelligence and Data Mining, Nashville, TN, 2009, pp. 324-331.

[8] X. Liu, J. Wu and Z. Zhou, "Exploratory Undersampling for Class-Imbalance Learning," in IEEE Transactions on Systems, Man, and Cybernetics, Part B (Cybernetics), vol. 39, no. 2, pp. 539-550, April 2009.

[9] M. H. Popel, K. M. Hasib, S. Ahsan Habib and F. Muhammad Shah, "A Hybrid Under-Sampling Method (HUSBoost) to Classify Imbalanced Data," 2018 21st International Conference of Computer and Information Technology (ICCIT), Dhaka, Bangladesh, 2018, pp. 1-7.

[10] T. Rong, X. Tian and W. W. Y. Ng, "Location bagging-based undersampling for imbalanced classification problems," 2016 International Conference on Wavelet Analysis and Pattern Recognition (ICWAPR), Jeju, 2016, pp. 72-77.

[11] Qianqian Huang and Xiaolong Zhang, "An improved ensemble learning method with SMOTE for protein interaction hot spots prediction," 2016 IEEE International Conference on Bioinformatics and Biomedicine (BIBM), Shenzhen, 2016, pp. 1584-1589.

[12] H. R. Sanabila and W. Jatmiko, "Ensemble Learning on Large Scale Financial Imbalanced Data," 2018 International Workshop on Big Data and Information Security (IWBIS), Jakarta, 2018, pp. 93-98.

A Framework to Identify Critical Design Parameters for Enhancing User's Satisfaction in Human-AI Interactions

J G Shin[1,2], J B Kim[1,2] and S H Kim[1,2,*]

[1]BrainKorea21[PLUS] Cultivating Creative Design Engineer Project Group
[2]Graduate school of Industrial Engineering, Kumoh National Institute of Technology, Gumi, Republic of Korea

[*]E-mail: kimsh@kumoh.ac.kr

Abstract. The advancements in artificial intelligence technology have made changes in how people interact with systems. Unique features and user requirements of Human-AI Interactions (HAII) need to be identified with respect to those of Human-Computer interactions (HCI). This study proposes a way to find critical parameters of interaction design for enhancing user's satisfaction when people interact with intelligent systems through voice user interfaces. We summarised distinguished user requirements for intelligent products identified from previous researches. Then match them with design parameters in terms of performance indexes that will make differences in the user's satisfaction. The interaction scenario was set as users ask simple questions with their own voices to the system and the system answer to the questions with synthesized voices after it got to the answer by AI function. The critical performance indexes derived are the number of trials to get the right answer for a question, response time to get to the next interaction, sentence structures of the answer, and pace of the answer. An experimental setup is ready to evaluate user's satisfaction among different levels of the above performance indexes by Wizard of Oz design method applied on a voice user interface we implement. We are going to validate the effects of performance indexes in HAII on the user's satisfaction, which will be measured in terms of verbal and non-verbal measures.

1. Introduction

Through the development of artificial intelligence technology, various intelligent products such as AI speakers (Google's Google Home®, Amazon's Echo®) and virtual assistant systems (Samsung's Bixby, Apple's Siri) on mobile devices are being released [1], resulting in interaction between users and artificial intelligence [2]. Those interactions are called as Advanced Human-Computer Interaction or Human-AI Interaction (HAII). There is a tendency to recognize and understand them as a new version of Human-Computer Interaction (HCI) [3]. But Terry Winograd (2006) argued they differ from the existing Human-Computer Interaction (HCI) [4]. The intelligent products require users to interact with them in new ways. Although HAII has not yet fully established, it is necessary to check whether the user will be satisfactory while interacting with these brand new systems. It is also to check whether there is a need for establishing new criteria to evaluate the goodness of HAII design rather than utilizing traditional way of HCI evaluation method [5]. In this study, a few design parameters that may affect the user's satisfaction are identified by analysing a user's requirements for intelligent products. Finally, we devise and implement an evaluation system for checking out the effect of those design parameters on user's satisfaction while interacting with the intelligent system through the VUI (Voice User Interface).

IOP Conf. Series: Journal of Physics: Conf. Series **1284** (2019) 012036 doi:10.1088/1742-6596/1284/1/012036

2. Method

The overall study flow is shown in Figure 1 below.

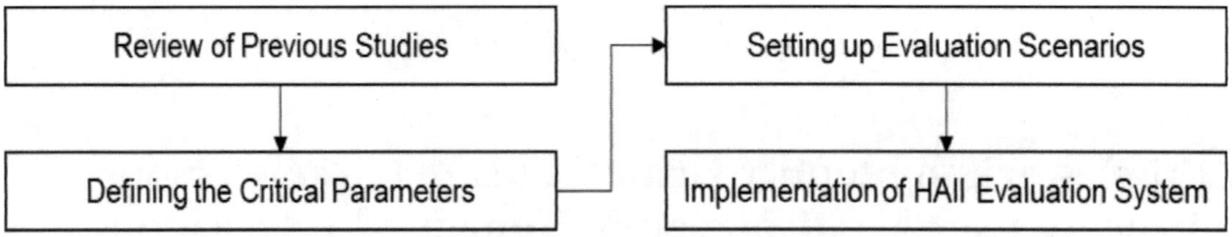

Figure 1. Flowchart for building HAII Evaluation System.

(1) In order to identify a few critical design parameters of intelligent products, user requirements were gathered by reviewing previous studies. Critical design parameters that can affect the satisfaction of intelligent product are drown out by mapping these user requirements with performance indexes. Commercialized intelligent products such as Samsung's Bixby or Apple's Siri were investigated as the source of variations for the design parameters.

(2) A model HAII system was designed and implemented to conduct experiments for evaluating the effects of identified design parameters. Since it is not easy to control the design parameters of the commercialized products, we took the WoZ (Wizard of Oz) evaluation method. It means that this model HAII system was designed as it can provide different interactions with modified design parameters in random order by researchers in the rear side, but the subjects are feeling and believing that they interact with the real AI system [6].

3. Result

3.1. Definitions for Critical Design Parameters

We collected user requirements in two previous studies related to user requirements for intelligent products. User requirements are shown in Table 1 below.

Table 1. User Requirements for HAII.

Paper	User Requirement	Selection
Laila Dybkjær and Wolfgang Minker 2008. [7]	• Modality appropriateness	
	• Input Recognition adequacy	
	• Naturalness of user speech relative to the task including coverage of user vocabulary and Grammar	
	• Output voice quality	V
	• Output phrasing adequacy	
	• Feedback adequacy	V
	• Adequacy of dialogue initiative relative to the task	V
	• Naturalness of the dialogue structure relative to the task	
	• Sufficiency of task and domain coverage	
	• Sufficiency of the system's reasoning capabilities	
	• Sufficiency of interaction guidance	V
	• Error handing adequacy	
	• Sufficiency of adaptation to user differences	

	• Number of interaction problems	
	• User satisfaction	
Valéria Farinazzo et al, 2010. [8]	• Appropriate modality	
	• Suitable Feedback	
	• User diversity and user perception	
	• Appropriate phases out	
	• Output Voice Quality	V
	• Proper entry recognition	
	• Appropriate dialog start out and adequate instruction about how to interact with the application	V
	• Help tool	
	• Error Prevention	V
	• Handling errors	V

With these chosen user requirements, we analyse the functional aspects of the intelligent systems in terms of design parameters so that we can map the user requirement and the design parameters. Eight user requirements were selected and then mapped with four design parameters as shown in Table 2 below.

Table 2. Critical Design Parameters Identified.

User Requirements	Critical Design Parameters
• Appropriate dialog start out	• Response time to get to next interaction
• Error Prevention • Error handling adequacy	• Number of trials to get the right answer for a question
• Output Voice Quality	• Pace of the answer
• Feedback Adequacy • Adequacy of dialogue initiative relative to the task • Naturalness of the dialogue structure relative to the task • Output phrasing adequacy	• Sentence structures of the answer

The four design parameters derived from user requirements are defined as follows:

- Response time to get to next interaction - The time it takes for an intelligent product to respond to a user's question.
- Number of trials to get the right answer for a question - The number of the trial that an intelligent product challenged to give the right answer.
- Pace of the answer - The pace of answers that users can easily recognize.
- Sentence structures of the answer - The amount of information in the answer presented by the intelligent product after the user requests the information.

In order to design satisfactory HAII, an evaluation system is to develop for identifying the effect and proper values of the design variables. Before starting the evaluation experiments we have to specify a range and levels for each design parameter considered as independent variables. On the basis of commercial system values and the previous studies, the range of design parameters to be tested are determined as shown in Table 3 below.

Table 3. Scope of Critical Design Parameters.

Critical Design Parameters	Scope
• Response time to get to next interaction	• Between 1 to 5 sec
• Number of trials to get the right answer for a question	• 1 trials • 2 trials • 3 trials
• Pace of the answer	• 4 Syllable/sec • 6 Syllable/sec • 8 Syllable/sec
• Sentence structures of the answer	• Answer only • Repeat Question then Answer • Repeat Question then Answer with Clear Reference Source

The response time to get to next interaction is 2 seconds on average for the released intelligent product (Bixby&Siri), and a range of 1 to 5 seconds was conceived as a good control group to test its effect on the user satisfaction. Regarding the number of trials to get the right answer for a question, based on a result from a study that VUI is excusable for up to two errors [9], three levels from first to third trials were considered as a good range for the test. Pace of the answer was set at 6 Syllable/sec, which is the average response speed of Bixby as middle levels and set 4 Syllable/sec and 8 Syllable/sec as slow and fast paces. Lastly, the sentence structure of the answer was set as three different cases: feeds the answer back only, repeat the given question before the answer to make the user sure that the system understood the question correctly, and repeat the given questions before the answer with clear and convincing reference source.

3.2. Design for HAII Evaluation System

The HAII evaluation system was designed and implemented so that the system can emulate varying interaction conditions by combining different design parameter values described in the previous section. The system was devised to be good at conducting mixed within-subject design since the time and cost of the testing experiments as well as the fatigue of subjects will be burdensome if all the design parameters are considered as within-subject variables. Response time to get to next interaction and pace of the answer were considered as between-subject variables, and the remaining two parameters (number of trials to get the right answer for a question, sentence structures of the answer) were considered as within-subject variables. By this design, any subject can finish their tasks for only 9 trials. As an experimental session begins, the subject is given a question represented on a computer screen. The subject needs to initiate the interaction by asking the question to the HAII with their own voices. Then the HAII system feeds the answer back depends on the experimental protocol. The overall flow of the interaction scenario is shown in Figure 2 below.

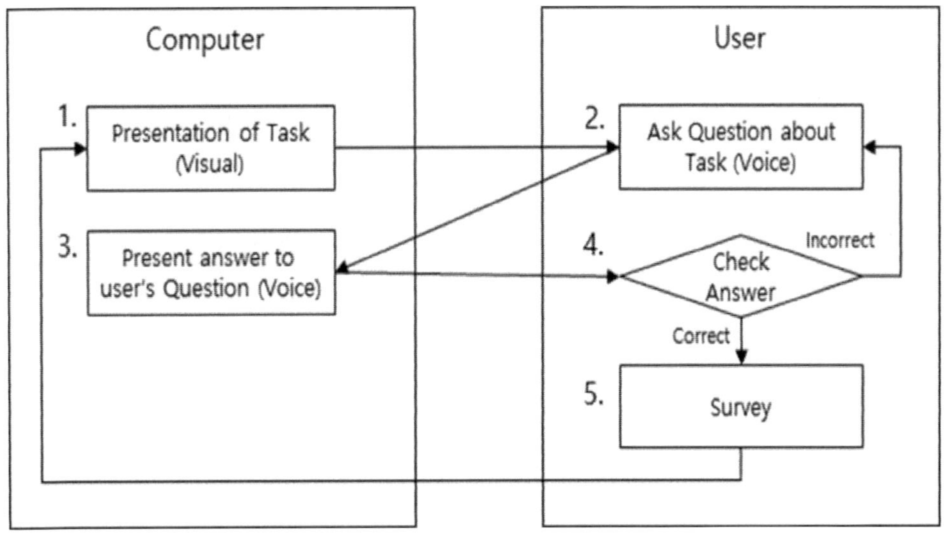

Figure 2. Flowchart for HAII Scenario.

Figure 3 illustrates the different scenarios or trials of experiments that a subject needs to get through for a given question.

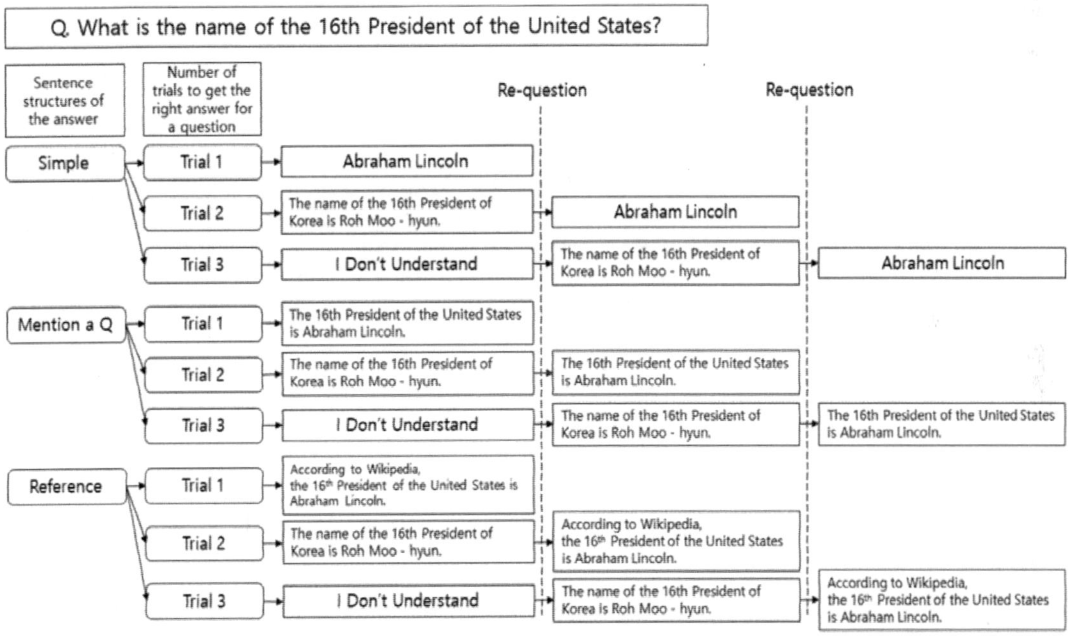

Figure 3. Example: Answer Design for Task.

In the given example above, a subject is given a question that "Who is the 16th President of the United States?" The subject asks to the evaluation system through the VUI for searching the answer for the question, and the evaluation system randomly feeds one of the nine scenarios back. The interface of the HAII evaluation system is shown in Figure 4 below.

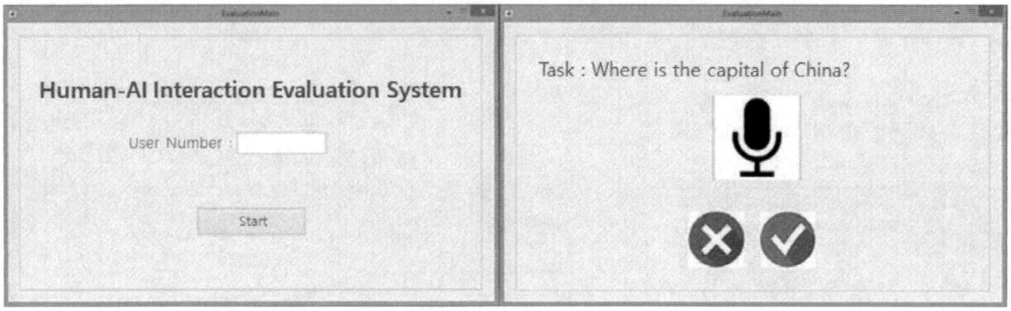

Figure 4. HAII Evaluation System Interface.
(Left: Start Screen, Right: Evaluation Screen)

4. Discussion

In this study, we derived four critical design parameters of the HAII that can affect user satisfaction during interacting under VUI environment. These design parameters were carefully identified by mapping the selected user requirements on functional performance indexes of HAII. We focused on 8 out of 25 user requirements that were dealt in previous studies. We took the 8 by eliminating those ones that are redundant or does not directly related to the quality of answers from the system. To get more thorough information about how to design satisfactory HAII, it is supposed to collect the user requirements more diversely in further studies including direct surveys on experienced users of the system.

The current version of HAII evaluation system implemented in this study considered the 4 design parameters with specified ranges, but the system is devised as flexible for the changes of parameters as well as their ranges. Since we adapt the Wizard of Oz method, it is easy to add or remove a set of design parameters having a variety of ranges. The evaluation system was developed as an embedded software so that it has another advantage of portability makes it free to apply on a variety of environments using VUI.

To validate the criticality of the four parameters in the HAII design, it is mandatory to proceed with further studies regarding real evaluation of user satisfaction. User satisfaction is a subjective matter and that is used to measure through verbal responses by rating scales or semantic differential method [10]. Non-verbal evaluation protocols utilizing bio-signals such as EEG, heart rate, facial expressions are suggested recently as alternative ways to supplement the verbal protocols [11]. In the next study, we are going to utilize both protocols to find the effects and the proper values of the four parameters derived in this study.

Acknowledgments

This research was supported by the Brain Korea 21 Plus Project (Dept. of Industrial Engineering, Kumoh National Institute of Technology).

References

[1] Gustavo L, Luis Q and Luis A G 2017 *Alexa vs. Siri vs. Cortana vs. Google Assistant: A Comparison of Speech-based Natural User Interfaces*. (Int. Conf. on Applied Human Factors and Ergonomics) pp 241-250

[2] Jinbae K, Saleh S A A S and Hyunsoo L 2018 *Pattern Recognition and Classifier Design of Bio Signals collected in HAiI* (Proc. of KIIS Fall Conf. Vol 28 Issue 3) pp 9-10

[3] Johnson H *Relationship between User Models in HCI and AI* 1994 (IEE Proc. Computer Digital Techniques Vol 141 No 2)

[4] Terry W 2006 *Shifting Viewpoints: Artificial Intelligence and Human–Computer Interaction* (Artificial Intelligence Vol 170 Issue 18) pp 1256-1258

[5] Jonggyu S, Ingwon J and Sangho K 2018 *A Study on the Difference between Human-AI*

IOP Conf. Series: Journal of Physics: Conf. Series **1284** (2019) 012036 doi:10.1088/1742-6596/1284/1/012036

Interaction and Human-Computer Interaction (Proc. of Ergonomics Society of Korea Fall Conf.)
[6] Bella M and Hanington B 2012 *Universal Methods of Design: 100 Ways to Research Complex Problems, Develop Innovative Ideas, and Design Effective Solutions* chapter 99 p 204
[7] Laila D and Wolfgang M 2008 *Recent Trends in Discourse and Dialogue* (Berlin: Springer-Verlag)
[8] Valeria F, Martins S, Andre L S K and Joao S O N 2010 *An Empirical Approach for the Evaluation of Voice User Interfaces (User Interfaces)* ed Rita M (Hungary: Eötvös Loránd University) Chapter 10
[9] Myeongji L and Kihyung H 2011 *Design and Implementation of a Usability Testing Tool for User-oriented Design of Command-and-Control Voice User Interfaces* (Phonetics and Speech Sciences Vol 3 No 2) pp 79-87
[10] Sangho K, Sunah K, Jongkyu S and Jeongyoon A 2016 *A Human Sensibility Ergonomic Design for Developing Esthetically and Emotionally Affecting Glass Panels of Changing Colors* (Journal of the Ergonomics Society of Korea Vol 35 Issue 6) pp 535-550
[11] Dongmin S, Dongil S and Dongkyoo S 2015 *Research of Real-Time Emotion Recognition Interface Using Multiple Physiological Signals of EEG and ECG* (Journal of Korea Game Society Vol 15 Issue 2) pp 105-114

IOP Conf. Series: Journal of Physics: Conf. Series **1284** (2019) 012037 doi:10.1088/1742-6596/1284/1/012037

Application of Data Mining Technology in Evaluating Real Estate Investment Plan Based on GRA-AHP

ZHUANG MEIER, WEN-TSAO PAN*, SHI ZHUOHONG, ZHOU YINGYING and ZHONG ZUCHANG

Guangdong University of Foreign Studies, Guangzhou, Guangdong, China

*E-mail: teacherp0162@126.com

Abstract. Data mining and information technology is increasingly important in the era of big data. Many researches have been discussing the application of data mining and information technology in all walks of life. To further explore the practical value of data mining technology in economic and business field, this paper studies its application in evaluating real estate investment plans. R software is applied as the statistical computing console and relative program code is created to implement grey relational analysis-analytical hierarchy process(GRA-AHP), which establishes an improved evaluation model of decision-making. Firstly, the related data of the research is divided into several layers, namely the target layer, criteria layer and plan layer. In the criteria layer, the processing method is different from previous research because in this paper the indicator value of the plan layer to the criteria layer is calculated based on grey relational analysis rather than the pairwise comparison judgment matrix proposed by previous scholars, which makes this analysis more scientific and accurate. Then the plan layer single sort weight and the total sort weight of each plan can be calculated to provide reference for the final investment decision. This paper aims to propose a better evaluation method in the real estate investment and promote the application of data mining technology in economic fields.

1. Introduction

As an up-to-date analytical method, data mining technology has been frequently used in various fields in the modern research. Among them, the combination of data mining technology and real estate industry is an increasingly important research hotspot [1-2].

So far, plenty of researchers have proposed all kind of related theories in the field. Jovanovic conducted sensitivity analysis for investment projects under risk conditions [3]; Borgonovo and Peccati discussed the overall situation of investment decisions under uncertain conditions [4]; Kollmann and Kuckertz analysed venture capitalists' risk investment decision-making processes [5]; Ke Xiaoling established a different evaluation system from several aspects [6].

Different from previous studies, in this paper data mining technology will be applied to analyse the real estate investment. The specific calculating methods used in the study are grey relational analysis (GRA) and analytic hierarchy process (AHP). According to the GRA theory, the real estate investment decision-making problem is a grey system where both known and unknown information are included. The indicators of real estate investment decisions are not independent of each other. [7-8]. As one of the analytical methods of data mining technology, AHP is a useful tool for decision-making and evaluation of complex problems proposed by American operations researcher T. L. Saaty [9]. It is suitable for analysing decision-making problems of complex systems composed of many interrelated factors. [10]. AHP has the advantages of system and flexibility and is widely applied in all sorts of fields such as economical management and public administration [11].

Therefore, to further explore the practical value of data mining technology in the real estate investment decision-making field, this paper combines GRA and AHP to construct the evaluation model of the real estate investment development plan. The association hierarchy analysis and evaluation system will be proposed in the paper, as well as the real estate investment evaluation model method based on GRA-AHP. In brief, what we focus on is the application of data mining technology in a real estate industry and how it spurs the development of this industry.

2. Research Methods

2.1. Analytic Hierarchy Process(AHP)

AHP is to decompose related elements into goals, criteria, programs and other levels, and conduct qualitative and quantitative analysis on this basis. AHP first proposes a general goal based on the nature and requirements of the problem. Then the problem is decomposed into layers, and the factors in the same level are determined by the pairwise comparison method to determine the respective weight coefficients relative to the upper layer target. This layer-by-layer analysis continues until the last level, giving a ranking of all factors relative to the overall goal.

2.2. Grey Relational Analysis(GRA)

GRA is a direct and comprehensive evaluation method when conducting data mining. The degree of association is essentially the degree of difference in geometry between curves. Therefore, the difference between the curves can be used as a measure of the degree of association. The research object of this method is the non-dimensionalized sequence, which is divided into a parent sequence and a subsequence. In most cases, the parent sequence, also called reference sequence, is a standard sequence in the grey relational analysis, reflecting the behaviour characteristics of the system. On the other hand, the subsequence, also named comparison sequence, is a sequence of objects in the gray relational analysis, which is a data sequence composed of factors affecting system behaviour.

2.3. Specific Steps of AHP based on GRA

First, establish a hierarchical structure model. Next, construct all judgment matrices in each level. Finally, conduct the hierarchical single sorting, hierarchical total sorting and consistency checking.

For each upper element, consider the lower elements that have a logical relationship with them, and make a pairwise comparison between them. The computing results of the judgment are given in quantitative numbers and are represented in a certain matrix, which is called a "judgment matrix".

GRA-AHP is to use the grey relational degree calculation to determine the weight of the relative importance ranking of each element in each level, list the mutual factors of the criterion layer mutual factor weight table and the program layer to the criterion layer. Through the comprehensive evaluation of each level, the total ranking weight of each scheme for the target layer is given.

3. Application of Investment Plan Evaluation Model Based on GRA-AHP

3.1. Annual Measurement Index

Here is a real estate development enterprise that is to carry out investment development planning. There are four investment options to choose from, and the results of the annual indicators of each program are shown in the following table summarized from a previous study [12]. (see table 1)

To improve the accuracy of analysis and reduce the subjectivity of decision-making, certain qualitative indicators such as expected sales situation and customer satisfaction have been graded according to uniform standards, and then processed into quantitative indicators through fuzzy analysis.

Table 1. Annual Indicators of Four Plans.

Index		Plan T_1	Plan T_2	Plan T_3	Plan T_4
Economic benefit	Financial internal rate of return (%)	16.3	15.97	15.16	15.65
	Financial net present value (Ten thousand yuan)	9579	10300	7931	9682
	Economic internal rate of return (%)	9.03	9.21	8.38	8.93
	Economic net present value (Ten thousand yuan)	2925.6	4120.5	2225.1	3461.2
Degree of risk	Investment loss (Ten thousand yuan)	96.8	110	97.9	110
	Project risk (%)	50	60	50	60
Investment returns	Total investment (Ten thousand yuan)	3100.2	3563.4	3369.8	3647.3
	Payback period(Year)	4.56	7.64	7.88	7.71
Market effect	Expected sales (%)	60	70	50	60
	Customer satisfaction (%)	50	60	60	50

3.2. Establish an Evaluation Index System

According to the fundamental theory of analytic hierarchy process, the evaluation index system is established in such a way as follows. The corporate real estate development investment plan is set as the target level A, which is the problem needed to be solved. The target layer A includes four sub-targets B1, B2, B3, and B4, each of which is a preferred criterion for a real estate development investment plan, so this layer is called a criterion layer and is represented by B. Each criterion layer contains several indicators, and the values of these indicators are applied to the total target A through their subordination principle B. The four options for development investment become the layer of the program to be evaluated, denoted by T.

The following figure illustrates how the evaluation system works. (see figure 1)

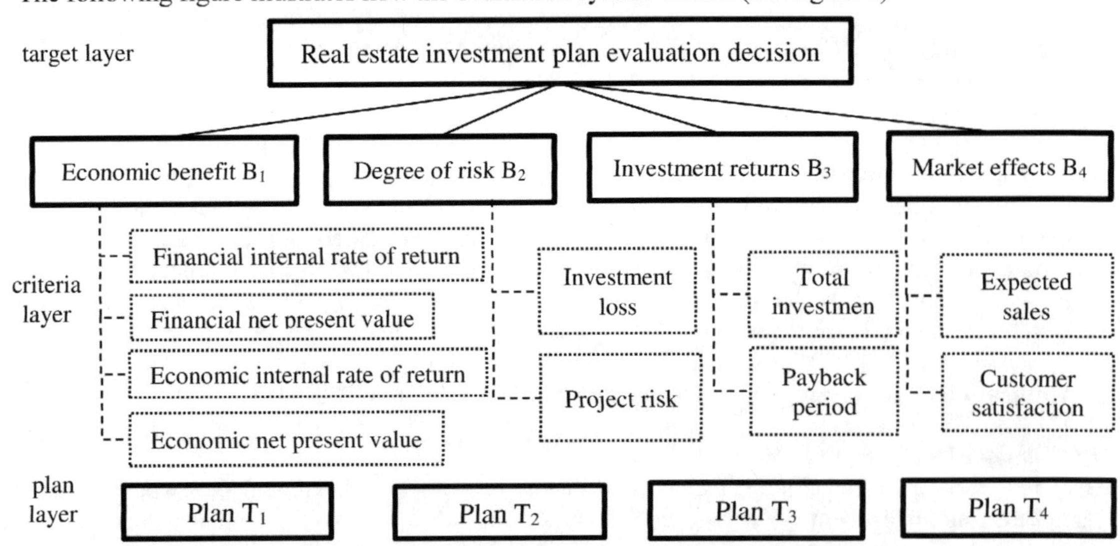

Figure 1. Evaluation Model Based on GRA-AHP.

3.3. GRA-AHP Empirical Analysis Steps

This paper is based on GRA-AHP, which is implemented in the R software using the GRA program.

Since each of the overall indicators in the criterion layer is composed of several subordinate indicators, the GRA of the subordinate indicators in the criterion layer is performed to obtain the overall indicators of the matching criterion layers of each project which needs selecting.

First, according to each subordinate index value below the criterion layer, the standard sequence and the comparison sequence of the subordinate index are calculated, and the gray correlation program is used to obtain the overall index value B of each scheme T relative to each criterion.

Specifically, the GRA is implemented on the four subordinate indicators of financial internal rate of return, financial net present value, economic internal rate of return and economic net present value, with the aim to obtain the corresponding economic benefit index values of each program; Similarily, the GRA is carried out on the investment loss amount and the project risk degree, and then the index value of the risk degree is obtained as well. In the same way, we use GRA to implement the data mining process on the total investment amount and the investment recovery period in order to obtain the investment income index value; It is nearly the same way to calculate the data of the sales situation and the customer satisfaction, and therefore, the market effect indicator value is obtained.

After the gray relational analysis processing, the overall index values of the corresponding criterion layers of each scheme are shown in the following table. (see table 2)

Table 2. Indicator Value of the Plan Layer to the Criteria Layer Based on GRA.

Target layer A	Plan T_1	Plan T_2	Plan T_3	Plan T_4
Economic benefit B_1	3/4	1	2/3	6/7
Degree of risk B_2	2/3	1	2/3	1
Investment returns B_3	2/3	1	5/6	1
Market effects B_4	1	1	2/3	1

Second, calculate the weight of the relative importance order of each element in the criterion layer. Specifically, after determining the index values of the corresponding criterion layers in the solution layer, the grey relational analysis is again used, and the index values obtained in the previous step are brought into the GRA program, and each scheme corresponds to the overall of the criteria. The index value is used as a comparison sequence, and the corresponding maximum value in each comparison sequence is used as a reference sequence, thereby calculating the weight of the relative importance ranking of each element in the criterion layer, and the results are shown in table 3.

Table 3. Weight Value of the Criteria Layer.

Criteria layer	Weights
Economic benefit B_1	5/7
Degree of risk B_2	4/7
Investment returns B_3	1/2
Market effects B_4	1

Then, according to the formulation principle of the judgment matrix in the analytic hierarchy process and the above-mentioned data, the weight values of the overall indicators in the criterion layer with respect to the target layer A are calculated, and the criterion layer B mutual factor weight table is listed. As it is shown in the table below. (see table 4)

Table 4. Criteria Layer B Mutual Factor Weight.

A	Economic benefit B_1	Degree of risk B_2	Investment returns B_3	Market effects B_4
Economic benefit B_1	1	4/5	2/3	7/5
Degree of risk B_2	5/4	1	7/8	7/4
Investment returns B_3	3/2	8/7	1	2
Market effects B_4	5/7	4/7	1/2	1

From the above calculation, it is convenient for us to work out the weights of the scheme layer to each criterion layer with the help of data mining technology, according to the all those known data. Hence, the scheme layer to the criterion layer mutual factor can be listed smoothly, which is illustrated in table 5.

Table 5. Scheme Layer to the Criterion Layer Mutual Factor.

Economic benefit B_1	Plan T_1	Plan T_2	Plan T_3	Plan T_4
Plan T_1	1	4/3	8/9	8/7
Plan T_2	3/4	1	2/3	6/7
Plan T_3	9/8	3/2	1	9/7
Plan T_4	7/8	7/6	7/9	1
Degree of risk B_2	Plan T_1	Plan T_2	Plan T_3	Plan T_4
Plan T_1	1	3/2	1	3/2
Plan T_2	2/3	1	2/3	1
Plan T_3	1	3/2	1	3/2
Plan T_4	2/3	1	2/3	1
Investment returns B_3	Plan T_1	Plan T_2	Plan T_3	Plan T_4
Plan T_1	1	3/2	5/4	3/2
Plan T_2	2/3	1	5/6	1
Plan T_3	4/5	6/5	1	6/5
Plan T_4	2/3	1	5/6	1
Market effects B_4	Plan T_1	Plan T_2	Plan T_3	Plan T_4
Plan T_1	1	1	2/3	1
Plan T_2	1	1	2/3	1
Plan T_3	3/2	3/2	1	3/2
Plan T_4	1	1	2/3	1

After constructing the judgment matrices, the analysis process is started using the procedure of the AHP. The numerical values in the judgment matrix obtained above are substituted into the AHP, and the results of the analytic hierarchy process data are illustrated in table 6.

Table 6. Data Mining Result of GRA-AHP.

Criteria layer B		Economic benefit B_1	Degree of risk B_2	Investment returns B_3	Market effects B_4	Total sort weight
Criteria layer weight value		0.2569610	0.2055688	0.1777249	0.3597453	
Scheme layer single sort weight	Plan T_1	0.2290909	0.2	0.1904762	0.2727273	0.2319459
	Plan T_2	0.3054545	0.3	0.2857143	0.2727273	0.2890514
	Plan T_3	0.2036364	0.2	0.2380952	0.1818182	0.2011641
	Plan T_4	0.2618182	0.3	0.2857143	0.2727273	0.2778386

Among the total ranking weights, the scheme T2>T4>T1>T3, which means T2 is the best investment plan and should be adopted or at least taken more consideration than others; T4 is the second appropriate selection, which can be regarded as an alternative for decision makers' reference; while T3 and T4 are relatively poor, they should be excluded. (To better present the data results and highlight the differences in the total ranking weights of the various schemes, the form of the decimals is restored here. In addition, after conducting GRA-AHP, the obtained value of CR is less than 0.1, which means it passed the consistency test and the results are reliable)

4. Conclusions

Data mining and information technology can greatly improve the effectiveness and efficiency of decision-making in the real estate investment plans. As a specific method of data mining technology, GRA-AHP exerts positive effects on the development administration of real estate industry, which is worthy of wide application.

Applying GRA-AHP helps us to comprehensively consider various factors affecting real estate investment decision-making, and organically combine qualitative and quantitative analysis, which can fully reflect the ambiguity of evaluation factors and evaluation process while minimizing the drawbacks caused by individual subjective judgments. Since the data mining technology is conducive to making full use of the data, the GRA-AHP can reflect the real estate development investment situation more realistically, facilitate the horizontal and vertical comparison of comprehensive evaluation, and provide a technical mathematical model for real estate development management. The analysis results based on data mining technology are more objective and practical than those ordinary evaluation methods such as scoring. Therefore, compared with previous evaluation methods, the data mining technology is much more scientific, thus the evaluation model established in this research are far more reliable and valuable.

Acknowledgments

The research is supported by Guangdong University of Foreign Studies Featured Innovation Program (project number: 299-GK18GS53) and National Natural Science Foundation of China (71673064), Innovation team project (Humanities and Social Science) of universities in Guangdong, China(2017WCXTD003).

References

[1] Luo Jing, Zhu Jianfeng. Empirical analysis of real estate price based on data mining and spatial measurement. *Research of Development*.2013, (5), p 128-131
[2] Sun Shanshan. Real estate price prediction based on data mining. *Modern Electronic Technique*. 2017,40(5), p 126-129
[3] Jovanovic P. Application of sensitivity analysis in investment project evaluation under uncertainty and risk. *International Journal of project management*. 1999, 17(4), p 217 -222
[4] Borgonovo E, Peccait L. Uncertainty and global sensitivity analysis in the evaluation of investment projects. *Journal of Production Economics*. 2006, 104(1), p 62-73
[5] Kollmann T, Kuckertz A. Evaluation uncertainty of venture capitalists' investment criteria. *Journal of Business Research*. 2010, 63(7), p 741-747
[6] Ke Xiaoling, Zhu Kejun, Diao Fengqin. Improved model for fuzzy comprehensive evaluation of real estate investment risk. *Reform and Strategy*. 2010, 26(4), p 43-45
[7] Zhang Ping, Ma Li, Zhai Bowen. The grey fuzzy comprehensive evaluation of real estate investment risk. *Journal of Technical Economics and Management*, 2018, (4), p 22-25
[8] Zhou Xiang, Jiang Genmou. Risk analysis of construction project based on combination weighting and improved grey interrelated. *Construction Management and Modernization*. 2018,103(6), p 78-81
[9] Saaty T L. How to make a decision: the analytic hierarchy process. *European Journal of Operational Research*,1990,48(1), p 9-26
[10] Deng Xue, Li Jiaming, Zeng Haojian, Chen Junyang, Zhao Junfeng. Research on computing method of AHP weight vector and its applications. *Mathematics in Practice and Theory*. 2012,42(7), p 93-100
[11] Liu Zhaoming, Han Bin. Analysis on regional venture capital environment based on AHP fuzzy comprehensive evaluation. *Journal of Southwest Jiaotong University*.2005, 40(3), p 379-384
[12] Sun Haitao. Fuzzy optimization model of real estate investment project based on AHP and its Application. *Enterprise Economy*. 2011,369(5), p 167-170

IOP Conf. Series: Journal of Physics: Conf. Series **1284** (2019) 012038 doi:10.1088/1742-6596/1284/1/012038

Truthful Mechanism for Task Assignment in Crowdsourcing with Social Network

Haiyan Qin[1*], Yonglong Zhang[2], and Bin Li[3]

[1]Guangling College, Yangzhou University, China
[2]College of Information Engineering Yangzhou University, China
[3]State Key Laboratory for Novel Software Technology Nanjing University, China

E-mail: 13092006508@163.com

Abstract. Macrotasks like document writing, product design, or web development, require expert skills and the collective contributions of the crowds. In fact, there are few studies give close attention to the workers with social network in crowdsourcing. Social networks provide an available platform for cooperation among social workers. In this paper, we discuss the complex task assignment in crowdsourcing where the workers are in a social network. A Truthful Mechanism for the task assignment in Crowdsourcing with Social Network (TMC-SN) is proposed to solve the challenges in this scenario. The problem is modeled as an auction, where the requesters are buyers, workers are sellers, and the crowdsourcing platform serves as an auctioneer. In order to find out the most appropriate team, TMC-SN measures Worker Fitness of the worker for the team from both Marginal Contribution and Team Cohesion. We demonstrate that TMC-SN has fine economic properties such as truthfulness, individual rationality, and budget balance. The evaluation results show that TMC-SN is advantageous in terms of Social Welfare and favorable for the workers.

1. Introduction

Crowdsourcing, coined by Jeff Howe in 2006, is defined as the method of outsourcing a specific task to an set of uncertain people instead of assigning it to designated employees. From then on, crowdsourcing is prevalent in many fields. In fact, the technique of crowdsourcing can be traced back to 1884 when Oxford English Dictionary depended on a crowd to catalog words[1]. Amazon Mechanical Turk is a functional crowdsourcing platform where employers release tasks such as translation and labelling etc and Internet workers like to complete tasks with small rewards. Despite micro payments, part-time or full-time crowds are willing to take different types of tasks which not only decreases overall expense of the requester but also helps to generate high-quality work.

It is worth noting that there are few studies that give close attention to the workers with social network in crowdsourcing. Conversely, the scientific collaboration network has been considered as a representative social network in previous studies, in which teamwork is an important issue in scientific and technological activities[2]. Obviously, some research projects are too complex to be performed by a single scientist. Therefore, scientific projects are necessarily conducted with the collaboration of many scientists who tend to form or have builded up communities in accordance with their research topics and interests. Social networks provide an available platform for cooperation among social workers. In crowdsourcing, TopCoder is a representative crowdsourcing marketplace for software development which needs the cooperation of a set of workers to generate the project. Actually, once the social network of the workers is introduced into crowdsourcing, it is triple-win for the workers, task requesters and the the platform. The workers would rather collaborate with those who are familiar with than those who need to be further run-in. Then, the task would be completed more outstandingly and efficiently. Task requesters are certainly glad to receive the finished task with higher-quality. Furthermore, the platform may hire less workers and then cut costs.

In this paper, we discuss the complex task assignment in crowdsourcing where the workers are connected. The problem is modeled as an auction, where the requesters are buyers, workers are sellers, and the crowdsourcing platform serves as an auctioneer. After the requester posts the task and the connected workers submit skills, the platform will make the assignment. In this scenario, there are several unsolved challenges. First of all, it has been proved that the complexity of task assignment in social networks remains NP-hard[3]. Thus, presenting an efficient mechanism is the most important issue. After that, the workers in social networks are connected with complex relationship. In that way, the mechanism should figure out the relations of the connected workers and then find out a suitable team. Last but not least, the self-serving workers may attempt to misreport their ask to gain a higher utility. Therefore, the mechanism has to be truthful.

A Truthful Mechanism for the task assignment in Crowdsourcing with Social Network (TMC-SN) is proposed to solve the challenges mentioned above. In TMC-SN, an efficient assignment within polynomial time complexity is proposed. In order to find out the most appropriate team, TMC-SN measures Worker Fitness of the worker for the team from both Marginal Contribution and Team Cohesion. Conventional solution ignored the self-centred players who may misreport their value in order to gain higher utility. However, auction is introduced into TMC-SN as a valid method to forbid the players from lying. TMC-SN satisfies all the properties of auction, truthfulness, individual rationality, and budget balance.

The remainder of this paper is structured as follows. The detailed system model and some definitions are described in Section 2. We present the design of TMC-SN in detail in Section 3 and then analyse related properties, specifically the truthfulness. Our simulation results are presented in Section 4, followed by conclusions in Section 5.

2. Problem Formulation

2.1. *System Model*

Social Networks: A social network, $SN = \langle A, E, D\rangle$ is an undirected graph, where $A = \{a_1, a_2, \ldots, a_m\}$ is the set of workers. $\forall (a_i, a_j) \in E$ indicates that there is a direct social interaction between a_i and a_j. Each pair of workers, $(a_i, a_j) \in E$ is associated with a cost $d(a_i, a_j) \in D$ incurred by the necessary communication and $d(a_i, a_j) = d(a_j, a_i)$. In other words, $d(a_i, a_j) \in D$ is the value of edge for $(a_i, a_j) \in E$.

Requester: The requester posts her task R with bid b to the platform. b is the requester's claimed value if the task is completed and $\tilde{b}$ is the truthful value. We assume that there are τ types skills $S = \{s_1, s_2, \ldots, s_\tau\}$ the requesters need. $S_R = \{s_r^1, s_r^2, \ldots, s_r^\tau\}$ is a τ-dimension vector, where $s_r^k = \{0, 1\}$ means that k-th skill is required or not in R. We assume that there is an uncovered skill vector $S_R^{'}$ as the members are added into the team and $S_R^{'}$ is the same as S_R at the beginning of process.

Worker: Each worker $a_i \in A$ submits a tuple $C_i = \{< s_i^k, c_i^k >_{k=1,2,\ldots,\tau}\}$ to the platform. Among C_i, $s_i^k = \{0, 1\}$ denotes whether a_i holds the k-th skill or not and c_i^k is the reward that a_i expects to be paid by contributing s_k. c_i^k is not necessary equal to $\tilde{c}_i^k$ which is the true cost of performing the s_k. $C = \{C_1, C_2, \ldots, C_m\}$ is the bid vector of all the workers. $Nei(a_i)$ is the set of agents who have the direct connection with a_i.

The worker may contribute one or several skills to the task. Therefore, we import overall cost oC_i to deal with the gathering skills. $H = \{h_1, h_2, \ldots, h_\tau\}$ represents the weight vector of S, where h_k means the workload of the skill s_k. If a_i has the skills that is required in $S_R^{'}$, the overall cost of a_i can be calculated as follows.

$$oC_i = \frac{\sum_{k=1}^{\tau} h_k s_r^k s_i^k c_i^k}{\sum_{k=1}^{\tau} h_k s_r^k s_i^k} \tag{1}$$

Platform: After gathering information about requesters and workers, the platform produces the outcome of the auction, including the chosen team members $T \subseteq A$ and the payment vector $P = \{p_1, p_2, \ldots, p_m\}$ of the workers in T.

Therefore, the utility of the requester is the difference between the value and the payment:

$$U_R = \begin{cases} \tilde{b} - \sum_{a_i \in T} p_i & \text{if } R \text{ is completed },\\ 0 & \text{otherwise.} \end{cases} \tag{2}$$

Similarly, the utility of winning worker $a_i \in T$ is the reward from the auctioneer minus the cost of supplying the skills:

$$U_i = \begin{cases} p_i - \tilde{oC_i} & \text{if } a_i \in T ,\\ 0 & \text{otherwise.} \end{cases} \tag{3}$$

Here, $\tilde{oC_i}$ is the true overall cost corresponding to oC_i.

2.2. Definition of Concepts

In this section, we introduce several economic properties we are desirable to achieve and some concepts we will use.

Definition 1. *(Truthfulness). An auction is truthful shows that neither buyers nor sellers can gain a higher utility by deviating from their true value or cost. In our auction, it implies for* $\forall a_i \in A$, *if* $oC_i = \tilde{oC_i}$, U_i *is maximized; for* R, *if* $b = \tilde{b}$, U_R *is maximized.*

Truthfulness is the most crucial property in auction theory. Since exposing true bid or ask procures the highest utility, no rational buyer or seller would cheat any more. As a result, the trade can be free from market manipulation. In addition, irrespective of other's behavior, players would simply apply their truth-telling strategy so that the system would reach an equilibrium. Myerson theorem is the most common theory used in auction which is shown as follows.

Theorem 1. *(Myerson theorem) An auction mechanism is truthful if and only if [4]:*

- *Monotone allocation: If* a_i *is allocated successfully by bidding* c_i^k, *he can also win by bidding* $c_k^{'} \leq c_k$.
- *Critical value: If* a_i *asks higher than the critical value, he will not win the auction.*

Definition 2. *(Individual Rationality). A winning seller is paid more than his cost and the winning buyer is charged less than her budget. In other words, if the trade is successful, for* $\forall a_i \in T$, $p_i \geq oC_i$; *for* R, $\sum_{a_i \in T} p_i \leq b$.

Definition 3. *(Budget Balance). Budget balance means that the budget from the requester is more than the total payment to the workers, i.e.,* $b - \sum_{a_i \in T} p_i \geq 0$.

Definition 4. *(Social Welfare). Social Welfare is calculated as the aggregate utility of all the winning buyers, sellers and the auctioneer. It is also defined as the difference between the sum of winning buyers' value and the sum of winning sellers' costs.*

3. TMC-SN Mechanism

3.1. Design Details

Many factors will influence the performance of a team. The requester prefers the workers who can make more contributions to the team. It is mentioned in Ref. [5] that group cohesion was positively related to overall performance. Therefore, considering both contribution and cohesion, we give the definitions as follows.

Definition 5. *(Marginal Contribution). If* a_i *is selected into the team* T, a_i*'s Marginal Contribution is the workload that* a_i *can take to the uncovered skills in* $S_R^{'}$, *i.e.,*

$$\Delta_i = \sum_{s_r^k \in S_R^{'}} s_r^k s_i^k h_k \tag{4}$$

Definition 6. *(Team Cohesion). Team Cohesion is calculated as the communication cost with his neighbours if* a_i *is selected to* T, *i.e.,*

$$\Gamma_i = \sum_{a_j \in T'} d(a_i, a_j) \tag{5}$$

Definition 7. *(Worker Fitness). Both Marginal Contribution and Team Cohesion are involved in the fitness of* a_i. *Then, the fitness of* a_i, $fitness\,(a_i)$, *is defined as*

$$fitness(a_i) = \alpha * nor\,(\Delta_i) + (1 - \alpha) * nor\left(\frac{1}{\Gamma_i}\right) \tag{6}$$

Here, $\alpha \in [0, 1]$ *is a tune parameter between Marginal Contribution and Team Cohesion and* $nor(\cdot)$ *is the normalization function.*

The TMC-SN mechanism is shown in Algorithm 1. At the beginning of the mechanism, we search for the neighbours of the teammates and add them into $Neis$. In allocation rule, the worker with the smallest cost per fitness is selected as the candidate, i.e., the lowest $\frac{oC_i}{fitness(a_i)}$. It's worth mentioning that the candidate workers are connected with each other. Only the neighbors of $a_i \in T$ will be added into the team. In other words, the winning workers form a connected subgraph at the end of the process. The payment rule follows the Myerson Theorem in order to make sure that the selected worker is paid the highest cost he can report or else he will not be selected. We rebuild the team without the participation of a_i. It select a_j from $\{Neis^{'} \setminus T\}$ with the smallest cost per fitness and a_j still has the connection with the existing workers. Therefore,

$$\begin{aligned} \frac{oC_i}{fitness\,(a_i)} &\leq \frac{oC_j}{fitness\,(a_j)} \\ oC_i &\leq \frac{oC_j}{fitness\,(a_j)} * fitness\,(a_i) \end{aligned} \tag{7}$$

In the case that $\frac{oC_i}{fitness(a_i)} > \frac{oC_j}{fitness(a_j)}$, a_j will be selected instead of a_i.

Repeat selecting a_j until the budget is run out or the skills are covered. Throughout the process, the largest value $\{\frac{oC_j}{fitness(a_j)} * fitness\,(a_i)\,, p_i\}$ is set as the critical value of a_i. Moreover, if the payment for a_i is less than remaining budget, a_i will be the winner. During the process, Δ_i and Γ_i are updated once a new worker is added into the team. In the end of the process, the trade is a success in the case that the skill requirements of R are covered.

Algorithm 1 TMC-SN

Input: SN, b, S_R, C.
Output: T, P.

$T \leftarrow \varnothing, S_R^{'} \leftarrow S_R, p_{1\sim n} \leftarrow 0$
while $A \neq \varnothing$ and $S_R^{'}$ hasn't been covered **do**
  if $T \neq \varnothing$ **then**
    $Neis \leftarrow T \cup \{\cup_{a_i \in T} Nei\,(a_i)\}$
  else
    $Neis \leftarrow A$
  end if
  $a_i \leftarrow argmin_{a_i \in Neis \setminus T} \frac{oC_i}{fitness(a_i)}$
  $Neis^{'} \leftarrow Neis \setminus \{T \cup \{a_i\}\}$
  $K \leftarrow T$
  while $b \geq p_i$ and $\Delta_j \neq 0$ **do**
    $a_j \leftarrow argmin_{a_j \in Neis' \setminus K} \frac{oC_j}{fitness(a_j)}$
    $p_i = max\{\frac{oC_j}{fitness(a_j)} * fitness\,(a_i)\,, p_i\}$
    $K \leftarrow K \cup \{j\}$
  end while
  if $b \geq p_i$ **then**
    $T \leftarrow T \cup \{a_i\}, b \leftarrow b - p_i$
    update the uncovered skills in $S_R^{'}$
  else
    $A \leftarrow A \setminus \{a_i\}$
  end if
end while
if the skills of S_R have not been covered **then**
  $T \leftarrow \varnothing, p_{1\sim n} \leftarrow 0$
end if

3.2. Analysis of Desirable Properties

Having given the detailed design of TMC-SN, we now prove that TMC-SN achieves truthfulness, individual rationality, and budget balance.

Theorem 2. *TMC-SN achieves truthfulness.*

Proof: The truthfulness of the requester is obvious. There are two possible reasons why R can not be allocated. One is that the skills of R cannot be covered by the remaining workers, and the other is that the budget is limited. If it's the first reason, the requester can not be assigned anyway. If a requester with a limited budget wants to be successfully allocated, the requester must raise the bid until it is higher than the asks of the team. However, it would result in a negative benefit for the requester. Therefore, the requester will not lie about his bid.

For the workers, it is obvious that TMC-SN has a monotone allocation. If a_i is selected by bidding c_i^k, he can also be selected by bidding $c_i^{k'} \leq c_i^k$ with a smaller $\frac{oC_i}{fitness(a_i)}$.

The payment to a_i is decided by selecting a team without a_i's participation. a_j is selected from the set $Neis\backslash\{K \bigcup a_i\}$ with the minimum $\frac{oC_j}{fitness(a_j)}$. In the case that $\frac{oC_i}{fitness(a_i)} > \frac{oC_j}{fitness(a_j)}$, a_j will be selected instead of a_i. The payment to a_i is temporarily equal to $p_i^{'} = \frac{oC_j}{fitness(a_j)} * fitness\,(a_i)$. Here, $p_i^{'}$ isn't the highest price which a_i can report because his competitors in $Neis\backslash\{K \cup a_i\}$ haven't been considered entirely and the skill requirements of a_i haven't been covered. Therefore, the process of iteration continues until all the skills are covered or the budget is exhausted. Finally, the payment is the critical value, i.e., $p_i = max\{\frac{oC_j}{fitness(a_j)} * fitness\,(a_i)\,, p_i\}$. Because of critical value, if a_i asks higher than p_i, a_i will be placed after the last selected worker, then a_i will not be selected as the winner. From above, the critical value has been proved.

Adhering to Theorem 1, the truthfulness of workers is proved. ■

Theorem 3. *TMC-SN achieves individual rationality.*

Proof: For the requester, $b \geq p_i$ is checked every time before including a new teammate and the budget is updated after a successful trade. In other words, remaining budget is always higher than the payment to the new joined workers. Thus, $U_R = b - \sum_{a_i \in T} p_i \geq 0$ is satisfied. For the workers, we have $oC_i \leq \frac{oC_j}{fitness(a_j)} * fitness\,(a_i) \leq max\{\frac{oC_j}{fitness(a_j)} * fitness\,(a_i)\,, p_i\} = p_i$. Therefore, $U_i = p_i - oC_i = p_i - o\tilde{C}_i \geq 0$ is satisfied.

In general, the individual rationality of TMC-SN is achieved. ■

Theorem 4. *TMC-SN achieves budget balance.*

Proof: According to the payment rule, the remaining budget is checked whether enough for the payment of the new joined workers. Hence, we have $b - \sum_{a_i \in T} p_i \geq 0$. The budget balance of TMC-SN achieves. ■

4. Performance Evaluation

In this section, we simulate a crowdsourcing platform where workers are in a social network. In the experiment, the truthfulness of the TMC-SN is verified. The TMC-SN mechanism is compared with the complex task mechanism design without social network structure, non-SN, in terms of number of transactions, utility, and social welfare.

4.1. Simulation Settings

The number of workers m varies from 10 to 100 with a default value of 50. The bid of requester is randomly distributing over $(0.5, 6.5]$. The ask of workers for a single skill is randomly selected in the interval $(0, 1]$. If $(a_i, a_j) \in E$, the distance between a_i and a_j, $d\,(a_i, a_j)$, is varied from 2 to 12. We assume that the requester demands 6 kinds of skills at most. The tune parameter α is randomly selected in the interval $(0, 1]$, and the default value is 0.5. Some indicated datum in the figures is the average result of 1000 independent instances in each setting. In each instance, there is only one requester post one task.

4.2. Truthfulness

In this section, the truthfulness of players in TMC-SN mechanism is analyzed. One winner and one loser are randomly selected, and then their utilities are examined when they ask or bid different value. As shown in Figure 1(a), a randomly selected winning requester bids truthfully with $b = \tilde{b} = 4.57$ and achieves utility 2.13. When the winning requester attempts to decrease his bid to gain a higher utility, he would fail and gain no utility. Similarly, the loser case in Figure 1(a) indicates that when a losing requester raises his ask to win the auction, he would gain a negative utility. The truthfulness of workers in Figure 1(b) can be explained in the same way.

4.3. Comparison of TMC-SN and non-SN.

In this section, non-SN is introduced to compare with TMC-SN. In the non-SN mechanism, the social relations among workers are neglected. In that case, the Worker Fitness is only related to Marginal Contribution, i.e., $\alpha = 1$. From Figure 2(a), it can be observed that with the increase of workers, the number of transactions of non-SN is always higher than that of TMC-SN. This is because that TMC-SN considers the Team Cohesion of workers. The more stringent the trading are, the lower the transaction rate will be. Figure 2(b) indicates that the utility of requester increases as the augment of workers while the utility of workers decreases. It can be seen that the utility of requesters in non-SN is better than that in TMC-SN, but the utility of workers in non-SN is worse.

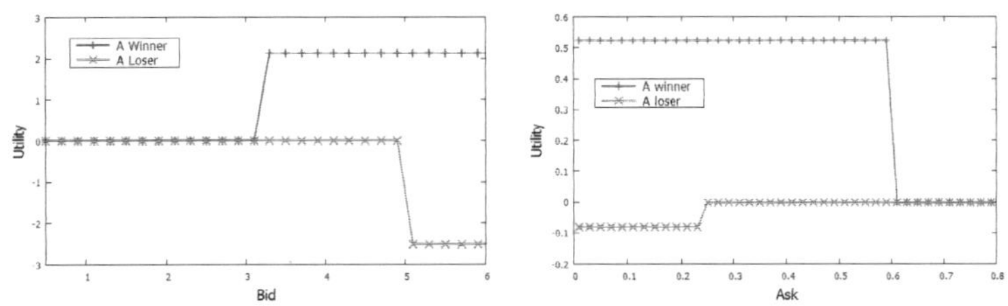

(a) Truthfulness for requesters. (b) Truthfulness for workers.

Figure 1: Truthfulness.

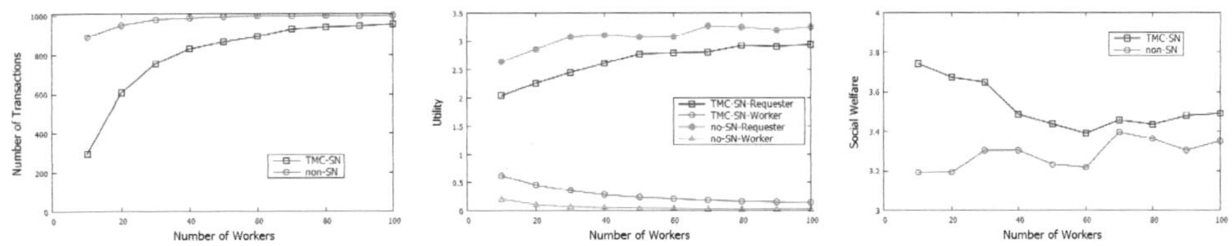

(a) Number of transactions. (b) Utility. (c) Social Welfare.

Figure 2: Comparison of TMC-SN and non-SN.

This demonstrates that the concern in Team Cohesion is profitable for workers but disadvantageous for requesters. Figure 2(c) shows that the Social Welfare of TMC-SN and non-SN is fluctuant. However, Social Welfare in TMC-SN performs better than non-SN as a whole. The concern in Team Cohesion is serviceable in teams of Social Welfare.

5. Conclusion

In this paper, we discuss the complex task assignment in crowdsourcing where the workers are in a social network. The TMC-SN mechanism is designed to solve the challenges in this scenario. In order to find out the most appropriate team, TMC-SN measures Worker Fitness of the worker for the team from both Marginal Contribution and Team Cohesion. The payment rule follows Myerson theorem. The theoretical analysis has proved that TMC-SN achieves truthfulness, individual rationality, and budget balance. In performance evaluation, the truthfulness of TMC-SN is further verified. We also compare the performance between TMC-SN and non-SN. The evaluation results show that TMC-SN is advantageous in terms of Social Welfare and favorable for the workers.

Acknowledgment

This work was funded in part by Open Project Foundation of Information Technology Research Base of CAAC (Grant No.CAAC-ITRB-201604); Natural Science Research Project of Guangling College of Yangzhou University (Grant No.ZKYB180020).

References

[1] Anand Inasu Chittilappilly, Lei Chen and Sihem Amer-Yahia, A Survey of General-Purpose Crowdsourcing Techniques, IEEE Transactions on Knowledge and Data Engineering, vol. 28, no. 9, pp. 2246-2266, April, 2016.

[2] Jordi Duch, Joshua S. Waitzman and Luis A. Nunes Amaral, Quantifying the Performance of Individual Players in a Team Activity, PlOS ONE, vol. 5, no. 6, p. e10937, June 2010.

[3] Mathijs de Weerdt, Yingqian Zhang and Tomas Klos, Distributed Task Allocation in Social Networks, in Proc. 6th AAMAS, Honolulu, May, 2007, pp. 500-507.

[4] R. Myerson, Optimal auction design, Mathematics of Operations Research, vol. 6, no. 1, pp. 58-73, February 1981.

[5] Heng-Li Yang and Jih-Hsin Tang, Team structure and team performance in IS development: a social network perspective, Information & Management, vol.41, no. 3, pp.335-349, 2004.

IOP Conf. Series: Journal of Physics: Conf. Series **1284** (2019) 012039 doi:10.1088/1742-6596/1284/1/012039

Construction and Application of MOOC Resources Database for Software Engineering Major Based on Integrable Ware

Jiujiu Yu[*], Jishan Zhang, Liqiong Pan, Ning Wu and Yuefang Liang

College of Computer Engineering, Anhui SanLian University, Hefei, China

[*]yjjyjL@163.com

Abstract. In recent years, the construction of MOOC (Massive Open Online Course) resources database has been paid more attention on MOOC research field. Based on the idea of integrable ware, MOOC elements are designed to form the corresponding MOOC learning knowledge points for some courses of software engineering major in a local university. They have the characteristics of openness, flexible and reusable. MOOC elements are fully applied in construction of the MOOC resources database on local campus for personalized learning effectively. The results show that teachers can select several different MOOC elements to form an important knowledge point for MOOC teaching process according to an actual course and reuse some existed MOOC elements to form another MOOC knowledge point for other courses. Different MOOC elements can be combined freely to form the related learning knowledge points for one course of software engineering major to stimulate and improve students' interests in personalized learning. Finally, further work is put forward. Firstly, compatible supported conditions on software and hardware are to be created based on big data. Secondly, capacities on integrating and reusing of various types of MOOC resources by teachers should be improved. Thirdly, more reusable MOOC elements based on integrable ware for the courses of other majors should be designed and constructed for personalized learning.

1. Introduction

Accelerating the process of education informatization, focusing on the development and application of high-quality education resources, constructing a modern network teaching resource system and introducing more high-quality teaching resources are effective ways to improve the quality of teaching level in universities and reduce the cost of higher education [1]. The improvement of construction of higher education teaching resources database have a direct impact on the formation and development of quality, healthy and comprehensive development of students after they participate in the work in the future.

MOOC (Massive Open Online Course) has been widely focused on in recent years. Unprecedented development and researches on MOOC in distance education field, and MOOC has played an active role in education fairness, teaching and learning at any time and any place, online interaction, and online evaluation [2]. Learning resources on MOOC are mostly designed according to the structure of the chapter (module)-section (unit), each section is connected with a complete knowledge point, and the whole of the MOOC resource is mostly concerned from the technical and application. MOOC is for the massive learning groups in society, which the contents of MOOC are complete and full opening, and it is difficult for MOOC resources to match the curriculum goals and the accumulation of existing knowledge of most social learners [3, 4]. Although construction activities on MOOCs are carried out in many local universities in China, researches on designing and innovation of MOOC resources are ignored. "Internet+" has provides much wider space for the construction of teaching resources on MOOC, but basic theoretical research such as knowledge attributes of MOOC resources is not enough yet [5, 6].

In this paper, MOOC resources are designed that based on integrable ware for some courses of software engineering, software testing, software architecture, object-oriented programming, and UML, and are fully implemented for MOOC construction of the software engineering major in a local university. The structure of the paper is organized as follows: Section 2 presents the relevant concepts and technologies of integrable ware and resources database. Section 3 and section 4 descript the construction and application of the MOOC resources database for the software engineering major. Section 5 concludes the paper and puts forward the further work.

2. Relevant Concepts and Technologies

2.1. Integrable ware

Integrable ware is also called micro-courseware, and the major principle is to use the unified coding standard to decompose the knowledge into tiny units [7]. These tiny units can be freely reorganized to form a knowledge section (unit) according to the actual learning needs for learners, and they are not limited to the contents of a particular course. Integrable ware has the characteristics of openness, universality and flexibility. The idea of integrable ware emphasizes that students are the main body of cognition, and the construction of the learning knowledge is completed by learners themselves finally. When the idea of integrable ware is applied to the development of MOOC resources, the MOOC resources can be minimized and integrable, thus fulfilling the requirements of MOOC's universality, openness and flexibility, and new design ideas for MOOC resource development can be provided [6].

2.2. Technology of DHTML dynamic web

DHTML supports the development of dynamic Web systems for resources database in the form of HTML and WWW access. It also supports Java, Script, CSS, DOM and other technologies for cooperating with each other, and it has the advantages of dynamic style, dynamic content and dynamic positioning [8]. DHTML can interact with the system for repository access through hardware devices such as keyboard and mouse, and it can be responded accordingly. It can make the interface operation by its own way or respond to the user's operation, and realize the movement of information such as text and pictures on the page.

2.3. Technology of XML database

XML database is a data persistence software system that allows data to be specified, and sometimes stored, in XML format [9]. This data can be queried, transformed, exported and returned to a calling system [9]. Data processing requirements can be implemented by XML database directly, and it also has a function of UI interface friendly. It provides a great convenience in designing, invocation, modification, and combination of integrable ware of MOOC resource.

2.4. Technology of message push

Message push technology is a technology for the field of web application development, which means that the server sends the information for the users' interests and purpose to the end users on time in an active manner according to the needs of the user. It searches and filters information based on users' interests and pushes them to users on regular time to help users to discovering valuable information efficiently [10].

The message push function on the platform of PC or smart phone can be well implemented by some network protocols of MQTT, RSMB, and XMPP and so on. Of course, there are also some third-party push platforms in domestic that can provide the message push service by means of payment.

3. Construction of MOOC Resources Database

3.1. Design of integrable ware for MOOC resources

A knowledge point of MOOCs for learning is needed to be divided into several MOOC integrable wares. Integrable ware is the smallest unit on MOOC resources, which is also known as the MOOC element that cannot be decomposed [6]. Users can freely combine these MOOC elements to form the corresponding resources for one MOOC or several MOOCs according to the requirement for local learning. All MOOC elements will be coded uniformly and stored in the database of the MOOC platform by a certain file format to form a MOOC resources database.

Different from the idea on construction of traditional MOOC resources, flexible and reusable are the most characteristics on MOOC integrable ware.

For the MOOC integrable ware, each MOOC element should be followed the designing criteria for MOOC resources. The description of a MOOC element is allowed in the form of text, image, animation, video, and so on. Teachers can choose the most appropriate description form to design each MOOC element according to the actual teaching situation, but the picture and its accompanying essay for the description should be both excellent.

As a case of an important MOOC element of "Class Diagram" for software engineering major, the framework of the information on a MOOC element for designing is shown as Table 1.

Here, MOOC elements cover most of the important knowledge points on the courses of software engineering, software testing, software architecture, object-oriented programming, and UML for software engineering major. The list of the MOOC elements is shown as Table 2. Some new MOOC elements could be added during the learning process, and the list of the MOOC elements is often updated.

Table 1. Framework of the information on a MOOC element.

No.	Name	Text Description	Image Description	Animation Description	Video Description	Others
19	Class Diagram	*Describe the definition and characteristics of a class diagram(.PPT)	*Describe the relationships between classes (.JPG/.PNG)	*Describe the formation process of a class diagram (.SWF/.FLV)	*Describe the applications of class diagram (.MPG4<=5mins)	

Table 2. The list of MOOC elements.

No.	Name	Knowledge Point	Comments
……			
19	Class Diagram	UML Diagram	
20	Sequence Diagram	UML Diagram	
……			
250	Code Review	White Testing	Unit Testing by Coders
……			
489	Boundary Value Analysis	Black Box Testing	
……			

3.2. Development of subsystem for combination of MOOC integrable ware

The main function of the subsystem is to provide teachers with the ability to combine MOOC elements freely. For one hand, teachers can select the corresponding MOOC integrable wares to form a MOOC learning knowledge point according to the actual teaching of a course. On the other hand, learners can also combine the MOOC integrable wares freely according to the learning interests to complete the personalized MOOC learning.

The subsystem for combination of MOOC integrable ware is developed by JSP, which is based on MyEclipse 2015. SQL-Server 2010 is the database management tool. The interface on this subsystem is beautiful and easier to operate. Each MOOC element is stored in the database of the subsystem in the form of a standard encoded file, to convenient for users to export and retrieve by keywords. As the registered and authorized learners on local campus, the elements which are extracted by users can be displayed in the resource list, and they are free to be selected and combined for different learners. Then, a complete MOOC knowledge point is generated for displaying finally. Due to limits on length of the paper, here, only the functions of the subsystem are briefly described, but the development process is omitted.

3.3. Construction of MOOC resources database

The construction of MOOC resource database needs to be closely integrated with the daily teaching content to achieve integrated management [11]. On one hand, it is necessary to complete the functions of updating, accessing, and publishing of teaching resources in a unified system. On the other hand, openness, shareable, and usability are the guiding principle on construction of MOOC resource database.

The author's experimental university relies on a lightweight MOOC platform (abbreviated as E-MOOC platform) in domestic to construct of MOOC resources database for some courses of software engineering major, and organizes students on senior grades from the college of computer engineering on local campus to carry out learning activities on these MOOC courses actively, such as software engineering, software testing, software architecture, object oriented programming, and UML.

Technology of DHTML dynamic web, XML database, and message push are fully applied in E-MOOC platform, and enable to provide functions of MOOC course management, exam management, interactive topic management, score management, MOOC learning and management, participation in interaction, message push, and so on. As a new and important part, the subsystem for combination of MOOC integrable ware is integrated into the E-MOOC platform. The new function of MOOC integrable ware database is added to improve the function of MOOC course management. It provides great convenience on the generation, reorganization and reuse of MOOC resources on local campus.

4. Application of MOOC Resources Database

The application of MOOC resources construction based on integrable ware for actual teaching is also the driving force for the development of MOOC resources database on local universities. Development and construction of combinable and reusable MOOC resources, teaching method on network in the period of "Internet+" can be innovated, and a MOOC resources database is formed initially which is corresponding to the software engineering subject knowledge system.

Such as Table 1, more than five hundred MOOC elements has been designed that based on integrable ware are imported to the subsystem of E-MOOC platform for the last two years, and nearly two hundred learning knowledge points on about six to eight courses for software engineering major are covered. Different MOOC elements can be combined freely to form the related learning knowledge points for one course, and they are shown as Table 3 as a case. Teachers can select several different MOOC elements to form an important knowledge point for MOOC teaching process according to an actual course, and reuse some existed MOOC elements to form another MOOC knowledge point for other courses.

More than 80 percent of MOOC elements are provided with clear animation demos or videos for learners. More than four hundred students of computer and software major on local campus completed learning process on MOOC by combining the MOOC elements freely and purposeful according to own interests in personalized learning.

Table 3. Learning knowledge points for MOOC elements combination.

No.	MOOC Element	Order of Combinations	Knowledge Point	Course
……				
19	Class Diagram	①	Static View of UML	Software Engineering
22	Object Diagram	②		
……				
287	Integration Strategy	③	Integration Testing Process	Software Testing
288	Stub Module	①		
280	Driven Module	②		
……				
75	Inheritance	③	Design of a Class	Object Oriented Programming
76	Polymorphism	④		
77	Encapsulation	⑤		
19	Class Diagram	①		
22	Object Diagram	②		
……				

5. Conclusions and Further Work

In this paper, MOOC elements are based on integrable ware which are designed and constructed for some courses of software engineering major, and they could be freely combined to form all sorts of MOOC learning knowledge points to the MOOC resources database as a subsystem of E-MOOC platform.

Furthermore, in big data era, the construction of teaching resource database is also the basis on the integration of teaching resources in universities in the future [1]. Further work will be done is the following.

Firstly, all sorts of MOOC resources based on big data are needed to be developed and applied continuously to expand and extend the depth and scope of teaching resources using on local universities. Compatible supported conditions on software and hardware should be created to promote the integration of teaching resources based on big data and maximize the utilization of MOOC resources in universities [1].

Secondly, teachers' comprehensive application capabilities based on big data need to be improved. Integration and reusing of various types of MOOC resources based on integrable ware by teachers in big data environment is essential, teachers should optimize the teaching idea and integrate the big data teaching theory with information technology in the MOOC teaching practice.

Thirdly, more and more reusable MOOC elements based on integrable ware for the courses of other majors should be designed and constructed for personalized learning on local campus, to improve the level of teaching information.

Acknowledgments
The work in this paper was financially supported by Foundation of "Excellent Young Talent Support Project of Anhui Province University" in year of 2019 (Grant No. gxyq2019138) and "Massive Open Online Course on Software Engineering" of Anhui Province (Grant No. 2015mooc104).

Moreover, this paper was also supported by Foundation of "Teaching Team of Software Engineering Course" of Anhui SanLian University (Grant No. 15zlgc029), "Project of Quality Engineering of Anhui Province University" (Grant No. 2017jxtd131) and "Project of Quality Engineering of Anhui SanLian University" (Grant No. 16zlgc031).

In the end, as the corresponding author of this paper, I would like to express my heartfelt gratitude to my colleagues of Jishan Zhang, Liqiong Pan, Ning Wu, Yuefang Liang, and all the authors of the references which are listed at the end of this paper.

References

[1] Y.L. Yan 2018 Research on the Integration of Teaching Resources in Colleges and Universities under the Background of Big Data *Journal of China Adult Education.* **24** 37-9

[2] J.T. Luo, M. Sun and X.Q. Gu 2014 The Innovation Research of MOOC from the Perspective of Blended Learning: Case Study of SPOC *Journal of Modern Educational Technology.* **7** 18-24

[3] J.L. Fan and H.P. Fang 2015 Reflections on Curriculum Teaching Reform of Universities in China during Post-MOOC Period *Journal of Changchun University.* **12** 89-92

[4] S.D. Jia and Y. Yang 2014 Enlightenment of MOOC Mode to Open Courses in China *Journal of China Education Info.* **3** 6-10

[5] G.R. Zheng and H.J. Wang 2018 Analysis of Current Situation and Strategy Research on Construction of SPOC Teaching Resources under the Background of "Internet +" *Journal of Education Modernization.* **44** 208-11

[6] H.B. Gui and J.M. Zhang 2017 Design of MOOC Teaching Resource Database System Based on Integrable Ware Theory *Journal of Mudanjiang University.* **6** 135-7

[7] G.J. Lv, T.X. Rui and L.Y. Chen 2014 on the Technology of Integrable Ware Based on Flash *Journal of Jiangsu Normal University (Natural Science Edition).* **4** 55-8

[8] J.P. Zeng 2015 Research and Realization of network teaching resource database based on CDIO *A Thesis for the Degree of Master (Hunan University)*

[9] Information on *https://en.wikipedia.org/wiki/XML_database*

[10] Information on *https://www.cnblogs.com/xueshui20/p/4151405.html*

[11] J.P. Li 2018 Construction and Application of the Network Teaching Resource Base under the Thought of Product *Journal of Microcomputer Applications.* **10** 94-7

IOP Conf. Series: Journal of Physics: Conf. Series **1284** (2019) 012040 doi:10.1088/1742-6596/1284/1/012040

Improved Biclustering Algorithm Based on Cuckoo Search and Firefly Algorithm

FAN Zhenhao[1] and OU Ling[2]

College of Computer and Information Science, Southwest University, Chongqing 400700,China

E-mail: [1] fan_swu@163.com, [2] ouling@swu.edu.cn

Abstract. Biclustering the gene expression data, which focuses on finding a subset of genes and a subset of experimental conditions simultaneously, is a good method to help researchers understanding the data. It can be regarded as optimization problem. Unfortunately, the biclustering result is always fall into local optimal area. To deal with this problem, this paper propose CFB biclustering algorithm. That is, combining Cuckoo Search and Firefly Algorithm for biclustering. The experimental results show that CFB inherits not only the global optimization ability of Cuckoo Search but also the fast convergence speed of Firefly Algorithm, and is more powful to find the biclusters of the gene expression data.

1. Introduction

With the rapid development of biotechnology, people can investigate thousands of genes by measureing their expression levels under different experimental conditions. Usually, these data are organized as a matrix, where each row corresponds to a gene and each column corresponds to a sample or an experimental conditions. The struct of gene expression data is showing in Figure 1

	condition 1	condition 2	condition 3	condition 4	condition 5
gene1	1.2	-3.4	2.6	1.9	-0.2
gene2	0.5	-3.5	-3.4	5.6	8.2
gene3	1.3	2.7	2.0	-1.2	1.3
gene4	0.1	-2.2	3.6	3.9	-6.9
gene5	3.8	1.1	2.4	-1.6	-5.5

Figure 1. Struct of gene expression data

Clustering techniques became very important tools for analysising gene expression data. Traditional clustering algorithm allow researchers to improve the understanding of the functions of the genes from an organism, like functional annotation, tissue classification and motif identification[1–3]. However, there are some drawbacks, such as, each gene or experimental condition is assigned to only one cluster, and each gene or experimental condition must be assigned to some cluster. Thus, a traditional clustering method may not be convincing enough.

In order to overcome the limitations, Cheng and Church[4] propose the concept of bicluster firstly. The biclustering algorithm is actually a multi-objective optimization local search algorithm. It's unreasonable that put the complicated gene data into the biclustering algorithm directly. This paper propose CFB, which combining CS and FA for biclustering.

2. Related work

In this section, we will elaborate the related work about biclustering algorithm and swarm intelligence algorithm.

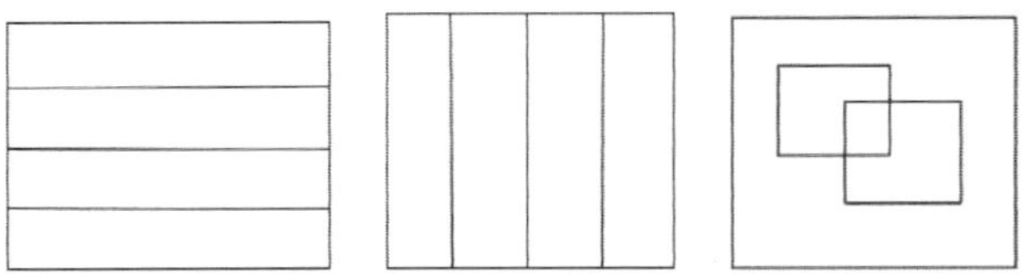

(a) gene-wise clustering **(b)** condition-wise clustering **(c)** biclustering

Figure 2. Traditional clustering and biclustering

As Figure 2 shows, unlike traditional clustering method that treat similarity as a function of pairs of genes or pairs of conditions, the bicluster model measures coherence within the subset of genes and conditions. So that we can find out the biological information hidden in local data.

As for a gene matrix $A(X,Y)$, $B(I,J)$ is a submatrix of A, which I is a subset of rows(genes), and J is a subset of columns(conditions). The $B(IJ)$ is called a bicluster, as long as its *MSR (mean squared residue)* is smaller than a given threshold σ . The *MSR* design formulas is:

$$MSR(I,J) = \frac{\sum\limits_{i \in I, j \in J} (a_{ij} - a_{iJ} - a_{Ij} + a_{IJ})^2}{V_{IJ}} \tag{1}$$

where a_{ij} is valid element of $B(I,J)$, and, a_{iJ} , a_{Ij} and a_{IJ} are the row average, the column average and the matrix average.

Recent years, there are many algorithms for biclustering. Ayadi[5] classified the biclustering algorithms into systematic search algorithms and stochastic search algorithms. Among the reprentative systematic search algorithms are **CC**[4], **BiMine**[6], **OPSM**[7], **FLOC**[8], **SAMBA**[9], **OP-Cluster**[10], **Motifs**[11], **Bimax**[12], and **Paid**[13], etc. The stochastic searching algorithms have **SEBI**[14], **SSB**[15], **CMOPSOB**[16], and **SAB**[17].

Each of these algorithms has its own advantages and disadvantages. Mostly, facing the complex gene data, the systematic search algorithms are easily catched into local optimal solutions. Although the stochastic search algorithms can avoiding local optimal solutions,they are not so stable. So, how to slove this problem is the current research hotspots.

There are some researchers apply swarm intelligence algorithms, Such as **PSO**[18],**FA**[19], **GA**[20], **CS**[21], to biclustering. Swarm intelligence algorithms can be seen as the stochastic search algorithms essentially. But, different algorithm hao its own brilling and drawback. The properly combining can get a better result. In this paper, we choose CS and FA. The former has many outstand properties, like fewer parameters and greater global optimize ability. And FA can converge faster.

3. CFB

3.1. Encoding

Given $A(X, Y)$, a bicluster $B(I, J)$ can be present by $x_p = (g_1, \ldots, g_i, \ldots, g_m, s_1, \ldots, s_j, \ldots, s_n)$, $p = 1, \ldots, N$. where N, m, n are the population number, the number of genes and the number of samples of A. When the ith gene or jth sample in A is selected as B, $g_i = 1$ or $s_j = 1$, otherwise, $g_i = 0$ or $s_j = 0$. $1 \leq i \leq m$ and $1 \leq j \leq n$.

The original swarm intelligence algorithm solutions (nests, fireflies) are multidimensional continuous values that need to be mapped to corresponding bit strings before they can be used to represent double clusters.The usual practice is to set the upper limit to 1, the lower limit to 0, and then determine whether it is greater than 0.5, and map the real value to a bit value. As Figure 3 shows:

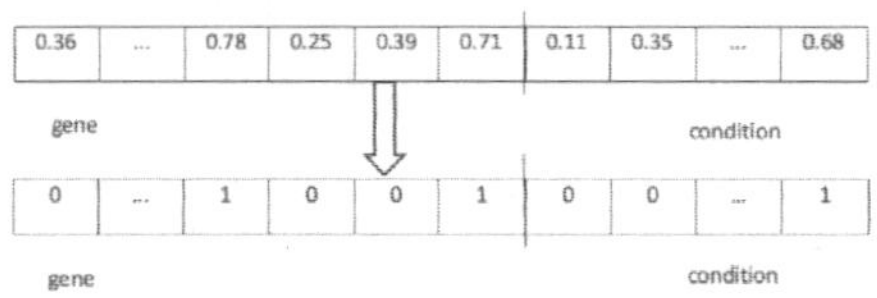

Figure 3. Continuous solution mapping to bicluster

3.2. Fitness Function

For swarm intelligence algorithms, the design of the fitness function has a decisive influence on the final result. Generaly, there are *MSR*, *GV (Gene Volume)*, *CV (Condition Volume)* and *Var (Row Variance)* for evaluating the bicluster. *MSR* represents the consistency of the fluctuations in the expression of genes, and the smaller the value, the more consistent the changes in the expression levels of genes. *Var* represents the range of fluctuations in the expression of genes, and the greater the range of fluctuations, the higher the credibility associated with genes. Mostly, the smaller the GV and CV, the smaller the *MSR* and *Var*; the larger the GV and CV, the larger the *MSR* and *Var* become. However, if the volume of the bicluster is too large or too small, the meaning of the bicluster itself is lost, so the algorithm should be able to find the *MSR* minimum solution under reasonable GV and CV. For a bicluster $B(I, J)$, its fitness value is:

$$f(B) = MSR(B) + \frac{\lambda}{GV(B)} + \frac{\mu}{CV(B)} + \frac{\omega}{Var(B)} \tag{2}$$

$$MSR(B) = \frac{1}{k \times l} \sum_{i \in I, j \in J} (a_{ij} - a_{iJ} - a_{Ij} + a_{IJ})^2 \tag{3}$$

$$Var(B) = \frac{1}{k \times l} \sum_{i \in I, j \in J} (a_{ij} - a_{iJ})^2 \tag{4}$$

where a_{iJ}, a_{Ij}, a_{IJ} are the i-th row, the j-th column and the entire average of $B(I, J)$. $k = |I|$, $l = |J|$ are the volume of the gene and experimental conditions in B, i.e. GV and CV. λ, μ, ω is to solve the problem of different dimensions. The larger the λ, μ and ω, the greater the influence of GV, CV and *Var* on the fitness value, and the value depends on the data set.

The goal of the algorithm is to try to find a solution with a small fitness value, that is, *MSR* is as small as possible, and *Var* are as large as possible.

3.3. Algorithm Describe

Algorithm divide original matrix by CS first, then perform FA clustering on the divided sub-matrix. Its flow chart as Figure 4 shows.

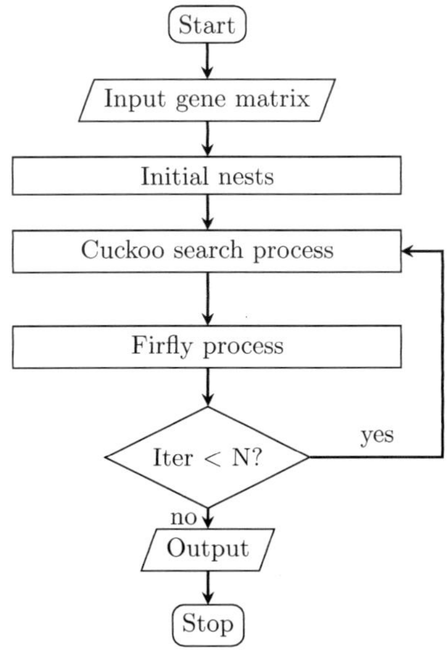

Figure 4. Flow Chart

4. Experiment

4.1. Dataset

In order to evaluate the four algorithms, we did experiments on four datasets. They are Leukaemia dataset, Yeast Cell Cycle dataset, DLBCL (Diffuse Large B-cell Lymphoma) dataset and Breast Cancer dataset. The detail information are showing in Table 1.

Table 1. Datasets for experiment

Dataset	GeneVolume	ConditionVolume	λ	μ	ω
Leukaemia	7128	72	5.0E02	1.0E01	3.0E-02
Yeast Cell Cycle	2884	17	1.0E06	1.0E03	3.0E05
DLBCL	12625	21	2.0E07	1.0E04	3.0E07
Breast Cancer	13666	117	2.0E03	3.0E02	1.0E-01

4.2. Results

Figure 5 are the fitness curves of each algorithm on four datasets. From the figures, we can concluse that, CFB inherits not only the global optimization ability of CS but also the fast convergence speed of FA.

Table 2 to Table 5 are the evaluation indexes of the bicluster results of each algorithm on each dataset. It can be seen from the tables that, CFB is good at discovering biclusters with small MSR and large volume on most datasets.

5. GO analysis

For the study of gene expression data, the final problem is to find its biological significance. Therefore, we need to verify whether the gene set in the bicluster participates in some biological processes. GO (Gene Ontology)[22] can be used to investigate if a group of genes belonging to a bicluster presents significant enrichment about a specific GO term. GO provides functional

IOP Conf. Series: Journal of Physics: Conf. Series **1284** (2019) 012040 doi:10.1088/1742-6596/1284/1/012040

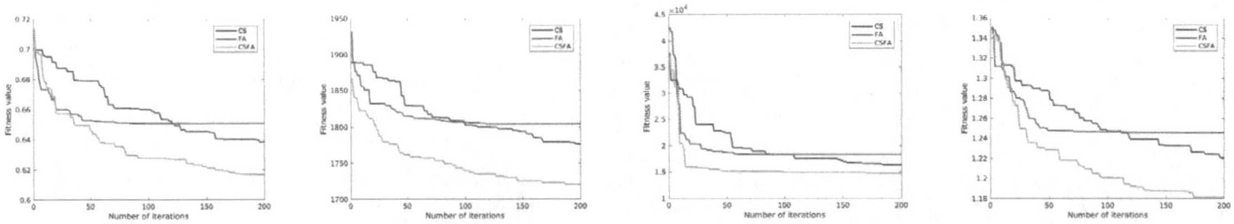

(a) Leukaemia (b) Yeast Cell Cycle (c) DLBCL (d) Breast Cancer

Figure 5. Fitness curves on each dataset

Table 2. Evaluation indexes on Leukaemia

Algorithm	MSR	Volume	GV	CV	Var
CS	0.1769	231488	3617	64	0.1797
FA	0.1776	214937	3643	59	0.1807
CFB	0.1769	269352	3741	72	0.1794

Table 3. Evaluation indexes on Yeast Cell Cycle

Algorithm	MSR	Volume	GV	CV	Var
CS	578.8211	10822	1546	7	735.4996
FA	640.3960	11025	1575	7	777.6675
CFB	529.3312	9972	1662	6	709.6392

Table 4. Evaluation indexes on DLBCL

Algorithm	MSR	Volume	GV	CV	Var
CS	9139	56637	6293	9	10308
FA	11752	57942	6438	9	13256
CFB	5019	39210	6535	6	6025

Table 5. Evaluation indexes on Breast Cancer

Algorithm	MSR	Volume	GV	CV	Var
CS	0.3128	683793	6907	99	0.3172
FA	0.3118	631358	6938	91	0.3164
CFB	0.3189	787384	6968	113	0.3229

annotation enrichment analysis of biological processes, cell component, and molecular function. Uploading the biclustered gene set to GO and selecting the species category, GO will returns the gene set and enrichment analysis between each biological process, cell component and molecular function. GO uses P-value to measure the degree of enrichment. When the P-value is less than 0.05, it indicates that the biclustered gene set is enriched in the corresponding GO function category.

In this paper, the enrichment analysis of the biclustering results of three algorithms on DLBCL is carried out in Table 6. Only the most significant terms are shown. For example to the bicluster of CFB, the genes are mainly involved in catalytic activity. The tuple (n=1910,p=3.28E-31) represents that out of 1482 genes in bicluster of CFB, 1910 genes belong to catalytic activity function, and the statistical significance is given by the P-value of p=3.28E-31.

Table 6. GO analysis

Algorithm	GV	Biological Processes	Molecular Function	Cell Component
CS	6293	lymphocyte activation (n=185,p=1.22E-14)	kinase binding (n=326,p=1.44E-18)	transferase complex (n=269,p=3.03E-06)
FA	6438	response to steroid hormone (n=187,p=3.14E-18)	hormone binding (n=51,p=1.63E-04)	ribonucleoprotein complex (n=282,p=9.89E-05)
CFB	6968	regulation of gliogenesis (n=68,p=1.41E-07)	catalytic activity (n=1910,p=3.28E-31)	pigment granule (n=59,p=2.05E-06)

6. Conclusion

Swarm intelligence techniques are based on collective intelligence of groups of simple agents. A few algorithms are very efficient and they are popular tools for solving real-world problems. In this paper, we propose CFB biclustering algorithm which combining the CS and FA for biclustering. And the experiment results show that the algorithm is more suitable for biclustering than CS and FA.

References

[1] Pablo A Jaskowiak, Ricardo JGB Campello, and Ivan G Costa Filho. Proximity measures for clustering gene expression microarray data: a validation methodology and a comparative analysis. *IEEE/ACM Transactions on Computational Biology and Bioinformatics (TCBB)*, 10(4):845–857, 2013.

[2] Pablo A Jaskowiak, Ricardo JGB Campello, and Ivan G Costa. On the selection of appropriate distances for gene expression data clustering. In *BMC bioinformatics*, volume 15, page S2. BioMed Central, 2014.

[3] Amos Tanay, Roded Sharan, and Ron Shamir. Discovering statistically significant biclusters in gene expression data. *Bioinformatics*, 18(suppl_1):S136–S144, 2002.

[4] Yizong Cheng and George M Church. Biclustering of expression data. In *Ismb*, volume 8, pages 93–103, 2000.

[5] Wassim Ayadi, Mourad Elloumi, and Jin-Kao Hao. A biclustering algorithm based on a bicluster enumeration tree: application to dna microarray data. *BioData mining*, 2(1):9, 2009.

[6] Wassim Ayadi, Mourad Elloumi, and Jin-Kao Hao. A biclustering algorithm based on a bicluster enumeration tree: application to dna microarray data. *BioData mining*, 2(1):9, 2009.

[7] Amir Ben-Dor, Benny Chor, Richard Karp, and Zohar Yakhini. Discovering local structure in gene expression data: the order-preserving submatrix problem. *Journal of computational biology*, 10(3-4):373–384, 2003.

[8] Yang Jiong, Wang Haixun, Wang Wei, and P. Yu. Enhanced biclustering on expression data. In *Third IEEE Symposium on Bioinformatics and Bioengineering, 2003. Proceedings.*, pages 321–327, March 2003.

[9] Amos Tanay, Roded Sharan, and Ron Shamir. Discovering statistically significant biclusters in gene expression data. *Bioinformatics*, 18(suppl_1):S136–S144, 2002.

[10] Xiaobo Li, Huijuan Lu, and Mingjun Wang. A hybrid gene selection method for multi-category tumor classification using microarray data. *International Journal Bioautomation*, 17(4), 2013.

[11] TM Murali and Simon Kasif. Extracting conserved gene expression motifs from gene expression data. In *Biocomputing 2003*, pages 77–88. World Scientific, 2002.

[12] Amela Prelić, Stefan Bleuler, Philip Zimmermann, Anja Wille, Peter Bühlmann, Wilhelm Gruissem, Lars Hennig, Lothar Thiele, and Eckart Zitzler. A systematic comparison and evaluation of biclustering methods for gene expression data. *Bioinformatics*, 22(9):1122–1129, 2006.

[13] Laura Lazzeroni and Art Owen. Plaid models for gene expression data. *Statistica sinica*, pages 61–86, 2002.

[14] Federico Divina and Jesus S Aguilar-Ruiz. Biclustering of expression data with evolutionary computation. *IEEE transactions on knowledge and data engineering*, 18(5):590–602, 2006.

[15] Juan A Nepomuceno, Alicia Troncoso, and Jesús S Aguilar-Ruiz. Biclustering of gene expression data by correlation-based scatter search. *BioData mining*, 4(1):3, 2011.

[16] Junwan Liu, Zhoujun Li, Xiaohua Hu, and Yiming Chen. Biclustering of microarray data with mospo based on crowding distance. In *BMC bioinformatics*, volume 10, page S9. BioMed Central, 2009.

[17] Kenneth Bryan, Pádraig Cunningham, and Nadia Bolshakova. Application of simulated annealing to the biclustering of gene expression data. *IEEE transactions on information technology in biomedicine*, 10(3):519–525, 2006.

[18] Yuhui Shi and Russell C Eberhart. Empirical study of particle swarm optimization. In *Proceedings of the 1999 Congress on Evolutionary Computation-CEC99 (Cat. No. 99TH8406)*, volume 3, pages 1945–1950. IEEE, 1999.

[19] Xin-She Yang. Firefly algorithm, lévy flights and global optimization. In Max Bramer, Richard Ellis, and Miltos Petridis, editors, *Research and Development in Intelligent Systems XXVI*, pages 209–218, London, 2010. Springer London.

[20] Lawrence Davis. Handbook of genetic algorithms. 1991.

[21] Yang Xin-She and Deb Suash. Cuckoo search via lvy flights. In *2009 World Congress on Nature Biologically Inspired Computing (NaBIC)*, pages 210–214, Dec 2009.

[22] Michael Ashburner, Catherine A Ball, Judith A Blake, David Botstein, Heather Butler, J Michael Cherry, Allan P Davis, Kara Dolinski, Selina S Dwight, Janan T Eppig, et al. Gene ontology: tool for the unification of biology. *Nature genetics*, 25(1):25, 2000.

IOP Conf. Series: Journal of Physics: Conf. Series **1284** (2019) 012041 doi:10.1088/1742-6596/1284/1/012041

Identifying Significant Components of Complex Software for Improving Reliability: Using Invocation Relationships and Component Characteristics

Lixing XUE, Zhan ZHANG and Decheng ZUO

School of Computer Science and Technology, Harbin Institute of Technology, Harbin 150001, China

zz@ftcl.hit.edu.cn

Abstract. The scales of software systems are getting larger and larger, which may make the system reliability become low. To improve their reliability, a set of important components that are able to strongly influence the system reliability are usually determined first. However, the existing determining approaches only consider the components which are often called by other components as significant ones, but overlook the components which often call others also have a strong effect on the system. Besides, the approaches all require component invocation probabilities which cannot be obtained easily in large-scale systems. To attack the problems, we propose a novel approach for identifying the significant components in complex systems. This approach includes two component ranking algorithms, which take into account not only the components that are frequently invoked, but also the components which often invoke others. The two algorithms, which do not require component invocation probabilities, can either make significant component ranking only based on the component invocation relationships or consider not only the component invocation relationships but also the component characteristics to achieve results. The significant components are selected according to the two ranking results. Extensive experiments are provided to evaluate the approach and draw comparisons with existing methods.

1. Introduction

Nowadays, various forms of computers are everywhere in our daily life -- desktop computers or laptops we use, smart phones, electronic control units in cars and embedded systems in intelligent electrical appliances, and this offers tremendous gains in convenience. However, any failure of these computers may bring great impact to people, leading the reliability to be a particular concern. As an indispensable part of a computer, software is becoming large scale and rather complex. Unfortunately, a large quantity of components and their complicated invocation relationships incur a huge threat to the system reliability and cause system failures easily. To find out an efficient method for improving the reliability of complex software has been an urgent and practical problem.

Improving software reliability is not a thoroughly new topic. There are four main methods in traditional software reliability engineering: fault prevention, fault removal, fault tolerance, and fault forecasting [1]. However, due to the massive number of components in a large-scale software system, any of these methods can be applied to all components. It will incur too much cost and it is not efficient. In fact, there is no need to do so, because the 80-20 rule also acts in software reliability. Microsoft pointed out that 80% of failures and crashes in Windows and Office can be ascribed to 20%

of the most reported bugs [2]. Therefore, identifying the significant components that impact the system reliability much more than others, is the first job to optimize the reliability of complex software.

For small-scale software, performing sensibility analysis based on a Markov model can be employed to locate the significant components that impact the system reliability much more than others [3-5]. But utilizing this classic approach to a large-scale system will lead to the state-space-explosion problem. Researchers have attempted to attack this issue from other perspectives [6-10]. FTCloud [6], which is one of the ideas, performs component ranking by invocation frequencies and then find out a set of important components which are able to impact the system reliability much more than other components. FTCloud2 [7], a refined work of FTCloud, adds attention on the component characteristics, namely components are divided into noncritical components or critical ones. However, there are still two issues in theses component ranking approaches. On one hand, they take into account only components that are frequently invoked by other components, but ignore components that frequently invoke other components. On the other hand, the approach requires stable usage profiles (i.e. component invocation probabilities), which need to be recorded during a long period and cannot be obtained when building new complex software.

To address the problems, we put forward a new component ranking approach for identifying significant components in complex software. The approach utilizes the invocation relationships as well as the component characteristics to performing component ranking twice. One performs the ranking by the impact of components that are frequently called, and the other does component ranking by the effect of components that often invoke others. According to the two ranking results, a set of the significant components that can greatly influence the system reliability are determined. Compared to the existing approaches, the proposed method which does not require component invocation probabilities takes into account the component characteristics and is able to achieve better results. This method also can be used to locate the most significant components when designing the architecture.

The rest of the paper is organized as follows. Structures of software systems are modelled in Section 2. Section 3 details the two component ranking algorithms and discusses how to determine significant components. Massive experiments are provided to evaluate our approach within Section 4. At last, we conclude the paper.

2. Component Invocation Graph

The structure of a software system, i.e. the component invocation relationships, can be modelled by a directed graph $G=<C,E>$, where an element $c_i \in C$ denotes a component in the software, and a directed link $e_{ij} \in E$ from a node c_i to another node c_j represents a component invocation relationship, i.e. c_i invokes c_j.

Each component invocation graph containing n components is corresponding with an $n \times n$ adjacent matrix $A=\left[a_{ij}\right]_{n\times n}$, where each entry a_{ij} indicates whether there is a directed link from node c_i to node c_j. If the directed link exists, the value of a_{ij} is set to be 1, otherwise it is 0. This can be formulized as:

$$a_{ij}=\begin{cases}1 & if\ e_{ij}\in E\\ 0 & else\end{cases} \tag{1}$$

A directed link ending at node c_i is called an incoming link of c_i and the number of incoming links of c_i is named its indegree $\deg^+(c_i)$. $\deg^+(c_i)$ is able to be calculated by:

$$\deg^+(c_i)=\sum_{k=1}^{n} a_{ki} \tag{2}$$

A directed link starting from node c_i is named an outgoing link of c_i and the number of outgoing links of c_i is called its outdegree $\deg^-(c_i)$. $\deg^-(c_i)$ can be computed by:

$$\deg^-(c_i) = \sum_{k=1}^{n} a_{ik} \tag{3}$$

For a component c_i, the number of its parent nodes is $\deg^+(c_i)$, indicating that a total of $\deg^+(c_i)$ component will invoke it, while the number of its children nodes is $\deg^-(c_i)$, showing that it will invoke a total of $\deg^-(c_i)$ components in the software system.

3. Ranking and Significant Component Determining

Section 3.1 details one of the algorithms and Section 3.2 introduces the other one. Then determination of significant components which are able to greatly impact on the system reliability is discussed in Section 3.3.

3.1. Component Ranking Inspired by PageRank

In a component invocation graph, some components may have larger indegrees than other components. We intuitively feel that these components may be frequently invoked by others and therefore they are able to influence the global reliability of this software much more than other components [3]. In terms of the principle of PageRank [11], an algorithm is put forwarded to score the significance of the software components.

Considering a system which has n components, each component c_i $i = 1,2,L\ ,n$, with $n \in \mathbf{N}$ has a correspoding a significance value $VP(c_i)$ in the range of [0,1] . $VP(c_i)$ can be calculated by:

$$VP(c_i) = (1-\alpha) f(c_i) + \alpha \sum_{e_{ji} \in E} \frac{VP(c_j)}{\deg^-(c_j)} \tag{4}$$

where $f(c_i)$ is the basic significance value of c_i itself, while $\sum_{e_{ji} \in E} \frac{VP(c_j)}{\deg^-(c_j)}$ is the significance value derived from other components which are invoked by c_i. The parameter α $(0 \le \alpha \le 1)$ is used to balance the derived value and the basic value, which is usually set as 0.85 [6, 7].The basic significance value $f(c_i)$ is defined as

$$f(c_i) = \begin{cases} \dfrac{\beta}{|C|} & \text{if } c_i \text{ is a critical component} \\ \dfrac{1-\beta}{|NC|} & \text{if } c_i \text{ is a noncritical component} \end{cases} \tag{5}$$

In Formula (5), $|C|$ and $|NC|$ denote the quantity of critical components and noncritical components, respectively. The parameter β ($\frac{|C|}{n} \le \beta \le 1$) balances the weight of the two category of components. If $\beta = \frac{|C|}{n}$, this approach does not consider the component characteristics at all and the ranking result is only determined by invocation relationships.

Formula (4) can be written in vector form:

$$\begin{bmatrix} VP(c_1) \\ \vdots \\ VP(c_n) \end{bmatrix} = (1-\alpha)\begin{bmatrix} f(c_i) \\ \vdots \\ f(c_n) \end{bmatrix} + \alpha T \begin{bmatrix} VP(c_1) \\ \vdots \\ VP(c_n) \end{bmatrix} \tag{6}$$

where the matrix $T = \left[t_{ij}\right]_{n\times n}$ is defined by:

$$t_{ij} = \begin{cases} \dfrac{1}{\deg^{-}(c_j)} & if\ e_{ji} \in E \\ 0 & else \end{cases} \tag{7}$$

Formula (6) can be solved via calculating the eigenvector with eigenvalue 1.

This algorithm ranks the components only based on the component invocation relationships if the critical components cannot be determined. Ones they are provided, the algorithm is able to achieve more accurate ranking results. The algorithm outputs the significance values of all the components and a component that has a larger value has the power impacting the global reliability much more than the one with a small value.

3.2. Component Ranking Inspired by TrustRank

Opposite to the condition discussed in Section 3.1, a component invocation graph may have larger outdegrees than other components. These components seem to call others frequently and their failures may have influence on many subsequent components. Therefore, the components should also be considered to be significant. According to the thought of inverse PageRank in TrustRank [12], we propose another method to estimate the significance of the software components.

Assuming that a software system has n components, a significance value $VT(c_i)$ is defined for each component C_i, which is in the range of $[0,1]$. $VT(c_i)$ can be obtained by:

$$VT(c_i) = (1-\alpha) f(c_i) + \alpha \sum_{e_{ij}\in E} \frac{VT(c_j)}{\deg^{+}(c_j)} \tag{8}$$

where $f(c_i)$ is the basic significance value of c_i itself, while $\sum_{e_{ji}\in E} \frac{VP(c_j)}{\deg^{-}(c_j)}$ is the significance value derived from other components which are invoked by c_i. The definitions of α and $f(c_i)$ are the same as and those in Section 3.1.

Formula (8) can be written in vector form:

$$\begin{bmatrix} VT(c_1) \\ \vdots \\ VT(c_n) \end{bmatrix} = (1-\alpha)\begin{bmatrix} f(c_i) \\ \vdots \\ f(c_n) \end{bmatrix} + \alpha U \begin{bmatrix} VT(c_1) \\ \vdots \\ VT(c_n) \end{bmatrix} \tag{9}$$

where the matrix $U = \left[u_{ij}\right]_{n\times n}$ is defined by:

$$u_{ij} = \begin{cases} \dfrac{1}{\deg^{+}(c_j)} & if\ e_{ij} \in E \\ 0 & else \end{cases} \tag{10}$$

In this way, it can be solved by calculating the eigenvector with eigenvalue 1.

Similarly, this algorithm is able to rank the components only using the component invocation relationships as well as to achieve more accurate ranking results when adding the information of component characteristics. The algorithm also gives the significance values of all the components as the output and a component that has a larger value has more influence on the system reliability.

3.3. Significant Components Determining
On the basis of the above two algorithms, two series of significance values of all the components in the software are able to be obtained. One focuses on the components which are frequently invoked and the other pays attention on the components that frequently calls others. The two ranking results are both effective and valuable.

Therefore, Top-$\frac{K}{2}$% ($0 < K \leq 100$) components are respectively selected from the two ranking results and a total of $K\%$ most significant components are determined.

4. Experiments
A lot of experiments are provided in this section in order to validate the proposed approach, estimate the impact of the parameters and draw performance comparisons between our approach and FTCloud / FTCloud2.

4.1. Experimental Setup
To validate the effectiveness and efficiency of the approach proposed in this paper, we compare seven approaches, which are as follows:

- RandSel. We randomly choose K percent components and then improve their reliability by applying fault-tolerance strategies.
- OurID1. Top-K percent most important components are identified by the proposed method, which only utilizes the invocation relationships. Then their reliability is improved.
- OurID2. Top-K percent most important components are identified by the proposed method, which adds the consideration on the component characteristics. Then their reliability is improved.
- FTCloud. Top-K percent most important components are by FTCloud. Then their reliability is improved.
- FTCloud2. Top-K percent most important components are by FTCloud2. Then their reliability is improved.

A graph whose degree distribution exhibit a power law is called a scale-free graph. A lot of previous works show that not only many networks but also the program structures, such as MySQL and Linux Kernel, appear to be approximate scale-free properties [13, 14]. Similarly to [6, 7], we also apply Pajek [15], which is network analysis software, to randomly draw scale-free directed graphs as component invocation graphs for the experimental studies and draw performance comparisons. In the section, three graphs which have 500, 1000 and 2000 nodes respectively are generated.

In is section, *intf* denotes the internal failure probability of the components, including two initial settings -- 0.05 and 0.1. For fair comparisons, we suppose that the internal failure probability (*intf*) can be reduced by 30 percent after improving the reliability.

In OurID1, OurID2, FTCloud and FTCloud2, the parameter α is used to balance the weight of the derived significance and the basic significance the component has itself. Similar works [6, 7] have proved that 0.85 is an optimal value. Thus, in this section, we also set the parameter α to be 0.85.

In Section 4.2, OurID1 and RandSel are compared to validate the identification by the proposed method. The impact of Top-K and component internal failure probability is discussed in Section 4.3 and 4.4 respectively. Performance comparison between OurID1 and FTCloud is drawn within Section 4.5. In this comparison, the component characteristics are not involved. In Section 4.6, OurID2 and FTCloud2 are compared when considering the component characteristics.

4.2. Validity of Identification
To validate the significant components identified by our approach have greater power impacting on the system, RandSel and OurID are compared in each graph respectively with different K values. Table 1 lists the experimental results of software reliability.

Table 1. Validity of Identification.

Node Numbers	Methods	Component intf=0.05				Component intf=0.1			
		Top2%	Top6%	Top10%	Top20%	Top2%	Top6%	Top10%	Top20%
500	RandSel	0.7913	0.7934	0.7988	0.8109	0.6281	0.6312	0.6397	0.6591
	OurID1	0.8072	0.8221	0.8424	0.8558	0.6543	0.6786	0.7120	0.7343
1000	RandSel	0.7625	0.7634	0.7642	0.7678	0.5870	0.5882	0.5893	0.5947
	OurID1	0.8007	0.8237	0.8308	0.8438	0.6476	0.6837	0.6950	0.7159
2000	RandSel	0.6964	0.6984	0.6972	0.7059	0.4988	0.5025	0.5003	0.5164
	OurID1	0.7206	0.7322	0.7390	0.7494	0.5450	0.5681	0.5817	0.6029

Table 1 shows that OurID1 provides better reliability performance than RandSel, in all experimental settings. This phenomenon shows that improving the reliability of significant components identified by OurID1 can get higher system reliability than enhancing the reliability of randomly selected components. Obviously, OurID1 is able to find out significant components that have stronger effect on the global reliability.

As the K value grows, the reliability provided by OurID1 increases monotonously, while the reliability provided by RandSel may or may not increase. This result indicates that it is effective to apply OurID1 to locate important components.

With the growths of node number, OurID1 is able to consistently provide better the reliability performance. This observation suggests that OurID1 is capable of effectively identifying significant components in different sizes of software systems.

When the component internal failure probability changes from 0.05 to 0.1, OurID1 can consistently provide higher reliability than RandSel, indicating that OurID1 is free from component internal failure probabilities.

4.3. Impact of K

To demonstrate that the parameter K has no impact on the identification results, we compare RandSel and OurID1 under multiple K values and record the reliability results in Table 2. The node number is 1000.

Table 2. Impact of Top-K% on System Reliability.

Componet intf	Methods	Values of Top-K%							
		Top10%	Top20%	Top30%	Top40%	Top50%	Top60%	Top80%	Top100%
0.05	RandSel	0.7642	0.7678	0.7809	0.7930	0.8380	0.8602	0.8680	0.8718
	OurID1	0.8308	0.8438	0.8508	0.8561	0.8604	0.8640	0.8687	0.8718
0.1	RandSel	0.5893	0.5947	0.6142	0.6320	0.6951	0.7413	0.7559	0.7627
	OurID1	0.6950	0.7159	0.7274	0.7359	0.7430	0.7488	0.7567	0.7627

When the K value is set as not less than 80, using OurID1 can obtain better reliability performance than using RandSel regardless of the component internal failure probabilities. When K reaches 100, there is no difference between the two methods, since all the components in either methods are selected. This observation demonstrates that K values do not have impact on OurID1.

4.4. Impact of Component Internal Failure Probability

To show that OurID1 is free from the component internal failure probability (*intf*) settings, OurID1 and RandSel are compared respectively, under component internal failure probability values ranging from 0.01 to 0.1 with a step value of 0.01. In this experiment, the node number is 1000 too, and four K values (2, 6, 10 and 20) are used. The results are depicted Figure 1.

As shown in Figure 1, employing OurID1 can get higher reliability than using RandSel, no matter what the K value is and how *intf* changes. This result suggests that different component internal failure probability settings have no impact on OurID1.

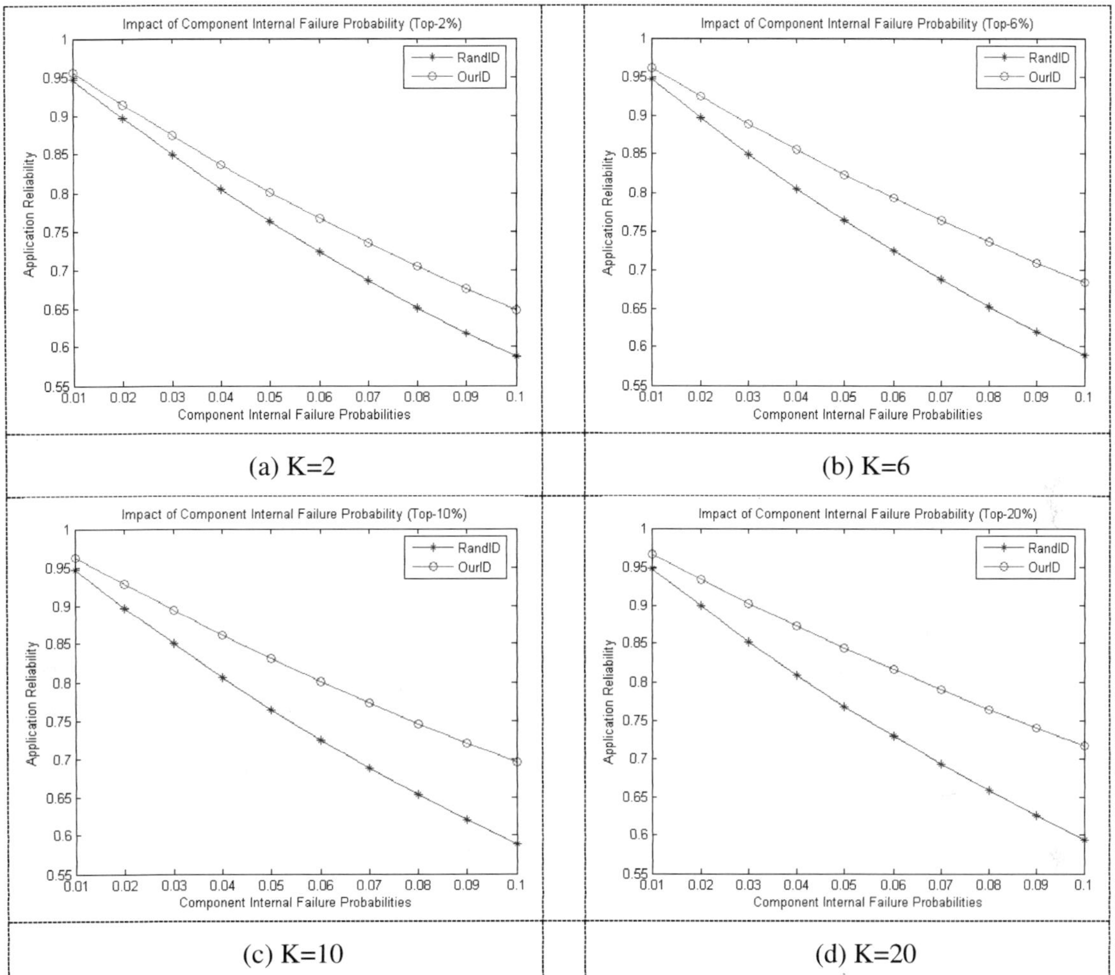

(a) K=2 (b) K=6

(c) K=10 (d) K=20

Figure 1. Impact of Component Internal Failure Probability.

4.5. Performance Comparison without FTCloud

FTCloud requires the invocation relationships as well as the component invocation probabilities while our approach does not need the component invocation probabilities and ranks the components. Furthermore, our approach takes into account both the components which are frequently invoked and the ones which frequently invoke other components, while FTCloud takes into account only the former.

The comparative experiments are conducted on three settings of node numbers (500, 1000 and 2000) respectively with various K values. The initial component internal failure probability is set as 0.1 ($intf = 0.01$). The experiment results of software system reliability are illustrated in Table 3.

Table 3. Performance Comparison with FTCloud.

Node	Methods	Top2%	Top4%	Top6%	Top10%	Top20%	Top30%	Top40%	Top50%
500	OurID1	0.6543	0.6696	0.6786	0.7120	0.7343	0.7516	0.7619	0.7711
	FTCloud	0.6544	0.6696	0.6753	0.6955	0.7106	0.7218	0.7385	0.7639
1000	OurID1	0.6476	0.6746	0.6837	0.6950	0.7159	0.7274	0.7359	0.7430
	FTCloud	0.6476	0.6745	0.6804	0.6857	0.7013	0.7084	0.7178	0.7261
2000	OurID1	0.5450	0.5579	0.5681	0.5817	0.6029	0.6047	0.6066	0.6083
	FTCloud	0.5363	0.5468	0.5579	0.5683	0.5855	0.5902	0.5960	0.6001

When the node number is 500, OurID1 provides better reliability performance than FTCloud under all the Top-K% settings except 2%. When the node number grows to 1000, the reliability performance of OurID1 is better than that of FTCloud under all the Top-K% settings. When the node number reaches 2000, OurID1 consistently outperforms FTCloud. This observation indicates that OurID1 is more efficient than FTCloud in larger software systems.

When the node number is smaller and the Top-K% value is set to be small, that is only a few (may be less than 30) significant components are selected, the performance of OurID1 is a little worse than FTCloud, but the difference is very small. Nevertheless, it is acceptable, because OurID1 requires less information and is able to be applied more widely. FTCloud requires stable component invocation probabilities and can only be used to optimize the systems which have been running after a period of time while OurID1 works only by employing component invocation relationships and can be applied as long as system structure information is provided.

4.6. Performance Comparison with FTCloud2

Compared to FTCloud, FTCloud2 introduces the component characteristics, but still ignores the impact of the components that frequently call others on the system reliability. The method put forward in this paper allows users to rank components taking into account the component characteristics and requires less information than FTCloud2.

The comparisons between our method and FTCloud2 are performed on three settings of node numbers (500, 1000 and 2000) respectively using multifarious K values. The initial component internal failure probability is set as 0.1 ($intf = 0.1$), too. The parameter β is set as 0.7 [7]. The experiment results of Table 4 records the experiment results of the system global reliability.

Table 4. Performance Comparison with FTCloud2.

Node	Methods	Top2%	Top4%	Top6%	Top10%	Top20%	Top30%	Top40%	Top50%
500	OurID2	0.634	0.6494	0.6582	0.6917	0.7139	0.7315	0.7411	0.7511
	FTCloud2	0.63411	0.6493	0.6551	0.675	0.6903	0.7016	0.7181	0.7431
1000	OurID2	0.6341	0.6611	0.6703	0.6818	0.7022	0.7141	0.7227	0.7399
	FTCloud2	0.634	0.6609	0.6669	0.6725	0.688	0.6953	0.7048	0.713
2000	OurID2	0.5367	0.5495	0.5598	0.5735	0.5947	0.5965	0.5985	0.6003
	FTCloud2	0.5279	0.5385	0.5497	0.5601	0.5772	0.5819	0.588	0.5921

Similarly to the results in Section 4.5, OurID2 outperforms FTCloud2 in all the experimental settings except the node number and the Top-K% are both rather small. OurID2 employs only the invocation relationships to achieve better results than FTCloud2 which utilizes not only the invocation relationships but also the component invocation probabilities.

5. Conclusion & Future Work

We introduce a ranking-based approach for identifying the significant components in complex software. This approach includes two component ranking algorithms proposed from two different perspectives. One regards the components that are often invoked by others as significant components. The other views the components that frequently call others as significant components. The components in software are ranked twice by employing the two algorithms and then the top K percent most important ones are selected in terms of the two ranking results. The effectiveness and efficiency of the approach are validated by a lot of experiments. Compared with FTCloud and FTCloud2, our approach requires less information but is able to provide better performance in most cases, and is able to be employed when designing the system architecture.

The determination of significant components in this approach treats the two ranking results equally, but there may be a better method to select the important components. This will be studied in the future. The future work also includes more experimental analysis of actual software systems and component ranking considering more factors.

References

[1] Lyu M R 1996 *Handbook of Software Reliability Engineering* (New York: McGraw-Hill)
[2] Rooney P 2002 Microsoft's CEO: 80-20 Rule Applies to Bugs, Not Just Features *ChannelWeb*
[3] Cheung R C 1980 A User-Oriented Software Reliability Model *IEEE Transactions on Software Engineering* **SE-6** 118-125
[4] Cortellessa V and Grassi V 2007 A Modeling Approach to Analyze the Impact of Error Propagation on Reliability of Component-Based Systems *2007 Int. Symp. on Component-Based Software Engineering* (Springer Berlin Heidelberg) 140-156
[5] Pham TT and Xefago D 2012 Reliability Prediction for Component-Based Systems: Incorporating Error Propagation Analysis and Different Execution Models *2012 12th Int Conf. on Quality Software* (IEEE) 106-115
[6] Zheng Z, Zhou TC, Lyu MR and King I 2010 FTCloud: A Component Ranking Framework for Fault-Tolerant Cloud Applications *2010 IEEE 21st Int. Symp. on Software Reliability Engineering* (IEEE) 398-407
[7] Zheng Z, Zhou T C, Lyu M R and King I 2012 Component Ranking for Fault-Tolerant Cloud Applications *IEEE Transactions on Services Computing* **5** 540-550
[8] Arianyan E, Ahmadi M R and Maleki D 2016 A Novel Taxonomy and Comparison Method for Ranking Cloud Computing Software Products *International Journal of Grid and Distribution Computing* **9** 173-190
[9] Naji-Hasan A H and Shu G O 2014 A Service Profile Based Similarity Framework for Web Service Discovery *Telkomnika (Telecommunication Computing Electronics and Control)* **12** 1079-1087
[10] Lu G, Hai Y and Sun Y 2014 A Reliable Web Services Selection Method for Concurrent Requests *Telkomnika (Telecommunication Computing Electronics and Control)* **12** 1053-1063.
[11] Page L, Brin S, Motwani R and Winograd T 1998 The PageRank Citation Ranking: Bringing Order to the Web *Stanford InfoLab Report number:1999-66*
[12] Gyongyi Z, Garcia-Molina H and Pedersen J 2004 Combating Web Spam with Trustrank *Proc. of the 13th Int. Conf. on Very Large Databases* (ACM) 576-587
[13] de Moura A P S, Lai Y C and Motter A E 2003 Signatures of Small-World and Scale-Free Properties in Large Computer Programs *Physical Review E* **68** 017102
[14] Myers C R 2003 Software Systems as Complex Networks: Structure, Function and Evolvability of Software Collaboration Graphs *Physical Review E* **68** 046116
[15] de Nooy W, Mrvar W and Batagelj V 2011 *Exploratory Social Network Analysis with Pajek* (New York: Cambridge University Press)

Time-Series Data Mining of Minimum Design Height for River Bridge Deck Using Seasonal Trend Decomposition

Haiyan Xie and Xi Hu

Department of Technology, Illinois State University, Turner Hall 5100, Normal, IL 61790-5100, USA

Email: hxie@ilstu.edu;

Abstract. Using the traditional calculation of minimum design height (MDH) could not provide the effective decision support for engineers under complex situations. This research proposes a time-series data mining system for the decomposition and analysis of seasonal trends of hydrologic fluctuations, using stage-discharge curves, Log-Pearson frequency curves, and plotting positions. The research purpose is to provide the decision support based on precise time series for bridge design and feasibility analysis. Prior to conducting the gage-height design for bridges, this study investigates the construction site of the bridge at Kinston Mines of Illinois River in the United States and processes the correlation tests between gage heights and hydrologic discharge. In addition, the study simulates and verifies the system outcomes using Mann-Kendall Mutation, stability tests, and non-linear regressions. The results show the correlation coefficient of 0.9719 for the discharge trends from 1994 to 2017. In addition, the study selects the best-fitted theoretical frequency curve based on the most appropriate convergence to the empirical frequency curve. Overall, this research achieves satisfactory results for the MDH design and has substantial influence for engineers and designers to meet design requirements and save project costs. The data mining method sheds light upon data analysis for engineering design.

1. Introduction

The minimum design height (MDH) of bridge deck is always an inevitable factor in the feasibility study of a new river bridge [1, 2]. The traditional methods to calculate the MDH are usually based on the statistical means of multiple observations on hydraulic data [3-7]. However, the current practice of bridge maintenance requires the considerations of the hydrological fluctuations of rivers [5, 6]. Particularly, hydrologic discharge is the volume of water flowing past a given point in the stream in a given period of time which is an important indicator for river changes [5, 6, 8]. It is named after stream-flow by fluvial hydrologists and plays a critical role in the study of hydrogeological cycles. The most common form of discharge is water flow derived from net precipitation and melting glaciers [8]. Bridge design and maintenance should consider gage height or stage, which is the water height in the stream above a reference point [5, 6, 8, 9]. Empirically, there is a quantitative relation between gage height and discharge in a specific cross-section of river, which can be manifested by Stage-Discharge Curve [10]. However, in the real world, transport systems connect geographical isolation by highway, railway, and airlines. Hence, it's undoubted that bridges, which connect two points over two sides of rivers or valleys at specific heights, are significant components of the transportation systems. In the planning and design phases of river bridges, engineers and planners should always consider flooding argumentation in order to make sure that this newly-constructed bridge meet the specific flooding control requirement. This ensures that it will not be submerged by a once-in-a-century flood [10, 11]. Based on a design flood, bridge engineers should define the flood discharge for the particular project according to historical time-series data. Then they the graph the stage-discharge curve correspondingly to find the conforming gage height. In addition, when deciding on the minimum vertical height from the lowest point of the bridge center and surface water, engineers and planners should take into account the navigable clearance height if this waterway is eligible for river transportation [12]. For example, on Illinois River in the United States, the minimum design height of bridge deck consists of designed flood level and navigable clearance height [3].

Researchers noticed the knowledge gap between the traditional methods of bridge design and the current needs of cost control on maintenances and repair. Engineers, designers and planners need to have improved understanding on how to estimate and predict the possible river flow changes. For example, Amirreza and Arash stated briefly that the relationship between stage and discharge was one of the methods used to measure the continuous river flow [5]. Researchers proposed ideas for the estimation of stage-discharge curve in open channels [5, 6]. For instance, the model of Ahmadi et al. introduced the rating curve taking advantage of the discharge information at a referenced water level [6]. Another method was suggested by Fread DL using numerical solutions of the dynamic, unsteady, non-uniform flow equations [10, 11]. Recent studies proposed inverse methods based on Manning's equation to estimate the discharge from remotely sensed observations of water surface elevation, width and slope [12, 13]. Researchers discussed that useful discharge estimates could be retrieved for certain types of river reaches using satellite images. The estimates provided information on the river widths and the log-linear relationships of at-a-station hydraulic geometries (AHG) [14]. The log-Pearson type III distribution is the upgraded transformation of Pearson type III distribution and currently proposed to do flood-frequency analysis by U.S. federal agencies as described by B17.IACWD-1982 [15]. The method of moment' s estimator in log space might be suitable to estimate the parameters of the distribution [7]. Nevertheless, researchers devoted much time and effort to update the methods to the accuracy of stage-discharge curve and estimation of distribution parameters. Furthermore, few scholars gathered those two methods in practice though both are functional tools for flooding control or flood argumentation [12, 13]. In this study, the authors consider using the combination of these two technical means to determine the minimum design river bridge height and explicitly explain the feasibilities.

2. Literature Review

In hydrology, a rating curve is a graph of discharge versus stage for a given spot on the stream at a gouge station [9]. Generally, the rating curve is plotted as stage on Y-axis versus discharge on X-axis [8]. At a specific gauge station, the stream discharge is measured, and the stage can also be acquired by water level gauge. Theoretically, the curve scope diminishes gradually with a level off to 0 [8, 9].

Log-Pearson III distribution (LP3) is the updated transformation of Pearson III distribution (P3) [13, 14, 16]. In term of P3, there are shape and location parameters α, β and ε with the probability density function in Equation (1):

$$f(x) = \frac{\beta^{\alpha}}{\Gamma(\alpha)}\left(\frac{x-\varepsilon}{\beta}\right)^{\alpha-1} e^{-\beta(x-\varepsilon)} \quad (1)$$

where α > 0, (x-ε)/β > 0, Γ(α) = complete gamma function.

The density function can be confirmed if these three parameters defined. The parameters of the P3 distribution are functions of the first three population moments ($\bar{x}$, σ_x and γ_x): $\alpha = \frac{4}{\gamma_x^2}$; $\beta = \frac{\sigma_x \gamma_x}{2}$; $\varepsilon = \bar{x}(1 - 2\frac{\sigma_x}{\gamma_x})$, in which, $\bar{x}$, σ_x and γ_x are sample mean, the variance coefficient and skewness coefficient(Cs) respectively. Similarly, the LP3 distribution has probability density function:

$$f(q) = \frac{1}{|\beta| q\Gamma(\alpha)}\left(\frac{\ln q-\varepsilon}{\beta}\right)^{\alpha-1} \cdot e^{-\left(\frac{\ln q-\varepsilon}{\beta}\right)} \quad (2)$$

Where, ln q = x has a P3 distribution with parametersα, β and ε. The density function is defined for α > 0, and 0 < q < exp [ε] for $\gamma_x < 0$, and exp [ε]≤q for $\gamma_x > 0$. When $\gamma_x = 0$, the PL3 distribution

IOP Conf. Series: Journal of Physics: Conf. Series **1284** (2019) 012042 doi:10.1088/1742-6596/1284/1/012042

converges to a LN distribution.

However, the differences between LP3 and P3 are not limited to the density functions. As far as variance coefficient (Cv) or σ_x for LP3, it should be calculated as follow:

$$\sigma_x = \sqrt{\frac{\sum_{i=1}^{n}(K_i-1)^2}{n-1}} \quad (3)$$

In which, K_i is modulus coefficient; n is sample size. In term of skewness coefficient σ_x (Cs), it can be estimated by Cv. For instance, Cs = (2~3) · Cv. In hydrological computation, when a specific frequency P given, the random response variable x_p is acquired through an analysis of frequency density curve. In practice, the complicated process of calculation has been simplified in order to make it more efficient. When conducting frequency computation, based on the calculated Cv value, a hypothetical Cs should be given in order to find the Ø with a specific frequency value of P. Then, values of Ø and Cv can be substituted into formula 5 to get the according $\ln x_p$.

$$\emptyset = \frac{\ln x - \ln \bar{x}}{C_v \cdot \ln x} \quad (4)$$

$$\ln x_p = \ln \bar{x} \cdot (1 + C_v \cdot \emptyset) \quad (5)$$

where, Ø is standardized value, named coefficient of deviation mean.

3. Method Design

In this research, the following steps are used for the calculation and analysis of time series of the seasonal trends of hydrological data. The steps can help to calculate the minimum design height of river bridge deck minus navigable clearance.

- Rank the observed values of discharge data from high to low first, calculate the empirical frequency of each discharge item and graph empirical curve thereafter.
- Define hydrological frequency distribution curve (Log-Pearson III Frequency Curve).
- Work out the values of $\ln\bar{x}$, Cv and Cs by using Central Moment Method and suppose them as the subjects for first attempt. Due to large error comes out when calculating Cs by Central Moment, it is generally estimated by 2 - 3 times of Cv value.
- The identified $\ln\bar{x}$, Cv and Cs help to calculate Ø. Plot $\ln\bar{x}$ as y-axis versus P as x-axis, Log-Pearson frequency curve is graphed.
- Change the Cs then find the decent Log-Pearson frequency curve which fits the empirical curve well. Then, take advantage of those parameters of this well-fitted curve as population statistics for practical applying.
- Calculate the design discharge value with a designated frequency.

4. Data Collection and Analysis

All the data used in this study is derived from USGS. A gauge station on the stream at Kinston Mines recorded monthly data of stage and discharge of Illinois River from October 1993 to July 2018. Due to malfunction of stage measurement instrument or tele-transmission failure, especially before 2002, there were some deletions of monthly data of gauge height. But the stage data is sufficient for this study because the lost data only stands for a small proportion.

Prior to graph stage-discharge curve, it's necessary to test the correlation between gage height and discharge to ensure that there is a certain relation between these two factors. The monthly average data of gage height and discharge is used to do correlation test; the result of correlation test indicates that the correlation coefficient is 0.9719.

Mann-Kendall (M-K) test (Mann 1945; Kendall 1975), a non-parametric way, in which is widely utilized to evaluate the statistically significant trends in hydrological and climatological time series. So as to make sure whether there are extreme values of discharge data and if the time series discharge data are acceptable within a specific interval, M-K test is conducted. With a significant level $\alpha = 0.05$ and $U_{0.05} = \pm 1.96$, figure 1 (a) shows that the discharge of Illinois River at Kinston Mines has a distinct uptrend according to both UFk and UBk curves. In addition, there are three UFk excess values which are assigned with the year 2011, 2016 and 2017 respectively. The difference between UFk and upper limit $U_{0.05} = 1.96$ is most significant. Hence, it's affirmative that there is at least one monthly average extreme discharge data in 2017 and 2017 should be selected as typical year for subsequent definition of design water level.

Due to some lost data points of gage height, of which some assigned discharge data is not paired, so these discharge data points were not undertaken in drawing the stage-discharge curve. In the end, the data points of gage height and discharge used for drawing following curve (as shown in figure 1 (b) and (c)) were one-to-one correspondence (the data will be available upon request). Based on the observed values of discharge, the data is ranked from high to low and serial numbers i (i=1, 2, …24) are the newly-ranked data. Figure 1(d) shows the empirical frequency calculated by following formula:

$$Pi= [i/(n+1)]*100\% \tag{6}$$

In this research, three different attempts of plotting position are carried out. The results are shown in figures 2 (a) – (f) respectively. Figures 2(g), 2(h) and 3 combine the outcomes of those three attempts to analyse which curve with a set of Cv and Cs is best-fitted and shrinks towards the empirical frequency curve. The outcome of first attempt is the most ideal, so the Cv and Cs are the estimated parameters of $\ln x_p = ln\,\bar{x} \cdot (1 + C_v \cdot \emptyset)$. Furthermore, this curve will be used to do flood forecasting at Kinston Mines of Illinois River. The $ln\,\bar{x}$ is a constant value of 10.49, calculated from the average of annual maximum discharge values within 24 years.

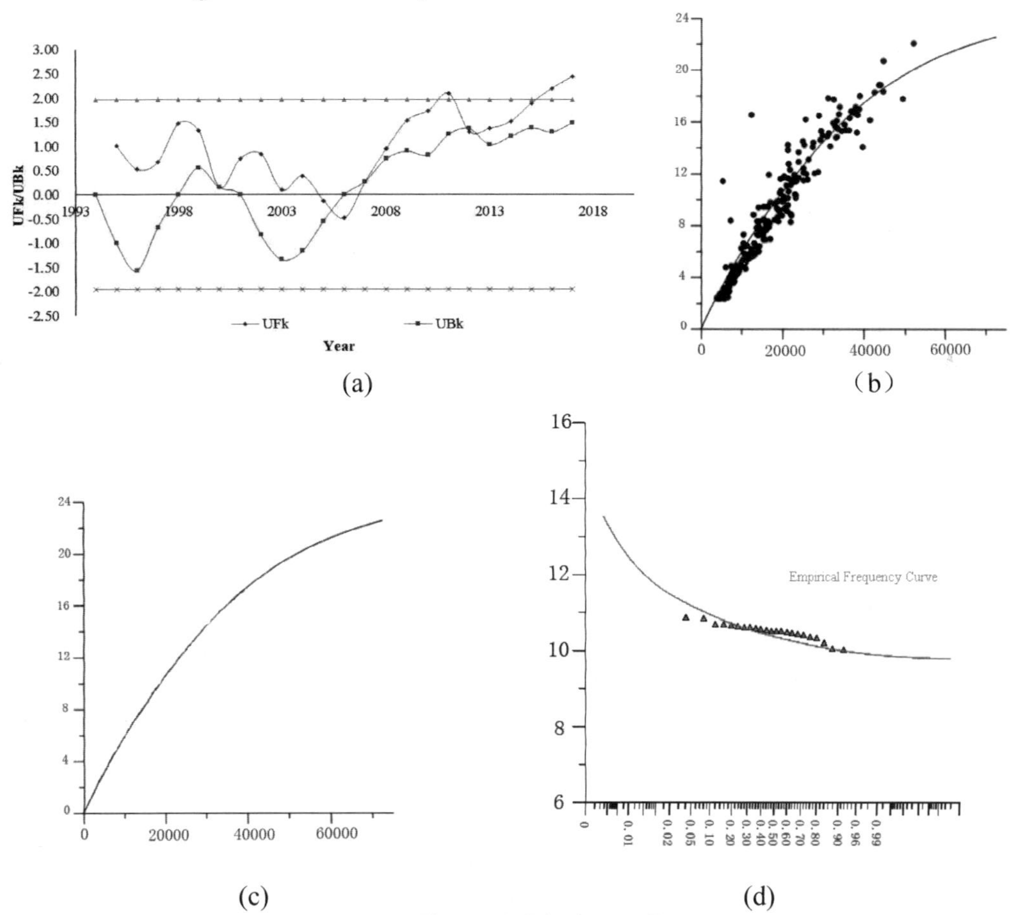

Figure 1. Discharge Curves

IOP Conf. Series: Journal of Physics: Conf. Series **1284** (2019) 012042 doi:10.1088/1742-6596/1284/1/012042

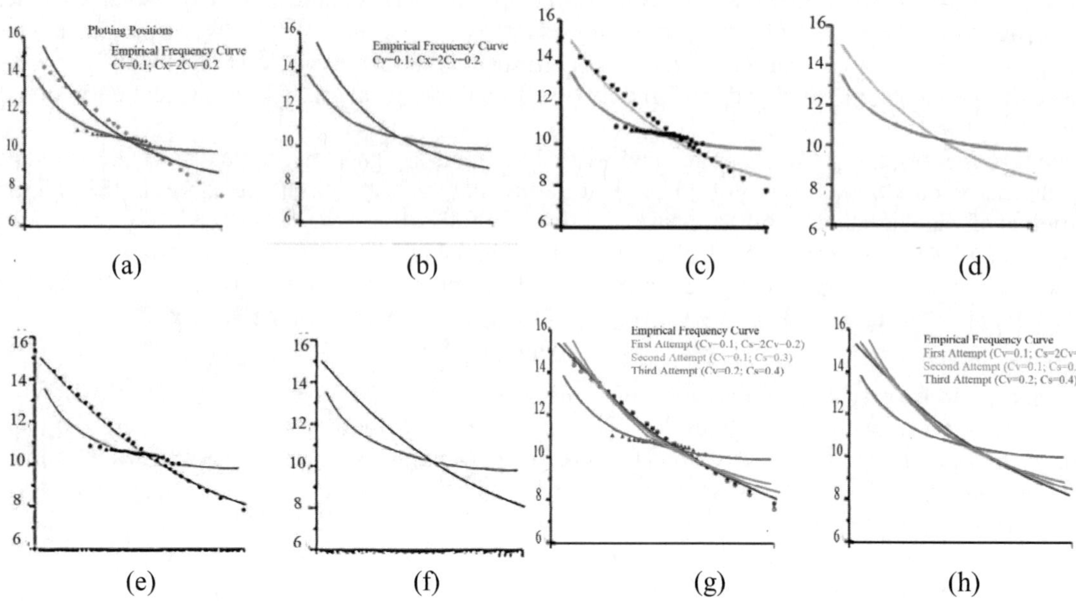

Figure 2. (a)-(f) Three Attempts of Plotting Positions; (g) and (h): Combined Plotting Positions

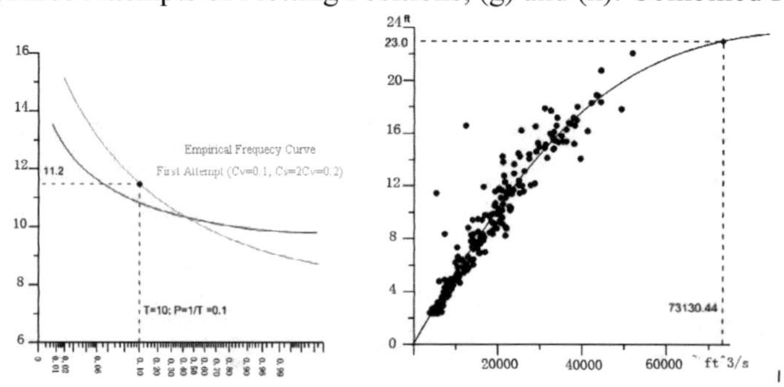

Figure 3. Regression Charts

5. Conclusions

This study discusses the minimum design height of a river bridge. In terms of the correlation test between gage height and discharge, the correlation coefficient is significantly closed to 1. In other words, it means the stage-discharge curve is drawn based on reliable data points. The M-K test was well applied to verify the annual discharge variation trend. Precisely, the discharge of Illinois River at Kinston Mines has an upward tendency from 1994 to 2017, which means the according gage height will also goes up to some extent. Hence, when doing estimation of design gage height at Kinston, Mines, the design discharge should be amplified slightly greater. Compared Pearson Frequency Curve with Log-Pearson Frequency Curve, the principles of each other is largely identical but with minor differences. Log-Pearson Frequency Curve is satisfyingly used to do plotting positions.

The research shows that design discharge with a designated return period can be deduced by the theoretical frequency curve, and then the design gage height is worked out through stage-discharge curve. The design discharge with a 10-year return period corresponds to a 23-ft gage height at Kinston Mines of Illinois River. Furthermore, when considering the minimum height of a hypothetical river bridge, it can be defined by navigable clearance and design gage height.

This hypothetical study is for the minimum design height of river bridge deck and for the first time, shows the possibility to combine the function of stage-discharge curve, Log-Pearson III Frequency

Curve or Pearson III frequency curve. In general, this study is an exploration to connect hydrological system and transport, which inspires more researchers to step forwards an interdisciplinary domain. There are still several limitations in this study. For instance, the study subject is merely a hypothetical object, so there is no data material collected about navigation system, such as the ship height. In addition, in the process of plotting positions, it's inevitable to generate some personal errors due to the curves were delineated manually. In future research, the authors plan to collect navigable clearance data to complete the study.

Acknowledgments
Authors wishing to acknowledge the financial support from the Sustainability Grant of Illinois State University (10400-01-510000000).

References
[1] Cheng Y-M, Leu S-S. Integrating data mining with KJ method to classify bridge construction defects. Expert Systems with Applications. 2011;38(6):7143-50.
[2] Wong K-Y. Design of a structural health monitoring system for long-span bridges. Structure and Infrastructure Engineering. 2007;3(2):169-85.
[3] IDOT. Standard Specifications for Road and Bridge Construction. Illinois Department of Transportation. http://www.idot.illinois.gov/Assets/uploads/files/Doing-Business/Manuals-Guides-&-Handbooks/Highways/Construction/Standard-Specifications/Standard%20Specifications%20for%20Road%20and%20Bridge%20Construction%202016.pdf2016. p. 1225.
[4] AASHTO L. AASHTO LRFD bridge design specifications. Transportation (Amst) American Association of State Highway and Transportation Officials, Inc: Washington, DC. 2007.
[5] Kavousizadeh A, Ahmadi A. High-performance approach for estimating stage-discharge curves in the open channels. Journal of hydrology. 2018; 565:197-213.
[6] Maghrebi MF, Ahmadi A. Stage-discharge prediction in natural rivers using an innovative approach. Journal of hydrology. 2017; 545:172-81.
[7] Haan CT. Statistical methods in hydrology: The Iowa State University Press; 1977.
[8] Kennedy E. Discharge ratings at gaging stations: US Geological Survey Techniques of Water-Resources Investigations, book 3, chap. A10. 1984;59.
[9] Herschy RW. Hydrometry: principles and practices. 1999.
[10] Fread D. A dynamic model of stage-discharge relations affected by changing discharge. 1973.
[11] Fread D. COMPUTATION OF STAGE-DISCHARGE RELATIONSHIPS AFFECTED BY UNSTEADY FLOW 1. JAWRA Journal of the American Water Resources Association. 1975;11(2):213-28.
[12] Durand M, Neal J, Rodríguez E, Andreadis KM, Smith LC, Yoon Y. Estimating reach-averaged discharge for the River Severn from measurements of river water surface elevation and slope. Journal of Hydrology. 2014; 511:92-104.
[13] Garambois P-A, Monnier J. Inference of effective river properties from remotely sensed observations of water surface. Advances in Water Resources. 2015; 79:103-20.
[14] Gleason CJ, Smith LC. Toward global mapping of river discharge using satellite images and at-many-stations hydraulic geometry. Proceedings of the National Academy of Sciences. 2014;111(13):4788-91.
[15] Pierson TC. Hyperconcentrated flow—transitional process between water flow and debris flow. Debris-flow hazards and related phenomena: Springer; 2005. p. 159-202.
[16] Turban E, Sharda R, Delen D, Efraim T. Decision support and business intelligence systems: Pearson; 2014.

Design and Implementation Method for Supporting Complex CAE Software System Module Development

Xiaohui Wang[1,a], Xiaohua Nie[1,b], Liang Chang[1,c] and Shihui Duan[1,d]

[1]Aircraft Strength Research Institute of China, Xi'an Shanxi 710065, China

Corresponding author: [a]wxhd100@163.com; [b]392292400@qq.com; [c]changliang4201@aliyun.com; [d]duansh63@163.com

Abstract. The use of customizable or combined modular software products to meet the growing personalization needs of users has become one of the major development directions of the software market in the future. Based on the modular design idea, a highly open CAE software architecture for HAJIF software system is proposed, and a special modular design method is clarified. According to the analysis process of aircraft structural strength, a spatial surface interpolation method based on RBF is proposed for the problem of distributed load transfer to finite element load on space curved structure. The problem is solved well and modularized. The modular design of CAE software is clear, and it can design the CAE system comprehensively, improve the development environment, shorten the development cycle, and have important significance for accelerating CAE software customization.

1. Introduction

With the rapid development of information technology, the use of finite element analysis technology for strength analysis and assessment has become one of the standard processes of aircraft structural design process. Among them, CAE software plays a very important role in the finite element analysis process as a powerful auxiliary analysis tool. In recent years, due to the launch of large-scale domestic aircraft and the replacement of various new types of aircraft, designers have become more and more high-tech tactical indicators for aircraft structures, and the finite element analysis technology in aircraft design has been continuously improved. The traditional load model analysis gradually transitions to the refined model analysis, and the analysis methods and means also undergo great changes. In the aircraft structure design process, in addition to the mainstream large-scale commercial general CAE software system, some host scientific research institutes have more urgent requirements for special feature modules, such as pneumatic load handling, test evaluation analysis, strength check, and engineering database. Therefore, the use of customizable or configurable software modular products to meet the growing personal needs of software products will be one of the major development directions for software companies in the future. Although the commercial software provides the secondary development function, due to the blockade of the underlying core code, the software function expansion is insufficient, and it is difficult to meet the actual needs of the project. In order to better design and analyze the aircraft structure, it is necessary to integrate these special tool modules and technologies into the CAE system software to achieve a high degree of uniformity in the CAE software analysis process.

IOP Conf. Series: Journal of Physics: Conf. Series **1284** (2019) 012043 doi:10.1088/1742-6596/1284/1/012043

With the occurrence of the ZTE incident, the domestic intellectual property rights of independent software are becoming more and more important, and the demand and development of special software are becoming more and more rapid. The Aviation Structure Analysis System (HAJIF), as an aerospace industry strength, has developed a civil aviation strength analysis and optimization system with independent intellectual property rights in the domestic CAE field. In recent years, with the development of computer hardware and software technology and researchers for many years Throughout the research, HAJIF computing system has greatly improved the solution of scale efficiency [1,2,3] and image visualization [4], and further established its position in the domestic CAE software industry. At the same time, HAJIF system has a highly open CAE software architecture and independent intellectual property rights, which has inherent advantages for the integration of dedicated functional modules. Therefore, based on the open software architecture of HAJIF system, this paper adopts component development technology and modular design idea in integrated development environment [5], and completes the implementation of pneumatic load processing module and integration with structural strength general analysis platform to better serve. It is analyzed and analyzed in aircraft structure, and it is convenient to provide users with customized development and service.

2. Open software architecture

Based on the modular design idea, with the goal of open, easy to expand and easy to maintain, the HAJIF software system interface, the underlying knowledge resources, the data communication and the external interface (the expansion of new modules and new algorithms) are layered and designed. A highly open software architecture as shown in Figure 1.

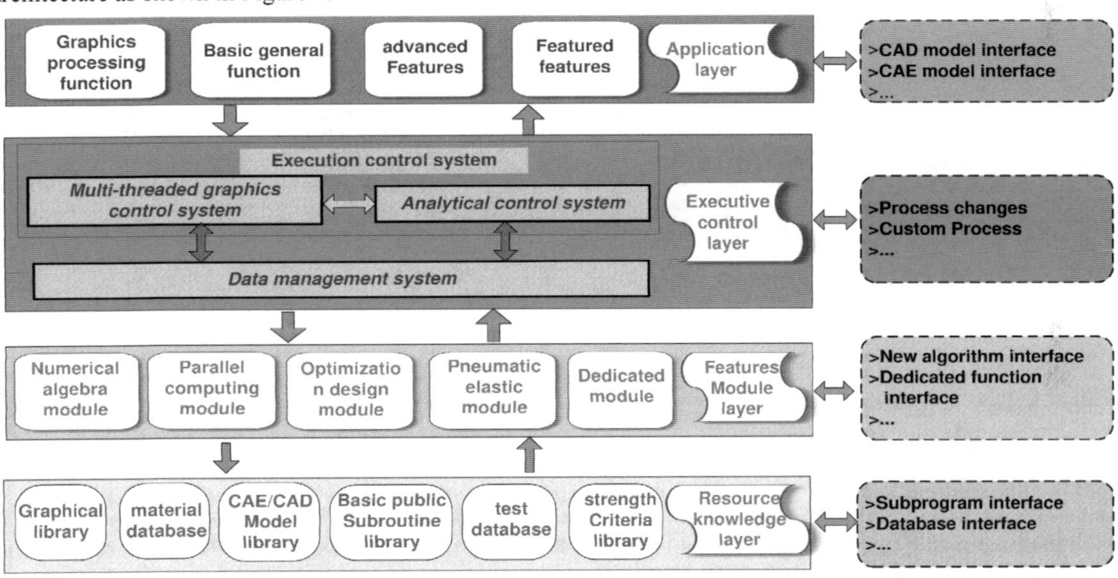

Figure 1. Open software architecture of the HAJIF system

The HAJIF software architecture is logically divided into four layers from top to bottom, namely the functional application layer, the execution control layer, the functional module layer and the resource knowledge layer.

- Function application layer: It includes graphic visualization function, general function, advanced function and special function, which is an application function that the user intuitively feels.
- Execution control layer: Contains the execution control system and data management system. The execution control system is composed of a multi-thread graphics control system and an analysis execution control system, and data communication and storage between the modules are managed by the data management system.

- Functional module layer: It contains functional modules such as numerical algebra module, parallel computing module, optimization design module, aero module and special module. It can solve the most basic and single mechanics and mathematics problems and consists of the basic resources of the resource layer.
- Resource knowledge layer: It contains the underlying support database such as graphic kernel library, material library, model library, basic common subroutine library, test database and strength criterion library, and provides interfaces for upper layer calls.

Compared with the traditional CAE software architecture, the most prominent feature of HAJIF's software architecture is its high degree of openness. It has a variety of open interfaces, including database interfaces, algorithm interfaces, dedicated module interfaces, process control interfaces and dedicated development interfaces. Data exchange with the outside world, program addition, and new software extensions.

3. System function modular design ideas

In view of the openness of the framework, the whole software programming consists of a main program and multiple basic service components. The entire build process follows object-oriented and modular software construction techniques, and forms a series of basic service components through dynamic link libraries. Different components work together by defining interfaces to complete data transfer [4] as shown in Figure 2.

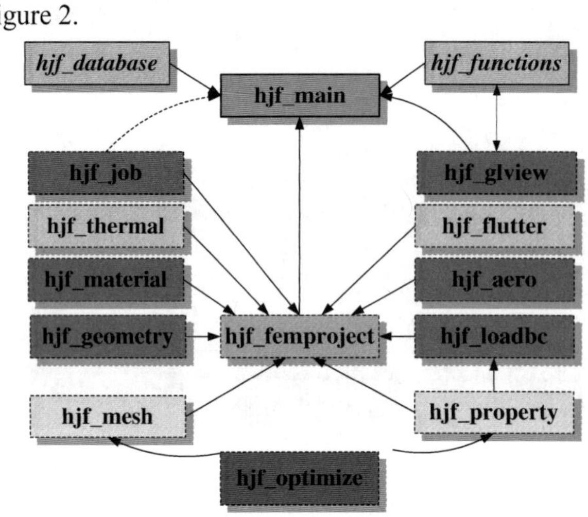

Figure 2. Component-based software programming method

3.1 Development environment

The development of the software system is mainly developed under the framework of Visual Studio 2010 using object-oriented programming ideas, specifically using the following tools:

- Computer operating system: Windows 7 64-bit system;
- Programming languages: C++, Qt, OpenGL, Python;
- Programming development platform: Visual Studio 2010.

Among them, Qt is a cross-platform C++ graphical user interface application framework. Because of its good cross-platform performance and interactive communication mechanism, it has become the main tool for data interaction between functional modules and system platforms [6].

3.2 Modular design process

In terms of software coding, the coding tools used vary from the functional requirements of each module, but the integration of each module and the basic platform can be achieved through a unified data interface. The specific implementation process as shown in Figure 3.

The general solver module is mainly implemented by Fortran and C language programming, the purpose is to make the efficiency of the solution algorithm and facilitate the management of software and hardware;

The front and rear modules are mainly in the Visual Studio 2010 environment, using C++, Qt and OpenGL mixed programming to achieve graphical interaction and display functions [4];

The solver for the dedicated module is mostly implemented by Python script file, compiling the dynamic library or executing program that can be called.

Based on the modular idea, on the software front and rear platform, the interactive Qt graphics framework is adopted to give full play to its superiority in data communication and interface linking, complete UI design and data interaction, and realize its module special performance [7].

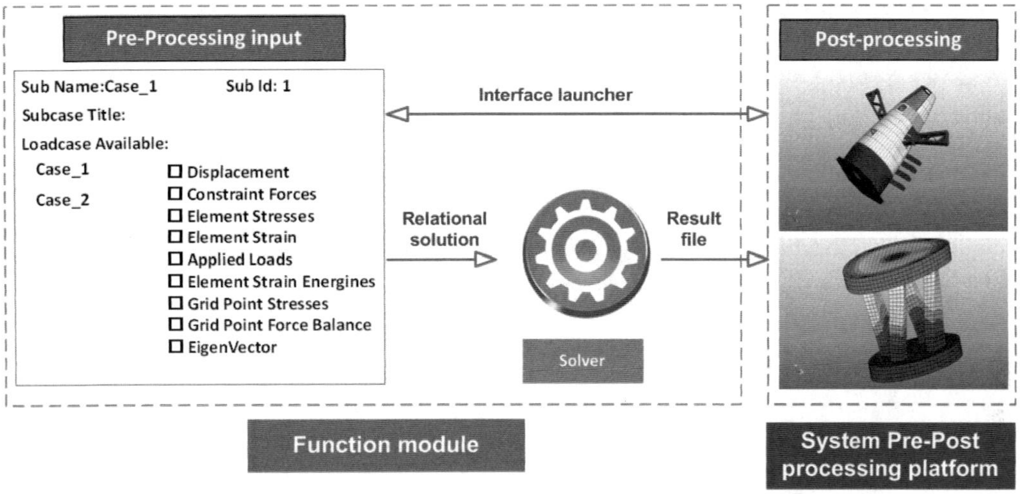

Figure 3. Software modular design development process

The HAJIF system now has a mature expansion interface and a basic general platform, and its functions are perfect, especially in the integration of new solving algorithms. The new functional modules have the inherent advantages of source code, so deep integration of algorithms and functions can be achieved. Realize the function of the algorithm and ensure the stability of the system performance. The following is verified by the development of the specific aerodynamic load processing module.

4. application examples

According to the idea of modular software development, in the process of strength analysis, the problem of equivalent transformation of aerodynamic load to finite element model load, through module interface development, integrates new algorithm, solves new problems, and expands the new function of HAJIF system.

4.1 Background knowledge

In order to analyze the high-speed aircraft affected by the aerodynamic load, the aerodynamic load and inertial load of the aircraft structure need to be effectively and reasonably applied to the finite element model for analysis [8]. Most of the radomes are hyperbolic structures with large curvature, which cannot be processed by two-dimensional problems. The traditional three-point method, four-point method and multi-point method cannot be used [9]. In this paper, by summarizing the technical characteristics of different load processing methods, the radial basis function has the advantages of high efficiency, convenient storage and simple operation in dealing with multivariate problems [10], and a local radial basis function (Radial Basis Function) is proposed. The load handling method of RBF) solves the load handling of special shape structures such as the fuselage radome. Here, through the software modular design idea, the new algorithm is integrated into the HAJIF software to better serve the structural strength analysis process.

4.2 Surface structure distribution load to finite element load conversion method

The method described here is also used for the airfoil structure. The radome or canopy load is typically provided as a distributed load in a set of spatial points and the lift coefficient thereon, ie a set of $\{(x_i, y_i, z_i), f_i\}_{i=1}^{N}$, intended to be applied to the finite element model. For this problem, we use the method of radial basis function.

The radial basis function was originally applied to the approximation theory and is considered to be the most accurate and stable method for solving discrete data interpolation [10]. Radial basis function interpolation is defined as: for a given multivariate scattered data $\{X_i, F_i\}_{i=1}^{N} \in R^n \times R$, the radial function Φ: $R^+ \to R$ is selected, and use the translation $\{\phi\|X - X_i\|\}_{i=1}^{N}$, where the length of the vector V is expressed $\|V\|$, ie $\|V\| = \sqrt{x^2 + y^2 + z^2}$, and the interpolation function *S(X)* is found as determined by the equation (1).

$$S(X) = p(X) + \sum \lambda_i \phi\|X - X_i\| \tag{1}$$

Where $$P(X)=c_1x+c_2y+c_3z+c_4 \tag{2}$$

The interpolation condition is satisfied as by the equation (3)

$$S(Xi)=fi \tag{3}$$

And the orthogonal conditions as determined by the equation (4).

$$E \sum_{i=}^{N} \lambda_i = \sum_{i=}^{N} \lambda_i x_i = \sum_{i=}^{N} \lambda_i y_i = \sum_{i=}^{N} \lambda_i z_i = 0 \tag{4}$$

Solving the following equations (5) yields the coefficients λ_i and c_1, c_2, c_3, c_4.

$$\begin{bmatrix} \phi(\|X_1 - X_1\|) & \phi(\|X_1 - X_2\|) & \cdots & \phi(\|X_1 - X_N\|) & x_1 & y_1 & z_1 & 1 \\ \phi(\|X_2 - X_1\|) & \phi(\|X_2 - X_2\|) & \cdots & \phi(\|X_2 - X_N\|) & x_2 & y_2 & z_2 & 1 \\ \cdots & \cdots & \cdots & \cdots & \cdots & \cdots & \cdots & \cdots \\ \phi(\|X_N - X_1\|) & \phi(\|X_N - X_2\|) & \cdots & \phi(\|X_1 - X_N\|) & x_N & y_N & z_N & 1 \\ x_1 & x_2 & \cdots & x_N & 0 & 0 & 0 & 0 \\ y_1 & y_2 & \cdots & y_N & 0 & 0 & 0 & 0 \\ z_1 & z_2 & \cdots & z_N & 0 & 0 & 0 & 0 \\ 1 & 1 & \cdots & 1 & 0 & 0 & 0 & 0 \end{bmatrix} \begin{bmatrix} \lambda_1 \\ \lambda_2 \\ \cdots \\ \lambda_N \\ c_1 \\ c_2 \\ c_3 \\ c_4 \end{bmatrix} = \begin{bmatrix} f_1 \\ f_2 \\ \cdots \\ f_n \\ 0 \\ 0 \\ 0 \\ 0 \end{bmatrix} \tag{5}$$

According to actual needs, we use the basis function $\phi(r) = r^3$. Substituting the obtained coefficients λ_i and c_1, c_2, c_3, c_4 into the equation (1), the radial basis function *S(X)* is obtained, and all the required values can be obtained by using *S(X)*. For the convenience of description, the formula (5) is abbreviated as formula (6):

$$AX=F \tag{6}$$

According to most of the papers on radial basis functions, the basis function $\phi(r)$ is chosen appropriately. The coefficient matrix ***A*** is symmetric positive definite, but in fact this statement does not hold. For example, all points are in the same plane, and the determinant of ***A*** is near zero. The number is large; for example, z_i is all zero, one row and one column of matrix ***A*** is zero, and the determinant of ***A*** is zero. To this end we use a pseudo inverse matrix or ***moore-Penrose*** inverse to overcome this problem.

The matrix ***G*** satisfying the equation (7) is a pseudo inverse matrix of ***A***, and is represented by the symbol ***A+***.

$$AGA=A, GAG=G, (AG)T=AG, (GA)T=GAb \tag{7}$$

Using the singular value decomposition algorithm (SVD), *A+* is obtained and ***X=A+F*** is obtained.

The implementation process of the RBF program is as follows:

1) Construct a linear equation formula (5) using the coordinates of the aerodynamic load distribution point and the lift coefficient;
2) Solve the linear equations (5). If the determinant of $\boldsymbol{A}$ is not zero, solve it directly. Otherwise, use the pseudo inverse matrix to obtain the coefficients λ_i and c_1, c_2, c_3, c_4.
3) The coefficients λ_i and c_1, c_2, c_3, c_4 are substituted into equation (1) to obtain the radial basis function *S(X)*;
4) For all finite element cycles that need to handle the load, use the radial basis function *S(X)* to obtain the load distribution lift coefficient of the element node or centroid point;
5) End the aerodynamic load point cycle and output the load card required for structural analysis.

4.3 Software Implementation

Based on the software modular design development process, the implementation of the pneumatic load processing module corresponds to the functional module layer in Figure 1, according to the development process as shown in Figure 3, the implementation process of the pneumatic load processing module is shown in Figure 4.

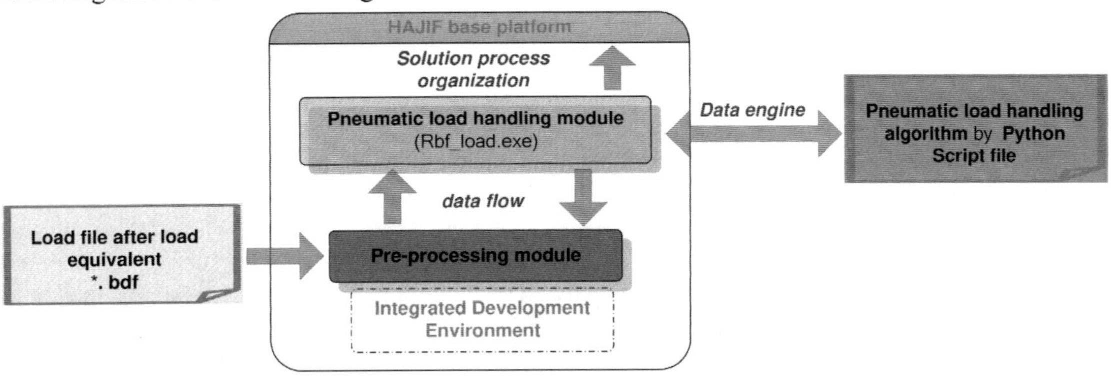

Figure 4. Pneumatic load processing module implementation process

Step 1, use the Python script file to implement the RBF processing algorithm in Section 4.2, and compile and generate the executable file rbf_load.Exe;

Step 2, use the Qt graphics framework to complete the interface organization and design of the aerodynamic load processing module in the software integrated development environment (Visual Studio 2010), including the input of the load file. Selection of equivalent methods, equivalent direction, and selection of target objects;

Step 3, based on the modular design, uses the QT communication mechanism to open the input interface of the pneumatic load processing module through signal transmission. The program is implemented as follows:

Connect(actAerodynamic, SIGNAL(triggered()), this, SLOT(activePneumaticLoadMenu()));

Through the component calling technology, the pneumatic load processing module inputs the information, starts the rbf_load.Exe through the data engine, passes the input data, completes the data connection between the module and the system platform, and the program is realized as follows:

QProcess myProcess; *(create process)*

myProcess.start(exeFile,param); *(Submit input parameters and execute the launcher)*

myProcess.waitForFinished(); *(wait for the process to abort)*

Step 4, after the pneumatic load module is processed, the data interface is defined according to the file input format of the HAJIF pre-module, and the real-time display of the load processing can be realized as shown in Figure 5.

The function module can be operated independently or integrated with the analysis platform, which makes the software modules reusable, independent and combinable. It has been used for load handling of multiple models.

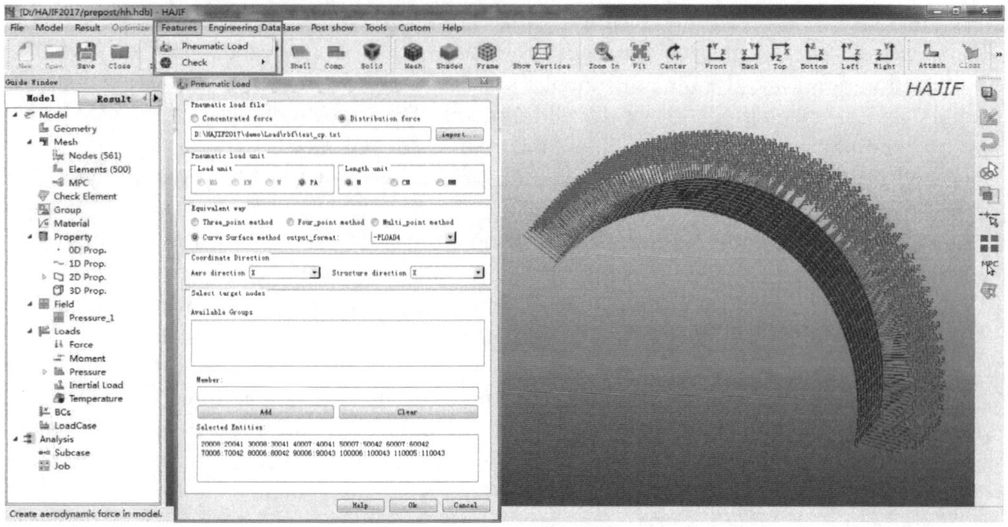

Figure 5. Aerodynamic load handling module based on HAJIF system

5. Conclusion

Firstly, by introducing modular design ideas, this paper proposes a highly open software architecture of HAJIF software system, which facilitates the seamless integration of new algorithms and special function modules, improves the software system functions, and improves the software integration efficiency. Secondly, combined with engineering application, a special method of aerodynamic load based on radial basis function (RBF) is studied for the special hyperboloid structure of radome, and through the modular design idea, from the function input, output and platform link The feasibility division is carried out, and the algorithm is successfully integrated into the HAJIF software system. The implementation of the dedicated module is completed, which fully verifies the important value of the modular design in the design and development process of the complex CAE software system.

References

[1] Chang Liang, Duan Shihui, Wang Likai et al. Large-scale parallel analysis and optimization application of aviation structure[J]. *Computer Aided Engineering*. 2017, 26(3): 45-50.

[2] Luo Lilong, Chang Liang, Wang Likai et al. Research on the solution technique of hyperscale variable structure sensitivity[J].*Mechanical Science and Technology*.2017, 36(5):816-820.

[3] Wang Likai, Guo Yuchao, Luo Lilong. Implementation of large-scale optimization design in HAJIF system[J]. *Engineering and Experimentation*. 2016, 56(3): 91-93.

[4] Wang Xiaohui, Huang He, Guo Yuchao et al. Design and implementation of pre- and post-processing modules based on HAJIF[J]. Computer Engineering and Design. 2017, 45(3): 558-564.

[5] Rao Tiangui, Li Kai, Li Yufeng. Modular design of locomotive display application software [J]. *Locomotive electric drive*.2016, 1:71-74.

[6] Jing Yazhi. Research on architecture design in software development [J]. *Technology Guide*. 2014, 21: 142-143.

[7] Lu Wenzhou. Qt5 Development and Examples [M], Second Edition. Beijing: Publishing House of Electronics Industry, 2015.

[8] Lei Li, Han Qing, Zhong Xiaoping. Comparative Analysis of the Load Transfer Method of Airfoil Finite Element Nodes[J]. *Progress in Aeronautical Engineering*, 2014, 3(5): 383-391.

[9] Jin Jinli. Chen Quanli et al. Research on aerodynamic load treatment method [J]. *Science and Technology Innovation and Application*, 2015, 26: 44.

[10] Hu Chaobin et al. Research on Local Radial Basis Function Interpolation Algorithm[J]. *Mechani-cal & Electrical Technology*, 2014,2:43-45.

Cloud Server Oriented FPGA Accelerator for Long Short-Term Memory Recurrent Neural Networks

Jiasheng Wang[1,a], Yu Zhou[1,b], Yuyang Sun[1,c], Keyang Li[1,d] and Jun Liu[1,e]

[1]Center for Data Science, Beijing University of Posts and Telecommunications, Beijing, China, 100876

E-mail: [a]wangjiasheng@bupt.edu.cn, [b]zyicy@bupt.edu.cn, [c]sunyuyang@bupt.edu.cn, [d]2015302580280@whu.edu.cn, [e]liujun@bupt.edu.cn

Abstract. Long Short-Term Memory network(LSTM), which is the most widely used and representative recurrent neural network architecture, plays an important role in language modeling, machine translation, image captioning, etc. However, due to its recurrent nature, general-purpose processors like CPUs and GPUs cannot achieve high parallelism, not to mention their high power consumption. FPGA accelerators can outperform them by flexibility, energy-efficiency and more delicate optimization in each phase of the algorithm. In this paper, we present a cloud-oriented FPGA accelerator for LSTM based on OpenCL. Different from previous works which are designed for embedded systems, our FPGA accelerator performs multiple time series predictions in parallel. We provide a general matrix optimization model to optimize the computation of LSTM in the cloud environments. The performance of our implementation beats both the CPU implementation and other previous hardware implementations. We present and analyze the performance results of our work.

1. Introduction

In recent years, the great progress of inference accuracy of deep learning makes it a hot topic in computer science. Deep Neural Networks (DNN) provides a practical method for computers to learn knowledges from massive labeled data. Recurrent neural network (RNN) is a class of Deep Neural Networks which processes sequential data and performs time series predictions. RNN has been proved to be successful in various time series prediction related applications like language modeling [1] [2], machine translation [3], and image captioning [4]-[6], etc. A number of RNN models are created to promote the performance of the traditional RNN. Among these models, Long Short-Term Memory network (LSTM) is the most commonly used recurrent neural network architecture. It implements a learned memory controller to improve the prediction accuracy of RNNs.

From model aspect, LSTM focus on how to improve the prediction accuracy while paying little attention to the throughput, respond time or the power consumption of the application. As the LSTM applications are becoming more and more mature, they are increasingly deployed in cloud environments. In this kind of situation, high throughput, short response time, and low power consumption are very important. However, owing to the recurrent nature, LSTM implementation on general-purpose processors like CPUs and GPUs cannot achieve high parallelism and needs high power consumption.

To solve this problem, a number of works have been done to utilize FPGA to accelerate LSTM [7] [8] [9]. FPGA accelerators can offer more opportunity and ability to optimize the applications delicately, given that they can perform the computation through the process of pipelining and the hardware structure can be customized to adapt to specific applications when being synthesized. In addition, FPGA accelerators are known to achieve low power consumption. However, most previous hardware implementations are only optimized on embedded platforms. As for the cloud environment, the host server usually receives multiple input streams in parallel. Traditional LSTM algorithms are largely composed of vector-based manipulations, which read vectors sequentially, reducing the overall throughput of the system.

In this paper, we design and implement a cloud server oriented FPGA accelerator of LSTM. The design and implementation are highly optimized for the LSTM applications in cloud environments. We use OpenCL to develop our accelerator. OpenCL is an open, cross-platform parallel programming language that can be used in CPU, GPU and FPGA developments, supported by many vendors [10]. Xilinx SDAccel SDK implements the OpenCL standard based on its high-level synthesis tools.

Our implementation transforms the vector manipulations into matrix operations, providing a general matrix optimization model that optimizes the matrix computation in cloud environments. We process multiple input streams in parallel, ensuring the high throughput of our system. Given that we predict multiple data sequences in a deep pipeline way instead of creating copies of processing hardware, increasing the number of input streams makes no impact on the resource usage. Our work achieves high performance in cloud environments. We measure the throughput by the number of predictions it can produce per second. The peak throughput of our design is 7573462.59 predictions per second, which means that the average execution time of one prediction per input stream is 132.04ns.

The rest of this paper is structured as follows. In Section II, we compare our implementation with related work. Section III discusses the details of our design and implementation. Section IV presents and analyzes the results of our work. In Section V, we conclude and point out future works.

2. Related Work

LSTM was first designed in [11], which has some updated variations [1] [12] [13]. The LSTM with peephole introduced by [14] is a famous variation that the cell memory influences the forget, input and output gates. As a result, the model peeps into the memory cell before deciding whether to memorize or forget. All those variations presented above have similar performance as shown in [15], which surveys on the most commonly used variations, discovering that the variations have few effects on overall prediction accuracy. Our work implements the LSTM version without peepholes [16].

LSTM network has been widely used in such various areas as language modeling, machine translation, image captioning, etc. Most software-implemented LSTM applications are paying too much attention on prediction accuracy, while ignoring the throughput and power consumption, which are vital in real production scenarios. Efforts have rarely been made to implement a hardware accelerator. In [7], Chang et al. implements a two layer LSTM network with 128 neurons each, which is regarded as the earliest actual hardware implementation for LSTM. However, it does not achieve high parallelism. Meanwhile, its performance is not pretty well compared to some latest implementations. Work in [8] optimizes the LSTM network both in computation and communication phase. The authors use pipelining and unrolling to improve the computation performance. Work in [9] focuses on a delicate optimization for LSTM. Then the authors compare their implementation with [7] directly, proving that their work in [9] has a significant speed-up over [7] and the software implementations.

Previous works like [7], [8] and [9] all aim at embedded systems.

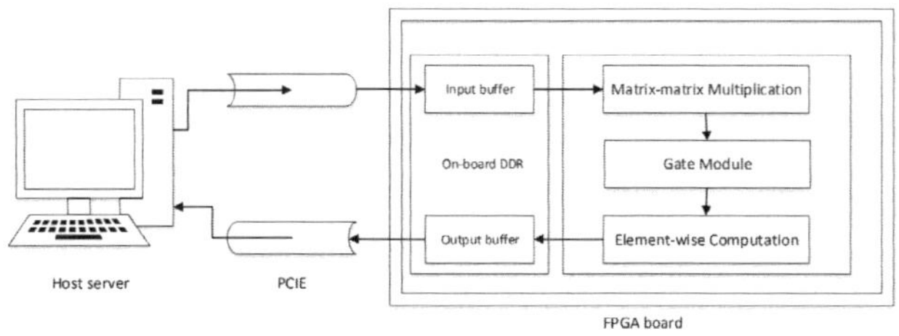

Figure 1. System Architecture

However, we could hardly find any LSTM hardware implementations that are optimized for cloud environments. In cloud environments, host servers usually receive multiple input streams in parallel. We make some modifications on traditional LSTM algorithms and implement a cloud environment hardware system. Since our work is similar to [9], we compare our results to those in [9], proving that the on-chip computation performance of our work outperforms the software implementation and hardware implementation in [9].

3. Design and Implementation

In this section, we will first show the whole view of the system architecture of our FPGA accelerator. Secondly, we will demonstrate the data-flow between each component of our system, describing how they interact with each other. Then we will discuss our design in detail, providing a general matrix optimization model to analyze the core principles and methods of our optimization. Finally we will present and illustrate the specific implementation of our accelerator.

3.1. Architecture

Figure 1 shows the system architecture of our work. Since our accelerator is designed for cloud environments, our system architecture is quite different from that in embedded environments. It includes not only the FPGA board but also the host server, which is a universal server equipped with CPUs to receive requests from remote clients and send response to them. The host server controls the data transmission and schedules the computation-intensive tasks for the FPGA board. It transfers data with the FPGA board through PCIe bus. The DDR4 DRAMs persisted on the FPGA board receive input vectors from PCIe bus and save output vectors generated by the FPGA chip, functioning as a connector between FPGA chip and host server. The data transmissions between the DDR4 DRAMs and the FPGA chip are done on the AXI4 bus or the AXI4Lite bus. The three modules in the FPGA chip are our core computation components, which implement three different phases of the LSTM network. The on-chip modules are also connected to AXI4 bus or AXI4Lite bus. Since LSTM network is a recurrent neural network which needs multiple time steps to generate final output vectors, the computation of the three on-chip modules and the data transmission among them need to loop through several time steps. In each time step, output vectors from the previous module function as the input vectors of the next module. What's more, The output vectors generated by the element-wise computation module from the previous time step are regarded as recurrent state, which are also feeded to the matrix-matrix multiplication module as input data. The following sections will discuss the design principles and implementation details of three on-chip modules mentioned above.

3.2. Matrix Optimization Model

The LSTM algorithm can be characterized by the following equations:

$$i_t = \sigma(W_{xi} * x_t + W_{hi} * h_{t-1} + b_i) \tag{1}$$

$$f_t = \sigma(W_{xf} * x_t + W_{hf} * h_{t-1} + b_f) \tag{2}$$

$$\tilde{c}_t = tanh(W_{xc} * x_t + W_{hc} * h_{t-1} + b_c) \tag{3}$$

$$o_t = \sigma(W_{xo} * x_t + W_{ho} * h_{t-1} + b_o) \tag{4}$$

$$c_t = f_t \odot c_{t-1} + i_t \odot \tilde{c}_t \tag{5}$$

$$h_t = o_t \odot tanh(c_t) \tag{6}$$

where $\odot$ is element-wise multiplication, σ is the *sigmoid* function, x is the input vector, W is the corresponding weight matrix, b is the corresponding bias vector, c is memory cell activation, and h is the layer output vector. $i, f, \tilde{c}, o$ represent the input gate , forget gate, candidate memory cell gate, and output gate respectively. The subscript t denotes the corresponding time step.

(1) (2) (3) (4) (5) (6) indicate that the traditional LSTM algorithm is based on vector manipulation. Previous LSTM hardware accelerators like [7] [8] [9] focus on optimization for this vector-based model, including the matrix-vector multiplication and some kinds of element-wise vector computation. However, the defective vector-based model mainly lies in its low throughput, because the LSTM network can only process one input stream at a time. Provided that in cloud environments we must handle multiple input streams, the traditional vector-based model is not longer suitable for our accelerator. Our implementation transforms the vector-based model to a matrix-based model, grouping multiple input vectors from various input streams into an input matrix. The LSTM implementation of our work can be characterized by the following equations:

$$I_t = \sigma(X_t * W_{xi} + H_{t-1} * W_{hi} + B_i) \tag{7}$$

$$F_t = \sigma(X_t * W_{xf} + H_{t-1} * W_{hf} + B_f) \tag{8}$$

$$\tilde{C}_t = tanh(X_t * W_{xc} + H_{t-1} * W_{hc} + B_c) \tag{9}$$

$$O_t = \sigma(X_t * W_{xo} + H_{t-1} * W_{ho} + B_o) \tag{10}$$

$$C_t = F_t \odot C_{t-1} + I_t \odot \tilde{C}_t \tag{11}$$

$$H_t = O_t \odot tanh(C_t) \tag{12}$$

As (7) (8) (9) (10) (11) (12) show, after grouping multiple input vectors together, our implementation is largely based on matrix computations, including matrix-matrix multiplication, matrix addition, element-wise matrix multiplication, element-wise matrix *sigmoid* function, and element-wise matrix *tanh* function. In order to optimize the computation phase of our system, we put forward a general matrix optimization model that optimize all these matrix computations by using unrolling and pipelining manners. All of three on-chip modules presented in Figure 1 use this model to implement the LSTM network.

We notice that these different types of matrix computations have similar structures, while the main differences are matrix operators. That is to say, when performing matrix computations, each element in two input matrices would perform a certain type of basic matrix operation. Figure 2 shows five matrix operators, which are MAC, $+$, $*$, *sigmoid* and *tanh*, referring to the matrix-matrix multiplication, the matrix addition, the element-wise multiplication, the element-wise matrix sigmoid function and the element-wise matrix *tanh* function respectively.

Figure 3 can be considered as the expanded view of Figure 2, presenting the details of our matrix optimization model. *Input Stream Count* denotes the number of input streams which need to be processed, while *LSTM Size* denotes the scale of the LSTM model, which is determined by the size of the weight matrix.

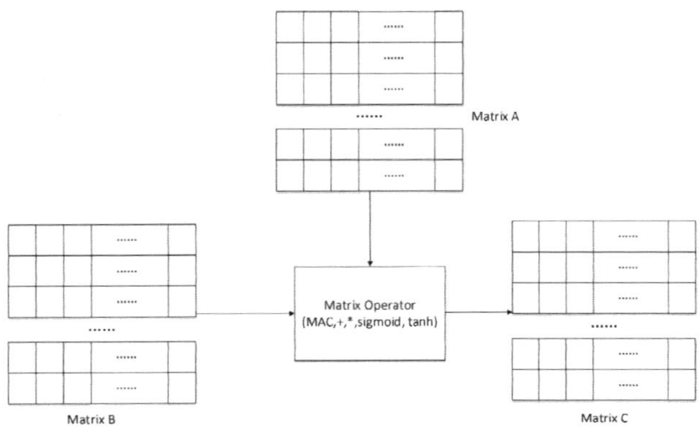

Figure 2. Matrix Computations

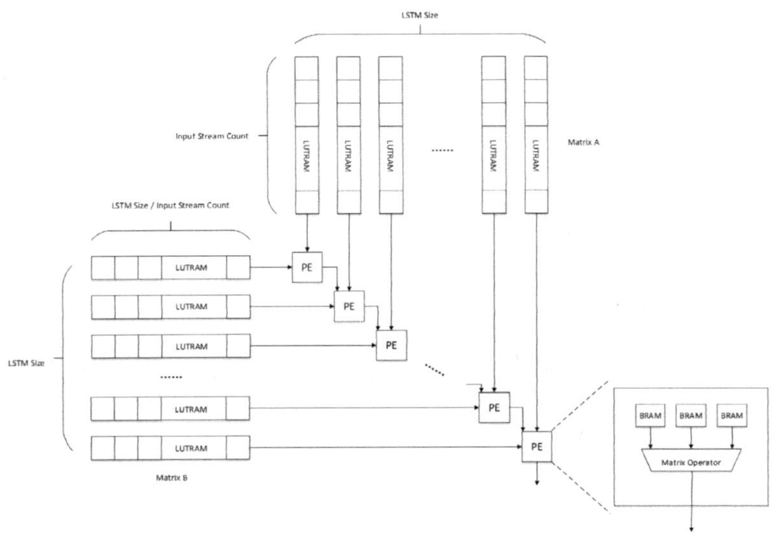

Figure 3. Matrix Optimization Model

We assume that the size of the input vector is less than the *LSTM Size*. In Figure 3, *Matrix A* represents the input matrix in on-chip computation modules, while *Matrix B* represents the weight matrix in the matrix-matrix multiplication module or the input matrix in the other two modules. When performing matrix computation, we always unroll the loop of iterating the columns of input matrix or the rows of the weight matrix. The PE(Processing Element) in Figure 3 shows the unrolling logic. The number of PEs is *LSTM Size*. Every PE fetches one value from the column of the input matrix or the corresponding row of the weight matrix to its inside BRAM, where matrix computation with specific operator are processed parallel. If the matrix B is the weight matrix in the matrix-matrix multiplication module, the previous PE also has to write the output value to the next PE, satisfying the need of *MAC* operator.

All PEs are running in a deep pipeline way. Different PEs are processing different input streams simultaneously, and even a single PE is processing multiple input streams. For instance, it can read one value of input stream 3 from Matrix A, while reading one value of input stream 2 from Matrix B and performing the computation of the matrix operator of input stream 1 in parallel. In conclusion, all PEs are processing different phases of the computation of different input streams. This pipeline mechanism keeps the hardware working all the time and improves performance significantly.

In order to maximize the pipeline performance, we must ensure the PEs carry no dependencies. Considering that we have $LSTM\ Size$ PEs, which could access at most $LSTM\ Size$ columns of the input matrix and at most $LSTM\ Size$ rows of the weight matrix in parallel, we store the matrices in LUTRAM. The input matrix is partitioned by columns while the weight matrix is partitioned by rows, constructing N-ports memory which can provide N parallel accesses. After the partition, the PEs carry no dependencies and the initiation interval of the pipeline is 1.

The following subsections in this chapter will present the implementation details of the three on-chip computation modules, based on the matrix optimization model.

3.3. Matrix-Matrix Multiplication

Algorithm 1 Matrix-matrix multiplication

Input:

$A = [a_{ij}]_{1 \leqslant i \leqslant MAX\_STREAMS, 1 \leqslant j \leqslant MAX\_LSTM\_SIZE} \rightarrow$ The input matrix
$W = [w_{ij}]_{1 \leqslant i,j \leqslant MAX\_LSTM\_SIZE} \rightarrow$ The weight matrix

Output:

$B = [b_{ij}]_{1 \leqslant i \leqslant MAX\_STREAMS, 1 \leqslant j \leqslant MAX\_LSTM\_SIZE} \rightarrow$ The output matrix

for j = 1 **to** $real\_lstm\_size$ **do**
    #pragma pipeline
    for i = 1 **to** $stream\_count$ **do**
        Initialize Temp to zero
        #pragma unroll by factor of MAX_LSTM_SIZE
        for k = 1 **to** MAX_LSTM_SIZE **do**
            $Temp = a_{i,k} * w_{k,j} + Temp$
        end for
        $b_{i,j} = Temp$
    end for
end for

According to the (7) (8) (9) (10) in Section 3.2, we can see that matrix-matrix computation is the first task of the LSTM network. The pseudocode of the matrix-matrix multiplication is presented in Algorithm 1. The input matrix A can be the layer input matrix X_t or the recurrent state matrix H_{t-1} from the previous time step and the weight matrix can be any of the $W_{xi}, W_{xf}, W_{xc}, W_{xo}, W_{hi}, W_{hf}, W_{hc}, W_{ho}$. We define $MAX\_STREAMS$ as an upper bound of the number of rows of the input matrix and the output matrix, deciding the maximum input streams we can process together. We also define MAX_LSTM_SIZE as an upper bound of the number of rows of the weight matrix and columns of the input matrix, the weight matrix and the output matrix. LSTM applications can use this model if weight scale is no more than MAX_LSTM_SIZE. We assume that the size of input vector are less than the LSTM weight size. The matrix-matrix multiplication is performed by three nested loops. $real\_lstm\_size$ denotes the actual weight size of the LSTM layer and $stream\_count$ denotes the number of input streams. The innermost loop generate one element of the output matrix and the rest two outer loop traverse the output matrix in row-major order.

We completely unroll the innermost loop and pipeline the second loop to achieve high parallelism. We can see that Algorithm 1 is based on the matrix optimization model and the matrix operator is MAC. The two inner loops implement the matrix optimization model while the outermost loop performs the calculation sequentially. In this way, we can predict multiple data sequences in parallel and achieve high throughput.

3.4. Gate Module
After the matrix-matrix multiplication, we can see that the rest of work in (7) (8) (9) (10) are matrix addition, element-wise *sigmoid* function and element-wise *tanh* function. Given that we construct the data in the form of matrices rather than vectors, we can keep benefiting from the deep pipeline way. Provided that the *sigmoid* function and the *tanh* function are non-linear which are hard to be implemented in FPGA, we use the Polynomial Approximations [9] to implement approximate linear functions. We adopt the same polynomial coefficients which are also used in [9]. The maximum approximation are $1.408X10^{-3}$ for the *sigmoid* function and $1.21X10^{-2}$ for the *tanh* function.

Algorithm 2 Gate Module

Input:
$A = [a_{ij}]_{1 \leqslant i \leqslant MAX\_STREAMS, 1 \leqslant j \leqslant MAX\_LSTM\_SIZE}$ → The input matrix 1
$B = [b_{ij}]_{1 \leqslant i \leqslant MAX\_STREAMS, 1 \leqslant j \leqslant MAX\_LSTM\_SIZE}$ → The input matrix 2
$bias = [v_i]_{1 \leqslant i \leqslant MAX\_LSTM\_SIZE}$ → The bias vector

Output:
$C = [c_{ij}]_{1 \leqslant i \leqslant MAX\_STREAMS, 1 \leqslant j \leqslant MAX\_LSTM\_SIZE}$ → The output matrix

```
#pragma pipeline
for i = 1 to stream_count do
    #pragma unroll by factor of MAX_LSTM_SIZE
    for j = 1 to MAX_LSTM_SIZE do
        Temp = a_{i,j} + b_{i,j} + v_j
        c_{i,j} = approximate linear function(Temp)
    end for
end for
```

Algorithm 2 presents the details of matrix addition and element-wise activation functions. The A matrix is the output matrix from $X_t * W_{xi}, X_t * W_{xf}, X_t * W_{xc}, or X_t * W_{xi}$, the B matrix is the output matrix from $H_{t-1} * W_{hi}, H_{t-1} * W_{hf}, H_{t-1} * W_{hc}, or H_{t-1} * W_{ho}$, $bias$ can be any weight bias vector of b_i, b_f, b_c, b_o, and the C matrix can be one of $I_t, F_t, \tilde{C}_t, O_t$ respectively. We can see that Algorithm 2 is based on the matrix optimization model, using +, *sigmoid* and *tanh* as the matrix operators.

3.5. Element-wise Computation
The rest work of our accelerator is implementing the last two Equations (11) and (12), which mainly consist of element-wise multiplication, matrix addition and element-wise *tanh* function. The optimization methods are similar to Algorithm 2 and the only difference is the matrix operators. This module is also based on the matrix optimization model, using the $*$, $+$, and *tanh* as matrix operators.

4. Results
We first introduce our experiment environment, including the hardware and software platforms as well as experimental methods. Then we analyze the results of different-sized synthesized networks and various numbers of input streams.

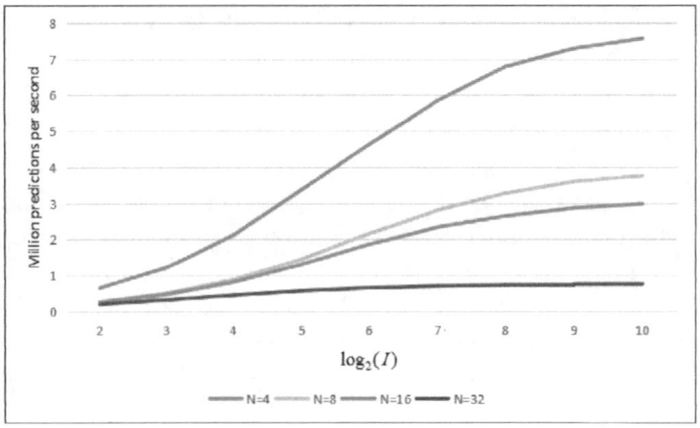

Figure 4. Millions of predictions per second according to the number of input streams I

4.1. Experimental Environment

We implement an OpenCL-based FPGA accelerator with the assistance of a high-level synthesis design tool, Xilinx SDAccel v2017.1. Our hardware system consists of a Xilinx Kintex UltraScale KCU1500 Acceleration Development Board and a general host server with Intel(R) Pentium(R) CPU G2030. The host server connects with FPGA board using PCIe Gen2(5 GT/s) with 8 lanes.

We use Keras with TensorFlow backend to train and test the software LSTM network. The same weight parameters and input vectors are used as the input of our hardware implementation. The output vectors of the hardware implementation are compared with the software LSTM network, ensuring that two implementations have the same results within a certain error range because of the non-linear activation functions explained in 3.4.

We test our system with different-sized synthesized networks and various numbers of input streams to examine the performance. We define the size of synthesized network as N, representing for the scale of the weight matrix, which can be 4, 8, 16 and 32. The number of input stream is defined as $I \in \{4, 8, 32, 64, 128, 512, 1024\}$. The different results of various N and I are presented and analyzed in the next subsection. We run our application 1000 times for each combination of N and I.

4.2. Performance

To show the benefits of our implementation, we compare the results to a similar FPGA accelerator as presented in [9]. The [9] focuses on on-chip computation performance. The [9] also implement a software LSTM layer in python. We compare the average computation time per input stream of our accelerator with the performance of hardware implementation and python implementation in [9]. The comparison result is reported in Table 1. The results prove that the performance of our implementation is better than the [9] in all of the different-sized synthesized networks.

The results presented in Table 1 are our best results when $I = 1024$.

Table 1. Comparison with Previous Implementation

	OUrs	**Refs.[9] hardware**	**Refs.[9] python**
N=4	132.04 ns	259.12 ns	65 μs
N=8	265.77 ns	317.52 ns	72 μs
N=16	335.60 ns	461.336 ns	96 μs
N=32	1.33 μs	1.586 μs	185 μs

Table 2. Clock Frequency

	OUrs	**Refs.[9]**
N=4	$300MHz$	$158MHz$
N=8	$280MHz$	$154MHz$
N=16	$250MHz$	$140MHz$
N=32	$190MHz$	$102MHz$

Figure 4 demonstrates that the throughput improves when I grows. Especially when I is small, the throughput grows linearly. It is because of the high parallelism that our matrix model performs better.

As Section 3.2 explains, we process multiple input streams in a deep pipeline way, predicting multiple data sequences in parallel. When I is small, the pipeline queue is always empty, leading to the hardware idleness. For example, if one PE can process four input streams in parallel and there are totally 16 PEs(N is 16), we must process at least 64 input streams together to make the pipeline full, keeping the hardware busy all the time. We achieve the highest throughput when $I = 1024$, and when the I keep growing, the impact on throughput is so little that it can be ignored.

Table 2 presents different clock frequencies when N changes. When N grows, the clock frequency decreases to avoid timing violations and guarantee the accelerator functions properly. We can see that the clock frequencies of our implementation outperform the previous implementation [9]. We do not present the results of different I because I hardly influences the clock compared to N, and we can ignore it. In fact, as for each N, we synthesize the network which can process at most 1024 input streams, then we can change I from 4 to 1024 to test the performance.

4.3. Resource Utilization and Power Consumption

In Table 3, we can see that BRAM, DSP, FF and LUT all scale up to a 2X factor when N changes from 4 to 32. Our accelerator needs LUT most because of the memory partition as explained in Section 3.2. In order to achieve high performance, our resource utilization is pretty high and we cannot synthesize the network when $N >= 64$. Given that we process multiple input streams in a deep pipeline way instead of unrolling, the increment of I would not consume hardware resource, so that our accelerator can process plenty of input streams together. Table 3 also presents the power consumption of each N. As one would expect, the smaller the network is, the less power is consumed.

5. Conclusion and Future Work

We have presented an OpenCL-based FPGA accelerator for long short-term memory recurrent neural networks. We optimize our work in cloud environments, processing multiple input streams in parallel. We provide a general matrix optimization model, transforming the vector-based manipulations to matrix-based operations, which improves the throughput. When the number of input streams is big enough, we obtain high throughput while still retaining the same resource consumption. Additionally, it is also worthwhile to mention that our work still have potential for further research and improvements. First, we only focus on the on-chip computation performance in this implementation while the communication phase need more optimization. Second, we used 32 bit floating point operations instead of fixed point operations, which would exert negative impact on performance and resource consumption. Third, we will add matrix tiling to make our work more scalable.

Table 3. Resource Utilization and Power Consumption

	BRAM	DSP	FF	LUT	POWER
N=4	3%	7%	9%	20%	3.6W
N=8	6%	14%	15%	28%	4.3W
N=16	11%	28%	27%	48%	5.5W
N=32	56%	55%	41%	89%	7.0W

References

[1] H. Sak, A. Senior, and F. Beaufays, "Long short-term memory recurrent neural network architectures for large scale acoustic modeling," in Proc. of the Annual Conf. of INTERSPEECH, 2014.
[2] Alan Graves, Abdel-rahman Mohamed, and Geoffrey Hinton, "Speech recognition with deep recurrent neural networks," in Acoustics, Speech and Signal Processing (ICASSP), 2013 IEEE International Conference on, pages 6645-6649. IEEE, 2013.
[3] Ilya Sutskever, Oriol Vinyals, and Quoc VV Le, "Sequence to sequence learning with neural networks," in Advances in neural information processing systems, pages 3104-3112, 2014.
[4] O. Vinyals, A. Toshev, S. Bengio, and D. Erhan, "Show and tell: A neural image caption generator," CVPR, 2015.
[5] J. Mao, W. Xu, Y. Yang, J. Wang, and A. L. Yuille, "Explain images with multimodal recurrent neural networks," arXiv preprint arXiv:1410.1090, 2014.
[6] H. Fang, S. Gupta, F. Iandola, R. Srivastava, L. Deng, P. Doll'ar, J. Gao, X. He, M. Mitchell, J. Platt, et al., "From captions to visual concepts and back," arXiv preprint arXiv:1411.4952, 2014.
[7] Andre Xian Ming Chang, Berin Martini, and Eugenio Culurciello, "Recurrent neural networks hardware implementation on FPGA," arXiv preprint arXiv:1511.05552, 2015.
[8] Yijin Guan, Zhihang Yuan, Guangyu Sun, Jason Cong, "FPGA-based Accelerator for Long Short-Term Memory Recurrent Neural Networks," 2017 22nd Asia and South Pacific Design Automation Conference (ASP-DAC), 2017.
[9] Joao Canas Ferreira, Jose Fonseca, "An FPGA implementation of a long short-term memory neural network," 2016 International Conference on ReConFigurable Computing and FPGAs (ReConFig), 2016.
[10] Khronos, "The open standard for parallel programming of heterogeneous systems", 2015.
[11] S. Hochreiter and J. Schmidhuber, "Long short-term memory," Neural Computation, vol. 9, no. 8, pp. 1735-1780, 1997.
[12] K. Cho, B. Van Merriënboer, C. Gulcehre, D. Bahdanau, F. Bougares, H. Schwenk, and Y. Bengio, "Learning phrase representations using rnn encoder-decoder for statistical machine translation," EMNLP, 2014.
[13] S. Otte, M. Liwicki, and A. Zell, "Dynamic cortex memory: Enhancing recurrent neural networks for gradient-based sequence learning," in Artificial Neural Networks and Machine Learning-ICANN 2014, Springer, 2014, pp. 1-8.
[14] F. Gers, J. Schmidhuber, et al., "Recurrent nets that time and count," in Neural Networks, 2000. IJCNN 2000, Proceedings of the IEEE-INNS-ENNS International Joint Conference on, vol. 3. IEEE, 2000, pp. 189-194.
[15] K. Greff, R. K. Srivastava, J. Koutník, B. R. Steunebrink, and J. Schmidhuber, "Lstm: A search space odyssey," arXiv preprint arXiv:1503.04069, 2015.
[16] A. Graves and J. Schmidhuber, "Framewise phoneme classification with bidirectional lstm and other neural network architectures," Neural Networks, 2005, 18(5-6): 602-610.

PHASE: An Environment for Parallel High-performance Agent-based Simulating

Jiamin Guo[1], Bo Li[1,*], Yingliang Zhao[1], Yuan Yu[1] and Zuolin Chen[2]

[1] The School of Electronic and Information Engineering, Xi'an Jiaotong University, Xi'an, Shaanxi, 710049, P.R. China
[2]People's Liberation Army 63626 troops, China.

[*]boblee@xjtu.edu.cn

Abstract. Agent-based simulation models have become an increasingly popular research and management tool in many different fields. However, most of the platforms are foreign, and there is no relevant public parallel simulation platform available in China. So it is necessary to construct a public high-performance parallel computing platform for Agent-based Simulation. This paper uses X10 language to design a highly scalable Agent-based simulation platform by using an asynchronous partitioned global address space model, and describes the construction principle and implementation process of the simulation platform from four core modules, the data organization and partitioning mechanism, Agent communication mechanism, simulation promotion mechanism and data collection mechanism. And we verified the correctness and feasibility of the simulation platform by running the classic GameOfLife model. Our work fills the gap that there is no public high-performance parallel simulation platform in China and contributes to the future research work.

1. Introduction

Our goal in this work is to design an Agent-based modeling and simulation platform with high versatility and high parallelism. Our approach is to use the high parallel X10 programming language and the Asynchronous Partitioned Global Address Space (APGAS) model to complete the related design and implementation of the platform. We are not the first to use the X10 language as the language of the development platform (see Japan XASDI [1]), but we believe that there are many unique ideas in our approach, especially in different core parts using different and effective methods.

1.1. ABMS

ABMS (Agent Based Modeling System, ABMS) is a scientific research method, which can be used to study the autonomous agent with interactive behaviors to describe the system abstractly and to study the emergence phenomenon of the system through bottom-up modelling [2,3]. ABMS allows the modeler to specify individual agent behavior rules for each agent, describe the environment in which the individual is located, and then execute to determine possible system results according to rules [4].

1.2. X10

The X10 parallel programming language is a typical implementation of the APGAS model [5]. It can improve programming efficiency by using X10 to write programs, because it does not need to pay too much attention to the heterogeneous details of the underlying computing resources, and so it is also easy to write and the X10 language can be used to write multi-thread distributed programs suitable for execution in heterogeneous cluster environments.

The core concepts in the X10 language are the Place and the Activity. The Place can be thought of as a virtual shared-memory multi-processor, a unit of computation. And the calculation of the X10 language is done by executing the Activity, which can be considered a lightweight thread.

1.3. APGAS Model

In the process of developing X10 language, IBM first proposed the Asynchronous Partitioned Global Address Space (APGAS) model [6]. The APGAS model extends the implementation framework of PGAS and provides more asynchronous support for the program, making the model not limited to programs that express SPMD types [7].

2. Related Work And Motivation

2.1. Related Work

As for the research of parallel multi-agent simulation and simulation platform, in China, most researchers carry out some simulation experiments through the simulation platform, one part is to improve the performance, and few people research the development of the platform. At present, the foreign popular multi-agent social simulation platforms mainly include: the Repast HPC platform developed jointly by The University of Chicago and ANL based on Repast [8], the FLAME (Flexible Large-scale Agent Modelling Environment) platform jointly developed by Rutherford Appleton Laboratory and the University of Sheffield [9], the SASSY (Scalable Agent-Based Simulation System) platform developed by Maria Hybinette et al [10], XASDI (X10-based Agent Simulation on Distributed Infrastructure) platform based on ZASE developed by IBM Research-Tokyo [1], D-Mason simulation platform based on Mason developed by George Mason University [11], while in China, YU-SUPE platform developed by National University of Defense Technology.

The Repast HPC platform is a parallel extension of the Repast platform, implemented in the C++ language. For the data storage and management of the Agent, the platform uses the two data structures Context and Projection. And the Context is used to store the Agent object, the Projection is the mapping between the Agent and the model environment. For communication, the platform proposes a package communication mechanism, which uses the MPI in the Boost library to call the explicit ***Send*** and ***Receive*** functions to complete the communication. For the advancement of simulation, it uses the discrete event scheduling strategy.

FLAME is implemented in C language. For data storage and management and communication, the platform uses a Message Board mechanism, that is, the data is designed as a set of message boards, and then the Agent can obtain information and interact by accessing to the message boards. For the advancement of the simulation, the platform uses the execution of the Agent's behavior and the transition of the state to advance.

SASSY can be seen as a middleware between Agent-based API and PDES emulation kernel. For data storage and management, the Agent and simulation environment are divided into several areas, which would be allocated to different processes. For communication, SASSY uses the Interest Management (IM) mechanism to interact. For the advancement of the simulation, the platform uses parallel discrete event technology, and uses event-driven mode combined with optimistic synchronization strategy for orderly simulation.

The XASDI platform is implemented in the X10 language. The platform has established a high-performance multi-agent simulation platform for traffic conditions in Tokyo, Japan. For data storage and management and communication, the X10 language's own mechanism and data structure are implemented, such as ***DistArray***, ***Block***, ***at(P) S***, Team toolkit, etc. For the advancement of the simulation, global clock control is implemented through Place0 node, and the simulation is advanced in parallel with the X10 communication primitive.

The D-MASON platform is a distributed version of MASON. For data partitioning and storage, before the simulation starts, the simulation environment is divided into several areas. For communication, D-MASON uses a Publish-Subscribe mechanism. Who subscribes to my message, who will I send a message to. For the advancement of the simulation, the platform considers ABM as a step-by-step calculation, which is to calculate Agent behaviors in successive simulation steps. While in each simulation step, it is divided into three phases: communication, synchronization and simulation.

2.2. Motivation

First of all, through the above analysis, the Repast HPC platform uses a tight synchronization time management mechanism, and this mechanism does not further optimize the synchronization mode, and the parallel simulation execution performance is not fully demonstrated. The topology of the Agents in the FLAME platform is not clear enough in the simulation modeling process, which makes the simulation efficiency not high enough [12]. SASSY is mainly suitable for with local interaction intensive models, and so for the frequent interactions models between nodes it obviously will have communication bottlenecks. The XASDI platform is mainly used for traffic simulation, and its applicability is not extensive enough. The data division in D-MASON is static, and load balancing cannot be guaranteed during the simulation process. YU-SUPE is a platform for internal research projects in schools in China. It is not public, and the platform is not very versatile and easy to use.

Secondly, when many non-computer people are working on Agent modeling in his fields, if there is no theoretical basis, they have to implement the modeling process of the entire project from start to finish, which is time-consuming and labor-intensive. Although the model environment may not be the exactly same in different fields, it is similar or even repetitive in some basic simulation coding work. Therefore, if there is a general parallel agent simulation platform, the simulation modeling cycle in different fields can be greatly reduced, thereby the costs will be saved.

In order to solve the above problems comprehensively, this paper mainly combines the X10 high parallel programing language and the discrete event-based simulation advancement mechanism to develop the Parallel High-performence Agent-based Simulating Environment (PHASE) with APGAS model.

3. PHASE Platform Design And Implementation

The PHASE platform designed and developed in this paper consists of four core parts—data organization and partition mechanism, Agent communication mechanism, parallel simulation advance mechanism, and data collection mechanism.

3.1. Data Organization And Partition Mechanism

Agent data organization and partition generally have two ways: one is based on environmental, and the other is based on Agent. And Agent environments can generally be divided into four types: Grids, Networks, Irregular Polygons, and Geographic Information Systems (GIS). In this paper, the data organization model combines two division methods, which is to use environment-based partitioning and building through Grids at the beginning of simulation, and to use Agent-based partitioning in the simulation process. The reasons are as follows:

- Grids is a two-dimensional spatial topology with a simple structure and can be seemed as a checkerboard and be suitable for most models. Figure 1 shows two partitioning ways based on grid.
- We can obtain Agent and other operations according to unique categories or two-dimensional coordinates. It is not necessary to assign each agent a unique identification number ID like other simulation platforms (such as ReapstHPC). At the same time, various entities in the environment are also abstracted into different types of Agent, which is quite scalable by supporting the custom functions in the Agent and the environment.
- Store all agents using a distributed array (***DistArray***). Since each agent is distinguished from other agents by category and grid coordinates. In some models consisted of multi types of

Agent, each type of agent can be stored using the ***DistArray*** data structure in X10. When the program runs, all data in the ***DistArray*** is separated into several blocks by the largest average, and then these blocks are assigned to each Place in some way, which ensures that the static balance of the load is partitioned before the simulation.

- We use the Agent-based partitioning method in the simulation process. However, this method is not global, but a local load migration. For example, in the simulation process, if some nodes were overloaded, the system will move from the overload node to the underload node. The migration of the agent is performed, so that the load of each computing node is roughly in a dynamic balance state, and the parallel performance of the computer can be fully utilized.

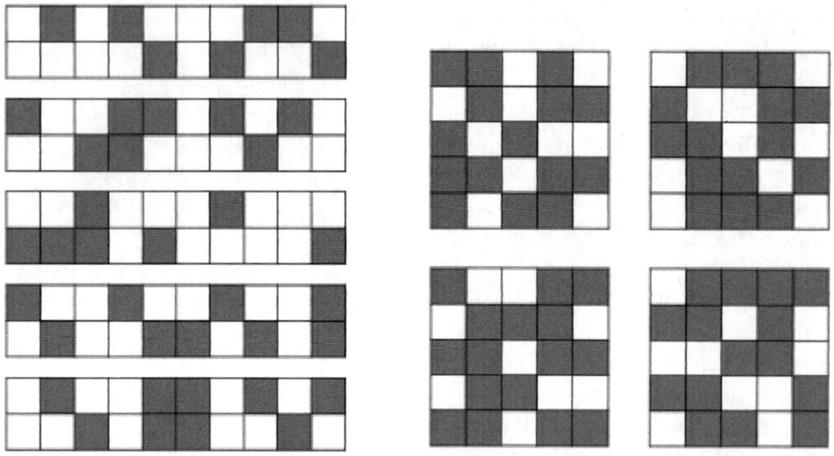

Figure 1. Grid-based band and block partitioning in the GameOfLife model

In addition, we have also tried some other ways of partitioning. Such as, Cluster based partitioning, that is, we divided the distributed-intensive Agents into the same Place. See [13].

3.2. Agent Communication Mechanism

The Agent communication mechanism adopts a combination of global communication and local communication.

- Global communication. The basic operation ***at ()*** in the X10 allows direct access from one compute node to another compute node. For example, execute ***at (place(2)) s*** in Place1; this statement indicates that the Activity executed in the current Place1 will pause and execute the S statement in Place2. Therefore, it is easy to complete the operation of mutual access operation in multiple Places. However, the time overhead of such operations is too large because these calculations belong to communication operations across compute nodes.
- Local communication. If all communication modes use global communication, it will inevitably lead to frequent cross-Place communication, and the computational overhead will be extremely large, which will affect the iterative speed of the simulation system. In order to reduce the computational overhead, the locality principle is used for redundancy calculation [14], and the number of global communication is reduced by storing the nonlocal data into the local (current computing node) redundant area in advance, which also reduces the correlation between the computing nodes. So that the calculation no longer relies on frequent global communication to obtain nonlocal data, and then can be executed concurrently to the maximum extent.

As shown in Figure 2, the dark part is the redundant area, the light part is the local sub-model data block, and a number of redundant areas are added on the outside of the sub-model data block in each place for storing. When updating their state, the Agents can obtain the nonlocal data in local area.

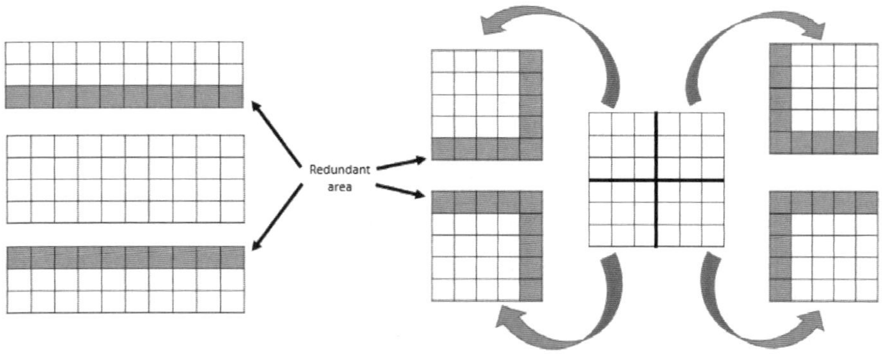

Figure 2. Concept of the redundant area

As shown in Figure 2, the dark part is the redundant area, the light part is the local sub-model data block, and a number of redundant areas are added on the outside of the sub-model data block in each place for storing. When updating their state, the Agents can obtain the nonlocal data in local area.

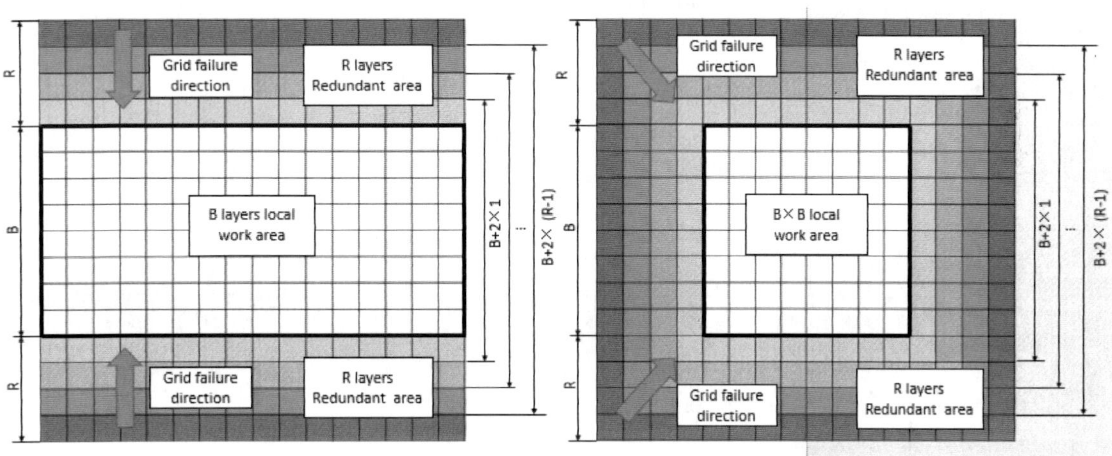

Figure 3. Redundant area

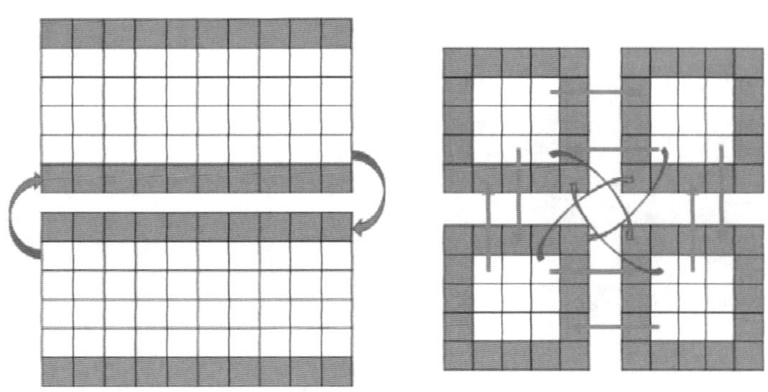

Figure 4. Source in Redundant area

In addition, in order to further improve communication efficiency, we have also made various optimizations for communication methods, such as optimization of redundant area layers, optimization of minimum traffic, and optimization of communication forms. As shown in Figure 3, the optimization of the number of redundant areas means the selection of the value of R. the minimum traffic

IOP Conf. Series: Journal of Physics: Conf. Series **1284** (2019) 012045 doi:10.1088/1742-6596/1284/1/012045

optimization means that when updating the status information of the agents in each Place, the Agents in the work area does not need to be acquired from Place0, that is, it does not need to acquire the all data, and only need to obtain the overlap area data from Place0, thus reducing Inter-node communication from $(B + 2R) \times (B + 2R)$ to $((B + 2R) \times (B + 2R) - B \times B)$. As shown in Figure 4, the optimization of communication form means that each Place gets data from Place0, which will increase the communication load pressure of Place0. In order to reduce the pressure, data can be exchanged directly between the Places to update local information.

3.3. Parallel Simulation Advancement Mechanism

We design the discrete event advancement mechanism based on conservative strategies similar to the SASSY. As shown in Figure 5, the global controller in Place0 controls the global simulation time. Each Place also has its own local simulation time. The synchronization of the Place is adjusted by the global controller in Place0. Each Place has its own local event table and scheduler for storing and managing local events and determining the execution time and order of local events. The global controller in Place0 selects event with the global timestamp minimum from all Places. When the simulation starts, the scheduler in each Place continuously and cyclically selects event with the local timestamp minimum event from the local event table, then the global controller selects the event with the global timestamp minimum and then executes the event. And the simulation will end until executing the end event. In addition, we combine ***finish***, ***clock***, ***when*** and other primitives in the X10 language to achieve message synchronization among Places.

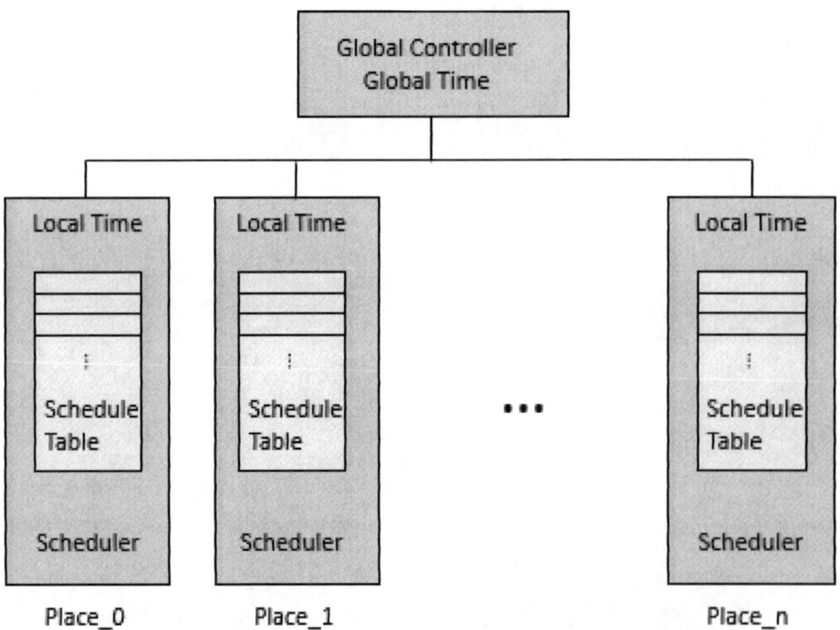

Figure 5. Schematic diagram of Simulation advancement mechanism

3.4. Data Collection Mechanism

Data collection in Parallel Agent simulation is timely and completely storage and management of the data generated during the parallel agent simulation process, so as to provide the in-depth relevant details for the simulation participants in the simulation or after the simulation, which is an extremely important step to understand the simulation process. This platform uses "Decentralized" collection at any location mechanism. It is to define a class (DataCollection), and then the user can directly call the corresponding function in its model code. As a result, the relevant data will be captured and stored in the data collecting file.

4. Experimental Environment And Results Analysis

4.1. Experimental Environment

The hardware and software environment are as shown in table 1 and table 2.

Table 1. Hardware environment

Hardware	**hpms-1**	**hpms-2**
Architecture	X86_64	X86_64
CPU(s)	16	8
On-line CPU(s) list	0-15	0-7
Number of threads per core	2	1
Number of cores per socket	4	8
Socket(s)	2	1
NUMA node	2	2
CPU MHz	2200.000	800.000
L1d cache	16K	64K
L1i cache	64K	64K
L2 cache	2048K	512K
L3 cache	6144K	5118K
NUMA node0 CPU(s)	0-7	0-3
NUMA node1 CPU(s)	8-15	4-7
Company ID	AuthenticAMD	AuthenticAMD

Table 2. Software environment

Software	version
Linux system	Ubuntu-14.04.4
MPI	Mpich-3.2
SSH	3.6
JDK	Jdk-8u111-linux-x64
ANT	Apache-ant-1.9.7
X10dt	X10-2.6

4.2. GameOfLife Model Experiment And Result Analysis

In order to verify the correctness and feasibility of the PHASE simulation platform, we simulated the classic model of ABMS-GameOfLife model on the PHASE. The relative steps for adding the GameOfLife model to the parallel simulation platform are as follows:

- ***Agent.*** Add all the attributes of the entity in the Agent.x10 file. The existing two-dimensional coordinate position and status information are available. Others can be added after the comment "AddCode";
- ***Grid.*** The grid parameters are set in the Grid.x10 file. In this example, the grid is initially set

to 360 in both landscape and portrait orientation.

- ***Event.*** The simulation event is designed in the EventSequence.x10 file. In this example, we set four events, which are game initialization (*init ()*), Agent neighbor information acquisition (*AgentNeighborState()*), and game evolution according to rules (*Interactive()*), Agent information display (*Display()*). And at the same time they should be written to the corresponding step according to the scheduling order.
- ***Model.*** Event scheduling is designed in the Model.x10 file to run the specified event at the specified time. It should be noted that an end event is designed to stop the simulation.
- ***DataCollection.*** Data collection in the DataCollection.x10 file can occur in any given event or at any time. If the latter, you need to customize the data collection event.
- ***Run.*** Run main function and count the simulation time in the Run.x10 file. And configure the simulation properties in the Setting.x10 file. In this example, the initial number of agents, the iteration step, and the directory where the data collection files are stored.

In this simulation experiment, according to the model, we added some methods such as location acquisition, state acquisition, and neighbor acquisition. In addition, for each type of Agent that needs to communicate across the Place, a redundant area is designed in every Place by using the ***DistArray*** data structure, and the nonlocal data is required in advance into the local boundary buffer by the asynchronous operation (***async()***) before the calculation , thereby Communication could be hided. In addition, the model initial parameters set are as follows: The environment scale is a 360 × 360 grid environment, the number of compute node (the Place) is 2, the initial number of Agent is 20000, and the total steps is 200. The number of alive Agent changes as shown in Figure 5.

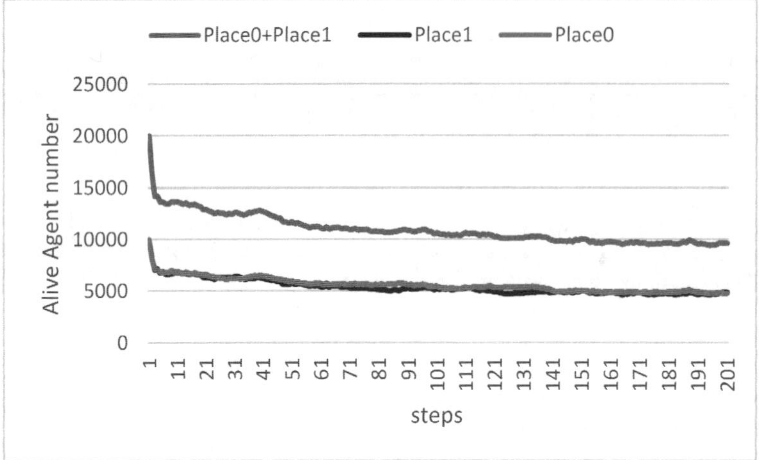

Figure 6. The change of the number of Agents in the simulation

As shown in the Figure 6, the number of alive Agent decreases rapidly at the simulating initial time. This is because the Agent is randomly distributed, and then according to the simulation rules, many Agents could die, and a few steps later, the changing speed of alive Agents number will gradually slow down, and finally maintain a certain number of ups and downs in a dynamic balance. The conclusion of this experiment is a good description of the simulation process of the GameOfLife model. Also the number of death and rebirth of the Agent in each step both can be viewed in the logging file of collecting data.

5. Summary And Future Work

This paper describes the design principle and implementation process of the simulation platform from four core modules: data organization and division mechanism, Agent communication mechanism, parallel simulation advance mechanism and data collection mechanism, and verify the correctness and feasibility of the basic functions of the platform by simulating the classic model of GameOfLife on the platform. The related design and experiment details of the platform has been publicly released on the Github platform (Github: https://github.com/hpmsxjtu/PHASE) for studying and using by relevant scientific researchers. And at the same time, it fills the gap that there is no public high-performance parallel simulation platform in China. Moreover, the platform is more versatile, not just for traffic simulation.

The PHASE simulation platform makes full use of the advantages of APGAS and has an ideal adaptability to models based on grid space. However, it is a topic worthy of further studying that how to extend the platform from the grid structure to the other topology structures such as network and geographic information system. In addition, the PHASE platform still has several problems to be solved in practical applications, such as dynamic load balancing and further communication optimization. At present, we have integrated the simulation advance mechanism and data collection mechanism into a toolkit for the other simulators to use, and the related data organization and division mechanism and communication mechanism toolkit have just been completed initially. And the basic functions of the platform have been realized and there are many extra functions that need to be further refined, and it is expected to form a more completed high-performance Agent simulation platform in the future.

Reference

[1] Suzumura T, Kato S, Imamichi T, et al. X10-based massive parallel large-scale traffic flow simulation[C]// *Acm Sigplan X10 Workshop*. 2012.

[2] Klügl F, Bazzan AL. Agent-based modeling and simulation[J]. *AI Magazine*, 2012, **33 (3)**: 29-41.

[3] North M J , Macal C M . Managing Business Complexity: Discovering Strategic Solutions With Agent-Based Modeling and Simulation[J]. *Oxford University Press*, 2007, **35(6)**:1120-1122.

[4] Dubitzky W , Kurowski K , Schott B . Repast HPC: A Platform for Large-Scale Agent-Based Modeling[C]// *Wiley-IEEE Press*, 2012.

[5] Charles P , Grothoff C , Saraswat V A , et al. X10: an object-oriented approach to non-uniform cluster computing.[C]// *Acm Sigplan Conf on Object-oriented Programming. ACM*, 2005.

[6] Yelick K, Dan B, Chen W Y, et al. Productivity and performance using partitioned global address space languages[C]// *Parallel Symbolic Computation*, Pasco, International Workshop, July, University of Western Ontario, London, Ontario, Canada. 2007.

[7] Hao Zhijun. Research on Cluster Resource Management and Optimization Technology for Asynchronous Partitioned Address Space Programming Model [D]. 2014.

[8] Collier N, North M. Parallel Agent-based simulation with Repast for High Performance Computing[J]. *Simulation*, 2013, **89 (10)**: 1215-1235.

[9] Kiran M , Richmond P , Holcombe M , et al. FLAME: simulating large populations of Agents on parallel hardware architectures[C]// *International Conf on Autonomous Agents & MultiAgent Systems*: *Volume*. International Foundation for Autonomous Agents and MultiAgent Systems, 2010.

[10] Hybinette M , Kraemer E , Xiong Y , et al. SASSY: A Design for a Scalable Agent-Based Simulation System using a Distributed Discrete Event Infrastructure[C]// *Proceedings of the Winter Simulation Conf WSC 2006*, Monterey, California, USA, December 3-6, 2006. DBLP, 2006.

[11] Cordasco G , Chiara R D , Mancuso A , et al. A Framework for Distributing Agent-Based Simulations[C]// *International Conf on Parallel Processing.* Springer-Verlag, 2012.

[12] Dubitzky W, Kurowski K, Schott B. Large-Scale Computing Techniques for Complex System Simulations[M]: John Wiley & Sons, 2012.

[13] Wang Y, Lees M, Cai W, et al. Cluster based partitioning for agent-based crowd simulations[C]. *Winter Simulation Conf*, 2009: 1047-1058.

[14] Chen Li, Zhang Zhaoqing, Feng Xiao-bing. Redundant computational segmentation of optimized communication in distributed storage systems[J]. *Journal of Computer*, 2003, **26(2)**.

IOP Conf. Series: Journal of Physics: Conf. Series **1284** (2019) 012046 doi:10.1088/1742-6596/1284/1/012046

Large-scale Agent Data Partitioning Based on DensityRepel-K_medoids

Lingjuan Wu*, Jie Liang and Bo Li

The School of Electronic and Information Engineering, Xi'an Jiaotong University, Xi'an, Shaanxi, 710049, P.R. China

*wulingjuan0910@stu.xjtu.edu.cn

Abstract. Large-scale Agent data partitioning is the premise of parallel distributed computing in the process of ABMS (Agent-based Modeling and Simulation) . Based on the distance-based K_medoids clustering algorithm, this paper proposes an improved K_medoids algorithm (DensityRepel-K_medoids), which is implemented by high-performance programming language X10 and applied to large-scale Agent data partitioning simulation based on distance interaction. The DensityRepel-Kmedoids algorithm first determines the density value and the repulsion value of each Agent data in the Agent data set, and secondly pre-selects the cluster center according to the density value and the repulsion value of each Agent data, and finally uses the pre-selected cluster centers as the initial cluster centers for continuous iterative clustering until convergence. The algorithm avoids the defects of K_means clustering algorithm sensitive to outliers, and avoids the shortcomings of K_medoids clustering algorithm for large-scale data processing. By comparing and analyzing the simulation experiments of Agent data sets of different scales, the algorithm presents a better performance.

1. Introduction

ABMS (Agent-Based Modelling and Simulation) uses interactive, autonomous, and social agent to describe complex distributed intelligent systems. As the number of agents increases, the communication pressure and coordination difficulty between agents will also increase. Therefore, parallel distributed computing is important in this issue, and large-scale Agent data partitioning is the key to realize parallel distributed computing in the process of ABMS. On the current stage [1], the research on large-scale Agent data partitioning is still in the development periord about the computer simulation field. At present, there are many researches on Agent data partitioning based on K_means clustering algorithm and K_mdoids clustering algorithm, but they all have certain defects: K_means algorithm is more sensitive to outliers. Compared with the K_means algorithm, the K_medoids algorithm has lower sensitivity to outliers, but the calculation is much larger than Kmeans, which is generally suitable for small-scale data.

Therefore, this paper proposes an improved K_medoids algorithm (DensityRepel-K_medoids).Compared to K_means algorithm and K_medoids algorithm, the DensityRepel-K_medoids algorithm not only reduces the sensitivity to outliers, greatly reduces the amount of computation . It makes a reasonable and stable partitioning of large-scale Agent data, at the same as, it greatly improves the efficiency of large-scale Agent data partitioning.

2. K_medoids algorithm

The K_medoids algorithm is a derivative of the K_means algorithm. Compared to K_means, the K_medoids algorithm is greatly reduced by the influence of outliers, but the consequent drawback is that calculation is much larger than K_means.In the following step, K_means and K_medoids are compared [2-4], and the main differences, running speed, robustness, and data scale are analyzed, as shown in Table1.

Table 1. Comparison of K_means algorithm and K_medoids algorithm

Agorithm	K_means	K_medoids
Algorithm step	a. K centers were randomly selected. b. Mark the clusters with centers, Calculate the distance from each agent data point to the center. c. Divide the data points into the cluster of the nearest center marked, Form K clusters. d. According to the classified cluster, the center is recalculated in each cluster: the average of all the data points in the cluster is selected as the new center. e. The iteration is repeated until the deviation is less than the set value.	a. K centers were randomly selected. (Centers must be some data point). b. Mark the clusters with centers, Calculate the distance from each agent data to the center. c. Divide the data points into the cluster of the nearest center marked, Form K clusters. d. According to the classified cluster, the center is recalculated in each cluster: the data point with the smallest sum of Manhattan distances from all the data points in the cluster is selected as the new center. e. The iteration is repeated until the deviation is less than the set value.
Main differences	The centroid is the average of all data points and may be a point that does not exist in the data points.	The centroid must be a data point that exists in the data points.
Running speed	Time complexity is O(n), n is the number of data points, It is more efficient and scalable for larger data.	Time complexity is O(n^2), n is the number of data points . Running slower because the time to update the centers is larger.
Robustness	Poor, affected by the outliers.	Better.
Data scale	Can handle large scale data.	More suitable for small sample data, the time complexity is larger.

3. DensityRepel-K_medoids

Based on the distance-based K_medoids clustering algorithm, this paper proposes an improved K_medoids algorithm (DensityRepel-K_medoids). The DensityRepel-Kmedoids algorithm first determines the density value and the repulsion value of each Agent data in the Agent data set, and secondly pre-selects the cluster center according to the density value and the repulsion value of each Agent data, and finally uses the pre-selected cluster centers as the initial cluster centers for continuous iterative clustering until convergence.

First, determine the density value D_i and the repel value R_i of agent data i in the Agent data set. The density value D_i of the data point i is [5-6]:

$$D_i = \sum_j \chi(d_{ij} - d_c) \quad (1)$$

agent data j and agent data i are not equal, and both belong to the data set; χ(x)is a function that represents x<0, χ(x)=1; x>=0, χ(x)=0; d_{ij} represents the distance from agent data j to agent data i; d_c represents the distance threshold; That is, D_i represents the number of agent datas whose distance from the agent data i is less than d_c.

Appropriate distance threshold d_c:

d_c is a distance threshold and play an important role in the implementation of the algorithm. This parameter is used to calculate the density value D for each agent data. During the process of large-scale Agent data partitioning, We need to determine the number of partitions based on the number of processors available. To ensure load balancing across processors and reduce communication interactions,This paper uses a method to select the appropriate d_c. Assume that the number of available machines is k, data set range is N*M. The d_c is:

$$d_c = \min(N,M)/k \quad (2)$$

min(N,M) represents the minimum of N and M; k represents the number of processors available.

The repel value R_i of the data point i is [5-6]:

$$R_i = \begin{cases} \min(d_{ij}), & \exists j: D_j > D_i \\ \max(d_{ij}), & \forall j: D_j < D_i \end{cases} \quad (3)$$

agent data j and agent data i are not equal, and both belong to the data set; ∃agent data j, agent data j has a density value D_j greater than the density value D_i of agent data i, R_i represents the closest distance between agent data i and the agent data with a density value greater than agent data i's density value D_i; ∀agent data j, its density value D_j is smaller than the density value D_i of agent data i, and R_i represents the largest distance between the agent data i and the agent data from the data set.

Secondly, pre-select the cluster center according to the density value D and the repulsion value R of each agent data. Constructor $DR_i=\rho*D_i+\delta*R_i$, Select agent datas with larger density value and larger repel value, that's DR_i is larger. The value of ρ+δ is 1, ρ and δ are based on a large number of experiments, and the experimental results are shown in Figure 1.

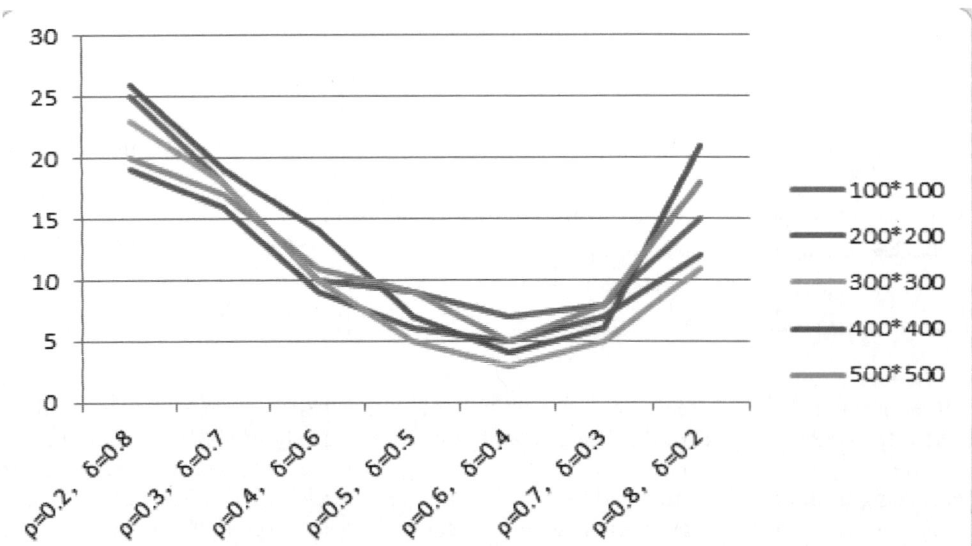

Figure 1. Effect of ρ and δ values on the number of iterations.

Finally, the pre-selected cluster centers are used as the initial cluster centers for continuous iterative clustering until convergence.

Algorithm steps:

1) Data initialization
   - Setting the number of clusters according to the number of available processors;
   - Calculating the density value D_i and the repel value R_i of each agent data i from the data set;
   - Construct $DR_i=\rho*D_i+\delta*R_i$, where ρ and δ are the influence factors, and select agent datas that D_i and R_i are both larger(DR_i is larger) served as pre-selected cluster centers from data set;
   - The pre-selected clusters centers are taken as the initial cluster centers.
2) Forming the initial cluster
   - The non-centrographic agent datas are divided into cluster of the center marked that is the most closest to them, and the loop operation is performed to complete the first division of all non-centrographic agent datas.
3) Update the cluster centers
   - Recalculate the new center in each of the formed clusters. And within the cluster, choose the agent data that smallest sum of Manhattan distances between it and agent data of all agent datas in the cluster as the new center.
   - Divide all non-centrographic agent datas into cluster of the center marked that is the most closest to them.
   - Loop operation until all the centers that have updated are the same as all the centers of the previous one.

4. Experiment

The cellular automaton model is a commonly used model in simulation. It is essentially defined in a cell space composed of discrete, finite state cells, and evolves in discrete time dimensions according to certain behavioral rules. It is widely used in the field of ecology, for example, wolf-sheep-grass ecosystem; in the field of physics, for example, the formation of dendrites, simulated snowflakes; in the field of transportation science, for example, the study of urban traffic flow; in the field of computer science and informatics , for example, the proliferation of network information.

The experiment uses the cellular automaton model as the research object.Compare K_modoids and DensityRepel-K_medoids.The experimental environment is Inter® CoreTM i5-3230M CPU@2.60 GHz, 4GB RAM, 400GB hard drive, 24 core, X10-dt development environment. The initial cell (agent data) is generated using a random number.The algorithm clustering results were evaluated by the number of iterations and simulation time, as shown in Table 2.

Table 2. Comparison of two clustering algorithms K_medoids and DensityRepel-K_medoids.

Data set range	Valid data set	Number of iterations		Running time/ms	
		K_medoids	DensityRepel-K_medoids	K_medoids	DensityRepel-K_medoids
100*100	6000	9	7	87	79
200*200	32000	7	5	762	432
300*300	54000	7	3	2017	1455
400*400	96000	6	4	4205	3761
500*500	150000	8	5	8437	5645

It can be seen from Table2 that compared to K_medoids, the DensityRepel-K_medoids algorithm presents a better performance, which reduces the number of iterations and reduces the running time.

5. Conclusions and prospects

The DensityRepel-K_medoids algorithm, that proposed in this paper, compared to the K_means algorithm, it reduces the sensitivity of outliers; compared with the K_medoids algorithm, it reduces the number of iterations, which reduces the running time and weakens the shortcomings of K_medoids' slow processing of large-scale data. The algorithm is applied to the partitioning of large-scale agent data, which improves the efficiency of partitioning and explains the feasibility of the DensityRepel-K_medoids algorithm.

The current partitioning algorithm is just a master machine single-thread to complete the division of all agent data. From the perspective of algorithm optimization, when the number of agents gets larger and larger, the time and space overhead of the clustering-based partitioning process will become longer and longer, and the partitioning process can be parallelized in subsequent work.

Acknowledgments

In the process of completing the paper, I got technical guidance of my teacher Bo Li's and senior school mate Jie Liang's. As the same as, my classmates around me also provide great help.Thank you very much for the guidance of the teacher, the help of the senior school mate and the help of my classmates around me.

References

[1] Wang Y, Lees M and Cai W. 2009 Cluster based partitioning for agent-based crowd simulations[C]. *Winter Simulation Conf*, 1047-1058.
[2] Madhulatha T S . Comparison between K-Means and K-Medoids Clustering Algorithms[J]. 2011.
[3] Meng Y, Luo K, Liu J H. An Improved K-Medoids Clustering Algorithm[J]. *Advanced Materials Research*, 2012, **562-564**:2106-2110.
[4] Arora P , Deepali D , Varshney S . Analysis of K-Means and K-Medoids Algorithm For Big Data ☆[J]. *Procedia Computer Science*, 2016, **78**:507-512.
[5] Alex R and Alessandro L 2014 Clustering by fast search and find of density peaks[J]. *Science*, **344(1492)**: 1492-1496
[6] Zheng J , Sun H , Zhang M , et al. Research on Optimization of Clustering by Fast Search and Find of Density Peaks[C]. *Web Information Systems & Applications Conf.* IEEE, 2017.

Analysis of Chinese Zither Modes Based on Finite Element Analysis

Hu bing[1] and Li Haoran[2]

[1]Xi'an University of Technology, Shaanxi, Xi'an, 710054.
[2]Shaanxi Heavy Duty Automobile Co.Ltd, Shaanxi, Xi'an, 710200.

1301288807@qq.com

Abstract. This paper makes a modal analysis of Chinese zither by using the finite element analysis method. By combining traditional Chinese national musical instruments with computer simulation technology, it studies the modal structure of Chinese zither according to the method of "data model - establishing FEM model - assigning material parameters - establishing constraints - solving the case". Through modal analysis, it not only verifies the validity of computer simulation method, but also provides data support and theoretical support for the improvement of Chinese zither structure and the structural research of traditional Chinese plucked instruments.

1. Introduction

Chinese zither is the representative of the cultural heritage of the China. As a traditional Chinese plucked instrument, it has more than two thousand years history. The external form of zither is simple, but structure is extremely complex[1]. Although most of the modern methods of making zither can be made by machine, they still cannot be separated from the manual method. Therefore, the selection of timber, shape and the production planning of musical instruments are based on the experience of the zither maker. They are well-made and well-used materials, so the production process of each zither factory also demonstrates different style.

2. The structural characteristics of the zither

Chinese zither has mellow timbre, long lasting charm and wide gamut, and has become the leader of traditional plucked instruments. These characteristics of tone colour and timbre are closely related to their structural characteristics[2]. Chinese zither consists of panel board, baseboard（or rear panel）, sound hole, bulkhead（zither inside）, curb plate, front and rear YueShan, perforation, string, zither code (Yanzhu), zither box, string nail and other parts. As shown in the Figure 1:

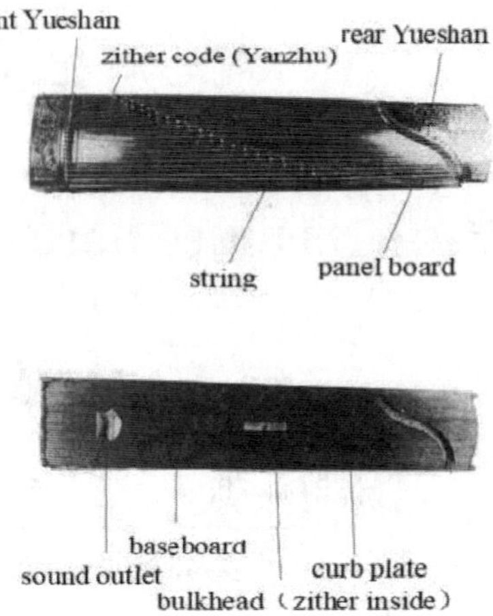

Figure 1. zither structure

There are one or more sound holes of different shapes on the baseboards, usually three. The shape and position of the sound hole play a certain role in the change of resonance, volume and tone quality of the zither. The hollow part between the panel boards and the baseboards has a partition, which plays a supporting role in the structure of the zither and makes its structure stable. In order to decorate, the four sides of zither are often made of precious wood, such as rosewood, red sandalwood and broad-leaved yellow sandalwood[3].

3. Finite element model description of zither structure

3.1. Instruction of finite element analysis of Chinese zither

With the rapid development of finite element technology, modal simulation analysis method has been widely concerned and applied in engineering structures such as aircraft, ships and automobiles. This method of analysis combined with structural vibration has great advantages in solving structural vibration and analyzing the relationship between structure and sound. The structure of zither is complex, so it is difficult to get its vibration characteristics through theoretical analysis. It is usually analyzed by finite element method in engineering application[4]. In the process of analysis, the structure of the zither is simplified properly. It uses the finite element analysis software to establish the finite element mesh model of the three-dimensional structure of the zither, reasonably chooses the wood material parameters, and calculates the structural response of the zither at various frequencies. The pre-processing software uses HyperMesh 14.0, the solver uses Optistruct, and the post-processing software uses HyperView 14.0.

3.2. Modal analysis finite element method

3.2.1 Structure discretization. This step is exactly the same as static analysis. Because of their different analysis contents, they may have different requirements for grid form. For example, in static analysis, it is required to refine the mesh in the stress concentration area, but in dynamic analysis, because the natural frequency and the main mode of vibration are mainly related to the mass and stiffness distribution of the structure, it requires the whole structure to adopt the uniform mesh form as far as possible.

3.2.2 Element analysis. The task of element analysis is still to establish element characteristic matrix and form element characteristic equation. In dynamic analysis, the element characteristic matrix includes not only stiffness matrix, but also mass matrix and damping matrix. The virtual displacement principle is used to establish the element characteristic matrix. Under the action of dynamic load, the virtual displacement of element nodes is assumed for any instantaneous moment $\{\delta q\}^e$ .Then the corresponding virtual displacement $\{\delta d\}$ and virtual strain $\{\delta\varepsilon\}$ are generated in the element. At this time, the virtual strain energy generated in the element is:

$$\delta U = \iiint_V \{\delta\varepsilon\}^T \{\sigma\} dV \tag{1}$$

$\{\sigma\}$ is the stress of the node, dV is the element of volume.

In this case, the element is not only subjected to dynamic loads, but also to inertial forces $-\rho\{\ddot{d}\}dV$ and $-\nu\{\dot{d}\}dV$, ρ is the material density and ν is the linear damping coefficient. Therefore, the virtual work done by external forces is as follows:

$$\delta W = \iiint_V \{\delta d\}^T \{P_V\} dV + \iint_A \{\delta d\}^T \{P_s\} dA + \{\delta d\}^T \{P_c\} - \iiint_V \rho \{\delta d\}^T \{\ddot{d}\} dV - \iiint_V \nu \{\delta d\}^T \{\dot{d}\} dV \tag{2}$$

$\{P_V\}$ is the dynamic force acting on the element.

$\{P_s\}$ is the dynamic surface force acting on the element.

$\{P_c\}$ is the dynamic concentrated force acting on the element.

V element volume;

A element area;

$\{d\} = [N]\{q\}^e$, $\{\varepsilon\} = [B]\{q\}^e$

$[N]$ shape function matrix, which is only a function of coordinates, has nothing to do with time.

$[B]$ strain matrix, in which each non-zero element is a constant determined by the node coordinates:

$$\left.\begin{aligned} \{\dot{d}\} &= [N]\{\dot{q}\}^e \\ \{\ddot{d}\} &= [N]\{\ddot{q}\}^e \\ \{\delta d\} &= [N]\{\delta \dot{q}\}^e \\ \{\delta\varepsilon\} &= [B]\{\delta \dot{q}\}^e \end{aligned}\right\} \tag{3}$$

According to the principle of virtual displacement, there is the following formula

$$\delta U = \delta W \tag{4}$$

By substituting and sorting (1), (2) and (3) into (4), the equation of motion of the element can be obtained.

$$[m]^e\{q\}^e + [C]^e\{q\}^e + [k]^e\{q\}^e = \{R(t)\}^e \tag{5}$$

$$[k]^e = \iiint_V [B]^T [D][B] dV \tag{6}$$

$$[k]^e = \iiint_V [B]^T [D][B] dV \tag{7}$$

$$[c]^e = \iiint_V [N]^T \nu [N] dV \tag{8}$$

$[D]$ elastic matrix is determined by elastic modulus and Poisson's ratio, which is independent of coordinates. $[k]^e$, $[m]^e$, $[c]^e$ are the stiffness matrix, mass matrix and damping matrix, which are the characteristic matrices that determine the dynamic performance of the element.

$$\{R(t)\}^e = \iiint_V [N]^T \{P_v\} dV + \iint_A [N]^T \{P_s\} dA + [N]^T \{P_c\} \tag{9}$$

IOP Conf. Series: Journal of Physics: Conf. Series **1284** (2019) 012047 doi:10.1088/1742-6596/1284/1/012047

(9) is called the element node load array. It is the result of the displacement of the physical force, surface force and concentrated force acting on the element to the element node.

3.3. Global matrix integration
The task of global matrix integration is to assemble the characteristic matrices of each element into the characteristic matrices of the whole structure so as to obtain the global equilibrium equation.

$$[M]\{\ddot{q}\}+[C]\{\dot{q}\}+[K]\{q\}=\{R(t)\} \tag{10}$$

In the formula, $\{q\}$ is an n-order array composed of displacement components of all nodes.

n the total degree of freedom of the structure.

$\{R(t)\}$ the node load array;

$$\{R(t)\}=\sum_{i=1}^{n}\{R_i(t)\} \tag{11}$$

i the number of nodes:

$[K]$ the stiffness matrix of the structure.

$[M]$ the mass matrix of the structure.

$[C]$ the damping matrix of the structure.

The stiffness matrix, mass matrix and damping matrix of the structure are all composed of the corresponding matrix of the element.

3.4. The modal extraction method of Radioss
In general, we choose the first four methods: Subspace method, Block Lanczos method, Power Dynamics method and Reduced method. Unsymmetric method, Damped method and QRDamped method are only used in special cases. Modal extraction methods mainly depend on the size of the model (relative to the computing power of the computer) and specific application occasions. Comparing these four methods synthetically, the finite element model of the zither contains more nodes and units, and combined with the model and computer hardware configuration, this paper chooses the block Lanczos method for modal analysis[5].

3.5. Chinese zither mode
Chinese zither is modeled by 2nd-order tetrahedron element, the basic element size is 3mm. The string is simulated by spring element. The zither has a finite element model, the number of model elements is 612767, and the number of nodes is 162264, as shown in Figure 2.

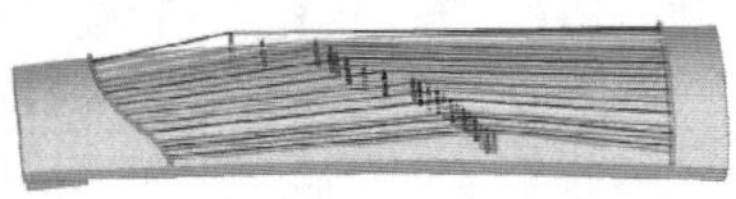

Figure 2. Chinese zither finite element model

Paulownia density is 2.52X10-10T/mm^3, elastic modulus is 5000MPa and poisson's ratio is 0.49, as shown in Table 1. The total mass of Chinese zither is 7.579Kg.

IOP Conf. Series: Journal of Physics: Conf. Series **1284** (2019) 012047 doi:10.1088/1742-6596/1284/1/012047

Table 1. Material parameters of Chinese zither finite element model

Material	Elastic modulus (MPa)	Poisson's ratio	Density (T/mm^3)
Paulownia	5000	0.49	2.52X10-10

4. Chinese zither modal result

Modal analysis is a common method used by many scholars at home and abroad to study musical instruments. By calculating the free mode of the zither, the modal results within the first 400 Hz of the model are obtained. The first 12 order modal results are shown in Figure 3-5.

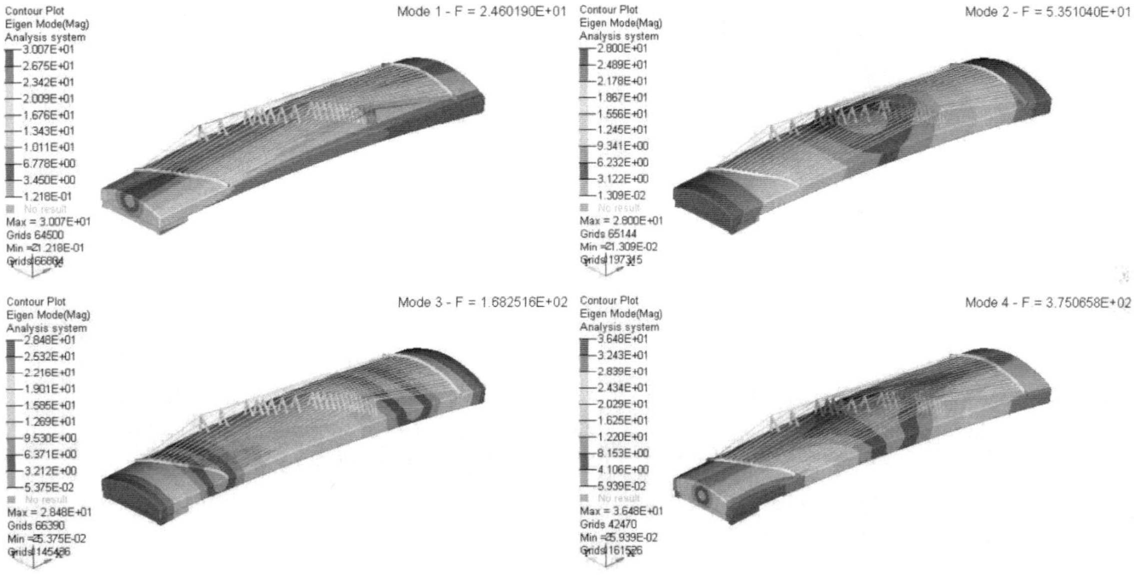

Figure 3. 1 ~ 4 mode

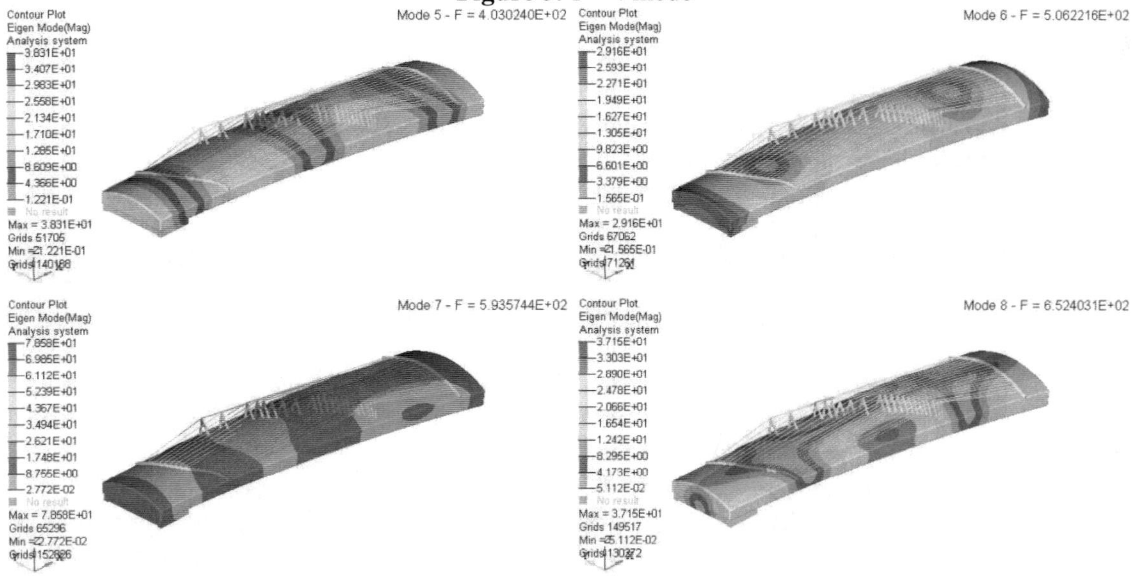

Figure 4. 5~8 mode

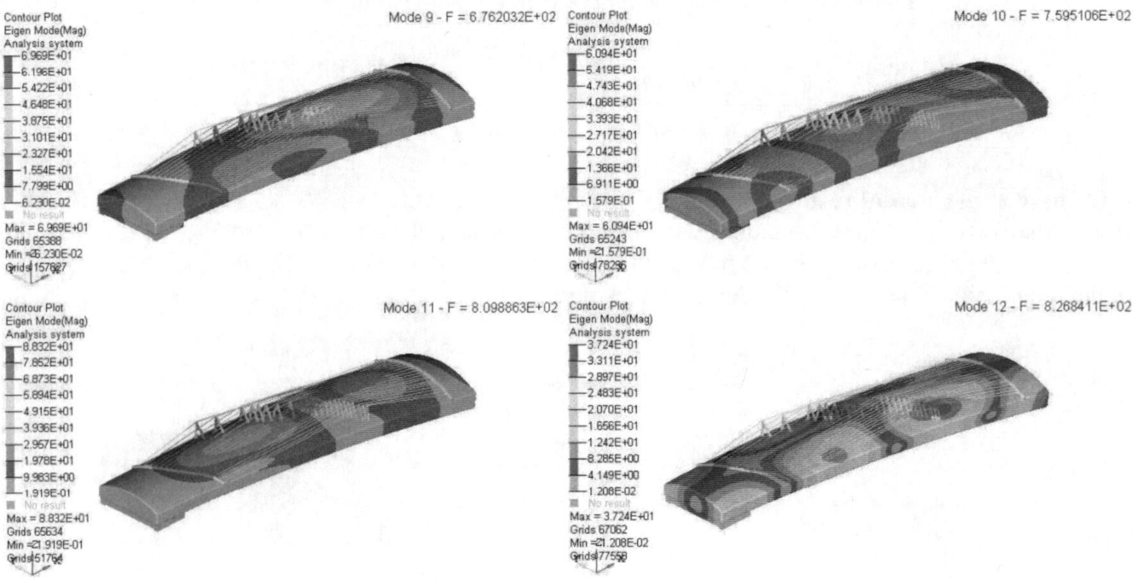

Figure 5. 9~12 mode

5. Conclusion

Using the method of computer virtual simulation, this paper uses Hypermesh software to analyze the modal of the finite element model of the zither and to explore the vibration characteristics of the musical instrument scientifically, which is the first step in the simulation study of the national musical instrument. Vibration modal analysis of zither is only one part of the problem of frequency domain characteristics of vibration. The simulation analysis of noise of the zither needs to be further explored.

References

[1] Hutchins C.M. A history of violin research [J]. The Journal of the Acoustical Society of America, 1983, 73(5): 1421-1432

[2] MCINTYRE M E, WOODHOUSE J. The acoustics of stringed musical instruments[J]. Interdisciplinary Science Reviews, 1987, 3(2): 157-173

[3] Han Baoqiang, Instrument Acoustics System and Space Music Acoustics[J]. Performing Arts Equipment and Technology. 2007(2): 64-68.

[4] Wang Zhonglin, Fan Wenbo, Chinese zither structure, principle and design modeling[J]. Journal of Dong Hua University(Natural Science) 2012. Vol 38, No.5, 564-571

[5] Fu Zhifang, Hua Hongxing. Modal Analysis Theory and Application[M]. ShangHai ShangHai Shanghai Jiaotong University Press. 2000

IOP Conf. Series: Journal of Physics: Conf. Series **1284** (2019) 012048 doi:10.1088/1742-6596/1284/1/012048

Forecasting of Maritime Traffic Accident based on Residual Error Unbiased Grey Forecast Model

Sun Shuang[1,*], Chen Yan[1], Zhang Jinsong[1] and Li Taoying[1]

[1]School of Maritime Economics and Management, Dalian Maritime University, Dalian, China

*470582807@qq.com

Abstract. In order to enhance the safety of maritime transportation and improve the accuracy of maritime traffic accident prediction, an unbiased grey forecast model based on residual error is applied to predict maritime traffic accident. Based on the historical data of maritime traffic accidents from 2008 to 2017, the traditional unbiased grey model prediction and residual error unbiased grey model prediction are carried out, and the fitting curves of actual value and predicted value of the two models are drawn. The results show that the prediction accuracy and fitting curve of residual error unbiased grey model are better than those of traditional unbiased grey model, which can truly reflect the development trend of comprehensive safety of maritime traffic, and the prediction results have certain reliability and practicability.

1. Introduction

Grey theory was first put forward by Deng Julong in 1982.[1] It is a new method to study the problems of minority data, poor information and uncertainty. This theory takes the uncertain system of "small sample and poor information" with "part information known and part information unknown" as the research object. Through the generation and development of some known information, it extracts valuable information, and realizes the correct description and effective monitoring of the system's operation behaviour and evolution law. In grey forecasting model, GM (1,1) model of time series forecasting is the most commonly used model. It has been widely used in industry and data mining forecasting. [2] However, when using GM (1,1) model to model and predict, it sometimes has big deviation, and has some defects, such as low learning ability to data and low prediction accuracy.[3]

The basic idea of maritime traffic accident prediction is to reasonably infer and analyse the possible maritime traffic accidents based on the statistics, analysis and processing of the accident data and the causes and rules of the accidents. [4] Common forecasting methods include regression analysis method, time series forecasting method, grey forecasting method, BP neural network forecasting method, etc. [5] Practice shows that each prediction method has its own characteristics and can basically meet the expected requirements, but at the same time there are corresponding defects. [6] For example, the grey model uses the accumulated new data to model, ignoring the randomness of the predicted data. [7] Although BP neural network has high prediction accuracy, it needs very complete data, otherwise the network extension trained will be very poor.[8]

IOP Conf. Series: Journal of Physics: Conf. Series **1284** (2019) 012048 doi:10.1088/1742-6596/1284/1/012048

2. Model establishment

2.1. grey GM(1,1) model

1) Selecting n-year maritime traffic accident volume as the initial time series $X^{(0)}$ of grey model, a set of original time series samples of water traffic accident are obtained,

$$X^{(0)} = (x^{(0)}(1), x^{(0)}(2), \cdots, x^{(0)}(n)) \tag{1}$$

In formula (1), $x^{(0)}(k) \geq 0, k = 1,2,3,\cdots,n$, the sequence reflects the changing process of maritime traffic accidents over time.

2) In order to weaken the randomness and volatility of water traffic accidents, a new time series is generated by 1-AGO accumulation of the original series $X^{(0)}$,

$$X^{(1)} = (x^{(1)}(1), x^{(1)}(2), \cdots, x^{(1)}(n)) \tag{2}$$

In formula (2), $x^{(1)}(k) = \sum_{i=1}^{k} x^{(0)}(i), k = 1,2,3,\cdots,n$.

3) The smoothness and quasi-exponential test of the generated new time series $X^{(1)}$ are carried out. When $k > 3$, if $\rho(k) < 0.5$ and $1 < \sigma(k) < 1.5$, the data satisfy the rules of smoothness and quasi-exponential, the GM (1,1) model can be established. The formula for testing smoothness and quasi-exponential law is as follows:

$$\rho(k) = x^{(0)}(k) / x^{(1)}(k) \tag{3}$$

$$\sigma^{(1)}(k) = x^{(1)}(k) / x^{(1)}(k-1) \tag{4}$$

4) Generate a series of adjacent mean values. Adjacent mean values of Sequence $X^{(1)}$:

$$Z^{(1)} = (z^{(1)}(1), z^{(1)}(2), \cdots, z^{(1)}(n)) \tag{5}$$

The formula is $z^{(1)}(k) = \frac{1}{2}(x^{(1)}(k) + x^{(1)}(k-1)), k = 2,3,\cdots,n$.

5) Determine parameters of weighted GM(1,1) model by least square method for $X^{(1)}$:

$$\hat{\boldsymbol{a}} = [a,u]^{\mathrm{T}} = (\boldsymbol{B}^{\mathrm{T}}\boldsymbol{B})^{-1}\boldsymbol{B}^{\mathrm{T}}\boldsymbol{Y} \tag{6}$$

In formula (6), $\boldsymbol{B}^{\mathrm{T}} = \begin{bmatrix} -z^{(1)}(2) & -z^{(1)}(3) & \cdots & -z^{(1)}(n) \\ 1 & 1 & \cdots & 1 \end{bmatrix}$; $\boldsymbol{Y}^{\mathrm{T}} = \begin{bmatrix} x^{(0)}(2) & x^{(0)}(3) & \cdots & x^{(0)}(n) \end{bmatrix}$.

6) By substituting the parameters a and u into the prediction model derived from albino differential equation, the accumulative series of predictions is obtained.

$$\hat{x}^{(1)}(k) = [x^{(0)}(1) - \frac{u}{a}]e^{-a(k-1)} + \frac{u}{a} \tag{7}$$

The initial value of cumulative reduction is: $x^{(0)}(k+1) = x^{(1)}(k+1) - x^{(1)}(k)$.

2.2. Residual error unbiased grey GM(1,1) model

1) According to the grey GM(1,1) model, the parameters a and u are obtained, from which the parameters b and A of unbiased grey GM (1,1) model can be obtained: [9]

$$b = \ln\left((2-a)/(2+a)\right) \tag{8}$$

$$A = 2u/(2+a) \tag{9}$$

2) By calculating the values of b and A, the fitting model of unbiased grey GM (1, 1) model is:

$$\overline{x}^{(0)}(1) = x^{(0)}(1), \overline{x}^{(0)}(k+1) = Ae^{bk} \quad k = 1,2,3,\cdots,n \tag{10}$$

According to the model, a group of predictions can be listed as follows:

$$\overline{x}^{(0)}(t)=\left\{\overline{x}^{(0)}(1),\overline{x}^{(0)}(2),\overline{x}^{(0)}(3),\cdots,\overline{x}^{(0)}(n)\right\} \tag{11}$$

3) The difference between the original series and the predicted series is $e^{(0)}(t)=x^{(0)}(t)-\overline{x}^{(0)}(t)$, then the series of residuals is listed as:

$$e^{(0)}(t)=\left\{e^{(0)}(1),e^{(0)}(2),e^{(0)}(3),\cdots,e^{(0)}(n)\right\} \tag{12}$$

4) If $e^{(0)}(t)$ series has the negative number, it should be positively processed, and the corresponding absolute value should be taken, after which a new series can be obtained:

$$e^{(1)}(t)=\left\{e^{(1)}(1),e^{(1)}(2),e^{(1)}(3),\cdots,e^{(1)}(n)\right\} \tag{13}$$

5) The GM (1,1) model of $e^{(1)}(t)$ is built and the results are as follows:

$$e^{(1)}(t+1)=\left[e^{(0)}(1)-\frac{\mu_e}{a_e}\right]e^{-a_e t}+\frac{\mu_e}{a_e} \tag{14}$$

The initial value of cumulative reduction is:

$$\overline{e}^{(0)}(t+1)=\left[e^{(0)}(1)-\frac{\mu_e}{a_e}\right](e^{-a_e(t+1)}-e^{-a_e t}),(t=1,2,\cdots,n-1) \tag{15}$$

6) Then the unbiased grey prediction model of residual error is as follows:

$$\overline{x}^{(0)}(k+1)=\overline{x}^{(0)}(t+1)+\eta\overline{e}^{(0)}(t+1),(t=1,2,\cdots,n-1) \tag{16}$$

Among it, the correction factor is:

$$\eta=\begin{cases}1 & x^0(k)-\overline{x}^0(k)\geq 0\\ -1 & x^0(k)-\overline{x}^0(k)<0\end{cases} \tag{17}$$

7) Markov state transition matrix is used to judge the sign of residual prediction value when $k\geq n$.The state transition matrix P is introduced.

The steps are as follows: Determine + 1 and - 1 states; Find out the state transition matrix P; Determination of initial state vector $\pi(0)$;The result of the state transition in the t period is obtained according to $\pi(t)=\pi(0)\cdot P^t$ (P^t represents the state transition matrix in the t period) and the probabilistic state is taken. If the probability of positive and negative signs is the same, the symbols determined in the previous period are generally taken. [10]

2.3. Test of Forecast Model

Posterior error test and small error probability test are commonly used in grey forecast model test.

2.3.1. Posterior error test. The mean $\overline{X}$ *and variance* S_1 *of the original series* $X^{(0)}$ *are calculated:*

$$\overline{X}=\frac{1}{n}\sum_{k=1}^{n}X^{(0)}(k)\ ;\ S_1^2=\frac{1}{n}\sum_{k=1}^{n}(X^{(0)}(k)-\overline{X})^2 \tag{18}$$

The mean $\overline{X}$ and variance S_1 of the original series $e^{(0)}(t)$ are also calculated:

$$\overline{e}=\frac{1}{n}\sum_{k=1}^{n}e^{(0)}(k)\ ;\ S_2^2=\frac{1}{n}\sum_{k=1}^{n}\left(e^{(0)}(k)-\overline{e}\right)^2 \tag{19}$$

Calculating mean square deviation ratio: $C=S_2/S_1$ (20)

2.3.2. Small error probability test.

$$p=p\left\{\left|e(k)-\overline{e}\right|<0.6745S_1\right\} \tag{21}$$

Generally speaking, the accuracy level of the model is divided into the following table1: [11]

Table 1. Accuracy criteria of grey forecast model.

Precision grades	C	P	Precision grades	C	P
Excellent(Level 1)	0.35	0.95	Qualified(Level 3)	0.65	0.70
Good(Level 2)	0.50	0.80	Poor(Level 4)	0.80	0.60

3. Model validation and analysis

Taking the total number of maritime accidents in China from 2008 to 2017 published by the Ministry of Communications as the research object (see Table 2). From the historical statistics of maritime traffic accidents in recent 10 years, it can be seen that the number of maritime traffic accidents in China is on the decline as a whole.

Table 2. 2008-2017 Distribution table of maritime traffic accidents.

year	2008	2009	2010	2011	2012	2013	2014	2015	2016	2017
accidents	342	358	331	298	270	262	260	231	196	196

3.1. Establishment of unbiased grey forecast model

Taking the number of maritime traffic accidents from 2008 to 2017 as the original sequence, $X^{(0)}$ =(342,358,331,298,270,262,260,231,196,196). The original sequence is accumulated once and according to formula (6), it can be obtained that:

$$[a,u]^{\mathrm{T}}=(\boldsymbol{B}^{\mathrm{T}}\boldsymbol{B})^{-1}=\begin{pmatrix}0.078402161\\395.8989012\end{pmatrix}$$, So we can get: $b=-0.07844236,\ A=3.809646742$.

The unbiased grey GM (1,1) forecast model is obtained by introducing b and A into formula (10):

$$\begin{cases}\overline{x}^{(0)}(1)=x^{(0)}(1)\\ \overline{x}^{(0)}(k+1)=3.809646742e^{-0.07844236k},(k=1,2,3,\cdots,n)\end{cases}\tag{22}$$

The predicted values and residuals are obtained by formula (22) as shown in table 3.

Table 3. Results of unbiased grey forecast model for maritime traffic accidents.

Year	Actual Value	Prediction Value	Residual Error	Year	Actual Value	Prediction Value	Residual Error
2008	342	342	0	2014	260	238	22
2009	358	352	6	2015	213	220	-7
2010	331	326	5	2016	196	203	-7
2011	298	301	-3	2017	196	188	8
2012	270	278	-8	2018	-	174	-
2013	262	257	5	2019	-	161	-

According to table 3, although unbiased grey prediction fits the overall distribution characteristics of maritime traffic accidents and shows an exponential downward trend, it fails to fully reflect the random and fluctuating characteristics of accident volume. According to formula (18) - (21), $C=0.1564$, $p=1$ is obtained.

3.2. Establishment of unbiased grey forecast model based on residual error

Taking the residual error of maritime traffic accidents prediction results as the original sequence, $e^{(0)}(t)$=(0,6,5,-3,-8,5,22,-7,-7,8). Based on the modelling theory described above, the unbiased grey model of residual error can be obtained that: $[a,u]^{\mathrm{T}} = (\boldsymbol{B}^{\mathrm{T}}\boldsymbol{B})^{-1} = \begin{pmatrix} -0.063662856 \\ 5.883508929 \end{pmatrix}$.The predicted values and residuals are obtained and shown in table 4.

Table4. Results of residual error unbiased grey forecast model for maritime traffic accidents.

Year	Actual Value	Prediction Value	Residual Error	Year	Actual Value	Prediction Value	Residual Error
2008	342	342	0	2014	260	246	14
2009	358	358	0	2015	213	211	2
2010	331	333	-2	2016	196	194	2
2011	298	294	4	2017	196	198	-2
2012	270	271	-1	2018	-	185	-
2013	262	265	-3	2019	-	173	-

According to formula (18) - (21), $C = 0.1549, p = 1$ is obtained. The comparison results of the two models' accuracy are shown in Table 5.

Table 5. The comparison results of the two model's accuracy.

Model	Relative Error E	Mean square deviation ratio C	Small error probability P
unbiased grey forecast model	0.0285	0.1564	1
residual error unbiased grey forecast model	0.0118	0.1549	1

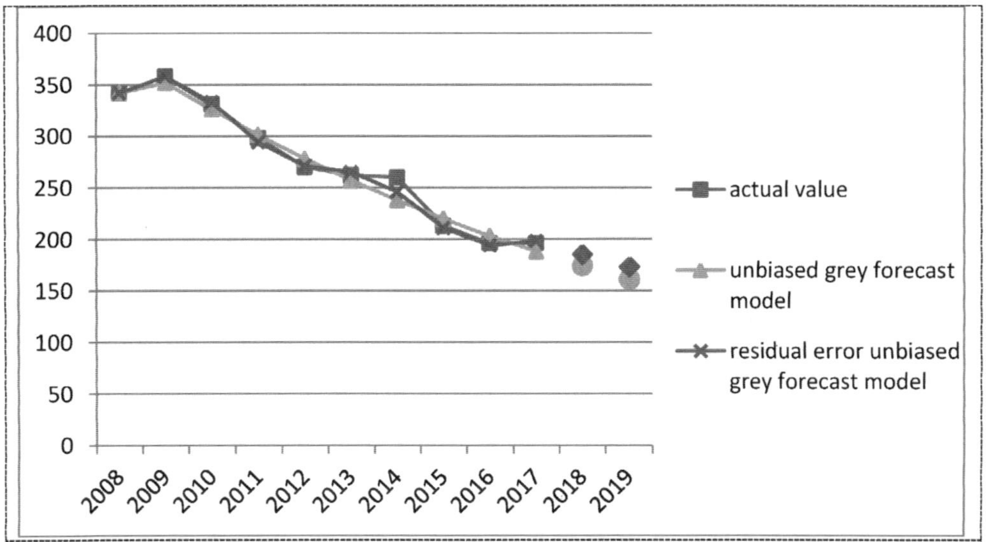

Figure 1. Comparison of two forecast models and real values.

The prediction value of unbiased grey forecast model and residual error unbiased grey forecast model are compared with the actual value. The results are shown in figure 1.From figure 1, it can be seen that the residual error unbiased grey model improves the prediction accuracy of unbiased grey model, and the prediction results are close to the actual values. From Table 5, it can be seen that the prediction accuracy of the unbiased grey prediction model with residual error is much higher than that of the unbiased grey prediction model, and the average relative error is reduced by 58.60%. For the posterior difference ratio, the unbiased grey prediction model and the unbiased grey prediction model with residual correction both reach the first level, but the value with residual error is smaller. Therefore, it is better to establish the residual error grey forecast model to predict maritime traffic accidents, and the method is feasible.

4. Summary

The residual error unbiased grey forecast model is used to predict maritime traffic accidents, and the results are satisfactory. The prediction accuracy of residual error unbiased grey forecast model is obviously higher than that of the unbiased grey forecast model. It shows that the residual grey prediction model has certain practicability and effectiveness in the forecasting of maritime traffic accidents. In addition, in the process of correcting unbiased grey prediction model, the residual error model can be superimposed many times in order to obtain higher accuracy.

Acknowledgments

The authors are very grateful to the editors and reviewers for their valuable comments and suggestions. This work was supported by the National Natural Science Foundation of China (Grant No. 71271034), National Social Science Foundation of China (Grant No.L15CGL031).

References

[1] Ju-Long D 2002 *THE BASIS OF GREY THEORY.*

[2] Xiong-Bin W 2009 Application of residual error gray forecast Model in traffic accident forecast *Technology & Economy in Areas of Communications* **11**(1):33-4.

[3] Li K, Lu L, Zhai J, Taghi M K, Li T 2016 The improved grey model based on particle swarm optimization algorithm for time series prediction *Engineering Applications of Artificial Intelligence* **55**(C):285-91.

[4] Lang-Xiong G , Bao-Gang Z , Yuan-Zhou Z and Jia-Min P 2008 Application of the grey model theory to forecast maritime traffic accident *Ship & Ocean Engineering* **37**(6):99-102.

[5] Yang Z Q 2011 Marine traffic accident prediction based on particle swarm optimization-based RBF neural network *International Conference on Computer Research & Development*.

[6] Zhao L 2013 Traffic accident prediction based on equal dimension and new information unbiased grey Markov model *Computer Engineering & Applications* **49**(7):35-383.

[7] Qi W and Zhiming W 2013 Forecasting of maritime traffic accidents based on the improved SCGM(1,1)_c—Markov Model *Navigation of China* **36**(4):119-24.

[8] Gang L, Huang T, Yan H, Huang X 2009 Grey residual error model of highway traffic accident forecast Journal of Traffic & Transportation Engineering 9(5):88-93.

[9] Ji P R, Huang W S, Hu X Y 2000 An Unbiased Grey Forecasting Model *Systems Engineering & Electronics* **22**(6):6-8.

[10] Lin Y , Chen S , Chen Y and Zhai J 2013 Grey-Markov model and algorithm for traffic accidents forecasting *Journal of Wuhan University of Technology* **37**(5):924-8.

[11] Changyuan C, Ran D, Jijun F , Xingwang Y and Jie Z 2017 Improved grey model applied in maritime traffic safety index prediction *Navigation of China* **40**(1):58-63.

A Fair TDMA Scheduling Based on Weight of Flows

Xiaoying Shuai[1]

[1] College of Computer Science and Technology, Taizhou University, Taizhou, China

E-mail: xyshuai@163.com

Abstract. Nodes in wireless ad hoc networks transmit data to 1-hop neighbour nodes through broadcast channel. Conflicts occur when multiple nodes within the communication range send data at the same time or when one node receives data sent by multiple nodes at the same time. For broadcast scheduling problem (BSP), many researchers used TDMA (Time Division Multiple Access) scheme based on SVC (Sequential Vertex Colouring) in ad hoc networks to achieve conflict-free scheduling. However, SVC is related to the order of nodes. To improve the fairness of network, the slots of nodes are adjusted by changing the order of nodes. In this paper, we proposed a fair scheduling algorithm for TDMA, which takes into account the number of flows, the size of the flows and the bandwidth of each node. The simulation and analysis results show that this scheduling algorithm can enhance the fairness of the ad hoc compared with the traditional algorithm based on SVC.

1. Introduction

Wireless ad hoc is temporary dynamic topology structure using shared wireless channel to interconnect many nodes, and it don't rely on any fixed network infrastructure. Nodes can transmit directly with 1-hop neighbors. When a target node is located far away, a sequence of intermediate nodes will store and forward packets to the destination. Since nodes share broadcast channel, collision will emerge when a node transmits with its 1-hop or 2-hop neighbors simultaneously, as shown by figure1.

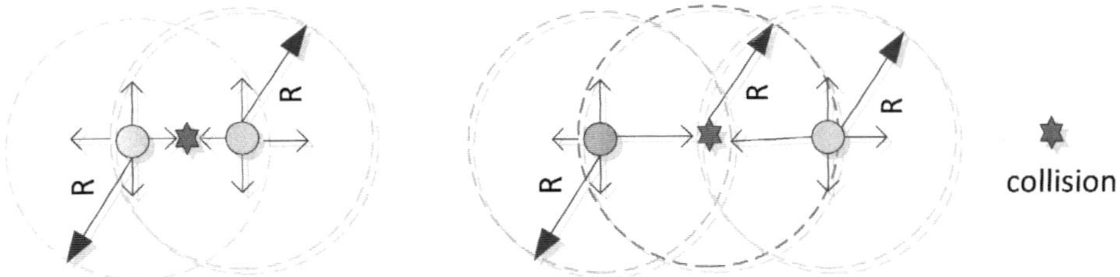

Figure 1. Collision in ad hoc.

To collision-free broadcast schedules, TDMA is widely used in ad hoc. BSP is referred as the TDMA scheduling problem to minimize time slots of a frame [1]. BSP to schedule at least one slot for each node and maximize the throughput in an ad hoc is a NP-complete problem. For BSP, papers [2][3] proposed many TDMA schemes. The result of scheduling is related to the order of nodes. Therefore, different order will result in different slots of nodes, which will lead to fairness in ad hoc.

2. Related work

2.1 SVC algorithm

Vertex coloring method is efficient algorithm to deal with BSP [4]. An undirected graph is represented as $G=(V, E)$. Nodes u, $v \in V$ are 1-hop neighbors, then undirected edge $e=(u, v) \in E$. If vertex u is assigned color C_u, vertex v is C_v, then $C_u \neq C_v$. The slots schedule in TDMA is considered as vertex color allocation. So, ad hoc achieves high channel utilization without collision. SVC orders node by specified criterion.

```
SVC algorithm
  Order nodes as 1,2,…,N
  Color i=1;
  Repeat
    For vertex=1 to N do
      If vertex can color with i then color
    Color i++
  Untill all vertices colored
```

An efficient algorithm for BSP based on SVC is proposed in paper3. VCP (Vertex Coloring Problem) is NP-hard. The algorithm includes two phases. The first phase assigned only one slot to a node to minimize the frame length. To maximize the total number of slots in a frame, the algorithm in the phase 2 allocated an exist available slot to other nodes that are not 1-hop and 2-hop neighbors. This algorithm has the property of greedy selection, the fairness of which is related to the order of nodes. Aggeliki Sgora et al. [5][6] proposed fair TDMA scheduling based on the active flows in ad hoc networks.

2.2 Model

$G=(V, E)$ represents an ad hoc. If nodes can hear each other, we consider they are 1-hop neighbor. The connectivity table A_{nn} as follow, where n represents the number of nodes.

$$A_{ij} = \begin{cases} 1 & \text{if node i and j can communicate directly} \\ 0 & \text{otherwise} \end{cases} \tag{1}$$

If a common neighbor of node i and j is node k is, then $A_{ik}=A_{kj}=1$. And if $A_{ij}=0$, 2-hop neighbor of node i is node j. The matrix B_{nn} as follow:

$$B_{ij} = \begin{cases} 1 & \text{if node i and j have a common neighbor} \\ 0 & \text{otherwise} \end{cases} \tag{2}$$

The compatibility matrix D_{nn} as follow:

$$D_{ij} = A_{ij} \vee B_{ij} = \begin{cases} 1 & \text{if node i and j are within 2 hops} \\ 0 & \text{otherwise} \end{cases} \tag{3}$$

Let S represent the slots of per frame. Let $s : S \times N \mapsto \{0,1\}$ to express a transmission schedule, is defined by

$$s_{ij} = \begin{cases} 1 & \text{if slot i assigned to node j} \\ 0 & \text{otherwise} \end{cases} \tag{4}$$

So, TDMA scheduling in ad hoc can be famulated as multi-objective optimization problem [7].

Minimize S

$$\text{Maximize} \sum_{i=1}^{S} \sum_{j=1}^{N} s_{ij} \tag{5}$$

$$s.t. \sum_{i=1}^{S} s_{ij} \geq 1 \qquad i = 1, 2, \cdots, S \tag{6}$$

$$\sum_{i=1}^{S} \sum_{j=1}^{N} \sum_{k=1,k\neq j}^{N} s_{ij} d_{jk} s_{ik} = 0 \tag{7}$$

Equation (6) shows each node must broadcast at least once. Equation (7) prevents collisions, means that nodes within 1-hop or 2-hop can't transmit in the same slot.

3. TDMA scheduling based on weight of flows

A new TDMA scheduling scheme is proposed to improve fairness. Different bandwidth or size of the flow will lead to the actual data transmission time of the node is different. The node has lager number of flows or data will give priority to more slots. Nodes are ordered according to three parameters: the number of flows, the size of each flow and bandwidth of a node. Firstly, the weight of the flow at the nodes are calculated based on these three parameters. Then, the nodes are sorted according to the weights to minimize the frame length. Finally, maximize the total number of slots in a frame.

3.1 Weight of flows

Let F_{ij} represent the weight of the flow j in node i.

$$F_{ij} = \begin{cases} sf_{ij}/(BW_i * tslot) + 1 & \text{if } sf_{ij} > 0 \\ 0 & \text{otherwise} \end{cases} \tag{8}$$

Where sf_{ij} is the size of F_{ij}, BW_i represent the bandwidth of node i, *tslot* is unit time of the slot. Weight vector W is follow, where T is the number of flow in the node i.

$$W_i = \sum_{j=1}^{T} F_{ij} \tag{9}$$

3.2. Minimize frame length

This stage minimizes the frame length and ensures that each node with data has at least one slot. First, the nodes are ordered by descending according to the weights. Then, an available slot is searched to assign to each node until all nodes are allocated.

Algorithm minimize frame length as follow

```
Compute W;
Sort node by Wj;
Different slots (m1) are allocated to the first node and its 1-hop neighbors;
i=1, j=m1+1, S=m1;
While (j≤N) //the remaining nodes
    If Wj≠0
        If slot i can be allocated to node j
        Then sij=1, j++, i=1;
        Else i++;
    Else j++;
    If i>S then S=i;
```

3.3 Maximize throughput

In this stage, a slot is allocated to as many collision-free nodes as possible to maximize channel utilization. Because the scheduling scheme is related to the order of nodes, the phase 2 sorts the nodes by the weights. It is possible for those nodes that need more slots to get more slots. The algorithm improves the fairness of the system.

Algorithm maximize throughput as follow

```
Sort node by W_j;
j=1;
While (j≤N)
    If W_j≠0
        i=1
        While(i≤S)
            If slot i can be assigned to node j
            Then s_ij=1;
            i++;
    j++;
```

4. Simulation

A simulation C program has been created to compute and analyze the fairness of my algorithm. By simulations and analysis, we compared our algorithm with the algorithm in paper [3][5]. The simulation network topology of 15 nodes is introduced by Wang and Ansari, shown in figure2 [8]. The bandwidth of each node in the ad hoc can randomly select the following data: 1M,2M,5M,10M, etc. Nodes randomly produce several flows and the size of each flow.

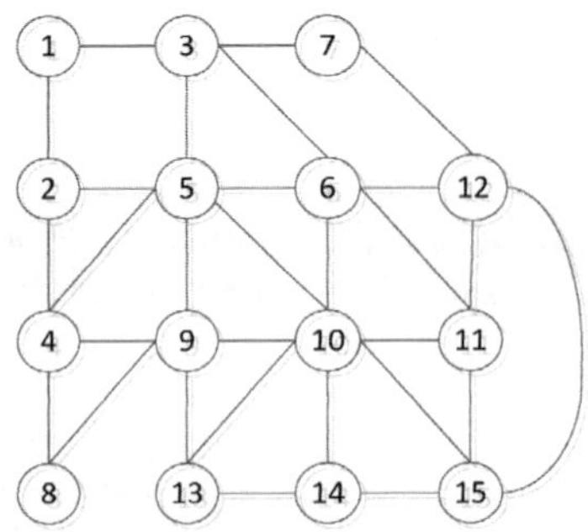

Figure 2. Topology of 15-nodes network.

4.1 Fairness index

The fairness is a key factor in ad hoc. Unfair channel allocation may lead to starvation, waste of resources and others, thus reducing channel utilization. Therefore, many researchers have studied how to alleviate the unfairness in ad hoc [9][10]. The proposed algorithm hopes to provide TDMA scheduling in a fair way. Consider not only the number of actual flows in the ad hoc, but also the length of each flow. To measure fairness, the Jain's Fairness Index [11] was adopted in this paper, defined as:

$$f(x) = \frac{(\sum_{i=1}^{n} x_i)^2}{n\sum_{i=1}^{n} x_i^2} \tag{10}$$

Where x_i represents the amount of slots allocated to node i, n is the number of nodes.

4.2 Result analysis

After more than 100 simulations, the average simulation result shows that the proposed scheduling algorithm surpassed the algorithm ordered only by the number of neighbor nodes or by flows in fairness. The result is shown as figure3.

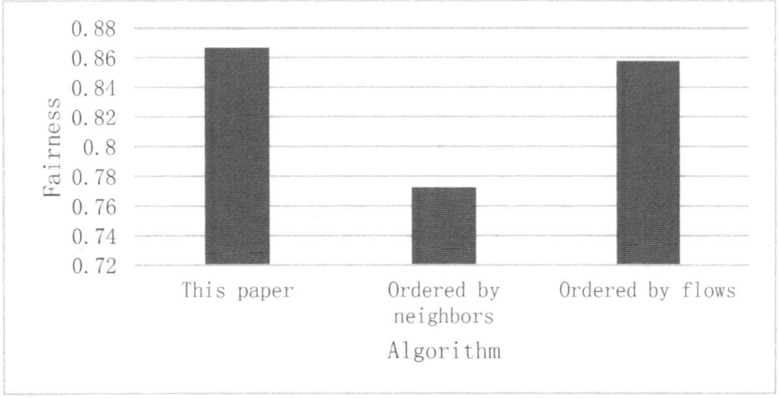

Figure 3. Average fairness of different algorithms.

5. Summary
An ad hoc network is a self-organizing and dynamic topology wireless network without infrastructure. Nodes in the communication range transmit data through broadcast channel. To achieve collision-free scheduling, TDMA is widely used in ad hoc. BSP on TDMA ad hoc networks is a multi-objective optimization problem, that is, the shortest frame length, maximizing channel utilization. TDMA based on SVC is closely related to order of nodes. This paper proposed a fair scheduling algorithm for TDMA, which calculated the weights of each node's flow. Then the nodes are ordered according to its weight. The algorithm improves the fairness of ad hoc network.

The time complexity of my algorithm is $O(n^3)$. In the future, we will enhance the performance of the scheduling algorithm in the following: reduce computation time and optimize slot allocation.

Acknowledgement
This research was supported by the Taizhou University Foundation for the Talents (QD2016035).

References
[1] Clayton W. Commander and Panos M. Pardalos 2009 A combinatorial algorithm for the TDMA message scheduling problem *Computational Optimization and Applications* **43** p 449
[2] A. Ephremides and T. V. Truong 1990 Scheduling broadcast in multi-hop radio networks *IEEE Transactions on Communications* **38** p 456
[3] Jaehyun Yeo, Heesoo Lee and Sehun Kim 2002 An efficient broadcast scheduling algorithm for TDMA ad-hoc networks *Computers & Operations Research* **29** p 1793
[4] Enrico Malaguti and Paolo Toth 2009 A survey on vertex coloring problems *International Transactions in Operational* **17** p 1
[5] Aggeliki Sgora, Dimitrios J. Vergados and Dimitrios D. Vergados 2007 Efficient Fair TDMA Scheduling in Wireless Multi-Hop Networks *Proceedings of the 4th IFIP International Conference on Artificial Intelligence Applications and Innovations* (USA) **247** p 279
[6] Dimitrios J. Vergados, Aggeliki Sgora and Dimitrios D. Vergados et al 2010 Fair TDMA scheduling in wireless multihop networks *Telecommunication Systems* **30** p1
[7] Clayton W. Commander 2007 *Broadcast Scheduling Problem* (Boston: Springer Press)
[8] Gangsheng Wang and Nirwan Ansari 1997 Optimal broadcast scheduling in packet radio networks using mean field annealing *IEEE Journal on Selected Areas in Communications* **15** p 250
[9] Jun He and Hung Keng Pung Fairness of medium access control protocols for multi-hop ad hoc wireless networks *Computer networks* **48** p 867
[10] Huaizhou SHI, R. Venkatesha Prasad and Ertan Onur et al 2014 Fairness in Wireless Networks Issues, Measures and Challenges *IEEE Communications Surveys & Tutorials* **16** p 5
[11] Jain Raj 1991 *The art of computer systems performance analysis: techniques for experimental design, measurement, simulation, and modeling* (New York: John Wiley & Sons Inc)

Stochastic Analysis of Downlink Heterogeneous Cellular Networks: Based on Spatial Hybrid Model

Pan Ziyu[1], Yang Jie[1] and Guo Nan[2]

[1] Department of Communication Engineering, Nanjing Institute of Technology, Nanjing, 211167, China
[2] Department of Mathematics and Physics, Nanjing Institute of Technology, Nanjing, Jiangsu, 211167, China

panziyu@njit.edu.cn

Abstract. In this paper, a hybrid model for heterogeneous cellular networks (HCNs) is proposed based on the differences in the distribution characteristics, path loss models and transmit power of macro and micro base stations (BSs). The macro base stations in HCNs are modelled as 2-D PPP which obeys certain parameters, and the small base stations are modelled as 3-D PPP. According to the stochastic geometry theory and the relevant knowledge of probability theory, the interference characteristics in HCNs are analysed, and the downlink coverage probability of the target users in the network are derived. Compared the hybrid interference model proposed in this paper with the traditional model, simulation shows that the hybrid interference model can illustrate the distribution characteristics of each kind of base station in HCNs more accurately, and the network performance based on model proposed in this paper is more accurate.

1. Introduction

As cellular network architecture tends to isomerization, the randomness of communication nodes in spatial distribution becomes more significant. Therefore, stochastic geometric theory plays an increasingly important role in modelling and analysing HCNs [1]. In [2], the authors analyse the multi-tier HCNs by assuming that each tier of the BS obeys the 2-D PPP and all BSs share the same spectrum resources. On this basis, the author deduces the probability of users accessing all BSs and the load of each BS, and further estimates the downlink coverage probability and transmission rate. Although this model is relatively simple, it pioneered the stochastic geometric analysis of HCNs, and laid the foundation for further research. Most of the subsequent research results are based on this. The load balancing problem of HCNs is studied in [3]. In [4], the partial frequency reuse technology in HCNs is studied. In [5], the inter tier spectrum allocation scheme and the BS access mechanism in the two level family base station network are studied. In [6], the cell access strategy in HCNs is analysed. In [7], the interference management and space time association in HCNs are studied.

In addition to the above PPP based HCN performance analysis, many scholars adopt non PPP modelling of HCNs in order to obtain more accurate performance analysis results. Starting from the correlation between BSs, the authors describe the distribution of micro BSs for macro BS edge coverage and hot spot coverage with two non PPP respectively [8]. The literature [9-10] uses poisson clustering process to describe HCNs, the macro and micro BSs are assumed to be the parent processes and sub processes in the poisson cluster process respectively. Considering the minimum distance between adjacent base stations, a cellular network analysis method based on poisson hard core process is proposed [11]. All the above point process analysis methods are more suitable for cellular networks than PPP and grid methods, and can better describe the spatial characteristics of BSs distribution in real networks. However, the biggest weakness of these point processes is that it is difficult to analyse and is difficult to obtain closed instruction with guiding significance, which limits their application in modelling and analysing interference in wireless cellular networks.

In this work, we propose a new interference model on the basis of the different characteristics of macro and micro BSs in HCNs. According to the theories of PPP and probability, the interferences in HCNs are analysed and downlink coverage probability is obtained. Simulation shows that the network performance index estimated by this model is 1-2dB accurate than the traditional method.

2. System Model

In the urban central area with a large number of buildings, macro BSs are usually deployed at the top of the buildings to achieve wide coverage and plane distribution, while the micro BSs are deployed inside the building to meet the indoor coverage and are stereoscopic, shown in Figure 1. In the hybrid model, the macro BSs are modelled as 2-D PPP Φ_1 with parameter λ_1, whereas the micro BSs are modelled as the 3-D PPP Φ_2 with parameter λ_2. It is worth noting that for the different distribution dimension of macro and micro BS, the units of λ_1 and λ_2 are different (m^{-2} and m^{-3} respectively). P_1 and P_2 indicate the transmitting power of macro and micro BS respectively; β_1 and β_2 respectively represent the SINR threshold of macro and micro BS.

Due to the deployment of micro BSs indoors, the path loss of the indoor wireless channel fading model $G\cdot x^{-(\alpha+1)}$ described in [12] is adopted, where G is constant. The general path loss model $\mathrm{PL}(x)=x^{-\alpha}$ is adopted for macro BS. Without loss of generality, the target user is assumed at the origin of the coordinates [13].

When a target user communicates with a macro BS, its received SINR is

$$\mathrm{SINR}_1(x)=\frac{P_1h_{1,0}x^{-\alpha_1}}{\sum\limits_{i\in\Phi_1\backslash B_{1,0}}P_1h_{1i}\left|Y_{1i}\right|^{-\alpha_1}+\sum\limits_{i\in\Phi_2}P_2Gh_{2i}\left|Y_{2i}\right|^{-(\alpha_2+1)}+\sigma^2},\tag{1}$$

while communicates with a micro BS, the SINR is represented as

$$\mathrm{SINR}_2(x)=\frac{P_2Gh_{2,0}x^{-(\alpha_2+1)}}{\sum\limits_{i\in\Phi_1}P_1h_{1i}\left|Y_{1i}\right|^{-\alpha_1}+\sum\limits_{i\in\Phi_2\backslash B_{2,0}}P_2Gh_{2i}\left|Y_{2i}\right|^{-(\alpha_2+1)}+\sigma^2},\tag{2}$$

where h represent shadow fading and Y is the distance from the interfering macro or micro BS to the target user.

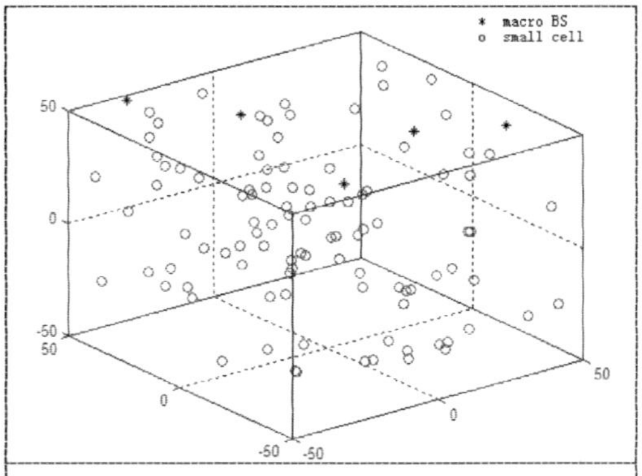

Figure 1. Two-tier HCN spatial hybrid model

3. Coverage Probability

This section discusses the downlink coverage probability of HCNs based on hybrid distribution model. In this section, we first discuss the access problem of user in HCNs, including the probability of association between target user and macro or micro BSs, and the average number of associated users per macro or micro BS. Then we derive the PDF of the distance between the target user and its associated BS, and finally deduce the target user's downlink coverage probability.

Lemma 1: The probability that a target user associate with the macro and micro BS is

$$A_1 = 2\pi\lambda_1 \int_0^\infty r \exp\left\{ -\frac{4}{3}\pi\lambda_2 \left(\frac{P_2 G}{P_1}\right)^{3/(\alpha_2+1)} r^{3\alpha_1/(\alpha_2+1)} - \pi\lambda_1 r^2 \right\} dr, \quad (3)$$

and

$$A_2 = 4\pi\lambda_2 \int_0^\infty r^2 \exp\left\{ -\pi\lambda_1 \left(\frac{P_1}{P_2 G}\right)^{2/\alpha_1} r^{2(\alpha_2+1)/\alpha_1} - \frac{4}{3}\pi\lambda_2 r^3 \right\} dr. \quad (4)$$

Proof: Denote symbol n as an index of the BS tier that a typical user communicate with, n will be 1 or 2.

If the received power from the nearest macro BS $P_{r,1}$ is larger than that of the nearest micro BS, the user will communicate with macro BS. In this way n=1; otherwise n=2.

$$A_1 = \mathrm{P}[n=1] = \mathrm{E}_{R_1}\left[\mathrm{P}\left[P_1 R_1^{-\alpha_1} > P_2 G R_2^{-(\alpha_2+1)}\right]\right] = \int_0^\infty \mathrm{P}\left[R_2 > \left(\frac{P_2 G}{P_1}\right)^{1/\alpha_2+1} r^{\alpha_1/\alpha_2+1}\right] f_{R_1}(r)\, dr, \quad (5)$$

where

$$\mathrm{P}\left[R_2 > \left(\frac{P_2 G}{P_1}\right)^{1/(\alpha_2+1)} r^{\alpha_1/(\alpha_2+1)}\right] = \exp\left\{ -\frac{4}{3}\pi\lambda_2 \left(\frac{P_2 G}{P_1}\right)^{3/(\alpha_2+1)} r^{3\alpha_1/(\alpha_2+1)} \right\} \quad (6)$$

and

$$f_{R_1}(r) = \frac{d\left[1-\mathrm{P}\left[R_1 > r\right]\right]}{dr} = e^{-\pi\lambda_1 r^2} 2\pi\lambda_1 r. \quad (7)$$

Combining (5), (6) with (7), we can get (3). Similarly, we get (4).

Lemma 1 reveals a basic rule, that is, users tend to associate with those BSs with larger distribution density and transmit power, and are unwilling to associate with the BSs with sparse distribution and low transmitting power. This indicates that when we deploy HCNs, we need to consider the transmission power and distribution density of different types of BSs. This enables users to choose associated BSs in macro and micro BSs instead of accessing the same type of BSs uniformly, and ultimately achieves the purpose of balancing network load and improving network efficiency.

Lemma 2: When the target user communicated with macro BS, the pdf of the distance between the target user and its associated macro BS will be

$$f_{X_1}(x) = \frac{2\pi\lambda_1}{A_1} x \exp\left\{ -\frac{4}{3}\pi\lambda_2 \left(\frac{P_2 G}{P_1}\right)^{3/\alpha_2} x^{3\alpha_1/\alpha_2} - \pi\lambda_1 x^2 \right\}, \quad (8)$$

and that between the target user and its associated micro BS is

$$f_{X_2}(x) = \frac{4\pi\lambda_2}{A_2} x^2 \exp\left\{ -\pi\lambda_1 \left(\frac{P_1}{P_2 G}\right)^{2/\alpha_1} x^{2(\alpha_2+1)/\alpha_1} - \frac{4}{3}\pi\lambda_2 x^3 \right\}, \quad (9)$$

Proof: The probability that $X_1 > x$ can be given as

$$\mathrm{P}[X_1 > x] = \mathrm{P}[R_1 > x \mid n=1] = \frac{\mathrm{P}[R_1 > x, n=1]}{\mathrm{P}[n=1]}, \tag{10}$$

the numerator can be expressed as

$$\begin{aligned} \mathrm{P}[R_1 > x \mid n=1] &= \mathrm{P}\left[R_1 > x, P_{r,1}(R_1) > \max P_{r,2}\right] \\ &= \int_x^{\infty} \mathrm{P}\left[R_2 > \left(\frac{P_2 G}{P_1}\right)^{1/(\alpha_2+1)} r^{\alpha_1/(\alpha_2+1)}\right] f_{R_1}(r)\,dr \\ &= 2\pi\lambda_1 \int_x^{\infty} r \exp\left\{-\frac{4}{3}\pi\lambda_2 \left(\frac{P_2 G}{P_1}\right)^{3/(\alpha_2+1)} r^{3\alpha_1/(\alpha_2+1)} - \pi\lambda_1 r^2\right\} dr \end{aligned}, \tag{11}$$

while the denominator can be determined by Lemma 1,

So,

$$\mathrm{P}[X_1 > x] = \frac{2\pi\lambda_1}{A_1} \int_x^{\infty} r \exp\left\{-\frac{4}{3}\pi\lambda_2 \left(\frac{P_2 G}{P_1}\right)^{3/(\alpha_2+1)} r^{3\alpha_1/(\alpha_2+1)} - \pi\lambda_1 r^2\right\} dr. \tag{12}$$

The pdf of X_1 is

$$f_{X_1}(x) = \frac{d\left[1 - \mathrm{P}[X_1 > x]\right]}{dx} = \frac{2\pi\lambda_1}{A_1} x \exp\left\{-\frac{4}{3}\pi\lambda_2 \left(\frac{P_2 G}{P_1}\right)^{3/(\alpha_2+1)} x^{3\alpha_1/(\alpha_2+1)} - \pi\lambda_1 x^2\right\}. \tag{13}$$

Similar, we can get (9).

In HCNs composed of macro and micro BSs, the downlink coverage probability of target user *Cov* is defined as the probability that the target user's instantaneous SINR in the network is larger than the predefined threshold. Therefore, we can obtain the coverage probability of the whole heterogeneous network through the total probability formula [13]

$$Cov = \sum_{k=1}^{2} \mathrm{cov}_k A_k \tag{14}$$

where A_k indicates the association probability which is determined by lemma 1 and cov_k is the coverage probability associated with macro or micro BS.

Theorem 1: When the target user communicated with macro BS, its coverage probability will be

$$\begin{aligned} Cov_1 = \frac{2\pi\lambda_1}{A_1} \int_0^{\infty} x \exp\{&-\frac{Tx^{\alpha_1}}{SNR} - \pi\lambda_1 \int_{T^{-2/\alpha_1}}^{\infty} \frac{x^2 T^{2/\alpha_1}}{1+u^{\alpha_1/2}} du \\ &-\frac{4}{3}\pi\lambda_2 \int_0^{\infty} \frac{\left(x^{\alpha_1} P_2 G P_1^{-1}\right)^{3/(\alpha_2+1)}}{1+v^{(\alpha_2+1)/3}} dv - \frac{4}{3}\pi\lambda_2 \left(P_2 G P_1^{-1}\right)^{3/(\alpha_2+1)} x^{3\alpha_1/(\alpha_2+1)} - \pi\lambda_1 x^2\} dx \end{aligned}, \tag{15}$$

and that with micro BS is

$$\begin{aligned} Cov_2 = \frac{4\pi\lambda_2}{A_2} \int_0^{\infty} x^2 \exp\{&-\frac{Tx^{(\alpha_2+1)} G^{-1}}{SNR} - \pi\lambda_1 \int_0^{\infty} \frac{\left(x^{(\alpha_2+1)} P_2^{-1} G^{-1} P_1 T\right)^{2/\alpha_1}}{1+u^{\alpha_1/2}} du \\ &-\frac{4}{3}\pi\lambda_2 \int_{T^{-3/(\alpha_2+1)}}^{\infty} \frac{x^3 T^{3/(\alpha_2+1)}}{1+v^{(\alpha_2+1)/3}} dv - \pi\lambda_1 \left(P_2^{-1} G^{-1} P_1\right)^{2/\alpha_1} x^{2(\alpha_2+1)/\alpha_1} - \frac{4}{3}\pi\lambda_2 x^3\} dx \end{aligned}. \tag{16}$$

Therefore, the downlink coverage probability of the entire HCN is

$$Cov = \mathrm{cov}_1 A_1 + \mathrm{cov}_2 A_2. \tag{17}$$

Proof: When communicate with macro BS, the coverage probability is

$$Cov_1 = \mathrm{E}_x\left[\mathrm{P}\left[\mathrm{SINR}_1(x) > T\right]\right] = \int_0^\infty \mathrm{P}\left[\mathrm{SINR}_1(x) > T\right] f_{X_1}(x)dx, \tag{18}$$

where $f_{X_1}(x)$ is given in Lemma 2.

Rewrite the SINR of target user as $\mathrm{SINR}_1(x) = \frac{h_{1,0}}{x^{\alpha_1} P_1^{-1}\left(I_r + \sigma^2\right)}$, where $I_r = \sum_{i\in\Phi_1\backslash B_{1,0}} P_1 h_{1i}\left|Y_{1i}\right|^{-\alpha_1} + \sum_{i\in\Phi_2} P_2 G h_{2i}\left|Y_{2i}\right|^{-(\alpha_2+1)}$ is interference. Therefore, the probability $\mathrm{P}\left[\mathrm{SINR}_1(x) > T\right]$ in (18), that is the CCDF of received SINR, can be expressed as

$$\mathrm{P}\left[\mathrm{SINR}_1(x) > T\right] = \mathrm{P}\left[h_{1,0} > x^{\alpha_1} P_1^{-1} T\left(I_r + \sigma^2\right)\right] = e^{-x^{\alpha_1} P_1^{-1} T\sigma^2} \mathcal{L}_{I_r}\left(x^{\alpha_1} P_1^{-1} T\right). \tag{19}$$

According to the definition of Laplace transform, the Laplace transform in (19) can be expanded as

$$\begin{aligned}\mathcal{L}_{I_r}\left(x^{\alpha_1} P_1^{-1} T\right) &= \mathrm{E}\left[\exp\left(-x^{\alpha_1} P_1^{-1} T\left(\sum_{i\in\Phi_1\backslash B_{1,0}} P_1 h_{1i}\left|Y_{1i}\right|^{-\alpha_1} + \sum_{i\in\Phi_2} P_2 G h_{2i}\left|Y_{2i}\right|^{-(\alpha_2+1)}\right)\right)\right] \\ &= \mathrm{E}\left[\prod_{i\in\Phi_1\backslash B_{1,0}} \exp\left(-x^{\alpha_1} T h_{1i}\left|Y_{1i}\right|^{-\alpha_1}\right)\right]\cdot \mathrm{E}\left[\prod_{i\in\Phi_2} \exp\left(-x^{\alpha_1} P_1^{-1} P_2 G T h_{2i}\left|Y_{2i}\right|^{-(\alpha_2+1)}\right)\right]\end{aligned} \tag{20}$$

Following the probability generating functional (PGFL) of 2-D and 3-D PPP [13] and h ~exp(1)

$$\begin{aligned}\mathcal{L}_{I_r}\left(x^{\alpha_1} P_1^{-1} T\right) = &\exp\left\{-2\pi\lambda_1 \int_x^\infty \left(1 - \frac{1}{1 + x^{\alpha_1} T y^{-\alpha_1}}\right) y dy\right\}\cdot \\ &\exp\left\{-4\pi\lambda_2 \int_0^\infty \left(1 - \frac{1}{1 + x^{\alpha_1} P_1^{-1} P_2 G T y^{-(\alpha_2+1)}}\right) y^2 dy\right\}\end{aligned}. \tag{21}$$

Employing variable substitutions $u = \left(x^{\alpha_1} T\right)^{-2/\alpha_1} y^2$ and $v = \left(x^{\alpha_1} P_1^{-1} P_2 G T\right)^{-3/(\alpha_2+1)} y^3$ to the two exponentiation expressions in formula results in (21), we can get

$$\mathcal{L}_{I_r}\left(x^{\alpha_1} P_1^{-1} T\right) = \exp\left\{-\pi\lambda_1 \int_{T^{-2/\alpha_1}}^\infty \frac{\left(x^{\alpha_1} T\right)^{2/\alpha_1}}{1 + u^{\alpha_1/2}} du\right\} \exp\left\{-\frac{4}{3}\pi\lambda_2 \int_0^\infty \frac{\left(x^{\alpha_1} P_1^{-1} P_2 G T\right)^{3/(\alpha_2+1)}}{1 + v^{(\alpha_2+1)/3}} dv\right\}. \tag{22}$$

Combining (22) and (19), setting $\frac{T}{SNR} = \frac{T\sigma^2}{P_1}$, we can get (15). In the similar way, formula (16) can be obtained. Furthermore, the downlink coverage probability in the whole network can be obtained by plugging (15) and (16) into (17) .

4. Numerical Results

In this section, we simulate and analyse the coverage probability of two-tier HCN downlink system composed of macro and micro BSs. In addition to special instructions, the simulation environment is set as follows: the radius of a central spherical area R =100m; the transmission power of macro and micro BS P_1=63dBm, P_2=33dBm, and the density of macro and micro BSs $\lambda_1=10^{-5}/m^2$, $\lambda_2=2\times10^{-4}/m^3$; the path loss factor of macro and micro BS α_1 = 3.8, α_2 = 4.5; the additive Gauss white noise power σ^2=-104dBm.

Comparisons of the coverage performance between the hybrid interference modelling method proposed in this paper and the traditional analysis method based 2-D PPP in case of noiseless and noisy environments are shown in Figure 2 and Figure 3 respectively. The distribution of macro and micro BSs in the actual environment is taken as a reference. The abscissa coordinates the SIR/SINR threshold, and the ordinate represents the coverage probability.

From Figure 2 and Figure 3, we can find that the simulation results based on the hybrid model proposed in this paper are very close to the simulation results of the actual base station distribution scenario, which is more accurate than the traditional analysis based on 2-D PPP 1-2dB. Obviously, the hybrid model in this paper is more suitable for describing the distribution of base stations in HCNs, and can make more accurate prediction of network performance indicators. The reason is that, macro and micro BS appear different spatial distribution characteristics and transmission characteristics in actual environment. In the traditional analysis process, they are uniformly described as 2-D PPP distribution, but there are many differences in parameters such as base station density, path loss factor and transmission power. This paper uses different model to describe the spatial distribution of macro and micro BSs respectively, describes the path loss of macro and micro BSs with different fading models, and achieves good results.

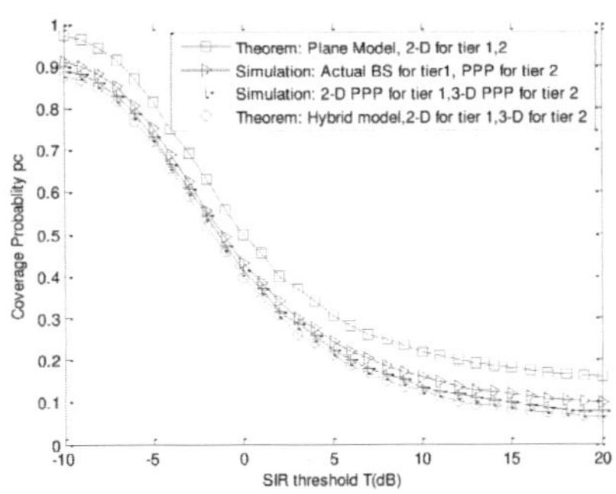

Figure 2. Coverage probability in no noise environment

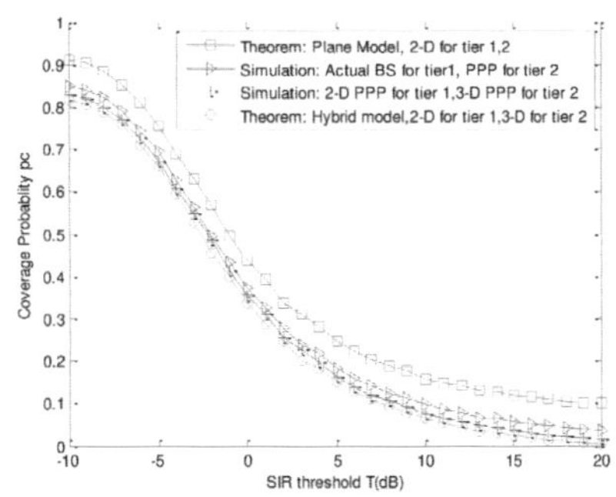

Figure 3. Coverage probability in noisy environment

Figure 4 presents the coverage probability for different values of α in the no noise and noisy environment. Compared the line of no-noise with noisy environments, we note that the additive noise has little influence on the coverage probability and the heterogeneous cellular networks are interference limited. In addition, we can find that the coverage probability increases when the path loss exponent α increases, whether or not the noise exists. The reason is that, when α increases, the received signal and the interference will be reduced, but the interference decreases more significantly, so the SINR and coverage probability increase.

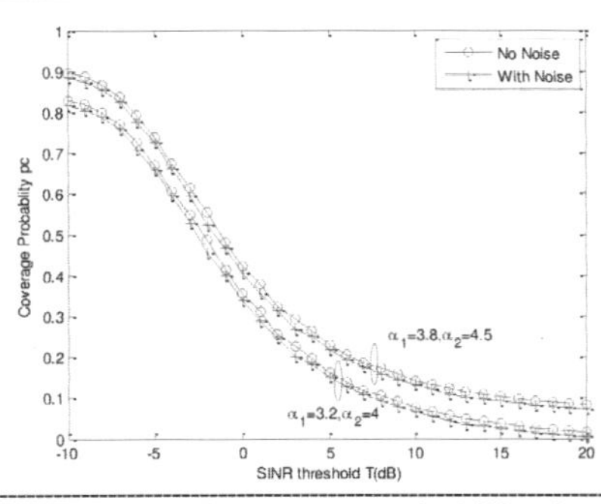

Figure 4. Coverage probability in a two-tier HCN with and without thermal noise

IOP Conf. Series: Journal of Physics: Conf. Series **1284** (2019) 012050 doi:10.1088/1742-6596/1284/1/012050

5. Conclusions

Through analysing the different characteristics of macro and micro BSs in two-tier HCNs, this paper proposes a hybrid model for interference modelling of HCNs based on 2-D and 3-D PPP, and analyses the network performance based on the new model. Simulations show that the proposed new model in this work appears closer to the reality than traditional 2-D PPP in terms of accuracy.

Acknowledgments

This work is supported by the Natural Science Foundation of the Jiangsu Higher Education Institutions of China(18KJB510017), the Foundation of Nanjing Institute of Technology (ZKJ201801, PTKJ201603), the Young Scientists Fund of the National Natural Science Foundation of China (61701221) and the Young Scientists Fund of Natural Science Foundation of Jiangsu Province, China(BK20160781) .

References

[1] Andrews J G, Claussen H, Dohler M, Rangan S and Reed M C 2012 *IEEE Journ. Sel. Areas in Commun.*, **30** 497

[2] Dhillon H S, Ganti R K, Baccelli F and Andrews J G 2012 *IEEE Journ. Sel. Areas in Commun.* **30** 550

[3] Dhillon H S, Ganti R K and Andrews J G 2013 *IEEE Trans. Wireless Commun.* **12**, 1666

[4] Novlan T D, Ganti R K, Ghosh A 2012 *IEEE Trans. Wireless Commun.* **60** 2029

[5] Wang C C, Quek T Q S, Kountouris M2012 *IEEE Journ. Sel. Areas in Commun.* **30** 561

[6] Yi Z, Zhang W 2013 *IEEE Trans. Wireless Commun.* **61** 3016

[7] Li C, Zhang J, Haenggi M 2015 *IEEE Trans. Wireless Commun.* **63** 1419

[8] Deng N, Zhou W, Haenggi M 2015 *IEEE Journ. Sel. Areas in Commun.* **33** 2167

[9] Suryaprakash V, Møller J, Fettweis G. On the Modeling and Analysis of Heterogeneous Radio Access Networks Using a Poisson Cluster Process[J]. IEEE Transactions on Wireless Communications, 2015, 14(2):1035-1047.

[10] Wang Y, Zhu Q 2017 *IEEE Commun. Lett.* **21** 576

[11] Haenggi M. 2011 *IEEE Commun. Lett.* **15** 792

[12] Devasirvatham D M J, Banerjee C, Murray R R 1991 *IEEE Global Communications Conference* 1282

[13] Sung N C, Dietrich S, Wilfrid S K and Joseph M 2013 Stochastic Geometry and Its Applications (3rd Edition)(U.S.: John Wiley & Sons Inc.)

Beamspace MUSIC Algorithm based on Space-time Array Signal Model

B Yang[1], B Yang[2], C Wang[2] and H X Shen[3]

[1] Zhengzhou Information Science and Technology Institute, Zhengzhou, 450001, China

[2] National Digital Switching System Engineering and Technology Research Center, Zhengzhou, 450002, China

[3] Graduate School of Chinese Academy of Social Sciences, Beijing, 102488, China

E-mail: forestsyoung@outlook.com

Abstract. This paper is devoted to the direction-of-arrival (DOA) algorithm in the moving array scenario. A beamspace MUSIC algorithm based on space-time signal model is proposed to cope with this scenario. First, the DOA information contained in Doppler shift is used to combine with the array spatial information, thus constructing a space-time signal model. Second, allowing for the considerable dimension and complexity of the space-time model, we transform the array sensor data into beamspace to reduce the computation load. The introduction of Doppler information leads to significant enhancement in algorithm performance. Simulation results demonstrate the promotion of proposed algorithm.

1. Introduction

Direction-of-arrival (DOA) estimation using an antenna array is an important problem across a number of applications. Its performance mainly depends on the array aperture. In the modern communication system, the array is always disposed on mobile carriers such as ships, aeroplanes and satellites to obtain better manoeuvrability [1]. Because of the limitation of the carrier size, the array aperture is restricted which will degrade the DOA performance [2].

To improve DOA estimation performance, researchers have put forward various modified methods. Literature [3-5] improve DOA performance by making use of non-circularity, cyclostationarity and fourth-order cumulant of the signal, respectively. However, in essence, these techniques merely use the properties of signals to improve DOA performance, but not solve the conflict between the array aperture and DOA performance in moving array scenario.

Except for above methods, DOA estimation utilizing Doppler information has drawn increasing attention. One of the representative methods is synthetic aperture (SA) technique, which includes ETAM [6-7] and FFTSA [8] algorithms. It has been widely used for moving array and provides improved performance. However, this technique has strict demand for the motion attitude of the array, so it is only suitable for specific scenarios. Another type of method is space-time signal processing. This technique exploits the DOA information in space and time domain simultaneously and constructs a space-time signal model [9]. Algorithm based on this signal model has been proposed in [10] and demonstrates good performance. However, to the best of our knowledge, existing space-time processing methods mainly focus on the scenario where a static array detects a moving target [11]. Methods intended for moving array have not been addressed often.

IOP Conf. Series: Journal of Physics: Conf. Series **1284** (2019) 012051 doi:10.1088/1742-6596/1284/1/012051

According to the moving array scenario assumed in this paper, we construct a space-time signal model based on existing methods, so that the spatial and Doppler information has been combined and embedded in the space-time manifold vector. To achieve high accuracy and resolution estimation, we adopt MUSIC algorithm. Because of the accumulation of temporal data, the dimension of space-time model grows manyfold compared with that of conventional signal model, which causes heavy computation load. Thus we utilize beamspace MUSIC algorithm to reduce the dimension of the subspace without degradation of performance. Finally, numerical simulations are presented to illustrate the reliability and effectiveness of proposed algorithm.

2. Scenario description and problem formulation

2.1. Scenario description and assumptions

Assume that an M sensors uniform linear array with an inter-element spacing d is equipped on a moving platform, which moves along its baseline at velocity v. P far-field narrowband signals of DOA θ_p, $1 \le p \le P$, is received by the array. The carrier frequencies of signals are f_0, which is known to the receiver. The Doppler shift caused by array motion can be expressed as

$$f\left(\theta_p\right) = f_0 v \sin\theta_p / c \tag{1}$$

where c is the velocity of electromagnetic wave.

Then we can represent the received signal of the mth sensor at time t as

$$\begin{aligned} x_m(t) &= \sum_{p=1}^{P} s_p e^{j2\pi f_0\left(1-\frac{v\sin\theta_p}{c}\right)\left(t-\frac{(m-1)d}{c}\sin\theta_p\right)} + n_m(t) \\ &= \sum_{p=1}^{P} s_p e^{j2\pi f_0\left(t-\frac{v\sin\theta_p}{c}t-\frac{(m-1)d}{c}\sin\theta_p+\frac{v(m-1)d}{c^2}\sin^2\theta_p\right)} + n_m(t) \end{aligned} \tag{2}$$

where s_p is the complex envelope of the pth signal and $n_m(t)$ is the additive white noise.

According to the fact that $v \ll c$ and f_0 is a prior knowledge, we can simplify equation (2) as

$$\boldsymbol{x}(t) = \sum_{p=1}^{P} \boldsymbol{a}(\theta_p) s_p e^{-j2\pi f_0 \frac{v\sin\theta_p}{c}t} + \boldsymbol{n}(t) \tag{3}$$

where $\boldsymbol{a}(\theta_p) = \left[1, e^{-j2\pi f_0 \frac{d}{c}\sin\theta_p}, \cdots, e^{-j2\pi f_0 \frac{(M-1)d}{c}\sin\theta_p}\right]^T$ is the spatial manifold vector, $\boldsymbol{n}(t)$ is the noise vector. We assume that the signals are independent.

2.2. Problem formulation for space-time model

From the signal model in equation (3), we observe that Doppler shift contains DOA information, which is useful for DOA estimation. Besides, the time domain samples constitute a geometric sequence, which is similar to spatial manifold. Thus, we consider constructing a space-time signal model to make full use of Doppler information and acquire better DOA estimation performance.

Assuming that the total sample number is K, we can divide the samples into Q equal intervals and every interval has N ($N = K/Q$) samples. Then during the qth interval, the array output vector at the nth sample moment can be expressed as

$$\begin{aligned} \boldsymbol{x}_n(q) &= \sum_{p=1}^{P} s_p \boldsymbol{a}(\theta_p) e^{-j2\pi f_0 \frac{v\sin\theta_p}{c}[(q-1)N+(n-1)]T_s} + \boldsymbol{n}_n(q) \\ &= \sum_{p=1}^{P} s_p l_p(q) \boldsymbol{a}(\theta_p) e^{-j2\pi f_0 \frac{v\sin\theta_p}{c}(n-1)T_s} + \boldsymbol{n}_n(q) \\ &= \sum_{p=1}^{P} \beta_p(q) \boldsymbol{a}(\theta_p) e^{-j2\pi f_0 \frac{v\sin\theta_p}{c}(n-1)T_s} + \boldsymbol{n}_n(q) \end{aligned} \tag{4}$$

where $l_p(q) = \exp\left(-j2\pi f_0 \frac{v\sin\theta_p}{c}(q-1)NT_s\right)$ and $\beta_p(q) = s_p l_p(q)$. T_s represents sample interval.

From equation (4), it can be observed that the phases of the samples in an interval can form a vector as

$$\boldsymbol{b}(\theta_p) = \left[1, \exp\left(-j2\pi f_0 \frac{v\sin\theta_p}{c} T_s\right), \cdots, \exp\left(-j2\pi f_0 \frac{v\sin\theta_p}{c}(N-1)T_s\right)\right]^T \tag{5}$$

The vector $\boldsymbol{b}(\theta_p)$ has similar structure with the spatial manifold vector $\boldsymbol{a}(\theta_p)$, so it can be defined as temporal manifold vector. Then we use the array output vector at every moment in an interval to construct an MN-dimensional vector $\boldsymbol{y}(q) = \left[\boldsymbol{x}_1^T(q), \boldsymbol{x}_2^T(q), \cdots, \boldsymbol{x}_N^T(q)\right]^T$. $\boldsymbol{y}(q)$ is the space-time observed data and its expression is

$$\begin{aligned} \boldsymbol{y}(q) &= \sum_{p=1}^{P} \left[\boldsymbol{b}(\theta_p) \otimes \boldsymbol{a}(\theta_p)\right] \beta_p(q) + \mu(q) \\ &= \sum_{p=1}^{P} \boldsymbol{g}(\theta_p) \beta_p(p) + \mu(q) \end{aligned} \tag{6}$$

where $\otimes$ represents Kronecker product. $\boldsymbol{g}(\theta_p) = \boldsymbol{b}(\theta_p) \otimes \boldsymbol{a}(\theta_p)$ is the space-time manifold vector and $\mu(q) = \left[\mu_1^T(q), \mu_2^T(q), \cdots, \mu_N^T(q)\right]^T$ is the space-time noise vector. For simplicity, we can rewrite equation (6) as

$$\begin{aligned} \boldsymbol{y}(q) &= (\boldsymbol{B} \circ \boldsymbol{A}) \beta(q) + \mu(q) \\ &= \boldsymbol{G}\beta(q) + \mu(q) \end{aligned} \tag{7}$$

where $\circ$ denotes Khatri-Rao product. $\boldsymbol{A} = [\boldsymbol{a}(\theta_1), \boldsymbol{a}(\theta_2), \cdots, \boldsymbol{a}(\theta_P)]$, $\boldsymbol{B} = [\boldsymbol{b}(\theta_1), \boldsymbol{b}(\theta_2), \cdots, \boldsymbol{b}(\theta_P)]$ and $\beta(q) = [\beta_1(q), \beta_2(q), \cdots, \beta_P(q)]^T$. $\boldsymbol{G} = [\boldsymbol{g}(\theta_1), \boldsymbol{g}(\theta_2), \cdots, \boldsymbol{g}(\theta_P)]$, and it is an $MN \times P$-dimensional space-time manifold matrix.

In next section, we will come up with an algorithm to estimate the DOA based on the proposed signal model.

3. Beamspace MUSIC algorithm for DOA estimation

In this section, we consider using MUSIC algorithm to estimate the DOAs of signals. However, equation (7) indicates that the dimension of proposed signal model is N times as that of conventional model, which will increase the complexity of algorithm and computation load. With this in mind, we switch observed signal to beamspace to reduce the complexity [12].

The principle of beamspace processing is demonstrated in figure 1. The array sensors are passed through beamformers, then the beam data is used to estimate DOAs [13].

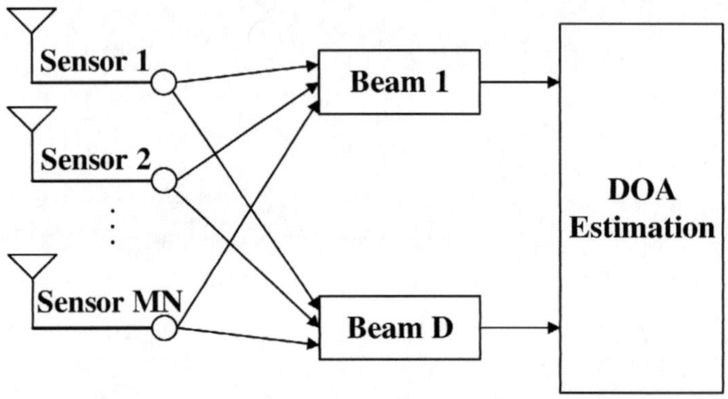

Figure 1. Principle of beamspace DOA estimation

From section 2, we know that the space-time manifold vector $g(\theta_p)$ is

$$\begin{aligned} g(\theta_p) = \Big[& 1, e^{-j2\pi f_0 \frac{d\sin\theta_p}{c}}, \cdots, e^{-j2\pi f_0 \frac{(M-1)d\sin\theta_p}{c}}, \cdots, e^{-j2\pi f_0 \frac{vT_s\sin\theta_p}{c}}, \\ & e^{-j2\pi f_0\left(\frac{d\sin\theta_p}{c}+\frac{vT_s\sin\theta_p}{c}\right)}, \cdots, e^{-j2\pi f_0\left(\frac{(M-1)d\sin\theta_p}{c}+\frac{vT_s\sin\theta_p}{c}\right)}, \cdots, e^{-j2\pi f_0 \frac{(N-1)vT_s\sin\theta_p}{c}}, \\ & e^{-j2\pi f_0\left(\frac{d\sin\theta_p}{c}+\frac{(N-1)vT_s\sin\theta_p}{c}\right)}, \cdots, e^{-j2\pi f_0\left(\frac{(M-1)d\sin\theta_p}{c}+\frac{(N-1)vT_s\sin\theta_p}{c}\right)} \Big]^T \end{aligned} \tag{8}$$

According to the form of $g(\theta_p)$, we can define a beamforming weight vector as

$$\begin{aligned} w(u) = \Big[& 1, e^{-j2\pi u f_0 \frac{d}{c}}, \cdots, e^{-j2\pi u f_0 \frac{(M-1)d}{c}}, \cdots, e^{-j2\pi u f_0 \frac{vT_s}{c}}, \\ & e^{-j2\pi u f_0\left(\frac{d}{c}+\frac{vT_s}{c}\right)}, \cdots, e^{-j2\pi u f_0\left(\frac{(M-1)d}{c}+\frac{vT_s}{c}\right)}, \cdots, e^{-j2\pi u f_0 \frac{(N-1)vT_s}{c}}, \\ & e^{-j2\pi u f_0\left(\frac{d}{c}+\frac{(N-1)vT_s}{c}\right)}, \cdots, e^{-j2\pi u f_0\left(\frac{(M-1)d}{c}+\frac{(N-1)vT_s}{c}\right)} \Big]^T \end{aligned} \tag{9}$$

where $u = \sin\theta$, and the interval $-1 \le u \le 1$ is corresponding to the angular interval $-90^\circ \le \theta \le 90^\circ$. $w(u)$ represents the beamformer of the beam whose main lobe points at θ. Based on this, D adjacent beamforming vectors of the structure in (9) with respective pointing angles equispaced by $\Delta = 2/MN$ can make up a $MN \times D$ beamforming matrix

$$W = \frac{1}{\sqrt{MN}}\left[w\left(r\frac{2}{MN}\right), w\left((r+1)\frac{2}{MN}\right), \cdots, w\left((r+D-1)\frac{2}{MN}\right)\right] \tag{10}$$

Note that $w(u)$ does not possess Vandermonde structure, hence the columns of W are not orthonormal. We can orthogonalize W and obtain the transform matrix as

$$H = W\left(W^H W\right)^{-\frac{1}{2}} \tag{11}$$

Utilizing matrix H, the observed data $y(q)$ can be transformed into

$$z(q) = H^H y(q) \tag{12}$$

It is noteworthy that when $H = I$, the beamspace data $z(q)$ is equivalent to sensor data $y(q)$. Besides, equation (12) also indicates that the dimension of observed data is reduced from $MN \times 1$ to

$D \times 1$. The corresponding covariance matrix is

$$\begin{aligned} \boldsymbol{R}_{zz} &= \boldsymbol{H}^H \boldsymbol{R}_{yy} \boldsymbol{H} \\ &= \boldsymbol{H}^H \boldsymbol{G} \boldsymbol{R}_{\beta\beta} \boldsymbol{G}^H \boldsymbol{H} + \sigma^2 \boldsymbol{I} \end{aligned} \quad (13)$$

where $\boldsymbol{R}_{yy}$ and $\boldsymbol{R}_{\beta\beta}$ are the covariance matrices of $\boldsymbol{y}(q)$ and $\beta(q)$, respectively. σ^2 is the power of noise. In light of the procedure of MUSIC algorithm, the noise subspace $\tilde{\boldsymbol{E}}_\mu$ can be obtained through eigen decomposition of $\boldsymbol{R}_{zz}$. Then the beamspace MUSIC spatial spectrum is formulated as

$$\hat{\boldsymbol{S}}_{MUSIC}(\theta) = \left[\boldsymbol{g}^H(\theta) \boldsymbol{H} \tilde{\boldsymbol{E}}_\mu \tilde{\boldsymbol{E}}_\mu^H \boldsymbol{H}^H \boldsymbol{g}(\theta)\right]^{-1} \quad (14)$$

Finally, θ can be estimated by searching the maximum of $\hat{\boldsymbol{S}}_{MUSIC}(\theta)$ as follow

$$\hat{\theta} = \arg\max_{\theta} \hat{\boldsymbol{S}}_{MUSIC}(\theta) \quad (15)$$

As is well known, the main computation load of MUSIC algorithm is centered on eigen decomposition. The complexity of original MUSIC algorithm for space-time signal model is $O\left(M^3 N^3\right)$, while beamspace MUSIC algorithm requires only $O\left(D^3\right)$, which is significantly reduced.

4. Numerical results

In this section, numerical simulations of the proposed algorithm (BMUSIC-ST) are presented and compared with conventional spatial MUSIC (MUSIC-spatial), space-time MUSIC (MUSIC-ST) and ETAM algorithm. In simulations, the resolution and root mean square error (RMSE) are exploited to evaluate the performance of algorithm.

In the first scenario, we consider an 8 sensors uniform linear array with half-wavelength inter-element spacing. The array moves at velocity 200m/s.

Figure 2 demonstrates the comparison versus signal-to-noise ratio (SNR) with 900 signal samples based on 300 Monte Carlo trials. We assume that two narrow-band signals arrive from 20° and 23°, respectively. Their center frequencies are 300MHz and the dimension of vector $\boldsymbol{b}(\theta)$ (temporal manifold vector) is 18 (N = 18). From the figure, we can observe that BMUSIC-ST and MUSIC-ST can absolutely distinguish two signals at -4dB and 6dB, respectively, while spatial MUSIC algorithm cannot achieve this until 14dB. It can be concluded that algorithms based on space-time model possess better resolution than those based on conventional model. Besides, obviously, the resolution of BMUSIC algorithm is superior to that of MUSIC algorithm.

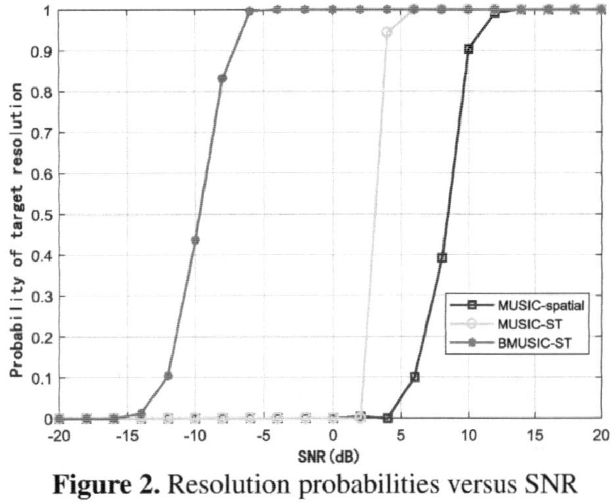

Figure 2. Resolution probabilities versus SNR

In figure 3, we present the performance comparison versus SNR amongst four algorithms to examine the accuracy of proposed algorithm. Assume that a narrow-band signal impinges on the array from 15°. Other conditions remain the same as those in figure 2. In particular, we introduce sensor errors in figure 3(b), where the gain and phase errors of each sensor are drawn from the interval [−0.5dB, 0.5dB] and [−20°, 20°], respectively. In figure 3(a), it can be seen that the performances of BMUSIC-ST and MUSIC-ST almost coincide, and the two algorithms outperform the other comparison algorithms. As the SNR increases, the RMSEs of BMUSIC-ST and MUSIC-ST approaches the associated CRB. In figure 3(b), we can observe that the performances of all the algorithms are degraded by the presence of sensor errors. However, BMUSIC-ST and MUSIC-ST still exhibit better accuracy and robustness than other algorithms under the influence of sensor errors.

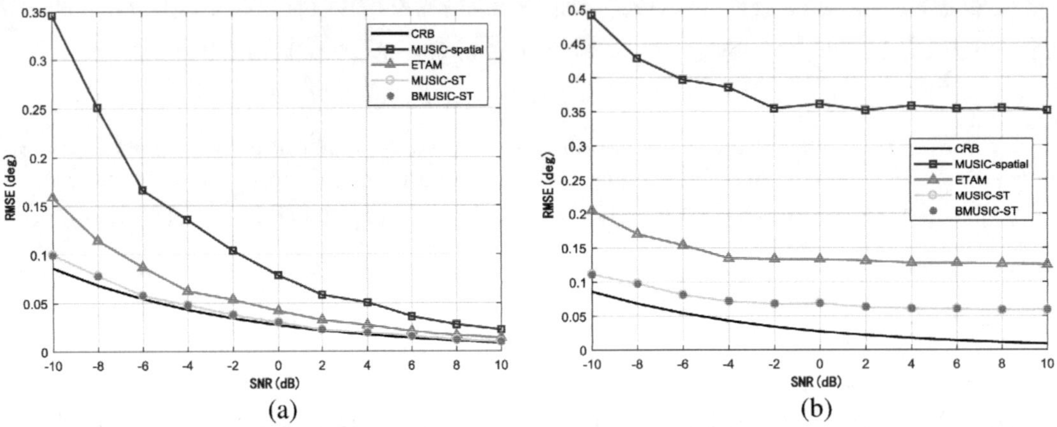

Figure 3. RMSEs and CRB versus SNR: (a) without sensor errors, (b) with sensor errors

Next, we investigate the impact of temporal manifold vector dimension N on the performance of proposed algorithm. Consider a 9 sensors uniform linear array, which moves at 200m/s. In figure 4, the DOAs of two narrow-band signals are set to be 20° and 23°, respectively. For each SNR, the probabilities are calculated by 300 Monte Carlo runs for 1200 snapshots. Other conditions stay unchanged. We can find that the resolution is improved as N increases. In figure 5, we simulate a signal impinging from 15°, and other conditions are the same as those in figure 4. It can be seen that the RMSE reduces as N increases. From figure 4 and 5, we can conclude that the estimation accuracy and resolution can be enhanced when temporal manifold vector dimension increases.

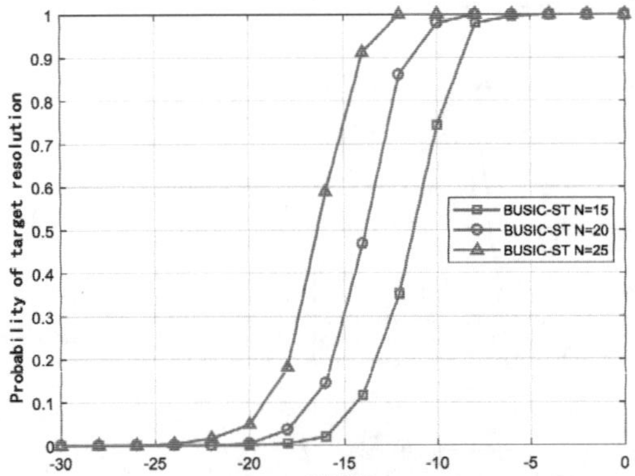

Figure 4. Resolution probabilities versus SNR for different temporal manifold vector dimensions

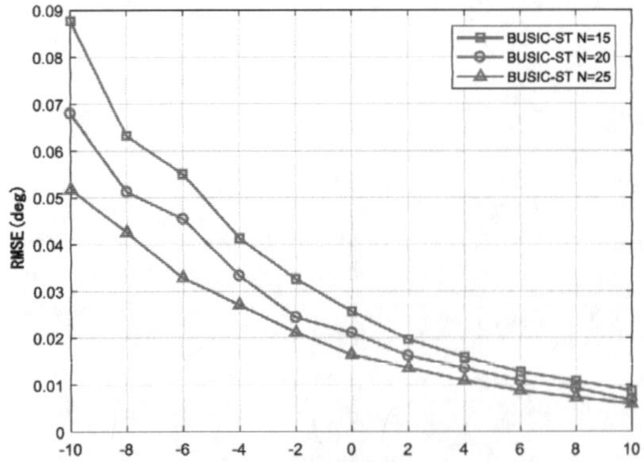

Figure 5. RMSEs versus SNR for different temporal manifold vector dimensions

IOP Conf. Series: Journal of Physics: Conf. Series **1284** (2019) 012051 doi:10.1088/1742-6596/1284/1/012051

5. Conclusions

In this paper, we propose a beamspace MUSIC algorithm based on moving array. First, a space-time signal model is established, which combines spatial information with temporal information and makes it possible to acquire acceptable DOA estimation with limited array aperture. Then the beamspace MUSIC algorithm is proposed for fast and high-accuracy DOA estimation. Numerical results demonstrated that the algorithms based on space-time model outperform other contrastive algorithms. Besides, beamspace MUSIC algorithm shows similar accuracy and superior resolution compared with original MUSIC algorithm.

Acknowledgments

Authors wishing to acknowledge the National Natural Science Foundation of China (Grant No. 61801514) and the China Postdoctoral Science Foundation (Grant No. 2016M592989) for their assistance.

References

[1] Zhang C, Pang K and Ma L 2015 Interpolated airborne MIMO antenna array *IEEE Antennas Wirel. Propag. Lett.* **14** 72-75
[2] Gu J F, Zhu W P and Swamy M N S 2015 Joint 2-D DOA estimation via sparse L-shaped array *IEEE Trans. Signal Process.* **63** 1171-1182
[3] Han S N, Dong-Sheng L I and Yong A X et al. 2015 DOA estimation method by sparse representation of fourth-order cumulants *J. Signal Process.* **31** 314-318
[4] You G, Jiang B and Qin H et al 2018 A novel cyclic correntropy MUSIC algorithm of cyclostationary signal based on UCA in impulsive noise *Proc. IEEE Chinese Automation Congress* (Jinan: China)
[5] Xu L Y and Wen F Q 2017 Fast noncircular 2D-DOA estimation for rectangular planar array *Sensors* **17** 840
[6] Oudompheng B, Nicolas B and Lamotte L 2015 Passive synthetic aperture array to improve noise mapping of a moving ship *Proc. IEEE Oceans* (Genoa: Italy)
[7] Kendra J R 2017 Motion-Extended Array Synthesis-Part I: Theory and Method *IEEE Trans. Geosci. Remote Sens.* **99** 1-17
[8] Stergiopoulos S and Urban H 2002 A new passive synthetic aperture technique for towed arrays *IEEE J. Oceanic Eng.* **17** 16-25
[9] Claudio E D D, Parisi R and Jacovitti G 2017 Space time MUSIC: Consistent signal subspace estimation for wide-band sensor arrays *IEEE Trans. Signal Process.* **66** 2685-2699
[10] Jia Y K and Bao Z 1997 The Cramer-Rao bound of DOA estimation for signals with temporal-spatial 2-dimension model *Acta Electronica Sin.* **25** 15–19
[11] Zeng W J, Li X L and So H C 2013 Direction-of-arrival estimation based on spatial–temporal statistics without knowing the source number *Signal Process.* **93** 3479–3486
[12] Vasylyshyn V I 2017 DOA estimation using beamspace root-music based estimator bank *Proc. Int. Conf. on Antenna Theory & Techniques* (Kiev: Ukraine)
[13] Wang Y L, Chen H, Peng Y N and Wan, Q 2004 *Theory and Algorithms of Spatial Spectrum Estimation* (Beijing: Tsinghua University Press) pp 117–118

Multi-modulation Recognition Using Convolution Gated Recurrent Unit Networks

TAN Jiyuan[1], ZHANG Limin[2], ZHONG Zhaogen[3] and YU Wenlong[4]

[1,2,3]Naval Aviation University , Yantai, Shandong 264001, China
[4]Naval Air Force 91213

E-mail address:173160056@qq.com

Abstract. To solve the problems of long modulation and recognition time and complexity of convolutional neural networks, a modulation method based on CNN-GRU (Convolutional Neural Networks and Gated Recurrent Unit) is proposed. Firstly, the spatial features of the signal are extracted by CNN convolution operation, then the timing correlation of the signal is extracted by GRU and finally the recognition probability is output by using the softmax layer to achieve the purpose of multi-modulation recognition. The experimental results show that the recognition performance of the method is further improved under the condition of no prior information such as channel and noise, and 11 modulation categories such as 16QAM and 64QAM can be effectively identified, and the complexity of the method is low, which greatly saves. The training identification time has good engineering application value.

1. Introduction

Signal modulation recognition technology has always been a research hotspot in the field of wireless communication. It is widely used in military and civilian fields, such as electronic countermeasures, signal monitoring and information acquisition. It is the basis of cognitive radio and perceptual spectrum [1][2][3]. In recent years, modulation recognition technology has been greatly developed. Domestic and foreign scholars have done a lot of research on the identification of different modulation modes of signals and the debugging and identification of signals under different channel and noise conditions, and have obtained some practical results, including the largest. Likelihood method [4] [5], high-order cumulant method [5] [6], cyclic spectrum algorithm [7] and KS detection algorithm [1] [8].

With the development of communication technology, the types of modulation methods of communication signals are also diverse. How to effectively distinguish multiple modulation methods under different signal-to-noise ratios and different channel conditions is still a difficult problem in research. At present, the literature in this aspect is relatively more difficult. less. N.Ghani proposed to classify the difference between the square spectrum peaks of the equal signals and the power spectrum [9]. This method can identify 10 modulation modes, but only for the AWGN channel. In 2007, K.Maliatsos et al. first introduced wavelet transform [10], firstly reduced the influence of noise by wavelet transform, and then extracted features for modulation recognition, which realized the recognition of multiple modulation methods of PSK, FSK, ASK and QAM. At the same time, the computational complexity is reduced, but the performance of the algorithm when the signal-to-noise ratio is lower than 10 dB is not analyzed. In recent years, deep learning has gradually been applied to the field of modulation recognition. Compared with traditional machine learning, deep learning does not need to know information about modulated signals, and has the advantage of automatically learning features directly from original data. In 2016, Timothy J O'Shea et al. proposed the use of convolutional neural networks to classify 11 modulated signals [11]. The results show that this method is superior to traditional identification methods, and has a variety of recognition types and accuracy. Improve, but the recognition rate of QPSK is lower, and the algorithm takes longer to operate. In 2017, et al. proposed a Hierarchical Deep Neural Networks (H-DNN)-based recognition method [12], which solved the problem of low recognition rate of WBFM in the literature [11], but produced new ones. Problem, the recognition rate between 4-QAM and 16-QAM has been reduced. In 2018, Rundong Li et al. proposed robust automatic VHF based on deep convolutional neural network for modulation recognition [13]. This method improves the anti-noise performance and the robustness of the algorithm, but the recognition type is reduced and the recognition is common. Seven modulation methods have been added, and the network complexity has increased. For the first time in literature [14], convolutional long-term and short-term deep neural networks combined with CNNs and LSTM are applied to the field of modulation identification. Compared with CNNs identification network, CLDNN has improved recognition performance due to the increase of network depth.

In this paper, the convolutional neural network is combined with the gated loop unit to propose a signal multi-modulation method for the convolution-gated loop unit neural network. For the first time, this method combines CNN and GRU in the field of modulation identification, and reduces the number of convolutional layers. It is constructed into a simpler neural network to achieve the purpose of modulation recognition. The CNN-GRU method proposed in this paper has the following advantages: (1) the number of network layers and training parameters are reduced, and the recognition time is greatly reduced; (2) the prior identification information such as channel and noise is not needed, and the modulation recognition rate is improved.

2. network description

2.1. Convolutional Neural Network

CNNs is a feedforward neural network. Convolutional neural networks have the characteristics of translation invariance in image processing, so they have always been one of the core algorithms in the field of image recognition. Standard CNNs generally consist of an input layer, a convolutional layer, a pooled layer, a fully connected layer, and an output layer [15][16], as shown in Figure 1.

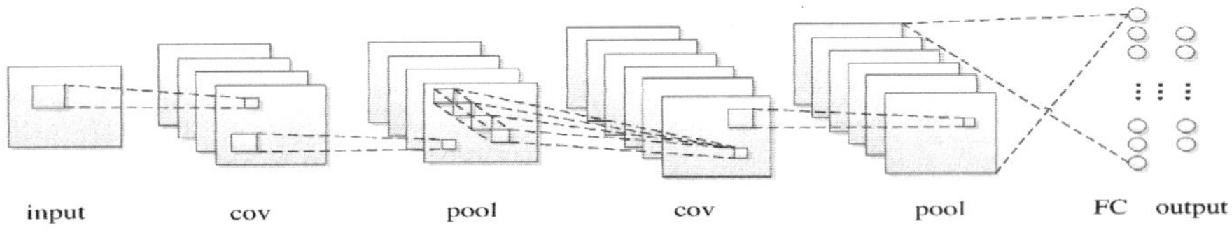

Figure 1. Standard Convolutional Neural Network

2.2. Gated Recurrent Unit

GRU is a special kind of Recurrent Neural Networks (RNN), which can use the previous input of prediction information together with several inputs later, so that the prediction will be more accurate. The network structure of the GRU is shown in Figure 2 below.

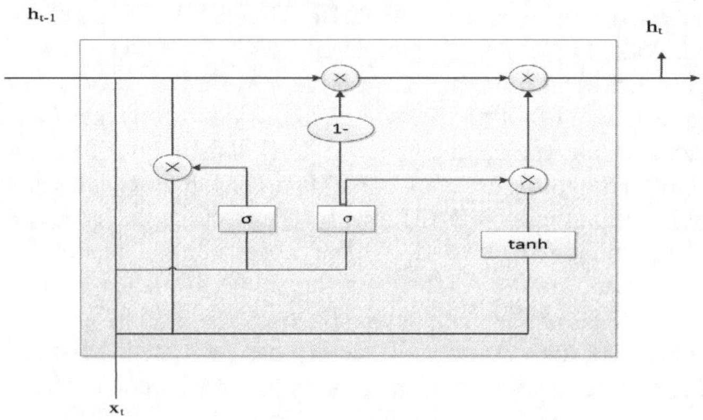

Figure 2. GRU network structure

2.2.1. Forward propagation of GRU

$$\begin{aligned} r_t &= s(w_r \cdot [h_{t-1}, x_t]) \\ z_t &= \sigma(w_z \cdot [h_{t-1}, x_t]) \\ \tilde{h}_t &= \tanh(w_{\tilde{h}} \cdot [r_t * h_{t-1}, x_t]) \\ h_t &= (1 - z_t) * h_{t-1} + z_t * \tilde{h}_t \\ y_t &= \sigma(w_o \cdot h_t) \end{aligned} \tag{1}$$

Where [] indicates that two vectors are connected, and * indicates the product of the matrix.

2.2.2. Training process of GRU

From the formula in the forward propagation process, it can be seen that the parameters to be learned are w_r, w_z, w_h, w_o. The first three parameters are spliced (because the latter vectors are also spliced), so you need to split them during the training:

$$\begin{aligned} w_r &= w_{rx} + w_{rh} \\ w_z &= w_{zx} + w_{zh} \\ w_{\tilde{h}} &= w_{\tilde{h}x} + w_{\tilde{h}h} \end{aligned} \tag{2}$$

The input of the output layer is $y_t^i = w_o h$, and the output of the output layer is $y_t^o = \sigma(y_t^i)$ .After the final output is obtained, the loss of network transmission can be written. The loss of a single sample at a certain moment is $E_t = \frac{1}{2}(y_d - y_t^o)^2$, and the loss of a single sample at all times is $E = \sum_{t=1}^{T} E_t$. The backward error propagation algorithm is used to learn the network, and the loss function is first obtained. Partial guidance for each parameter:

$$\begin{aligned} &\frac{\partial E}{\partial W_o} = \delta_{y,t} h_t, \quad \frac{\partial E}{\partial W_{z,x}} = \delta_{z,t} x_t, \quad \frac{\partial E}{\partial W_{z,h}} = \delta_{z,t} h_{t-1}, \quad \frac{\partial E}{\partial W_{\tilde{h}x}} = \delta_t x_t \\ &\frac{\partial E}{\partial W_{\tilde{h}h}} = \delta_t (r_t \cdot h_{t-1}), \quad \frac{\partial E}{\partial W_{rx}} = \delta_{r,t} x_t, \quad \frac{\partial E}{\partial W_{rh}} = \delta_{r,t} h_{t-1} \end{aligned} \tag{3}$$

Wherein, each intermediate parameter is $\delta_{y,t} = (y_d - y_t^o) \cdot \sigma'$

$$
\begin{aligned}
&\delta_{h,t} = \delta_{y,t} W_o + \delta_{z,t+1} W_{zh} + \delta_{t+1} W_{\tilde{h}h} \cdot r_{t+1} + \delta_{h,t+1} W_{rh} + \delta_{h,t+1} \cdot (1 - z_{t+1}) \\
&\delta_{z,t} = \delta_{t,h} \cdot (\tilde{h}_t - h_{t-1}) \cdot \sigma' \\
&\delta_t = \delta_{h,t} \cdot z_t \cdot \phi' \\
&\delta_{r,t} = h_{t-1} \cdot [(\delta_{h,t} \cdot z_t \cdot \phi') W_{\tilde{h}h}] \cdot \sigma'
\end{aligned} \tag{4}
$$

After calculating the partial derivatives of each parameter, the parameters can be updated, and the loss convergence is obtained by sequentially iterating.

3. Experimental data and CNN-GRU network

3.1. experimental database

The database of modulation recognition is derived from the RML2016.04C data sets published by Timothy J O'Shea[17]. The data set is composed of I/Q two-way data and is one that includes each sampled data. In-phase and quadrature components, the dimension is 2×128 . The data sets are composed of 220,000 modulated data samples, including 11 modulation categories, including AM-DSB, AM-SSB, WBFM three analog modulations and BPSK, 8PSK, CPFSK Eight digital modulations such as GFSK, 4QAM, 16QAM, 64QAM and QPSK. These data are approximately evenly distributed at 2dB intervals from SNR = -20dB to SNR = 18dB. These modulated data samples are obtained using real speech and text signals under channel conditions such as fading, multipath and additive white Gaussian noise.

3.2. Network structure

The CNN-GRU network used in this paper consists of two convolutional layers, one GRU layer and one fully connected layer. The two convolutional layers use the modified linear unit ReLU as the activation function, and the full connection layer uses the soft maximum function SoftMax as the activation function to classify the 11 kinds of modulation data, while the convolutional layer and the GRU layer use the missing output Dropout to prevent over-fitting. phenomenon. At the same time, Adam is selected as the gradient descent optimization algorithm. The specific network structure is shown in Figure 3.

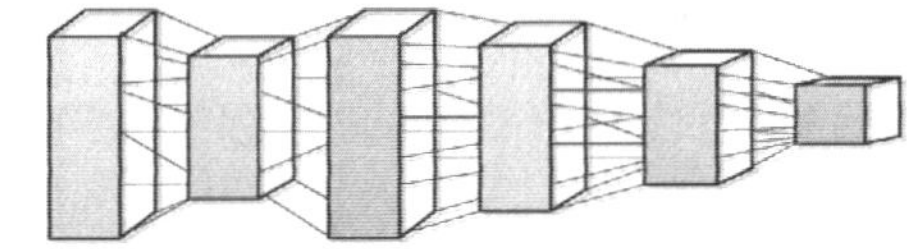

Figure 3. CNN-GRU network structure

The input layer is the modulation data 2×128 of the I/Q two-way dimension. The size of the first layer convolutional layer convolution kernel is 1×3, and the number is 128. The size of the second convolutional layer convolution kernel is 2×3, and the number is 32. The number of neurons in the GRU layer is 128, and the number of neurons in the last layer of the fully connected layer is 11, corresponding to the 11 modulation type.

3.3. Network implementation

First, the input data is processed, and the Reshape function in Keras is used to convert the 2×128 dimension modulation data into (None, 1, 2, 128) 4D tensors, which in turn represent (sample number, image height, image width, image channel number). As an input to the convolution layer. After convolution of the two convolutional layers, the output is (None, 128, 1, 124) and the dimension is the 4D tensor of 32×124 . Since the input requirement of the GRU layer is 3D tensor, the Reshape function is used again to convert the 4D tensor to (None, 128, 124), 32 as the dimension of the GRU network input sequence, and 124 as the length of the GRU network input sequence to the GRU network. After the two-way operation of equations (1), (2), and (3), a 2D tensor (None, 128) is output. It is then passed to the last fully connected layer.

3.3.1. Learning of convolutional layers. In the convolutional layer, it is assumed that the input of the convolutional layer is $X \in R^{A \times B}$, where A represents the number of features of the input signal and B represents the number of input signals. Let $x = [x_1\ x_2 \cdots x_B]$, where $\boldsymbol{x}_b$ represents the eigenvector of signal b. The activation of the convolutional layer can be calculated as:

$$h_{j,k} = \theta(\sum_{b=1}^{s} w_{b,j}^{T} x_{b+k-1} + a_j) \tag{5}$$

Where $h_{j,k}$ represents the convolutional network output of the jth feature map, $w_{b,j}^{T}$ is the weight parameter of the jth convolver, s is the size of the convolver, and a_j is the offset of the jth feature map. After the input and convolver are weighted averaged, the output node value of the network is calculated by the nonlinear activation function θ. Through the formula (6), the convolution layer realizes the convolution of the feature map, and the corresponding output features are obtained. The nonlinear activation function θ used in this paper is ReLU, and the specific expression is shown in formula (7).

$$\mathrm{ReLU}(x) = \max(0, x) \tag{6}$$

The number of input channels of the convolutional layer is determined by the number of channels that input 4D tensors. The number of channels for the output tensor is determined by the number of output channels of the convolutional layer. The height and width of the output tensor are calculated as follows.

$$\begin{aligned} height_{out} &= (height_{in} - height_{kernel} + 2 * padding) / stride + 1 \\ width_{out} &= (width_{in} - width_{kernel} + 2 * padding) / stride + 1 \end{aligned} \tag{7}$$

Among them, $height_{out}$ represents the height of the convolutional layer output tensor, $height_{in}$ represents the width of the convolutional layer input tensor, $height_{kernel}$ represents the height of the convolution kernel, $padding$ represents the scanning mode, and the selection in this paper is valid. $stride$ represents the number of convolution kernels, $width_{out}$ represents the width of the output tensor, and $width_{in}, width_{kernel}$ represents the width of the convolution layer input tensor and the width of the convolution kernel, respectively.

3.3.2. softmax layer. The full connection layer uses the softmax function, which can be used in the multi-classification process to map the inputs of multiple neurons to (0, 1), and each mapping can be viewed as an output probability for multi-classification. Assuming that the output of the previous layer is $\{z_1, z_2, \cdots, z_n\}$, the softmax layer operation formula is as follows.

$$k_i = \frac{e^{z_i}}{\sum_{1}^{n} e^{z_i}} \tag{8}$$

Where k_i represents the probability that z_i represents the predicted result. Converting the output probability into an image display is a confusion matrix, which can intuitively reflect the recognition of each modulation mode.

4. Experimental results and analysis

4.1. Experimental environment

The experiment runs under Windows 7 system, the CPU is Intel(R) Core(TM) i5-4210U, without GPU participation, the memory is 12G. The deep learning library built by the network uses Keras, and the back end selects TensorFlow. The network training is done 100 times. The EarlyStopping function is used to set the training loss function of the network to the monitoring value. When the monitoring value exceeds 5 times, the training will stop and the training result will be output. 50% of the data under the respective signal-to-noise ratios of the data sets are used as training data, and the remaining 50% of the modulated data is used as the detection data.

4.2. Recognition accurary comparison

It can be seen from the comparison of the recognition rates from figure 4 that the recognition rate of the network proposed in this paper is higher than that of the CNN network. It can be seen from the identification confusion matrix of figure 5, figure 6, figure 7 and figure 8 that the improvement of the network recognition rate proposed in this paper is mainly due to the improvement of the recognition rate of 8PSK.

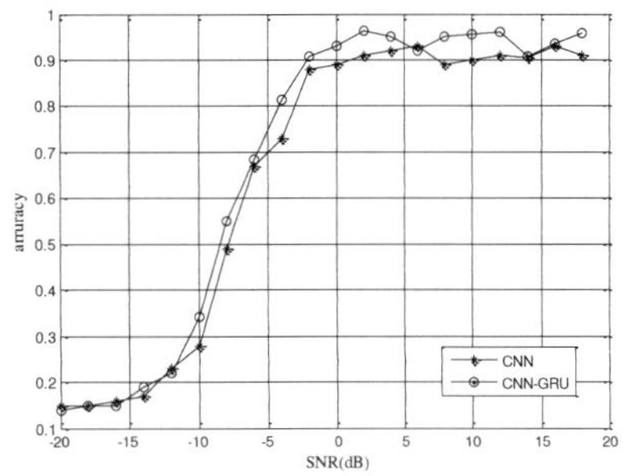

Figure 4. Comparison of CNN-GRU and CNN2 recognition rates under RML2016.04C

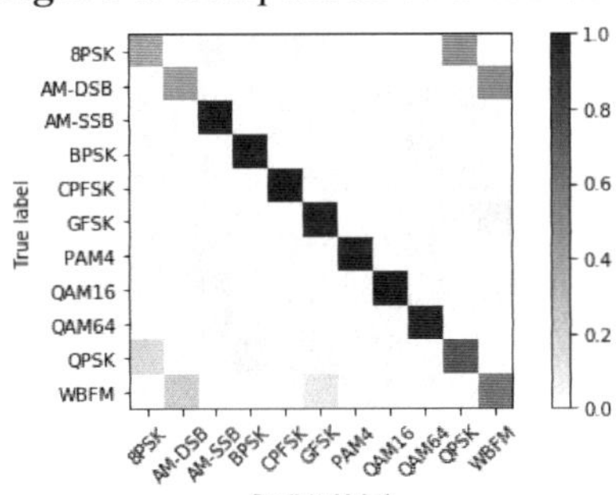

Figure 5. CNN confusion matrix when SNR=0dB

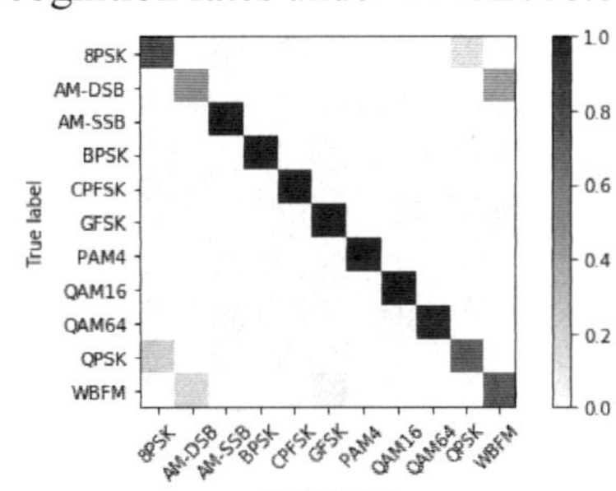

Figure 6. CNN-GRU confusion matrix when SNR=0dB

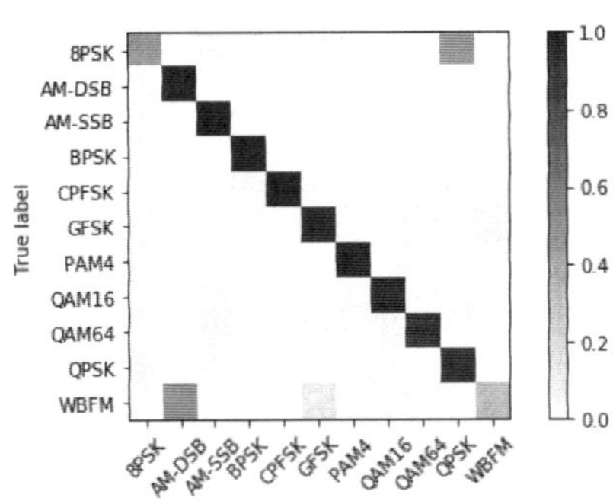

Figure 7. CNN confusion matrix when SNR=18dB

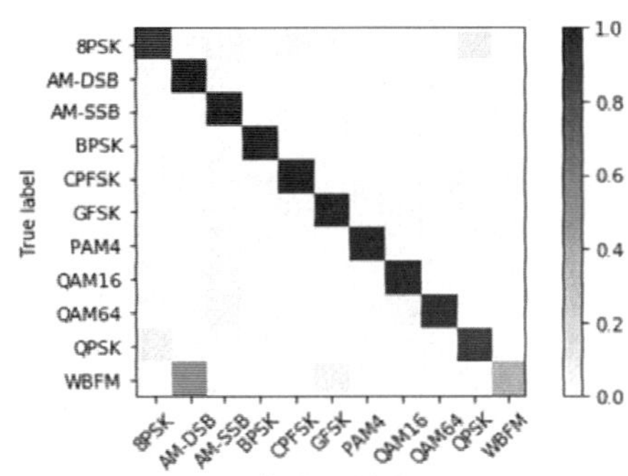

Figure 8. CNN-GRU confusion matrix when SNR=18dB

5. Conclusion

Based on the published data sets of RML2016.04C, this paper studies the identification of 11 modulation methods. Based on convolutional neural network and bidirectional long-term and short-term memory neural network, the proposed method establishes the CNN-GRU modulation recognition network, making full use of CNN's advantages in image processing and GRU's time-series correlation, and achieves 11 different modulation modes. the goal of. The experimental results show that it greatly reduces the network complexity and the modulation recognition performance has been improved. The next step will be to study how to effectively improve the recognition performance of the algorithm when the signal-to-noise ratio is lower than -14dB.

Reference

[1] Wang F, Dobre O A, Chan C, et al. Fold-based Kolmogorov–Smirnov Modulation Classifier *J. IEEE Signal Processing Letters*, 2016,23(7):1003-1007.

[2] Marey M and Dobre O A. Blind Modulation Classification Algorithm for Single and Multiple-Antenna Systems Over Frequency-Selective Channels *J. IEEE Signal Processing Letters*, 2014, 21(9):1098-1102.

[3] Dobre O. Signal identification for emerging intelligent radios: classical problems and new challenges *J. Instrumentation & Measureme-nt Magazine IEEE*, 2015, 18(2):11-18.

[4] Hameed F, Dobre O A and Popescu D C. On the likelihood-based approach to modulation classification *J. IEEE Transactions on Wireless Communications*, 2009, 8(12):5884-5892.

[5] Abu-Romoh M, Aboutaleb A and Rezki Z. Automatic Modulation Classifcation Using Moments And Likelihood Maximization *J. IEEE Communications Letters*, 2018:1-1.

[6] Abdelbar M, Tranter W H and Bose T. Cooperative Cumulants-Based Modulation Classification in Distributed Networks *J.* 2018, PP(99):1-1.

[7] Wu T and Di Wa Huang G. MPSK Signal Modulation Recognition Algorithm Based on Joint Frequency Offset Estimation and Recurring Moment *J. Instrument Technology and Sensors*, 2014, 117 (12): 102-104.

[8] Wu H C, Saquib M and Yun Z. Novel Automatic Modulation Classification Using Cumulant Features for Communications via Multipath Channels *J. IEEE Transactions on Wireless Communications*, 2008, 7(8):3098-3105.

[9] Ghani N and Lamontagne R. Neural networks applied to the classification of spectral features for automatic modulation recognition *C// Military Communications Conference*, 1993. MILCOM '93. Conference record. Communications on the Move. IEEE. IEEE, 1993:111-115 vol.1.

[10] Zhou H. Research on blind identification of digital modulated signals under low SNR *D. University of Electronic Science and Technology of China*, 2010.

[11] O'Shea T J, Corgan J and Clancy T C. Convolutional Radio Modulation Recognition Networks *C// International Conference on Engineering Applications of Neural Networks*. Springer International Publishing, 2016:213-226.

[12] Karra K, Kuzdeba S and Petersen J. Modulation recognition using hierarchical deep neural networks *C// IEEE International Symposium on Dynamic Spectrum Access Networks*. IEEE, 2017:1-3.

[13] Li R, Li L, Yang S, et al. Robust Automated VHF Modulation Recognition Based on Deep Convolutional Neural Networks *J. IEEE Communications Letters*, 2018, PP(99):1-1.

[14] Liu X , Yang D and Gamal A E . Deep Neural Network Architectures for Modulation Classification *J.* 2017.

[15] Li Y, Zhang T, Shan C, et al. Deep learning Convolutional neural networks from entry to proficiency *M.* Mechanical Industry Press, 2018: 4-5.

[16] Tensorflowai. Deep learning using Keras: (vi) LSTM and two-way LSTM explanation and practice [OL]. https://blog.csdn.net/fendouaini/article/details/80198994, 2018.12.

[17] T. J. O'Shea, https://radioml.com/datasets/radioml-2016-04-dataset/.

Research on Routing Algorithm Based on Reinforcement Learning in SDN

Chuntao Fang, Cheng Cheng, Zhongyun Tang and Chuanhuang Li

School of Information and Electronic Engineering, Zhejiang Gongshang University, Zhejiang Hangzhou 310000, China

E-mail: 1147197814@qq.com

Abstract. Reinforcement learning is an unsupervised learning method which has been used in many fields. Actually, the essence of Reinforcement learning is a decision-making problem. It constantly tries to interact with the environment. Each interaction process will get a different feedback value, and then it adjusts each trial strategy through feedback. In this paper, we apply the Reinforcement learning technology to software defined network routing algorithm, and propose the routing algorithm based on Q-learning. Through the combination of Reinforcement learning and neural network, which means the Q-table in Q-learning is replaced by neural network, we present routing algorithm based on Deep Q-learning.

1. Introduction

In recent years, mobile Internet and Internet of Things are rapidly developing. New applications and fast-growing data traffic require better quality of service (QoS). Software defined network (SDN)[1][2][3] separates the infrastructure layer and the control layer of the network device to realize the intelligent management of network traffic. Usually, SDN controller [4][5][6] forwarding the packet bases on Dijkstra algorithm. If forwarding of all packets only relies on the shortest path algorithm, it may cause a serious problem that all data streams are brought together by selecting the same forwarding path, which greatly reduces the utilization of network link and leads to network congestion. Therefore, we need a better strategy to manage the routing of the SDN network [7]. At present, there is no fixed routing algorithm applied in SDN network routing. Designing a routing algorithm under the SDN network and finding a good forwarding path for the data packet is a hot spot in the current SDN network research.

Hoanghuu V [8] proposed a dynamic routing algorithm based on the maximum correlation pair model to find a path from the starting point to the ending point in the unknown environment. Compared with the traditional dynamic routing algorithm, the experimental results show that the algorithm has higher search efficiency. Jie Sun et al. [9] proposed the SAS (Scheduling According to Stickness) algorithm. The SAS algorithm implements the QoS requirements of the link based on the link sticking value. Generally, the data stream with smaller link sticking value will enjoy the right to forward preferentially. The importance of the data flow is measured by the link sticking value. So the data stream with the higher priority is forwarded first, and the data stream with the lower priority is forwarded later. Huaidong Zhang et al. [10] modeled the path selection of elephant flow based on multi-commodity flow problems. Compared with the ECMP (Equal Cost Multi Path) [11] algorithm, network bandwidth can be fully utilized to improve network utilization.

Mentioned above, SDN network does not consider more about the requirements of link's quality of service. Therefore, ensuring service deployment and optimal utilization of network links in the SDN network architecture, using certain QoS requirements as the criterion for judging the path pros and cons, reinforcement learning is applied to SDN network routing in this paper. Reinforcement learning can continuously explore the surrounding environment through agents without prior knowledge, familiar with the entire environment after a number of training cycles, and make the right choice finally. Different from the Dijkstra algorithm [12] using in traditional network routing, we propose and implement a routing algorithm based on Q-learning. Also, because of the feature that routing algorithm Q-learning needs the defects of artificial design, using the neural network to replace the Q-table in the traditional network, the routing algorithm based on Deep Q-learning is proposed.

2. Model research

2.1. Research on routing algorithm based on Q-learning

Q-learning as a no-model advanced version of reinforcement learning, relying on Q-value iteration to achieve the balance with exploration and utilization. During the iterative process of the Q-learning algorithm, people generally select the direction in which Q-value is the largest. However, not every iteration will follow this principle. Under the results of multiple experiments, Q-learning algorithm will use a certain probability of which the distribution will be carried out in the direction with the largest possible that Q-value is the largest. There are several common terms in Q-learning, states, reward, and actions. The goal of Q-learning is to achieve the maximum reward value by Q-value iteration.

When designing the routing algorithm, a numerical iterative solution method is used to approximate the optimal value. The complex network environment in which new packets are located is a finite state Markov process. When a new packet selects an action from the action set, the network environment accepts the action and transitions state, giving the decision a reward value at the same time. The greater the reward value that a packet gets from an action, then the probability that the packet will select this action again will increase. At a certain time t, the state of the packet is s_t, the data packet selects an action a_t, the network environment state is transferred from s_t to s_{t+1}, and the reward value given by the environment system is r_t, then the routing decision process in the environment is shown in equation (1).

$$r_t = r(s_t, a_t, s_{t+1}) \quad (1)$$

In the learning process, the optimal value function is approximated by iteratively calculating $Q(s, a)$. The basic transfer rules of Q-learning algorithm are shown in equation (2) and equation (3). s, a respectively represent the current state and action of the packet, s' and a' respectively represent the next state and action of the current state, and the learning parameter γ is a constant satisfying $0 > \gamma > 1$.

If the learning parameter tends to 0, indicating that packet routing decision cares more about the reward value of the current action. If the learning parameter tends to 1, packet routing decision will take into account the reward value of the future action.

$$Q(s,a) = R(s,a) + \gamma \max_{a}\{Q(\tilde{s},\tilde{a})\} \quad (2)$$

$$Q(s,a) = (1-\lambda_t)Q(s,a) + \lambda_t\{r + \gamma \max_{a\in A} Q(s',a')\} \quad (3)$$

Without supervision, the packet only study relying on the current experience, and it constantly moves from one state to another to explore until it reaches the target state. Each exploration of a packet is called an episode. When the packet arrives at the target state from the initial state, it represents the end of an episode.

The basic steps based on the Q-learning routing algorithm can be summarized as follows:

(1) Give the learning parameter γ and the bonus value matrix R.

(2) When the Q-matrix is initialized, set the value of all elements in the matrix to 0.
(3) For each episode:
(3.1) Select a state *s* randomly from state set as an initial state.
(3.2) Before reaching the target state, the following steps will be performed:
(3.2.1) Select an action *a* among all possible actions of the current state *s*.
(3.2.2) Obtain the next state *s'* with the selected action *a*.
(3.2.3) Calculate *Q(s, a)* according to the algorithm transfer rule mentioned above.
(3.2.4) Let $s = s'$.

Once the Q-matrix training is completed, the newly transmitted data packet can easily find a forwarding path from the arbitrary state s_0 to the target state from the trained Q-matrix. The specific process is as follows:
(1) Let the current state $s = s_0$.
(2) Determine the action *a* in the current state, which satisfies $Q(s, a) = max \{Q(s, a')\}$.
(3) Let the current state $s = s'$ (s' denotes the next state corresponding to *a*).
(4) Repeat (2) and (3) until *s* becomes the target state.

The iterative process for every route attempt using of packet which is based on the Q-learning routing algorithm is shown in figure 1. S_0 represents the position of the start state of the packet in one iteration process, and Sn represents the n state of one iteration. The direction indicated by the arrow is the direction in which the packet state transitions. It can be seen that based on the Q-learning routing algorithm, each state transition of the packet occurs between two adjacent states, so such a Q-learning algorithm is also called a single-step Q-learning algorithm. The single-step Q-learning algorithm takes up a relatively large amount of memory and at the same time converges slowly.

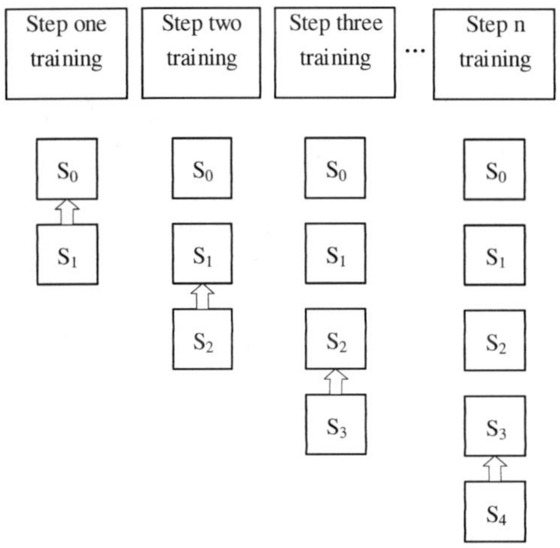

Figure 1. Iterative process based on Q-learning routing algorithm

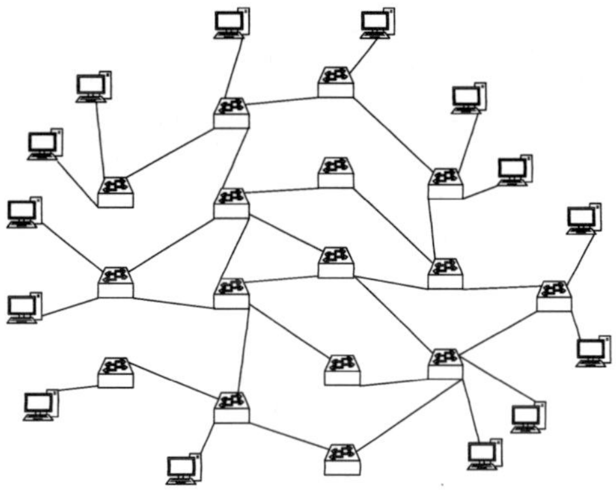

Figure 2. Algorithm test simulation experiment environment model diagram

2.2. Research on routing algorithm based on deep Q-learning

The Q-learning algorithm can learn from the surrounding environment, but it needs to manually design the corresponding features to train a convergent Q-matrix, while neural networks have good processing power for massive data. So replaced the Q-matrix in Q-learning with a neural network, it is the Deep Q-learning algorithm. A Deep Q-learning algorithm model is a combination of a multi-layer neural network model and a reinforcement learning model. Give a state *s*, output a vector of action values Q (s, *; θ), θ is the parameter of the network. And neural network as a function can map the vector from an n-dimensional space to an m-dimensional action space. The Deep Q-learning algorithm has two main points, one is the target network, and the other is experience replay. Equation (4) is the target network update equation.

IOP Conf. Series: Journal of Physics: Conf. Series **1284** (2019) 012053 doi:10.1088/1742-6596/1284/1/012053

$$Y_t^{DQN} \equiv R_{t+1} + \gamma \max_a Q(S_{t+1}, a; \theta_t^-) \tag{4}$$

For experience replay in reinforcement learning, the observed state transitions are first stored. After the samples have accumulated to a certain extent, then sampling randomly to update the network. Experience replay is a very important part of deep reinforcement learning, which can greatly improve the system performance of deep reinforcement learning.

There are two key elements based on the Deep Q-learning routing algorithm: one is the calculation of the loss function, and the other is the extraction of the training samples in the neural network training process.

Calculation of the loss function: The relationship between the target Q-value and the old Q-value in Q-learning coincides with the relationship between the output value and the result value in the neural network learning, so the loss function is calculated as shown in equation (5).

$$loss = (target\_q - q)^2 \tag{5}$$

Training sample extraction: Based on the concept of experience replay, the reason for randomly sampling is that a sample obtained after a packet exploring by the surrounding environment is a time-associated sequence with temporal correlation. Because of the time correlation, if the data is directly used to train and update the Q-table, the system convergence will be greatly affected. Therefore, we adopt random sampling to solve the problem of time correlation.

The complete process based on the Deep Q-learning routing algorithm is as follows:

(1) Initialize packet randomly as one state, initialize the memory pool, and set the observation value

(2) Loop traversal:

(2.1) Packet policy selects an action a.

(2.2) After select an action a. Obtain the corresponding bonus value r, the state s' after the action ends, and the value done which judges whether the entire search ends.

(2.3) Save the relevant parameters s, a, r, s', *done* in the first two steps to the memory pool.

(2.4) Determine the amount of data stored in the memory pool (whether the number of data in the memory pool exceeds the set observation value). If not, go to step (2.5). If the data is sufficient, perform the following steps:

(2.4.1) Take a part of the data randomly from the memory pool as samples for training.

(2.4.2) Take sample state s' which is taken randomly as training samples, and obtain the Q-table corresponding to state through training.

(2.4.3) Calculate *target_q* value table corresponding to Q-table according to equation (6).

$$Q(s,a) = r + \gamma^* MaxQ(s', all\_actians) \tag{6}$$

(2.4.4) Train the neural network by q and *target_q*.

(2.5) Determine if the search process is over.

(2.5.1) Find the end, set a state to s randomly.

(2.5.2) If the search is not completed, the current state s is updated to s'.

3. Experimental simulation design

3.1. Routing algorithm simulation environment

The tests of the two algorithm models are based on the network topology shown in figure 2. In this experiment, the link is graded by a certain QoS standard. The higher the level, the better the link QoS, and the stronger the trend of selecting the link. The ultimate goal of the experiment is to select a forwarding path with better QoS performance for the packet.

3.2. Routing Algorithm Status Description

The main points of reinforcement learning are to find out the state of the environmental system and the corresponding executable actions. According to the simulation experiment environment model diagram above, all the state sets under the adopted topology can be described as follows: S = [S1, S2, S3… S15, L*_1, L*_2, L*_3]. S1 ~ S15 respectively represents 15 switches in the experimental environment, * placeholder represents the links of any two switches. In order to fully represent all the possible link conditions in the entire network environment, each link performance is divided into three levels according to the QoS standard, _1, _2, _3. The link performance corresponding to each level is shown in Table 1. The higher the percentage of link QoS performance, the better the link status. Since there are two or more links on a switch, each link is divided into three different levels, so one link of a switch usually corresponds to an actual state. All states can be materialized according to three different levels of link state. Take switch S3 as an example. There are three links connected to it: L2, L5, and L8. So the three links correspond to three states, and each link can be divided into three states according to the link performance level. The three links of a switch actually correspond to nine specific states. As shown in table 2.

Table 1. Link performance corresponding to different link levels

Link level	Link QoS performance
L*_1	0%~30%
L*_2	31%~60%
L*_3	61%~100%

Table 2. All state sets of switch S3

S1=[3,L2_1]	S2=[3,L2_2]	S3=[3,L2_3]
S4=[3,L5_1]	S5=[3,L5_2]	S6=[3,L5_3]
S7=[3,L8_1]	S8=[3,L8_2]	S9=[3,L8_3]

S1 = [3, L2_1] represents the state s1, the switch in which it is located is S3, the link is L2, and the link level is _1. After all the states are determined, the optional set of actions in the experimental environment is determined. An optional action refers to moving from one state to another. In this paper, the optional state is defined as A = {s[i], s[t]}, which means to transfer from one state *i* to another state *t*.

3.3. Routing Algorithm Reward Function Definition

The purpose of the routing algorithm is to select a link with better comprehensive performance from among the many optional links. In this paper, the network QoS standard is taken as the main consideration. So the purpose of the routing algorithm is to select a path with better QoS synthesis. Since the ultimate goal of reinforcement learning algorithm we adopted is to select the link with the larger reward value as the better link after training, the setting of the bonus value R is particularly critical.

Since the performance of the link has been divided into three levels, the reward value can be designed based on the performance level of the link. Generally speaking, we will choose the link with higher link performance value as the better link, so according to the three levels of link performance, the set bonus value is as follows: {[link QoS performance: 0%~30 % Reward value R1 = 50], [Link QoS performance: 31% to 60% Reward value R2 = 100], [Link QoS performance: 61% to 100% Reward value R3 = 150]}.

3.4. Deep reinforcement learning algorithm parameter setting

Deep Q-learning uses neural network to replace the Q-table in Q-learning. In this paper, we design a simple three-layer neural network. After setting the parameter debugging and operating, if the loss function value continues to decrease, as shown in figure 3, confirm that the neural network can be normal operation. The loss function, which can also be called the error function, represents the closeness of the sample's predicted value to the actual value. Learning rate, the number of layers, and the number of neurons in neural network can be adjusted according to the experimental results. In the

IOP Conf. Series: Journal of Physics: Conf. Series **1284** (2019) 012053 doi:10.1088/1742-6596/1284/1/012053

experiment, after debugging the learning rate several times, find that setting the value to 0.01 is appropriate.

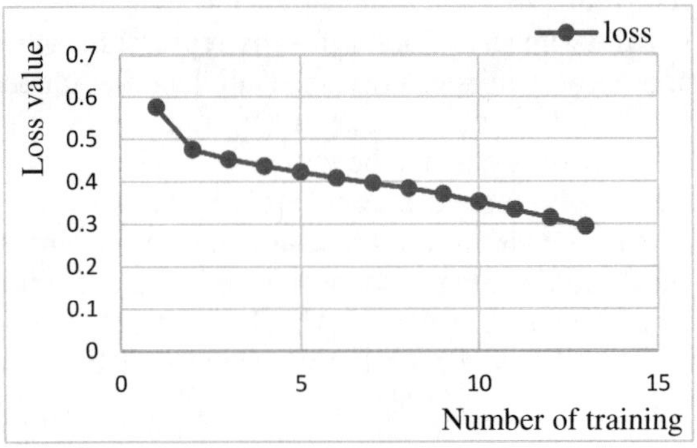

Figure 3. Neural network loss function value changes with training period

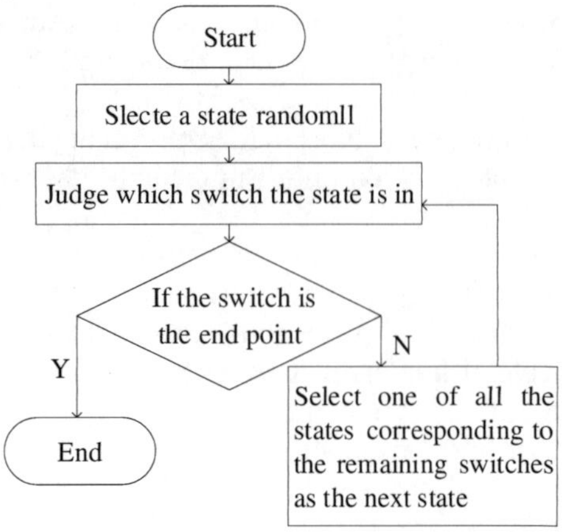

Figure 4. Routing loop problem solving strategy flow chart

3.5. Routing Loop Problem

In the above experimental scheme design, there may be a routing loop problem. And traditional RIP protocol has four main methods to solve the routing loop problem [13]: trigger update, horizontal split, toxicity reversal and suppression timing method. We adopt a kind of horizontal splitting method: if a packet passes through a switch node, set a variable Node = [S1, S2, S3...Sn] to record the switch that the packet has already experienced. The optional path of the next hop of the packet can only be selected from the remaining switch nodes, but cannot be returned to the node that has already arrived. The strategy flow chart for solving the loop problem is shown in figure 4.

4. Experimental results and analysis

4.1. Routing Algorithm Experimental Results

Table 3. Q-matrix based on Q-learning training

0	0	0	0	252	252	0	0	0	0	0	0	0	0	0
0	0	0	0	201	0	0	0	0	0	0	0	0	0	0
0	0	0	0	0	0	252	0	0	0	0	0	0	0	0
0	160	0	0	252	0	0	252	0	0	0	0	0	0	0
201	0	0	201	0	252	0	0	316	316	0	0	0	0	0
201	0	0	0	252	0	252	0	0	316	316	0	0	0	0
0	0	201	0	0	252	0	0	0	0	0	316	0	0	0
0	0	0	201	0	0	0	0	0	0	0	0	316	0	0
0	0	0	0	252	0	0	0	0	0	0	0	0	0	0
0	0	0	0	252	252	0	0	0	0	0	0	0	396	0
0	0	0	0	0	252	0	0	0	0	0	0	0	396	0
0	0	0	0	0	0	252	0	0	0	0	0	0	396	0

0	0	0	0	0	0	0	252	0	0	0	0	0	0	0
0	0	0	0	0	0	0	0	316	316	0	0	316	0	496
0	0	0	0	0	0	0	0	0	316	316	316	0	0	496
0	0	0	0	0	0	0	0	0	0	0	0	0	496	0

The final result of the Q-learning algorithm is to train a convergent Q-matrix, through which the link from a given starting point to the end point can be directly selected. The results of the Q-matrix solved by the Q-learning algorithm in this experiment are shown in Table 3. Through the trained Q-matrix, it is easy to get the full path results from any starting point to the ending point 15 as shown in Table 4. Based on the same topology design, the Deep Q-learning routing algorithm can also easily find the complete path from any starting point to the ending point 15. The partial search results after the model training is completed are shown in Table 5.

Table 4. Arbitrary start-to-end routing results

Start	Path	Start	Path
0	0->4->8->13->15	1	1->3->4->8->13->15
2	2->6->11->14->15	3	3->4->8->13->15
4	4->8->13->15	5	5->9->13->15
6	6->11->14->15	7	7->12->13->15
8	8->13->15	9	9->13->15
10	10->14->15	11	11->14->15
12	12->13->15	13	13->15
14	14->15		

4.2. Analysis of experimental results of routing algorithm

Based on the given network topology environment, the number of steps required to reach the end point in the training process based on the Q-learning routing algorithm is analysed. The results of partial cycle convergence steps are shown in table 6. It can be easily seen from figure 5 that based on the Q-learning-based routing algorithm, the convergence speed near the end state does not change the convergence speed of the current state.

Table 5. Partial routing results of deep Q-learning routing algorithm

Start	Path
0	0->4->8->13->15
1	1->3->4->8->13->15
2	2->6->11->14->15
3	3->4->8->13->15

Table 6. Q-learning convergence steps

Cycle	10	20	30	40	50	60	70
QL	104	73	26	9	7	8	6

In terms of algorithm accuracy, the Q-learning belongs to unsupervised learning. Each training period in the early stage is in the process of trying and training the Q-matrix, so the accuracy rate is very low, but as the number of training increases, the Q-matrix gradually converges. During the process, it can be found that the accuracy of training is rising rapidly as shown in Table 7. And it is easy to see from figure 6 that the accuracy of the Q-learning can reach a stable value faster. The environment used in this training process is not complicated, and the final accuracy of the algorithm can reach almost 100%.

Table 7. Q-learning accuracy rate

Cycle	10	20	30	40	50	60	70	80	90	100	110
QL	0.72	0.88	0.985	0.9957	0.9984	0.9999	0.9998	1	1	1	1

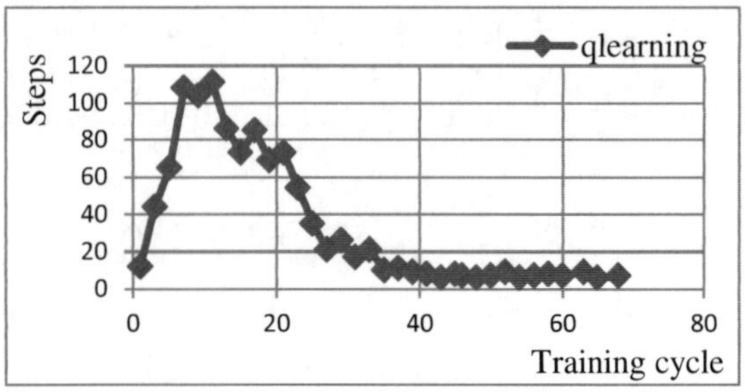

Figure 5. the average number of steps to the end point changes with the training period

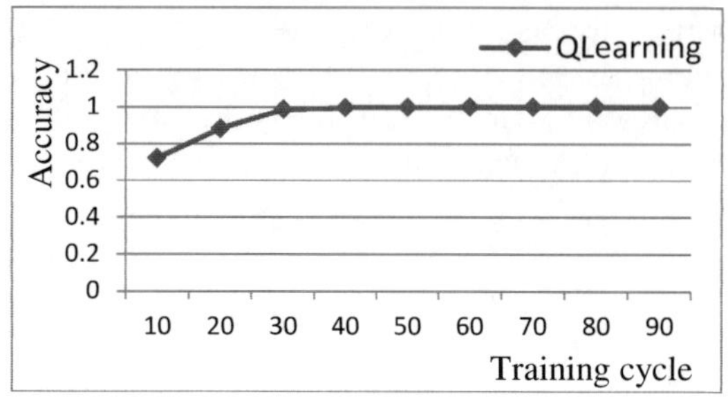

Figure 6. Accuracy rate as a function of training cycle

5. Conclusion

In this work, we mainly use reinforcement learning to SDN network routing, and design a routing algorithm based on Q-learning. In the design process of this algorithm, the network link QoS standard is mainly considered. Obviously, this algorithm can make full use of the network link. From the experimental test results, the algorithm can find the route with better QoS performance with nearly 100% accuracy after a certain period of training. There is a disadvantage in the Q-learning-based routing algorithm: the corresponding features need to be manually designed. However, in the actual application environment system, the features are often much complicated. It is very difficult to extract the design features manually, and the effect is not satisfactory. Therefore, we combine neural network with reinforcement learning, replace the Q-table with an approximate function trained by neural networks, and proposes a routing algorithm based on deep reinforcement learning. The experimental results show that the routing algorithm based on Deep Q-learning also shows good performance.

Acknowledgments

This work was supported in part by a grant from the National Key Research and Development Program of China (No.2017YFB0803202), the Natural Science Foundation of China(61871468), the Zhejiang Provincial Natural Science Foundation of China (LY18F010006), the Zhejiang's Key Project of Research and Development Plan (2017C03058), the Zhejiang Provincial Key Laboratory of New Network Standards and Technologies (No.2013E10012), Scientific Research Projects of Zhejiang Education Department (Y201737967).

References

[1] Foundation O N 2012 Software-defined networking *The New Norm for Networks*.
[2] Shunmiao Z and Fumin Z 2013 Survey on software defined network research *Application Research of Computers* **30** pp 2246-51.
[3] Hongyi Z, Peyman K, George V and Nick M 2014 Automatic test packet generation *IEEE/ACM Transactions on Networking* **22** pp 554-566.
[4] Dixit A, Hao F, Mukherjee S, Lakshman T V and Kompella R 2013 Towards an elastic distributed SDN controller *Acm Special Interest Group on Data Communication* **43** pp 7-12.

[5] Uhlig R, Neiger G and Rodgers D 2005 Intel virtualization technology *IEEE Computer* **38** pp 48-56.
[6] Bosshart P, Gibb G, Kim H S, Varghese G, McKeown N, Izzard M, et al 2013 Forwarding metamorphosis: fast programmable match-action processing in hardware for SDN *ACM SIGCOMM Computer Communication Review* **43** pp 99-110.
[7] Yingya G, Zhiliang W, Xia Y, Xingang S and Jianping W 2014 Traffic engineering in SDN/OSPF hybrid network *2014 IEEE 22nd International Conference on Network Protocols*, Raleigh NC USA pp 563-568.
[8] Hoanghuu V, Sanghyeok A and Taechoong C 2011 Extended dyna-q algorithm for path planning of mobile robots *Journal of Measurement Science and Instrumentation* **2** pp 283-287.
[9] Jie S, Li L and Subin S 2016 A routing strategy based on qos and dynamic load balancing *Computer Technology and Development* **26** pp 188-194.
[10] Huaidong Z and Qingwei D 2016 A method of data center network dynamic traffic scheduling based on SDN *Computer and Modernization* pp 80-86.
[11] Hopps C E 2000 Analysis of an equal-cost multi-path algorithm *Journal of Allergy & Clinical Immunology* **109** S265.
[12] Bauer R, Delling D, Sanders P, Schieferdecker D, Schultes D and Wagner D 2008 Combining hierarchical and goal-directed speed-up techniques for dijkstra's algorithm *WEA'08 Proceedings of the 7th international conference on Experimental Algorithms* pp 303-318.
[13] Fan Y 2012 RIP routing protocol analysis and configuration brief *Silicon Valley* pp 16-17.

Mobility Analysis of Mobile VLC with Optical Zoom Antenna

Shanshan Hu, Yuee Chang, Xingkui Wu and Yi Tang*

School of Optics and Photonics, Beijing Institute of Technology, Beijing 100081, China

*tangyi4510@bit.edu.cn,

Abstract. This paper specifically analyse the mobility performance in optical zoom receiving mobile VLC. Optical zoom antenna adjust the optical gain by changing the FOV and then effect the signal to noise ratio (SNR) at the receiver. Firstly, with same transmit power, we analysis the communicable area between fixed focus VLC systems at different FOV. Besides, we contrast the mobility between fixed-focus and optical zoom VLC. Comparing with fixed focus VLC, the mobility can be increased by 90% with optical zoom receiving at 800Mbps. As data rate increasing, fixed focus system is completely disrupted, while the optical zoom antenna still ensure well mobility. According to these analysis, optical zoom antenna can guarantee a stable and high speed communication over a wide range in mobile VLC.

1. Introduction

Smart home need all intelligent devices indoor could access Internet [1]. For most indoor intelligent devices, they are dispersed. Thus, ensuring all devices can communicate well is an urgent issue. VLC is a technology combines the transmission task and illumination role with indoor infrastructure, offering advantageous such as low cost, long lifetime and a large number of available spectrum resources [2,3,4]. The spectrum energy distribution marches to Gaussian distribution, leading to a non-uniform communication performance within the room domain. Even worse, some region of the devices could not receive the signal [5].

Previously,researchers have been dedicated to high-speed VLC. Mobility is another research focus [6]. In 2018, an LED implementation approach based on hyperheuristic evolutionary algorithm (HypEA) show that mobility coverage optimized by 13.23% in compared with uniform deployment and 70.67% compared with centralized arrangement [7]. Later, the approached HypEA successfully investigated optimal LED deployment for fully mobile indoor VLC system at a height of 0.85m above the ground [8]. The uniformity of illumination was improved from 0.55 to 0.86 in a room with the size of 7.8 m×6 m×3m by a LED lamp arrangement algorithm, ensuring almost equal SNR effects for all locations indoor. There are 2 × 3 lamps distribute on the ceiling [9]. From the aspect of transmitter, in 2014, a linear programming (LP) method is proposed to optimize transmit power of LEDs, decreasing the fluctuation of received power [10]. In 2015, Zhang Jian optimized the distribution of SNR from the aspects of LED lamp layout and power optimization based on Genetic Algorithm [11]. These methods find the optimal solution to guarantee the mobility, while also limit the basic equipment in the house. When the size of the room changes, it need to find a new solution to ensure mobility. In 2016, optical zoom antenna was proposed to keep the stability and mobility signal in VLC [12]. However, the data analysis about optical zoom antenna is not detailed.

In our work, we sufficiently analyse the mobility in optical zoom receiving VLC. In section two, average mobility area ratio (M) is introduced to measure the comparable area of indoor VLC. Besides, we describe the SNR calculation formula with optical zoom antenna. Later, we set up a four LED lamp VLC system for the following simulation. Finally, we adequately analyse the mobility of in fixed focus and optical zoom receiving mobile VLC with MATALB. Mobility is a novel research direction in VLC. As the mobility improved, the communication reliability of smart home network can be guaranteed.

2. Average mobility area ratio (M) and SNR in mobile VLC

2.1. Average mobility area ratio (M) in VLC

The average mobility area ratio (M) is used to measure the mobility of indoor VLC, indicating the overall mobility achieved during the movement [7]. If the value of M is 0, it indicates the full communication interruption, whereas the value of R is 0, it indicates complete coverage the mobility of the receiver:

$$M_{average} = \frac{S_{SNR>13.6}}{S_{total}} \tag{1}$$

For on-off keying (OOK) modulation, The BER is:

$$\mathrm{BER} = \mathrm{Q}(\sqrt{SNR}) \tag{2}$$

Where $\mathrm{Q(x)} = \frac{1}{\sqrt{2\pi}}\int_x^{\infty} e^{-y^2/2}\,dy$. According to the relationship between BER and SNR, when the BER is less than 10^{-6}, the SNR must be more than 13.6dB. In this way, based on Monte Carlo (MC) algorithm, $M_{average}$ depend on every ray.

$$M_{average} = \frac{\sum APD_{SNR>13.6\mathrm{dB}}}{APD_{total}} \tag{3}$$

2.2. Signal to noise ratio (SNR) in VLC with optical zoom antenna

MC algorithm has a widespread application in VLC [13,14]. In this work, we use MC algorithm for describing the randomness of photos. It is assumed that four 10cm×10cm LED lamps are installed on the ceiling, and each LED lamp has a built-in four-color LED module. Any point on the illuminating surface is a point source (n_{x},n_{y},n_{z}) with an intensity of I_0.

$$\begin{cases} n_{\mathrm{x}} = x_0 + a * sin\theta \\ n_{\mathrm{y}} = y_0 + b * sin\varphi \\ n_{\mathrm{z}} = 3 \end{cases} \tag{4}$$

Where a and b is Side length of the LED lamp. θ and φ are both random number:

$$\varphi = 2\pi r_3 \tag{5}$$

$$\theta = 2\pi r_4 \tag{6}$$

Where r_3 and r_4 are both uniform random numbers between 0 and 1, generated by the rand function. LED acts as a Lambert radiation source, randomly sampling the exit direction of the light can uniquely determine a random light. Based on the MC algorithm, the luminous intensity of the random light is :

$$I = I_0 \cdot cos\gamma \tag{7}$$

Where I_0 is the luminous intensity of LED source. γ is the angle between the emitted light and the surface normal. Simplify,we assume that the normal direction is perpendicular to the ceiling and is (0, 0, -1).

Optical gain is defined as the ratio between the optical power received on the photosensitive surface of a detector when it's equipped with and without an optical antenna [15]. For fixed-focus VLC system, the gain of the optical antenna can be expressed as:

$$G(\psi) = \begin{cases} \frac{n^2}{sin^2\Psi} & \psi < \Psi \\ 0 & \psi > \Psi \end{cases} \tag{8}$$

where n is internal refractive index, Ψ is FOV of optical antenna. As shown in figure 1, the acceptance angle of light A is smaller than FOV of APD and light A can be received. Inversely, light B can't be received.

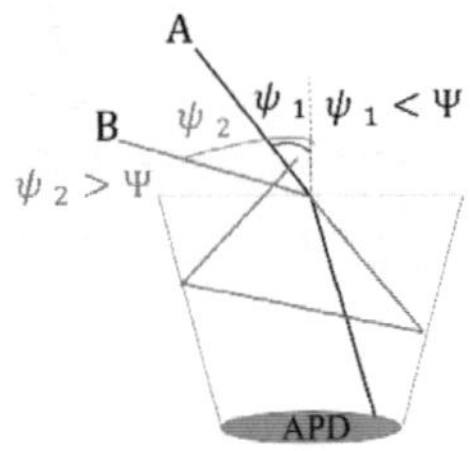

Figure 1. Simple model of APD

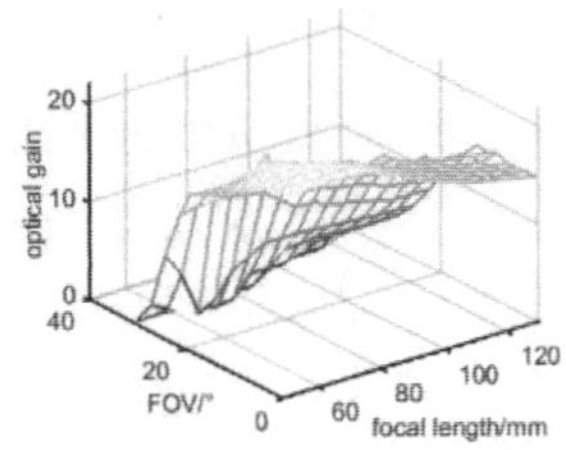

Figure 2. Optical gain versus FOV and focal length

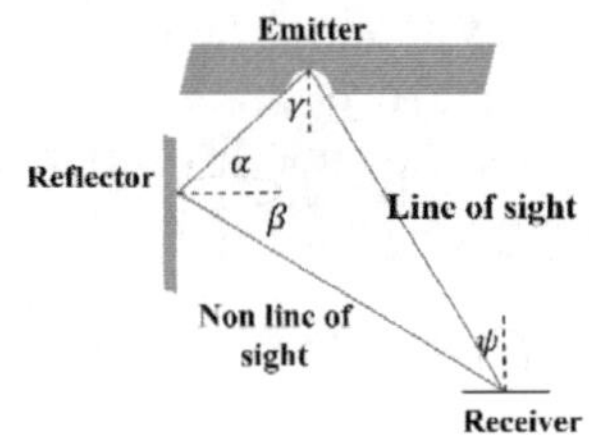

Figure 3. Channel model for NLOS VLC.

The process of optical zoom antenna is to change the focal length of the antenna through the movement of optical lens. As shown in figure 2, we simulate the optical gain using the laboratory's existing zoom antenna. The relationship between FOV and optical gain is inverse. For optical zoom receiving VLC system, the optical gain can be expressed as:

$$G(\psi, f) = \begin{cases} \frac{n^2}{sin^2\Psi(f)} & \psi < \Psi \\ 0 & \psi > \Psi \end{cases} \tag{9}$$

Considering the channel model in figure 3, we consider the effects of wall reflection and the reflection coefficient of wall is 0.8. For the design of an NLOS optical link, the received power $P_{r,i,j}$ at receiver i from ray j can be expressed as [2]:

$$P_{r,i,j} = (H_{d,i,j} + H_{ref,i,j}) \times P_t \tag{10}$$

Where $H_{d,i,j}$ and $H_{ref,i,j}$ are gain on direct path and reflect path respectively. P_t is the transmit power of single random ray.

$$H_{d,i,j}(\psi, f) = \begin{cases} \frac{(m+1)A_R}{2\pi d^2} cos^m(\gamma) T(\lambda, \psi) G(\psi, f) \cos(\psi) & 0 \le \psi \le \Psi \\ 0 & \psi \ge \Psi \end{cases} \tag{11}$$

$$dH_{ref,i,j}(\psi, f) = \begin{cases} \frac{(m+1)A_R}{2\pi d_1^2 d_2^2} \rho dA cos^m(\gamma) \times \cos(\alpha)\cos(\beta)\, T(\lambda, \psi) G(\psi, f) \cos(\psi) & 0 \le \psi \le \Psi \\ 0 & \psi \ge \Psi \end{cases} \tag{12}$$

where d is the distance between the transmitter and receiver, dA is the surface element of reflection wall, $T(\lambda, \psi)$ is the transmission function of an optical filter, γ is the angle of the emergence, ψ is the angle of incidence, Ψ and f represent the FOV and the focal length of the optical zoom antenna.

According to the channel characteristics of indoor visible light communication,after free space propagation, the received power within Ψ can be expressed as equation (13)

$$P_{r,total} = \sum_{rays} \left\{ H_{d,i,j} \times P_t + \int_{s-wall} \times P_t dH_{ref,i,j} \right\} \tag{13}$$

For VLC system, AC power upload information and DC power correspond to average intensity. The SNR of APD at receiving surface is:

$$SNR_i = \frac{\left(P_{AC,i} \cdot i_{so,i}\right)^2}{crosstalk_{i,j} + \sigma_{excess}^2 + \sigma_{dark}^2 + \sigma_{thermal}^2} \tag{14}$$

Where $P_{AC,i}$ and $P_{DC,i}$ are AC power and DC power of LED i respectively, $P_{DC,j}$ is the DC power of LED . σ_{excess}^2, σ_{dark}^2, $\sigma_{thermal}^2$ are the excess noise, dark current noise, and thermal noise of the APD caused by photocurrent. $crosstalk_{i,j}$ is spectral crosstalk for LED i form LED j [16]:

$$crosstalk_{i,j} = \sum_{j=1, j\ne i}^{n} \left(P_{AC,j} \bullet i_{c0,i,j}\right)^2 \tag{15}$$

in indoor VLC systems, user's slow moving changes not only the propagation distance, but also the angle of incidence and irradiance. Both distance and angle are key factors for the capacity in VLC systems.

3. Indoor VLC geometric distribution model

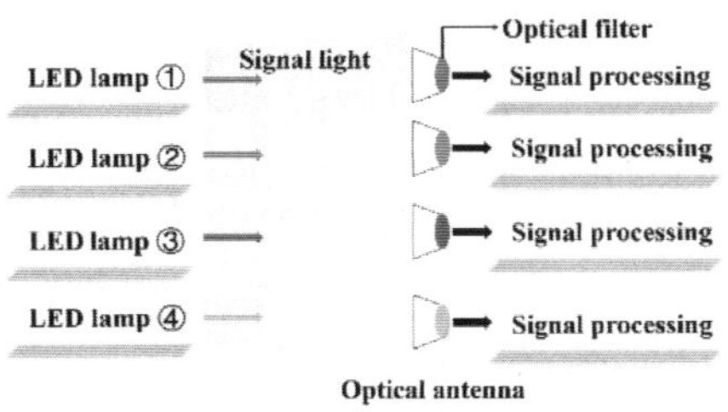

Figure4. Multi-spectrum VLC block diagram

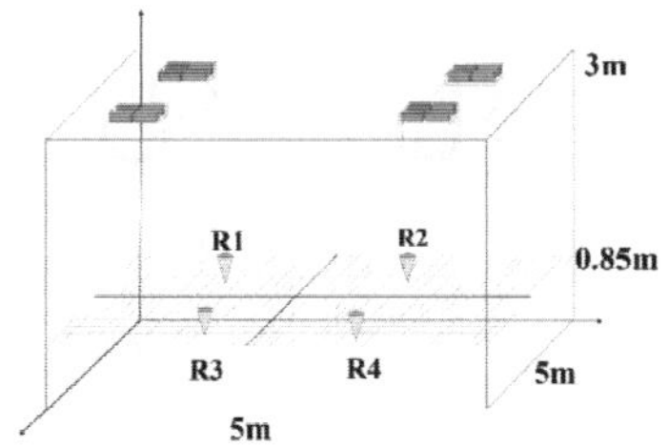

Figure 5. The room model.

As shown in figure 4, signals are loaded on sub-communication channels using different colour LEDs. We assume each LED lamp adopts a unique random sequence to transmit information. Thus, spectral crosstalk is the most serious interference problem. APD need to select the maximum instantaneous SNR among the four LED lamps. So each LED lamp is responsible for a region, like in figure 5.

The indoor VLC system is set up in a room with a size of 5 m×5 m×3 m. There are four LED lamp are located at (1.2, 1.2, 3), (1.2, 3.6, 3), (3.6, 1.2, 3), (3.6, 3.6, 3) on the ceiling and every LED lamp is 10 watts. The semi-angle at half power $\Phi_{1/2}$ is 60°. Each white light LED lamp consist of four different colour LED. The central wavelength is 463nm (blue), 520nm (green), 595nm (orange), and 640nm (red), respectively. The semi-angle at half illuminance of the LEDs is 60° (m=1). According to the principle of mixed light, the coefficient ratio for mixing white light is about 1:1:1:1, so we design every colour of LED is 2.5 watts [17]. To ensure better forecast accuracy, we suppose each LED emits 50,000 random rays and there are 100 APDs in the receiving at the height of 0.85m above the ground level. The parameters of VLC system are given in table 1.

Table 1. System parameters

Parameters	Value
Order of the Lambert emission m	$m = ln2/\ln(\Phi_{1/2})$
Active area of APD A_R	3 mm^2
FOV of optical zoom antenna	Between 16° and 40°
Filter transmittance	80% within the bandwidth, 0% out the bandwidth
Filter FWHM	10 nm

4. Mobility simulation results

4.1. Mobility in fixed focus VLC

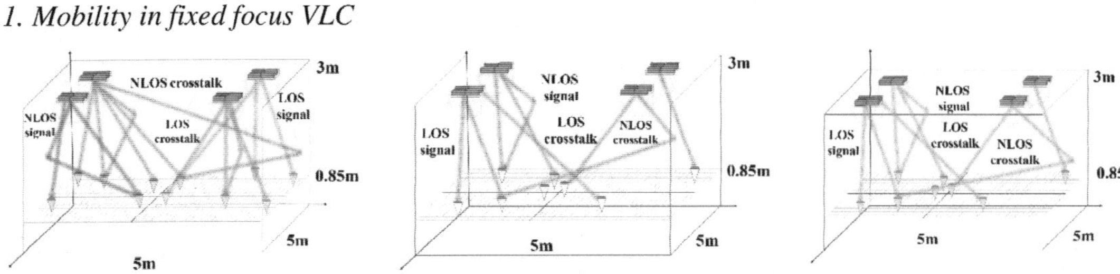

(a) (b) (c)

Figure 6. Scene diagram of indoor VLC. (a) Red monochrome VLC, (b) Green monochrome VLC, (c) RGBY four-spectrum VLC

Monochrome VLC system is that there is one spectrum signal channel in the VLC system, others are all hold the role of supplementing white light illumination, like Figure 6(a) and Figure 6(b). Multi-spectrum VLC use multi-colour LEDs and each colour of LED lamp is responsible for a corresponding region, like in Figure 6(c). Taking red LED lamp and green LED lamp as examples, we compare the communicable area between monochrome fixed focus VLC and multi-spectrum fixed focus VLC at the receiving plane. The transmit power of every LED lamp is the same.

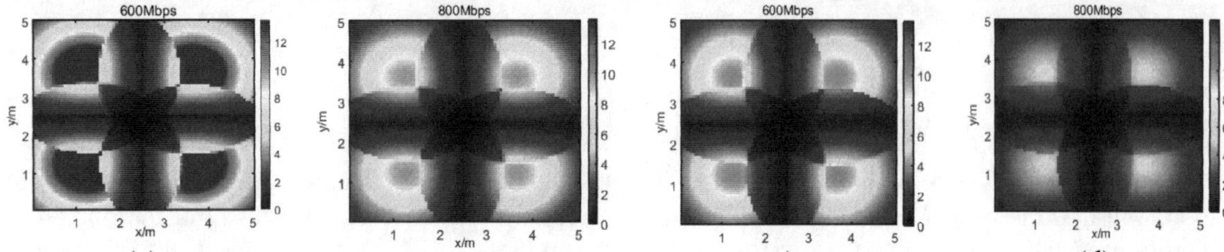

(a) (b) (c) (d)

Figure 7. (a) and (b) are comparable area of red monochrome fixed focus VLC with 50° FOV ; (c) and (d) are comparable area of green monochrome fixed focus VLC with 50° FOV.

As simulation results in figure 7 and figure 8, compared to monochrome VLC, the communicable area of RGBY four-spectrum VLC is in a higher proportion. For multi-spectrum VLC, the effect of the optical filter is more prominent, and the crosstalk is reduced, leading to a more stable communication. The problem is that with data rate increasing, the communication still suffer from mobility limitation.

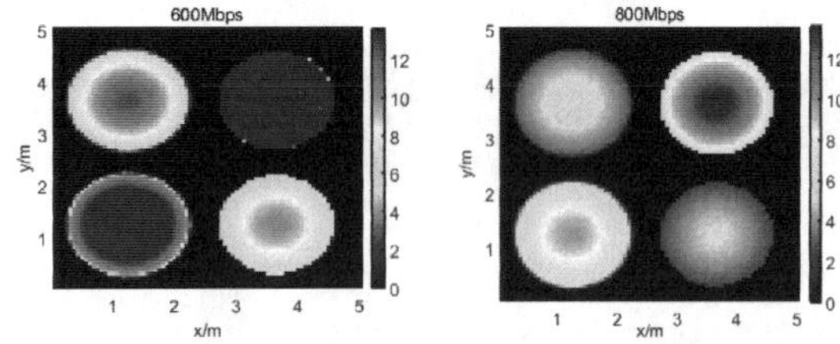

(a) (b)

Figure 8. Comparable area of RGBY four-spectrum fixed focus VLC with 50° FOV

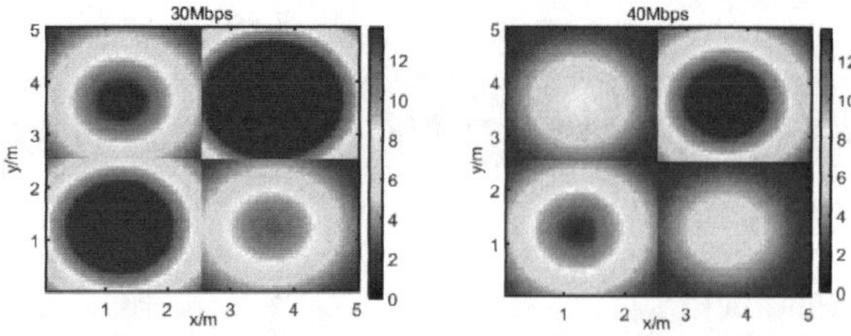

(a) (b)

Figure 9. Comparable area of RGBY four-spectrum fixed focus VLC with 90° FOV

In the following analysis, we compared the mobility between different FOV VLC systems. According to the results in figure 8 and figure 9, the optical antenna with a small FOV is easy to achieve a high speed communication, but the mobility is limited. Contrary, for lager FOV VLC, the receiver could receive a wider field of light. A big communicable area could be achieved, but the communication capacity is limited due to the interference of crosstalk channel.

4.2. Mobility in optical zoom antenna VLC

Optical zoom antenna can change the optical gain by adjusting the FOV of receiver as shown in figure 3. The FOV of the optical zoom antenna we use can change between 16° and 40° and the maximal system gain is 20.66. The communicable area of optical zoom antenna VLC system show in figure 10. According to the calculation of M, at the speed of 800Mbps, with same transmit power, the mobility performance of RGBY optical zoom four-spectrum VLC can be improved by 90% compared with RGBY fixed focus four-spectrum VLC with 50° FOV. When the data rate is more than 1Gbps, the value M of fixed focus VLC is almost 0, meaning the communication outage. Value M of red ,green and blue VLC with 50° FOV, RGBY fixed focus four-spectrum VLC with 50° FOV and RGBY optical zoom four-spectrum VLC is shown in figure 11.

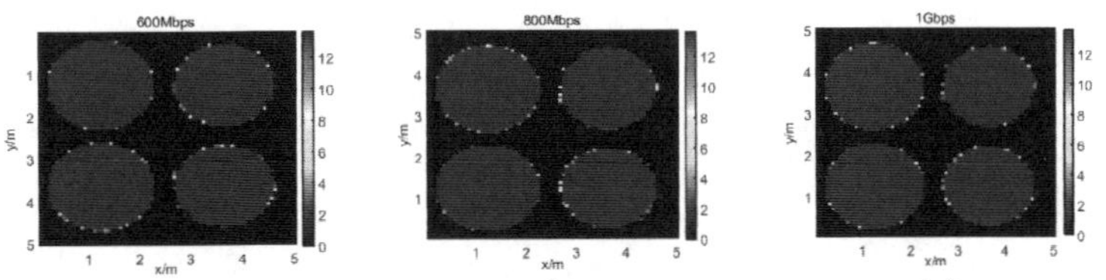

(a) (b) (c)

Figure 10. SNR distribution of RGBY four-spectrum optical zoom VLC

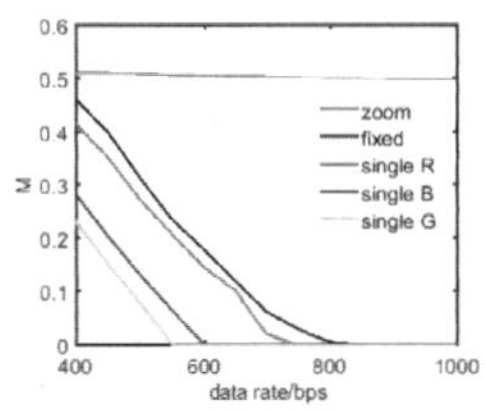

Figure 11. Value of M versus data rate

5. Conclusion

Mobility is an urgent problem to VLC system. In our work, we add focal length into the SNR formula and analyse the mobility of VLC with optical zoom antenna. We contrast the average mobility area ratio (M) between fixed focus and optical zoom antenna VLC. According to the calculation results, the optical zoom antenna effectively provide optical compensation in mobile VLC. The optical zoom antenna receiving VLC can support a more stable mobile communication, especially at high speed transmission.

References

[1] Tadokoro, S., Jia, Q. S., Zhao, Q., Darabi H., Huang G., Becerikgerber, B.. (2014). Smart building technology [tc spotlight]. *IEEE Robotics & Automation Magazine,* **21**(2), 18-20.

[2] Komine, T., Nakagawa, M.. (2004). Fundamental analysis for visible light communication system using led light. *IEEE Transactions on Consumer Electronics,* **50**(1), 100-107.

[3] Cui, K., Chen, G., Xu, Z., Roberts, R. D.. (2010). Line-of-sight visible light communication system design and demonstration. *International Symposium on Communication Systems Networks & Digital Signal Processing*.

[4] Pathak P H, Feng X, Hu P Visible Light Communication, Networking, and Sensing: A Survey, Potential and Challenges[J]. *IEEE Communications Surveys & Tutorials*, 2015, **17**(4):2047-2077.

[5] Ohno, Yoshi. (2005). Spectral design considerations for white led colour rendering. *Optical Engineering,* **44**(11), 111302.

[6] Burton, A., Minh, H. L.. (2012). A Study of LED Lumination Uniformity with Mobility for Visible Light Communications. *International Workshop on Optical Wireless Communications*. IEEE.

[7] Sharma, R., Kumari, A. C., Aggarwal, M., & Ahuja, S.. (2018). Hyperheuristic-based analysis and optimization of a mobile indoor visible light communication system. *Transactions on Emerging Telecommunications Technologies,* **29**(9), e3274.

[8] Sharma R, Kumari A C, Aggarwal M. Optimal LED deployment for mobile indoor visible light communication system: Performance analysis[J]. *AEU-International Journal of Electronics and Communications*, 2018, 83: 427-432.

[9] Hao H G, Zhang D D, Tang S. Analysis of the LED Lamp Arrangement for Uniformity of Illumination in Indoor VLC System[J]. *Journal of the Optical Society of Korea*, 2014, **18**(6):663-671.

[10] Yingrui D, Lihua L I. LED Power Optimization and Angle Parameters Analysis for Indoor Visible Light Communication Systems[J]. 2014.

[11] Jian Z, Hui W. An improved SNR uniformity optimization scheme for VLC system[J]. *Journal of Chongqing University of Posts and Telecommunications (Natural Science Edition),* 2015.

[12] Xuebin Zhang, Yi Tang, Lu Cui. Design of a Spherical Optical Zoom Antenna Using for Mobile Visible Light Communication[J]. *Acta Optica Sinica*, 2016, **36**(1):0106004.

[13] Dong, L., Wenbo, W., Yonghua, L.. (2009). Monte Carlo Simulation with Error Classification for Multipath Rayleigh Fading Channel. *International Conference on Wireless Communications*. IEEE Press.

[14] Ramirez-Aguilera, A. M., Luna-Rivera, J. M., Guerra, V., Rabadan, J., & Lopez-Hernandez, F. J.. (2016). Multi-wavelength modelling for VLC indoor channels using Monte carlo simulation. *2016 10th International Symposium on Communication Systems, Networks and Digital Signal Processing (CSNDSP)*. IEEE.

[15] Bonakdar, A., Mohseni, H.. (2013). Hybrid optical antenna with high directivity gain. *Optics Letters,* **38**(15), 2726.

[16] Lu, C., Yi, T., Qing-Wei, Z., Jia-Bin, L., & Shan-Shan, H.. (2016). Analysis of channel crosstalk in muliti-spectrum visible light communication system. *Acta Physica Sinica*.

[17] Lu, C., Yi, T., Qingwei, Z., Heqing, H., Xuebin, Z.. (2016). A multispectral mixed white light wdm-vlc system combining color domain with time domain. *Opto-Electronic Engineering*.

IOP Conf. Series: Journal of Physics: Conf. Series **1284** (2019) 012055 doi:10.1088/1742-6596/1284/1/012055

Analysis of Desired Receive Signal Impact in Wireless Co-time Co-frequency Full Duplex

Juan Zhou[1], **Chao Chen**[1,*], **Yajuan Xue**[1], **Ying Shen**[2] **and Meng He**[2]

[1] College of Communication Engineering, Chengdu University of Information Technology, Chengdu, 610225, China;

[2] National Key Laboratory of Science and Technology on Communications, University of Electronic Science and Technology of China, Chengdu, 610054, China;

[*] Email: roosterchen@163.com

Abstract. With self-interference cancellation (SIC), a wireless full duplex terminal is allowed to transmit and receive signal simultaneously in the same frequency band. This paper focuses on the analysis of desired receive signal impact on the radio frequency (RF) SIC performance. The full-duplex system model is first given, and the principle of radio frequency self-interference cancellation is introduced. Then, from the perspective of the convex function, the influence of the desired signal on the RF SIC is analyzed when the total power of the signal after RF SIC is taken as the objective function. It is found that when there is no desired signal, the objective function is a convex function, and the residual self-interference power can converge to the minimum value by the gradient descent method. If the desired signal exists and the channel is constant, the objective function is also a convex function, and the residual self-interference power can also converge to the minimum value. However, in the presence of the desired signal and the channel fluctuates, the target is not a convex function, and the residual self-interference power cannot be converged to the minimum value by the gradient descent method.

1. Introduction

Self-interference problem arises and becomes one of the biggest practical impediments to co-time co-frequency full duplex (CCFD) operation, since a wireless terminal is allowed to transmit and receive simultaneously in the same frequency band [1–3]. To suppress the self-interference, three methods are always considered, including antenna separation, RF cancellation, and digital cancellation [4].

As an important SIC method, the RF cancellation attracts worldwide research interest [5–9]. A novel optical approach to implement RF self-interference cancellation for CCFD communication using phase modulation and optical sideband filtering is proposed in [5]. In [6], a fully integrated technique for wideband cancellation of transmitter (TX) self-interference (SI) in the RF domain is proposed. From [7], it is found that the amplitude and phase errors degrade the SIC performance.

As has been shown in the existing literatures, RF SIC is limited by hardware imperfections and SI modeling inaccuracy. The residual SI after RF cancellation is typically assumed to be well above the desired receive signal power, and therefore the impact of the desired receive signal is normally neglected. However, with a well designed RF SIC scheme, compared to the residual SI, the desired receive signal can be higher and the impact on SIC performance is inevitable.

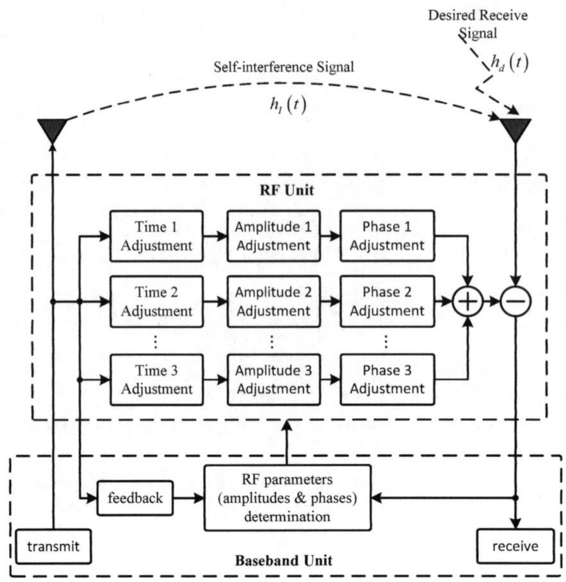

Figure 1. Block diagram of RF self-interference cancellation in full-duplex system.

In this paper, the desired signal impact on the RF SIC performance is analyzed mathematically. Compared with the RF SIC without desired signal, the RF SIC performance with the desired signal is limited by the fluctuation of desired receive signal power.

The system model and the impact on SIC of the RF chain isolation are explained in Section II. In Section III, the proposed scheme for reducing the RF chain isolation impact is proposed. The numerical results are presented in Section IV and Section V concludes the paper.

2. System Model

The system model is shown in figure 1, both the desired signal from the remote terminal and the self-interference signal from the local transmit antenna are received. The RF self-interference signal is estimated with multi-path structure, and then subtracted from the receive signal. Different path has different time delay, phase, and amplitude, which are determined by the baseband unit.

Without loss of generality, local transmit signal $x(t)$ can be described as

$$x(t) = \mathrm{Re}\{b(t)e^{j\omega_c t}\} \tag{1}$$

where $b(t)$ is the complex baseband signal expressed in the RF domain, w_c is the carrier frequency, and $\mathrm{Re}(\alpha)$ denote the real part and the imaginary part of α. Correspondingly, the received self-interference can be expressed as

$$r_x(t) = \sum_{i=1}^{N} h_i \mathrm{Re}\{b(t-\tau_i)e^{j\omega_c(t-\tau_i)}e^{j\phi_i}\} \tag{2}$$

where h_i, τ_i, and ϕ_i denote the self-interference channel gain, transmission delay, and phase offset of the i-th path, respectively. Let $\mathbf{G} = [h_1 e^{-j\phi_1} h_2 e^{-j\phi_2} \cdots h_N e^{-j\phi_N}]$ and $\mathbf{D} = [b(t-\tau_1)e^{-j\omega_c\tau_1} b(t-\tau_2)e^{-j\omega_c\tau_2} \cdots b(t-\tau_N)e^{-j\omega_c\tau_N}]$, (2) can be rewritten as

$$r_x(t) = \mathrm{Re}\{\mathbf{D}^T\mathbf{G}e^{-j\omega_c t}\} \tag{3}$$

The RF self-interference signal is estimated for SIC, which can be described as

$$\hat{r}_x(t) = \sum_{k=1}^{K} \hat{h}_k \mathrm{Re}\{b(t-\hat{\tau}_k)e^{j\omega_c(t-\hat{\tau}_k)}e^{j\hat{\phi}_k}\} \tag{4}$$

where $\hat{h}_k$, $\hat{\tau}_k$, and $\hat{\phi}_k$ denote the estimated self-interference channel gain, transmission delay, and phase offset of the k-th path, respectively. $\hat{h}_k$ and $\hat{\phi}_k$ are determined at the baseband unit in an adaptive way, and $\hat{\tau}_k$ is usually manually adjusted by the change of cable length. With $\mathbf{G_r} = [\hat{h}_1 e^{-j\hat{\phi}_1} \quad \hat{h}_2 e^{-j\hat{\phi}_2} \cdots \hat{h}_N e^{-j\hat{\phi}_N}]$ and $\mathbf{D_r} = [b(t-\hat{\tau}_1)e^{-j\omega_c\hat{\tau}_1} \quad b(t-\hat{\tau}_2)e^{-j\omega_c\hat{\tau}_2} \cdots b(t-\tau_N)e^{-j\omega_c\hat{\tau}_N}]$, (4) can be rewritten as

$$\hat{r}_x(t) = \mathrm{Re}\{\mathbf{D_r}^T \mathbf{G_r} e^{-j\omega_c t}\} \tag{5}$$

Similarly, the remote transmit signal $s(t)$ can be written as

$$s(t) = \mathrm{Re}\{b_s(t)e^{j\omega_c t}\} \tag{6}$$

Then the desired receive signal can be given by

$$r_s(t) = \sum_{m=1}^{M} \lambda_m \mathrm{Re}\{b_s(t-\beta_m)e^{j\omega_c(t-\beta_m)}e^{j\theta_m}\} \tag{7}$$

where λ_m, β_m, and θ_m denote channel gain, transmission delay, and phase offset the of desired signal in the m-th path, respectively.

The RF SIC mechanism is shown in figure 1, the RF self-interference is rebuilt and subtracted from the whole receive signal. Hence, the residual signal after RF SIC can be obtained by

$$\begin{aligned} \mu(t) &= r_x(t) + r_s(t) - \hat{r}_x(t) + n(t) \\ &= I(t) + r_s(t) + n(t) \end{aligned} \tag{8}$$

where $n(t)$ is the White Noise and the residual self-interference $I(t)$ is

$$I(t) = \mathrm{Re}\{(\mathbf{D}^T\mathbf{G} - \mathbf{D_r}^T\mathbf{G_r})e^{-j\omega_c t}\} \tag{9}$$

Correspondingly, the residual signal power is

$$P_r = \mathrm{E}(\|\mu(t)\|^2) \tag{10}$$

Since the self-interference signal, the desired signal and the white noise are independent of each other, (10) can be written as

$$P_r = P_I + P_s + P_n \tag{11}$$

where the residual self-interference power P_I is

$$P_I = \mathrm{E}(\|I(t)\|^2) \tag{12}$$

the desired receive power P_s

$$P_s = \mathrm{E}(\|r_s(t)\|^2) \tag{13}$$

and the white noise power is $P_n = \mathrm{E}(\|n(t)\|^2)$.

The purpose of RF SIC is to minimize the residual self-interference power P_I. In other words, $\mathbf{D}^T\mathbf{G}$ should be closely approximated by $\mathbf{D_r}^T\mathbf{G_r}$. In practice, we first determine the transmission delay $\hat{\tau}_k$ of each path, which is fixed later in the SIC operation. Then the amplitude $\hat{h}_k$ and phase offset $\hat{\phi}_k$ are estimated in real time at the baseband unit, so as to minimize the residual self-interference power P_I. However, the self-interference and the desired receive signal are mixed together. Hence,P_I is replaced by the residual receive signal power P_r, meaning that $\hat{h}_k$ and $\hat{\phi}_k$ are determined based on the minimization of P_r.

3. Impacts on RF SIC from Desired Signal

As shown in figure. 1, the RF self-interference signal is first estimated and then subtracted from the receive signal, so as to get the desired signal. P_r is used as the criterion for the determination of RF parameters. Based on the minimization of P_r, the impacts on RF SIC from the desired signal are analyzed as follows.

1) No Desired signal.

It means $P_s = 0$ and $P_r = P_I + P_n$. Based on (9), the residual self-interference signal power can be expressed as

$$P_I = E\{\|\mathbf{D}^T\mathbf{G} - \mathbf{D}_\mathbf{r}^T\mathbf{G}_\mathbf{r}\|^2\} \tag{14}$$

For the convenience of analysis, let

$$\mathbf{D}^T\mathbf{G} = \mathbf{CY} \tag{15}$$

where $\mathbf{C} = [\mathrm{Re}\{\mathbf{D}^T\} + j\mathrm{Im}\{\mathbf{D}^T\} \quad -\mathrm{Im}\{\mathbf{D}^T\} + j\mathrm{Re}\{\mathbf{D}^T\}]$ and $\mathbf{Y} = [\mathrm{Re}\{\mathbf{G}\} \quad \mathrm{Im}\{\mathbf{G}\}]^T$. $\mathrm{Im}(\alpha)$ denotes the imaginary part of α.

Then the residual self-interference signal power can be expressed as

$$\begin{aligned} P_I(\mathbf{Y}) =& E\{\|\mathbf{CY} - \mathbf{D}_\mathbf{r}^T\mathbf{G}_\mathbf{r}\|^2\} \\ =& \frac{1}{2}E\left\{\left(\mathbf{CY} - \mathbf{D}_\mathbf{r}^T\mathbf{G}_\mathbf{r}\right)^H\left(\mathbf{CY} - \mathbf{D}_\mathbf{r}^T\mathbf{G}_\mathbf{r}\right)\right\} \\ =& \frac{1}{2}\left(P_0 + \mathbf{Y}^H\mathbf{RY} - 2\mathrm{Re}\{\mathbf{ZY}\}\right) \end{aligned} \tag{16}$$

where $P_0 = E\{(\mathbf{D}^T\mathbf{G})^H(\mathbf{D}^T\mathbf{G})\}$ is the power of the receive self-interference signal, $\mathbf{R} = E\{\mathbf{C}^H\mathbf{C}\}$, and $\mathbf{Z} = \mathbf{G}^H E\{\mathbf{D}^*\mathbf{C}\}$

Since $\mathbf{R}$ is autocorrelation and positive semidefinite matrix, $P_I(\mathbf{Y})$ is convex on $\mathbf{Y}$ [10]. $\mathbf{Y}$ denotes the amplitude and phase offset parameters, hence, $P_I(\mathbf{Y})$ is convex on the amplitude and phase parameters. With $P_s = 0$, the minimization of P_I is equal to the minimization of P_r. It means that the optimal amplitude and phase offset can be obtained by the minimization of P_r.

In conclusion, when $P_s = 0$, with the optimal amplitude and phase parameters $\mathbf{Y}^*$, the residual self-interference signal power can be minimized as $P_{I,m} = P_I(\mathbf{Y}^*)$.

2) With Desired Signal.

It means $P_s \neq 0$ and $P_r = P_I + P_s + P_n$. If the channel gain λ of desired signal from remote node is time-invariant, the signal power P_s is constant based on (12) and (7). The optimal amplitude and phase parameters can also be achieved by the minimization of P_r. However, in practice the wireless channel is usually time-variant, which means

$$P_r(\mathbf{Y}, t) = P_I(\mathbf{Y}) + P_s(t) + P_n \tag{17}$$

If time t is fixed and $P_s(t)$ is constant, $P_r(Y, t)$ is still convex on Y. However, t is variant during the RF SIC process, and the convexity of $P_r(Y')$ can not be preserved. For the convenience of analysis, gradient descent method is applied to solve the minimization problem. A natural choice for the search direction is the negative gradient. Suppose the RF SIC starts at t_1 and ends at t_2, and the $i-$ dimension of $\mathbf{Y}$ is $\mathbf{Y}_i$. The search direction of $\mathbf{Y}_i$ is calculated as

$$\begin{aligned} D(t) \approx& \frac{P_r(\mathbf{Y} + \kappa\mathbf{Y}_i, t_1) - P_r(\mathbf{Y}, t_2)}{\kappa} \\ =& \frac{P_I(\mathbf{Y} + \kappa\mathbf{Y}_\mathbf{i}) - P_I(\mathbf{Y}) + P_s(t_1) - P_s(t_2)}{\kappa} \\ =& \frac{\Delta P_I(\kappa) - \Delta P_s(t_2 - t_1)}{\kappa} \end{aligned} \tag{18}$$

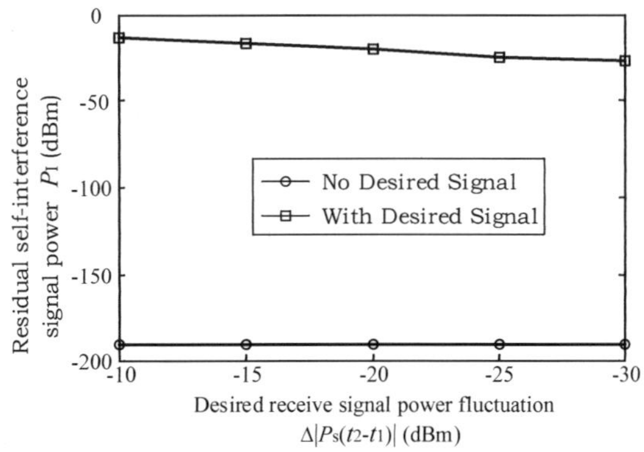

Figure 2. Impact of desired receive signal in single-path self-interference channel.

where κ is the step size, and $\Delta P_s(t_2 - t_1)$ is the variance power of the desired receive signal. The impact of $\Delta P_s(t_2 - t_1)$ on RF SIC is analyzed as follows.

In the first period of RF SIC, since the residual self-interference signal power is much larger than the desired receive signal power, $P_I(\mathbf{Y}) \gg |\Delta P_s(t_2 - t_1)|$. With the suitable κ, $|\Delta P_I(\kappa)| > |\Delta P_s(t_2 - t_1)|$ and the sign of the $D(t)$ is not changed. Hence$P_I(\mathbf{Y})$ can be decreased with the right direction. However, with the decrease of $P_I(\mathbf{Y})$, $|\Delta P_I(\kappa)|$ becomes smaller with the same κ. When $|\Delta P_I(\kappa)|$ is approximated with $|\Delta P_s(t_2 - t_1)|$, $|\Delta P_I(\kappa)| > |\Delta P_s(t_2 - t_1)|$ can not be preserved and the sign of $D(t)$ can be reversed. In this case, the wrong search direction is used and $P_I(\mathbf{Y})$ can not be decreased, meaning that the minimum $P_I(\mathbf{Y})$ is $|\Delta P_s(t_2 - t_1)|$.

With the descriptions above, the minimum residual self-interference signal power can be given as

$$P_{I,s,m} = \max\{|\Delta P_s(t_2 - t_1)|, \quad P_{I,m}\} \tag{19}$$

where $P_{I,m}$ is the minimum residual self-interference power without the desired signal described above. When $P_{I,m} > |\Delta P_s(t_2 - t_1)|$, the desired signal impact can be neglected. However, if the desired signal transmitter is near to the local receiver and $P_{I,m} < |\Delta P_s(t_2 - t_1)|$, the RF SIC performance is limited by $|\Delta P_s(t_2 - t_1)|$.

4. Numerical Results

To confirm the theoretical analysis, the simulation platform is developed as follows. 3GPP LTE protocol with 20MHz bandwidth is applied in the CCFD system. The amplitude of desired signal is variant with time. The received self-interference signal power and the received desired signal power are 0dBm and -10dBm, respectively.

The self-interference signal channel is first supposed as the single-path channel, and then $P_{I,m} = 0$ under the ideal self-interference channel reconstruction. Figure 2 shows the desired receive signal impact on the residual self-interference signal power. It is found that the residual self-interference signal power can be as less as -190dBm without the desired signal. However, in the case of desired receive signal exists, the RF SIC performance is restricted by the desired signal power fluctuation $|\Delta P_s(t_2 - t_1)|$. The residual self-interference signal power P_I is larger than $|\Delta P_s(t_2 - t_1)|$. When $|\Delta P_s(t_2 - t_1)| = -30$dBm, $P_I \approx -25$dBm.

Figure 3 show the desired receive signal impact on the RF SIC performance, considering the multi-path self-interference signal channel. The multi-path signal channel is established based on the experimental power delay profile, with the delay spread as 200ns. In the RF signal reconstruction, 4-path channel model is used. Since reconstruction channel model is estimated with inevitable error, the self-interference signal can not be cancelled completely even though no desired receive signal. It is found that the residual self-interference signal power $P_I = -36$dBm

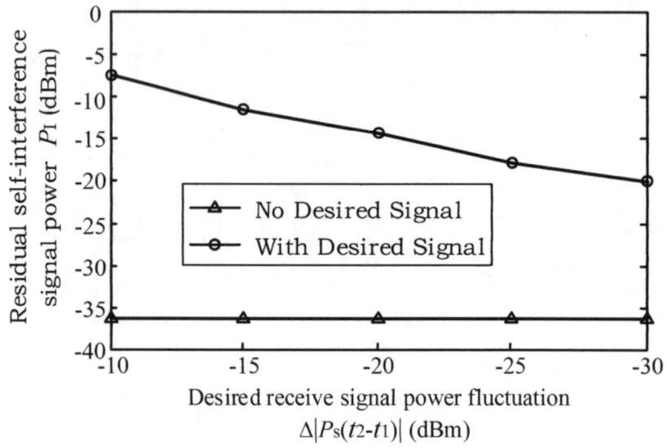

Figure 3. Impact of desired receive signal in multi-path self-interference channel.

without desired receive signal. However, when the desired receive signal exists, the performance is damaged apparently. When the desired signal power fluctuation $|\Delta P_s(t_2 - t_1)| = -10\text{dBm}$, $P_I \approx -7.5\text{dBm}$.

5. Conclusion

In this paper, it is first described that desired receive signal damages the self-interference cancelation performance in CCFD wireless communication system. Through the mathematical analysis, we find that the residual self-interference signal power is limited by the desired signal power fluctuation. In practice, it is usually larger than the desired signal power fluctuation. And it suggests that the study on the method to reduce the desired signal impact is an interesting topic for future research.

Acknowledgments

This work was supported by the National Natural Science Foundation of China under Grant No.61601064.

References

[1] Y. Hua, P. Liang, Y. Ma, A. C. Cirik, and Q. Gao, "A Method for Broadband Full-Duplex MIMO Radio," *IEEE Signal Process. Lett.,*, vol. 19, pp. 793-796, Dec. 2012.

[2] A. Sahai, G. Patel, C. Dick, and A. Sabharwal, "On the Impact of Phase Noise on Active Cancelation in Wireless Full-Duplex," *IEEE Trans. Veh. Commun.,*, vol. 62, pp. 4494-4510, Nov. 2013.

[3] A. Sabharwal, P. Schniter, D. Guo, D. W. Bliss, S. Rangarajan, and R. Wichman, "In-Band Full-Duplex Wireless: Challenges and Opportunities," *IEEE J. Sel. Areas Commun.,*, vol. 32, pp. 1637-1652, Sep. 2014.

[4] M. Duarte, C. Dick, and A. Sabharwal, "Experiment-Driven Characterization of Full-Duplex Wireless Systems," *IEEE Trans. Wireless Commu.,*, vol. 11, pp. 4296-4307, Dec. 2012.

[5] X. Han, B. Huo, Y. Shao, C. Wang, and M. Zhao "RF Self-Interference Cancellation Using Phase Modulation and Optical Sideband Filtering," *IEEE Photonics technology letters,*, vol. 29, no. 11, pp. 917-920, Jun. 2017.

[6] J. Zhou, T. Chuang, T. Dinc, and H. Krishnaswamy "Integrated Wideband Self-Interference Cancellation in the RF Domain for FDD and Full-Duplex Wireless," *IEEE Journal of solid-state circuits,*, vol. 50, no. 12, pp. 3015-3031, Dec. 2015.

[7] Z. He, S. Shao, Y. Shen, C. Qing, and Y. Tang, "Performance Analysis of RF Self-Interference Cancellation in Full-Duplex Wireless Communications," *IEEE Wireless Commu. Lets.,*, vol. 3, pp. 405-408, Aug. 2014.

[8] Y. Shen, J. Zhou, and Youxi Tang. "Digital Self-Interference Cancellation in Wireless Co-time and Co-frequency Full-Duplex System," *Wireless Pers. Commun.*, vol. 82, pp. 2557-2565, Feb. 2015.

[9] E. Ahmed and A. M. Eltawil. "Self-Interference Cancellation Technique for Full-Duplex Systems," *IEEE Trans. Wire. Commun.,*, vol. 14, pp. 3519-3532, Jul. 2015.

[10] S. Boyd and L. Vandenberge. *Convex Optimization.* Cambridge University Press, 2004.

Optimization Design of Networked Control System with Communication Interference

Liangliang Niu [1], Haiping Wei[1, *] and Lina Jin[1]

[1] School of Computer and Communication Engineering, Liaoning ShiHua University, Liaoning, China

[*]416089317@qq.com

Abstract. For network control systems with long delay and packet loss, firstly, it is based on the effectiveness of model-based control in network systems with long delay, and it also establish a feedback model that simulates delay as a system component. Considering the possible modeling errors of the feedback model, a design of the traditional linear system is proposed to design the controller by using the communication interference observer to compensate for the delay characteristics. Then, the steady-state error of the characteristic equation of the system is obtained and prove the feasibility and robustness of the method, and carried out the equivalent analysis of packet loss. Finally, the effectiveness of the controller is demonstrated by the simulation of communication interference data from comparing with the Smith predictor.

1. Introduction

Network Control System (NCS) refers to a closed-loop feedback control system formed by a communication network connection. The main components are controllers, controlled objects, sensors and actuators, etc. The module composed of this system component is used as a network node to access the network, so it forms a new closed loop. The network control system has the advantages of less wiring and lower cost compared with the traditional point-to-point wired control system. Even in the face of complex environments, it is easy to install, maintain, and expand [1-2]. However, due to the introduction of communication networks, some unstable factors of the network will also be brought into the control system [3-5].

Due to the introduction of the network, it is necessary to avoid and reduce the impact on the control system. In recent years, the academic community has also done a lot of research on some problems about the network control system [6-8]. These studies have reduced key issues such as network inducing delay and packet loss in different directions [9]. Document uses Markov-based stochastic process to describe the characteristics of system packet loss. A time-varying controller based on the principle of packet loss and capable of compensating for the steady-state error of the system is designed [10]. Literature uses state augmentation to overcome the impact of delay on network control system analysis. Then consider the random packet loss situation and convert the augmentation system into a switching system. Based on this, a network robust controller design method is proposed.

This paper adopts a model-based network control system to compensate for long delay and packet loss. The problem of existence of model feedback errors and simultaneous time delay and packet loss is studied. Design controllers by using the traditional linear system, design method by using the Communication Disturbance Observer (CDOB) [11]. Then separate the controller and model error feedback module on both sides of the network. The research on communication interference shows that the method has good robustness. The final data simulation also verified that the controller design was effective [12].

2. Questions and main content

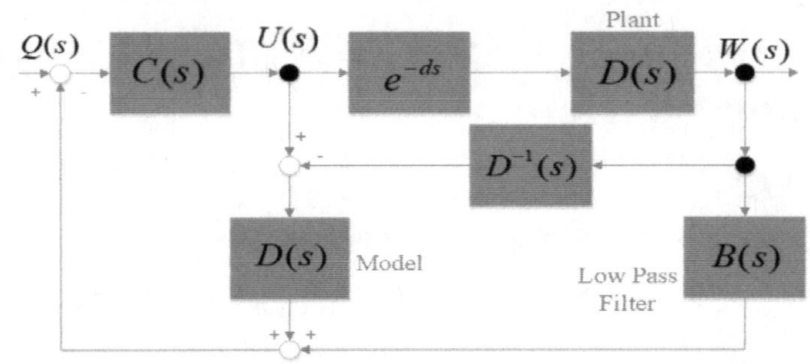

Figure 1. System structure about feedback model.

Consider designing a Communication Disturbance Observer (CDOB) to eliminate the effects of time delays on the system, and this communication interference observer can also effectively compensate for time delays. According to Figure 1, important conclusions can be drawn. The transfer function $G_{cl}(s)$ from $Q(s)$ to $W(s)$ is as shown in equation (1).

$$G_{cl}(s) = \frac{C(s)D(s)e^{-ds}}{1+C(s)D(s)} \tag{1}$$

Where $C(s)$ is the controller at the near end of the network, d is the actual time delay, and $D(s)$ is simply a separate transfer function of the remote controlled object of the network. Assume that $B(s)=1$, the low-pass filter converter in the block diagram, has a high frequency cutting efficiency. Since the compensation of CDOB is obtained, and it can be seen from equation (1) that the time delay element e^{-ds} exists only in the numerator of the fraction, and because there is no time variable in the characteristic polynomial, the design of the conventional linear system is adopted here. The controller $C(s)$ method is designed for the controllers in this section.

3. Assume the impact of the error

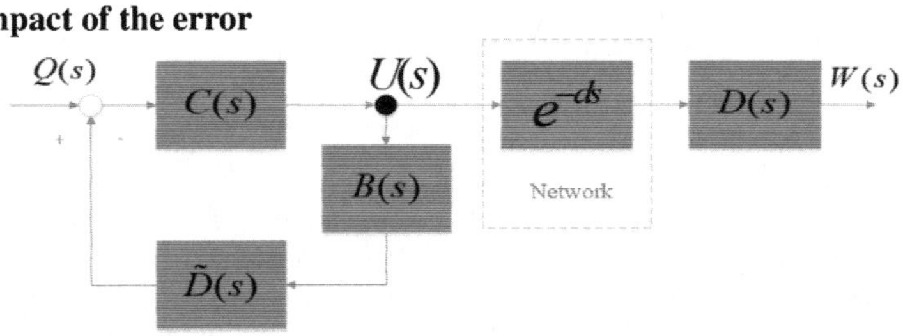

Figure 2. Schematic diagram of system model error feedback structure.

Considering that modeling errors may affect the communication disturbance observer. When there is a modeling error in the inverse model of the structure of the CDOB in Figure 1. The basic structure of Figure 1 will be converted into the structure shown in Figure 2. Where $\tilde{D}(s)$ is the key model, its role is to identify the controlled object and feedback information to the controller. The closed-loop transfer function $G_{cl}(s)$ from the data input $Q(s)$ to the information output $W(s)$ in Figure 2 is obtained as in equation (2).

$$G_{cl}(s)=\frac{C(s)D(s)e^{-ds}}{1+C(s)\tilde{D}(s)} \quad (2)$$

$$E(s)=Q(s)-G_{cl}(s)Q(s)=\frac{C^{-1}(s)+(\tilde{D}(s)-D(s)e^{-ds})}{C^{-1}(s)+\tilde{D}(s)}Q(s) \quad (3)$$

In the equation (3), $E(s)$ is the deviation between the target data $Q(s)$ at the time of system input and the information $W(s)$ at the time of system output.

Let $Q(s)=1/s$, in the equation (4) obtain the steady-state error of the control input.

$$\lim_{s\to 0} sE(s)=\lim_{s\to 0} s\frac{C^{-1}(s)+(\tilde{D}(s)-D(s)e^{-ds})}{C^{-1}(s)+\tilde{D}(s)}\frac{1}{s}=\lim_{s\to 0}\frac{C^{-1}(s)+(\tilde{D}(s)-D(s))}{C^{-1}(s)+\tilde{D}(s)} \quad (4)$$

Assuming the system is stable and not modeled correctly and without errors, i.e. $\tilde{D}(s)=D(s)$, and $C(s)$, $D(s)$ or $\tilde{D}(s)$ has a pole at the origin when there is no step input, then $\lim_{s\to 0} sE(s)=0$, i.e. the system does not have a steady-state error. When $\tilde{D}(s)\neq D(s)$, even if $C(s)$ has a pole at the origin, the system will interrupt the feedback due to the presence of steady-state error. As shown in Figure 2, the controller cannot receive any feedback from the state of the object. Therefore, even if the output value of the object has a large deviation from the target value, the offset compensation cannot be adjusted to the controller.

4. Compensation method and steady state error

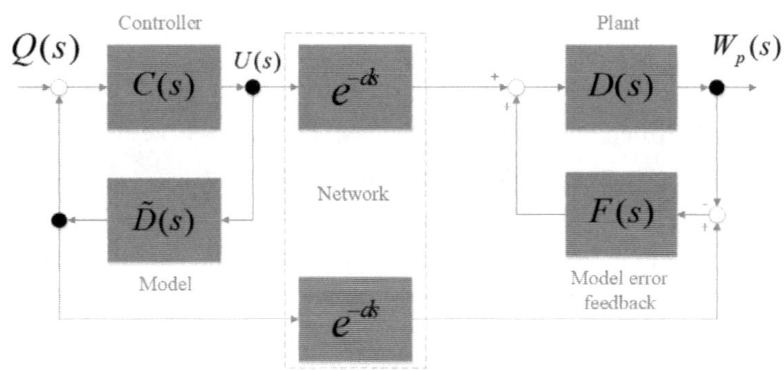

Figure 3. System model under the new controller.

As shown in Figure 3, it is the system model after setting up the new controller. Where $\tilde{D}(s)$ is the established model and $F(s)$ is the model error feedback module. Even if there is an undetermined steady-state deviation between the object transfer function $D(s)$ and the $\tilde{D}(s)$ G of the feedback model.

In order to detect the stability of the method, simulate the interference of time delay on the closed-loop system by inputting a step signal to the target value $Q(s)$. The transfer function of the system for obtaining $Q(s)$ to $W_p(s)$ simulation data from equation (5)

$$\frac{W_p(s)}{Q(s)}=\frac{C(s)D(s)e^{-ds}(1+F(s)\tilde{D}(s))}{(1+C(s)\tilde{D}(s))(1+F(s)D(s))} \quad (5)$$

There is no unknown function time delay element e in the denominator of equation (5). Therefore, the network near-end controller $K(s)$ and the network far-end feedback model $F(s)$ can be designed using the traditional linear time-invariant design method. Now suppose that the object transfer function $D(s)$ and the system model $\tilde{D}(s)$ have poles at the origin, namely $D(s)=D'(s)/s$ and $\tilde{D}(s)=\tilde{D}'(s)/s$. Therefore, Equation (5) is transformed into equation (6).

$$\frac{W_p(s)}{Q(s)}=\frac{K(s)D'(s)e^{-ed}(s+F(s)\tilde{D}'(s))}{(s+K(s)\tilde{D}(s))(s+F(s)D(s))} \quad (6)$$

The input set in this system is step input, i.e. $Q(s)=1/s$, the steady state error can be obtained in equation (7).

$$\lim_{s\to 0} s(Q(s)-W_p(s))=\lim s(1-\frac{W_p(s)}{Q(s)})Q(s)=\lim_{s\to 0} s\left(1-\frac{C(s)D'(s)e^{-ds}(s+F(s)\tilde{D}'(s))}{(s+C(s)\tilde{D}'(s))(s+F(s)D'(s))}\right)\frac{1}{s} \quad (7)$$

If the object transfer function $D(s)$ and the feedback model $\tilde{D}(s)$ have no poles at the origin. At the time of design, the network near-end controller $K(s)$ and the network far-end feedback model $F(s)$ will have poles at the origin, namely $C(s)=C'(s)/s$, $F(s)=F'(s)/s$.

$$\lim_{s\to 0} s(Q(s)-W_p(s))=\lim_{s\to 0} s\left(1-\frac{C'(s)D(s)e^{-ds}(s+F'(s)\tilde{D}(s))}{(s+C'(s)\tilde{D}(s))(s+F'(s)D(s))}\right)\frac{1}{s}=0 \quad (8)$$

Therefore, as can be seen from equation (8), the steady-state error of the system becomes zero at this time.

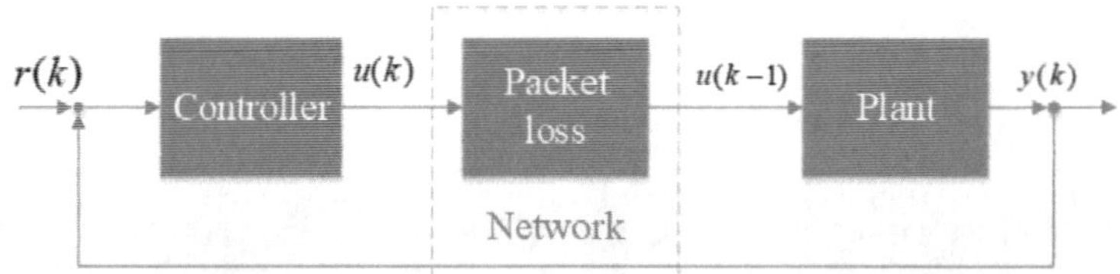

Figure 4. Structure of the impact of packet loss on system adjustment.

As shown in Figure 4, the environment is assumed to be a continuous input of the system for a certain period of time. The input mode can be zero. Assuming that time is discrete, continuous loss of data occurs when information is present in the network. At this time, the information $u(k)$ input to the controlled object becomes the $u(k-1)$ of the previous moment. And if the system input $u(k)$ is out of communication interference at time k , the communication interference is considered as $u(k)-u(k-n)$. However, the input value of the system object at time k is still $u(k-1)$. It can be seen that the input and output of the system in these two cases is equivalent. Even if continuous packet loss occurs, the communication interference at the system input $u(k)$ can be regarded as $u(k)-u(k-1)$. The above discussion shows that packet loss is equivalent to communication interference in the CDOB, so the packet loss is compensated.

5. Numerical simulation

This section will use data simulation to verify the proposed method. For the consideration of the model parameters, the model error is simulated by the special deviation of the DC gain, so that the gain obtained by D_m is larger than the gain obtained by D_p.

$$D_m=\frac{240}{152s(s+780)}, \quad D_p=\frac{140}{152s(s+780)}$$

Since the controller $K(s)$ and the error feedback model $F(s)$ of the method are designed as separate units. Therefore $K(s)$ and $F(s)$ are set to be implemented as follows:

$K(s)=45+10s, \quad F(s)=45+s$

IOP Conf. Series: Journal of Physics: Conf. Series **1284** (2019) 012056 doi:10.1088/1742-6596/1284/1/012056

In order to evaluate the compensation effect of the proposed method for long delay, the Smith predictor was added to the simulation for comparison. So the sampling period for the simulation is set to $T = 0.001s$. Set a longer time delay $d(k) = 4s$. The simulation results are shown in Figure 5.

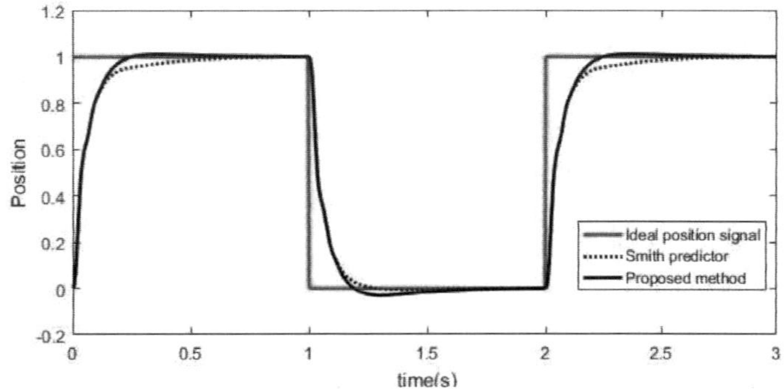

Figure 5. Feedback delay curve of time delay on system state.

A clear conclusion can be drawn from Figure 5. When the controller input signal is issued and the system network has a delay. In this paper, the controller design method can successfully reach the signal position. Moreover, it can be seen that this method can achieve the ideal target value faster than the Smith predictor. Although there are some very slight deviations after reaching the target value, the system is able to correct itself very quickly. So in some cases this method has better features.

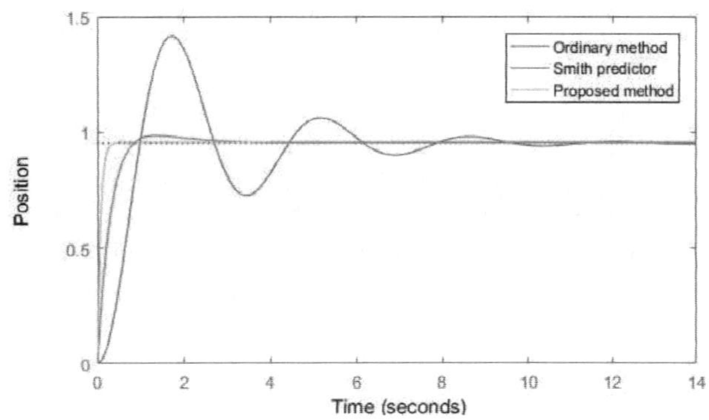

Figure 6. Status trajectory of the system after communication interference.

In order to detect whether the proposed method is robust in the case of system packet loss. Assume that the time delay is evenly distributed, i.e. $d(k) = 0.12s$, and the upper and lower limits of the boundary value are set.

$$d_{max}(k) = 3.50\text{sec}, \quad d_{min}(k) = 3.20\text{sec}$$

Since the time delay is set to be evenly distributed, it is not considered for the time being.

$$K(s) = 9s + 450; \quad F(s) = 9s + 450$$

Figure 6 shows the impact of packet loss testing on system stability. Where the yellow line is the curve of the proposed method. It can be clearly seen that good stability can be maintained even in the case of a serious packet loss rate of 10% in a short time. The above simulation experiments verify the robustness of the proposed method to packet delay, network packet loss and model error.

6. Conclusion

This paper is aimed at network control systems with long delay, packet loss and model establishment. Considering the error caused by modeling, a method of adding communication interference observer (CDOB) to the system is proposed to reduce the influence of communication interference. The method uses a dual feedback mode, which to build a model on the remote controller side. This model feeds back information to both the near-end controller and the remote object; Model error feedback is established on the far-end object side to compensate for steady-state errors. Simulation experiments show that the controller model designed in this paper has good stability under various communication interferences.

References

[1] You K Y and Xie L H. Survey of recent progress in networked control systems *J.* Acta Automatica Sinica, 2013,39(2): 101-118.

[2] Zhang Dongmei, Yu Li and Zhou Minghua. Stabilization of Networked Control Systems withFast Time-varying Delay and Packet Loss *J.* Control Theory and Applications, 2008, 25(3).

[3] Halder K, Das S and Dasgupta S. Controller design for networked control systems —An approach based on L2 inducednorm *J.*Nonlinear Analysis Hybrid Systems, 2016,19: 134-145.

[4] Schenato L. Optimal estimation in networked control systems subject to random delay and packet drop *J.* IEEE Trans on Automatic Control, 2008, 53(5): 1311-1317.

[5] Turner J R. Tow ards a theory of project management: The functions of project management *J* . Int J of Project M anagement, 2006, 24(3): 187-189.

[6] C. Tan, L. Li and H. Zhang: "Stabilization of networked control systems with both network-induced delay and packet dropout", Automatica, Vol.59, pp.194–199 (2015)

[7] Gao H J, Meng X Y and Chen T W. Stabilization of networked control systems with a new delay characterization *J.*IEEE Trans on Automatic Control, 2008, 53(9): 2142-2148.

[8] Wang J F and Yang H Z. Exponential stability of a class of networked control systems with time delays and packet dropouts *J.* Applied Mathematics and Computation, 2012,218(17): 8887-8894.

[9] Cai Yunze, Pan Ning and Xu Xiaoming. H∞ robust filtering for networked control systemswith long delay and packet loss *J.* Control and Decision, 2010, 25(12).

[10] Zhang T Y and Liu G P. Tracking control of wheeled mobile robots with communication delay and data loss *J.* Journal of Systems Science & Complexity, 2018(4):1-19.

[11] Liu Yicai, Liu Bin and Zhang Yong. Stability Analysis of Networked Control Systems with Bilateral Random Delay and Packet Loss *J.* Control and Decision, 2017(9).

[12] Mao Z, Jiang B and Shi P. *H*∞fault detection filter design for networked control systems modelled by discrete Markovian jump systems *J.* IET Control Theory and Applications, 2007, 115(5): 1336-1343.

IOP Conf. Series: Journal of Physics: Conf. Series **1284** (2019) 012057 doi:10.1088/1742-6596/1284/1/012057

White Light Transmitting Antenna for Four-spectrum Mobile WDM-VLC

Donglei He[1], Yi Tang[2], Yuee Chang[2] and Pingan Li[2]

[1.] Beijing Institute of Spacecraft System Engineering,Beijing 100051, China
[2.] School of Optics and Photonics, Beijing Institute of Technology, Beijing 100081, China.

E-mail:tangyi4510@bit.edu.cn

Abstract. In this paper, we propose a mixing white light antenna with a homogenizer and an amplification system, and demonstrated its utilization in a four-spectrum mobile VLC. The central element is the homogenizer, which is a square integrating rod that mixes four-spectrum light into white light. Behind it, there are two Fresnel lenses to amplify the light-mixing area. The system is compact with a length of 80 mm and maximal diameter of 58.5 mm. We determined the initial structural parameters of the antenna by theoretical analysis, and then we determine the best parameters by simulation. Simulation results indicate that a 0.75 m×0.75 m area can be illuminated at 1.7 m distance. The detection surface has a high uniformity (76.27%), which can meet illumination and mobile communication requirements simultaneously.

1. Introduction

As a new generation of illumination sources, light emitting diodes (LEDs) have many advantages, such as environmental protection, high efficiency, low cost, and modulated performance. An LED-based VLC can illuminate a room and transmit data simultaneously, relieving the tension of a wireless spectrum. An indoor VLC has no radiation interference and is considered to be one of the most promising devices for use in wireless communication [1].

With the development of modulation and equalization technology, the single-channel VLC can already transfer data at Gbps rates. Devices have gradually developed into multi-channel VLC devices by using WDM technology [2-6]. Transmitting antennas, whether in a monochrome VLC system or a multicolor VLC system, are generally used to improve transmission distance or acquire suitable color temperature, illumination value, and high uniformity to meet the dual requirements of mobile communication. Until 2015, Wang Yiguang et al. [7] applied a reflection cup with 60° divergence angle to decrease the beam angle of RGB-LEDs for longer transmission distance. When combined with CAP modulation and equalization, an aggregate indoor free space data transmission rate of 4.5 Gbps was demonstrated over 1.5 m. In the same year, Wang et al. [8] used the same transmitting antenna, but the light source was changed to RGBY-LEDs. Then the data rate improved to 8 Gbps at 1 m. However, this optical antenna produces a color ring with low uniformity. With the development of multi-channel VLCs in 2016, Cui et al. [9] showed that a WDM-VLC system can transmit in up to 10 channels by theoretical analysis, i.e., the number of spectrum in an LED can be 10. In multi-colour WDM-VLC systems, the transmitting antenna plays a significant role in improving transmission distance, mixing color into an appropriate white light, and emitting highly uniform illumination for mobile communication and indoor illumination simultaneously.

This paper proposed a white light transmitting antenna for a mobile VLC system based on RGBY-LEDs, which consists of an integrating rod and two Fresnel lenses. The simulation and experimental results all show that our antenna achieved a suitable color temperature of 5583 K and a high uniformity of 73%. The structure of the system is compact and cheap, with 80 mm length and 58.5 mm maximum diameter.

2. Model of white light transmission antenna

2.1. Introduction of white light transmitting antenna

Like in Figure. 1, an indoor VLC system based on RGBY-LEDs includes Uplink and Downlink. Uplink, that is, a transmitter, mainly includes an LED light source and a transmitting antenna. The transmitting antenna forms the uplink terminal and plays an important role in indoor illumination and mobile communication.

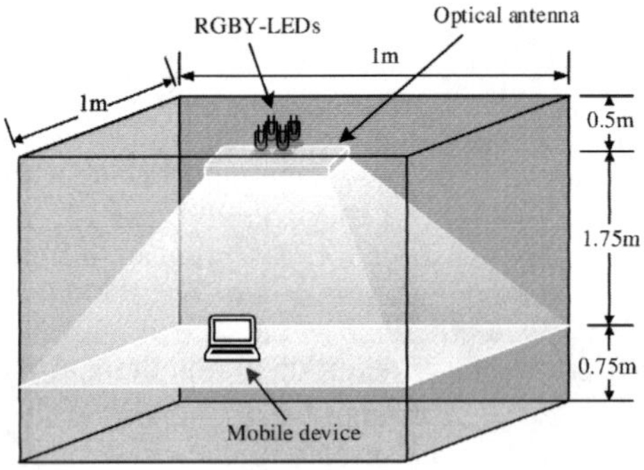

Figure 1. Indoor VLC model system based on RGBY-LEDs.

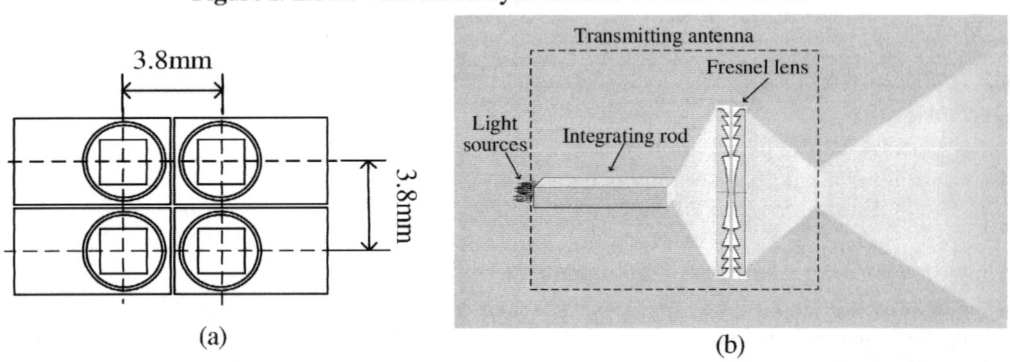

Figure 2. (a) Arrangement and spacing of light sources and (b) white light transmitting antenna system

According to indoor four-color WDM-VLC application demands and international illumination standards, we proposed the design standards of this white light transmitting antenna in table 1. The illumination uniformity and chromatic aberration should both be considered for illumination demands and signal to noise ratio (SNR) stability in mobile communication.

Table 1. White light transmitting antenna technical standards for a four-color VLC system

Parameter	Technical data specification

waveband range	400 nm–760 nm
illumination uniformity	≥70%[10]
chromatic aberration	<0.005[11]
communication distance	≥1.75 m
illumination area	>0.5×0.5 m2[10]
illumination value	200-800 lx[10]

2.2. System model and theoretical analysis of the transmitting antenna

At present, integrating rods and micro lens arrays are the most commonly used optical elements for reshaping a light beam for uniform illumination [12, 13]. A micro lens array can narrow the volume of the system, but incident light rays must be parallel, and the cost is high. The size of an integrating rod can be controlled. Incident parallel light rays are not required, and the cost is low. The light sources in our VLC system are LUXEON series Red (LXM2-PD01-0050), Yellow (LXML-PH01-0060), Green (LXML-PM01-0100) and Blue (LXML-PB02) (Philips) with 627 nm, 597 nm, 530 nm, and 470 nm peak wavelengths, respectively. These LEDs were arranged in a 2×2 array, as shown in figure 2 (a). Each LED has a similar large divergence angle. Thus, we used a square integrating rod.

The light mixing results from the integrating rod only appear in the output plane. If VLC system want to illuminate the room and transmit data at the same time, we should image the output plane in to obtain uniform illumination in a large area on the ceiling. This paper proposes a Fresnel lens design scheme. Compared with a traditional lens, Fresnel lenses is lightweight, and it facilitate system miniaturization. They also have large relative aperture, which can ensure energy efficiency. The entire system shown in figure 2 (b) can project a large illumination area at a far distance.

*2.2.1. Integrating-rod model.*The integrating rod can be either hollow or solid [14] and both rods will project beams with high efficiency and uniformity. A homogeneous N-BK7 solid rod was used in this study. The ray in the integrating rod travels under total internal reflection.

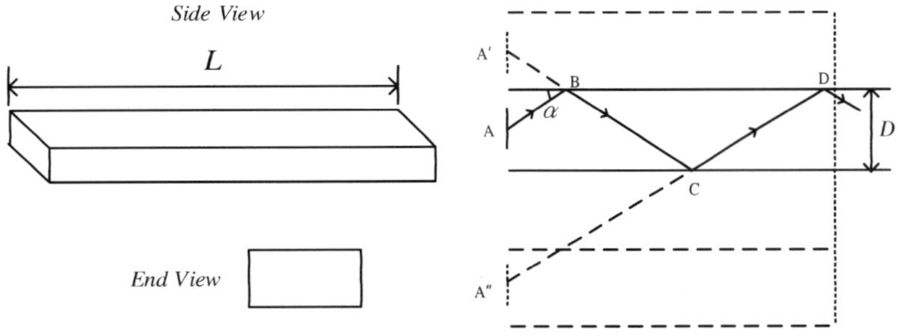

Figure 3. Integrating rod and internal light path diagram.

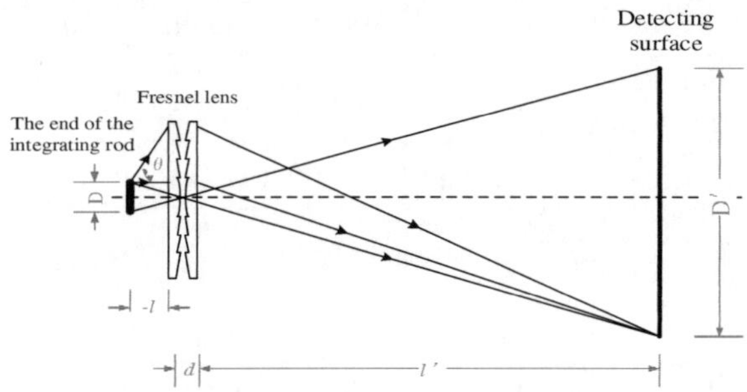

Figure 4. Schematic diagram of Fresnel lenses.

As shown in Figure. 3, by using the image method, each reflection is equivalent to illumination by a virtual source. The color and illumination of the true and virtual sources are superposed on the output end of the rod, thus we can realize color mixing and high illumination uniformity. The total internal reflection times N is related to uniformity and can be expressed as

$$N = \text{int}[\frac{Ltan\alpha}{D}] \tag{1}$$

where D and L are the diameter and length of the integrating rod, respectively, and α is the refraction angle at the input end of the integrating rod.

N must be at least 5 if more than 70% illumination uniformity is desired [15]. A greater number of reflections results in better uniformity at the output end. Generally speaking, we can only increase N by increasing the length L, but light absorption losses will also increase. The value of L cannot be too large as a long rod inhibits minimization.

Relatively speaking, the value of D has a greater impact on energy utilization than uniformity. In non-imaging optics, the energy transmission between systems is typically described by the Etendue value. This parameter can be defined as:

$$Etendue = n^2 \iint cos(\theta)dAd\Omega \tag{2}$$

The Etendue of the light source (RGBY-LEDs) is defined as:

$$E_{source} = K\pi n_S^2 sin^2\alpha(3A_{S_1} + A_{S_2}) \tag{3}$$

In Eq. (3), n is the internal refractive index of the light source, and α is LED's divergence half-angle. RGY-LEDs have the same luminous area AS1, and the luminous area of the blue LED is A_{S_2}.

We can also calculate the Etendue of the rod's output end using:

$$E_{rod} = \pi A_r n_r^2 sin\beta \tag{4}$$

where n_r is the rod's refractive index, A_r is the area of output end, and β is the effective ray incidence angle at the input end (sinβ=sinα/nr).

The design is reasonable if $E_{rod} > E_{source}$, and the energy utilization is proportional to E_{rod}. In fact, the value of D is limited due to its influence on uniformity. Fresnel lenses are Etendue-limited systems, and they can only work effectively within a certain range of NA. Considering all these limitations, we found that D falls in the range 6 mm<D<10 mm. Equation (1) was used to determine the range of L to be L>44 mm.

2.2.2 Fresnel-lens model. According to the standard room model, we designed an amplification system using two identical Fresnel lenses, and the schematic diagram is shown in Figure 4.

The distance between the rod and the Fresnel lens l and the focal length of Fresnel lens f' can be calculated to be 14.16 mm<|l|<23.60 mm, and 14.05 mm<f'<23.39 mm, respectively, according to Eq. (5):

$$\begin{cases} \frac{1}{l'} - \frac{1}{l} = \frac{1}{f'} \\ \frac{-l}{l'} = \frac{D}{D'} \end{cases} \tag{5}$$

IOP Conf. Series: Journal of Physics: Conf. Series **1284** (2019) 012057 doi:10.1088/1742-6596/1284/1/012057

The transmission distance l' is 1.77 m, and the length and width of the illumination area D' are both 0.75 m.

We can now calculate that the focal length range of one Fresnel lens must be 27.60 mm<f1'<46.07 mm using Eq. (6)

$$\frac{1}{f'} = \frac{2}{f_1'} - \frac{d}{f_1'^2} \tag{6}$$

Where d = 1 mm is the distance between two Fresnel lenses.

3. Design and optimization of white light transmitting antenna

3.1. Design of the white light transmitting antenna

According to the above values and the aforementioned standards, we optimized the design parameters for the white light transmitting antenna by ZEMAX. We found that an antenna with a square rod showed the best performance. We also designed an antenna with a hexagon rod as an example for comparison. The details are presented in Table 2.

Table 2. Parameter list of the white light transmitting antenna system

	parameters	values	
Integrating rod	Shape of the end	square	hexagon
	Side-length of input end (mm)	7	4.35
	Side-length of output end (mm)	7	4.35
	Length (mm)	60	60
Fresnel lens	Material	ACRYLIC	
	Groove/inch	232	
	Diameter (mm)	50.8	
	Focal length(mm)	51	
	Thickness (mm)	1.5	

In our design, each LED is placed next to the input end of the integrating rod, which can improve the efficiency and narrow the volume of the system. The distance between the rod and the Fresnel lens is 13.5 mm. The maximum diameter of the transmitter is 50.8 mm, and the total length is 80 mm.

3.2. Simulation of the transmitting antenna

According to the ANSI-NAMP IT 7.228-1997 illumination standard, we used 9 points method to calculate the uniformity degree with the following equations:

$$E_{ave} = \frac{\sum_{i=1}^{n} E(i)}{n} \tag{7}$$

$$N_1 = \left[1 - \frac{E_{max} - E_{ave}}{E_{ave}}\right] \times 100\%$$
$$N_2 = \left[1 - \frac{E_{min} - E_{ave}}{E_{ave}}\right] \times 100\% \tag{8}$$

$$N = min[N_1, N_2] \tag{9}$$

If we substitute the data shown in Figure. 5 into Eq. (7) to (9), the value of N_{square} was calculated to be 76.27%. Because of the limits of the simulation software, we cannot calculate an accurate value for the H-rod system. We can visually see that its uniformity is lower than that of the S-rod system.

The efficiency of the S-rod and H-rod systems were 45.87% and 43.03%, respectively. The practical illumination value can be adjusted based on different demands.

4. VLC experiments with white light transmit-ting antenna

4.1. Illumination experiment with transmitting antenna

As shown in Figure 6, according to the above simulation design, we built a model machine for the white light transmitting antenna.

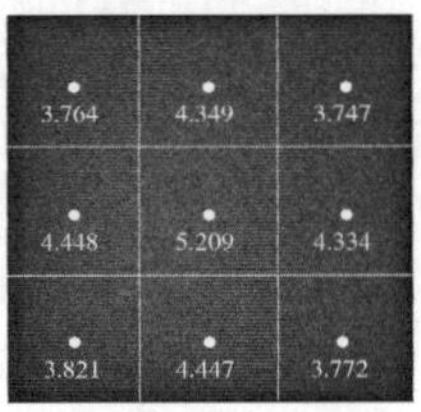

Figure 5. Detection surface: illumination distribution (lm)

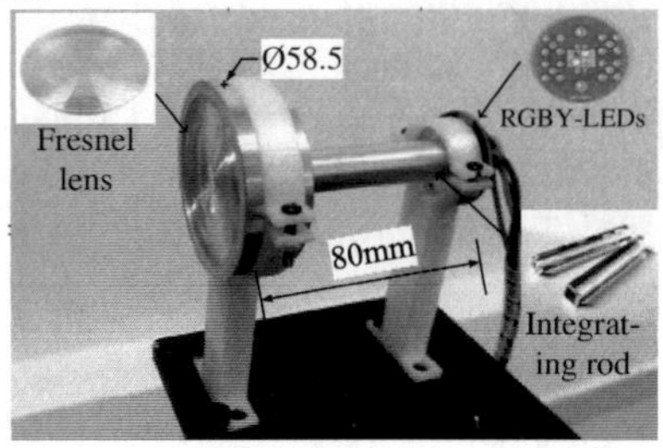

Figure 6. Model machine of the white light transmitting antenna

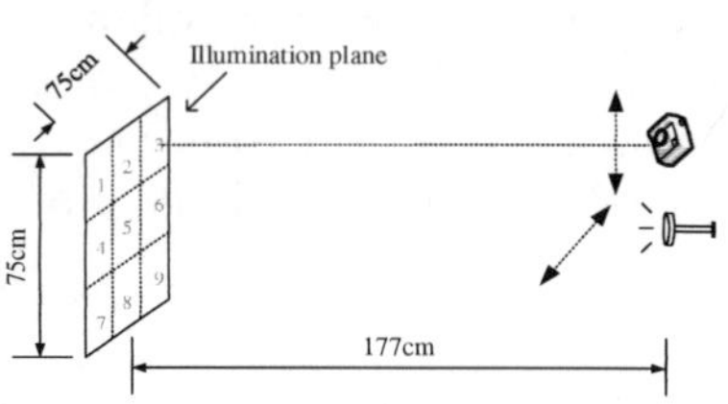

Figure 7. schematic diagram of 9 points method.

We illuminated a square region with 75 cm side length and a hexagon region with 46 cm side length at a distance of 1.77 m in a darkroom just as shown in Figure 7. The experimental results are shown in Figure 8, which shows the true-color graph of the square integrating rod system (S-rod system) and the hexagon integrating rod system (H-rod system) by Monte Carlo ray tracing method.

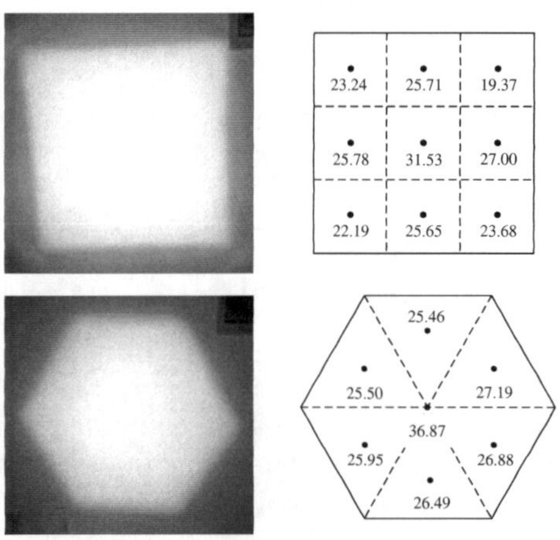

Figure 8. Actual illumination surface and illumination distribution (lx)

We can calculate total illumination E and the uniformity degree N using Eqs. (7) to (9). The S-rod system showed the best performance. The experimental results agree well with the simulation results.

$$E_{square} = 224.15lx, E_{hexagon} = 194.34lx \quad (10)$$

$$N_{square} = 73.37\%, N_{hexagon} = 67.18\% \quad (11)$$

The color temperature of the detection surface is 5583 K. Compared with a white color coordination value of 5500 K (CIE color coordination system), the chromatic aberration is 0.0021, which is less than eye resolution (<0.005). Thus, the device is suitable for indoor illumination.

4.2. Mobile VLC experiment with transmitting antenna

The setup of the monochromatic VLC system is shown in Figure 9. We used a blue LED as an example, other LEDs' results are similar with this blue LED because of the high uniformity of this system. We set up a VLC experimental link shown in Figure. 9(b) with a 1.77 m communication distance. An arbitrary sine wave generator (Agilent N9310A) with frequency set to 170 MHz was used to set the data transfer rate to 340 Mbps. 25 points are evenly selected in the 9 equal-sized areas in order to verify the mobile VLC performance. In the block diagram, we moved the transmitting terminal using translation stages.

As shown in Figure 10, all signal are the same. As shown in Figure 11, the maximum voltage is 2.08 V. The minimum value is 1.86 V.Thus all values meet the 3 dB attenuation requirement.

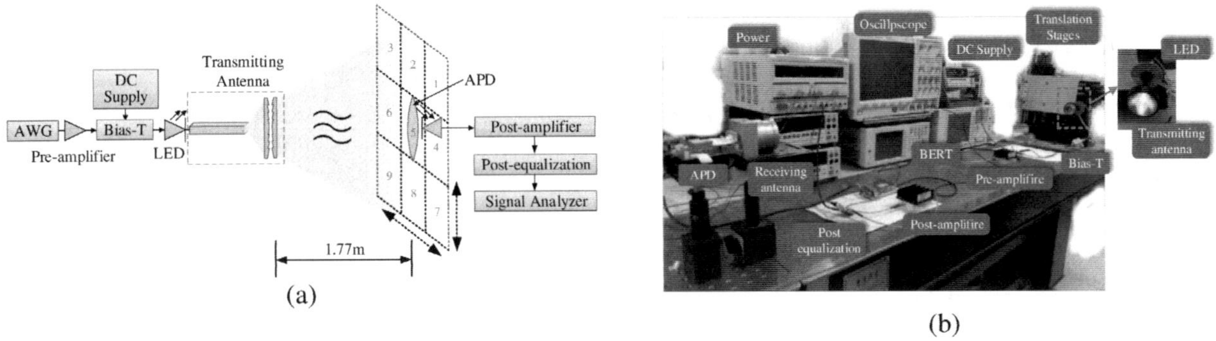

Figure 9. (a) Block diagram for mobile communication testing; (b) VLC experimental link.

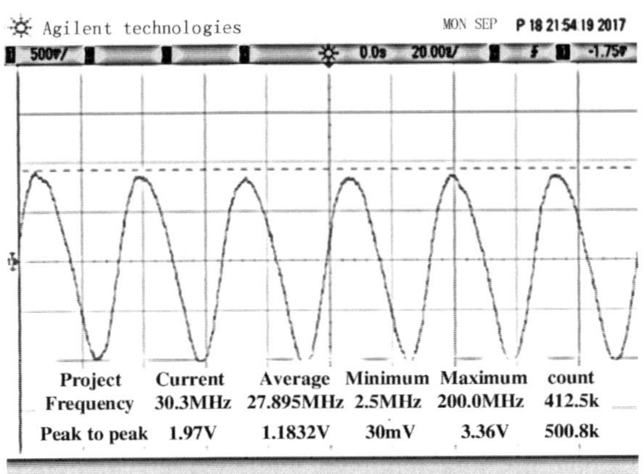

Figure 10. Waveform graph

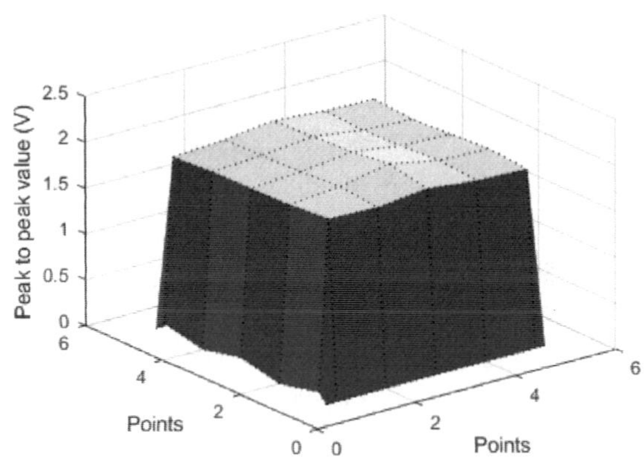

Figure 11. Peak to Peak value for each waveform

5. Conclusion

We presents a white light transmitting antenna based on an indoor four-spectrum VLC system, which included a square integrating rod and a Fresnel lens system. This antenna mixes beams from an array of RGBY-LEDs into white light with high uniformity (73.37%) and projects the mixed beam into an illumination area of 0.75 m×0.75 m at 1.77 m transmission distance. The CCT of the detection surface is 5583 K which is less than the value for an eye (<0.005).

Acknowledgments

The National Key Basic Research Program 973 Project (No.2013CB329202); National Natural Science Foundation of China (No. 61571067).

IOP Conf. Series: Journal of Physics: Conf. Series **1284** (2019) 012057 doi:10.1088/1742-6596/1284/1/012057

References

[1] Dong, S., Wan, J., Jia, Z., Yu, Y., Tao, W., & Song, W.. (2016). A voice transmission system based on visibllight communication. IEEE Global Conference on Consumer Electronics. IEEE.
[2] M. S. Gujar, S. Velankar, and A. Chavan, "Data Transmission Using Visible Light Communication," Journal of Engineering Science 2263 (2016).
[3] 750mb/s monochromatic led-based real-time visible light communication system employing a low-complexity cascaded post-equalizer. Chinese Optics Letters(12), 55-59. (2016).
[4] Huang, X., Wang, Z., Shi, J., Wang, Y., & Chi, N.. (2015). 16 Gbit/s phosphorescent white led based vlc transmission using a cascaded pre-equalization circuit and a differential outputs pin receiver. Optics Express, 23(17), 22034.
[5] Z. Wang, F. Chen, and J. Weng, "PAPR Reduction of OFDM Signal for Visible Light Communication by Using Hartley Transform and Companding Techniques," National Conference on Electrical, Electronics and Computer Engineering 2015.
[6] G. Zhang, et al. "Low-complexity frequency domain nonlinear compensation for OFDM based high-speed visible light communication systems with light emitting diodes," Optics Express 25(4) (2017): 3780-3794.
[7] Wang, Y., Huang, X., Tao, L., Shi, J., & Chi, N.. (2015). 4.5-gb/s rgb-led based wdm visible light communication system employing cap modulation and rls based adaptive equalization. Optics Express, 23(10), 13626.
[8] Wang, Y., Tao, L., Huang, X., Shi, J., & Chi, N. (2015). 8-Gb/s RGBY LED-based WDM VLC system employing high-order CAP modulation and hybrid post equalizer. IEEE Photonics Journal, 7(6), 1-7.
[9] Lu Cui ; Yi Tang ; Huiping Jia ; Jiabin Luo ; Bruce Gnade., "Analysis of the Multichannel WDM-VLC Communication System," Journal of Lightwave Technology 34(24) (2016): 5627-5634.
[10] European standard EN 12464–1: Lighting of indoor work places, (2003).
[11] Perduijn, A., Krijger, S. D., Claessens, J., Kaito, N., & Okada, S.. (2003). 43.2: light output feedback solution for rgb led backlight applications. SID Symposium Digest of Technical Papers, 34(1).
[12] Zhao, X. , Fang, Z. , Cui, J. , Zhang, X. , & Mu, G. . (2007). Illumination system using led sources for pocket-size projectors. Applied Optics, 46(4), 522-6.
[13] W. S. Sun, et al., "Compact LED projector design with high uniformity and efficiency," Applied Optics 53(29) (2014): p. H227-32.
[14] B. A. Jacobson, R. D. Gengelbach, and J. M. Ferri, "Beam shape transforming devices in high-efficiency projection systems," SPIE 3139 (1997): 141-150.
[15] Dimas, C. F., Read, S., & Kuta, J. J.. (2002). Integrating rod homogeneity as a function of cross-sectional shape. International Symposium on Optical Science & Technology. International Society for Optics and Photonics.

Supportability Evaluation of Radar Equipment Based on Cloud Model and Improved RS-G1

SHI Duanyang[1, 2, *], HU Bing[1], CHEN Jiaxun[1] and DONG Ruijie[1]

[1] Air Force Early Warning Academy, Wuhan, China
[2] 95174 PLA Troops, Wuhan, China

[*]Corresponding author: 644481846@qq.com

Abstract. In the context that factors affecting the supportability of radar equipment are numerous and of diverse dimensions, the existing index systems of supportability evaluation of radar equipment can hardly satisfy the continuously improving requirements of radar equipment support. First, aiming at the problem of transformation between quantitative description and qualitative concept, this paper converts the evaluation index set into an evaluation cloud model based on the cloud theory, generates the digital characteristics of the cloud model of the expert scoring by using the reverse cloud generator and the supportable comprehensive evaluation cloud drop chart of radar equipment by using forward cloud generator. Second, to make up for the shortcomings of the existing single weighting method, the Rough Set (RS) weighting method is improved with the introduction of the Order Relationship Analysis method (G1). Then this paper proposes the RS-G1 synthetic weighting method, and designs radar equipment supportability cloud evaluation based on this method. In the end, example shows that this method can evaluate the supportability level of radar equipment scientifically and effectively.

1. Introduction

The supportability of radar equipment refers to the capability that the design characteristics and planned support resources of radar equipment can meet the requirements of peacetime combat readiness and wartime use [1]. At present, evaluation methods of radar equipment supportability mainly include set pair analysis method [2], fuzzy evaluation method [3], grey correlation theory [4], etc. However, these methods consider the fuzziness and randomness of evaluation indexes separately, or quantify the fuzziness by using the fuzzy set method, or quantify the randomness by using the probability method, without taking these two uncertainties into comprehensive consideration. Cloud model [5] can use the form of cloud droplets to directly reflect the level of supportability evaluation of radar equipment, which is an effective method to study the supportability evaluation of radar equipment. However, in the cloud model, how to reasonably calculate the index weight is a very important issue.

At present, the calculation methods of weight coefficient can be divided into two basic categories according to their characteristics: subjective weighting methods and objective weighting methods [6]. But both of them have their own advantages and disadvantages. It is an effective way to solve the defects of single weighting method by combining the two weighting methods, making use of the advantages of them and drawing on each other's strength to realize synthetic weighting.

In this paper, combining the cloud theory with the improved RS-G1 comprehensive weighting method, this paper gives full play to the advantages of the two methods, and uses the comprehensive weighting method to settle the issue of determination of the index weight value in the supportability evaluation of radar equipment based on cloud model. A more accurate supportability level can be obtained through the cloud model, which can provide an auxiliary decision-making basis for the product upgrade of the developer and the equipment procurement of the user.

2. Construction of Radar Equipment Supportability Evaluation Index System

According to the definition of supportability of radar equipment, supportability requirements are divided into comprehensive supportability requirements, design requirements related to supportability and supportability resource requirements [7]. Therefore, the evaluation index of supportability of radar equipment can be divided into three parts: comprehensive supportability, design supportability and resource supportability.

Comprehensive supportability is proposed based on the capability of the equipment to complete and maintain specified tasks in the expected peacetime and wartime use conditions, which reflects the overall expectation of the military for the equipment supportability, and its indexes mainly include readiness, availability, mission sustainability, life cycle cost, etc.

Design supportability reflects the characteristics of radar equipment related to supportability, which directly affects the design of radar equipment and is an important requirement for the radar equipment to be easily support. Its indexes mainly include reliability, maintainability, testability, security and electromagnetic compatibility, etc.

Resource supportability refers to the resource requirements planned for the smooth implementation of equipment use and support in peacetime and wartime. It determines the variety and quantity of support resources, involving support personnel, support spare parts, support equipment, technical data, etc.

3. Algorithm Introduction

3.1 The improved RS-G1 comprehensive weighting method

3.1.1 The improved RS weighting method. In rough set theory, when the evaluation problem is expressed as a decision table, the weight of the index is the attribute importance, and the index weight can be determined by calculating the attribute importance. To calculate the importance degree of an index, you need to remove it from the table and examine how classification changes after the attribute is removed. If this attribute is removed, the classification will be changed accordingly, indicating that the importance of this attribute is high, otherwise the importance is low [8]. The information system decision table and attribute importance are defined as follows in [8,9].

Definition 1: The decision table of the comprehensive evaluation information system [9] is $S=\{U,R,V,f\}$, in which U represents the collected importance rating score data of each index given by several experts, called the theory domain. R represents a set of attributes; $V=\bigcup_{r\in R} V_r$ represents a set of attribute values; V_r represents the range of attribute value; $f:U\times R\rightarrow V$ represents an information function that specifies the various attribute values of each object x in U , the specific expression is $\forall r\in R, x\in U, f\left(x,r\right)\in V_r$.

Definition 2 [9]: For the decision table $S=\{U,R,V,f\}$, $R=C\bigcup D$,condition attribute $c_i\in C(i=1,2,\cdots,n)$,the C positive domain of D is $pos_C(D)$, and the importance degree of c_i to decision attribute D is $sig(c_i)$.

$$pos_C(D)=\{Y\in U/C \,|\, Y\subseteq D\} \tag{1}$$

$$sig(c_i) = \gamma_C(D) - \gamma_{C-c_i}(D) = \frac{card(pos_C(D)) - card(pos_{C-c_i}(D))}{card(U)} \tag{2}$$

The weight of each index can be obtained by normalizing the importance of each attribute:

$$W_{RS}(c_i) = \frac{sig(c_i)}{\sum_{i=1}^{n} sig(c_i)} \tag{3}$$

Thus, the greater the attribute important $sig(c_i)$ of an index c_i is, the greater its weight W_i will be. However, when the importance degree of an attribute is calculated by formula (2), the importance degree of an individual attribute may be zero, resulting in the weight value of this index being zero. Each index in the index system has its own significance. If the weight value of the index is zero, it is not in line with the actual situation. It can be seen that this method has shortcomings.

In order to ensure that the weight value of each index is greater than zero, it is necessary to ensure that the attribute importance of this index is greater than zero. Here, the concept of conditional entropy is introduced:

Definition 3 [10]: In the decision table $S = \{U, R, V, f\}$, the conditional entropy of decision attribute $D = \{D_1, D_2, \cdots, D_k\}$ relative to the conditional attribute $C = \{c_1, c_2, \cdots, c_n\}$ is:

$$I(D|C) = \sum_{i=1}^{n} \frac{|c_i|^2}{|U|^2} \sum_{j=1}^{k} \frac{|D_j \cap c_i|}{|c_i|}\left(1 - \frac{|D_j \cap c_i|}{|c_i|}\right) \tag{4}$$

The improved attribute importance is:

$$sig'(c_i) = I(D|C - \{c_i\}) - I(D|C) + \frac{\sum_{a\in C}|a(x)| - \sum_{a\in C-\{c_i\}}|a(x)|}{\sum_{a\in C}|a(x)|} \tag{5}$$

Then the improved weight of index c_i is:

$$W_{RS}{}'(c_i) = \frac{sig'(c_i) + I(D|\{c_i\})}{\sum_{a\in C}\{sig'(a) + I(D|\{a\})\}} \tag{6}$$

The attribute importance calculated by formula (5) is always greater than zero, so that the weight value of the index is always greater than zero. Therefore, by introducing the concept of conditional entropy, the problem of index weight value being zero in the original rough set weighting method can be avoided.

3.1.2 The G1 weighting method. In this paper, the order relation analysis method [11] is used to calculate the subjective weight of each index.

If the importance of the evaluation index x_i is greater than x_j, denoted as $x_i \succ x_j$ (symbol "$\succ$" indicates the superior relationship). If the evaluation indexes $x_1, x_2, \cdots, x_n$ have $x_1^* \succ x_2^* \succ \cdots \succ x_n^*$ based on an evaluation criterion, then order relationship between the evaluation indexes $x_1, x_2, \cdots, x_n$ is established according to "$\succ$". x_i^* is the evaluation index of item i $(i = 1, 2, \cdots, n)$ after $\{x_i\}$ is sorted in order relationship "$\succ$".

The ratio of importance between the evaluation indexes x_{k-1} and x_k is given as w_{k-1}/w_k, denoted as r_k:

$$r_k = w_{k-1}/w_k, k = n, n-1, \cdots, 3, 2 \tag{7}$$

If r_k given by the experts meets the following conditions:

$$r_{k-1} > \frac{1}{r_k}, k = n, n-1, \cdots, 3, 2 \tag{8}$$

Then the weight W_{G1} is:

$$W_{G1} = \left(1 + \sum_{k=2}^{n} \prod_{i=k}^{n} r_i\right)^{-1} \tag{9}$$

$$w_{k-1} = w_k r_k, k = n, n-1, \cdots, 3, 2 \tag{10}$$

3.1.3 The comprehensive weight. By using the method of multiplication synthesis normalization, the comprehensive weighting model is constructed as follows:

$$W_i = \frac{W_{iRS}{}' W_{iG1}}{\sum_{i=1}^{n} W_{iRS}{}' W_{iG1}} \tag{11}$$

3.2 The supportability cloud evaluation of radar equipment

The cloud model [12] is a transformation model between qualitative linguistic values and quantitative numerical domains, which can better describe the correlation between linguistic randomness and fuzziness.

The cloud model is determined by three numerical characteristics: expectation (Ex), entropy (En) and hyperentropy (He), denoted as $C(Ex, En, He)$. The cloud model is generated by forward cloud or reverse cloud generator algorithm. The forward cloud generator generates cloud droplets satisfying the normal cloud distribution rule according to the known digital characteristics $C(Ex, En, He)$. The generated cloud droplets constitute the whole cloud, thus the qualitative concept is expressed quantitatively. The reverse cloud generator, on the contrary, generates the corresponding digital characteristics through a given set of cloud droplets conforming to a normal cloud distribution rule as samples.

If the rating of radar equipment supportability is divided into five grades: excellent, good, average, poor and extremely poor, then the evaluation set is $V = \{excellent, good, average, poor, extremely\ poor\}$. Since it is a qualitative concept, a one-dimensional normal cloud can be used to describe each comment. For a comment with a bilateral constraint $[\lambda_{\min}, \lambda_{\max}]$, the expectation can be taken as the median of the constraint condition and the comment can be approximated by the cloud in the bilateral constraint region. The digital characteristics are calculated by the following formula:

$$\begin{cases} Ex = (\lambda_{\min} + \lambda_{\max})/2 \\ En = (\lambda_{\max} - \lambda_{\min})/6 \\ He = k \end{cases} \tag{12}$$

In the supportability evaluation index system of radar equipment, the m indexes are evaluated by n experts, and the matrix $S = (s_{ij})_{n \times m}$ is obtained. Then, the cloud model of the j index is as follows:

$$\begin{cases} Ex_j = \frac{1}{n}\sum_{i=1}^{n} s_{ij} \\ En_j = (\frac{\pi}{2})^{1/2}\frac{1}{n}\sum_{i=1}^{n}\left|s_{ij} - Ex_j\right| \\ Sx_j^{\ 2} = \frac{1}{n-1}\sum_{i=1}^{n}(s_{ij} - Ex_j)^2 \\ He_j = (Sx_j^{\ 2} - En_j^{\ 2})^{1/2} \end{cases} \tag{13}$$

According to formulas (13), the numerical characteristics of a certain level of indexes are obtained, and then the cloud model of the upper level indexes is calculated. Assuming that a certain upper level index includes n_k subordinate indexes, the cloud model of the upper level index is:

$$\begin{cases} Ex' = \dfrac{\sum_{j=1}^{n_k} Ex_j En_j w_j}{\sum_{j=1}^{n_k} En_j w_j} \\ En' = \sum_{j=1}^{n_k} En_j w_j \\ He' = \dfrac{\sum_{j=1}^{n_k} He_j En_j w_j}{\sum_{j=1}^{n_k} En_j w_j} \end{cases} \tag{14}$$

Calculate the membership of the topmost cloud droplet x_i according to the following formula:

$$\xi_j = \exp(-\frac{(x_i - Ex_j)^2}{2En_j^{\ 2}}) \tag{15}$$

Following the principle of maximum membership, the comprehensive evaluation level is the interval of the evaluation cloud corresponding to ξ_j with the largest membership degree.

4. Example analysis

Take a certain type of radar equipment as an example to evaluate its supportability. The grade of supportability evaluation of radar equipment is divided into five levels of I-V. According to the rating scale standard in literature [7], the scores of each level are: level I $(0.8,1]$ is excellent, level II $(0.6,0.8]$ is good, level III $(0.4,0.6]$ is average, level IV $(0.2,0.4]$ is poor, level V $[0,0.2]$ is extremely poor. According to the radar equipment supportability evaluation index system and the cloud evaluation model based on the improved RS-G1, the following calculation and analysis were performed.

4.1 Improved RS method to determine the objective weight

Under the improved RS weighting method, the weight of each evaluation index in the supportability evaluation index system of radar equipment is:

$W_{RS}' = (0.0709, 0.0837, 0.0558, 0.1007, 0.0516, 0.0965, 0.0324, 0.1157, 0.0773, 0.0901, 0.0965,$
$0.0580, 0.0709)$

IOP Conf. Series: Journal of Physics: Conf. Series **1284** (2019) 012058 doi:10.1088/1742-6596/1284/1/012058

4.2 G1 method to determine the subjective weight

Under the G1 weighting method, the weight of each evaluation index in the supportability evaluation index system of radar equipment is:

$$W_{G1} = (0.0954, 0.0682, 0.0682, 0.0524, 0.0885, 0.0804, 0.0665, 0.0731, 0.0665, 0.0978, 0.0808, 0.0735, 0.0889)$$

4.3 Determine the comprehensive weight

After the objective weight is calculated according to the improved RS method and the subjective weight is calculated by the G1 method, the comprehensive weight can be calculated.

$$W = (0.0881, 0.0743, 0.0495, 0.0687, 0.0595, 0.1010, 0.0281, 0.1101, 0.0669, 0.1147, 0.1015, 0.0555, 0.0821)$$

4.4 Supportability evaluation of radar equipment based on cloud model

According to the grade standard of supportability evaluation of radar equipment and formula (12), the digital characteristics of the standard evaluation of radar equipment supportability based on cloud model can be obtained, see table 1.

Table 1. Digital characteristics of the cloud model for standard reviews.

Standard reviews	(Ex, En, He)
Excellent	(0.9,0.0333,0.013)
Good	(0.7,0.0333,0.008)
Average	(0.5,0.0333,0.005)
Poor	(0.3,0.0333,0.008)
Extremely poor	(0.1,0.0333,0.013)

Seven experts including radar manufacturers, research institution and using units are invited to score each evaluation index system according to the evaluation criteria.

The numerical characteristics of each index corresponding to the cloud model are calculated according to formulas (13). According to the weight values of all levels of indexes and formulas (14), the digital characteristics of the comprehensive evaluation cloud model of the type of radar equipment supportability is (0.7376,0.0350,0.0207). The generated comprehensive evaluation cloud is compared with the standard evaluation cloud, as shown in figure 1.

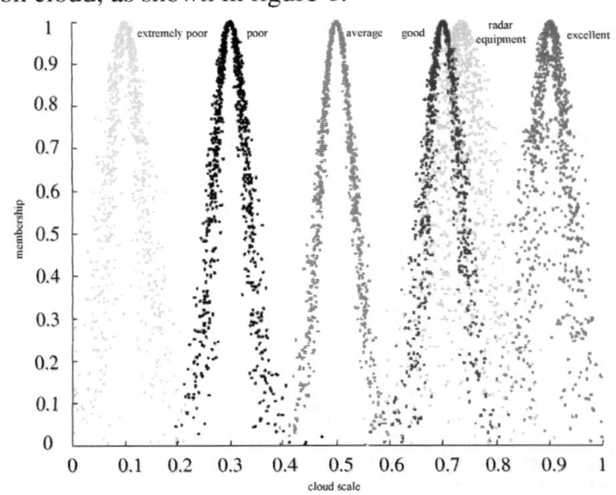

Figure 1. Comprehensive evaluation of radar equipment supportability cloud drop diagram

As can be seen from figure 1, most of the cloud droplets in the comprehensive supportability evaluation of radar equipment are in a "good" position, while the rest are in an "excellent" position. According to formula (15), the membership of evaluation are $\xi_{\mathrm{I}} = 6.85\times10^{-6}$, $\xi_{\mathrm{II}} = 0.53$, $\xi_{\mathrm{III}} = 8.81\times10^{-12}$, $\xi_{\mathrm{IV}} = 3.17\times10^{-38}$, $\xi_{\mathrm{V}} = 2.46\times10^{-80}$. Following the principle of maximum membership, the supportability evaluation of radar equipment level for II "good", which is consistent with the conclusion of cloud droplets diagram.

5. Conclusion

In this paper, according to the requirements of the definition of radar equipment supportability, the supportability evaluation index system of radar equipment is established. By improving the RS-G1 comprehensive weighting method, the deficiency of the single weighting method is made up, so that the calculation of index weight is more reasonable. According to the evaluation criteria of the supportability evaluation of radar equipment, the evaluation criteria are converted into the digital characteristics of the cloud model to better express the fuzziness and randomness in the evaluation process, and the evaluation process is more scientific. The example shows that the supportability evaluation method of radar equipment based on the cloud theory is feasible, which provides a new method for the supportability evaluation of radar equipment, and provides a basis for the manufacturers and users to promote the supportability of radar.

However, the supportability level of radar in this paper depends on the comparative analysis between the evaluation index cloud model and the evaluation index standard comments, and this comparative analysis relies on the knowledge and experience of experts, which is a problem worth studying in the next step.

References:

[1] Yang B X 2001 *Radar Integrated Technical Support Project* (Beijing: China Standard Press) p 17
[2] Cheng Y, Hu B and Dong W 2018 Supportability Evaluation of Radar Equipment Based on SPA-AHM *Modern Defence Technology* **46** pp 115-150
[3] Zhao Y, Zhai C S and Yan S Q 2009 Fuzzy Comprehensive Evaluation of Battle Effectiveness of New-type Radar Equipment *Journal of Air Force Radar Academy* **23** pp 324-326
[4] Cheng Y, Hu B and Dong Y H 2019 Supportability Evaluation of Radar Equipment Based on Grey-AHM *Modern Radar* **41** pp 9-13
[5] Li D Y and Du Y 2005 *Artificial Intelligence with Uncertainty* (Beijing: National Defence Industry Press) p 44
[6] Zhang F M 2018 *Comprehensive Evaluation of Basic Methods and Applications* (Beijing: Science Press) p 23
[7] Yang J P 2014 *Radar Equipment Support Science* (Beijing: Blue Sky Press) p 36
[8] Du D, Pang Q H and Wu Y 2015 *Modern Comprehensive Evaluation Methods and Case Selection* (Beijing: Tsinghua University Press) p 219
[9] Liu X W, Wang N and Li T R 2018 Improved Probabilistic Composite Rough Set Model and Attribute Reduction *Journal of Nanjing University(Natural Science)* **54** pp 958-966
[10] Bao X Z and Liu D 2009 A New Method of Ascertaining Attribute Weight Based on Rough Sets Theory *Chinese Journal of Management* **06** pp 729-732
[11] Wang W H, Huang R L and Zou B P 2018 Based On-sequence Relation Analysis Risk Assessment Method of Distribution Network *Computer and Digital Engineering* pp 1516-1519
[12] Liu B J, Cai L and Liu X Q 2017 Operational Effectiveness Evaluation of Air-to-ground Ammunition Based on Cloud Model and Fuzzy Analytic Hierarchy Process Entropy Weight Method *Fire Control & Command Control* **42** pp 80-87

Virtual Void-Based Marine Organism Protection Routing Protocol for Underwater WSNs

Hao Su[1], Yujie Qian[2] and Yuan Zhang[3]

[1] College of Internet of Things Engineering, HoHai University, 200 Jinling North Road, Changzhou, China

[2] HoHai University, Changzhou, 200 Jinling North Road, China

[3] College of Internet of Things Engineering, HoHai University, 200 Jinling North Road, Changzhou, China

E-mail: 20141923@hhu.edu.cn

Abstract. With the progress and development of the Internet of Underwater Things, underwater wireless sensor networks (UWSNs) have been showed as a promising technology to monitor and explore the oceans. Different from terrestrial WSNs, UWSNs adopts acoustic signals instead of radio frequencies for wireless communication. However, marine mammals' frequencies are heavily overlapped with the frequencies used by UWSNs, where the noise of underwater acoustic networks (UAN) made by human may interfere with or even harm marine organism. In order to protect these marine organism, it is very necessary and urgent to develop an environment-friendly UAN and cognitive acoustic (CA) is considered as a promising technique. Based on this, aiming at not only protecting marine mammals, but also attaining optimal network performance, in this paper, we propose a virtual void -based marine organism protection routing (VMOPR) protocol for UWSNs, which is a location-based flow transmission scheme. Simulation results show that the performance of the scheme is better than the corresponding EEIAR and VBF protocol in terms of packet delivery ratio, throughput and energy efficiency when taking marine mammals into account.

1. Introduction

Oceans cover more than 2/3 of this planet's surface, and numerous resources underneath. For decades, there has been a growing interest in monitoring aqueous environments. The corresponding UWSNs has a wide range of application, such as oceanographic data collection, pollution detection, underwater resource exploration, offshore infrastructure protection and target tracking [1], [2]. Acoustic communication is the only feasible method for underwater wireless communication in UWSNs [3]. The unique features of underwater acoustic communication bring many challenges such as temporary path loss, high noise, multipath fading, high communication energy cost and so on [4].

In addition to communication quality problems, it is worth noticing that the deployed area of UWSNs may be the same area where some marine creature appears. Due to the marine mammals' frequencies are heavily overlapped with the frequencies used by UWSNs [5,6], the noise of UWSNs made by human will have harmful impact on those marine creatures. Therefore, designing an environment friendly routing protocol with high network performance to protect marine organism is necessary.

In this paper, we propose a virtual void region-based marine organism protection routing (VMOPR) protocol for UWSNs that avoids interference with marine mammals' acoustic systems. VMOPR utilizes the location information of nodes and marine creatures to select the path. In order to effectively avoid channel interference. VMOPR adopts OFDM sub-channels at the MAC layer. Aiming at maximizing the throughput and improving the packet delivery ratio, flow routing is used [7]. In order to avoid interference with the marine mammals' acoustic system, the virtual void -based methodology is used in the proposed scheme, which means that the nodes around marine life will be set to idle state to avoid the interferences. And the nodes that may have impact on marine mammals' acoustic system will maintain the idle state until the marine life moves away.

The rest of the paper is organized as follows. In section 2, we introduce the network model of VMOPR. We describe the VMOPR with multiple constraints in section 3 and the simulation in section 4 is analyzed. Finally, section 5 concludes the paper and state our future works.

2. Network model

As shown in Figure. 1, we consider an UWSNs consisting of a sonobuoy, also named sink node, a group of forward nodes, a group of source nodes and several marine creatures. The sink node is fixed in the middle of the water surface, source nodes are anchored to the ocean bottom for underwater resource exploration, target tracking, or oceanography data collection. The forwarding nodes are randomly deployed in the water in the water. We assume that each sensor node knows its location and the location of marine mammals can be obtained by existing location techniques.

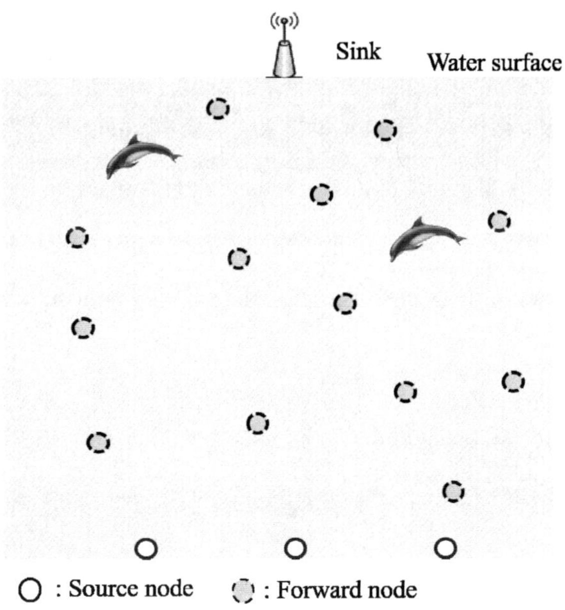

Figure 1: Network architecture of UWSN

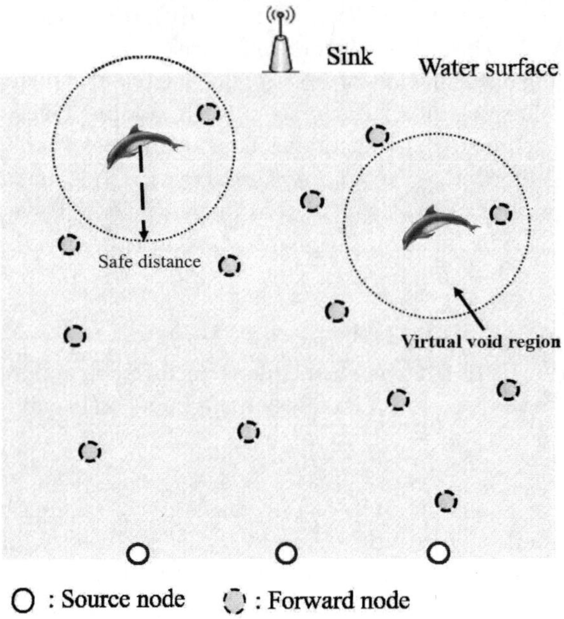

Figure 2: Virtual void node diagram

3. Virtual void-based marine organism protection routing protocol

In this section, a virtual void -based marine organism protection routing (VMOPR) protocol for UWSNs is elaborated. The available spectrum is divided into M OFDM sub-channels. In order to protect marine mammals, we define the safe distance D_s between marine life and sensor nodes, which indicates that marine organism will not be influenced by UWSNs acoustic signal from this distance. Based on these notations, the strategy of interference avoidance, selecting path and flow routing transmission is presented.

3.1. Strategy of selecting path

In order to protect marine mammals, we use the state $l_{i,j}$ to denote whether the node i links with node j:

$$l_{i,j} = \begin{cases} 1, i, j\text{links} \\ 0, \text{otherwise} \end{cases} \tag{1}$$

And the value of $l_{i,j}$ is a binary value. We use C_i to denote whether the node is in the biological interference range:

$$C_i = \begin{cases} 1, d_{im} \leq r \\ 0, \text{otherwise} \end{cases} \tag{2}$$

Where d_{im} is the distance between node i and the nearest marine mammal, R being the communication range of the sensor node as shown in Firgue. 2. If $C_i = 1$, it means that there are marine creatures around node iand node i is in the biological interference range, conversely $C_i = 0$. The nodes with $C_i = 1$ will not participate in forwarding and enter an idle state. With this strategy, the acoustic communication of UWSNs will not interfere with the marine creature. And the nodes of the idle state make the area centered the marine life a virtual void-region, which will not return to normal working state until the marine mammals swim away.

3.2. Route discovery and establishment
When the source node S have collecting data to transmit, it will initiate a route discovery through flooding. Firstly, it broadcasts a route request packet (REP) to all its neighbors, and each node with $C_i = 0$ receiving REP forwards REP to its neighbors. By analogy, multiple REPs will finally reach the destination –sink node. And sink node will choose the route that has the least hop count to be used as the data transfer path and enhance the selected path by replying route response packet (RRP). When the source node S receives the RRP, the data transfer path is established and enhanced.

3.3. Flow routing transmission
Aiming to maximize end-to-end throughput, VMPOR adopts flow routing. In flow routing, for each forward node, the total incoming data should be equal to the total outgoing data. Thus we have:

$$\sum \begin{matrix} i \neq source \\ \{i, j \in path\} \end{matrix} f_{ij} = \sum \begin{matrix} k \neq \text{sink} \\ \{j, k \in path\} \end{matrix} f_{jk} \tag{3}$$

Where f_{ij} denotes the flow arranged on the link from node i to node j and f_{ij} denotes the flow arranged on the link from node j to node k. In this constraint, node i, node j and node k present forwarding nodes in the selected path to sink node.

Then, for the source node, the total incoming data should be zero. Hence, the second constraint can be denoted as:

$$\sum_{\{s,j \in path\}} f_{js} = 0 \tag{4}$$

On the contrary, the total outgoing data of sink node should be zero.

$$\sum_{\{j,d \in path\}} f_{dj} = 0 \tag{5}$$

Since the forwarding nodes need to transmit upon receiving the data in flow routing, the VMPOR adopts OFDM at the MAC layer. The whole available spectrum can be divided into M OFDM sub-channels. One sub-channel is used as control packet delivery, and the other $M-1$ sub-channels are used as data transferring. In order to avoid collisions, each forwarder needs to utilizes two different sub-channels to transmit and receive data, respectively.

Table 1: The value of parameters.

Parameter	Value
`Network Size`	$4500m \times 4500m \times 2000m$
`Receive Power`	$0.756w$
`Forward Power`	$2w$
`Carrier Frequency`	$10kHz - 40kHz$
`Packet Size`	$160bytes$
`Number of Nodes`	$54 - 162$
`Data Generation Rate`	$0.16kbps - 0.96kbps$
`Safe Range`	$1000m$
`Initial Energy`	$100J$
`Transmission Rate`	$10kbps$

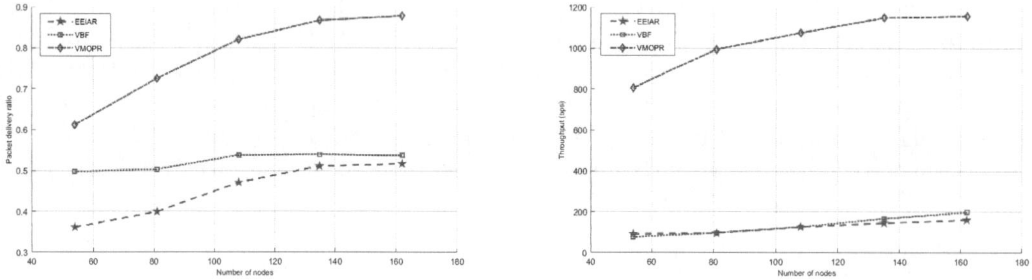

(a) Packet delivery ratio. (b) End-to-end throughput.

Figure 3: Simulation results with different number of nodes.

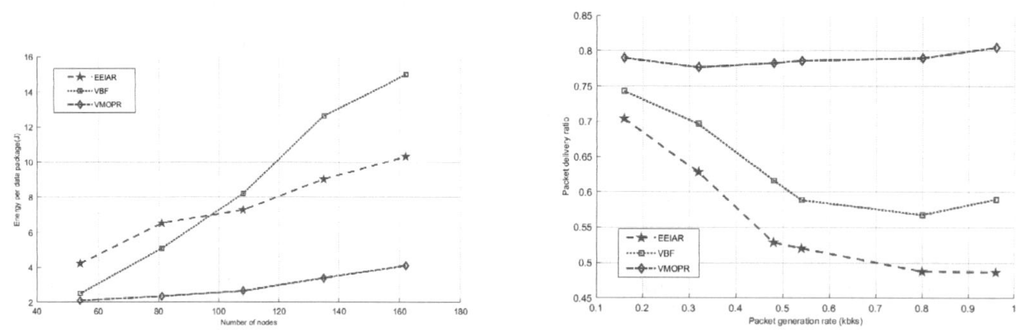

(a) Energy efficiency with different number of nodes. (b) PDR with different data generation rates.

Figure 4: Simulation results.

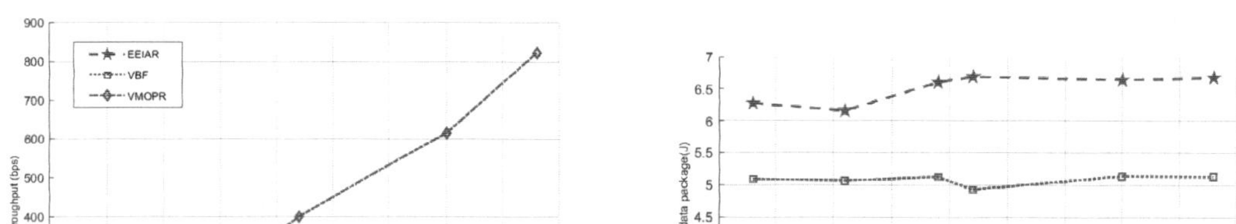

(a) End-to-end throughput. (b) Energy efficiency.

Figure 5: Simulation results with different data generation rates.

4. Performance evaluation

In this section, we compare our proposed protocol VMOPR with EEIAR (Energy Efficient Interference-aware Routing) [8] and VBF (Vector based Forwarding Protocol) [9] through extensive simulations on Matlab. Firstly, we simply explain the implementation of MAC protocol, and then we discuss the simulation results.

4.1. Implementation of MAC protocol
Because there is no standard protocol at the MAC layer, we adopt the simple MAC protocol based on CSMA in VBF and EEIAR. In this MAC protocol, the forward node senses the channel before forwarding data packets, if the channel is busy, it will utilize a back-off algorithm to wait until the channel is free [9]. Since VMOPR is a flow routing protocol, we implement a MAC protocol based on OFDM. To implement this MAC protocol, we divide the frequency into M sub-channels. What's more, there is no RTS/CTS and ACK in both MAC protocols.

4.2. Simulation setup
In our simulation, all nodes are deployed in the space of $4500m \times 4500m \times 2000m$. In order to make the distribution of nodes more reasonable, the region is divided into 27 sub-regions with size of $1500m \times 1500m \times 667m$. The number of source node is 9 and the number of forward nodes range from 27 to 162. We assume that the mobility of all sensor nodes can be ignored. The number of marine creature group is 2 and they can appear in the region randomly. Besides, the value of parameters are shown in Table 1.

First, we evaluate three protocols with number of forwarding nodes ranging from 54 to 162, and we want to know how they perform in terms of packet delivery rate, throughput, and energy consumption as the network density varies.

Figure. 3a shows the compassion results of the packet delivery rate among these three protocols. It can be seen from Figure. 3a that, as the forwarders increase, the packet delivery rate for all protocols improve in the overall trend. The performance of VMOPR with regarding to the packet delivery rate is better than two other protocols, because both VBF and EEIAR use vector forwarding to find the next hop, while VMOPR carries out the route discovery to find the most suitable path in the whole network through flooding. However, when the number of forwarding nodes exceeds 140, the he packet delivery rate almost reaches saturation for all protocols due to the packet loss caused by the channel conflict.

Figure. 3b shows the results concerning the throughput. VMOPR has a large advantage over VBF and EEIAR in the performance of throughput because of the use of flow routing methodology. Based on this, VMOPR can simultaneously transmit upon receiving the data.

Figure. 4a shows the results concerning the average energy consumption per received pcket. e can notice that these three protocols have little difference for low network density. However, as the number of forwarding nodes increases, the energy consumption of VMOPR has a very small rise, and the counterpart of VBF and EEIAR increases greatly. This happens because that the higher the network density, the more the forwarding nodes in the path involved in forwarding a packet successfully with VBF and EEIAR. Since VMOPR adopts a route discovery and establishment on demand method, it is almost unaffected by network density when the network model and transmission radius are constant.

Next, these three protocols are evaluated with the data packet generation rate ranging from $0.16kbps$ to $0.96kbs$. Figure. 4b shows the results of the packet delivery ratio. As shown in the plot, the performance of VMOPR is the best, compared with the other two protocol. It is because VMOPR will transmit the data packet successfully if there exists a path from source node to the sink. It happens more channel conflicts when more data packets need to be transmitted in VBF and EEIAR.

The comparison of throughput is shown in Figure. 5a. The throughput of our proposed scheme has proportional growth and outperforms other two protocols because it uses flow routing to maximize throughput.

Figure. 5b shows the results concerning the average energy consumption. As we set the number of forward nodes to 81, the ranking of energy consumption is consistent with Figure. 4a where the number of nodes is 81. As expected, the energy consumption performances of all protocols are basically stable.

5. Conclusion

This paper has presented a virtual void-based marine organism protection routing (VMOPR) protocol—an environmental friendly routing protocol for UWSNs. We create a virtual void region by setting the nodes interfering with marine creature to idle state which can only sense information during route discovery, and as a result these nodes will not be involved in route establishment ultimately. Therefore marine mammals' acoustic systems will not be interfered by the acoustic communication of UWSNs. Aiming to achieve high packet delivery ratio and end-to-end throughput, the data transfer is done based on flow routing method. The experimental simulation results show that the proposed VMOPR scheme can not only effectively avoid man-made noise causing fatal impact on marine mammals, but also have higher packet delivery ratio and end-to-end throughput compared with EEIAR and VBF protocol. Further study will take the environmental noise and the link capacity into consideration.

Acknowledgments

The work is supported by National Natural Science Foundation of China (No.11604077 and No.61601169).

References

[1] J.-H. Cui, J. Kong, M. Gerla, and S. Zhou, "The challenges of building mobile underwater wireless networks for aquatic applications," IEEE Netw., vol. 20, no. 3, pp. 12–18, May 2006.
[2] I. F. Akyildiz, D. Pompili, and T. Melodia, "Underwater acoustic sensor networks: research challenges," Ad Hoc Networks, vol. 3, no. 3, pp. 257–279, 2005.
[3] E. M. Sozer, M. Stojanovic, and J. G. Proakis, "Underwater acoustic networks," IEEE J. Ocean. Eng., vol. 25, no. 1, pp. 72–83, Jan. 2000.
[4] M. Stojanovic and J. Preisig, "Underwater acoustic com-munication channels: Propagation models and statistical characterization," IEEE Commun. Mag., vol. 47, no. 1, pp. 84–89, 2009.
[5] W. Yonggang, T. Jiansheng, P. Yue, and H. Li, "Underwater communication goes cognitive," in Proc. OCEANS, Quebec City, QC, Canada, Sep. 2008, pp. 1–4.
[6] Y. Luo, L. Pu, M. Zuba, Z. Peng, and J.-H. Cui, "Challenges and opportunities of underwater cognitive acoustic networks," IEEE Trans. Emerg. Topics Comput., vol. 2, no. 2, pp. 198–211, Jun. 2014.
[7] H. Y. Thomas and S. Yi, "Variable bit rate flow routing in wireless sensor networks," IEEE Transactions on Wireless Communications, v 6, n 6, pp. 2140-2148, June 2007.
[8] K.Anwar, J.Nadeem, A.Ihsan, A.-M.Hossein, R.-A.Ur, B.Naeem, Z.Muhammad and M.Hasan, "An Energy Efficient Interference-aware Routing Protocol for Underwater WSNs" in KSII transactions on internet and information systems Vol. 11, NO. 10, Oct. 2017, pp. 4844-4863.
[9] X. Peng, C. Jun-Hong and L. Li, "VBF:Vector-BasedForwarding Protocol for Underwater Sensor Networks," in uconn cse Technical Report: UbiNet-TR05-03, Feb.2006, pp. 1-20. [15] A. O. Bicen, A. B. Sahin, and O. B. Akan, "Spectrum-aware underwater networks: Cognitive acoustic communications," IEEE Veh. Technol. Mag., vol. 7, no. 2, pp. 34–40, Jun. 2012.

A New User Grouping and Power Allocation Scheme for Downlink Non-orthogonal Multiple Access Systems

Yin Wang*, Xiaohui Li, Wenwu Wang, Gongquan Zhang and Hongwei Zhang

Key Laboratory of Intelligent Computing and Signal Processing of Ministry of Education, Anhui University, Hefei 230039, China

*Email: wangyin950203@163.com

Abstract. In this paper, we propose a new user grouping and power allocation scheme based on beamforming for downlink non-orthogonal multiple access systems. The proposed user grouping scheme can effectively reduce the interference from the other user and other beams as well, and can effectively improve the weak user rate especially when the SNR is not large. In addition, a power allocation scheme that can maximize the sum capacity while satisfying a certain fairness index is proposed. From the simulation results, the user grouping and power allocation method proposed in this paper can not only improve the overall system throughput performance, but also improve the fairness of weak users.

1. Introduction

Due to the increasing number of communication devices, there is an increasing demand for various applications of wireless networks. 5G requires higher spectral efficiency, faster rate and larger capacity. Non-orthogonal multiple access technology is widely studied in 5G because of its fast rate and higher spectrum efficiency [1]-[3]. Compared to existing techniques, the NOMA technology distinguished by power dimension. In a NOMA system, Superposed signals from multiple users are sent by one base station. The user with good channel condition allocates a smaller power, and the user with better channel condition allocates a larger power.

Combining NOMA with MIMO technology and adopting zero-forcing beam precoding is a hot topic in recent years. Since the receiver uses successive interference cancellation technology, it can support multiple users. The number of receive antennas in the NOMA downlink model is often larger than the number of transmit antennas, which may cause the target user to be interfered by other users. We can divide the total number of user antennas into several different groups. The interference generated between different groups can be eliminated by beamforming techniques. Each cluster contains multiple users, that is one beam supports multiple users, which inevitably creates interference. Therefore, in order to reduce interference and increase the achievable rate as much as possible, reasonable clustering and power allocation methods are indispensable.

In recent years, many researchers have done a lot of work on user grouping and power allocation. In [4], a method of joint user clustering and dynamic power allocation was proposed, the scheme maximize the energy efficiency of the MIMO-NOMA system downlink. In [5], a joint clustering and precoding algorithm based on the eigenspace of the channel matrix was proposed, which can eliminate interference between groups and enhance the total capacity of the MIMO-NOMA system. The authors in [6] propose a user grouping algorithm with large SNR, and a power allocation method to maximize the sum rate when the minimum target rate is met. The authors in [7] performed a proportional fair based user grouping and power allocation method. Some of the above schemes do not consider the impact on user grouping when the value of SNR is very small, and some do not consider the fairness between users. In summary, this paper proposes a clustering method that selects strong users and weak users separately to improve sum capacity and to guarantee the fairness of the weak users.

The rest of this paper is organized as follows. The section II introduces our system model. The section III presents the user grouping and power allocation scheme. Numerical results and analyses are given in section IV. The section V is the summary of this paper. The symbol used are defined as follows, $()^{T}$, $()^{*}$, $()^{\dagger}$, $E()$ denotes matrix transpose, matrix conjugate transpose, matrix pseudo inverse, average value respectively.

2. System model

As shown in Figure 1, we considers a downlink multi-user MIMO-NOMA system. The base station is equipped with N antennas, and there are a total of K users in a cell. And each receiver have a single antenna. Different from the traditional MIMO-OMA technology, in order to increase the number of service users, each transmit antenna can support two or more users. In this case, up to N user groups are scheduled at the same time. For simplicity, we assume that there are two users in a cluster, and N beams can support 2N users. For two users in the same cluster, the user with large channel gain is called strong user, and the user with small channel gain is called weak user.

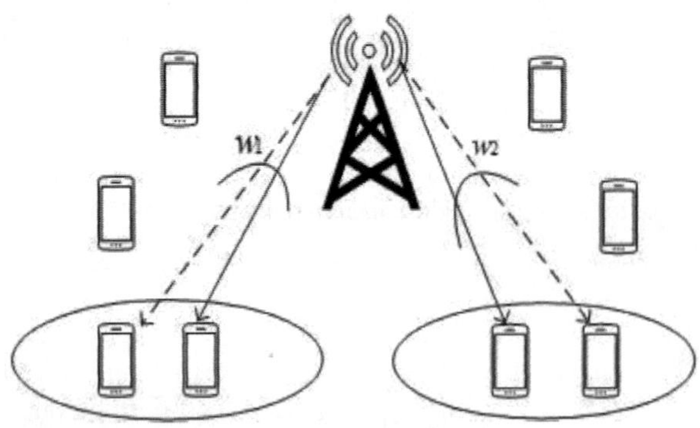

Figure 1. The proposed multi-user MIMO-NOMA system model

According to the above, the superposed signals of two users at the i-th cluster are expressed as

$$x_i = \sqrt{\alpha_{i,1}P_i}s_{i,1} + \sqrt{\alpha_{i,2}P_i}s_{i,2} \quad (1)$$

Where $E\left\{\left|s_{i,j}\right|^2\right\}=1$ for $j=1,2$. P_i is transmission power of the i-th cluster. $\alpha_{i,1}$ and $\alpha_{i,2}$ are power allocation coefficients assigned to strong users and weak users respectively, and satisfy $\alpha_{i,1}+\alpha_{i,2}=1$. In the n-th cluster, $y_{n,i}$ indicates the received signals of two users, given by

$$y_{n,i} = h_{n,i}\sum_{k=1}^{N} w_k x_k + n_{n,i} \qquad i=1,2 \quad (2)$$

Where $h_{n,i}$ is the $1\times N$ channel vector, which represents the channel coefficient of the i-th user of the n-th cluster. w_k is the $N\times 1$ precoding vector, and x_k is a superposed signal supported by a beam mentioned above, and $n_{n,i}$ is the additive complex Gaussian white noise with zero mean and a variance of σ_n^2. In this paper, we assume that the channel is a Rayleigh fading channel with zero mean and unit variance, and the transmitter knows the full channel state information. In this paper, the ZFBF is used for the precoding scheme. The BF vector w_n is calculated by combining all the strong user channels h_m in the user set, and satisfies the following relationship:

$$\frac{h_m}{|h_m|} w_n = \begin{cases} 0 & m \neq n \\ 1 & m = n \end{cases} \tag{3}$$

We set the channel matrix H to consist of strong user channel set, as follows:

$$\mathrm{H} = \left[h_{1,1}^T \dots h_{N,1}^T \right]^T \tag{4}$$

The precoding matrix W is obtained by

$$\mathrm{W} = [w_1 \dots w_N] = (\mathrm{H})^{\dagger} = (\mathrm{H})^* \left((\mathrm{H})(\mathrm{H})^* \right)^{-1} \tag{5}$$

Where w_n represents the precoding vector of the n-th cluster.

For a specific cluster, taking the n-th cluster as an example, the received signal of the strong user of the n-th cluster given by

$$y_{n,1} = h_{n,1} w_n \sqrt{\alpha_{n,1} P_n} s_{n,1} + h_{n,1} w_n \sqrt{\alpha_{n,2} P_n} s_{n,2} + h_{n,1} \sum_{k=1, k \neq n}^{N} w_k x_k + n_{n,1} \tag{6}$$

Where $h_{n,1} w_n \sqrt{\alpha_{n,1} P_n} s_{n,1}$ is the desired signal, $h_{n,1} w_n \sqrt{\alpha_{n,2} P_n} s_{n,2}$ is the interference from the other user, $h_{n,1} \sum_{k=1, k \neq n}^{N} w_k x_k$ is the interference from the other clusters to the n-th cluster. Since the SIC is used at the receiver, the interference of weak users to strong users can be eliminated by the SIC. We choose to generate BF vectors with strong user channels. According to the equation (3), it is known that $h_{n,1} w_k = 0$ for $1 < k < N$ and $k \neq n$, so the inter-cluster interference can also be eliminated. The rate achieved by the n-th cluster of strong user given by

$$R_{n,1} = \log_2 (1 + \frac{|h_{n,1} w_n|^2 \alpha_{n,1} P_n}{\sigma_n^2}) = \log_2 (1 + \frac{|h_{n,1}|^2 \alpha_{n,1} P_n}{\sigma_n^2}) \tag{7}$$

The received signal of the weak user is expressed as follows:

$$y_{n,2} = h_{n,2} w_n \sqrt{\alpha_{n,2} P_n} s_{n,2} + h_{n,2} w_n \sqrt{\alpha_{n,1} P_n} s_{n,1} + h_{n,2} \sum_{k=1, k \neq n}^{N} w_k x_k + n_{n,2} \tag{8}$$

Because weak user do not satisfy equation (3) and cannot use SIC to eliminate the interference from the other user, weak user received interference from the other user and other beams. The achievable rate of the weak user is given by

$$R_{n,2} = \log_2 (1 + \frac{|h_{n,2} w_n|^2 \alpha_{n,2} P_n}{|h_{n,2} w_n|^2 \alpha_{n,1} P_n + \left| h_{n,2} \sum_{k=1, k \neq n}^{N} w_k x_k \right|^2 + \sigma_n^2}) \tag{9}$$

We assume the transmit power of the base station transmitter to each cluster to be P_n .

3. Proposed user clustering and power allocation

In this section, a user clustering and power allocation scheme will be proposed. The simulation results show that the scheme has good performance in the maximum sum capacity and the fairness of weak users.

3.1. User clustering

The proposed user clustering scheme aims to reduce interference from the other user and other beams to maximize sum rate. In this paper, we select N strong users from K total users by using the SUS algorithm in [8], and then assign a weak user to each of the N clusters from K-N remaining users. Select weak users to meet:

$$\max\left(\frac{|h_{n,2}w_n|^2\alpha_{n,2}P_n}{|h_{n,2}w_n|^2\alpha_{n,1}P_n+\left|h_{n,2}\sum_{k=1,k\neq n}^{N}w_kx_k\right|^2+\sigma_n^2}\right) \tag{10}$$

Where SNR $\beta=\frac{P_n}{\sigma_n^2}$, equation (10) becomes

$$\max\left\{\alpha_{n,2}\left(\alpha_{n,1}+\frac{\left|h_{n,2}\sum_{k=1,k\neq n}^{N}w_k\right|^2}{|h_{n,2}w_n|^2}+\frac{1}{\beta|h_{n,2}w_n|^2}\right)^{-1}\right\} \tag{11}$$

We select users whose channel correlation with strong user are greater than ρ as weak users. When β is not large, $\frac{1}{\beta|h_{n,2}w_n|^2}$ has a significant impact on the SINR of weak users. Therefore, the user with the largest $|h_{n,2}w_n|^2$ value is selected as the weak user of the cluster among the users whose correlation is greater than ρ. The specific user clustering scheme steps are as follows:

Step 1 The set of total user $T=\{1,...,K\}$, and N strong users are selected into the set S according to the SUS algorithm, W represents a set of weak users, and R represents a user remaining after picking out N strong users from the total user set.

Step 2 Initialize $U=\varnothing, i=1$. For each user $k\in R$ in the set R, first calculate the correlation between all users in the R set and corresponding strong users:

$$Corr_{(i,k)}=\left\{\frac{|h_{i,1}h_k^H|}{\|h_{i,1}\|\|h_k\|}\right\} \tag{12}$$

When $Corr_{(i,k)}>\rho, U\leftarrow U\cup h_k$. If the user in the set R has calculated and $U\neq\varnothing$, and then go to Step 3; if the user in the set R has calculated and $U=\varnothing$, then

$\pi(k)=\arg\max\left\{\frac{|h_{i,1}h_k^H|}{\|h_{i,1}\|\|h_k\|}\right\}, W\leftarrow W\cup\pi(k), h_{i,2}=h_{\pi(k)}, i=i+1$. If $|W|<N$, repeat to Step 2, otherwise the user clustering scheme ends.

Step 3 Let $j=1,\ j\in U$.

$$\begin{aligned}&\pi(j)=\arg\max\left\{|h_jw_i|^2\right\}\\&W\leftarrow W\cup\{\pi(j)\}\\&h_{i,2}=h_{\pi(j)}\\&i=i+1\end{aligned} \tag{13}$$

If $|W|<N$, return to Step 2, otherwise the user clustering scheme ends.

3.2. Power allocation

After all users have completed the corresponding user grouping, the next step should be to allocate power to each user in the cluster, and maximize the sum rate of the cluster subject to that the fairness index achieve p.

Since there are two users in a cluster, we define the fairness index as [9]:

$$F = \frac{(\sum_{i=1}^{2} R_i)^2}{2\sum_{i=2}^{2} R_i^2} \tag{14}$$

which indicates the fairness of the system capacity of strong user and weak user in each cluster. The capacity for strong user and weak user gets close to each other when F gets close to 1.

The proposed power allocation scheme aims to maximize the cluster sum capacity under a fairness constraint. The optimal allocation problem can be formulated as

$$\max_{\alpha_{n,1},\alpha_{n,2}} imize \quad (R_{n,1} + R_{n,2}) \tag{15}$$

$$s.t. \; F \geq p \, , \alpha_{n,1} + \alpha_{n,2} = 1$$

p is a predefined adaptive parameter. We obtain the power allocation coefficients for two users in a cluster through exhaustive search. If none of the power allocation coefficients meet the fairness constraint, then assign all of the power to strong user.

4. Simulation results and analysis

The simulation diagrams below in this paper all take K=100 users and the number of antennas N=2 as an example. The noise energy $\sigma^2 = 1$, and the channel is a Rayleigh fading channel with zero mean and unit variance. Figure 2 is a simulation comparison of the correlations in the proposed user clustering scheme. It can be seen from the simulation results that the sum capacity is the highest when the correlation value is 0.95, so we all simulated with the correlation value of 0.95 in the following simulation experiments.

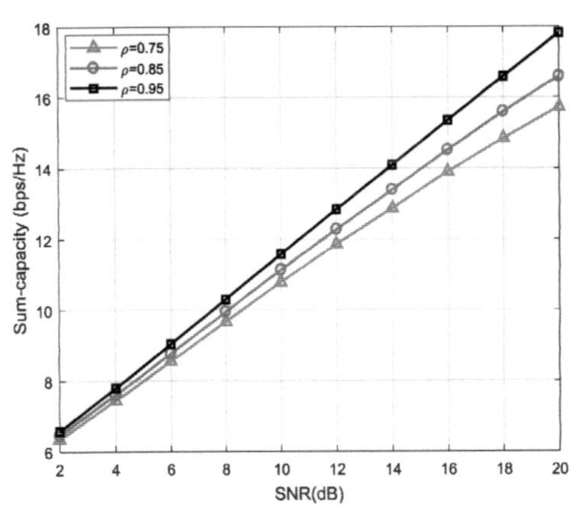

Figure 2. The relationship between the sum capacity and SNR of three different correlations at the fairness index p =0.8

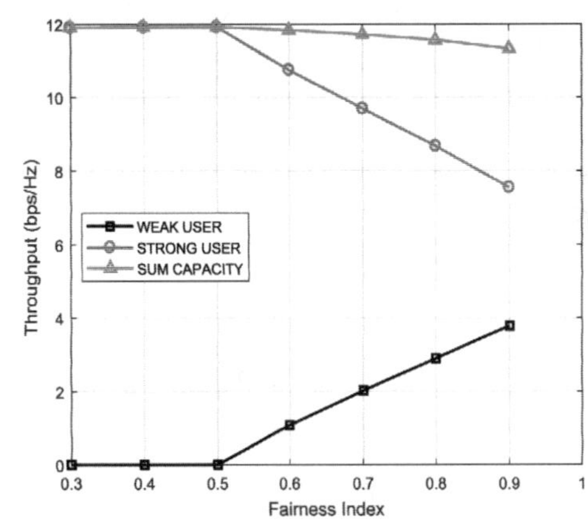

Figure 3. Capacity obtained by strong users and weak users with different fairness indexes

Figure 3 is a graph showing the relationship between weak users, strong users, and sum capacity and fairness index when SNR = 10 dB. It can be seen from the figure that the average achievable rate of weak users is almost zero when the fairness index is less than 0.5. When the fairness index is greater than 0.5, the average achievable rate of weak users increases with the increase of the fairness index and decreases with strong users. As the fairness index increases, the sum capacity decreases slightly.

IOP Conf. Series: Journal of Physics: Conf. Series **1284** (2019) 012060 doi:10.1088/1742-6596/1284/1/012060

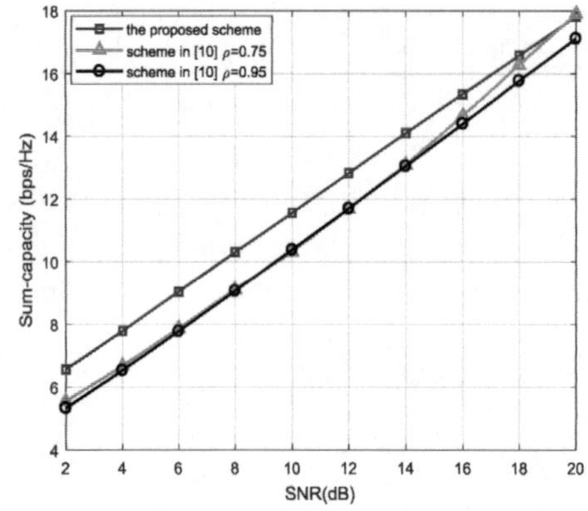

Figure 4. Comparison of the sum capacity between the proposed scheme and the scheme in [10]

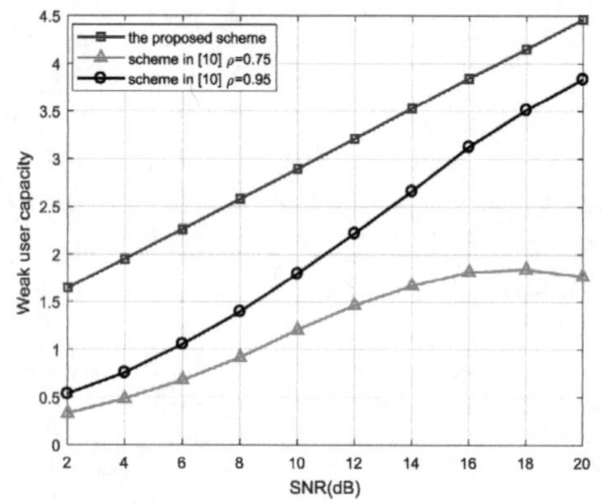

Figure 5. Weak user capacity with the proposed scheme and the scheme in [10] vs. the SNR(dB)

The authors in [10] performed user clustering according to the correlation of the user channel and the channel difference, and proposes a power allocation method that maximizes total system capacity while keeping the capacity of weak users equal to or greater than the capacity of a traditional multi-user BF system.When the fairness index is set to 0.8, it can be seen from Figure 4 that the sum capacity of the proposed scheme is greater than that in [10], especially when the SNR is small. This is because the smaller the SNR, the greater the impact of $|h_{n,2}w_n|^2$ on the weak user rate, and the proposed user clustering scheme can increase the weak user rate and thus increase the sum capacity. When the SNR is close to 20 dB, the scheme of the correlation of 0.75 in [10] is closer to the sum capacity of the proposed scheme.This is because the effect of $|h_{n,2}w_n|^2$ on the weak user rate is negligible as the SNR increases. But overall, the proposed scheme is slightly better.

The relationship between the weak user capacity and SNR is shown in Figure 5. When the fairness index is set to 0.8, it can be seen from Figure 5 that the sum capacity of weak users in proposed scheme is significantly larger than that in [10], especially when the SNR is not large. In addition, when ρ is high, the inter-cluster interference is reduced, thus enhancing the weak user's capacity.

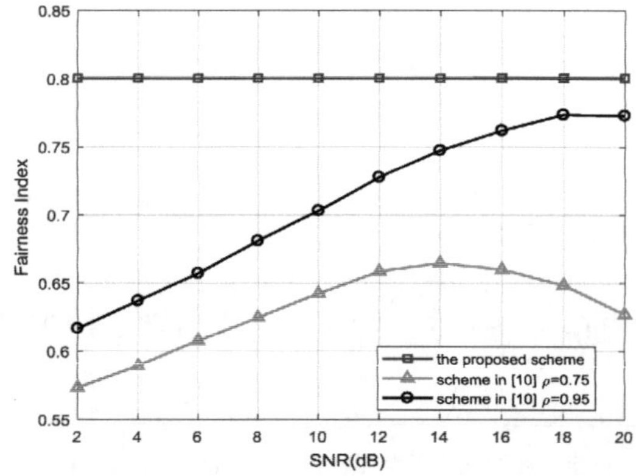

Figure 6. Fairness vs. number of users

The relationship between the fairness index and SNR is shown in Figure 6. From Figure 6, the fairness index of the proposed scheme is 0.8, and the fairness index with the correlation of 0.95 in [10] is close to the maximum at 18 dB. The maximum value of the fairness index with the correlation of

0.75 in [10] is also less than 0.7 with the increase of SNR. It is obvious that the fairness index of the proposed scheme is better than that of the scheme in [10]. This means that the fairness of weak users is guaranteed by the proposed power allocation scheme.

In summary, it can be seen that through the comparison of the above three simulation diagrams, the proposed scheme is superior to the scheme proposed in [10] whether sum capacity or weak user rate and fairness.

5. Conclusion

This paper proposed a user clustering and power allocation scheme for downlink multi-user MIMO-NOMA system. The proposed user grouping scheme not only effectively reduces the intra-cluster interference and inter-cluster interference, but also takes into account the impact on the weak user capacity when the SNR is not large, and increases the capacity of the weak user. The proposed power allocation scheme is based on the optimal power coefficient allocation of the fairness index, which can maximize sum capacity while guaranteeing certain fairness. The simulation diagram proves that compared with the existing methods, the scheme improves system capacity, weak user capacity and fairness.

References

[1] Linglong Dai, Bichai Wang, Zhiguo Ding, Zhaocheng Wang, Sheng Chen, Lajos Hanzo: A Survey of Non-Orthogonal Multiple Access for 5G[J]. IEEE Communications Surveys and Tutorials, 2018, 20(3): 2294-2323.

[2] Linglong Dai, Bichai Wang, Yifei Yuan, Shuangfeng Han, Chih-Lin I, Zhaocheng Wang: Non-orthogonal multiple access for 5G: solutions, challenges, opportunities, and future research trends[J]. IEEE Communications Magazine, 2015, 53(9): 74-81.

[3] Yuya Saito, Yoshihisa Kishiyama, Anass Benjebbour, Takehiro Nakamura, Anxin Li, Kenichi Higuchi.Non-Orthogonal Multiple Access (NOMA) for Cellular Future Radio Access[C]. VTC Spring 2013: 1-5.

[4] Sunil Chinnadurai, Poongundran Selvaprabhu, Moon Ho Lee. A novel joint user pairing and dynamic power allocation scheme in MIMO-NOMA system[C]. ICTC, 2017: 951-953

[5] Chin-Liang Wang, Jyun-Yu Chen, Siu-Hang Lam, Pei Xiao. Joint Clustering and Precoding for a Downlink Non-Orthogonal Multiple Access System with Multiple Antennas[C]. VTC Fall, 2016: 1-5.

[6] Chih-Chiang Chen, Shang-Ho Lawrence Tsai, Yuan-Pei Lin, Chia-Hua Lin.Resource allocation and minimum rate for precoded non-orthogonal multiple access[C]. APSIPA 2017: 813-817.

[7] Fei Liu, Petri Mähönen, Marina Petrova.Proportional fairness-based user pairing and power allocation for non-orthogonal multiple access[C].PIMRC 2015, IEEE, 2015: 1127-1131.

[8] Taesang Yoo, Andrea J. Goldsmith.On the optimality of multiantenna broadcast scheduling using zero-forcing beamforming [J]. IEEE Journal on Selected Areas in Communications, 2006, 24(3): 528-541.

[9] R.K. Jain, D.W. Chiu, and W.R. Hawe. "A Quantitative Measure of Fairness and Discrimination for Resource Allocation in Shared Computer System", Hudson, 1984, pp. 20-21

[10] Beomju Kim, Sungmook Lim, Hyungjong Kim, Sangwook Suh, Jonghyung Kwun, Sooyong Choi, Chungyong Lee, Sanghoon Lee, Daesik Hong. Non-orthogonal Multiple Access in a Downlink Multiuser Beamforming System[C]. MILCOM 2013, IEEE, 2013: 1278-1283.

Scale Adaptive Target Tracking Algorithm Based on Correlation Filtering

Fajun Lin[1], Haiping Wei[1,*], Yuanchen Qi[1] and Qinqin Li[1]

[1]Liaoning Shihua University, Liaoning, China

[*]1847749934@qq.com

Abstract. Focusing on the issue that the kernel correlation filter tracking method can not accurately adapt to the target scale change, a scale estimation method is added under the kernel correlation filter tracking frame, and a scale adaptive target tracking algorithm is proposed. The algorithm firstly uses the kernel function to obtain the kernel correlation filter, and estimates the position of the target for the candidate sample detection; then uses the scale estimation method to detect the target scale to further determine the optimal scale of the target to be tracked in the current frame; Finally, according to the occlusion judgment mechanism, the parameters of the target performance model and the scale model are updated online. The experimental results show that compared with kernel correlation filtering algorithm, the proposed algorithm can accurately target tracking process to adapt to changes in the target scale, and the center position error reduced by an average of 4.39 pixels.

1. Introduction

Target tracking has a wide range of applications in video surveillance [1], human motion analysis [2], human-computer interaction [2], and unmanned [3]. However, target tracking is still a challenging problem due to scale changes and target occlusion. The target tracking algorithm can be divided into a generation method and a discriminant method [4]. In recent years, due to the development of machine learning, the discriminant-based tracking method has achieved good tracking results, and the discriminant method trains a classifier based on the target and background information, and extracts the target from the background area [5].

Bolme et al. [6] introduced correlation filtering into the target tracking field for the first time, and proposed a Minimum Output Sum of Squared Error (MOSSE) based on correlation filtering. The algorithm used gray features to train the appearance model of the target area, and used Discrete Fourier Transform (DFT) to convert the similarity calculation between the target and all candidate areas to the frequency domain, which greatly enhanced the target tracking speed. Henrique et al. [7] proposed a Circulant Structure of Tracking by Detection with Kernels (CSK), the algorithm shifts training samples circularly, which can greatly increase the number of training samples. On the basis of this, Henrique et al. [8] proposed a Kernelized Correlation Filter (KCF) based on CSK, extracted the Histogram of Oriented Gradients (HOG) features to replace the former gray features [9], which improved the robustness of the algorithm.

Although the above algorithms have achieved good results in target tracking, these tracking algorithms do not take into account the scale changes of the target. Aiming at this problem, this paper proposes a method of target scale estimation, which realizes scale-adaptive target tracking. In order to verify the effectiveness of the proposed algorithm in this paper, the Matlab is employed to numerical simulations [10], further we verify the correctness of the algorithm theoretical derivation.

2. Tracking framework

The KCF algorithm trains the RLS classifier, solve the RLS classifier to obtain a filter template, and then detect the candidate samples to achieve tracking of the target [8].

2.1. *Training kernel correlated filter*

The purpose of training the Kernel Correlation Filter is obtained an objective function $f(x)=w^T x$ that the minimizes the squared error of all samples and their corresponding regression labels.

$$\min_{w} \sum_{i} (f(x_i)-y_i)^2+\lambda \| w \|^2 \tag{1}$$

The λ is a regularized parameter that controls overfitting. The w is a weight coefficient matrix, which is used to describe the relationship between the sample and the classifier.

To reduce the amount of computation, the input samples x are mapped onto a high dimensional feature space $\varphi(x)$, and a kernel function is defined $k(x,x')=\varphi^T(x)\varphi(x')$. Expressing the solution w as a linear combination of the samples [8], [11]:

$$w=\sum_{i} \alpha_i \varphi(x_i) \tag{2}$$

The α_i is classifier coefficients, the variables under optimization are thus α, instead of w. The solution to the kernelized version of Ridge Regression is given by [8]:

$$\alpha=(K+\lambda I)^{-1} y \tag{3}$$

Where I is an identity matrix, and each element of y is a regression target y_i. Since the equation (3) has a matrix inversion operation, the circulant matrix can be used to simplify the operation.

2.2. *Simplification of the circulant matrix*

Consider an $n \times 1$ vector representing a patch with the object of interest, denoted $x = (x_1, x_2, \cdots, x_{n-1}, x_n)$, We will refer to it as the base sample. The cyclic matrix X can be expressed as:

$$X=Fdig(\hat{x})F^H \tag{4}$$

Where F is a constant matrix, and $\hat{x}$ denotes the DFT of the x. From now on, we all always use a hat ^ as shorthand for the DFT of a vector. Equation (3) can be transformed into the Fourier domain, so that the calculation amount of the training is greatly reduced.

2.3. *Target position detection*

An image block of size z is extracted at the same center position of the current frame and the last frame. The element of matrix K^z is defined as $k^{xz}=k(x,z)$, and the matrix K^z is used to represent the kernel matrix between the training samples x and the candidate image patches z [8].

From equation (2), we can compute the regression function for all candidate patches get all the response:

$$\hat{f}(z)=\hat{k}^{xz} \odot \hat{\alpha} \tag{5}$$

Where $\odot$ denotes dot product. The target position of the current frame is estimated by the detected maximum value of the response. Then the classifier is updated as follows:

$$x_t=(1-\eta)x_{t-1}+\eta x \tag{6}$$

$$\alpha_t=(1-\eta)\alpha_{t-1}+\eta \alpha \tag{7}$$

Where η is the learning factor, x_t and x_{t-1} represent the target global appearance model of the current frame and the last frame. The α_t and α_{t-1} represent the classifier update parameters of the current frame and the last frame.

3. Scale adaptive target tracking

In this paper, the target scale estimation strategy and the model update strategy with occlusion judgment.

3.1. Adaptive scale estimation

In practice, the target has changed in size. The characteristics of the collected samples are inaccurate; therefore, it is necessary to solve the problem of scale change.

Assuming that the target size of frame $t-1$ is $M \times N$ and the scale filter size of frame $t-1$ is $S \times 1$. In the target position of the current frame, a total of $2M+1$ images of different scales are extracted [12], [13] and the scale formula can be written as:

$$l_i = 1 \pm \mu \cdot m \qquad m = 0,1,2,\ldots,M; i = 1,2,\ldots,2M+1 \tag{8}$$

Where $\mu \in (0,1)$ is the parameter for judging the scaling scale? When $\mu > 0$, $l_i > 1$, indicates the scale of magnification; when $\mu < 0$, $l_i < 1$ indicates the scale of reduction; when $l_i = 1$ ($m = 0$), the target scale remains unchanged. Image patches can be represented as $S_i = l_i M \times l_i N$. The Responses and updates of scale filters refer to Section 2.

3.2. Classifier update

In the target tracking, when the target is occluded, a method for occlusion judgment detection update is adopted.

In the assumption of forward-backward motion trajectory consistency [14], [15], the main factor that interferes with target tracking is that it is blocked by other objects. Therefore, it can be used as a basis for judging whether the target is occluded. According to this idea, this paper adopts a method of judging occlusion by using the Euclidean Distance between two trajectories.

Assuming that the $S = (m_t, m_{t+1}, \cdots, m_{t+k})$ is a sequence of images, forward tracking k step from t to $t+k$ can obtain a moving track $L_f = (x_t, x_{t+1}, \cdots, x_{t+k})$, f represents forward motion, k represents step size, x represents coordinate position of target center point. Then, a backward motion trajectory $L_b = (\hat{x}_t, \hat{x}_{t+1}, \cdots, \hat{x}_{t+k})$ is obtained by tracking k step backwards from $t+k$ to t. b represents backward motion, $x_{t+k} = \hat{x}_{t+k}$. Forward-backward error can be expressed by the Euclidean Distance d between two trajectories [15].

$$d = (|x_{t_x} - \hat{x}_{t_x}|^2 + |x_{t_y} - \hat{x}_{t_y}|^2)^{1/2} \tag{9}$$

Where x_{t_x}, x_{t_y} are the coordinates of the target central point? Normally, d is relatively small, indicating that the target is not occluded. Take the smaller threshold δ. If $\delta > d$, judge the target to be occluded. The update factor in the classifier and scale filter of the current frame is reduced by half. If $\delta < d$, the tracking is continued according to the original update method.

3.3. Algorithm steps

In this paper, target scale estimation method and the model update strategy with occlusion judgment, the steps of the algorithm are summarized as follows:

Input: An entire video sequence; the target position and scale of the previous frame.

Output: The position and scale of the current frame target.

① According to the position of the previous frame, the size of z image patch is extracted, the image patch and training sample are detected quickly. The maximum response value is obtained by the equation (5) to determine the position of the current frame.

② At the position of the target, a total number of $2M+1$ images of different scales are extracted, the images are quickly detected, and get the Best Scale.

③ When the target is occluded, to update classifiers and scale filters.

4. Experiments

In order to verify the effectiveness of the proposed algorithm, the experimental analysis was compared with the original algorithm using the evaluation method in reference [10].

4.1. Qualitative Analysis

In order to verify the effectiveness of the algorithm, the test video sequences are compared. The two algorithms use the same video sequence. Here, four video sequences are selected.

Experiment 1: For the case of scale change, the CarScale video sequence was selected. See Figure 1(a), Figure 1(b), the tracking frame size of the KCF algorithm is always unchanged, this algorithm can adapt to the change of the target scale well, and can reduce the relevant parameters of model updating in time under occlusion, such as 166th frame.

(a) KCF (b) Ours

Figure 1. Tracking results of CarScale (Frame 19, 139, 166 and 240).

Experiment 2: For the case of scale change and occlusion, the Girl video sequence was selected. There are obvious scale changes and occlusion in the target of the video sequence. As shown in Figure 2(a) and Figure 2(b), the scale of the tracking frame of the KCF algorithm does not change, While the size of the tracking frame of the algorithm in this paper keeps pace with the target scale.

(a) KCF (b) Ours

Figure 2. Tracking results of Girl (Frame 8, 156, 360 and 436).

Experiment 3: For the case of scale change, the Walking video sequence was selected. As shown in Figure 3(a) and Figure 3(b), comparing with this paper, the KCF algorithm has no scale change.

(a) KCF (b) Ours

Figure 3. Tracking results of Walking (Frame 12, 86, 208 and 356).

Experiment 4: For the case of scale change and occlusion, the Walking2 video sequence was selected. See Figure 4(a) and Figure 4(b), the KCF algorithm tracking is not accurate, this algorithm is more effective tracking.

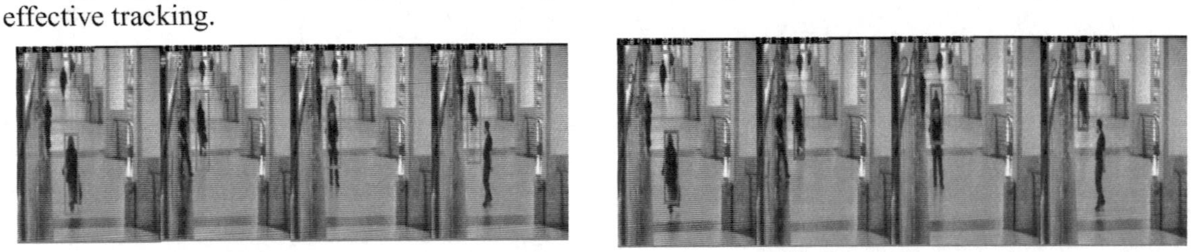

(a) KCF (b) Ours

Figure 4. Tracking results of Walking2 (Frame 6, 178, 204 and 287).

4.2. Quantitative analysis

In order to verify the effectiveness of the proposed algorithm, adopt Central precisionError (CLE) and Distance Precision (DP) as evaluation criteria. $CLE = [(x_t - x_g)^2 + (y_t - y_g)^2]^{1/2}$, it refers to the distance between the center coordinates (x_t, y_t) of the tracking target and the true center coordinates (x_g, y_g).

Table 1. 10 test videos for CLE (pixel).

Algo-rithm	CarScale	Girl	Singer1	David	Walking	Walking2	Car4	Dog1	Doll	Skating1
KCF	16.23	32.52	12.73	9.48	7.31	20.24	9.77	3.95	10.07	7.72
Ours	14.32	12.14	4.64	9.14	5.98	16.83	8.58	3.26	3.85	7.36

Table 1 shows the experimental results of 10 sets of center position errors. From Table 1, compared with KCF algorithm, CLE reduces by 4.39 pixels on average, which shows the effectiveness of the proposed algorithm.

DP is the ratio of the number of test video p frames to the total number of q frames, where p frames are the number of video frames whose CLE is less than a certain threshold (20 pixels in the experiment), obtaining $DP = p/q$.

Figure 5 is the DP curve of a partial test video, where the abscissa represents the threshold of the center position error, the ordinate represents the precision, i.e. the ratio of p to q. The higher the precision, the better the effect.

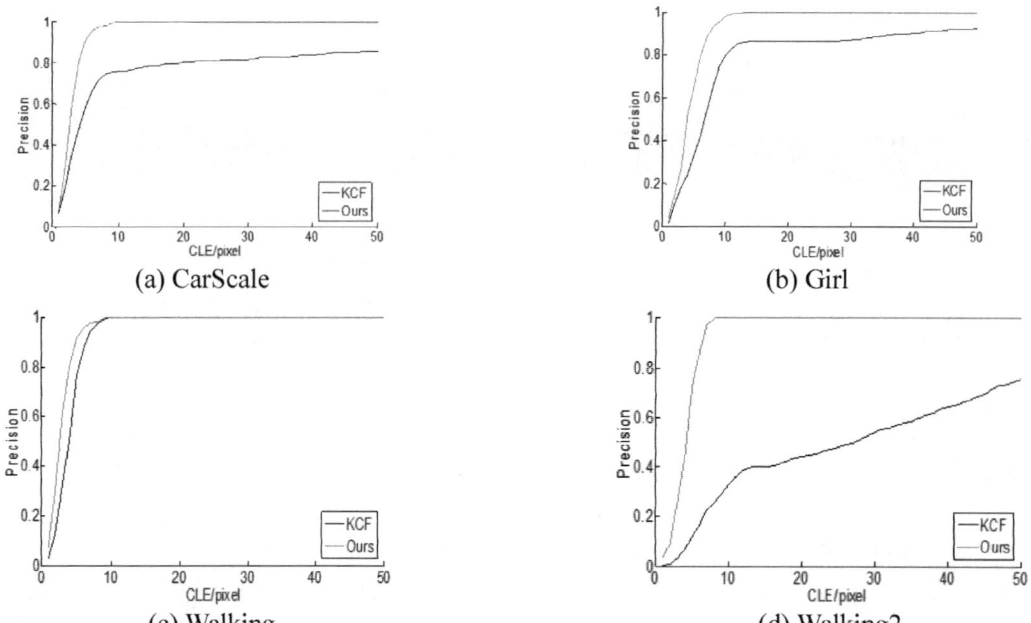

(a) CarScale (b) Girl

(c) Walking (d) Walking2

Figure 5. A part of the test videos DP curve.

5. Conclusion

In this paper, a correlation filter algorithm for scale adaptive tracking is proposed. By adding a target scale estimation strategy, the problem that KCF algorithm has no target scale change is solved. In addition, the algorithm also adds a model update strategy with occlusion judgment, which reduces the tracking error by reducing the model updating coefficient under occlusion, so as to improve the robustness of the algorithm. The experimental results on several groups of test videos show that the proposed algorithm greatly reduces the CLE and improves the DP. Therefore, compared with the traditional KCF algorithm, the proposed algorithm has good tracking effect in scale transformation and target occlusion.

References

[1] Smeulders A W M and Chu D M. Visual Tracking: An Experimental Survey *J*. IEEE Transactions on Pattern Analysis & Machine Intelligence,2014,36(7) :1442- 1468.
[2] Zhang K and Zhang L. Fast visual tracking via dese spatio-temporal context learning *C*. //Proc of the European Conference on Computer Vision, 2014.
[3] WANG N, SHI J and Yeung D Y. Understanding and diagnosing visual tracking systems C. //Proceedings of the IEEE International Conference on Computer Vision. New York: IEEE Press, 2015: 3101-3109.
[4] Liu Q, Zhao X and Hou Z. Survey of single-target visual tracking methods based on online learning *J*. IET Computer Vision, 2014, 8(5):419-428.
[5] Wei Q and Lao S Y. Review of visual target tracking based on correlation filter *J*. Computer Science, 2015, 43 (11): 1-5
[6] Bolme D S, Beveridge J R and Draper B A Visual object tracking using adaptive correlation filters *C. Los Alamition: Computer Vision and Pattern Recognition, 2010*
[7] Henriques J F, Caseiro R and Martins P Exploiting the ncirculant structure of tracking- by - detection with kernels *C*. New York: Europeon Comference on Computer Vision,2012
[8] Henriques J F, Caseiro R and Martins P High-speed tracking with kernelized correlation filters *J*. IEEE Transactions on Pattern Analysis and Machine Intelligence, 2015,37(3):583-596
[9] Dalal N and Triggs B Histograms of oriented gradients for human detection *C*. Santiago: Computer Vision and Pattern R ecognition, 2005
[10] Wu Y, Lim J and Yang M H Online Object Tracking: A Benchmark *C*. // Computer Vision and Pattern R ecognition. IEEE, 2013:2411-2418.
[11] Schölkopf B and Smola A Learning with kernels: Support vector machines, regularization, optimization, and beyond. MIT Press, 2002.
[12] Qian T H, Luo Z Q and Li G J Scale adaptive improvement of kernel correlation filter tracking algorithm *J*. Journal of Computer Applications, 2017, 37(3):811-816.
[13] Danelljan M, Häger G and Khan F S Accurate scale estimation for robust visual tracking *C*. // British Machine Vision Conference. 2014:65.1-65.11.
[14] Lv G J, Mao X and Hu Y J One Method to Detect Tracking Occlusions Based on Evaluation Forward and Backward Errors *J*. Infrared Technology, 2016.
[15] Zhao S S, Liu Xipeng and Shen J L The Affection of Thickness Error on THz Refractive Index *J*. Journal of Capital Normal University: Natural Science Edition, 2009, 30(2): 13-15

Application of Time-Weighted User Behavior Analysis Collaborative Filtering Algorithm in Short Video

Zhiqiang Li[1], **Yang Liu**[1,*] **and Difei Yu**[1]

[1]School of Computer and Communication Engineering, Liaoning Shihua University, Liaoning, China

[*]862720689@qq.com

Abstract. In the big data environment, the information overload is serious. The traditional search engine does not satisfy the user's demand for information. The user wants to obtain the user's acquired information conveniently and quickly, and the information producer wants to transmit the information to the target user. The intermediary that acts as both is the recommendation system. Recommendation system is an important way to deal with massive information. This technology brings great convenience to viewers in the field of short video. It is very difficult to extract features from missing short video data and from content information of short video. About this question. This paper proposes a time-weighted user behavior analysis collaborative filtering algorithm. This paper starts with the behavior of short video users. In order to compensate for date sparseness, the explicit and implicit behaviors of users are deeply analyzed and a user-video scoring model is established. The user's interest in video will change with time, so we use time weighting to adjust the time variable. Finally, the K most similar neighbors are extracted and recommended to the user in order from high to low. Experiments show that the proposed algorithm alleviates sparsity of date and improves the accuracy of recommendation.

1. Introduction

In recent years, with the rapid development of mobile internet, new media technology is supported by technology. With the support of related technologies such as data storage and video compression, short video has become the mainstream of network information dissemination from the previous mass information dissemination. Short videos are usually short, ranging from a few seconds to a few minutes, and have some innovative features. There are currently dozens of short video apps on the market. However, with the increasing number of short video uploaders and uploaded videos, the quality of short videos has become a huge problem. Nowadays, it has become the biggest problem for users to get interesting short videos from a large number of videos.

With the development of the recommendation system, many scholars and experts have proposed many different recommendation algorithms. The earliest recommendation algorithm for video is VideoRech's online video recommendation system, which is a real-time recommendation system proposed by Mei [1]. Personalize short videos based on the video the user is watching. But this system only considers some superficial information of the video itself. For example: title, author, release date. It does not build interest models for video users. However, short videos contain little or no textual information, and this algorithm does not work for short videos. It is not realistic for the tens of thousands of video data resources. Xing-yao Yang [2]. Add different attributes to the recommended project through artificial annotation. Further look for the relationship between user-attribute and attribute-video. Since most of the short video uploaders are ordinary Internet users and uploaded videos are short. The attribute of the video is blurred, and the accuracy of the recommendation is reduced. Jian-wei Niu [3] puts forward the idea of video recommendation based on emotional analysis. Recommend appropriate videos by capturing the current user's emotions. It is suggested that video recommendation is dynamic, but the user's emotion changes with time, mood and weather. This change is one that is extremely complex and unmeasurable. Wei-wei Song [4] proposed a time-weighted label algorithm, and he considered that time variation would have an impact on the recommended effect. But this is mostly recommended for movies and TV shows. Movies and TV shows contain more tags, such as actors, directors, movie types, 3D, and more. Short videos have very few tags, and most of the videos on short video apps are spontaneously uploaded by users. Most of these users do not have professional analysis of video content, and the content and features of the video may be unclear or even erroneous. It is very difficult to extract features from short video content. In addition, the user interest of the video may also change over time. The user may have different evaluations of the same video at different times. This paper uses a Time-Weighted User Behavior Analysis Collaborative Filtering Algorithm (TWUBCF) for these problems.

2. Definition of problems and basic methods

2.1. *User behavior information analysis*

The information of video user's behavior is stored in the form of logs on the mobile side. The definition of user behavior in the video recommendation system [5] usually includes five aspects: the behavior object, the user who generated the behavior, the content of the behavior and the content of the behavior, and the type of behavior. The behavior information of short video users mainly includes browsing time, browsing records, sharing friends, comments and scoring. These behaviors exist on the computer in the form of logs. According to the user's behavior information, there are two types: explicit behavior and implicit behavior. Explicit behavior requires the user to express his or her own interest in the video. Implicit behavior refers to the inability to directly see the like and dislike of the content.

2.2. *Modeling of user behavior information*

In order to find the intrinsic relationship between users and videos, this paper adopts the user behavior information modeling method. This paper establishes a model that can reflect user's interest in video content. Explicit behavior is used to judge whether the user is related to the video. In the implicit user behavior [6], we can set a reference value for the viewing time, the number of views, and the number of sharing to determine whether the user is related to the video content. Next, we use an algorithmic flow chart to verify the relevance of user and content. As shown in figure 1.

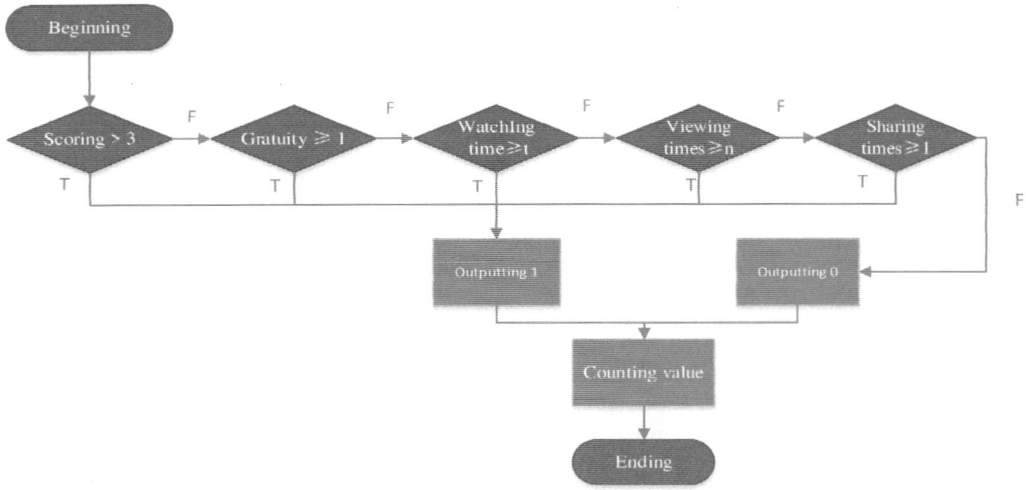

Figure 1. User-Video relationship judgment flow figure.

From Figure 1, we can see that the user likes the degree is 0, or is 1. 0 means that the user has no relationship with the video, and 1 means that the user has a relationship with the video. Based on the above, we create a scoring matrix M about user-video. As shown in Table 1, the matrix represents the user's rating of video, and it represents the user's preference for video.

Table 1. User-Video scoring matrix table.

User	Video			
	$Item_1$	$Item_2$	···	$Item_n$
U_1	$R_{1,1}$	$R_{1,2}$	···	$R_{1,n}$
⋮	⋮	⋮	···	⋮
U_s	$R_{s,1}$	$R_{s,2}$	···	$R_{s,n}$

In addition, in order to solve the problem of large data cold start, we let new users register an attribute table before using short video APP, which includes a series of information such as age, gender, region, occupation type, etc.

2.3. Calculating similarity of users

In this paper, the modified cosine similarity [7] is used. It treats user ratings as vectors in space. The modified cosine similarity formula can be expressed as equation (1).

$$\mathrm{Si}m(U_a,U_b)=\frac{\sum_{i_k\in I'}(R_{a,k}-\overline{R}_a)\times(R_{b,k}-\overline{R}_b)}{\left(\sum_{i_k\in I'}(R_{a,k}-\overline{R}_a)^2\right)^{1/2}\times\left(\sum_{i_k\in I'}(R_{b,k}-\overline{R}_b)^2\right)^{1/2}} \tag{1}$$

In the above equation (1), $R_{a,k}$ is the rating of the short video by the user U_a, $R_{b,k}$ is the rating of the short video by the user U_b, and $\overline{R}_a$ and $\overline{R}_b$ are the average of the ratings of the user U_a and the user U_b. Because everyone has different hobbies and experiences, different user ratings are also different. In order to cope with the above problem, the modified cosine similarity method first calculates the average value of the user for all videos, and then subtracts this average to weaken the difference in the user's scoring scale. The final result is that the value of $Sim(U_a,U_b)$ is within the interval $[0,1]$.

2.4. Improving the computation of similarity measure

Computing the similarity between short video users mainly depends on the short video they scored together. Assuming that there are fewer videos scored by video users together, there is some uncertainty in this similarity. Herlocker [8] et al. proposed adding a parameter to improve the accuracy of similarity. Based on this theory, we define I' as a video set that short video user U_a and short video user U_b score together, I' satisfies equation $I'=I_{U_a}\cap I_{U_b}$ .A threshold γ is set artificially and compared with $|I'|$.The improved similarity is shown in equation (2).

$$\mathrm{Si}m'(U_a,U_b)=\frac{\min(|I'|,\gamma)}{\gamma}\times\mathrm{Si}m(U_a,U_b) \tag{2}$$

From the algorithm of equation (2), we can get that the value range of the improved similarity quantity is still in [0, 1]. If the short video user U_a and the short video user U_b have a larger number of common scoring videos, according to $|I'|\geq\gamma$, we can get $\mathrm{Si}m'(U_a,U_b)=\mathrm{Si}m(U_a,U_b)$. If the short video user A and the short video user B have a small number of common scoring videos, according to $|I'|<\gamma$, we can get $\mathrm{Si}m'(U_a,U_b)<\mathrm{Si}m(U_a,U_b)$ .Through the above methods, we can improve the accuracy of user similarity.

3. Time-weighted user behavior analysis collaborative filtering algorithm

3.1. Calculating time weights

As time goes by, the user's interest will also change [9]. In a short period of time, the user's interest in watching video is basically stable, so the recently added user behavior has a great impact on recommendation prediction. Based on this problem, this paper proposes to add the factor of time to the user behavior analysis collaborative filtering algorithm. This is Time-weighted user behavior analysis collaborative filtering algorithm.

$$\log istic(t_{a,j}) = \frac{1}{1+e^{-t_{a,j}}} \tag{3}$$

In the above equation (3) $t \in [-1,1]$, $\log istic(t_{a,j}) \in (-1,1)$. $\log istic(t_{a,j})$ is a monotonically increasing function. With the increase of time t, $\log istic(t_{a,j})$ also increases and remains in the range of $(0,1)$. In this paper, we will map the variation range of time t to $[0,1]$ using a standardized method. The value of $\log istic(t_{a,j})$ changes almost linearly with time [10]. Based on the characteristics of this function, we will take this function as the weight of our time. Because short video has strong timeliness, in order to accurately calculate the user's nearest neighbor. On the basis of combining equation (2), we add the weight of time. The formula for adding time weights is given by equation (4).

$$Sim(U_a,U_b) = \frac{\sum_{i_k \in I'} (R_{a,k} \times \log istic(t_{a,j}) - \bar{R}_a) \times (R_{b,k} \times \log istic(t_{a,j}) - \bar{R}_b)}{\left(\sum_{i_k \in I'} (R_{a,k} \times \log istic(ta,j) - \bar{R}_a)^2\right)^{1/2} \times \left(\sum_{i_k \in I'} (R_{b,k} \times \log istic(ta,j) - \bar{R}_b)^2\right)^{1/2}} \tag{4}$$

3.2. Recommended results of k NN collaborative filtering

The k NN collaborative filtering recommendation algorithm is the most mature algorithm in data mining. The k NN approach relies primarily on a limited sample of users' proximity, rather than relying on short video types. For cross-over or overlapping sample sets such as video, the k NN method is more suitable than other methods.

By calculating the similarity between users, Time-weighted user behavior analysis collaborative filtering algorithm finds k neighbor for an unknown I_j and defines $S(U_a)$ and $|S(U_a)| = k$.

$$R_{a,j} = \bar{R}_a + \frac{\sum_{U_b \in S(U_a)} Sim'(U_a,U_b) \times (\bar{R}_{b,j} - \bar{R}_b)}{\sum_{U_b \in S(U_a)} Sim'(U_a,U_b)} \tag{5}$$

In the above equation (5), $R_{a,j}$ represents the predicted score of the user on the video B. The video is sorted in descending order according to the score value, and the top k items are selected as the recommendations.

3.3. Process of time-weighted user behavior analysis collaborative filtering algorithm

The first step: Modelling user behaviour. (Obtain the behaviour data of video users watching video and convert the behaviour data into user ratings)

The second step: Calculating user similarity by scoring matrix.

The third step: Weighting Time with Function $\log istic$.

The fourth step: Calculating the validity of time.

The fifth step: Choosing the appropriate k value.

The sixth step: Calculating k video sets of unknown score video after collaborative filtering.

IOP Conf. Series: Journal of Physics: Conf. Series **1284** (2019) 012062 doi:10.1088/1742-6596/1284/1/012062

4. Experimental results and analysis

4.1. Experimental data
The test data for the experiment is a data set given by Tubular Labs Research Centre, an American short video data organization. The data set contains approximately 1 million pieces of data. This date included 680 users watching 9343 short videos, and each user watched 17 short videos. The score matrix density of user-video is 15.7%. we can conclude that this data set is sparse.

4.2. Test the quality of the recommendation
In order to verify the validity of the time-weighted user behaviour analysis recommendation algorithm proposed in this paper. This paper uses *Precision* and Re*call* method to test it.

Table 2. Recommendation system evaluation rules table

	User favorite video	User dislike video
Method recommended video	TP	FP
Method do not recommended video	FN	TN

From the above Table 2, we can get some of the following equations.

$$Precision = \frac{TP}{(TP+FP)} \tag{6}$$

$$\mathrm{Re}call = \frac{TP}{(TP+FN)} \tag{7}$$

4.3. Experimental result
This paper analyses the data of sixty experimenters watching videos and compares Time-Weighted User Behaviour Analysis Collaborative Filtering Algorithm (TWUBCF) ,Time-Weighted Tab Collaborative Filtering Recommendation Algorithm（TBCF）, Uncertain Neighbors' Collaborative Filtering Recommendation Algorithm(UNCF), and Item-Based Collaborative Filtering Recommendation Algorithm (IBCF).

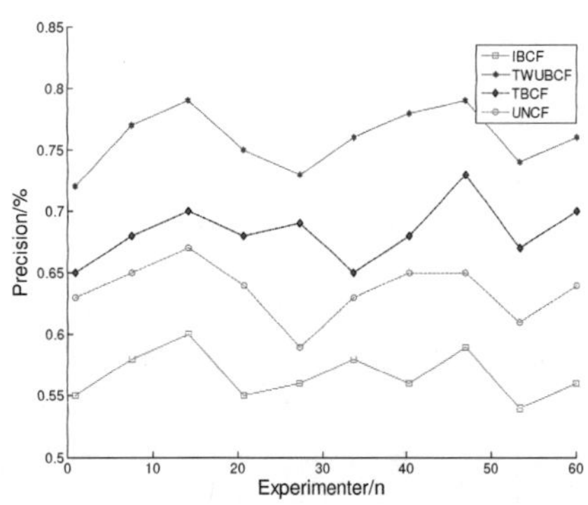

Figure 2. Comparing precision.

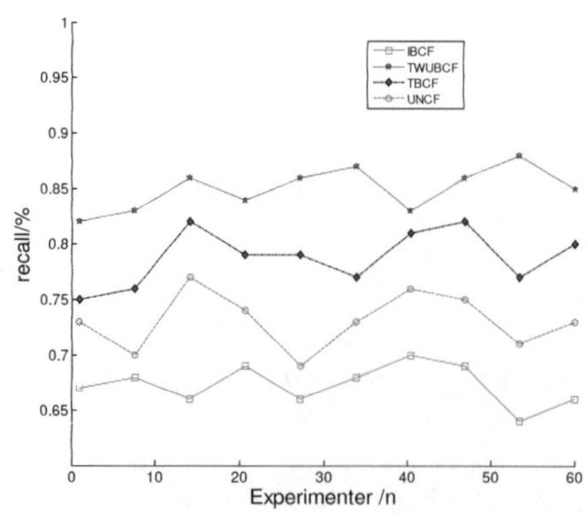

Figure 3. Comparing recall.

From the above Figure 2 and Figure 3, it is obvious that the TWUBCF proposed in this paper is far more comprehensive than the other three. The algorithm in this paper compares with TBCF, the former focuses on the relationship between users, while the latter focuses on the relationship between products. For short videos, it is difficult to find the relationship between videos, because short video information is less, quality is different, and it is difficult to distinguish. It's much easier to analyze the relationships between users, just to find similar user groups. UNCF is one of the classic algorithms. It balances the relationship between the item and the user. However, the score matrix of short video is sparse, and it does not fill the matrix in other ways, resulting in poor recommendation quality. TBCF A and TWUBCF are aware of the influence of time factors on the recommendation results, and weighted the time to improve the quality of recommendation. TBCF pays attention to artificially adding tags to videos, which is useful for traditional movies. Movies take a long time, are shot regularly, and are tagged by professionals. Make the tag fully express the characteristics of the movie itself.

5. Conclusion

Based on the characteristics of short video, TWUBCF is proposed in this paper. User behavior is deeply analyzed from the aspect of users. In order to avoid data sparsity, this paper not only analyses the explicit behavior of users, but also the invisible behavior of users. By improving the similarity, the relationship between users can be more accurately calculated. Time weighting can avoid the decline of recommendation quality caused by the change of user interest over time.

Reference

[1] Mei T, Yang B and Hua X VideoReach: An Online Video Recommendation System *C.//* International Acm Sigir Conference on Research & Development in Information Retrieval. ACM, 2007.
[2] Yang X, Jiong Y U and Ibrahim T. Collaborative Filtering Recommendation Model Based on Trust Model Filling *J.* Computer Engineering, 2015.
[3] Niu J, Zhao X and Zhu L. Affivir: An affect-based Internet video recommendation system *J.* Neurocomputing, 2013, 120(10): 422-433.
[4] Song Weiwei, Yang Degang and Zheng Min. Research on Collaborative Filtering Recommendation Algorithm Based on time-weighted labels *J.* Journal of Chongqing Normal University (Natural Science Edition), 2016 (5): 113-120.
[5] Jian D, Wang Y and Qi W. Football video recommendation system with automatic rating based on user behavior *C.//* International Congress on Image & Signal Processing. 2017.
[6] Quan S, Lu X and Shi Z. Analysis and Research of the Campus Network User's Behavior Based on k-Means Clustering Algorithm*C.//* International Conference on Digital Manufacturing & Automation. 2013.
[7] Hoppe A, Roxin A and Nicolle C. Dynamic, Behavior-Based User Profiling Using Semantic Web Technologies in a Big Data Context *C.//* Otm Confederated International Conferences "on the Move To Meaningful Internet Systems. 2013.
[8] Mclaughlin M R and Herlocker J L. A collaborative filtering algorithm and evaluation metric that accurately model the user experience *C.//* 2004.
[9] He L and Wu F. A Time-context-Based Collaborative Filtering Algorithm *C. //* IEEE International Conference on Granular Computing. IEEE, 2009.
[10] Zhang J, Lin Z and Bo X. An optimized item-based collaborative filtering recommendation algorithm *C //* IEEE International Conference on Network Infrastructure & Digital Content. 2009.

Wireless Sensor Network Deployment Optimization Using Reference-Point-Based Non-dominated Sorting Approach (NSGA-III)

Ali Metiaf[1,*] and Qianhong Wu[2]

[1]School of Electronics and Information Engineering, Beihang University, Beijing, 100191, China
[2]School of Cyber Science and Technology, Beihang University, Beijing, 100191, China

[*]E-mail: alimetiaf23@buaa.edu.cn

Abstract. Wireless sensor networks (WSN) nowadays have gained more interest, pushed by the growing necessity for data collection and transmission from both civilian and military domains. WSN is constructed from interconnected sensors and limited resource (battery), which requests great importance on the deployment strategy to increase the performance metrics for WSN (lifetime, coverage, QoS connectivity). Also, the deployment is considered as a fundamental issue in (WSNs) design, and it was taken from the perspective of the multi-objective optimization problem. Many of the existing deployment strategies are based on metaheuristics algorithms such as Genetic Algorithms (GAs) to resolve the deployment problem. In this article, we use and adopt one of the most attractive approaches for wireless sensor networks deployment (WSND) optimization based on metaheuristic searching which is named Non-dominated Sorting Genetic Algorithm-III (NSGA-III) In order to reach maximum coverage and minimize the consumption of energy to maximize the network lifetime under the connectivity constraint. The comparison results have proved the NSGA-III algorithm outperformed the Constrained Pareto-based Multi-objective Evolutionary Approach (CPMEA) that taken as a benchmark for this study. Those results encourage the application of NSGA-III to real-world deployment problems, and the importance of this approach is that it can be handled many objectives.

1. Introduction

WSNs are an increasing area of technology that has applications across many domains to keep pace with the developing technology of the Internet of Things (IoT). In general, the applications of WSNs can be classified into two types: monitoring and tracking. Monitoring is applied and used for analyzing; some of the examples are military, agriculture, health care, smart house, etc. Tracking is generally used for following the variation of an event; some of the examples are industry, public health, and military [1]. However, the WSN is not perfect. Indeed, given their low cost, and deployment in the often hostile or inaccessible area, which have some weaknesses including limited energy, thus a limited network lifetime, reduced bandwidth, low capture, processing, and communication. These limitations cause topology instabilities, degraded zone coverage, and network connectivity, and so on [2]. Therefore, to overcome all these critical limitations of the WSN, several research interests recently was issued in recent decades to enhance the performance of WSN, and the main ones concern the optimization of WSN deployment with the aim of improving the life-time of the network and coverage [3].

The deployment of the nodes in the interest zone is not necessarily predetermined. Thus, they can be a random or deterministic deployment in the interest zone. In situations where there is prior information of the interest zone, the deterministic deployment is preferably. While in other cases, because of the large size of the network or the inaccessibility of the interest zone, the use of random deployment is the only alternative [4]. As a result, the choice depends strongly on the sensor characteristics and the information of the terrain. Also, the deployment strategies may differ depending on the applications and objectives to be optimized, such as maximizing the coverage of the interest zone, ensuring good connectivity and maximizing the network lifetime. WSND is considered a multi-objective optimization (MOO) problem [3]. In order to reach maximum coverage and minimize the consumption of energy under the connectivity constraint in the same resolution model, many approaches were applied which formulated the problem as the form of a multi-objective function with constraints, have also been proposed in the literature. Such a problem modeled as a multi-objective function with constraints is generally proved to be an NP-complete problem.

2. Related Work

There are numerous problems related with several optimization aspects or criteria in the real world [5]. Numerous techniques based on optimization methods such as GA [6], heuristics and meta-heuristics [7], have been introduced for the resolution of NP-complete optimization problems in the WSND. In the following, we will present some of these solutions.

Deployment is sometimes considered a decision-making element is applied in some status where the user can place the sensors at specific positions. We find this approach in the study of Konstantinidis et al. [8]. where the authors present the problem DPAP (Deployment and Power Assignment Problem). Either a set of sensors of defined size, the objectives are to maximize the coverage (binary model) as well as to maximize the network lifetime. Lifetime depends to the moment when one of the sensors die (the batteries runs out). The decision variables are the sensor positions and the power allocated to the communication devices. To solve this problem, the author proposes a decomposition algorithm MOEA / D that he compares to NSGA-II. The authors redefine this problem in [9], where they integrate a more realistic energy model and are interested in dense deployments. MOEA / D previously used is modified and hybridized with heuristics specific to the problem; it proved to be more efficient than the previous methods. More accurate modeling of the area was discussed in [10], where the authors considered the obstacles, the variation of the detection radius and the inaccessible positions for the deployment. The authors have also developed a MOASA (Multi-objective Optimization Approach for Sensor Arrangement) algorithm, inspired by the SPEA2 algorithm to resolve these multi-objective problems: (i) maximizing the binary coverage; (ii) minimizing the redundancy of coverage; and (iii) minimize the deployed sensors number. Wei et al. [11]. developed a Multi-Objective GA (FD-MOGA) to resolve a three-dimensional deployment problem where the objectives are maximizing probabilistic coverage and maximizing detection levels, and minimizing energy consumption. The FD-MOGA algorithm proved to be more efficient than the MOGA algorithm. In the article of Oh et al. [12]. the authors used NSGA-II approach to optimize the following four objectives: (i) maximizing the binary coverage (taking into account several geometric coverage schemes) (ii) minimize the deployed sensors number, (iii) maximizing the user's preference, taking into account a classification of the types of sensors to be used and (iv) minimizing the distance between the nodes and the target. Chen et al. [13] suggested a hybrid GA based on the basic processes of genetic algorithms, merged with local search strategies. These hybrid methods divide the network nodes into a maximum disjoint coverings number that can cover every one of the targets. The purpose of their algorithm is to be able to resolve the problem, which is NP-complete by a detection technique comprehensive coverage and periodic switching of these disjoint covers. Thus, with such an approach, the network lifetime can be significantly improved. Chamam and Pierre proposed in [14] a centralized heuristic that allows dynamically calculating an almost-optimal set of sensors that can guarantee a predefined coverage rate while ensuring network connectivity in the situation where the communication radius (Rc) of each sensor is bigger than or equal to double its detection radius. However, it can be remarked that 100% coverage of the surveillance zone is not always guaranteed by their proposal. Cerulli et al. [15]. proposed a heuristic technique for life maximization in WSN via sensors with adjustable communication radius *Rc*. They implemented an approach based on a column generation technique and a heuristic incorporated into a local search pattern. This approach is tested in two different scenarios: the situation where the *Rc* of the sensors can be adjusted according to several levels of different powers, and the situation where the *Rc* of the sensors are fixed. They have shown that in the scenario where the sensors have adjustable *Rc*, the performance regarding lifetime is higher than the scenario where they have fixed *Rc* (the classic case). Also, Panag et al. [16]. proposed a new heuristic search scheme to find a maximum disjoint sensors number that absolutely cover a given surveillance area. To do this, they have implemented a method that enables applying a set of heuristics to search the solution space. The best-explored solution is used to update the next population. They have also shown, through tests done on applications requiring several coverage points, that the performance of their proposals in terms of optimality of the solution exceeds other existing proposals. Abdusy et al. [17]. used approach based on NSGA-II to get an optimal deployment that maintains full coverage and connectivity with a minimum sensor number. Authors defined a fitness function to maintain that each sensor is placed within the communication range of the other sensor and to restrict the sensors being near to each other in order to reach the optimal deployment. In addition, the authors apply this approach with the existence of the obstacles in [18].

The major aspect of this work aims to adapt the optimization algorithm NSGA-III introduced by Deb et al. [19]. for wireless sensor deployment problem the purpose is to maximize the coverage and the network lifetime and to compare our result with the state of the art approach CPMEA [3]. To the best of the author`s' knowledge, this strategy hasn't nevermore been suggested for WSND.

The remaining of the article is ordered as follows. In section III we provide the methodology for implementing the objective of the article. The simulation results are displayed in section IV while section V supplies a summary conclusion.

3. Proposed Deployment Methodology

This section provides the methodology used for WSND by adapting NSGA-III and the computation of the objective functions: coverage and the network lifetime.

3.1 Problem Formulation

We assume a 2-D static node in a rectangular sensing zone S, some similar sensors M with energy E and a static sink node K with infinite energy in the center of S. The sensors monitoring the zone of interest and reporting all issues to K. Each sensor i = 1...M has functional characteristics represented in sensing range (Rs) and communication range (Rc). However, these sensors make the communication with each other directly or via multiple hops through the adjacent sensors with K. The WSND problem is regarded from the aspect of the MOO problem due to its objectives. Therefore, for the formulation of the problem, we defined two contrasting objectives: the coverage zone and the network lifetime.

3.2 NSGA-III General Framework

The NSGA-III algorithm has been introduced in [19]. It is the last version of NSGA-II [20] one of the most used multi-objective algorithms, especially for its excellent performance, and its facility of adaptation to most continuous and combinatorial problems. The NSGA-III follows the pattern of a classic genetic algorithm, namely a reproduction cycle and a selection cycle; NSGA-III based on reference points instead of the crowd distance in NSGA-II to choose the individuals that will form the next generation in future, the main process is presented in Table 1. Order the population into Pareto fronts. The first front includes all the non-dominated individuals of the current population, and the second includes the individuals dominated by individuals of the first front. This distribution will intervene in the selection of the solutions until reached to the required size to form the next generation. If the last front to be inserted contain more individuals than is necessary, a selection must be made within this front based on reference points. Where the solutions that have the minimum perpendicular distance to the reference line has priority to be selected, the reference line is the line that starts from the center and passes on the reference point.

Table 1. Pseudo code of NSGA-III algorithm.

Algorithm 1 NSGA-III Main Procedure

Input: Reference points, Parent population, Number of Generation, Rc, Rs, M;
Output: Pareto front;
1: Create an initial population and evaluate objective values for every individual;
2: Repeat for the number of generations;
3: Sorting the population according to non-domination;
4: Use genetic operators to generate a child;
5: Create the combine populations of parent and child;
6: Realize the non-dominated population sorting;
7: Normalize the population members;
8: Assign points to the reference points and select the population with them;
9: End the run;

Finally, after we get the deployment solution that represents the positions of the sensors the computation of the objectives: coverage and lifetime, is provided in the following equations.

3.3 Coverage

Coverage: is a significant performance measure in the WSND. Let *S* be a region (zone) and let *G* $\{(x_1,y_1),(x_2,y_2)\ldots(x_G,y_G)\}$ be grids point of *S*. The coverage of region *S* by a set of sensors *M* mathematically defined in [3] by:

$$coverage=\frac{\sum_{all(x_G,y_G)}(x_G,y_G)}{N_G} \quad (1)$$

$$g(x_G,y_G)=\begin{cases}1 & \sqrt{\Delta x_{ij}^2+\Delta y_{ij}^2}\le R_s\\ 0 & otherwise\end{cases} \quad (2)$$

where $\Delta x_{ij}=x_i-x_j, \Delta y_{ij}=y_i-y_j, i=\{1,2,\ldots M\}, j=\{1,2,\ldots N\}$ *and* N_G represent the number of all grids.

3.4 Lifetime

Lifetime: is given by determining the necessity cycles for forwarding the messages until the first nodes die, and we divide the need cycles over the maximum number of cycles, which is computed according to the minimum distance among the distances between every two nodes. It calculated as following:

$$lifetime=\frac{\min\{T_{failure}\}}{T_{\max}} \quad (3)$$

4. Results and Discussion

For evaluating our adapting algorithm, a set of configurations has been adopted. As represented in Table 2 for taking more information concerning the metaheuristic searching optimization performance by modifying some parameters settings. The experimental researches were taken out in a MATLAB environment on a PC with 64-bit Windows 8.1 operating system, 8 GB RAM, and i5 CPU.

Table 2. Configuration set of the evaluation experiments.

Configurations Setting N°	N° of Generations	N° of Solutions	NSGA-III				CPMEA	
			Crossover Percentage	Mutation Percentage	Mutation Rate	N° of Division	Crossover Rate	Mutation Rate
C1	100	120	0.5	0.5	0.1	10	1	0.1
C2	250	120	0.5	0.5	0.1	10	1	0.5

C3	100	200	0.5	0.5	0.5	10	1	0.1
C4	250	200	0.5	0.5	0.5	10	1	0.5

We use each one of the configurations to produce results of our adaption (NSGA-III) and the benchmark (CPMEA). The Pareto front results presented in the following figures and compared between the two algorithms.

The results of our adaption (NSGA-III) have been compared with CPMEA results based on the configurations C1-C4. The Pareto front was plotted for each of the set of configurations, as presented in figure 1. The two axes represent the two main objectives of the deployment optimization: lifetime and coverage. From this figure, we can observe NSGA-III is superior to CPMEA in the most of the configurations.

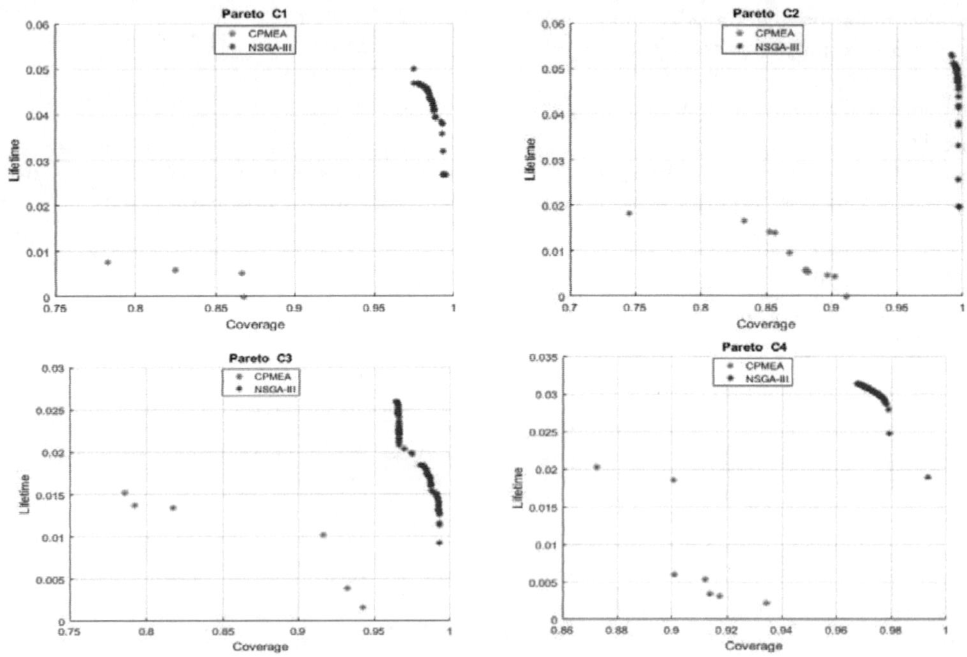

Figure 1. Pareto Front of NSGA-III vs. CPMEA.

5. Conclusion

In this article, we have proposed a formulation of the deployment problem taking into account the two objectives: coverage and network lifetime, a problem of multi-objective optimization is thus posed. NSGA-III was adapted for the goal of WSND. For evaluation, the approach was compared with the benchmark CPMEA. Four configurations of the parameters were employed for the evaluation, which allowed us to construct the optimal Pareto fronts. In the experiments, the NSGA-III found the optimal solutions for all configuration sets and stood out in the figure. Therefore, we can confirm the superiority of the NSGA-III compared to the benchmark (CPMEA) tested in the context of the wireless sensor network deployment problem.

References

[1] R. E. Mohamed, A. I. Saleh, M. Abdelrazzak, and A. S. Samra, "Survey on Wireless Sensor Network Applications and Energy Efficient Routing Protocols," *Wirel. Pers. Commun.*, 2018.

[2] S. M. Mohamed, H. S. Hamza, and I. A. Saroit, "Coverage in mobile wireless sensor networks (M-WSN): A survey," *Computer Communications*. 2017.

[3] M. Khalesian and M. Reza, "Wireless sensors deployment optimization using a constrained Pareto-based multi-objective evolutionary approach," *Eng. Appl. Artif. Intell.*, vol. 53, pp. 126–139, 2016.

[4] S. Abdollahzadeh and N. J. Navimipour, "Deployment strategies in the wireless sensor network : A comprehensive review," *Comput. Commun.*, vol. 0, pp. 1–16, 2016.
[5] A. Metiaf, Q. WU, and Y. ALJEROUDI, "Searching With Direction Awareness : Multi-Objective Genetic Algorithm Based on Angle Quantization and Crowding Distance MOGA-AQCD," *IEEE Access*, vol. 7, no. 1, pp. 10196–10207, 2019.
[6] U. Mehboob, J. Qadir, S. Ali, and A. Vasilakos, "Genetic algorithms in wireless networking : techniques , applications , and issues," *Soft Comput.*, 2016.
[7] Z. Muhammad and N. Saxena, "A REVIEW ON USE OF OPTIMIZATION APPROACHES FOR ENERGY EFFICIENT WIRELESS SENSOR NETWORKS," in *Proceedings of 102nd IASTEM International Conference, Seoul, South Korea*, 2018, no. January, pp. 18–20.
[8] A. Konstantinidis, K. Yang, Q. Zhang, and D. Zeinalipour-yazti, "A multi-objective evolutionary algorithm for the deployment and power assignment problem in wireless sensor networks," *Comput. Networks*, vol. 54, no. 6, pp. 960–976, 2010.
[9] A. Konstantinidis and K. Yang, "Multi-objective K-connected Deployment and Power Assignment in WSNs using a problem-specific constrained evolutionary algorithm based on decomposition," *Comput. Commun.*, vol. 34, no. 1, pp. 83–98, 2011.
[10] J. Lee, J. Seok, and J. Lee, "Multiobjective Optimization Approach for Sensor Arrangement in A Complex Indoor Environment," *IEEE Trans. Syst. Man, Cybern. Part C (Applications Rev.*, vol. 42, no. 2, pp. 174–186, 2012.
[11] L. Wei, C. Kang, J. Chen, and I. Engineering, "A Force-Driven Evolutionary Approach for Multi-objective 3D Differentiated Sensor Network Deployment," in *IEEE 6th International Conference on Mobile Adhoc and Sensor Systems*, 2009, pp. 983–988.
[12] S. C. Oh *et al.*, "Multiobjective Optimization of Sensor Network Deployment by a Genetic Algorithm," in *IEEE Congress on Evolutionary Computation (CEC 2007)*, 2007, pp. 3917–3921.
[13] C. Chen, C. Chuang, T. Lin, and C. Lee, "A Coverage-guaranteed Algorithm to Improve Network Lifetime of Wireless Sensor Networks," *Procedia Eng.*, vol. 5, pp. 192–195, 2010.
[14] A. Chamam and S. Pierre, "Dynamic Sensor Activation for Maximizing Network Lifetime under Coverage Constraint," in *International Symposium on Communications and Information Technologies (ISCIT2007)*, 2007, pp. 971–976.
[15] R. Cerulli, R. De Donato, and A. Raiconi, "Exact and heuristic methods to maximize network lifetime in wireless sensor networks with adjustable sensing ranges," *Eur. J. Oper. Res.*, vol. 220, no. 1, pp. 58–66, 2012.
[16] T. S. Panag and J. S. Dhillon, "Heuristic Search Algorithm (HSA) for Enhancing the Lifetime of Wireless Sensor Networks," *Int. J. Electr. Comput. Energ. Electron. Commun. Engeneering*, vol. 9, no. 8, pp. 822–828, 2015.
[17] A. Syarif, I. Benyahia, A. Abouaissa, L. Idoumghar, R. F. Sari, and P. Lorenz, "Evolutionary Multi-Objective Based Approach for Wireless Sensor Network Deployment," in *2014 IEEE International Conference on Communications (ICC)*, 2014, pp. 1831–1836.
[18] A. Syarif, A. Abouaissa, L. Idoumghar, R. F. Sari, and P. Lorenz, "Performance Analysis of Evolutionary Multi-Objective Based Approach for Deployment of Wireless Sensor Network with The Presence of Fixed Obstacles," in *In 2014 IEEE Global Communications Conference*, 2014, pp. 1–6.
[19] K. Deb and H. Jain, "An Evolutionary Many-Objective Optimization Algorithm Using Reference-Point-Based Nondominated Sorting Approach , Part I : Solving Problems With Box Constraints," *IEEE Trans. Evol. Comput.*, vol. 18, no. 4, pp. 577–601, 2014.
[20] K. Deb, A. Member, A. Pratap, S. Agarwal, and T. Meyarivan, "A Fast and Elitist Multiobjective Genetic Algorithm : NSGA-II," *IEEE Trans. Evol. Comput.*, vol. 6, no. 2, pp. 182–197, 2002.

Position and Velocity Estimation Using TOA and FOA Based on Lagrange Programming Neural Network

C Jia[1,2,*], J Yin[1,2], Z Yang[1,2] and L Zhang[1,2]

[1]National Digital Switching System Engineering and Technology Research Center, Zhengzhou 450002, P.R. China
[2]Zhengzhou Information Science and Technology Institute, Zhengzhou, Henan 450002, P.R. China

[*]E-mail: Kingsway0221@163.com

Abstract. This paper addresses the problem of estimating the position and velocity of a moving source utilizing the time-of-arrival (TOA) and frequency-of-arrival (FOA) measurements. Since the concerned estimation problem is highly non-linear and non-convex, we propose to utilize a novel neural circuit, namely the Lagrange programming neural network (LPNN) framework, to solve this problem. LPNN equips the abilities of fast convergence and the robustness of resisting high noise level, and thus these two advantages have drawn much attention for it recently. Since LPNN is able to solve the optimization problem with constraint, we first reformulate the original non-linear and non-convex maximum likelihood (ML) problem by introducing additional variables and constraints, and thus a neural network is built up based on the LPNN framework. Subsequently, the convergence and stability of the proposed neural network is mathematically proved and then verified by the results of numerical experiments. Different from the conventional numerical algorithms, the analog neural network can be utilized to fulfil the task of real-time calculation, especially when there are limited computation resources in some applications. The simulation results demonstrate that the proposed LPNN model equips the basic properties of convergence and stability, and also show the superior localization accuracy of the proposed method than other numerical algorithms.

1. Introduction

In recent years, passive source localization has attracted many attentions for its wide applications in communications, radar and others [1-3]. Among the positioning parameters used so far, the time-of-arrival (TOA) measurements can be utilized to locate a stationary radiation source with high precision [1, 2]. When there is relative movement between the receiving sensors and the radiation source, the frequency of arrival (FOA), combined with the TOAs, can be incorporated to estimate the source position and velocity. Theoretically, these two types of parameters can be obtained simultaneously when the carrier frequency of the signal can be obtained, and the clock synchronization between receivers and the radiation source can be achieved [2]. Under these two assumptions, the hybrid of TOA and FOA measurements can improve the localization accuracy [2]. Due to the possibility of jointing and its corresponding advantages, TOA and FOA joint positioning has received intensive investigations in geolocation based on communication satellites, direct position determination and sensor network positioning [1, 2].

It is not easy to solve the hybrid of TOA and FOA localization problem due to its high nonlinearity and non-convexity. Therefore, several types of methods have been utilized to tackle such nonlinear and non-convex problems [2-7]. The author in [3] proposes an iterative method based on Taylor-series to solve this problem. In general, the iterative methods are computationally expensive. To alleviate this shortcoming, algorithms with a closed form solution are developed based on the least squares criterion [4, 5]. However, when the measurement noise in the received signal is large, the second-order noise terms ignored in the two methods will significantly degrade the positioning performance of these algorithms. To obtain the optimal solution at a high noise level, the convex optimization technique is introduced to address this problem [6, 7]. But as shown in [7], the estimated result is sub-optimal if the relaxation applied to the original ML problem is not appropriate. More recently, a Monte Carlo importance sampling algorithm based on Pincus theorem is designed in [2] to generate an approximate global solution. However, this algorithm can only yield an optimal solution at moderate noise levels.

Unlike the numerical algorithms listed above, using neural circuit has attract many attentions since it can achieve real-time calculation through hardware implementation [8]. Among a series of neural networks proposed successively [8-10], the Lagrange programming neural network (LPNN) can solve a variety of optimization problems with constraints, and thus it has been applied in many fields [11-17], including source localization [13-17]. The authors in [13] and [17] first propose to use TOA (or TDOA) measurements based on LPNN framework for localization when only the measurement noise is considered. Then this method is employed to handle the outlier data [14] and a more sophisticated scenario in TOA positioning [15, 16]. These previous studies show the excellent performance of LPNN, but they do not consider the case where the radiation source and receivers are not stationary. Besides, the hybrid of TOA and FOA can be used to improve the positioning accuracy as mentioned before.

Therefore, based on the LPNN framework, we advocate to use TOA and FOA measurements to estimate the source position and velocity simultaneously. This is of great value for locating a moving source with high calculation speed, especially in applications that have limited computing resources, e.g., the satellite-based geolocation. Besides, note that the authors in [13, 14] simply ignore the inequality constraints in the process of reformulating the original ML problem. Instead, we advocate to introduce additional variables to convert inequality constraints into equalities as executed in [8, 16, 17], which can approximate the original maximum likelihood (ML) cost function more closely. Moreover, it is of great essential for a network of practical sense to equip the property of fast convergence, as well as the ability of maintaining its stable output when the neural network converges. Therefore, the convergence and stability of our model is analytically proved based on the analysis in [8]. These two properties of the proposed neural network, as well as its superior estimation performance, are verified by the simulation results.

The rest of the paper is organized as follows. First, the localization scenario and a TOA&FOA based model for moving source positioning are given in Section II. Section III constructs the proposed neural network and algorithm according to the LPNN framework, and the stability and computational complexity of the network is also analyzed. Then we conduct several simulation tests to examine the two basic properties and estimation performance of the proposed algorithm in Section IV. Finally, the conclusions are drawn in Section V.

2. Problem formulation

We consider the general problem of locating a moving source using TOA and FOA measurements in a sensor network. There are M sensors deployed in the D-dimensional $(D = 2 \text{ or } 3)$ space to determine the unknown positon $\boldsymbol{u}$ and velocity $\dot{\boldsymbol{u}}$ of a mobile source. The position and velocity of ith, $i = 1,2,\ldots,M$ sensor is accurately known, which is denoted by $\boldsymbol{s}_i$ and $\dot{\boldsymbol{s}}_i$. For simplicity, we assume line of sight (LOS) propagation. Thus the LOS TOA measurements with additive measurement noise e_i are given by

$$t_i = \frac{1}{c}\|\boldsymbol{u} - \boldsymbol{s}_i\| + e_i, \quad i = 1,2,\ldots,M, \tag{1}$$

where $\|\bullet\|$ is Euclidean norm and c represents the signal propagation speed. Multiplying the both sides of (1) by constant c, the range of arrival (ROA) measurements can be obtained as

$$d_i = \|\boldsymbol{u}-\boldsymbol{s}_i\| + n_i,\ \ i=1,2,\dots,M\ , \tag{2}$$

where $n_i = c\cdot e_i$. Taking the time derivation of (2) yields the FOA measurements:

$$\dot{d}_i = \frac{(\boldsymbol{u}-\boldsymbol{s}_i)^{\mathrm{T}}(\dot{\boldsymbol{u}}-\dot{\boldsymbol{s}}_i)}{\|\boldsymbol{u}-\boldsymbol{s}_i\|} + \dot{n}_i,\ \ i=1,2,\dots,M\ . \tag{3}$$

Without loss of generality, the measurement noises $\boldsymbol{n}=[n_1,n_2,\dots,n_M]^{\mathrm{T}}$ and $\dot{\boldsymbol{n}}=[\dot{n}_1,\dot{n}_2,\dots,\dot{n}_M]^{\mathrm{T}}$ are assumed to be uncorrelated. Moreover, these two types of noises follow the Gaussian distribution with zero means, whose covariance are given by $\mathrm{E}(\boldsymbol{n}\cdot\boldsymbol{n}^{\mathrm{T}})=\boldsymbol{Q}_\alpha$ and $\mathrm{E}(\dot{\boldsymbol{n}}\cdot\dot{\boldsymbol{n}}^{\mathrm{T}})=\boldsymbol{Q}_\beta$, respectively. Therefore, based on the above assumptions, the ML estimation of the source position and velocity can be directly computed as

$$\{\hat{\boldsymbol{u}},\hat{\dot{\boldsymbol{u}}}\} = \arg\ \min_{\boldsymbol{u},\dot{\boldsymbol{u}}} \frac{1}{2}(\boldsymbol{d}-\boldsymbol{d}^o)^{\mathrm{T}}\boldsymbol{Q}_\alpha^{-1}(\boldsymbol{d}-\boldsymbol{d}^o) + \frac{1}{2}(\dot{\boldsymbol{d}}-\dot{\boldsymbol{d}}^o)^{\mathrm{T}}\boldsymbol{Q}_\beta^{-1}(\dot{\boldsymbol{d}}-\dot{\boldsymbol{d}}^o) \tag{4}$$

where $\boldsymbol{d}=[d_1,d_2,\dots,d_M]^{\mathrm{T}}$ and $\dot{\boldsymbol{d}}=[\dot{d}_1,\dot{d}_2,\dots,\dot{d}_M]^{\mathrm{T}}$. Besides, $\boldsymbol{d}^o=[d_1^o,d_2^o,\dots,d_M^o]^{\mathrm{T}}$, where $d_i^o=\|\boldsymbol{u}-\boldsymbol{s}_i\|$, and $\dot{\boldsymbol{d}}^o=[\dot{d}_1^o,\dot{d}_2^o,\dots,\dot{d}_M^o]^{\mathrm{T}}$, where $\dot{d}_i^o=(\boldsymbol{u}-\boldsymbol{s}_1)^{\mathrm{T}}(\dot{\boldsymbol{u}}-\dot{\boldsymbol{s}}_1)/d_i^o$.

The problem of interest is to find the optimal estimates of $\boldsymbol{u}$ and $\dot{\boldsymbol{u}}$, which minimizes the ML cost function in (4) by making use of the available noisy measurements of TOA and FOA.

3. The proposed method

According to [8], the LPNN is designed to solve constrained optimization problems. This method defines two kinds of neurons, that is, the variable neurons and the Lagrange neurons. Moreover, these two types of neurons represent the optimization variables and the Lagrangian multipliers, respectively. Based on the original constrained optimization problem and the neurons defined, a Lagrangian function is firstly established. Then the differential dynamic equations of neurons are calculated from the Lagrangian function, and a self-feedback neural network model can be constructed based on these dynamic equations. During the working processing of the network, these equations will govern the network state until it reaches an equilibrium point. More detailed introduction of LPNN can be found in [8], and here we illustrate this method through the construction of the proposed neural network.

3.1. LPNN Model for TOA and FOA Based Localization.

When the measurement noise at different sensors are independent from each other, the measurement noise covariance matrices $\boldsymbol{Q}_\alpha = diag(\sigma_1^2,\sigma_2^2,\dots,\sigma_M^2)$ and $\boldsymbol{Q}_\beta = diag(\dot{\sigma}_1^2,\dot{\sigma}_2^2,\dots,\dot{\sigma}_M^2)$. Here $diag\{\cdot\}$ is the diagonal operator. Therefore, the cost function in (4) can be equivalently written as

$$\{\hat{\boldsymbol{u}},\hat{\dot{\boldsymbol{u}}}\} = \arg\ \min_{\boldsymbol{u},\dot{\boldsymbol{u}}} \sum_{i=1}^{M}\frac{1}{2\sigma_i^2}(d_i - \|\boldsymbol{u}-\boldsymbol{s}_i\|)^2 + \sum_{i=1}^{M}\frac{1}{2\dot{\sigma}_i^2}(\dot{d}_i - \frac{(\boldsymbol{u}-\boldsymbol{s}_i)^{\mathrm{T}}(\dot{\boldsymbol{u}}-\dot{\boldsymbol{s}}_i)}{\|\boldsymbol{u}-\boldsymbol{s}_i\|})^2\ , \tag{5}$$

In order to apply the LPNN framework, the above unconstrained problem need to be reformulated as a constrained one, and thus we introduce the dummy variable $g_i = \|\boldsymbol{u}-\boldsymbol{s}_i\|$ to the original problem in (5). Therefore, it is straightforward to obtain the following constrained optimization problem:

$$\begin{aligned} &\min_{\boldsymbol{u},\dot{\boldsymbol{u}}} \sum_{i=1}^{M}\left\{\frac{1}{2\sigma_i^2}(d_i - g_i)^2 + \frac{1}{2\dot{\sigma}_i^2}(\dot{d}_i - \frac{(\boldsymbol{u}-\boldsymbol{s}_i)^{\mathrm{T}}(\dot{\boldsymbol{u}}-\dot{\boldsymbol{s}}_i)}{g_i})^2\right\}, \\ &\text{s.t.}\ \ g_i^2 = \|\boldsymbol{u}-\boldsymbol{s}_i\|^2, \\ &\qquad g_i > 0,\ \ i=1,2,\dots,M. \end{aligned} \tag{6}$$

Remark 1: It is worthy to point out that in order to guarantee the validity of the objective function, g_i should not equal to 0, thus the original constraint $g_i \ge 0$ is further required to be $g_i > 0$. Hence, we have $g_i = \|\boldsymbol{u}-\boldsymbol{s}_i\| \neq 0$, which means that the location of source should not be the same as one of the sensor positions. In practice, this condition can be met in many localization scenarios.

As described in [8], the inequality constraints in (6) can be transferred into equality constraints by introducing additional variables y_i, $i=1,2,\ldots,M$. That is, to ensure that the non-negative property of g_i, we advocate to let $g_i = \alpha_i \cdot y_i^2$, α_i is a constant and usually $\alpha_i \geq 1$. Moreover, we define $\boldsymbol{g}=[g_1, g_2, \ldots, g_M]^{\mathrm{T}}$, and $\boldsymbol{y}=[y_1, y_2, \ldots, y_M]^{\mathrm{T}}$ is the additional variable vector. $\lambda=[\lambda_1,\lambda_2,\ldots,\lambda_M]^{\mathrm{T}}$ and $\mu=[\mu_1,\mu_2,\ldots,\mu_M]^{\mathrm{T}}$ are the Lagrange multipliers. Hence, the problem in (6) can be further recast as

$$\begin{aligned} \min_{\boldsymbol{u},\dot{\boldsymbol{u}},\boldsymbol{g}} \ & \sum_{i=1}^{M}\left\{\frac{1}{2\sigma_i^2}(d_i-g_i)^2+\frac{1}{2\dot{\sigma}_i^2}\Big(\dot{d}_i-\frac{(\boldsymbol{u}-\boldsymbol{s}_i)^{\mathrm{T}}(\dot{\boldsymbol{u}}-\dot{\boldsymbol{s}}_i)}{g_i}\Big)^2\right\}, \\ \text{s.t.} \ & g_i^2 = \|\boldsymbol{u}-\boldsymbol{s}_i\|^2, \\ & g_i = \alpha_i \cdot y_i^2,\ \alpha_i \geq 1,\ i=1,2,\ldots,M. \end{aligned} \tag{7}$$

Based on (7), a Lagrange function can be directly computed as

$$\mathcal{L}(\boldsymbol{u},\dot{\boldsymbol{u}},\boldsymbol{g},\boldsymbol{y},\lambda,\mu)=\sum_{i=1}^{M}\left\{\frac{1}{2\sigma_i^2}(d_i-g_i)^2+\frac{1}{2\dot{\sigma}_i^2}\Big(\dot{d}_i-\frac{(\boldsymbol{u}-\boldsymbol{s}_i)^{\mathrm{T}}(\dot{\boldsymbol{u}}-\dot{\boldsymbol{s}}_i)}{g_i}\Big)^2\right\}+\sum_{i=1}^{M}\left[\lambda_i(g_i^2-\|\boldsymbol{u}-\boldsymbol{s}_i\|^2)+\mu_i(g_i-\alpha_i\cdot y_i^2)\right] \tag{8}$$

However, our previous simulation experiments show that the LPNN model directly established by (8) is not satisfactory in terms of convergence speed and stability, as well as the final equilibrium output. Such kind of problem also appears in [12,14]. Therefore, we add penalty parameter C_0 and augmented term $\mathcal{A}(\boldsymbol{u},\boldsymbol{g},\boldsymbol{y})=(C_0/2)\{\sum_{i=1}^{M}[(g_i^2-\|\boldsymbol{u}-\boldsymbol{s}_i\|^2)^2+(g_i-\alpha_i\cdot y_i^2)^2]\}$ to the objective function in (7) for the sake of improving the convexity of it and further enhancing the stability of the LPNN model. Details can be found in [8].

Based on (7) and the augmented term, the LPNN framework can be applied since both the objective function and constraints are twice differentiable, and thus the augmented Lagrangian function is calculated as

$$\mathcal{L}_a(\boldsymbol{u},\dot{\boldsymbol{u}},\boldsymbol{g},\boldsymbol{y},\lambda,\mu)=\mathcal{L}(\boldsymbol{u},\dot{\boldsymbol{u}},\boldsymbol{g},\boldsymbol{y},\lambda,\mu)+\mathcal{A}(\boldsymbol{u},\boldsymbol{g},\boldsymbol{y}). \tag{9}$$

According to the Lagrangian function established (9) and the LPNN framework described in [8], the dynamics of variable neurons can be directly computed as

$$\frac{d\boldsymbol{u}}{dt}=-\frac{\partial\mathcal{L}_a}{\partial\boldsymbol{u}}=\sum_{i=1}^{M}\frac{1}{\dot{\sigma}_i^2\cdot g_i}\Big[\dot{d}_i-\frac{(\boldsymbol{u}-\boldsymbol{s}_i)^{\mathrm{T}}(\dot{\boldsymbol{u}}-\dot{\boldsymbol{s}}_i)}{g_i}\Big](\dot{\boldsymbol{u}}-\dot{\boldsymbol{s}}_i)+\sum_{i=1}^{M}2[C_0(g_i^2-\|\boldsymbol{u}-\boldsymbol{s}_i\|^2)+\lambda_i](\boldsymbol{u}-\boldsymbol{s}_i), \tag{10}$$

$$\frac{d\dot{\boldsymbol{u}}}{dt}=-\frac{\partial\mathcal{L}_a}{\partial\dot{\boldsymbol{u}}}=\sum_{i=1}^{M}\frac{1}{\dot{\sigma}_i^2\cdot g_i}\Big[\dot{d}_i-\frac{(\boldsymbol{u}-\boldsymbol{s}_i)^{\mathrm{T}}(\dot{\boldsymbol{u}}-\dot{\boldsymbol{s}}_i)}{g_i}\Big](\boldsymbol{u}-\boldsymbol{s}_i), \tag{11}$$

$$\begin{aligned}\frac{dg_i}{dt}=-\frac{\partial\mathcal{L}_a}{\partial g_i}=&-\frac{1}{\dot{\sigma}_i^2 g_i^2}\Big[\dot{d}_i-\frac{(\boldsymbol{u}-\boldsymbol{s}_i)^{\mathrm{T}}(\dot{\boldsymbol{u}}-\dot{\boldsymbol{s}}_i)}{g_i}\Big][(\boldsymbol{u}-\boldsymbol{s}_i)^{\mathrm{T}}(\dot{\boldsymbol{u}}-\dot{\boldsymbol{s}}_i)] \\ &-C_0[2g_i(g_i^2-\|\boldsymbol{u}-\boldsymbol{s}_i\|^2)+(g_i-\alpha_i\cdot y_i^2)]-\frac{1}{\sigma_i^2}(d_i-g_i)-2\lambda_i\cdot g_i-\mu_i,\end{aligned} \tag{12}$$

$$\frac{dy_i}{dt}=-\frac{\partial\mathcal{L}_a}{\partial y_i}=2C_0(g_i-\alpha_i\cdot y_i^2)\cdot\alpha_i\cdot y_i+2\mu_i\cdot\alpha_i\cdot y_i. \tag{13}$$

The dynamics of Lagrangian neurons can be obtained in the same way, and they are given by

$$\frac{d\lambda_i}{dt}=\frac{\partial\mathcal{L}_a}{\partial\lambda_i}=g_i^2-\|\boldsymbol{u}-\boldsymbol{s}_i\|^2, \tag{14}$$

$$\frac{d\mu_i}{dt}=\frac{\partial\mathcal{L}_a}{\partial\mu_i}=g_i-\alpha_i\cdot y_i^2. \tag{15}$$

By utilizing the LPNN framework, we can build up a neural network based on the above differential equations defined in (10)-(15). According to [10,13,14], the block diagram in Figure 1 shows the main structure of the neural network established. It can be noticed that the proposed neural model is a self-feedback network which consists of $2+4M$ function units and integrators. Besides, there are $2D+2M$ variable neurons to hold parameters $\boldsymbol{u}, \dot{\boldsymbol{u}}, \boldsymbol{g}$ and $\boldsymbol{y}$. Moreover, $4M$ Lagrangian neurons are employed to hold λ and μ. According to the analysis and circuit implementation approach given in [10], the working process of the network can be summarized as follows:

Step 1: Initialize the neurons $\boldsymbol{u}, \dot{\boldsymbol{u}}, \boldsymbol{g}, \boldsymbol{y}, \lambda, \mu$. Input the TOA and FOA measurements and set the integration time $t=0$

Step 2: The differential result of each neuron is first calculated and then transmitted to the integrator. The integrators will output the updated network states.

Step 3: Feedback the updated network states as inputs.

Step 4: Repeat steps 2 and 3 until the network converges to an equilibrium point.

Step 5: Take the steady output states of the network as the final estimation results.

The hardware realization of LPNN model is illustrated in [10,18]. For simplicity, as applied in [13-17], we advocate simulating the searching process of the designed neural network by employing the MATLAB ode solver to solve the differential equations (10)-(15). This is mainly because the purpose of our paper is to establish a TOA and FOA based neural network model for source localization, and further verify its stability and estimation performance. Therefore, the convergence and stability required for an asymptotical stable network are analyzed in the coming subsection.

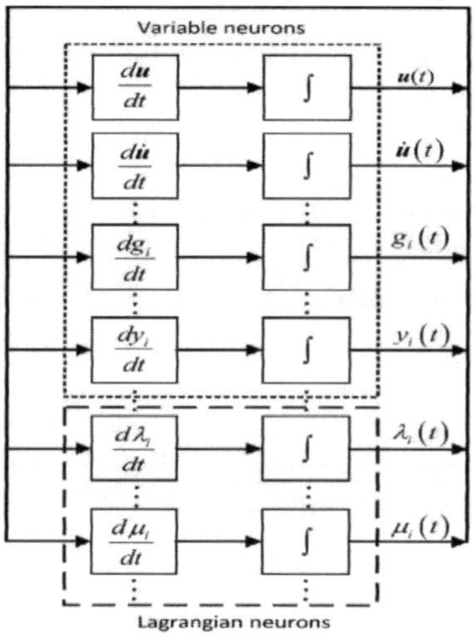

Figure 1. The structure of the proposed LPNN model.

3.2. Network convergence and stability analysis

To explain how the network can converge to the optimal solution, the saddle point property is employed to make an illustration in [8]. It is necessary to point out that our proposed LPNN model also satisfies the general analysis made in [8] since the problem considered meets the fundamental requirements of applying LPNN framework. Hence the convergence of network can be guaranteed. Then we prove that the designed neural network is asymptotically stable when the network converges to the minimum point (or equilibrium point), which means that the network can maintain the stable output state when it settles down at an equilibrium point from an arbitrary initial state within the attraction domain [8].

It is proved in [8] that when the gradient vectors of constraints in (7) are linearly independent and the Hessian matrix of the Lagrangian function in (8) is positive definite, then the equilibrium point $(\theta^{*\mathrm{T}}, \lambda^{*\mathrm{T}}, \mu^{*\mathrm{T}})^{\mathrm{T}}$ is asymptotically stable, where $\theta^{*} = (\boldsymbol{u}^{*\mathrm{T}}, \dot{\boldsymbol{u}}^{*\mathrm{T}}, \boldsymbol{g}^{*\mathrm{T}}, \boldsymbol{y}^{*\mathrm{T}})^{\mathrm{T}}$. Here the property of linear independence is first examined. The equality constraints in (7) can be rewritten in the form:

$$h_i(\theta) = g_i^2 - \|\boldsymbol{u} - \boldsymbol{s}_i\|^2, \quad h_{i+M}(\theta) = g_i - \alpha_i \cdot y_i^2. \tag{16}$$

The gradient vectors of above functions at the equilibrium point are given by

$$\begin{aligned}\left.\frac{\partial h_i(\theta)}{\partial\theta}\right|_{\theta=\theta^*} &= [-2(\boldsymbol{u}^* - \boldsymbol{s}_i^*)^{\mathrm{T}}, \boldsymbol{0}_{1\times 2}, \boldsymbol{0}_{1\times(i-1)}, 2g_i^*, \boldsymbol{0}_{1\times(M-i)}, \boldsymbol{0}_{1\times M}]^{\mathrm{T}}, \\ \left.\frac{\partial h_{i+M}(\theta)}{\partial\theta}\right|_{\theta=\theta^*} &= [\boldsymbol{0}_{1\times 4}, \boldsymbol{0}_{1\times(i-1)}, 1, \boldsymbol{0}_{1\times(M-i)}, \boldsymbol{0}_{1\times(i-1)}, -2\alpha_i\cdot y_i^*, \boldsymbol{0}_{1\times(M-i)}]^{\mathrm{T}}.\end{aligned} \tag{17}$$

It is easy to check that if $g_i = \|\boldsymbol{u} - \boldsymbol{s}_i\| \neq 0$, that is to say $\boldsymbol{u} - \boldsymbol{s}_i \neq 0$, the gradient vectors $\partial h_1(\theta^*)/\partial\theta^*$, $\partial h_2(\theta^*)/\partial\theta^*, \ldots, \partial h_M(\theta^*)/\partial\theta^*$, $\partial h_{M+1}(\theta^*)/\partial\theta^*$, $\partial h_{M+2}(\theta^*)/\partial\theta^*, \ldots, \partial h_{2M}(\theta^*)/\partial\theta^*$ are linearly independent. As stated in Remark 1, the sufficient condition that the source position is different from the sensor locations can be met in many localization scenarios. Thus, the linear independent property is established.

Then we show that the Hessian matrix of the proposed Lagrangian function is positive definite when the penalty parameter C_0 is taken sufficient large. According to (7)-(9), the augmented Lagrangian function can be re-expressed as

$$\mathcal{L}_a(\theta, \lambda, \mu) = \mathcal{L}(\theta, \lambda, \mu) + \frac{C_0}{2}\|\boldsymbol{h}(\theta)\|^2. \tag{18}$$

where $\boldsymbol{h}(\theta) = [h_1(\theta), h_2(\theta), \ldots, h_M(\theta), h_{1+M}(\theta), h_{2+M}(\theta), \ldots, h_{2M}(\theta)]^{\mathrm{T}}$. Therefore, at the equilibrium point, the Hessian matrix of (18) can be obtained as

$$\nabla^2_{\theta\theta}\mathcal{L}_a(\theta^*, \lambda^*, \mu^*) = \nabla^2_{\theta\theta}\mathcal{L}(\theta^*, \lambda^*, \mu^*) + C_0\cdot\nabla\boldsymbol{h}(\theta^*)\nabla\boldsymbol{h}(\theta^*)^{\mathrm{T}}, \tag{19}$$

where $\nabla^2_{\theta\theta}\mathcal{L}(\theta^*, \lambda^*, \mu^*)$ is the Hessian matrix of (8). From (19), we can easily observe that the local convexity of $\mathcal{L}(\theta, \lambda, \mu)$ around the equilibrium point can be remedied when the constant penalty parameter C_0 is set sufficiently large. Hence we have $\nabla^2_{\theta\theta}\mathcal{L}_a(\theta^*, \lambda^*, \mu^*) > 0$. Combine the above two proofs, we can reach the asymptotical stability of the neural network designed.

3.3. Complexity analysis

In this subsection, the computational complexity of the designed algorithm is analyzed. For comparison, we also include the analysis for other three classic numerical algorithms, such as Taylor-series (TS) method [3], weighted least square (WLS) algorithm extended from the algorithm compared in [5] and Monte Carlo importance sampling (MCIS) [2].

As put in [11,12,17], the main computational complexity of using neural network lies in the calculation of the time derivatives defined by the dynamic equations in (10)-(15). Typically, considering that there are M sensors in a K dimensional scenario, the complexity of calculating $\boldsymbol{Q}_\alpha^{-1}$ to obtain $1/\sigma_i^2$ is utmost $O(M^2)$. Similarly, the complexity of calculating $\boldsymbol{Q}_\beta^{-1}$ to obtain $1/\dot{\sigma}_i^2$ is $O(M^2)$. When obtaining the reciprocal of noise covariance, the complexity of computing the differential equations in (10)-(15) is roughly $MK^3 + MK^2 + MK + 20M$, and thus the computational complexity of our model is approximately $O(2M^2) + MK^3 + MK^2 + MK + 20M$ in a single iteration, as given in Table 1.

The main computational complexities of the other three classical numerical algorithms in a single iteration are also provided in Table 1. We can clearly see from Table 1 that the computational complexity of the TS method and the WLS algorithm is comparable, and the MCIS algorithm is computationally high due to the number of sampling points employed in this method. In contrast, the algorithm designed in this paper has a relatively low computational complexity compared with the other three algorithms. This is mainly due to the difference in methods, and the use of neural network has some certain computational advantages.

Table 1. Comparison of computational complexity.

Algorithms	Computational complexity
LPNN	$O(2M^2) + MK^3 + MK^2 + MK + 20M$
WLS	$O(2M^2) + O(K^3) + M^2 + 4M^2(K+2) + 3M(K+2)^2 + 5MK$

MCIS	$O(2M^2)+O(8K^3)+8K^2M+8KM^2+4KM+4K^2N+4KN+2M$, where N is the number of sampling points.
TS method	$O(2M^2)+O(8K^3)+8M^2K+8MK^2+9MK$

4. Simulation results

In this section, we first use noise-free and noisy TOA and FOA measurements to examine the convergence and stability of the designed network, and then we perform several simulations to test the localization performance of the network. In the simulation, we use Matlab ode solver to simulate the network dynamic process, which is governed by a group of differential equations defined in (10)-(15). The initial states of the variables are randomly generated between 0 and 10. Besides, the penalty parameter C_0 is set as 5, which is sufficient to guarantee the local convexity. The positive constants α_i, $i=1,2,\ldots,M$, are set as 100 to accelerate the convergence speed [8].

4.1 Verification of convergence and stability

In this subsection, we use four sensors deployed in the 2-dimension plane to locate a moving source at the position $\boldsymbol{u}=(600,-400)$ m with velocity $\dot{\boldsymbol{u}}=(30,-20)$ m/s. The positions and velocities of the four sensors are listed in Table 2. It is worthy to point out that our method can be directly generalized to a 3-dimension scenario, though a 2-dimension scenario is considered here for simplicity.

Table 2. Positions and velocities of four sensors.

Sensor no.	Sensor position (m)		Sensor velocity (m/s)	
	x	y	v_x	v_y
1	-300	200	10	-10
2	300	-200	20	10
3	300	200	-10	20
4	-300	-200	15	-15

When using the noise-free TOA and FOA measurements, the transient behavior of each neuron that represents the coordinates of source position and velocity after 50 independent runs is shown in Figure 2. It can be observed that after a short transient period from a random initial state, the neuron output state can eventually settle down at the true coordinates of source position and velocity. Besides, the output states can remain stable as the time elapses. The estimation results of using noise-free measurements exhibit that the proposed LPNN model is intrinsically to converge and remain stable.

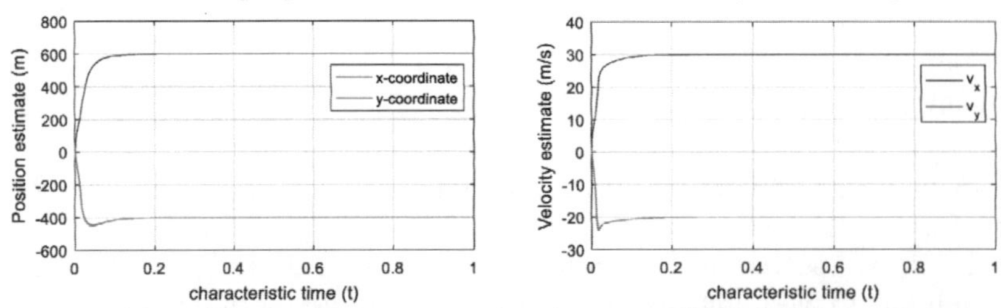

(a) position estimate (b) velocity estimate

Figure 2. Transient behaviors of using noise-free measurements.

Then we examine the proposed network with noisy TOA and FOA measurements since we can only obtain the contaminated positioning parameters in the non-ideal scenarios. For the sake of simplicity, assume that the measured noise power of different sensors is the same, namely $\sigma_1^2=\sigma_2^2=\ldots=\sigma_M^2=\sigma^2$ and $\dot{\sigma}_1^2=\dot{\sigma}_2^2=\ldots=\dot{\sigma}_M^2=\dot{\sigma}^2$. Besides, we let $\dot{\sigma}^2=0.1\sigma^2$ as claimed in [2,3]. Here we set covariance σ^2 as $1\,\mathrm{m}^2$ and $10\,\mathrm{m}^2$ respectively to achieve two different measurement noise levels. The transient behaviors of 50 independent runs for the neurons that of interest under two noise level are shown in Figure 3 and Figure 4. From the figures, we find that our neural network can converge and settle down after a short period of oscillation under both conditions. Moreover, it takes fewer time to converge when measurement noise is small, and the output estimates of source position and velocity are close to true values even when the measurement noise is large. In conclusion, our model equips the abilities of convergence and stability through the numerical experiment results.

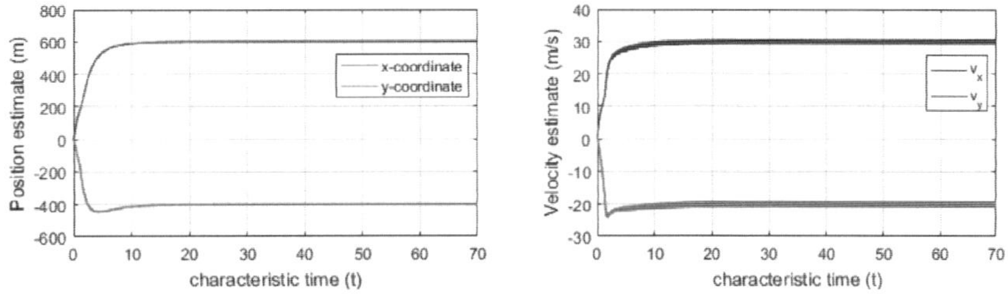

(a) position estimate (b) velocity estimate

Figure 3. Transient behaviors when σ^2=1 m^2.

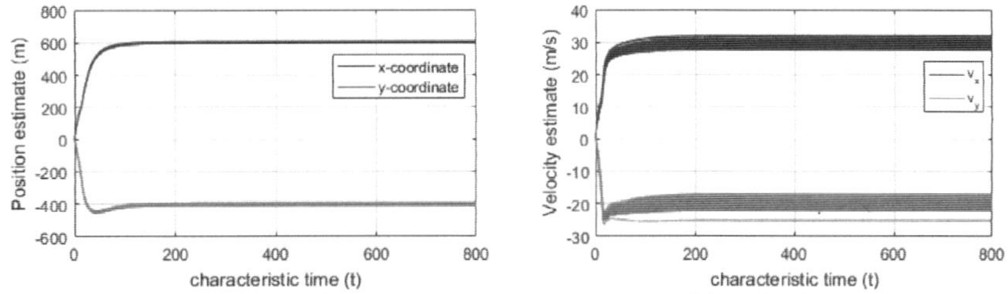

(a) position estimate (b) velocity estimate

Figure 4. Transient behaviors when σ^2=10 m^2.

4.2 Estimation performance

In this subsection, several simulations are conducted to evaluate the location and velocity estimation performance of the designed neural model. Meanwhile, the proposed method is compared with three classic numerical algorithms, namely the Taylor-series method (TS), weighted least square algorithm (WLS) and Monte Carlo importance sampling (MCIS) algorithm developed in [2]. Besides, the Cramér-Rao lower bound (CRLB) is also included as a benchmark.

We consider a 2-dimension scenario and four sensors are employed as listed in Table 1. The noise standard deviation σ is varied from 1 m to 17 m to achieve different noise level, and the localization performance is compared in terms of root mean square error (RMSE) through 1000 independent runs.

In the first test, the position of the moving source is $(600, -50)$ m with velocity $\dot{\boldsymbol{u}} = (30, -15)$ m/s. Figure 5 shows the RMSEs versus the noise standard deviation σ. It can be clearly seen that the method designed in this paper can reach the CRLB within a large noise variation range. However, the other three algorithms perform much worse than our method. As the measurement noise gradually grows large, the WLS algorithm suffers from the threshold effect, and thus its RMSE deviates from the CRLB immediately. The iterative Taylor series algorithm equips a higher threshold value than WLS, but the second-order term of noise ignored in this algorithm will cause a significant error when the measurement noise is large. Besides, the performance of the MCIS algorithm deteriorates seriously once the importance weight employed fail to modify the estimation in the importance sampling (IS) method at the high noise level.

In the second test, the true position and velocity of the moving source are randomly chosen from the 600×600 m and 60×60 square regions, respectively. Figure 6 shows the RMSEs versus the measurement noise. We can directly observe that the designed algorithm still outperforms the other three algorithms and can obtain the CRLB even when σ is large.

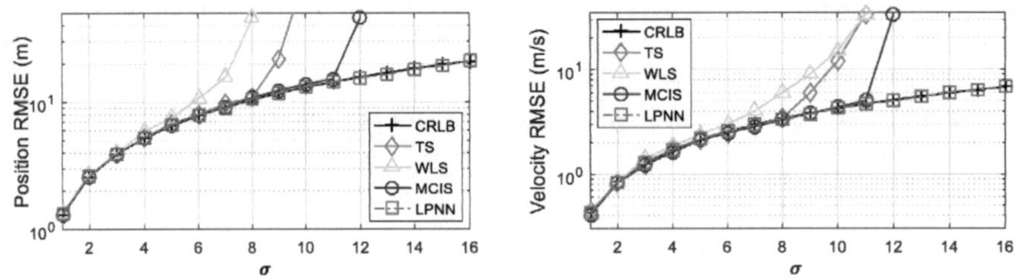

(a) position RMSE (b) velocity RMSE

Figure 5. RMSEs versus σ with $\boldsymbol{u} = (600, -50)$ m and $\dot{\boldsymbol{u}} = (30, -15)$ m/s.

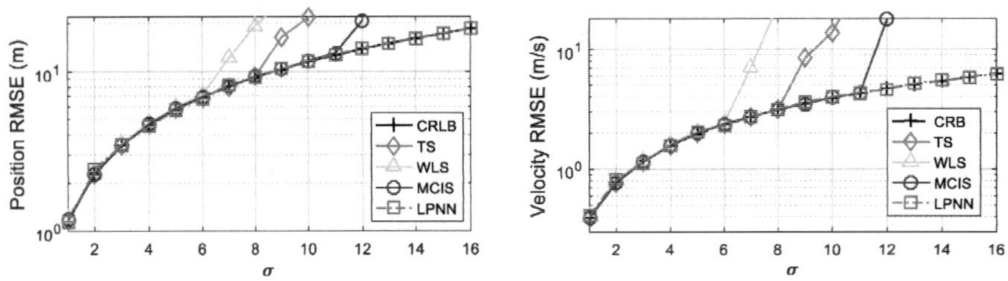

(a) position RMSE (b) velocity RMSE

Figure 6. RMSEs versus σ when the source position and velocity are randomly generated.

In the third test, since the non-ideal sensor geometry will greatly affect the positioning performance, we place the four sensors in a line to test the robustness of the proposed neural network. Here the moving source position is $(600, -50)$ m and $\dot{u} = (-30, 15)$ m/s, and the velocities of four sensors are the same as in Table 1, but the positions are changed to $(300, 300)$ m, $(100, 100)$ m, $(-100, -100)$ m and $(-300, -300)$ m. The RMSEs versus the measurement noise are shown in Figure 7 except for the RMSE of WLS algorithm. Because in this case, the WLS method has failed to estimate the source location and velocity, thus the estimation result of WLS is not given. Figure 7 exhibits the excellent performance of the proposed LPNN model even the other two types of algorithms deviate from CRLB consecutively under the non-ideal sensor localization geometry.

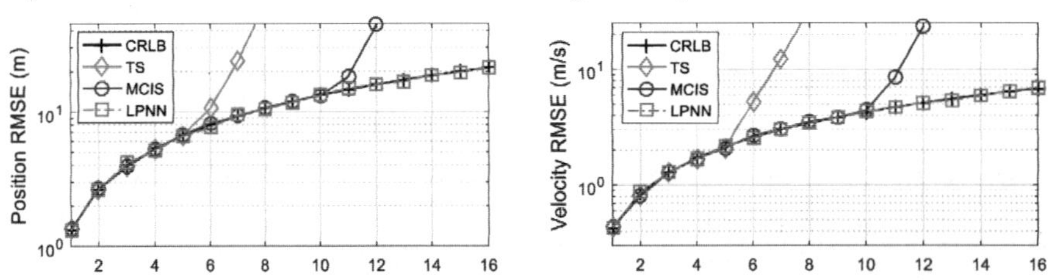

(a) position RMSE (b) velocity RMSE

Figure 7. RMSEs versus σ when the four sensors are placed in a line.

5. Conclusions

In this paper, we build up a neural network for mobile source localization based on LPNN framework, which uses TOA and FOA measurements. Moreover, the proposed analog neural model is proved to be asymptotically stable, and the stability of the network is also verified through numerical experiments. The simulation results demonstrate that our LPNN model has excellent estimation performance compared with the existing numerical algorithms and it can reach the CRLB even at the high noise level.

Acknowledgements
The authors acknowledge support from National Natural Science Foundation of China (Grant No. 61201381, No. 61401513 and No. 61772548), China Postdoctoral Science Foundation (Grant No. 2016M592989).

References
[1] Romero L A and Mason J 2015 Geolocation using TOA, FOA and altitude information at singular geometries *IEEE Trans. Aerosp. Electron. Sys.* **51** 1069–1078
[2] Liu R, Wang Y, Yin J, Wang D and Wu Y 2017 Passive source localization using importance sampling based on TOA and FOA measurements *Frontiers of Information Technology & Electronic Engineering* **18** 1167–1179
[3] Foy W H 1976 Position-location solutions by Taylor-series estimation *IEEE Trans. Aerosp. Electron. Syst.* **12** 187–194
[4] Ho K C, Lu X and Kovavisaruch L 2007 Source localization using TDOA and FDOA measurements in the presence of receiver location errors: Analysis and solution *IEEE Trans. Signal Process.* **55** 684–696
[5] Wang Y, Huang J, Yang L and Xue Y 2015 TOA-based joint synchronization and source localization with random errors in sensor positions and sensor clock biases *Ad Hoc Networks* **27** 99–111
[6] Zou Y and Wan Q 2106 Asynchrous Time-of-arrival-based source localization with sensor position uncertainties *IEEE Commun. Lett.* **20** 1860–1863
[7] Wang Y and Wu Y 2017 An efficient semidefinite relaxation algorithm for moving source localization using TDOA and FDOA measurements *IEEE Commun. Lett.* **21** 80–83
[8] Zhang S and Constantinides A G 1992 Lagrange programming neural networks *IEEE Trans. Circuits Syst. II, Analog Digit. Signal Process.* **39** 441–452
[9] Bouzerdoum A and Pattison T R 1993 Neural network for quadratic optimization with bound constraints *IEEE Trans. Neural Netw.* **4** 293–304
[10] Costantini G, Perfetti R and Todisco M 2008 Quasi-lagrangian neural network for convex quadratic optimization *IEEE Trans. Neural Netw.* **19** 1804–1809
[11] Liang J, So H C, Leung C S, Li J and Farina A 2015 Waveform design with unit modulus and spectra l shape constraints via Lagrange programming neural network *IEEE J. Sel. Topics Signal Process.* **9** 1377–1386
[12] Liang J, Leung C S and So H C 2016 Lagrange programming neural network approach for target localization in distributed MIMO radar *IEEE Trans. Signal Process.* **64** 1574–1585
[13] Leung C S, Sum J, So H C, Constantinides A G and Chan F K W 2014 Lagrange programming neural network approach for time-of-arrival based source localization *Neural Comput. Appl.* **12** 109–116
[14] Wang H, Feng R, Leung A C S and Tsang K F 2018 Lagrange programming neural network approaches for robust time-of-arrival localization *Cogn. Comput.* **10** 23–34
[15] Jia C, Yin J, Wang D and Zhang L 2018 Lagrange programming neural network for TOA-based localization with clock asynchronization and sensor location uncertainties *Sensors* **18** 2293
[16] Jia C, Wang D, Yin J, Chen X and Zhang L 2019 Joint multiple sources localization using TOA measurements based on Lagrange programming neural network *IEEE Access* **7** 3247–3263
[17] Han Z, Leung C S, So H C and Constantinides A G 218 Augmented Lagrange programming neural network for localization using time-difference-of-arrival measurements *IEEE Trans. Neural. Netw. Learn. Syst.* **29** 3879–3884

IOP Conf. Series: Journal of Physics: Conf. Series **1284** (2019) 012065 doi:10.1088/1742-6596/1284/1/012065

RCS Historical Data Screening Method Based on Spatial Observation Geometric Relationship

M X Hu*, J Z Lai and C Xu

Space Engineering University, Beijing, China

*E-mail: 1256064775@qq.com

Abstract. Selecting accurate Radar Cross Section (RCS) sequence historical data has an important impact on RCS-based spatial target attitude anomaly detection. RCS is related to factors such as satellite shape size, attitude and radar observation angle, and is a sensitive function of observation angle variation. In this paper, the influence of spatial observation geometric relationship on spatial target RCS sequence is analysed. Aiming at the orbital regression characteristics unique to spatial targets, a spatial target RCS historical data screening method based on dynamic time programming (DTW) distance is proposed. According to the relationship between the azimuth and the elevation angles of the radar line of sight, the orbit data with similar spatial geometry is found from the historical data. The orbital data of the system. This method can reduce the impact of Earth's rotation on data observation and establish a more efficient database. The variational mode decomposition (VMD) of RCS is carried out, and the fractal box dimension of each component is extracted as the data feature, and the BP neural network is used as the classifier for the anomaly detection. The simulation results prove that compared with the traditional historical data collection method, this method can improve the attitude detection rate of space targets and has better robustness.

1. Introduction

Today, satellites play an indispensable role in navigation and positioning, communications and other fields. The anomaly detection of space targets is mainly to quickly analyse the operational status of space targets such as satellites and spacecrafts, to confirm the possible abnormalities of the targets and the causes of the anomalies, and to provide a basis for rescue and emergency re-issue [1], which is an important part of ensuring the normal operation of the satellites.The attitude control mode of satellite can be the basis for judging the working state of the satellite. The literature [2-4] discriminates the attitude pattern of the satellite based on the simulated RCS data, and has achieved certain results. However, the influence of the radar observation angle change on the RCS value is not considered in the experiment. In this paper, based on the change of radar line-of-sight angle, the method of screening RCS historical data under similar spatial geometric relationship based on DTW distance is proposed. The detection process was analysed, and the improvement effect of the method combined with the traditional detection method was compared based on the simulation experiment.

2. Geometric relationship between ground-based monitoring radar and space targets

The RCS value is affected by the radar line of sight. The difference in spatial geometry makes the surface of the radar observation space target different, so that the RCS data obtained by the ground-based radar is different for the same spatial target.

IOP Conf. Series: Journal of Physics: Conf. Series **1284** (2019) 012065 doi:10.1088/1742-6596/1284/1/012065

2.1. Spatial geometric observation angle

Define the centroid orbit coordinate system O-XYZ of the space target. The origin of the coordinate system is in the centroid of the space target, the X axis points to the velocity direction, the Z axis points to the zenith direction, and the Y axis is orthogonal to the X and Z axes and satisfies the right-hand rule, which is presented in Figure 1. Under the O-XYZ, define the elevation angle between the radar line of sight and the +Z axis, and the azimuth angle between the projection of the radar line of sight at XOY and the +X axis, which is presented in Figure 2.

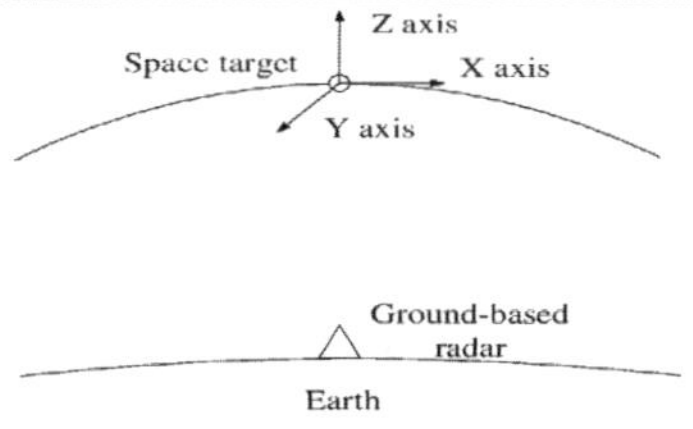

Figure 1. Space target centroid orbit coordinate system.

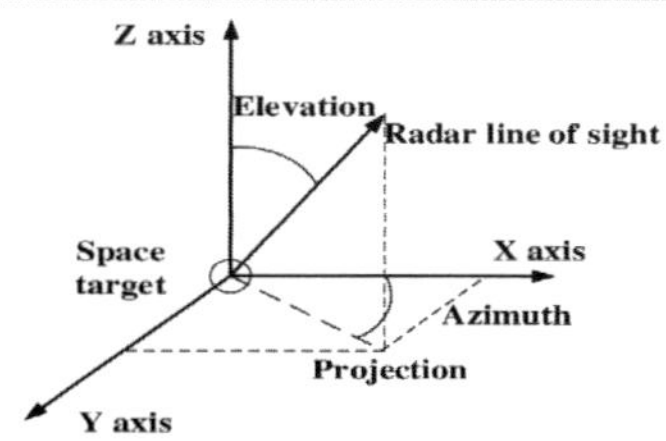

Figure 2. Spatial geometric relationship.

2.2. The influence of spatial geometry on RCS

Simulation was conducted to verify the influence of spatial geometry on RCS, and satellites with orbital roots as shown in table 1 were established in STK software.

Table 1. Orbital elements.

orbital elements	The parameter value
Semimajor axis	7378.14km
Eccentricity	0
Inclination	98°
Argument of periapsis	120°
Longitude of the ascending node	110°
Mean anomaly at epoch	0°

Radar station location set to 29.6° north latitude and longitude 100.5° , observation time is UTCG 00:00:00 on January 1, 2018, 000 and 2018 on February 1, 00:00:00. 000. The trajectory of sub-satellite points is shown in Figure 3.As can be seen from the figure 3, the rotation of the earth causes the deviation of the trajectory of the sub-satellite point, which causes the observation surface of the radar line of sight Angle to change, and the obtained RCS data changes in the numerical value and sequence length, stretching and translation in time and space.

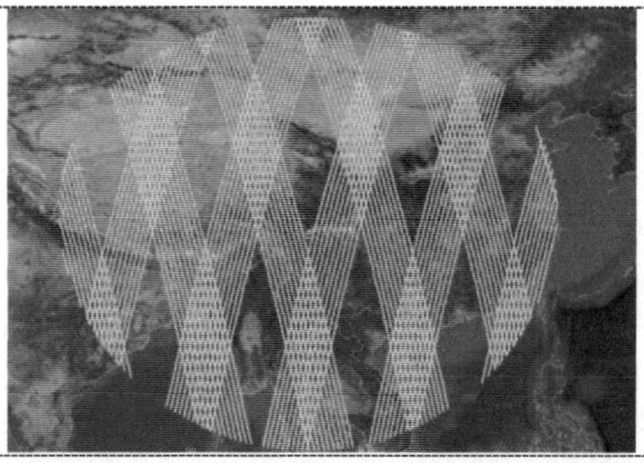

Figure 3. Sub-satellite trajectory change.

RCS under different observation angles of the same target are compared, as shown in Figure 4.

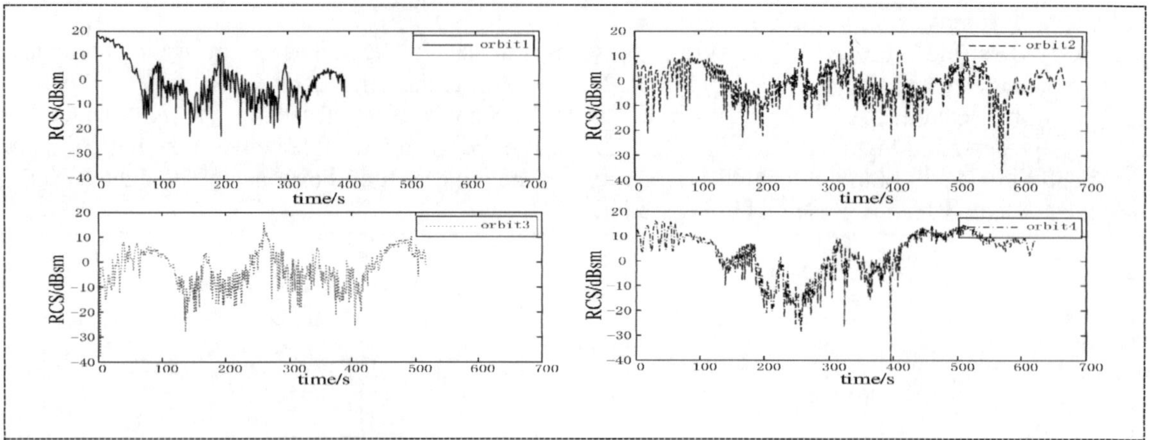

Figure 4. RCS with different spatial geometric relationships.

The traditional anomaly detection method ignores the influence of earth rotation on the RCS data in time and space. The influence can be reduced to a certain extent by screening RCS data with similar spatial geometric relations in historical data.

3. Theory of Dynamic Time Warping

The Dynamic Time Warping (DTW) algorithm was first used in the field of speech recognition. It is an algorithm for calculating distance by dynamic time planning and judging sequence similarity [5].The similarity between time series is generally judged by calculating distances, such as Euclidean distance and Markov distance [6], but there is a problem that only sequences of the same length can be calculated. The DTW algorithm is generated to deal with the problem of similarity between unequal long time series [7], and it is applicable to calculate the similarity between time series of spatial targets that vary due to the earth rotation.

There are two one-dimensional time series and, whose lengths are n and m respectively. The distance matrix D is calculated.

$$D(i,j) = \mathrm{d}(\mathrm{x}(i), \mathrm{y}(j)) \tag{1}$$

Where, d is called local distance, and Euclidean distance is used in this paper.

The regular path W is used to describe the mapping mode between two sequences.

$$\begin{cases} W = (w_1, w_2, \cdots, w_t) \\ w_t = d(\mathrm{x}(i), \mathrm{y}(j)) \end{cases} \tag{2}$$

Where, wt is the t element on the structured path W, and the length of the structured path W satisfies $t \in [\max(\mathrm{n}, \mathrm{m}), \mathrm{n}+\mathrm{m}-1]$.

In addition, the following conditions shall be met for the structured path:

- boundary: the two time series to be tested are corresponding at the beginning and end points, that is, the calculation of W starts from $(\mathrm{x}(1), \mathrm{y}(1))$ and ends from $(\mathrm{x}(n), \mathrm{y}(m))$.
- monotonicity and continuity: for the adjacent elements $w_t = d(\mathrm{x(i)}, \mathrm{y(j)})$, $w_{t-1} = d'(\mathrm{x(i)}', \mathrm{y(j)}')$ in W,which satisfy $|\mathrm{x}(i) - x(i)'| \in [0,1]$ and $|\mathrm{y}(j) - \mathrm{y}(j)'| \in [0,1]$, the distance between any point at any time and the point at the nearest time can only be calculated, and it should be carried out monotonically along the time axis, and the mapping between points at different time points cannot exist intersection.

Under the above conditions, the optimal path should be found so that the minimum value of the sum of the distances is the DTW distance, as shown in Figure 5.

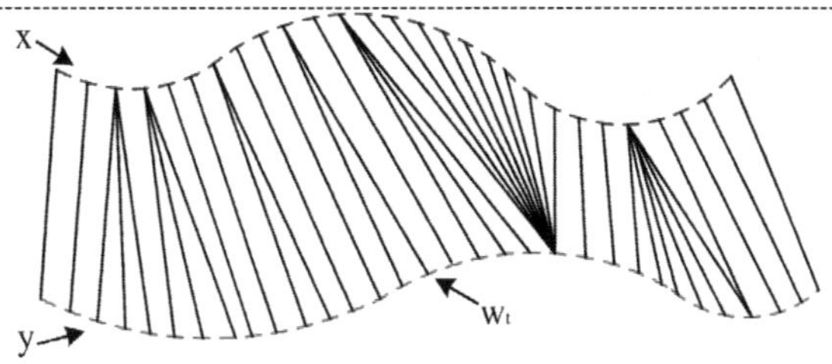

Figure 5. DTW distance diagram.

The distance accumulation matrix S is constructed to solve the DTW distance.

$$\begin{cases} S(1,1)=d(x(1),y(1)) \\ S(i,j)=d(x(i),y(j))+\min(S(i-1,j),S(i,j-1),S(i-1,j-1)) \end{cases} \quad (3)$$

4. Simulation and analysis

4.1. Method of screening RCS

Track parameters are set through STK toolbox, as shown in Table 1. The location of radar station is set at 29.6 degrees north latitude and 100.5 degrees east longitude. The observation time is from 00:00:00.000 on January 1, 2018 to 00:00.000 on February 1, 2018 at UTCG. Because the detection time of each circle of space target to ground station varies, the attitude motion information contained in too short RCS sequence is incomplete, so only RCS sequence whose length is longer than 300s is retained in this paper. The space target assumes *a* three-axis stable attitude, and the changes of the azimuth angle *a* and the elevation angle *e* under the space target centroid coordinate system O-XYZ are deduced from the time series three-dimensional coordinates. The screening process is as follows:

- Establish the historical observation arc database: including RCS sequence set, azimuth data set *a*, and elevation angle data set *e*.
- Measure the current observation target data: including RCS sequence, azimuth angle *a*' and elevation angle *e*' of the current observation target.
- Establish the database 2: The DTW distances between the azimuth angle *a*' of the current observation target and the azimuth angle of each observation arc in the historical data set *a* are calculated and sorted from small to large. The RCS sequence set arranged in the first K_1 historical observation arcs and the high and low angle data set of radar line of sight are selected to form the database 2. K_1 is the parameter that determines the size of the database 2.
- Establish the database 3: The DTW distances between the elevation angles of the current observation target *e*' and the high and low angles of each observation arc in database 2 are calculated and sorted according to the distance from small to large. The RCS sequence sets arranged in the first K_2 historical observation arcs are selected to form the database 3. K_2 is the parameter that determines the size of the database 3.

The results of DTW distance screening are shown in Figure 6. Through simulation, it can be found that RCS curves observed under similar spatial geometric relationships have high similarity, as shown in Figure 7.

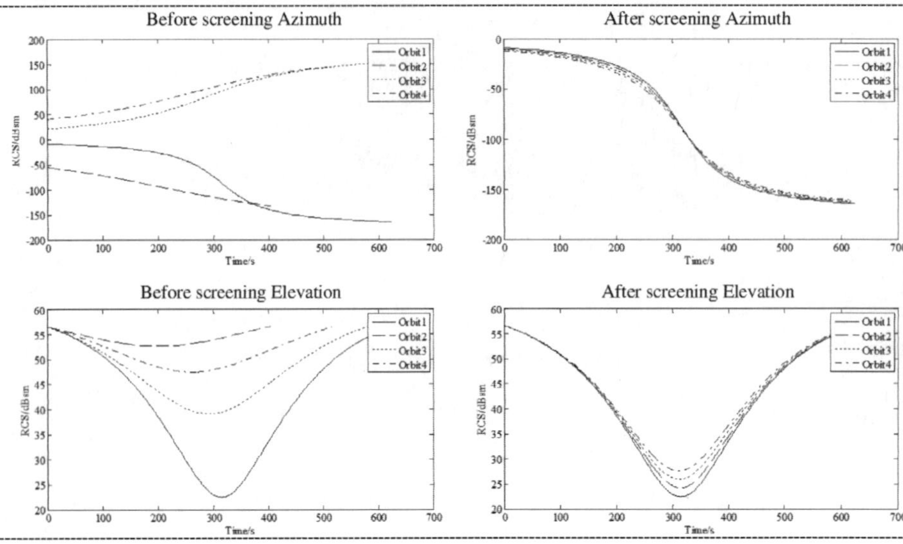

Figure 6. Azimuth, elevation angle before and after screening.

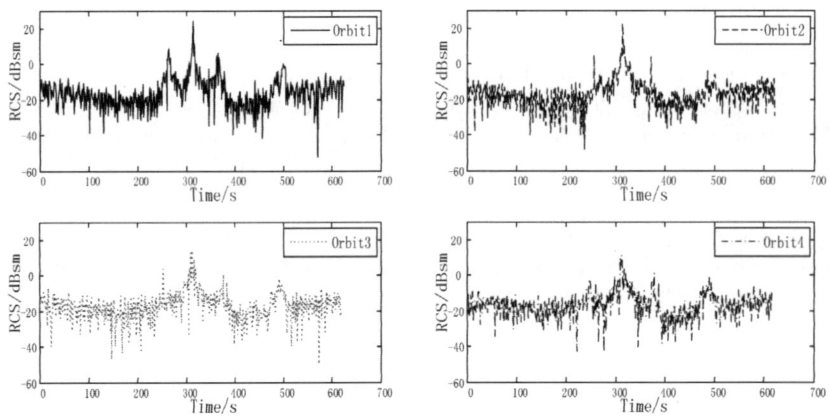

Figure 7. Comparison of RCS in similar spatial geometry.

4.2. Effect validation

4.2.1. Establishing target sample data. Simulation object space target model: A, B, C as shown in Figure 8.

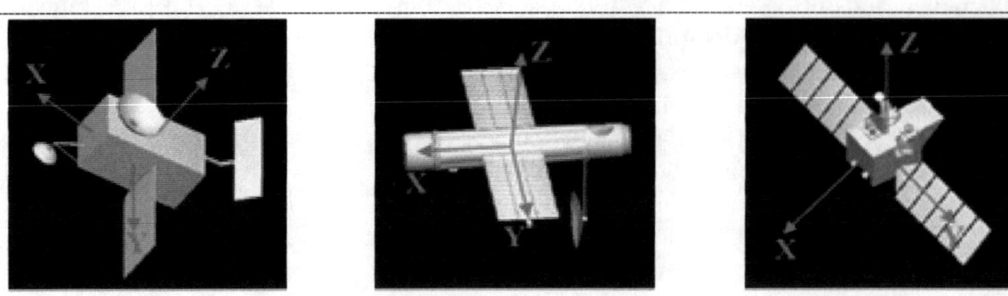

Figure 8. Spatial target model of the sample.

Simulation of four different motion postures:

- Rotate around Z axis at a rate of 3.5 revs/min.
- Rotate around Z axis at 6.5 revs/min.
- Rotate around Z axis at a rotation rate of 5 revs/min and precession rate of 5 revs/min.
- The Z-axis is aligned with the sky-bottom direction, and the X-axis is constrained by the normal direction of the orbit surface.

The number of orbital elements is shown in Table 1. The location of radar stations is set at 29.6 degrees north latitude and 100.5 degrees east longitude. There are few abnormal data in anomaly detection, so the observation time of three-axis stable attitude is from 00:00:00.000 on January 1, 2018 to 00:00.000 on January 1, 2019 at UTCG. The observation time of three different rotational speed rollover postures is from 00:00:00.000 on January 1, 2018 to 100:00.000 on March 1, 2018 at UTCG. The sampling frequency is 1 Hz, and the minimum height angle of radar observation is 15 degrees. Screening is carried out according to the flow shown in Figure 6. Choose 100 for K1 and 60 for k2.

After screening, the number of three-axis stabilized attitude orbits is 60, and the number of orbits is 27 from the simulated rollover attitude data, which constitute the test set 1 under the similar geometric relationship. In order to ensure that the base of the test set is the same, 60 three-axis stabilized attitude orbits are selected from the simulated data continuously, and nine of the three rollover attitude orbits are selected successively, respectively. The test set 2 under the conventional method is constituted.

4.2.2. Extraction feature. Variational mode decomposition (VMD) is a signal processing method that determines the bandwidth and center frequency of each intrinsic mode function (IMF) by seeking the optimal solution of the variational model and realizes adaptive signal decomposition. Fractal dimension is an important index to measure fractal, which can describe the complexity of signal. Box dimension is widely used because of its simple calculation and high efficiency. RCS is decomposed into variational modes to obtain multiple IMF components, and the box dimensions of each IMF component are obtained as features.

Among them, the initialization mode number K=8, penalty factor α and bandwidth τ use default values: α=2000 , τ=0 the box edge length is 1024. The box dimensions of K IMF components are calculated and the eigenvectors are formed.

4.2.3. BP Neural Network. The current mature pattern recognition toolbox of MATLAB neural network is used as classifier, in which the dimension of input parameters is used 10, the dimension of output is 1, the number of neurons is 10, the excitation function is sigmoid function, the learning method is elastic gradient descent method, the test set is 50% as training set, 20% as verification set, and 30% as test set. The remaining training parameters are shown in Table 2.

Table 2. BP neural network training parameter.

Parameter name	parameter values
Number of training cycles	1000 times
Training Accuracy Requirements	1e-4
Maximum number of failures	6 times
Minimum gradient requirement	1e-6
Maximum training time	Inf

Three different roll postures are regarded as abnormalities. The results of abnormal detection are shown in Figure 9.

Figure 9. Recognition result.

From the recognition results, it can be seen that the anomaly detection rate of RCS sequence after screening has been improved by 3.4%, 12.5% and 8.1%, respectively. Compared with the training model directly using historical data, the training of model after screening historical data according to the spatial geometric relationship of current observation data has better adaptability, which has practical physical and geometric significance. It has better robustness.

5. Conclusion

The RCS sequences of space targets with different space target models and different attitude motion modes are simulated. The historical data are screened based on the spatial geometry of the current observation data. The results are compared with the conventional methods using continuous historical data training model. The following conclusions can be drawn.

- The observation mechanism of RCS sequence of space target is studied. It is found that the selection of historical data in traditional methods does not take into account the influence of different spatial geometric relations. According to the motion characteristics of space target, a method of selecting historical data based on DTW distance is proposed and verified by simulation experiments. The results show that this method can improve the anomaly detection rate and provide a reference for fast detection of on-orbit target anomalies based on RCS data.
- RCS has a large amount of historical data, which requires more efficient screening methods. DTW algorithm has high complexity and time-consuming in large amount of data. Therefore, it is necessary to optimize the algorithm to improve the computational efficiency. Spatial target motion has regression characteristics. It is also a research direction to improve the efficiency of filtering by clustering the similar spatial geometric relations of historical data or calculating the regression period of spatial target.

References

[1] SINGLA P, SUBBARAO K, JUNKINS J L. Adaptive output feed- back control for spacecraft rendezvous and docking under measurement uncertainty[J]. Journal of Guidance Control & Dynamics, 2012, 29(4):892-902.

[2] Xu X, Zhao A.J, Ji S.S. Research on space target recognition based on RCS [J]. Firepower and command control, 2010, 35 (10): 134-136.

[3] Zhan W.P, Zheng Y.H, Wang J.X. Radar target recognition based on depth neural network model [J]. Modern radar, 2018 (1): 16-19.

[4] Bu Z.M, Li X. Y, Huang S.D. RCS characteristic analysis of space target based on wavelet power spectrum estimation [J]. Modern Radar, 2004, 26(2): 47-49.

[5] Li Z.X, Guo J.S, Wang Y, et al. DTW distance filtering search method [J]. Control and decision-making, 2018, 33 (7).

[6] Liu E.J, Yan Y, Yang Y.N. Spacecraft Orbital Anomaly Detection Based on TLE Data [C]// Sustainable Development of Space and Space Resources - Proceedings of the First China Conference on Space and Space Safety. 2015.

[7] Sakoe H. Two-level DP-matching--A dynamic programming-based pattern matching algorithm for connected word recognition[J]. IEEE Transactions on Acoustics Speech and Signal Processing, 1980, ASSP-27(6):588-595.

IOP Conf. Series: Journal of Physics: Conf. Series **1284** (2019) 012066 doi:10.1088/1742-6596/1284/1/012066

An Efficient Message Filtering Strategy Based on Asynchronous ADMM with L1 Regularization

Jiafeng Zhang

School of Computer Engineering and Science, Shanghai University, 200444, Shanghai, China

jiafengzhang@shu.edu.cn

Abstract. With the increment of data scale, distributed machine learning has received more and more attention. However, as the data grows, the dimension of the dataset will increase rapidly, which leads to the increment of the communication traffic in the distributed computing cluster and decreases the performance of the distributed algorithms. This paper proposes a message filtering strategy based on asynchronous alternating direction method of multipliers (ADMM), which can effectively reduce the communication time of the algorithm while ensuring the convergence of the algorithm. In this paper, a soft threshold filtering strategy based on L1 regularization is proposed to filter the parameter of master node, and a gradient truncation filtering strategy is proposed to filter the parameter of slave node. Besides, we update the algorithm asynchronously to reduce the waiting time of the master node. Experiments on large-scale sparse data show that our algorithm can effectively reduce the traffic of messages and make the algorithm reach convergence in a shorter time.

1. Introduction

In the era of big data, the scale of data is exponentially increasing, for machine learning, big data is not only reflected in the sample size of data, but also in the dimension of data. For example, in the problem of click-through rate prediction [1, 2] of online advertising, the dimensions of the dataset can reach more than 10 million dimensions, and the data is often stored in a sparse way. Under the large-scale data, it is difficult to optimize the traditional machine learning problem on a single machine. In this situation, distributed optimization [3, 4] is a better solution, and global optimization solution can be obtained by the cooperative computing and communication between nodes in a computing cluster.

ADMM [3] is a distributed optimization algorithm that can effectively solve global consensus optimization problems and has a wide range of applications in machine learning [5] and computer vision [6]. The ADMM algorithm accelerates the solution of machine learning problems by splitting the convex loss function $f(x)$ into N sub-functions $f_i(x_i)$ and computing in parallel. The update of the distributed ADMM algorithm can be described as:

$$
\begin{aligned}
x_i^{k+1} &= \underset{x_i}{\operatorname{argmin}} \left(f_i(x_i) + y_i^{kT}(x_i - z^k) + \left(\frac{\rho}{2}\right) \|x_i - z^k\|_2^2 \right) \\
z^{k+1} &= \frac{1}{N} \sum_{i=1}^{N} \left(x_i^{k+1} + \left(\frac{1}{\rho}\right) y_i^k \right) \\
y_i^{k+1} &= y_i^k + \rho\left(x_i^{k+1} - z^{k+1}\right)
\end{aligned} \tag{1}
$$

Where k is the number of iterations of the ADMM algorithm and ρ is the penalty term parameter. In the distributed ADMM algorithm, there is a master node responsible for updating the global variable z. The others are slave nodes, which solve the local objective function according to local dataset, obtain the local optimization solution x_i^{k+1} and y_i^{k+1}. In each iteration, the master node need to collect all the local optimization solution x_i^{k+1} and y_i^{k+1} from slave nodes to update global variable z. The ADMM algorithm achieves convergence in a certain number of iterations by alternating updates of the x, z and y.

In the distributed ADMM algorithms, each iteration, the master node need to communicate with all the slave nodes. However, when the dimension of the communication parameters is high, the communication time in the network will be greatly increased. In addition, because of the computing performance of each slave node is different, and the master node must wait the slowest node to complete computing for each iteration, which will make the algorithm waste a lot of time on waiting.

There are many researches to solve the optimizations problems of large-scale sparse data [7, 8]. The main idea of them is that each dimension of the model parameter has different convergence speed. In their approaches, each iteration, only partial valid dimensions are selected to update the model parameter, and the algorithms achieve faster convergence. In addition, recently, asynchronous update mechanism [4, 9, 10] is also proposed to reduce the waiting time of the distributed algorithm.

Motivated by the above researches, this paper proposes an efficient message filtering strategy based on ADMM algorithm to solve the communication problem of distributed optimization in large-scale sparse dataset. In our strategy, message will be filtered before the parameter is sent, which reduces the message traffic and communication time is significantly reduced. In addition, we also adopted the asynchronous way to update the algorithm, and the waiting time is reduced. The experiment based on sparse logistic regression shows that our algorithm achieves better convergence and lower communicate latency compared with the algorithm without message filtering.

This paper is organized as follow. Section 2 surveys related works. Section 3 describes the proposed algorithm, including the message filtering strategy and the asynchronous update mechanism. Section 4 carries on the distributed experiments to the algorithm with large-scale datasets. Section 5 concludes the paper.

2. Related work

In order to improve the computational efficiency of distributed algorithms, many researches have focused on reducing communication time and reducing the latency of computing network. In this paper, we will review the ADMM algorithm with asynchronously update and the distributed algorithm with message filtering.

In 2014, R.L. Zhang [9] proposed an asynchronous ADMM algorithm to reduce the waiting time while ensuring the convergence of the algorithm. In their approach, algorithm can update asynchronously with two control conditions: partial barrier and bounded delays. Reference [11] extends the asynchronous ADMM algorithm to solve the problem of non-convex optimization and gives a detailed. Reference [12] proposed an ADMM algorithm based on parameter server architecture, which divides the data according to the model, the service node is responsible for the aggregation of the model parameters, and the working node performs the asynchronous update calculation. Although this algorithm divides the data into dimensions, and reduces the traffic of communication, but it also increases the complexity of the algorithm.

Li Mu [13] proposed an asynchronous distributed block proximal gradient algorithm, which can filter the message of network, and the algorithm set a bounded delay to make the algorithm update asynchronously. In their approach, several message filter strategies is proposed to compress the data, and reduce the communication time of the network. In addition, for the optimization algorithm of large-scale sparse data, [8] proposed the FOBOS algorithm, the idea of the algorithm is to set a gradient truncation, at each iteration, the value of a dimension in the gradient is little than the gradient truncation, the dimension will be filtered. In this way, the algorithm achieves faster convergence. And [7] proposed a deep gradient compression algorithm to reduce the communication bandwidth for distributed training without losing accuracy

3. Algorithm Development

In this section, we will describe the message filtering strategy and the asynchronous update mechanism. Since the distributed ADMM update with a master node and several slave nodes. For master node, we propose a soft threshold filtering strategy to filter the message of global variable z. For slave node, we propose a gradient truncation update strategy to filter the message of local variable x_i and y_i. For the asynchronous control of the algorithm, we refer to [9]the implementation of the asynchronous ADMM algorithm.

3.1. message filtering strategy

In the distributed optimization with large-scale sparse dataset, the dimensions of the model parameters will be very high. Due to the different convergence speeds of different dimensions of the parameters, for those dimensions that have already converged, we can filter it before the message sent and reduce the communication traffic. In addition, when the large-scale sparse dataset is divided into several sub-datasets, some dimensions are invalid for a sub-dataset, and the value of these dimensions is 0 in the local model parameter. These dimensions have no benefit to the update of local model parameter, and we can also filter it to reduce communication.

Before introducing the message filtering strategy, we first define the message format for communication. The format of each message meta is defined as follows:

$$msg_i = \langle k, v\rangle \quad (2)$$

Where k denotes the index of each message in the parameter and v denotes the value corresponding to the kth dimension of the parameter. As shown in figure 1, before the message is sent, the valid elements of the parameter will be collected with the index marked. Then the collected elements with indexes will be sent to receiver. After receiving the filtered message, the receiver needs to aggregate the message and put the valid elements into the related index in the message cache.

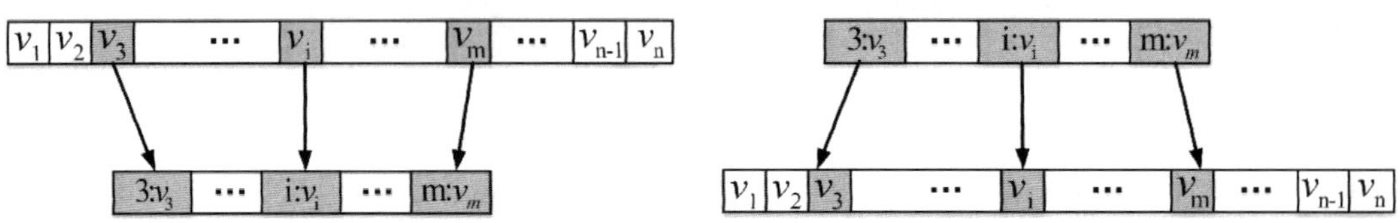

Figure 1. Message filtering and aggregation.

3.1.1. Soft threshold filtering strategy. In the distributed ADMM algorithm, the master node is responsible for the update of the global variable z, and the update formula of z is as shown in the equation (2), where $g(z)$ is generally a regularization term. When the ADMM algorithm contains the L1 regularization, $g(z) = \lambda\|z\|_1$, the update of z can be reformulate as:

$$z^{k+1} = \underset{z}{\operatorname{argmin}}\left(\lambda\|z\|_1 + \textstyle\sum_{i=1}^{N}\left(-y_i^{kT}z + (\rho/2)\left\|x_i^{k+1} - z\right\|_2^2\right)\right) \quad (3)$$

Since the update of z needs to collect the variables from all the slave node, and the update of z is quadratic equation, the update of z can be simplified as:

$$z^{k+1} = \underset{z}{\operatorname{argmin}}\left(\lambda\|z\|_1 + (\rho N/2)\left\|z - \bar{x}^{k+1} - (1/\rho)\bar{y}^k\right\|_2^2\right) \quad (4)$$

Where N denotes the number of slave nodes, $\bar{x}^{k+1}$ denotes the average value of the local variable x_i^{k+1}, and $\bar{y}^k$ denotes the average value of the dual variable y_i^k. Although the L1 regularization term is non-differentiable, we can still solve it by Sub-gradient differentiation:

$$z^{k+1} := S_{\lambda/\rho}\left(\bar{x}^{k+1} + (1/\rho)\bar{y}^k\right) \tag{5}$$

Where $S_{\lambda/\rho}$ denotes the soft threshold operator. We can get the final update formula of the global variable z:

$$\left(z^{k+1}\right)_j = \begin{cases} \left(\bar{x}^{k+1}\right)_j + (1/\rho)\left(\bar{y}^k\right)_j - \frac{\lambda}{\rho} & \left(\bar{x}^{k+1}\right)_j + (1/\rho)\left(\bar{y}^k\right)_j > \lambda/\rho \\ 0 & \left|\left(\bar{x}^{k+1}\right)_j + (1/\rho)\left(\bar{y}^k\right)_j\right| \le \lambda/\rho \\ \left(\bar{x}^{k+1}\right)_j + (1/\rho)\left(\bar{y}^k\right)_j - \frac{\lambda}{\rho} & \left(\bar{x}^{k+1}\right)_j + (1/\rho)\left(\bar{y}^k\right)_j < -\lambda/\rho \end{cases} \tag{6}$$

We use $\left(z^{k+1}\right)_j$ to denote the value of the jth dimension of the z^{k+1} parameter, which can be derived from the above update formula when $\left|\left(\bar{x}^{k+1}\right)_j + (1/\rho)\left(\bar{y}^k\right)_j\right| \le \lambda/\rho$, the value of $\left(z^{k+1}\right)_j$ takes as 0. In this situation, we can deem that $\left(z^{k+1}\right)_j$ have no help with update of slave nodes, so this dimension will be filtered.

The master node filters the updated global variable z according to the soft threshold strategy, and compress it according to the defined message format, and sends it to each slave node. After receiving the filtered message from master node, the slave node will decompress the message and store the global variable z^{k+1}. Then the local variable x_i^{k+1} and the dual variable y_i^{k+1} will update with the global variable z^{k+1}.

3.1.2. Gradient truncation strategy. In the distributed ADMM algorithm, each slave node is responsible for the update of the local variable x_i and the dual variable y_i. The dual variable y_i can be regarded as an auxiliary variable that keeps the global variable and the local variable consistent. According to the global consensus constraint, when the algorithm reaches convergence, the optimization solution (x_i^*, z^*) of the ADMM algorithm satisfies $x_i^* - z^* = 0$, and the update formula for dual variables can be simplified to:

$$y_i^{k+1} = y_i^k + \rho \Delta y_i \tag{7}$$

Where $\Delta y_i = x_i^{k+1} - z^k$, it can be seen as the gradient of the parameter y_i^{k+1}, when the algorithm approaches convergence, Δy_i approaches to 0.

However, with the different convergence of each dimension in the parameter, the value of each dimension in Δy_i is different. We set a truncation value θ of Δy_i, when the absolute value of a dimension in Δy_i is little than the truncation value θ, we consider that the dimension of the parameter has converged, and it will be filtered before parameter is sent. According to equation (1), we describe the update of x_i and y_i with message filtering as follow:

$$x_i^{k_i+\frac{1}{2}} = \underset{x_i}{\operatorname{argmin}}\left(f_i(x_i) + y_i^{k_i T}\left(x_i - z^k\right) + \left(\frac{\rho}{2}\right)\left\|x_i - z^k\right\|_2^2\right)$$

$$y_i^{k_i+\frac{1}{2}} = y_i^{k_i} + \rho\left(x_i^{k_i+\frac{1}{2}} - z^k\right)$$

$$\left(x_i^{k_i+1}\right)_j = \begin{cases} x_i^{k_i+\frac{1}{2}} & \left|\left(x_i^{k+\frac{1}{2}}\right)_j - (z^k)_j\right| \ge (1/\sqrt{k})\theta \\ 0 & \left|\left(x_i^{k+\frac{1}{2}}\right)_j - (z^k)_j\right| < (1/\sqrt{k})\theta \end{cases}$$

$$\left(y_i^{k_i+1}\right)_j = \begin{cases} y_i^{k_i+\frac{1}{2}} & \left|\left(x_i^{k+\frac{1}{2}}\right)_j - (z^k)_j\right| \geq (1/\sqrt{k})\theta \\ 0 & \left|\left(x_i^{k+\frac{1}{2}}\right)_j - (z^k)_j\right| < (1/\sqrt{k})\theta \end{cases} \quad (8)$$

Where $x_i^{k_i+\frac{1}{2}}$ and $y_i^{k_i+\frac{1}{2}}$ are intermediate result of x_i and y_i. $\left(x_i^{k_i+1}\right)_j$ and $\left(y_i^{k_i+1}\right)_j$ denote the jth dimension of $x_i^{k_i+1}$ and $y_i^{k_i+1}$. While [11]shows that ADMM algorithm achieves convergence with $O(1/\sqrt{k})$, so we adjust the truncation as $(1/\sqrt{k})\theta$ dynamically. At each iteration, slave nodes first perform the update of x_i and y_i, and parameters will be filtered by updating $x_i^{k_i+\frac{1}{2}}$ and $y_i^{k_i+\frac{1}{2}}$, then the parameter $x_i^{k_i+\frac{1}{2}}$ and $y_i^{k_i+\frac{1}{2}}$ will be sent to the master node.

3.2. Asynchronous update of the algorithm

In the synchronous ADMM algorithm, the master node needs to collect local variable x, y from all the slave nodes to update the global variable z. It means that, each iteration, the master node needs to wait the update completed of the slowest node, and much time is wasted on waiting. In the asynchronous ADMM algorithm, the master node only collects local variable from partial slave nodes, and the slave nodes can update asynchronously in a bounded delay. In this way, the asynchronous ADMM algorithm can achieves convergence in a shorter time. Next, we will introduce the update method of asynchronous ADMM algorithm.

3.2.1. Partial barrier. Partial barrier denotes that, each iteration, master node only collects local variable from partial slave nodes, and does not need to wait the slowest slave node to complete update. In this paper we use S to denote the partial barrier, and use Φ as a node set to record the partial slave nodes which the master node received message from. In each iteration, only when the count of message master collected reach to $S(|\Phi| \geq S)$, the master could to update global variable z. After completing update of z, the global variable z will be sent to the slave nodes which in Φ.

3.2.2. Bounded delay. Under the control of partial barrier conditions, the slave nodes can update asynchronously. However, if each node performs the update completely asynchronously, the convergence of the ADMM algorithm will not be guaranteed. Therefore, bounded delay is set to control the update of slave nodes. We use τ to denote the value of bounded delay. Bounded delay denotes that the master node must receive at least one message sent by each slave node every τ iterations. In this way the convergence of the ADMM algorithm can be guaranteed. Specifically, the master node sets a count variable τ_i for each slave node. When the master node receives the message from the node i, the value of the corresponding count variable τ_i in the master node is set to 1, when the z variable of the master has updated, the value of all the count variables τ_i is incremented by 1. When the delay of a node reaches τ ($\tau_i > \tau$), the other slave nodes must stop updating and wait for the node to complete the update.

We describe the message filtering strategy based on asynchronous update as shown in figure 2. In a network of one master node and three slave nodes, we set the partial barrier S to 2, and set bounded delay τ to 10. As shown in the figure, at iteration 0, the master node receives the message from the node 1 and the slave node 3, the partial barrier reaches to 2. Then the master node can update the variable z, and send the message to the node 1 and node 3 after message filtering. At this time, the delay count τ_1 and τ_3 will set to 1, and τ_2 will increment by 1 and set to 2. When τ_2 reaches to 10, slave node 1 and node 3 must stop update and wait slave node 2 to complete update. Based on the above analysis, we describe the asynchronous ADMM algorithm based on message filtering in algorithm 1.

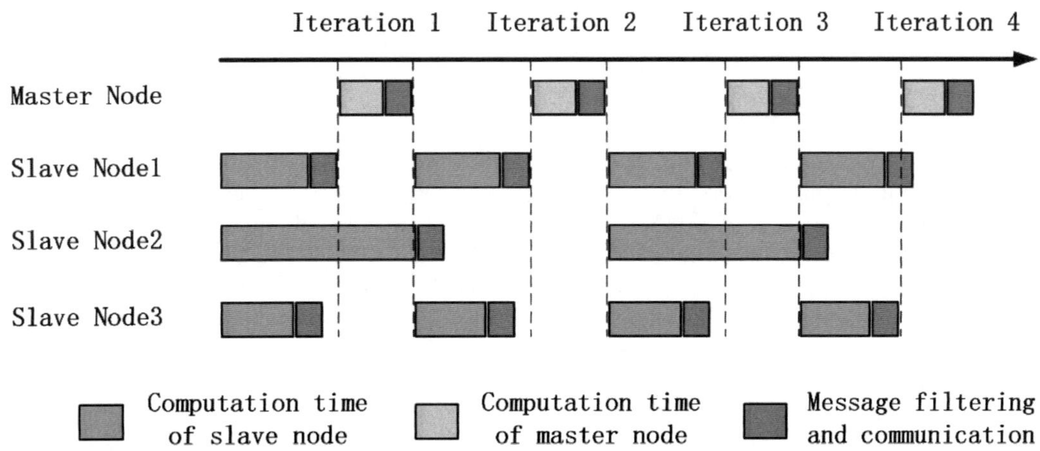

Figure 2. ADMM algorithm update asynchronously based on message filtering.

Algorithm 1 Asynchronous ADMM algorithm based on message filtering

Algorithm of the Master node:

Set $k = 0, \tau_i = 1$, initialize x_i, y_i, $i = 1,2, \dots N$;
repeat
  repeat
    Waiting to receive a message, and store the slave nodes in ϕ^k;
  until $|\phi^k| \geq S$ and $\max(\tau_1, \tau_2 \dots \tau_N) \leq \tau$
  for node $i \in \phi^k$: **do**
    Set $\tau_i = 1$;
    Decompress the message and aggregate the message into x_i and y_i;
  end for
  for node $i \notin \phi^k$: **do**
    Set $\tau_i = \tau_i + 1$;
  end for
  Update z and filter the message according to equation(6);
  for node $i \in \phi^k$: **do**
    Send the filtered message to node i;
  end for
  Set $k = k + 1$;
until satisfies the stopping condition;

Algorithm of the Slave node:

Initialize x_i, y_i and z, set $k = 0$;
repeat
  Waiting to receive message from master node, decompress the message and aggregate the message into z;
  Update x_i, y_i and filter the message according to equation(8);
  Send the filtered message to master node;
until satisfies the stopping condition;

4. Experiments

In this Section, we implement the ADMM algorithm based on message filtering strategy and use it to solve the sparse logistic regression problem under large-scale datasets. Logistic regression algorithm is a classic machine learning algorithm. It is widely applied in data mining, click through rate prediction and other fields. The problem of logistic regression can be formulated as:

$$L(W) = \sum_{i=1}^{N}[y_i(W \cdot x_i) - \log(1 + e^{W \cdot x_i})] + \frac{C}{2}\|W\|_2 \quad (9)$$

Where W is the mode parameter, C is the penalty parameter, x_i andy_i are different from the variable of ADMM algorithm. In this situation, x_i is the sample of dataset, y_i is the label of sample.

The algorithm is implemented in C++, and MPICH is adopted for distributed computing and communication. We tested the algorithm on the high-performance cluster of Shanghai University and used 65 cores for computing. We use one core to simulate the master node and the other 64 cores are used to simulate the slave nodes. Each core is allocated with 4GB memory, and the communication bandwidth between nodes is 5.6Gb/s. We use the trust region newton method [14] to solve the optimization of W, this method is seen to have a better performance than the method such as LBFGS [15] for large-scale logistic regression.

We perform experiment on 3 large-scale sparse datasets (kdd2010, avazu, url), which can be found in the Internet[1]. Among the datasets, kdd2010 has 19,264,097 samples and 1,163,024 dimensions, avazu has 40,428,967 samples and 1,000,000 dimensions, url has 2,396,130 samples and 3,231,961 dimensions.

We set the primal residual as $r^k = \sum_{i=1}^{N} x_i^k - z^k$ according to the paper [3], where k denotes the algorithm kth iteration, N is the number of slave nodes, and then the convergence stopping condition of the algorithm can be described as:

$$\|r^k\|_2 < \sqrt{N}\epsilon^{\text{abs}} + \epsilon^{rel}\max\left\{\left\|\sum_{i=1}^{N} x_i^k\right\|_2, \|z^k\|_2\right\} \quad (10)$$

Where $\epsilon^{\text{abs}} > 0$ is the absolute error, $\epsilon^{rel} > 0$ is the relative error, here we both set them to 10^{-3}.

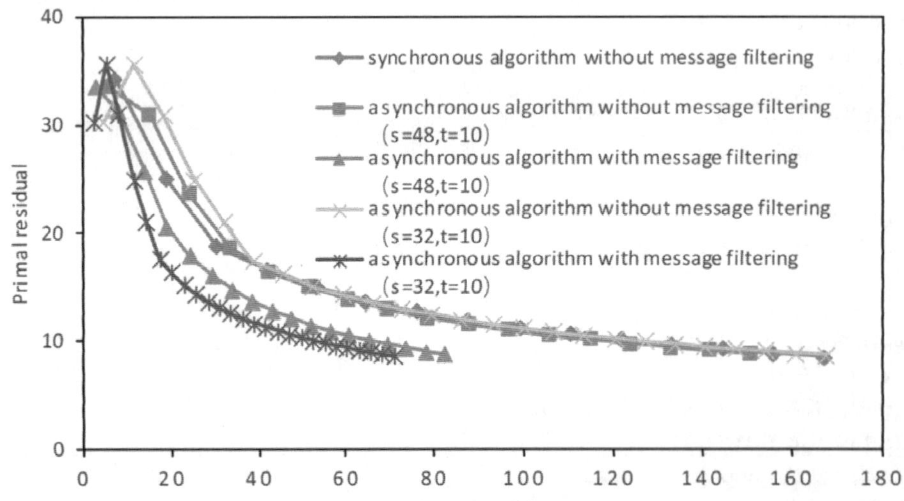

Figure 3. Compare the convergence of algorithm with different settings.

We first analyse the convergence performance of the algorithm with kdd2010 dataset. As shown in figure 3, we compare the primal residual of the algorithm under different settings. It can be seen from the figure that the convergence speed of the synchronization algorithm and the asynchronous algorithm are not much different when the message is not filtered, while the asynchronous algorithm adopting the message filtering strategy can achieve convergence in a shorter time. This is because of the message filtering strategy can significantly reduce the traffic of the message communication, and reduce the communication time of the algorithm. Next, we will analyse the communication performance of the algorithm in detail.

We perform the experiment on the proposed algorithm with 3 different datasets. In this experiment, the penalty parameter ρ of the ADMM is set to 2, and the penalty parameter λ of L1 regularization is set to 1. When the algorithm updates with message filtering strategy, the gradient truncation θ is set as 10^{-3}. When the algorithm updates asynchronously, partial barrier S is set to 32, and bounded delay τ is set to 10.

Table 1. The detailed result under different dataset.

Dataset	Synchronous algorithm without message filtering				Asynchronous algorithm with message filtering			
	Message traffic	C-time (s)	R-time (s)	Accuracy (%)	Message traffic	C-time (s)	R-time (s)	Accuracy (%)
avazu	1,000,000	91.45	122.38	80.17	11,797	24.45	56.34	78.90
kdd2010	1,163,024	131.57	182.64	88.65	35,612	5.8	77.6	88.70
url	3,231,961	607.35	827.22	98.01	48.231	11.16	104.03	97.81

Table 1 shows the detailed result of the proposed algorithm with different settings. Here, we use C-time to denote the communication time of algorithms, and use R-time to denote the run time of algorithms. As can be found in the table, the message traffic of each iteration can be greatly reduced while the algorithm adopting the message filtering strategy. Compared with synchronous algorithm, the Asynchronous algorithm with message filtering achieves faster convergence and has a better communication performance. When the algorithm is asynchronously updated, the algorithm has a lower accuracy. However, it is advisable to improve the performance of the algorithm with an appropriate accuracy loss.

5. Conclusion

This paper proposes a message filtering strategy based on asynchronous ADMM algorithm to solve the problem of communication in distributed optimization with large-scale sparse data. By using L1 regularity, a soft threshold filtering strategy is set on the master node, and gradient truncation is set by the update feature of the slave node. In addition, the algorithm adopts the asynchronous update mechanism to reduce the waiting time of the network. The related experiments show that the message filtering algorithm based on asynchronous ADMM can significantly reduce the communication time of the algorithm while ensuring the convergence of the algorithm. In the next work, the message filtering strategy will be considered to be applied to the ADMM algorithm of model parallel, and make efficiency of the algorithm will be further improved.

References

[1] Ho C H and Lin C J 2012 *Journal of Machine Learning Research* **13** 3323-3348
[2] Canini K, Chandra T, Ie E, McFadden J, Goldman K, Gunter M, Harmsen J, LeFevre K, Lepikhin D, LlinaresT *et al.* 2012 *Technical Talk* **1** 113
[3] Boyd S, Parikh N, Chu E, Peleato B, Eckstein J *et al.* 2011 *Foundations and Trends in Machine learning* **3** 1-122
[4] Li M, Andersen D G and Smola A 2013 Distributed delayed proximal gradient methods *NIPS Workshop on Optimization for Machine Learning* vol 3 p 3
[5] Lubell-Doughtie P and Sondag J 2013 Practical distributed classification using the alternating direction method of multipliers algorithm *2013 IEEE International Conference on Big Data* (IEEE) pp 773-776
[6] O'Connor D and Vandenberghe L 2014 *SIAM Journal on Imaging Sciences* **7** 1724-1754
[7] Lin Y, Han S, Mao H, Wang Y and Dally W J 2017 Deep gradient compression: Reducing the communication bandwidth for distributed training *31st Conference on Neural Information Processing Systems* pp 1-9

[8] Singer Y and Duchi J C 2009 Efficient learning using forward-backward splitting *Advances in Neural Information Processing Systems* pp 495-503
[9] Zhang R and Kwok J 2014 Asynchronous distributed admm for consensus optimization *International Conference on Machine Learning* pp 1701-1709
[10] Agarwal A and Duchi J C 2011 Distributed delayed stochastic optimization *Advances in Neural Information Processing Systems* pp 873-881
[11] Chang T H, Hong M, Liao W C and Wang X 2016 *IEEE Transactions on Signal Processing* **64** 3118-3130
[12] Zhu R, Niu D and Li Z 2018 *arXiv preprint arXiv:1802.08882*
[13] Li M, Andersen D G, Smola A J and Yu K 2014 Communication efficient distributed machine learning with the parameter server *Advances in Neural Information Processing Systems* pp 19-27
[14] Lin C J, Weng R C and Keerthi S S 2007 Trust region newton methods for large-scale logistic regression *Proceedings of the 24th international conference on Machine learning* (ACM) pp 561-568
[15] Nocedal J 1980 *Mathematics of computation* **35** 773-782

Network Traffic Anomaly Detection Based on Incremental Possibilistic Clustering Algorithm

YANG Tian-yi[1] and LIU Shi-yue[1,*] and LIU Jun-yi[1]

[1] School of Automation, Beijing Institute of Technology, Beijing 100081, P.R.China

*E-mail: xyyh@bit.edu.cn

Abstract. This paper proposed a Mahalanobis distance based Incremental Possibilistic Clustering (IPC) algorithm to detect abnormal flow. Firstly, the attributes of network flow is extracted by damped incremental statistics. Then the model of normal traffic will be generated by IPC algorithm. To extract the model of high-dimensional data without pre-known number of cluster centers, the algorithm gradually choose outliers as new clustering centers and merges the overlapping clustering centers. Finally, the data that doesn't belong to any normal model is regarded as abnormal data. By using the Mahalanobis distance instead of the traditional Euclidean distance, the defect that the possibilistic clustering tends to find the features of hypersphere is solved. The experiments show that this method can distinguish normal flow and abnormal flow effectively and reaches the detection rate of 98%.

1. Introduction

Due to the gradual development of internet technology and the popularity of internet of thing(IoT) devices, network security has become a crucial issue in cyberspace. As early as 2014, there were more than 40 million security incidents on the internet, resulting in significant economic losses and leakage of personal privacy [1]. With the accessing of IoT devices, including mobile phones, IP camera and other smart sensor, network traffic has become more and more complex.Although users' network reliance is increasing continued, users still lack awareness of network security [2]. In this case, a kind of high efficiency and low consumption detection method is required for IoT devices to detect network intrusion.

This paper proposes a Mahalanobis distance based Incremental Possibilistic Clustering (IPC) algorithm for anomaly detection. The attributes are extracted by damped incremental statistics instead of sliding window to reduce memory consumption. The IPC gradually detects outliers as new clustering centers, and uses possibility clustering to adjust and merge cluster centers. Finally the data that doesn't belong to any feature is regarded as abnormal data.

2. Related works

The anomaly detection model usually identify the attack by generate a detection model from normal traffic or the traffic combined with small percentage of abnormal packets. The detection model will divide the unknown network packets into different categories. The packet which has the characteristics of attack traffic will be assigned to the category of abnormal traffic or not fit into any category. Many researchers proposed unique detection model based on statistics, machine learning, data mining, information theory or deep learning to identify the

IOP Conf. Series: Journal of Physics: Conf. Series **1284** (2019) 012067 doi:10.1088/1742-6596/1284/1/012067

unknown attack traffic more accurately and efficiently. A survey [1] presents an in-depth analysis detection techniques. It proposed a generic framework for network anomaly detection, and list some detection model based on classification, statistics, information theory and clustering.

Machine learning is the most popular method of anomaly detection. Damopoulos et using four traditional machine-learning algorithms (Bayesian networks, radial basis function, K-nearest neighbors and random forest) [3] and reported that the detection rate has reached 99.8%. The deep learning is the newest favorite approach. Niyaz et designed a detection model by self-taught learning [4], Meidan et use deep auto-encoders to detect anomalous network traffic. The auto-encoders encode the packet and decode it, while only the normal packet will be restored correctly [5].

Clustering, as a common data mining method, is also suitable for anomaly detection. The paper [6] use K-means, k-medoid, EM clustering and KNN algorithm to detect unknown attack. The result shows that their detection accuracies is high while the anomaly detection module produces high false positive rate. Xie et designed an anomaly detection system based on fuzzy c-means [7], but this methods can't deal with high-dimensional data, and can't find the nonconvex structure datasets.

3. Naive clustering algorithm

K-means is a common Clustering algorithm which makes members in a same class have the minimum dispersion via reassigns class members in order to obtain the best clustering results. In K-means clustering, each sample belongs to the nearest cluster center. This method is time-saving and has rapid convergence rate, but the clustering result is not very satisfactory[8]. Based on Zadeh's fuzzy theory, Bezdek J C proposed fuzzy-C-means clustering algorithm(FCM) [9]. This method extends the membership degree of k-means clustering from 0,1 to the interval [0, 1], which makes the sample affect all clusters.

The following index is minimized in FCM:

$$J(U,V) = \sum_{i=1}^{c}\sum_{j=1}^{n} u_{ij}^{m} d_{ij}^{2}, \sum_{i=1}^{c} u_{ij} = 1 \tag{1}$$

where n is the total number of samples, c is the number of clusters, v_i and x_j is the $i^{th} and j^{th}$ vector of data and cluster center. And d is the distance from x_j to v_i, u is the membership from x_j to v_i, m is a default value of FCM, which will usually be 1.5 or 2.

However, FCM also has several obvious drawbacks such as sensitive to noise and need to pre-determine the number of cluster centers[10]. The most severe problem is that FCM can't find the unknown because every data point must belong to one cluster obtained from normal traffic[11]. In view of the above question, the paper proposes using possibilistic clustering instead. By relaxing the limit that the sum of membership degree must be one, we get possibility value from all samples. Then the objective function is defined as:

$$J_{FCM}(U,V) = \sum_{i=1}^{c}\sum_{j=1}^{n} u_{ij}^{m} d_{ij}^{2} + \sum_{i=1}^{c} \eta_i \sum_{j=1}^{n} (1 - u_{ij})^{m} \tag{2}$$

where U is the possibility partition matrix, u_{ij} means the possibility value that x_j belongs to v_i. To avoid getting an infinity cluster center, the function adds $\sum_{i=1}^{c} \eta_i \sum_{j=1}^{n} (1 - u_{ij})^{m}$ as penalty function. η_i is the pre-determined value which influence the clustering effect directly. The well-known value of η_i is $k\frac{\sum_{i=1}^{c}\sum_{j=1}^{n} u_{ij}^{m} d_{ij}^{2}}{\sum_{i=1}^{c}\sum_{j=1}^{n} u_{ij}^{m}}$ where $K > 0$ and usually is 1.

4. Anomaly detection based on IPC

In this section, we introduce the whole progress to detect abnormal network traffic. This section first introduces the method of incremental damping statistics, which can reduce the memory overhead of feature extraction. Then the unsupervised possibilities clustering method based on Mahalanobis distance is introduced. Finally, according to this method, an incremental possibilities clustering method is proposed to detect abnormal data.

4.1. Damped incremental statistics

Feature extraction is the first key problem in pattern recognition system. Generally, the behavior characteristic of network flow can be summarized as five fictions: source IP, destination IP, the size of packets, the sum number of packets and the time interval between each packet. There are three problems with extracting those features. Firstly, the packet arrival rate can be very high; secondly, there may be many different conversations at same time; thirdly, packets from different channels may be closely related [12, 13]. It is an ordinary method to maintain a window of packet from each conversation but this method will use a lot of memory and cannot be used on router.

Therefore, we use a framework named Damped Incremental Statistics which is proposed by Yisroel, Mirsky et. This framework uses incremental statistic maintained [14] over a damped window, makes the framework has a O(1) complexity.

Suppose $S = \{x_1, x_2, \cdots x_n\}$ is an unbounded stream, where x is the attribute of a packet (e.g. the size of packet). The mean, variance and standard deviation of S can be updated incrementally by maintaining the tuple $I_S = (N, L_S, S_S)$, where N, L_S, and S_S are the number, linear sum, and squared sum of the data stream. When a new data x is caught, I_S will update to $(N + 1, L_S + x, S_S + x^2)$.

In order to extract the current status of a data stream, old data must be thrown. The naive approach is to maintain a sliding window. While incremental statistics adds a decay function:

$$d_\lambda(t) = e^{-\lambda t} \tag{3}$$

where $\lambda \cdot t > 0$ is a default value and t is the time elapsed since the last observation from stream.

So the window of the framework adds a variable T_L which is the time last packet was caught. The window of damped incremental statistics will be $I_{Sd} = (n, L_S, S_S, T_L)$. The way to update the I_{Sd} is shown below[5]:

i) Compute decay factor $\gamma = d_\lambda \cdot (T_{now} - T_L)$
ii) Process decay $I_{Sd} = (\gamma n, \gamma L_S, \gamma S_S, T_L)$
iii) Insert new value $I_{Sd} = (n + 1, L_S + x, S_S + x^2, T_{now})$
iv) For variables that require two-dimensional statistics at the same time(for both income and outcome packets), compute $SR_{ij} = \gamma SR_{ij} + r_i r_j$
v) Return value

The statistcs which can be computed from the damped incremental framework is shown in Table 1.A total of 23 features can be extracted from a single time window(see Table 2). In this paper we extracts the same set of features from a total of five time windows : 0.1s,0.5s,1.5s,10s and 1min ($\lambda = 5, 3, 1, 0.1, 0.01$), thus totaling 115 features.

Table 1. Summary of the incremental statistics which can be computed from S_i and S_j

Type	Notation	Calculation
1D	weight(w)	w
	mean(μ)	L_S/w
	variance(σ)	$\sqrt{S_S/w - (L_S/w)^2}$
2D	size(O)	$\sqrt{\mu_i^2 + \mu_j^2}$
	radius(R)	$\sqrt{\sigma_i^4 + \sigma_j^4}$
	covariance(Cov_{ij})	$\frac{SR_{ij}}{w_i+w_j}$
	correlation coefficent(P_{ij})	$\frac{Cov_{ij}}{\sigma_i \sigma_j}$

Table 2. The features extracted from each time window

Value	Features	source
Size of outbound packets	μ, σ	Src_ip, Mac-ip,Channel, Socket
Number of packet	w	Src_ip, Mac-ip,Channel, Socket
Packet jitter	w, μ, σ	Channel
Size of both in/outbound packets	O, R, Cov_{ij}, P_{ij}	Channel

4.2. Incremental possibilistic clustering algorithm based on Mahalanobis distance

Under normal circumstance, the value of each attibute will stay in the boundary. When the exception, usually network attack, occurs, the value within each of the traffic features will surpass the range of threshold value. The question is, although we get a method which can improve the feature recognition performance, the clusters of normal traffic are not centralized and the shape of data may be an ultra-high dimensional irregular polyhedron. So we use an Incremental possibilistic clustering to get clusters of normal packets. Then we determine whether other packets are normal traffic. If a packet cannot satisfy any cluster extracted from normal traffic, the anomaly will be considered to happen.

4.2.1. Mahalanobis distance. The Mahalanobis distance is a measure of the distance between a point P and distribution D, introduced by P. C. Mahalanobis in 1936. It is the distance of the test point from the center of mass divided by the width of the ellipsoid in the direction of the point. Compared with Euclidean distance, it considers the relevance between each dimension and is not affected by the boundary of dataset[15]. It is obviously that the 115 features which are extracted from packet are highly correlated. Using Mahalanobis distance can make the clusters more realistic and reliable. This paper needs to compute the distance between each data to each cluster. While the shape of clusters and data can be considered as same, the function based on Mahalanobis distance is devised as:

$$d_{ij} = (x_j - v_i)\,\Sigma^{-1}\,(x_j - v_i)^T \tag{4}$$

where x_j is the data in the dataset $X = x_1, x_2 \cdots x_n$, v_i is one of the clusters, Σ is the covariance matrix of X. While the covariance matrix may be a singular matrix, the Mahalanobis distance cannot be calculated by the inverse matrix for these datasets. We carry out the pseudo inverse matrix instead by matrix factorization.

4.2.2. Unsupervised possibilistic clustering algorithm. By using Partition coefficient (PC) and Partition entropy (PE) as penalty function, the penalty function can be changed to

$\frac{\beta}{m^2\sqrt{c}}\sum_{i=1}^{c}\sum_{j=1}^{n}\left(u_{ij}\log u_{ij}^m - u_{ij}^m\right)$ where β is the co-variance of dataset[16]. In PCM based on Mahalanobis distance, it's recommanded to adjust the β to $\beta = \sum_{i=1}^{n} d\left(x_i, \overline{x}\right)/n$. Therefore, the result of clustering will be more accordant with division indicators.

Finally, the objective function of possibilistic clustering is defined as:

$$J_{PCM}\left(U,V\right) = \sum_{i=1}^{c}\sum_{j=1}^{n} u_{ij}^m d_{ij}^2 + \frac{\beta}{m^2\sqrt{c}}\sum_{i=1}^{c}\sum_{j=1}^{n}\left(u_{ij}\log u_{ij}^m - u_{ij}^m\right) \tag{5}$$

In order to minimize the objective function, the recursion formula will be:

$$u_{ij} = e^{-m\sqrt{c}d_{ij}^2/\beta} \tag{6}$$

$$v_i = \frac{\sum_{i=1}^{n} u_{ij}^m x_j}{\sum_{j=1}^{n} u_{ij}^m} \tag{7}$$

In (6), u_{ij} is determined by the distance between the cluster center to the sample. The range of u_{ij} is 0 to 1.This calculation method relaxes the restriction that the sum of membership degrees is 1, which means each cluster is calculated independently. Therefore, the true minimum number of objective function will be reached if and only if every cluster center comes together. Compared with FCM, PCM needs appropriate initial values, or the result will be really unsatisfied.

4.3. Increment possibilistic clustering

In PCM, to get the correct clustering result of dataset, the number of cluster center must be more than the right number, and initial cluster center must be with enough dispersivity. Generally, it's the defect of PCM. But for network traffic whose cluster number, shape and size are unpredictable, We design an Incremental possibilistic clustering algorithm (IPC) to get correct classifications, using PCM with excess number of initial cluster center which can make clustering accurately.

4.3.1. IPC principle. Suppose there is training set $X = \{x_1, x_2, \cdots, x_n\}$ and clustering centers $V = \{v_1, v_2, \cdots, v_c\}$. Then calculates the degree matrix of membership U and set a threshold u_{min}. When the membership between a data point to any cluster center is higher than the threshold, the point will be marked as a member of these clusters. Moreover, if the point cannot classified into any category, this point will be marked as outlier. This paper use normal flow to get the clusters and use these clusters with threshold to detect abnormal packets.

In order to find the cluster center and the threshold of normal packets, IPC use the iterative PCM. Firstly, use FCM to get the initial cluster centers $V = \{v_1, v_2, \cdots, v_c\}$ with an initial number c. Then put these cluster centers into the PCM. The extra centers from FCM will be gathered and some outliers will be appear. The outliers here may be the truly outliers of the dataset, or the point of undetected aggregates. Therefore, IPC put these outliers into the cluster centers' set to get a new set $V_{new} = \{v_1, \cdots, v_c, x_{i_1}, \cdots x_{i_m}\}$ and update the cluster number c into $c + m$. Finally, repeat the above actions for several times, the missing cluster centers will be retained, and fake centers from truly outliers will move to true centers.

4.3.2. realization of algorithm. Suppose there are training feature data $X = \{x_1, x_2, \cdots x_n\}$ extracted from normal network traffic. The steps for training are as follows.

a) Set initial value.
b) Using FCM to provide initial cluster centers:
c) Using PCM to cluster again and get the membership limit u_{min}:

i) Calculate the distance between x and v.
ii) Update the membership matrix by using (6).
iii) Update the clustering centers V by using (7).
iv) Repeat steps (i) to (iii) until the step reach the limit or V is stabilized.

d) Calculate the max membership degree of each data u_{max}.
e) Choose the lowest 3% point of u_{max} as u_{min} or set default u_{min}.
f) Add 1/10 of the data whose u_{max} is less than u_{min} to cluster centers V.
g) Re-compute the PCM again.
h) Repeat previous step for several times until the step reach the limit or every outliers has been added to V.
i) Merging Cluster Centers to get the final detection model .

5. Experiments and result Analysis

This paper uses three sets of data to compare with FCM algorithm. The first is iris datasets to qualitative analysis the influence of initial cluster centers. The second dataset is a self-made dataset to analysis the ability of finding arbitrary shape clustering. Finally the data from network flow is used to test the effect on anomaly detection.

5.1. Qualitative analysis

The first experiment used iris dataset. The parameter m is set to 1.5 and the number of initial clusters is the independent variables. Figure 1 shows that when cluster center is equal to the real cluster number, both FCM and IPC can perform well. With the increasing number of initial clusters, Figure 2 shows that the wrong clustering centers appeared by using FCM while the IPC still shows a Reliable result.

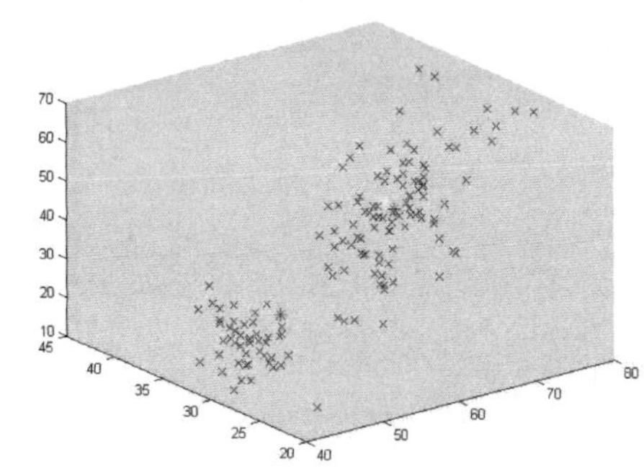

Figure 1. Iris dataset and the clustering result of FCM(red star) and IPC(yellow star) with 2 cluster center.

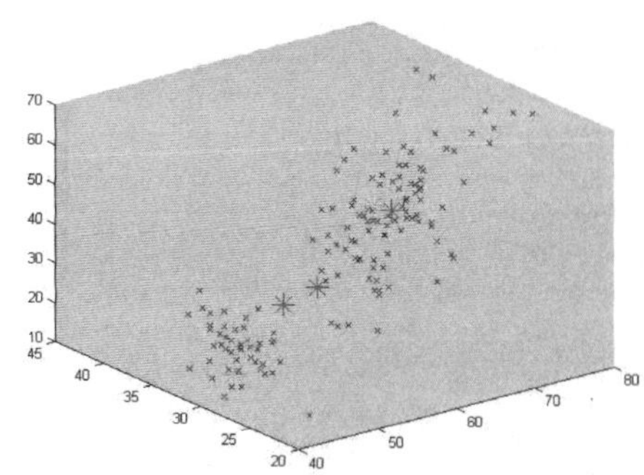

Figure 2. Iris dataset and the clustering result of FCM(red star) and IPC(green star) with 3 initial cluster center.

The second experiment used a self-made two-dimensional datasets that data points are random distributed on a ring ($r \in [40, 50]$). The number of initial cluster center is set to 10 and m=1.5. After clustering (Figure 3), the cluster center from FCM (red circle) cannot satisfied with the real data. The IPC detect 98 cluster centers and these centers reflect the true shape of data.

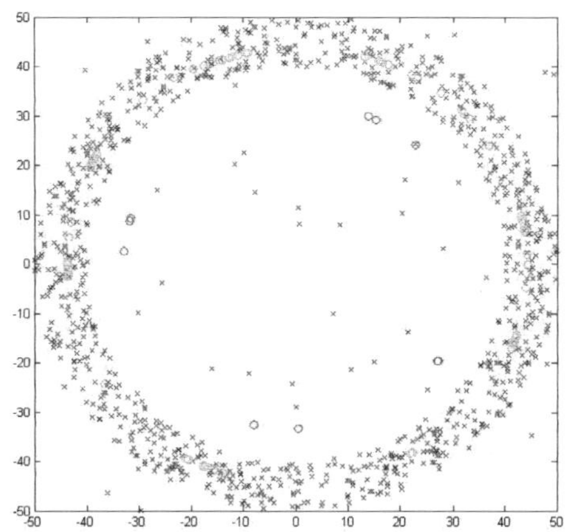

Figure 3. Self-made data and the cluster result of IPC and FCM.

5.2. IPC application to real datasets

The experiment used the dataset ISCX-Botnet 2014 created by Beigi E B et[17]. The dataset are composed of network flow from honeypot or through infecting computers with a given bot binary in a controlled environment. The dataset is including the abnormal packets and normal packets.

Firstly, we used damped incremental statistics to extract the attributes of normal network flow. 30 thousand of continued normal packets were used as training dataset. 10 thousand of packets combined with 47 normal packets and 9953 abnormal packets were used as testing dataset.This paper used MATLAB to run the IPC. Initial number of cluster center was set to 5.The parameter m was set to 1.5 and β was set to 180. Ceiling of the total steps was set to 100,000 and each turn of PCM cannot be more than 1/10 of the total steps. The FCM and PCM will stop until the movement of each point is less than 1×10^{-4} and the whole IPC will stop when all outliers have been chosen as the cluster center or the step reach the limit. The threshold to identify the outliers is set to the 3% lowest membership after the first turn of PCM.

Table 3. Result of the IPC anomaly detection

Type	Total Number	Detected	Precision
training set	40000	382	99.0%
Type	Total Number	Detected	Detection Rate
abnormal packet	9953	9837	98.8%
Type	Total Number	Detected	error rate
normal packet	10000	86	0.86%

After training, the IPC finally detect 141 cluster centers while FCM just find 5 overlapping clustering centers. the threshold to detect anomaly traffic $u_{min} = 1.9334 \times 10^{-5}$. In order to evaluate the effect of IPC in anomaly detection, we calculate the precision and detection rate for the test. The result is shown in the table3 .

From the result, we can see that model can detect 98.8% abnormal packets, while the error rate reached 1%. The algorithm has low precision when detect the abnormal packets from mixed network traffic. In this paper, IPC based on Mahalanobis distance and damped incremental

IOP Conf. Series: Journal of Physics: Conf. Series **1284** (2019) 012067 doi:10.1088/1742-6596/1284/1/012067

statistics can learn from normal packets and warn the researchers and internet users of coming network attack.

6. Conclusion

This paper introduces an anomaly detection method - incremental possibilities clustering. The method extract multi-dimensional features of network flow through damped incremental statistics that reduced memory overhead. Then using Mahalanobis distance to deal with the problem in dimension. The core algorithm can reduce the constraints of parameters and make it possible to deal with high-dimension data.

The experiment shows that IPC can find the pattern of training dataset and identify the abnormal data. But it means that the training dataset must be broad enough. In addition, the rule of merge clusters is the distance between each cluster centers. Density-based clustering algorithms can be added to IPC to get better clustering result.

References

[1] Mohiuddin Ahmed, Abdun Naser Mahmood, and Jiankun Hu. A survey of network anomaly detection techniques. *Journal of Network and Computer Applications*, 60:19–31, 2016.

[2] Eleonora Borgia. The internet of things vision: Key features, applications and open issues. *Computer Communications*, 54:1–31, 2014.

[3] Dimitrios Damopoulos, Sofia A Menesidou, Georgios Kambourakis, Maria Papadaki, Nathan Clarke, and Stefanos Gritzalis. Evaluation of anomaly-based ids for mobile devices using machine learning classifiers. *Security and Communication Networks*, 5(1):3–14, 2012.

[4] Ahmad Javaid, Quamar Niyaz, Weiqing Sun, and Mansoor Alam. A deep learning approach for network intrusion detection system. In *Proceedings of the 9th EAI International Conference on Bio-inspired Information and Communications Technologies (formerly BIONETICS)*, pages 21–26. ICST (Institute for Computer Sciences, Social-Informatics and . . . , 2016.

[5] Yair Meidan, Michael Bohadana, Yael Mathov, Yisroel Mirsky, Asaf Shabtai, Dominik Breitenbacher, and Yuval Elovici. N-baiot—network-based detection of iot botnet attacks using deep autoencoders. *IEEE Pervasive Computing*, 17(3):12–22, 2018.

[6] Iwan Syarif, Adam Prugel-Bennett, and Gary Wills. Unsupervised clustering approach for network anomaly detection. In *International Conference on Networked Digital Technologies*, pages 135–145. Springer, 2012.

[7] Linquan Xie, Ying Wang, Liping Chen, and Guangxue Yue. An anomaly detection method based on fuzzy c-means clustering algorithm. In *Second International Symposium on Networking and Network Security (ISNNS, 10) Jinggangshan, PR China*, pages 2–4. Citeseer, 2010.

[8] Yang Yan Xu Xiaodong and Li Gang. Network traffic anomaly detection based on k-means clustering. *Wireless Communication Technology*, 22(4):21–26, 2013.

[9] S Askari, N Montazerin, and MH Fazel Zarandi. Generalized possibilistic fuzzy c-means with novel cluster validity indices for clustering noisy data. *Applied Soft Computing*, 53:262–283, 2017.

[10] Liu Pengcheng Tang Chenghua and Tang shensheng. Anomaly instusion behavior detection based on fuzzy clustering and features selection. *Journal of Computer Research and Development*, 52(3):718–728, 2015.

[11] Wang Lina Luo Min and Zhang Huanguo. An unsupervised clustering-based intrusion detection method. *ACTA Electronica Sinica*, 31(11):1713–1716, 2003.

[12] K Munivara Prasad, A Rama Mohan Reddy, and K Venugopal Rao. Dos and ddos attacks: defense, detection and traceback mechanisms-a survey. *Global Journal of Computer Science and Technology*, 2014.

[13] Guo XJ, Zhou AP, and cheng G. Highspeed network traffic measurement method. *Journal of Software*, 25(1):135–153, 2014.

[14] Yisroel Mirsky, Tomer Doitshman, Yuval Elovici, and Asaf Shabtai. Kitsune: an ensemble of autoencoders for online network intrusion detection. *arXiv preprint arXiv:1802.09089*, 2018.

[15] Zu zhiwen and Li Qin. Km-fcm: A fuzzy clustering optimization algorithm based on mahalanobis distance. *journal of Hebei University of Science and Technology*, 39(2):159–165, 2018.

[16] Zuo Chuncheng Hu Yating and Qu Fuheng. Multi-scale possibilistic clustering algorithm. *Journal of Changchun University of Science and Technology(Natural Science Edition)*, 33(4):124–127, 2010.

[17] Elaheh Biglar Beigi, Hossein Hadian Jazi, Natalia Stakhanova, and Ali A Ghorbani. Towards effective feature selection in machine learning-based botnet detection approaches. In *2014 IEEE Conference on Communications and Network Security*, pages 247–255. IEEE, 2014.

IOP Conf. Series: Journal of Physics: Conf. Series **1284** (2019) 012068 doi:10.1088/1742-6596/1284/1/012068

Multi-focus Image Fusion Method based on Wavelet Transform

Linglong Tan*, **Yehui Chen and Wenju Zhang**

Electronic communication engineering college, Anhui Xinhua University, Hefei, 230088, China

*E-mail: tanlinglong@axhu.edu.cn;

Abstract. In this paper, the wavelet decomposition of multi-focus images is used to fuse the low-frequency components and high-frequency components obtained after decomposition. In order to make more information of the original image in the merged image, the method of maximizing the coefficient and the method of maximizing the variance of the region mean when the high frequency coefficient is selected, and combining the average and the selection in the selection of the low frequency coefficient. The simulation experiment results and the objective evaluation index values show that the modification method can make up for the defects of the two original images and obtain a relatively complete and clear image.

1.Introduction

Wavelet transform is a powerful tool for modern signal processing and analysis. The wavelet transform image can decompose the detail information in three different directions on each scale, and the transformed information has no redundancy compared with the original image information. Wavelet transform is widely used in signal processing, image analysis, computer applications and other fields in the field of image processing [1,2]. Graphic fusion is the synthesis of multiple images acquired by different sensors through the same target or the same scene. The synthesized image cannot be acquired by a single sensor. The composite image obtained after image fusion can maintain key information in multiple images, providing more precise conditions for target analysis. The traditional pixel-level image fusion method ignores the mutual relationship between pixels, resulting in poor image quality after fusion, and the visualization effect is not ideal. In this paper, wavelet transform is introduced into the image fusion algorithm, because the image can be separated by high-frequency components and low-frequency components after wavelet decomposition, and then the high-low frequency components are fused using different fusion rules in different frequency domains. This method can improve the target detection resolution and suppress the detection noise of different sensors at the same time[3].

2.Theoretical Basis

2.1 Discrete wavelet transform

The process of wavelet transform essentially refers to the signal being projected onto a set of wavelet functions, and then the signal is decomposed into a series of wavelet functions. Wavelet transform is an integral transform. The function f(t) in any $L_2(R)$ space is expanded under wavelet basis $\varphi_{\alpha,\tau}(t)$ as follows [4]:

$$WT_f(\alpha,\tau) = < f(t), \varphi_{\alpha,\tau}(t) > = \frac{1}{\sqrt{\alpha}} \int_R f(t)\, \overline{\varphi(\frac{t-\tau}{a})} dt \quad (1)$$

$$\varphi_{\alpha,\tau}(t) = a^{-1/2} \overline{\varphi(\frac{t-\tau}{a})} \quad , a > 0. \tau \in R \quad (2)$$

Among them, the scale factor and translation factor of the wavelet base are two important parameters of the wavelet base. The process of projecting a time function onto a two-dimensional time-scale phase plane is a wavelet basis expansion process. In wavelet transform, when analyzing low frequency signals, the time window is very large. When the high frequency signal is analyzed, the time window decreases.

In the continuous wavelet transform, the wavelet basis functions $\varphi_{\alpha,\tau}(t)$ have a large correlation under the continuous scale a and time τ values. This leads to redundant information in coefficients of continuous wavelet transform $WT_f(\alpha,\tau)$. Redundancy, in order to minimize the redundancy of the wavelet transform without losing the original signal information, the a, τ of the wavelet basis function is limited to some discrete points. The scale a is discretized in power series, taking $a_m = a_0^m$, where m is an integer. The time τ is uniformly and discretely taken to cover the entire time axis. Therefore, $\varphi_{\alpha,\tau}(t)$ is expressed as:

$$\varphi_{m,n}(t) = 2^{-\frac{m}{2}} \varphi(2^{-m}t - n) \qquad m, n \in Z \quad (3)$$

The wavelet transform of arbitrary signal f(t) can be expressed as:

$$WT_f(m,n) = \int_R f(t) \overline{\varphi_{m,n}(t)} dt \quad (4)$$

2.2 Fast Algorithm for Discrete Wavelet Transform

For any function $f(t) \in V_0$, it can be decomposed into detail part W_1 and large-scale approximation part V_1, and then the large-scale approximation part V_1 is further decomposed. By repeating this, the approximation part and the detail part at any scale can be obtained [5].

Let$f_s^i(t)$ be the profile signal at the j scale obtained after the function f(t) is projected onto the scale spaceV_j.

$$f_s^i(t) = \sum_k C_{j,k} \phi_{j,k} (2^{-j}t) = \sum_k C_{j,k} \phi_{j,k}(t) \quad (5)$$

Where $C_{j,k}$ is the scale expansion coefficient.

The function f(t) is projected onto the wavelet space $W_j$of different scales, and the detailed information $f_d^j(t)$ at different scales is obtained.

$$f_d^j(t) = \sum_k d_{j,k} \varphi_k (2^{-j}t) = \sum_k d_{j,k} \varphi_{j,k}(t) \quad (6)$$

Where $d_{j,k}$ is the wavelet expansion coefficient.

Expand $f(t) \in L^2(R)$ according to equation (7), then the function f(t) is expressed as equation (8).

$$L^2(R) = \sum_{j=-\infty}^{J} W_j \oplus V_j \quad (7)$$

$$f(t) = \sum_{j=-\infty}^{J} \sum_{k=-\infty}^{\infty} d_{j,k} \varphi_{j,k}(t) + \sum_{k=-\infty}^{\infty} c_{j,k} \varphi_{j,k}(t) \quad (8)$$

when $J \to \infty$.

Equation (8) is transformed into.

$$f(t) = \sum_{j=-\infty}^{J} \sum_{k=-\infty}^{\infty} d_{j,k} \varphi_{j,k}(t) \quad (9)$$

Equation (8) (9) is a comprehensive formula of discrete orthogonal wavelet transform.

3. Image Fusion Algorithms

The original image is decomposed into high-frequency and low-frequency regions by wavelet transform, and uses different fusion rules in different frequency domains to preserve the salient features of the original image in different frequency domains[6].

Wavelet transform is used in image fusion algorithm. firstly, multi-source images are geometrically accurately registered, and appropriate wavelet bases and decomposition layers are selected. The original image is decomposed into approximate coefficients and detail coefficients by multi-level wavelet decomposition. The fusion wavelet coefficients are fused by different wavelet fusion rules. Finally, the fused image is obtained by inverse transformation of the wavelet coefficients. The principle of the fusion algorithm is shown in Figure 1.

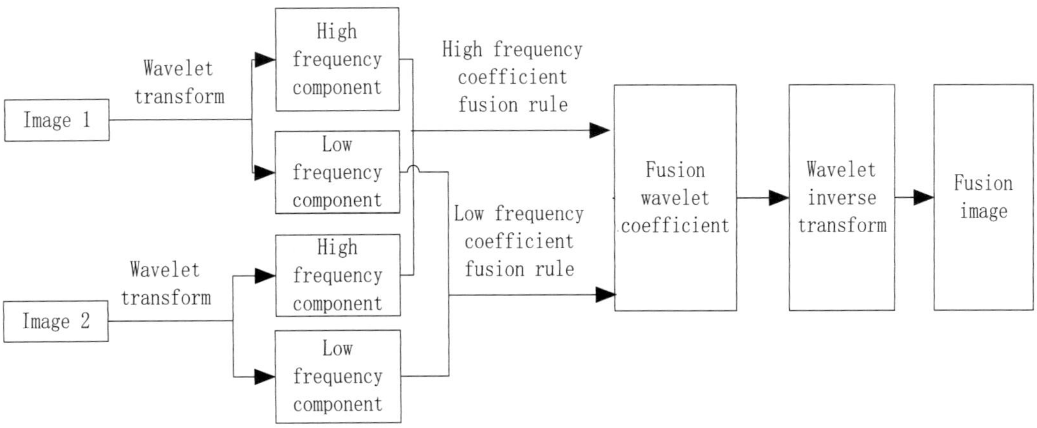

Figure 1. Fusion Algorithm

3.1 Wavelet high frequency coefficient fusion method

In this paper, large absolute coefficients and maximization of regional mean variance are used as fusion criteria for high frequency coefficients. Taking the fusion of two images as an example, the fused image of image 1 and image 2 is image 3. The wavelet coefficients of fused images are Formula 10.

$$\begin{cases} D_{N,3}^{i} = D_{N,1}^{i} & if\ |D_{N,1}^{i}| \geq |D_{N,2}^{i}| \\ D_{N,3}^{i} = D_{N,2}^{i} & else \end{cases} \tag{10}$$

In the formula, $D_{N,1}^{i}$ and $D_{N,2}^{i}$ represent the wavelet coefficients of image 1 and 2 in the direction i on the wavelet decomposition scale N.

In the middle decomposition layer, the wavelet coefficients of image 1 or image 2 are taken as the wavelet coefficients of fused image, that is, the mean variance of region 3×3 is the largest.

$$\begin{cases} D_{N,3}^{i} = D_{N,1}^{i} & if\ MSE1 \geq MSE2 \\ D_{N,3}^{i} = D_{N,2}^{i} & else \end{cases} \tag{11}$$

MSE1, MSE2 represents the local variance of image 1 and image 2 on the decomposition scale, and the mean variance MSE is defined as:

$$\mathrm{MSE} = \frac{1}{MN}\sum_{i=1}^{M}\sum_{j=1}^{N}(X_{i,j} - \bar{X})^2 \tag{12}$$

Where M represents the number of rows in the region, N represents the number of columns in the region,$X_{i,j}$ represents the gray value of the local region, and $\bar{X}$ represents the average gray value of the local region.

3.2 Wavelet low-frequency coefficient fusion method

In this paper, the selection of low-frequency scale coefficients is based on the combination of averaging and selection proposed by Burt, and SA(X, p) is used to represent the significance of the scale factor of image X at point p.

$$\mathrm{SA(X,p)} = \sum_{q\in Q} w(q)C_N^2(X,q) \tag{13}$$

Q represents the energy in a small area, w(q) is a weight, $C_N^2(X,q)$ is a wavelet coefficient of the image X at the q point, and the matching matrix R is defined as:

$$\mathrm{R(P)} = \frac{2\sum_{q\in Q} w(q)C_N(A,q)C_N(B,q)}{SA(A,P)+SA(B,P)} \tag{14}$$

The R value is close to 1 the image correlation is high, when the R value is larger, the weighted scale coefficient of the fused image two image scaling coefficient average, fusion function can be expressed as:

$$C_N(F,p) = W(A,p) \cdot C_N(A,p) + W(B,p) \cdot C_N(B,p) \quad (15)$$

4. Evaluation Indicators

In order to measure whether the image fusion method has good reliability and robustness, the quality and performance of the fused image should be evaluated.In addition to subjective fused image quality evaluation, this paper uses three objective evaluation indicators: Entropy (E), mutual information (MI) and average gradient (AG) are selected to evaluate the effect of fusion image.

Image entropy is an important index to evaluate the richness of image information. The size of the entropy reflects the size of the average information of the image. Let the gray distribution of the fused image be $p = \{p_1, p_2, \dots, p_i, \dots, p_n\}$, p_i be the ratio of the gray value of the first pixel to the gray value of the total pixel, n be the total gray level, and the entropy be defined as:

$$\mathrm{E} = -\sum_{i=1}^{n} p_i \log_2 p_i \quad (16)$$

The average gradient is a measure of image sharpness. The bigger the average gradient is, the higher the definition of the image is. The average gradient is defined as:

$$\mathrm{AG} = \frac{1}{(M-1)(N-1)} \sum_{i=1}^{M-1} \sum_{j=1}^{N-1} \sqrt{\frac{(F(i+1,j)-F(i,j))^2 + (F(i,j+1)-F(i,j))^2}{2}} \quad (17)$$

If the gray level distribution of the original image 1 is $q = \{q_1, q_2, \dots, q_i, \dots q_n\}$ and the joint probability density function of the two images is $\gamma_{i,j}$, the mutual information between the fused image and the original image 1 can be expressed as follows:

$$MI_1 = \sum_{i=1}^{n} \sum_{j=1}^{n} \gamma_{i,j} \log_2 \frac{\gamma_{i,j}}{p_i q_j} \quad (18)$$

The mutual information between the fused image and the two original images is represented by MI.

$$\mathrm{MI} = MI_1 + MI_2 \quad (19)$$

5. Experiment

To test the effectiveness of the proposed image fusion algorithm. In this paper, three groups of images with different focus are used for fusion experiments.The focus of each group of original images is distributed on different distant and near targets. The original images 1 and 2 in figure. 2, 3, 4 focus on the near and far respectively. The operation of two-dimensional wavelet decomposition, fusion and reconstruction is used to realize the fusion effect.

From the fusion effect of the three sets of images, the fusion algorithm used in this paper can perform better fusion processing on multifocal image. From the quality of the fused image, it can be seen that the image fusion method based on wavelet transform is better than that based on direct image fusion. Using wavelet transform to fuse the two original images can make up for the different defects of the two images, and obtains a relatively complete and clear image.

It can be seen from Table 1 that the fused image has a large entropy and an average gradient value, which indicates that the fused image contains rich information and has high definition. It can be seen from the mutual information value that the fusion image has a great correlation with the original image.

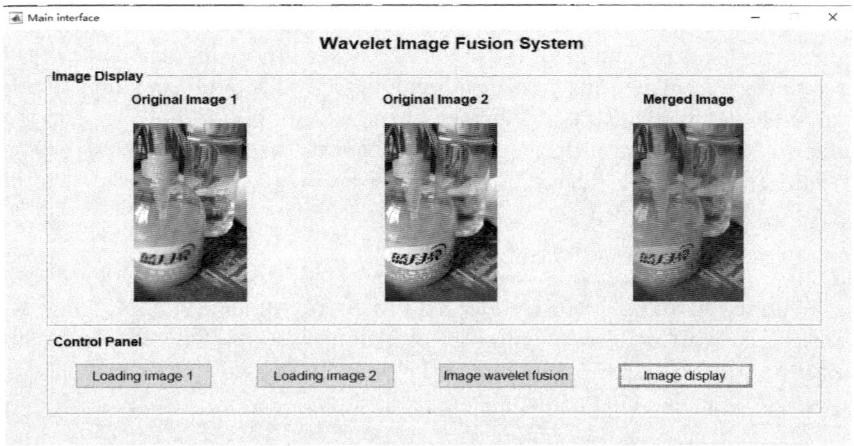

Figure 2. Image fusion experiment 1

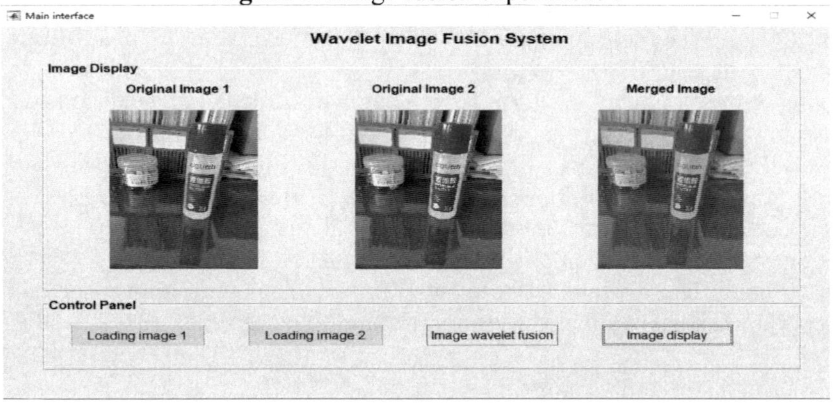

Figure 3. Image fusion experiment 2

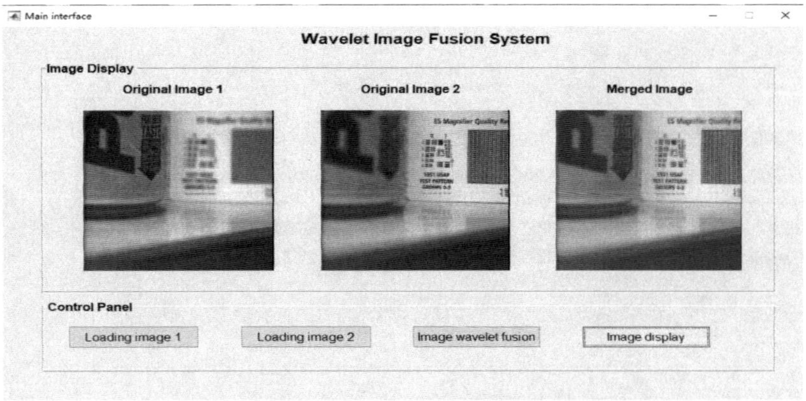

Figure 4. Image fusion experiment 3

Table 1. Evaluation results of fusion results

Evaluation index	E	MI	AG
Figure 2 fusion image	7.3101	6.2176	3.7805
Figure 3 fusion image	7.1772	6.5183	3.4415
Figure 4 fusion image	7.0561	6.3124	3.9451

6. Conclusion

Image fusion algorithm takes advantage of the human eye's sensitivity to local contrast transformation. According to certain fusion rules, some significant features are selected from multiple original images, and these features are retained in the final fused image. Different fusion algorithms are used for different decomposition layers and different frequency bands. The neighborhood-variance of wavelet coefficients is used to define the fusion factor, which can make good use of the time-frequency domain features of wavelet transform.

Acknowledgments

This work was supported by the Scientific Research Project of Anhui Xinhua College No. 2017zr003 and No. 2017zr001, Scientific Research Project of Anhui Natural Science Research Project No. KJ2017A628 and No. KJ2017A618.

References

[1] Ruan Xiaoyu, Han Chao. Image Fusion of Dual-Channel Pulse Coupled Neural Network Based on Wavelet Transform [J]. Journal of Pingdingshan University, 2019, 34 (02): 59-64.
[2] He Naiying, Li Lei, Wang Chuan-en. A Laplacian pyramid method for multi-focus image fusion [J]. Computer Age, 2019 (04): 18-20+25.
[3] Luo Xiaoqing, Xiong Mengyu, Zhang Zhengcheng. Multi-focus image fusion method based on joint convolution self-coding network [J/OL]. Control and decision-making: 1-8
[4] Kong Weiwei, Leiyang, Ren Cong, Zhao Wenyang. Multi-focus image fusion method based on improved CNN [J]. Journal of Zhengzhou University , 2019, 51 (02): 29-33.
[5] Zhao Xin, Cao Min. Multi-focus image fusion algorithm combining visual saliency [J]. Automation and instrumentation, 2019, 34 (03): 38-41.
[6] Chen Jie, Mao Jian, Zhang Jiemin. A multi-focus image fusion algorithm based on morphological components [J]. Journal of Jimei University (Natural Science Edition), 2019, 24 (01): 68-75.

IOP Conf. Series: Journal of Physics: Conf. Series **1284** (2019) 012069 doi:10.1088/1742-6596/1284/1/012069

Research on Image Compression Coding Technology

Linglong Tan, Yong Zeng* and Wenju Zhang

Electronic communication engineering college, Anhui Xinhua University, Hefei, 230088, China

*E-mail: 10312293033@qq.com

Absrtact. Image compression technology is the basis of all kinds of media compression and transmission, and its compression effect is directly related to the compression effect of media. In this paper, four commonly used image compression coding methods, namely, undistorted Huffman coding, run-length coding, finite distortion predictive coding and transform coding, are used to compress still images, and the efficiency of each coding method is analyzed. The experimental results show that the two lossless compression coding methods have remarkable compression effect, but the compression efficiency is low. Although the compression ratio of lossy compression coding method is high, the decompressed image has certain distortion.

1.Introduction

The contradiction between the large amount of data and the limited channel capacity of digital images reflects the importance of image compression, and usually there must be some correlation between the pixels of most images. The information contained between images belongs to image redundancy. In order to ensure image quality, data compression must eliminate image redundancy. Different compression methods are usually used to process images to eliminate different types of redundancy. There are many compression and coding methods for digital images. From the perspective of information theory, image compression and coding methods can be divided into two kinds: distortion-free compression and finite distortion coding. The principle of lossless compression coding is to reduce the data redundancy of image information source by variable length coding, which makes the encoding image data close to its information entropy and thus does not produce distortion. The principle of finite distortion coding is based on the unique visual characteristics of the human eye. Without forcing the image to be undistorted, the correlation of the image information source in space and time can be eliminated and the variance of the signal can be reduced by a certain signal transformation to achieve the requirement of compression coding [1-2].

2. Knowledge of compression coding theory

2.1 Huffman coding

In variable length coding method, information symbols with high probability are coded into short codewords, while information symbols with low probability are coded into long codewords. Huffman coding is a combination of Huffman algorithm and variable length optimal coding theorem. It is a coding algorithm that utilizes the probability distribution of information symbols. The specific coding method is as follows [3].

(1) First, the probability of source symbols is sorted in descending order.

(2) Then add up the two minimum probabilities and repeat the step, always arranging the high probabilistic branches on the right until they finally become probabilistic 1.

(3) Next, the path from probability 1 to each source symbol is marked, and the 0 and 1 along the path are recorded in sequence. Thus, the Hoffman codeword of the symbol is obtained.

(4) Finally, the left side of each pair of combinations is designated as 0, and the right side is designated as 1 (or vice versa).

The specific implementation of the encoding is shown in Table 1 and Table 2.

Table 1. Source reduction for Huffman coding

Initial source		Source reduction steps	
Symbol	probability	1	2
b_2	0.38	0.38	0.62
b_4	0.30	0.32	0.38
b_3	0.22	0.30	
b_1	0.10		

Table 2. Source assignment diagram of Huffman coding

Initial source			Assignment of reduced sources			
Symbol	probability	Codeword	1		2	
b_2	0.38	1	0.38	1	0.62	0
b_4	0.30	01	0.32	00	0.38	1
b_3	0.22	000	0.30	01		
b_1	0.10	001				

2.2 Run length coding

Usually, computer image is composed of a large number of color blocks of the same color. When storing the image, many consecutive scanning lines or continuous pixels on the same scanning line carry the same color value in the process of scanning the color blocks. Run-length coding is a coding method that stores a pixel value and the number of pixels with the same color instead of storing the same color blocks in the image one by one. Simply put, the color values and count values are used to represent the same color pixel values in a row.

Run-length coding has a high compression rate when an image contains a large number of the same color blocks. When the color value of the image is rich, the run-length value of the image is very short, and run-length coding will cause greater redundancy[4].

2.3 Predictive coding

The adjacent pixels of the image are correlated in gray level. Predictive coding uses this feature to predict the current signal through the previous signal, and then differential coding through the prediction error. Linear prediction is the most widely used prediction method, which is called Different Pulse Code Modulation.

DPCM encoder consists of linear predictor and quantizer.The output of the encoder is not the sample value of the image pixel, but the difference between the sample value and the predicted value, that is, the quantized value of the predicted error.According to the analysis of the statistical characteristics of the image signal, some appropriate prediction coefficients can be obtained, so that the distribution of prediction errors is mainly concentrated in the vicinity of "0". Through non-uniform quantization, the image data can be compressed by using fewer quantization layers[5,6].

2.4 Transform coding

Transform coding is to encode the transform coefficients of an image by making use of the gray correlation of all the pixels in the whole image. Because of the strong correlation of image data, image data will enter a new transform domain after special transformation. The original scattered image data can be centrally distributed in this new coordinate space. Because the transform coefficients of most images are small, compression coding can be realized by eliminating the coefficients approximating "0". After the image is reconstructed by inverse transform, the loss of irrelevant data will be negligible[7].

The effect of transform coding is related to the selection of transform method, image type and sub-image size. After the transformation method is decided, the choice of sub-image size will affect the amount of computation needed for transformation. Usually sub-images are divided into 8*8 or 16*16 sizes. The larger the sub-images used, the larger the computational complexity.

Image compression does not lie in transformation. The function of transformation is to map image data into a new spatial domain, so that the image can be compressed. Quantization process will cause errors in image data, and the effect of image compression mainly depends on the selection of quantizer.

Specific coding implementation steps are as follows:

(1) The first step is to segment the sub-image, generally dividing the N*n size image into (N/n) 2.

(2) Select the appropriate image transformation to transform the sub-image.

(3) Quantify the transform coefficients.

(4) Coding and inverse transformation.

3. Experimental Simulation

3.1 Huffman Coding Simulation

When Huffman coding is used to compress and encode image data, the probability statistics of the data in figure 1 (a) are carried out first, and the probability of each symbol is obtained. Then the probability is sorted and coded to realize image data compression. The compressed image is shown in figure 1(b).From the parameters in Table 3, the data after compression and decoding are consistent with the original data bits.

(a) original image (b) decompressed image

Figure 1. Huffman codec simulation results

Table 3. Huffman coding simulation results

Name	Size	Bytes	Class	Attributes
data	256x256	65536	uint8	
unzipped	1x65536	65536	uint8	
zipped	1x57712	57712	uint8	

3.2 Run-length coding simulation

Run-length encoding first binarizes figure 2 (a). The binarized image is shown in figure 2 (b). Then the binarized image is coded and decoded. The decompressed image is shown in figure 2 (c). The decompressed binary figure 2 (c) is basically undistorted compared with the original binary figure 2 (b). Compression parameters are shown in Table 4. The compression ratio of run-length encoding is very high, and there is no loss compared with the original image after encoding and decompression.

(a) original (b) binary image (c) compressed and decoded picture

Figure 2. Run-length codec simulation results

Table 4. Run-length coding simulation results

Name	Size	Bytes	Class	Attributes
data	256x256	65536	uint8	
unzipped	256x256	65536	uint8	
zipped	3543x2	7086	uint8	

3.3 Predictive coding simulation

Predictive coding achieves image compression by predictive error, and the effect of image compression is judged by predictive error. After reading gray figure 3 (a), the image is predicted and coded. As can be seen from figure 3 (b) and figure 4, the gray level of the image after predicted coding is greatly reduced and the image is obviously distorted.

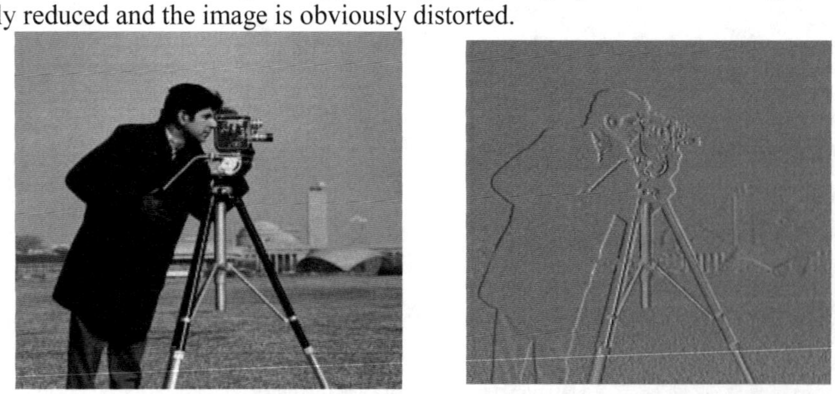

(a) Reverse of the original map (b) Prediction error map

Figure 3. Simulation results

IOP Conf. Series: Journal of Physics: Conf. Series **1284** (2019) 012069 doi:10.1088/1742-6596/1284/1/012069

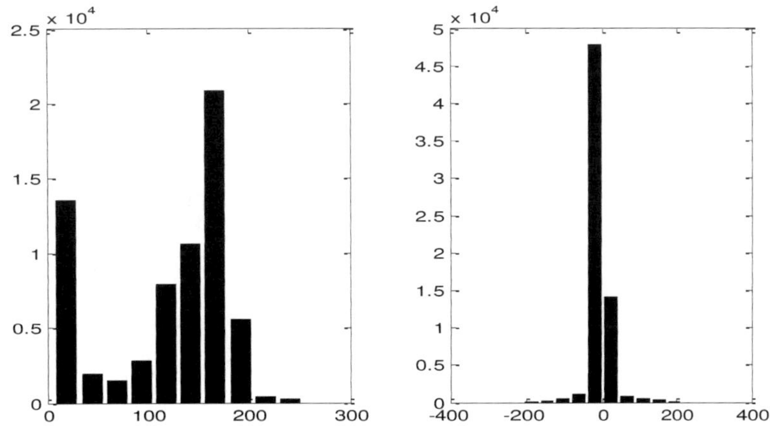

(a) Histogram of the original graph (b) Prediction error graph reversal

Figure 4. Predictive coding histogram

3.4 Transform Coding Simulation

Figure 5(a) is divided into 8 *8 blocks, and then two-dimensional DCT transform is performed on each block. Then each image is inversely transformed by DCT. Finally, figure 5 (b) is stitched together figure 5 of the simulation results shows that the image coded by DCT has obvious distortion compared with the original image, but it does not affect the overall visual effect of the image.

4. Conclusion

The simulation results show that the compressed image obtained by Huffman coding and run-length coding is almost the same as the original image after decompression. The coding algorithm is simple, but the coding compression ratio is relatively low. If the user does not require high quality of the image and pays little attention to the compression ratio, this coding method can be used. The main advantage of predictive coding is that it can achieve high coding compression ratio. The disadvantage of predictive coding is that its error resilience is too poor. When processing images, if there is a certain requirement for coding efficiency and low requirement for image quality, it can use predictive coding to process images. Transform coding mainly solves the problem of redundancy of correlation, but it leads to the decrease of image resolution, the increase of block effect and image noise, which will affect human visual effect.

(a) original (b) decompression

Figure 5. Simulation results

Acknowledgments

This work was supported by the Scientific Research Project of Anhui Xinhua College No. 2017zr003 and No. 2017zr001, Scientific Research Project of Anhui Natural Science Research Project No. KJ2017A628 and No. KJ2017A618 .

References

[1] Comparing and analyzing the influence of different image compression algorithms of Jiwenting. JPEG and JPEG2000 on image quality [J].Science and Technology Communication, 2019,11(07): 169-170+187.

[2] Yang Minzhu, Zou Zhipu, Han Changpei. Research on Lossless Compression Based on single-frame interferogram [J]. Modern electronic technology, 2019, 42 (07): 57-60.

[3] Zhang Songgang, Gaofeng, Yang Fulei, Liu Xiaowei. Research on UAV video compression based on compression algorithm [J]. Integrated circuit applications, 2019, 36 (04): 39-40.

[4] Luan Tingting, Zhu Jihua, Xu Siyu, Wang Jiaxing, Shixuan, Li Bichen, Huffman Coded Product Quantization for Image Hash Retrieval [J]. Chinese Journal of Image Graphics, 2019, 24 (03): 389-399.

[5] Zhang Leihong, Ye Hualong. Study on image transmission mechanism of correlation imaging algorithm based on Huffman coding [J]. Packaging Engineering, 2019,40 (05): 244-249.

[6] Run-length encoding and efficient compression of regular hexagonal raster data supported by Xinrui, Etinghua Gosper curves [J]. Journal of Surveying and Mapping, 2019, 48 (02): 226-237.

[7] Liu Aidong, Li Zhiyu, Wang Feng, He Linbo. A data compression algorithm supporting embedded calibration system [J]. Computer and Digital Engineering, 2018, 46 (12): 2607-2610.

IOP Conf. Series: Journal of Physics: Conf. Series **1284** (2019) 012070 doi:10.1088/1742-6596/1284/1/012070

Research of Coupled Inductance Three-Level Sepic Converter based on Parasitic Resistance

Zhen GUO [1, 2], Hong-xing ZHANG[1], Yun Liu[1], De-lin ZHAO[1], Lei YU[1], Feng-yu LIU[1] and Dai-ling YU[1]

[1] College of Information and Control Engineering, Qingdao University of Technology, Qingdao, 266520, China

[2] College of Information and Control Engineering, China University of Petroleum(East China), Qingdao, 266580, China

1446711839@qq.com

Abstract. All components are accompanied by some parasitic parameters during the DC converter experiment. These parameters have a certain effect on the voltage conversion rate and efficiency of the converter. In this paper, a three-level Sepic converter with high-voltage gain and low-voltage stress coupled inductor is studied to optimize the performance of existing DC converters. Firstly, constructing the differential equation of the converter according to the state space averaging method. Then, calculate the relationship between voltage gain and converter efficiency under the influence of parasitic parameters for analyzing the impact of each parasitic parameter on circuit performance. At last, through the Saber simulation, a correct result of the parasitic parameters affecting the converter is obtained.

1. Introduction

Parasitic parameters are not negligible during the design of the DC-DC converter. Therefore, a converter operating mode and performance are analyzed under the influence of parasitic parameters based on coupled inductor three-level Sepic converter. As shown in Figure 1, compared to the traditional Sepic basic circuit, the topology first adds switching transistors and capacitors through three-level conversion to maintain symmetry. Therefore, the voltage stress of the switching transistor and capacitor is reduced by half [1-3]. Then, the limit duty cycle can be avoided under the requirement of increasing voltage by introducing a coupling winding [4-6]. Finally, the diodes are connected in parallel to the coupled windings to reduce energy losses.

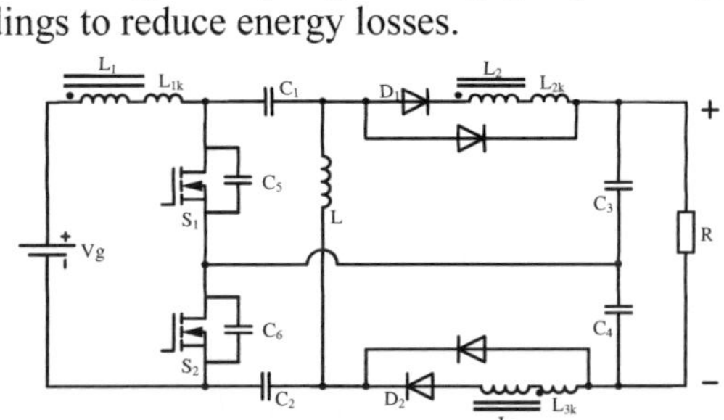

Figure 1. Coupled Inductor Three-Level Sepic Converter

IOP Conf. Series: Journal of Physics: Conf. Series **1284** (2019) 012070 doi:10.1088/1742-6596/1284/1/012070

2. Converter working process

As shown in Figure 2, considering the leakage inductance of the circuit, the working process of the coupled inductor three-level Sepic converter is mainly divided into 8 stages. The key waveform of the Saber simulation is shown in Figure 3.

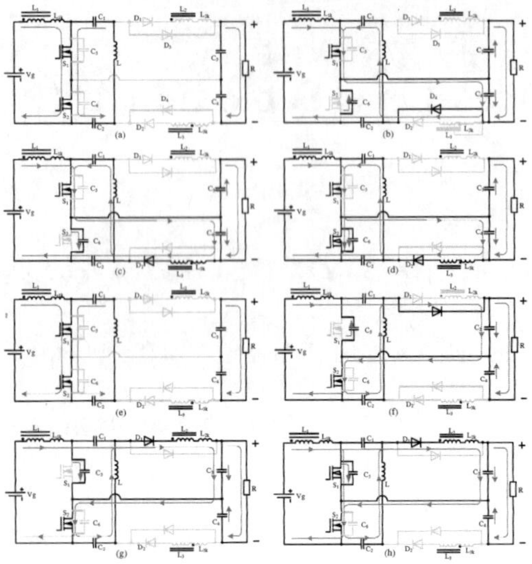

Figure 2. Operating mode of the converter

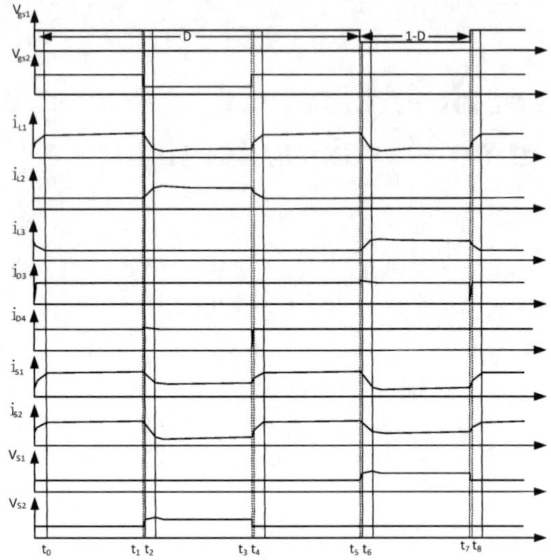

Figure 3. Key waveform of the converter

3. Steady-state analysis of converters and the influence of parasitic resistance

Steady-state analysis of four main working processes with consideration of some parasitic parameters: The two switching tubes S_1 and S_2 are simultaneously turned on. The switching tube S_1 is turned on, and the switching tube S_2 is turned off. The switching tubes S_1 and S_2 are turned on at the same time. The switch tube S_2 is turned on and the switch tube S_1 is turned off.

The converter is stable over a stable switching period. According to the volt-second balance law of the inductor, the change of the inductor current in the converter can be kept constant for one cycle. according to the capacitor's ampere-second balance rule, the change of the capacitor voltage of the converter can be kept constant in one cycle. Therefore, the converter has a static operating point and can be expressed as:

$$\begin{pmatrix} di_{L1}/dt \\ di_L/dt \\ dv_{C1}/dt \\ dv_{C2}/dt \\ dv_{C3}/dt \\ dv_{C4}/dt \end{pmatrix} = A \begin{pmatrix} i_{L1} \\ i_L \\ v_{C1} \\ v_{C2} \\ v_{C3} \\ v_{C4} \end{pmatrix} + B \begin{pmatrix} v_g \\ v_{D1} \\ v_{D4} \end{pmatrix} \tag{1}$$

According to the formula (7), the matrix A_1、B_1、A_2、B_2、A_3、B_3、A_4、B_4 can be obtained by analyzing the above four working states.

Due to:

$$A = A_1\left(D-\frac{1}{2}\right)*2 + A_2(1-D) + A_4(1-D),\ B = B_1\left(D-\frac{1}{2}\right)*2 + B_2(1-D) + B_4(1-D) \tag{2}$$

It can be obtained:

$$A=\begin{bmatrix} -\frac{\alpha}{(n+1)L_1} & -\frac{\beta}{(n+1)L_1} & -\frac{1-D}{(n+1)L_1} & -\frac{1-D}{(n+1)L_1} & -\frac{1-D}{(n+1)L_1} & -\frac{1-D}{(n+1)L_1} \\ -\frac{\chi}{L} & -\frac{\delta}{L} & \frac{1}{L} & \frac{1}{L} & 0 & 0 \\ \frac{1-D}{C_1} & -\frac{D}{C_1} & 0 & 0 & 0 & 0 \\ \frac{1-D}{C_2} & -\frac{D}{C_2} & 0 & 0 & 0 & 0 \\ \frac{1-D}{C_3} & \frac{1-D}{C_3} & 0 & 0 & -\frac{1}{RC_3} & -\frac{1}{RC_3} \\ \frac{1-D}{C_4} & \frac{1-D}{C_4} & 0 & 0 & -\frac{1}{RC_4} & -\frac{1}{RC_4} \end{bmatrix} \quad (3)$$

$$B=\begin{bmatrix} \frac{2nD-n+1}{(n+1)L_1} & -\frac{1-D}{(n+1)L_1} & -\frac{1-D}{(n+1)L_1} \\ 0 & 0 & 0 \\ 0 & 0 & 0 \\ 0 & 0 & 0 \\ 0 & 0 & 0 \\ 0 & 0 & 0 \end{bmatrix} \quad (4)$$

Where:

$$\begin{cases} \alpha=(2nD-n+1)\cdot(R_{S1}+R_{S2})+(2nD-n+1)\cdot R_L+(1-D)\cdot(R_{L2}+R_{L3}) \\ \beta=(2nD-n+D)\cdot(R_{S1}+R_{S2})+(1-D)\cdot(R_{L1}+R_{L2}) \\ \chi=D(R_{S1}+R_{S2})+(1-D)\cdot(R_{L2}+R_{L3}) \\ \delta=D(R_{S1}+R_{S2})+(1-D)\cdot(R_{L2}+R_{L3})+R_L \end{cases} \quad (5)$$

When the circuit is in steady state, the inductance and capacitance of the circuit should be balanced and charged. So the static working point by the system:

$$A\begin{pmatrix} i_{L1} \\ i_L \\ v_{C1} \\ v_{C2} \\ v_{C3} \\ v_{C4} \end{pmatrix}+B\begin{pmatrix} v_g \\ v_{D1} \\ v_{D4} \end{pmatrix}=0 \quad (6)$$

After substituting the simplification, the voltage gain and circuit efficiency of the topology can be obtained without considering the leakage inductance:

$$\begin{cases} B=\frac{V_o}{V_g}=\frac{(2nD-n+1)R-(1-D^2)(V_{D1}+V_{D4})R/V_g}{a(R_{S1}+R_{S2})+bR_L+cR_{L1}+dR_{L2}+eR_{L3}+R(1-D)} \\ \eta=\frac{V_o i_o}{V_g i_{IN}}=\frac{1-D}{D}\frac{(2nD-n+1)R-(1-D^2)(V_{D1}+V_{D4})R/V_g}{a(R_{S1}+R_{S2})+bR_L+cR_{L1}+dR_{L2}+eR_{L3}+R(1-D)} \end{cases} \quad (7)$$

$$\begin{cases} a=2nD^2-D^3+nD+3D-n \\ b=2nD+D^2-nD-D+1 \\ c=(1-D)(D-D^2+1) \\ d=(1-D)(D^2-3D+2) \\ e=1-D \end{cases} \quad (8)$$

The voltage gain and overall efficiency of the coupled inductor three-level Sepic converter with parasitic parameters can be analyzed by equations (7) and (8). Figure 4 is a comparison of voltage conversion capabilities under the influence of different parasitic parameters. It can be seen that the parasitic resistance of the coupled winding L1 has a great influence on the overall performance of the converter, and the remaining parasitic resistance has little effect on the circuit. Moreover, in the case of an open loop, the change in the load value causes a large change in the circuit. Overall, the trend of all the numerical curves is consistent with the sharp drop after reaching a certain duty cycle value. This indicates that the system should avoid excessive duty cycles to prevent the overall performance of the circuit from dropping rapidly.

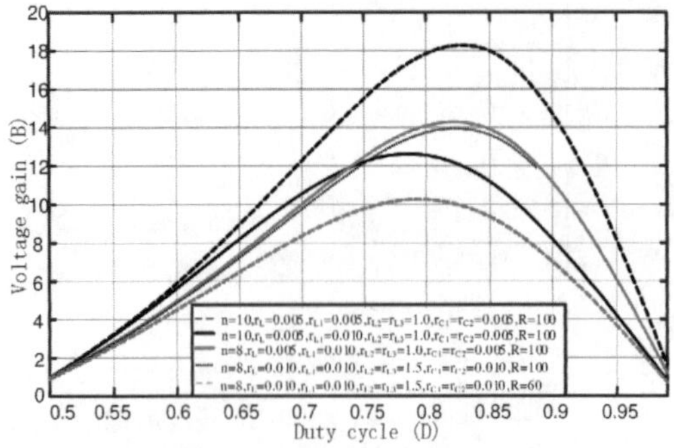

Figure 4. Voltage gain curve

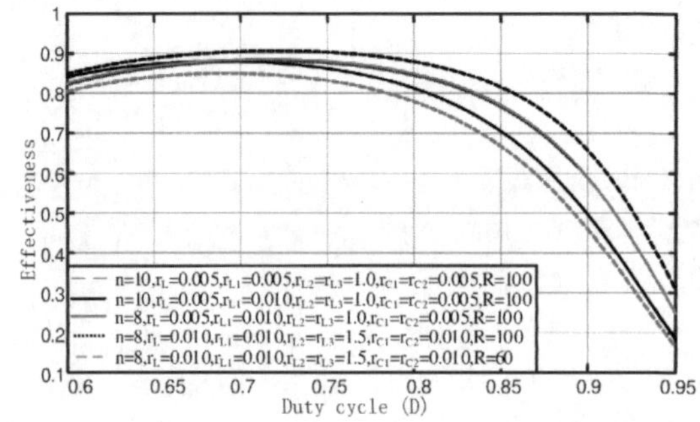

Figure 5. Efficiency graph

It can be seen from Figure 5 that under the influence of parasitic parameters of each device, the overall efficiency is always less than 1.The efficiency curve first increases and then decreases. When D reaches 0.85, the curve begins to drop sharply. The converter should avoid the occurrence of the extreme duty cycle, which not only ensures the safety of the experiment, but also ensures the efficiency of the experiment. When the coupling winding turns ratio n is increased and the parasitic resistance is reduced, it is helpful for the overall efficiency.

Figure 6 is a diagram showing the relationship between the voltage conversion capability and the overall efficiency of a coupled inductor three-level Sepic converter with parasitic resistances of different values.

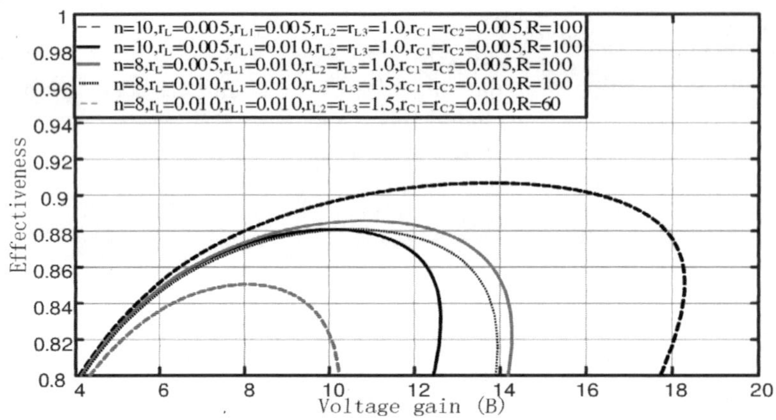

Figure 6. The relation curve of B-η

By analyzing the above three graphs of the important parameters of the coupled inductor three-level Sepic converter, It can be obtained: The parasitic parameters of the coupled winding L_1 have the greatest influence on the voltage gain and the overall efficiency of the converter. The load size also affects the system during open loop. Converter should focus on optimizing the parasitic parameters of the coupled winding L_1 to improve the overall efficiency and voltage gain. It is necessary to consider the influence of various factors such as parasitic parameters, duty cycle, voltage gain and parameter design efficiency, so that the converter can obtain the best experimental results to ensure its safety and efficiency.

4. Converter simulation analysis and verification

The main working process of the converter is verified by Saber software. Table 1 shows the parameters of the converter-related components during the simulation.

IOP Conf. Series: Journal of Physics: Conf. Series **1284** (2019) 012070 doi:10.1088/1742-6596/1284/1/012070

Table 1. The table of converter device parameters

parameter	Numerical value	parameter	Numerical value
L	100μH	C_1	470μF
L_1	95.8μH	C_2	470μF
L_2	366.6μH	C_3	1000μF
L_3	366.3μH	C_4	1000μF
n	2	V_g	12V
L_{1k}	0.05μH	f_S	20kHz
L_{2k}	0.1μH	D	0.75
L_{3k}	0.08μH	R_L	64Ω

Figure 7 is a schematic diagram of the open loop control simulation of the converter.

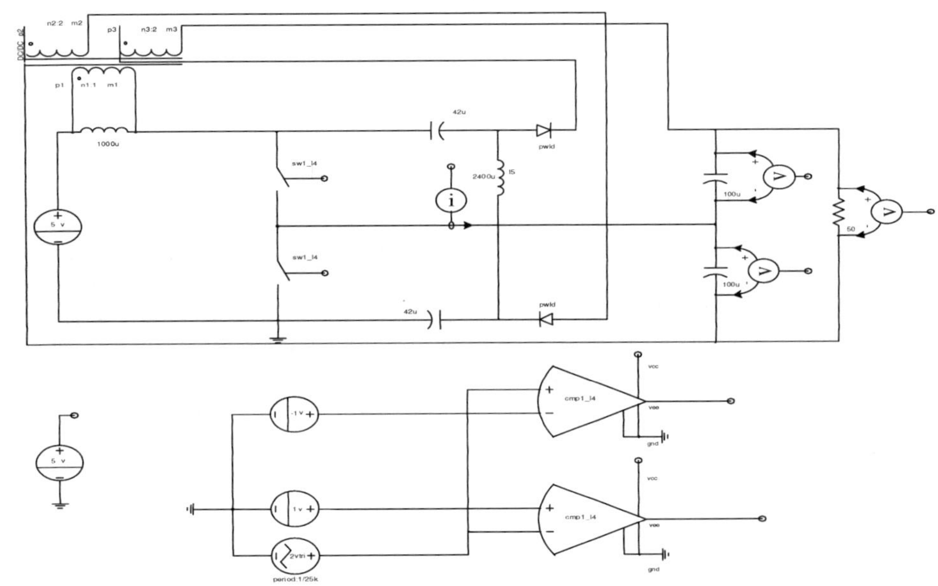

Figure 7. The principle diagram of the simulation

(a) and (b) in Figure 8 correspond to converter's input and output voltage simulation waveforms of the parasitic resistance $r_{L1} = 0.003\ \Omega$ and $r_{L2} = 0.015\ \Omega$.

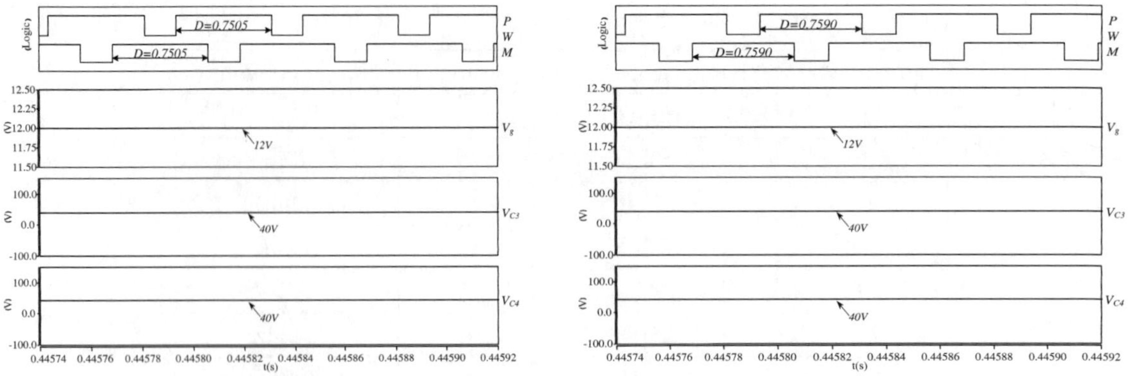

(a) r_{L1}= 0.003Ω (b) r_{L1} =0.015Ω

Figure 8. The simulation waveform of input and output voltage

5. Conclusion

This paper analyses the steady-state working process of the coupled inductor three-level Sepic converter under the influence of parasitic parameters. Establish a state space model and conduct simulation experiments, analyse the effects of various parasitic parameters on circuit performance. It can be concluded that the parasitic parameters of the coupled winding L_1 have the greatest impact on the converter voltage gain and converter efficiency.

References

[1] C.Yin,Y.Rao,K.Jin. Three-level full-bridge bi-directional DC-DC converter[A]. Advanced and Industry Research Center.Proceedings of the 2014 International Conference on Environmental Engineering and Computer Application (ICEECA 2014)[C].Advanced Science and Industry Research Center:,2014:4.

[2] Trillion Q.Zheng. A novel L-BOOST three-level DC-DC converter[A]. Intelligent Information Technology Application Research Association (IITA Association), Hong Kong、Shenzhen University, China.Proceedings of 2010 The 3rd International Conference on Power Electronics and Intelligent Transportation System(Volume 1)[C].Intelligent Information Technology Application Research Association (IITA Association), Hong Kong、Shenzhen University, 2010:4.

[3] Hisayuki Sugimura,Khairy Fathy,Sang-Pil Mun,Toshimitsu Doi,Tomokazu Mishima,Mutsuo Nakaoka. Three-Level Phase Shifted Soft Transition PWM DC-DC Power Converter with High Frequency Link for Arc Welders and Its Extended Version[A]. China Electrotechnical society、IEEE Power Electronics Society、National Science Foundation of China.2009 IEEE 6~(th) International Power Electronics and Motion Control Conference-ECCE Asia Conference Digests[C].China Electrotechnical society、IEEE Power Electronics Society、National Science Foundation of China:,2009:7.

[4] Xinping Ding, Dailing Yu, Yingjie Song,Bicui Xue. Switched-coupled inductor DC-DC converters[C]//Industrial Electronics for Sustainable Energy Systems (IESES), 2018 IEEE International Conference on. IEEE, 2018: 174-179.

[5] Mousa M., Ahmed, M.E., Orabi, M. New converter circuitry for high v applications using Switched Inductor Multilevel Converter[C]. IEEE 33rd International Telecommunications Energy Conference, Amsteram, 2011:1-8.

[6] Hwu, K.I., Yau, Y.T. Inductor-coupled KY boost converter [J]. Electronics Letters, 2010, 46 (24): 1624-1626.

Research of NCS Random Time Delay Prediction Based on FNN-PSO

Jian-He LEI, Jing-Xue LU, Bing-Xin XUE and Guang-Yi WANG

School of Information and Control Engineering, Qingdao University of Technology, Qingdao, 266520, China

1009856297@qq.com

Abstract. Networked control systems have been attracting considerable attention during the last decades, not only because of their theoretical value, but also of their widespread practical applications. In practice, networked control systems are inevitably affected by the random time delays. Aiming at the problem of system performance degradation caused by random time delay in networked control systems, a method of combining particle swarm optimization with fuzzy neural network is used to establish a networked control system time delay prediction model and predicts the future time delay. According to the simulation, we verify the effectiveness and feasibility of the proposed method. The results of simulation show that the fuzzy neural network optimized by particle swarm optimization has a high prediction accuracy for random time delay and is suitable for practical applications.

1. Introduction

The networked control system (NCS) refers to a closed-loop feedback control system formed by actuators, sensors and controllers in the control system to exchange information through the network. The addition of the network in the control loop makes the analysis of NCS complicated, especially the induced time delay in the network has an important impact on the performance of the system [1]. Therefore, accurate prediction of the network time delay is helpful to improve the performance of the NCS.

In recent years, a lot of research results have been reported on the network time delay prediction problem. In [2], a time-delay estimate algorithm for NCS based on least-square support vector machines is proposed. The authors in [3] use Auto-Regressive and Moving Average model to make forward time delay prediction and compare it with the prediction effect of Radial Basis Function neural network. In [4], a particle swarm optimization least square support vector machine method is used to establish an NCS time-delay prediction model and predict the future time-delay. For [5], the authors propose a novel method of NCS stochastic delay prediction based on the Autoregressive Intergrated Moving Average. By using this method, high prediction accuracy can be achieved.

Aiming at the random time delay in NCS, this paper uses the FNN [6-8] optimized by PSO to predict the time delay, which makes up for the shortcomings of the conventional BP algorithm, such as slow convergence speed and easy to fall into local extreme points, and ensures the accuracy of prediction.

2. Fuzzy neural network optimized by particle swarm optimization algorithm

2.1. Fuzzy neural network model

Fuzzy neural network is a combination of fuzzy set theory and neural network. It not only has the large-scale parallel processing ability, self-learning ability and self-adaptive ability of neural network, but also can process fuzzy information and complete fuzzy reasoning function with superior performance.

The prediction model is a four-layer network model with multiple inputs and single output, as shown in Figure 1. The four layers are the input layer, the fuzzification layer, the fuzzy inference layer, and the output layer.

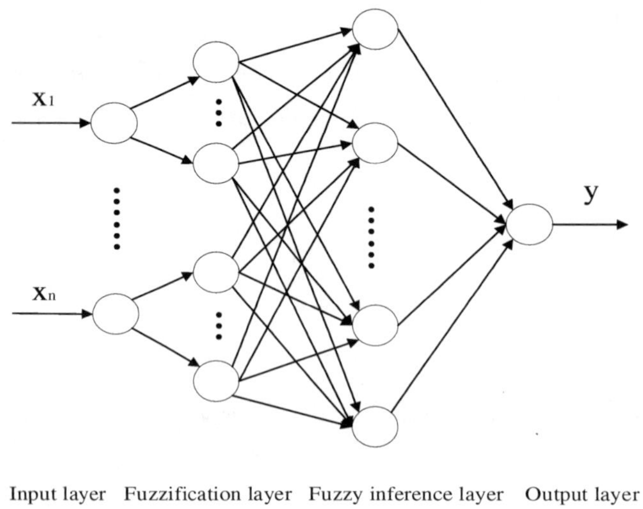

Figure 1. Fuzzy neural network structure.

The first layer is the input layer, the number of neurons in the input layer is the same as the number of input variables, and the input vector is $\boldsymbol{x} = (x_1, x_2, \cdots, x_n)^T$. y is the output layer variable. Since the output variable is the predicted value of future time delay and has only one, there is only one neuron in the output layer. The jth fuzzy rule of the system can be expressed as

$$R_j: if\ x_1 = X_{1j}, x_2 = X_{2j}, \cdots, x_n = X_{nj}, then\ y = Y_j \tag{1}$$

where Y_j is the result of the jth rule, j depends on the number of neurons in the fuzzy inference layer; X_{ij} is the fuzzy subset corresponding to the input variable, and $i = 1,2,\cdots,n$, j depends on the number of fuzzy subsets of each input variable.

The second layer is the fuzzification layer. In this layer, the membership degree of each input variable is calculated. The normal distribution function is used as the membership function, and the formula for calculating the ith input variable satisfying the fuzzy subset X_{ij} is as follows

$$\mu_{X_{ij}}(x_i) = exp\left[-\frac{(x_i - m_{ij})^2}{\sigma_{ij}^2}\right] \tag{2}$$

where m_{ij} is the expectation of normal distribution function, σ_{ij} is the standard deviation, and the membership degree of input variables is completely determined by parameters m_{ij} and σ_{ij}.

The third layer is the fuzzy inference layer. This paper defines the connection weights of the fuzzification layer and the fuzzy inference layer according to the fuzzy theory.

$$\mu_j = \mu_{X_{1j}}(x_1) \times \mu_{X_{2j}}(x_2) \times \cdots \times \mu_{X_{nj}}(x_n) = exp\left[-\sum_{i=1}^{n}\frac{\left(x_i - m_{ij}\right)^2}{\sigma_{ij}^2}\right] \tag{3}$$

where $\mu_{X_{ij}}(x_i)$ is the membership of the input variable x_i to the fuzzy subset X_{ij}.

The FNN performs the fuzzy decision function from the third to the fourth layer. Assuming that there are m rules in the system, this paper uses the weighted average method to defuzzify and obtains the output as

$$y = \frac{\sum_{j=1}^{m} w_j \mu_j}{\sum_{j=1}^{m} \mu_j} \tag{4}$$

where w_j is the connection weight between the fuzzy inference layer and the output layer.

To train the FNN, the expectation m_{ij} and standard deviation σ_{ij} of the normal distribution function and the connection weight w_j need to be determined. Gradient descent method is commonly used in neural networks. The algorithm has slow convergence speed and is easy to fall into local optimization problems, so it is difficult to get the global optimal solution. In order to solve this problem, the PSO algorithm with global optimization performance is used to optimize the parameters of the FNN.

2.2. *Particle swarm optimization algorithm*

2.2.1. Mathematical description of particle swarm optimization. If a target is found in a D dimensional space, the population is composed of m particles. Each particle in the population has its own position, and the position of the ith particle is represented by $\mathbf{x_i} = (\mathrm{x_{i1}, x_{i2}, \cdots, x_{iD}}), \mathrm{i} = 1,2,\cdots,\mathrm{m}$, and its velocity is also denoted by a D-dimensional vector $\mathbf{v_i} = (\mathrm{v_{i1}, v_{i2}, \cdots, v_{iD}})$. When each particle passes through a new position, it compares with all the positions that have passed before. When the ith particle flies to a certain position, the optimal position it passes is $\mathbf{p_i} = (\mathrm{p_{i1}, p_{i2}, \cdots, p_{iD}})$. By comparing the optimal positions of all particles in the particle swarm, the optimal positions of all particles are obtained and expressed as $\mathbf{p_g} = (\mathrm{p_{g1}, p_{g2}, \cdots, p_{gD}})$. Each particle changes its position according to the following formula.

$$v_{id}(t+1) = w v_{id}(t) + c_1 r_1 \big(p_{id} - x_{id}(t)\big) + c_2 r_2 \left(p_{gd} - x_{id}(t)\right) \tag{5}$$

$$x_{id}(t+1) = x_{id}(t) + v_{id}(t+1) \tag{6}$$

where i is the number of particles, $i = 1,2,\cdots,m$; d is the spatial dimension, $d = 1,2,\cdots,D$; w is the inertia weight and is non-negative; the acceleration factors c_1 and c_2 are defined as constants; r_1 and r_2 are random numbers and are evenly distributed between 0 and 1; $x_{id}(t) \in [-x_{max}, x_{max}]$, $v_{id} \in [-v_{max}, v_{max}]$, x_{max} and v_{max} are non-negative.

2.2.2. Algorithm flow chart. The particle swarm algorithm flow chart is shown in Figure 2. The position and velocity of the particle are randomly initialized, the particle fitness value is calculated, and the particle position is compared with the optimal position to determine whether the convergence criterion is satisfied. If satisfied, the output particle position is regarded as the optimal position of the particle swarm; if not, the position and velocity of the particle are updated, and the iteration continues.

2.2.3. Fitness function and parameter optimization steps. The PSO is used to optimize the connection weight and membership function of the FNN. That is to say, the element of the position vector $\mathbf{x_i}$ of the PSO is the connection weight of the FNN, the expectation and standard deviation of the membership function. The training of neural networks is mainly to minimize the parameter errors in the network. Therefore, the fitness function can be defined as

$$f = \frac{1}{N_s} \sum_{i=1}^{N_s} (Y_i - y_i) \tag{7}$$

where N_s is the number of training set samples; Y_i is the expected output; y_i is the actual output.

The steps of parameter optimization based on particle swarm optimization are as follows: 1) Initialization, randomly initialize the position and velocity of each particle, and determine parameters such as population number, inertia weight and acceleration factor; 2) Calculate the fitness of each initial particle, $\boldsymbol{p_i}$ is the optimal position of the current particle and $\boldsymbol{p_g}$ is the optimal position of the particle swarm. 3) Update the position and velocity of the particles according to formulas (5) and (6) to generate new particles; 4) Return to step 2) until the convergence criterion is satisfied, the iteration stops, and the global extremum is the optimal solution of the training problem.

3. Prediction model establishment and simulation analysis

3.1. Prediction model establishment

In this paper, the rolling prediction method is used to predict the network time delay. The first five delay values are taken as input sample 1 and the sixth delay value as output sample 1 to form a sample pair. The second to sixth delay values are taken as input sample 2 and the seventh delay value as output sample 2 to form the second sample pair. By analogy, 300 sample pairs are constructed as data for training neural networks. After that, the 301st to 305th delay values are taken as input samples, and input to the trained FNN as an input vector to obtain the first network delay prediction value. Then, the 302nd to 306th delay values are used as input samples to obtain the second network delay prediction value, and by analogy, 20 network delay prediction values are obtained. The prediction flow chart is shown in Figure 3, and the modeling steps are as follows.

(1) Establish and train fuzzy neural network. The structure of the neural network is determined by clustering algorithm, and the number of fuzzy membership degrees is selected by Kohonen network [9]. It is determined that each input variable is divided into two fuzzy subsets, so the structure of the fuzzy neural network is 5-10-32-1. Then the training data are used to train the fuzzy neural network.

(2) Optimize network parameters. Initialize the particle swarm optimization algorithm, set the population number to 30, the acceleration factor to $c_1 = c_2 = 2$, the fitness threshold to 0.001, and the maximum allowable number of iterations to 350.

(3) Model prediction. The test data is used to verify the model and the final prediction results are obtained.

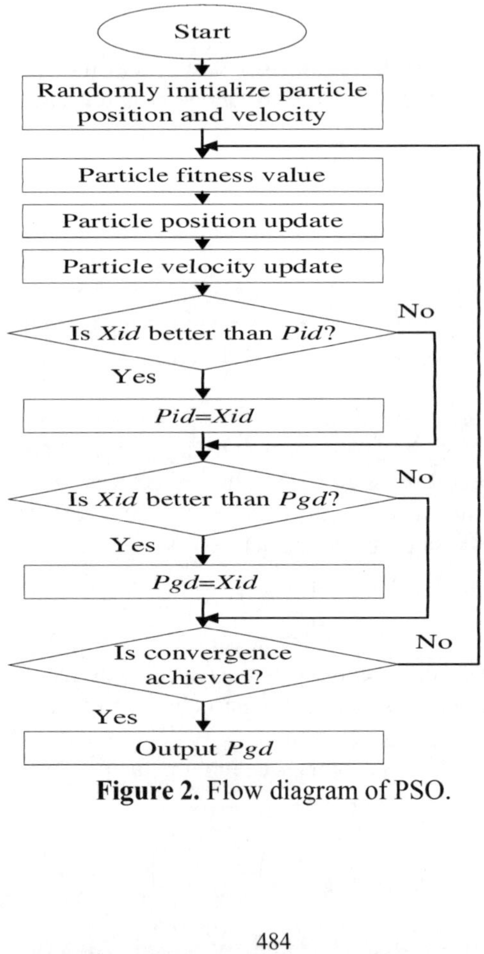

Figure 2. Flow diagram of PSO.

IOP Conf. Series: Journal of Physics: Conf. Series **1284** (2019) 012071 doi:10.1088/1742-6596/1284/1/012071

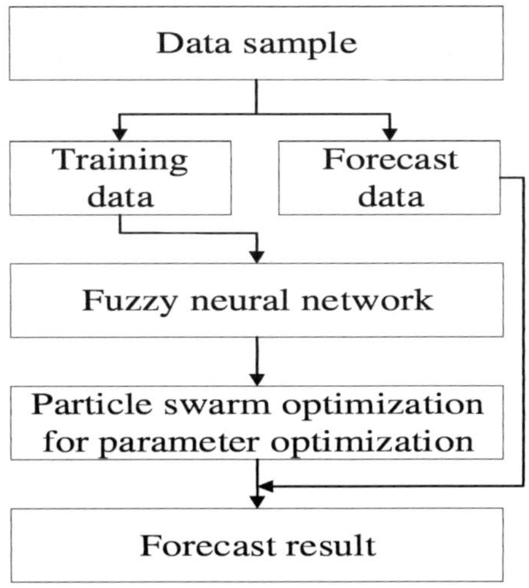

Figure 3. Flow diagram of prediction.

3.2. Simulation results and discussion

Two kinds of fuzzy neural networks are constructed. One uses gradient descent method to train the neural network parameters (FNN), and the other uses particle swarm optimization (FNN-PSO). The performance comparison of the two types of fuzzy neural networks is shown in Figure 4.

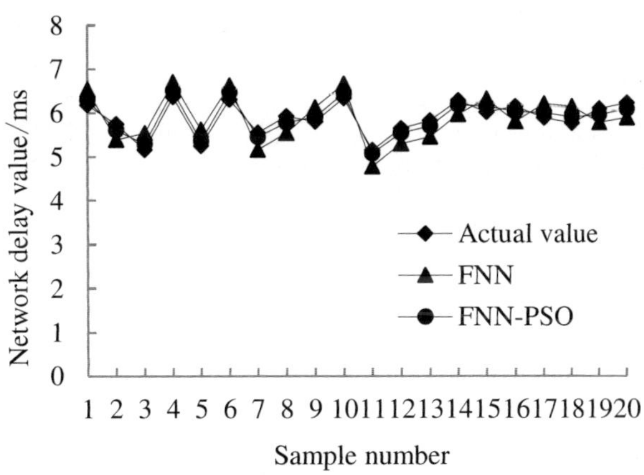

Figure 4. Performance comparison among measured value, FNN and FNN-PSO.

The root mean square error (RMSE), mean absolute percent error (MAPE) and mean absolute error (MAE) [10] are selected to analyze the error of the predicted data, and the error indexes are shown in Table 1.

$$R_{\mathrm{RMSE}} = \sqrt{\frac{1}{N}\sum_{i=1}^{N}(y_i - \hat{y}_i)^2} \tag{8}$$

$$M_{\mathrm{MAPE}} = \frac{1}{N}\sum_{i=1}^{N}\left|\frac{y_i - \hat{y}_i}{y_i}\right| \tag{9}$$

$$M_{\mathrm{MAE}} = \frac{1}{N}\sum_{i=1}^{N}|y_i - \hat{y}_i| \tag{10}$$

where y_i is the true value, $\hat{y}_i$ is the predicted value, and N is the total number of samples.

Table 1. Error indexes of the prediction results.

Prediction model	Error indexes		
	RMSE/%	MAPE/%	MAE/%
FNN	34.75	5.94	34.65
FNN-PSO	12.48	2.09	12.3

It can be seen from Table 1 that when predicting the network time delay value, compared with BP algorithm, such error indexes as RMSE, MAPE, and MAE of PSO algorithm are decreased by 22.27, 3.85, and 22.35 percentage points respectively. The error indexes prove the effectiveness of the PSO algorithm, and the prediction accuracy is higher than the conventional BP algorithm. According to Figure 4, it can be seen that the prediction model based on PSO and FNN has better fitting ability to actual data.

4. Conclusions

Aiming at the problem of random time delay affecting the performance of networked control system, a fuzzy neural network optimized by particle swarm optimization is proposed to predict random time delay. The simulation results show that the fuzzy neural network optimized by particle swarm optimization algorithm has higher accuracy for time delay prediction than the traditional BP algorithm, and reduces the computation time, which is more suitable for practical applications.

References

[1] Yan H C, Huang X H, Wang M. Delay-dependent stability of networked control systems with uncertainties and multiple time-varying delays[J]. Control Theory and Applications, 2008, 25(2): 303-306.

[2] Li C M, Xiao J, Zhang Y. Approach of adaptive prediction control on networked control systems based on least-squares support vector machines[J]. Journal of System Simulation, 2007, 19(15): 3494-3498.

[3] Xu P, Chen Q G, Ge Y, et al. Research of NCS forward time delay prediction based on ARMA model[J]. Computer Engineering and Applications, 2013, 49(20): 91-95.

[4] Tian Z D, Gao X W, Li K. Networked control system time-delay compensation algorithm based on predictive control[J]. Systems Engineering and Electronics, 2013, 35(10): 2165-5169.

[5] Xu W, Ge Y, Wang Y. Predicting NCS stochastic delay based on ARIMA[J]. Journal of Anhui Polytechnic University, 2016, 31(4): 72-76.

[6] JIANG S F, ZHANG C M, ZHANG S. Two-stage structural damage detection using fuzzy neural networks and data fusion techniques[J]. Expert Systems with Applications, 2011, 38(1): 511-519.

[7] Qian W, He Z X, Zhang D Y. Multi fire sensor characteristic parameters fusion algorithm based on fuzzy neural network[J]. Chinese Journal of Sensors and Actuators, 2017, 30(12): 1906-1911.

[8] Zhao J B, Xue B X. Fuzzy neural network for studying the prediction of dissolved oxygen in aquaculture ponds based on particle swarm[J]. Fishery Modernization, 2018, 45(4): 8-14.

[9] Gao F, Liu J, Li Y, et al. Identification of temperature measuring points for machine tools with Kohonen self-organizing competitive network[J]. China Mechanical Engineering, 2014, 25(7): 862-866.

[10] Huan J, Cao W J, Qin Y L, et al. Prediction of dissolved oxygen based on empirical mode decomposition and least squares support vector machine[J]. Fishery Modernization, 2017, 44(4): 37-43.

An Approach Combined the Faster RCNN and Mobilenet for Logo Detection

Terry Mudumbi, Naizheng Bian*, Yiyi Zhang, and Florian Hazoume

College of Computer Science and Electronic Engineering, Hunan University, Changsha, China.

*Email: 741565912@qq.com

Abstract. Although deep learning object detection tools such as Faster Recurrent Convolution Neural Network (Faster R-CNN) has demonstrated good performances in object detection, they also have a limited success rate for some applications. It is due to the lack of refinedness of feature maps for accurate localization, the insensitivity for small scale objects and fixed-window feature extraction in Region Proposal Network (RPN). In this paper, we performed a meticulous examination of both the proposal and the classification process by evaluating the adequacy of feature representations from different stages of the feature sequencing. We presented an approach to improve the Regional Proposal Network (RPN) by appropriate anchors selection, and proposed a modification by combining Faster R-CNN and MobileNet which influences higher-resolution feature maps for mobile devices. The results demonstrate that Faster R-CNN architecture with MobileNet has the best detection accuracy. The experiment result showed that we managed to achieve a final accuracy of 92.4% on a NVIDIA GeForce Gtx 1070 compare to previous work that achieved 90.8% of accuracy and found that our models performed well at the detection, with very low false positive rates possible for a fairly reasonably .

1. Introduction

Brand recognition is a very challenging topic with many useful applications in localization recognition, advertisement and marketing. Traditionally, logo detection is solved by defining key points and texture descriptors combined with statistical classifiers, such as Nearest Neighbour (NN) or Support Vector Machines (SVM) [1, 2]. In [3, 4] the images can be damaged by many artifacts and image distortions. In order to handle the limitations of the above models, in [5] they solved the problem by proposing Recurrent Convolution Neural Network (RCNN) and apply it for the classification task. Recent advances in hardware has increased multiple methods for detection using CNNs have been presented [6]. Through the Region-Based Convolutional Neural Network (RCNN) came out very good results compared to previous methods. In RCNN [7], regional proposals submitted using the proposal methods are implemented on CNN network.

The disadvantage of RCNN is its expensive computational cost, since each area is treated separately. By solving the expensive computational the Fast RCNN [8] improves the RCNN defect by using region proposal as detectors for a common feature map. Method [9] proposed the complex for object detection called Fast R-CNN that combines R-CNN and Spatial pyramid pooling networks (SPPnet). The network is formed by a set of convolutional layers, fully-connected layers, an external region proposal method typically Selective Search (SS) [10, 11, 12] and a Region of Interest (RoI) pooling layer. However, their method is still computationally heavy for real-time processing on an embedded platform. In the work of

Huang et al. [13], they showed the Single Short Multibox Detector (SSD) approach together with SquezeNet and MobileNet [14] as the backbone networks. Although SSD with SqueezeNet backbone results in a smaller model than MobileNet, [15, 16] says that the results are less accurate and its computation is slightly more expensive.

In this article, we want to improve the Faster R-CNN's mAP (mean average precision) by processing higher resolution layers objects which makes the feature map larger. Therefore, we proposed an approach based on Faster R-CNN and MobileNet for Logo detection and created implementations for a Faster R-CNN and MobileNet architectures which influence higher-resolution feature maps for mobile devices. The approach described the core layers that MobileNet is built as a good feature-extractor. Finally, we explored the networks that are able to locate and classify multiple logos in real time.

The rest of this paper is organized as follows: Section II analysis model and then discusses the details of our approach. Section III analysis the proposed approach and evaluates the performance through experiments. Finally, Section IV concludes the paper and points out the future research directions.

2. Proposed Approach

Our proposed architecture consists of 2 key steps: Firstly, we mathematically design the suitable sizes of anchor boxes with 6 aspect ratios {0.48, 0.65, 1.09, 1.28, 1.48, and 1.70} and perform the effectiveness of their choice. The RPN has proposed a set of bounding boxes with a trusted rating associated with potential logo. Secondly, we described detailed analysis of these fully convolution architectures by using MobileNet as a feature- extractor. To every proposal, the corresponding feature maps use the resolution in a fixed-size representation, many layers that are fully associated with presentations are classified within specific bounding box regression. On the features of different feature maps, the deep layers are likely to be able to provide better properties, which means that single activation processes for input stimuli are more specific than earlier layers. We establish that, features of the previous layers can provide performance for small objects that matches or even surpasses the performance of features from the deep layers. We rate our observations on the well-known FlickrLogos-32 dataset as an extension of the Faster R-CNN pipeline.

2.1 Regional Proposal Network

The main point of the RPN is to propose a set of bounding boxes with a trusted rating associated with potential logo. We modified the RPN to detect a logo with this configuration $B_a = 9$ anchors. Since each anchor box acts as a detector for sliding windows in a grouped image area, there are:

$$B_p = B_a \times \frac{w}{f_s} \times \frac{H}{f_s} \tag{1}$$

Where f_s corresponds to an anchor. The RPN structure conv1, conv2... , conv5. The general observation has been confirmed in part by the fact that increasing the efficiency of highway classification on the ImageNet. This applies to at least two common phase detection devices, such as Faster-RCNN and MobileNet. We used a network of regional presentations (RPN), consisting of two layers to locate the regions that can contain objects in feature maps (image). The network uses the RoI pool layer to reduce and resize resource maps based on proposals from that region. The maps use the new features of each region to select frame into three fully connected layers. In this work, MobileNet which took the layers as learning functions was used as a convolutional network the original feature extraction contains several layers and the first convolution stack structures acquired through transfer learning by using MobileNet.

2.2. Faster RCNN_Mobilenet

Our approach has two steps forming the current object detection such as: The first one consists on identifying ROI from images. These ROI can be considered as references in recommending some possible object location that are more carefully developed in the second step. As shown in figure 1 with 5 convolutional layers and 3 fully connected (FC) layers. While Faster R-CNN only uses the features of the last convolutional layer to localize and classify. In the first two convolution layers, after each successive layer and one Max-pooling layer, respectively. In the next three levels, just after each convolution layer, there is only one level of ReLU. In particular, on three levels, 3, 4 and 5, their outputs are also used as input data for the three levels of pooling of the ROI and the corresponding normalization levels. For each RPN anchor constituting a fully convolutional network, a degree is predicted which makes it possible to measure the probability of this anchor which contains the element of interest. In addition, the RPN provides the acceleration and measurement coefficients for each anchor that is part of the peripheral regression mechanism, thereby improving the position of the object.

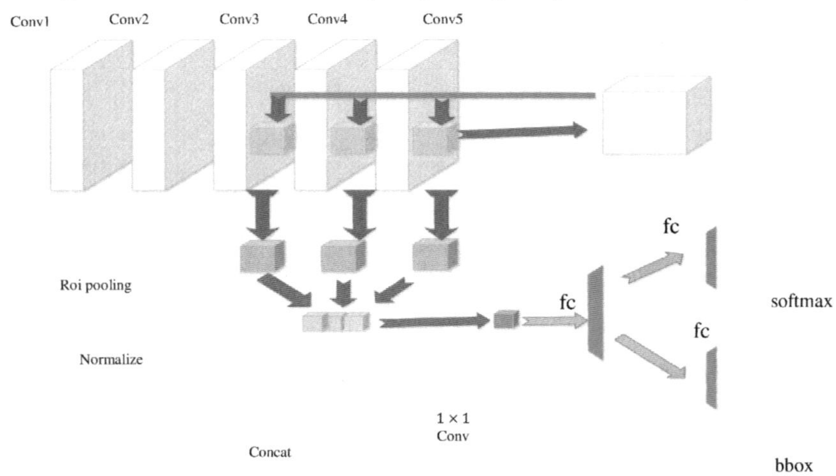

Figure 1. Our proposed Approach with 5 convolutional layers and 3 fully connected (FC) layers.

To properly illustrate this problem, considering the situation in figure 2a: we supposed that a second-order ground truth bounding box A_2 is delimited by a side length x_2 and a square anchor box A_1 of side lengthx_1.

$$t \leq IoU(A_2, A_1) = \frac{|A_2 \cap A_1|}{|A_2 \cup A_1|} = \frac{x_1^2}{x_1^2} \qquad (2)$$

In general, an anchor is considered a positive example if it contains an IoU greater than 0.5 for a ground truth objects.

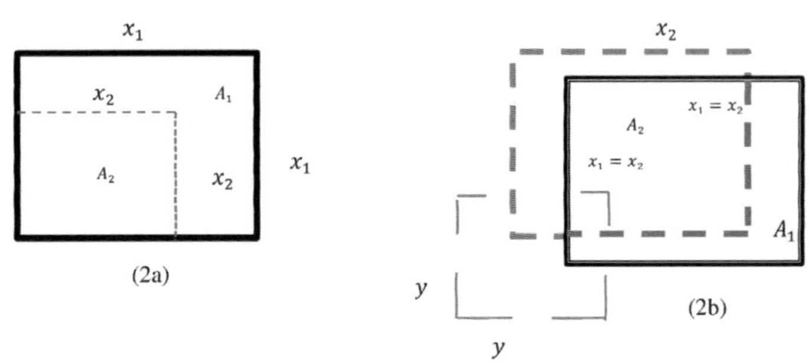

Figure 2. (2a) IoU can be expressed as the ratio of the areas of the limit box in the case of limits of the same proportion. (2b) shows two bounding boxes of the same size when the anchors are selected in stride y

To classify the anchor box as a positive example, ask the IoU to exceed a certain threshold t. It follows that for $> \sqrt{t}^{-1}$, the anchor cannot cover the field of truth sufficiently enough to be classified as a positive example. The same thing applies to non-quadrature anchors, provided that the ratio of groundtruth boxes and anchor boxes correspond to each other. For the above considerations, we suppose that there is an attachment point where the corner of an anchor is perfectly aligned with the ground truth example. In practice, this is not the case, because the network performance map based on RPN is usually much smaller than the original image. The reduction factor $y - 1$ between the source image and the object map effectively creates a network of anchors with stride y. To examine the effect of the characteristic resolution of the card on the potential RPN to determine the state of small objects, considering the situation in figure 2b, we suppose the case of quadratic atoms of phase A_2 and the existence of an anchor box A_1 of scale and the corresponding form factor. In the worst case, each box is moved a distance of $\frac{y}{2}$. The IoU between these boxes can be represented by:

$$IoU(A_2, A_1) = \frac{\left(x_2 - \frac{y}{2}\right)^2}{\left(x_2 - \frac{y}{2}\right)^2 + 2\left(2\frac{y}{2}\left(x_2 - \frac{y}{2}\right) + \frac{y^2}{4}\right)} \quad (3)$$

The initiative learning rate is 3e-3 and the stride size is set to, $x = 16$. Assuming that t = 0.5, this gives the minimum size of the detectable object. This indicates that for a small fraction of our size distribution, we need an object map at a higher resolution.

Second one consists of a depthwise decomposable approach to integrate the local context from each selected scale of the feature maps and then add it again. To factorize the convolution, a depthwise convolution a removable winding is used, as this can help to significantly reduce calculations and parameters. In CNN, convolution filters extract objects from input property maps through a sliding window. Different size of pixels will be extracted by the other size of filters, so that they can be utilized as context extractor's tools. By using these tools, we propose an end-to-end convolution approach with these context extractors that can be subdivided to give context to a local context. With a separable convolution depthwise, the computation is reduced by a factor of 8-9 times and our detection speed rate can be increased effectively. We define the convolution depth ratio in percentage $M = (\text{only } 0.25, 0.50, 0.75, 1)$ and adjust t = 0.5. If Object is in a group of C classes, L is the group of basic objects L_c ($c \in C \in$Cand N and there is a set of sentences of objects, then we can estimate the performance of this class's RPN, that its average value is Avg(c) expressed by:

$$Avg(c) = \frac{1}{|L_c|} \sum_{l \in L_c} \max_{n \in N} IoU(l, n) \quad (4)$$

3. Experiment and Result

To investigate the effects of RPN performance on the object size, different versions of the FlickrLogo-V2 dataset have been designed, and applied the following algorithm to each image. We began by choosing a point where the maximum distance between the two unbounded square boxes. This point specifies two axes that can divide the image between them into four parts. We ensure that the axes of division do not break any other point of truth. If no separation is found, the image will be discarded. For each resulting layer, which contains more than one main element, this process is applied repeatedly. After applying this algorithm, there is only one instance of the object in each image, which is measured to match the desired target size. To track the performance of RPNs at different levels, we create 3 RPN which are: RPN conv3, RPN conv4 and conv5 RPN. These networks use functions of level 2, 2 and 4 levels conv3 to predict the proposals of objects. Objects are transmitted through the normalization level, which normalizes the activation to change the mean and unit of measurement. We normalized the activation processes with respect to the training package then we put a standard RPN on the top of this normalization function, which is materialized by a 3×3 convolution using a similar number of channels as the previous layer.

Our approach improved by appropriate anchors selection in the Regional Proposal Network (RPN), had negative mining and non-maximum suppression. Our approach integrates a process that can be described as follows:

- **Regional proposal network** uses a RPN network that incorporates the regional generation network into a framework
- **Feature extraction:** includes many layers of conv+relu, and the first few convolution batch structures can be obtained through transfer learning. Faster R-CNN with mobilenet feature

extractor for 450k iteration, batch 8, learning rate 3e-03. A supplemental conv+relu layer is added to export the feature. According to each image position, 9 possible filter windows are considered, which are called anchors.

- **Classification and box regression:** At each point, the cls_score output keys have 9 anchor points for the forward and backward probabilities, the window regression prediction layer (bbox pred) displays each position and the 9 anchor matching window must move the hyper-parameters. Previous initializing hyper-parameters can be established as follows:
  - The selected appropriate scales and aspect ratios of the anchors, making it more efficient to detect object. For the normal Faster R-CNN (with inception_v1, inception_V2, etc...) used 9 anchors with 4 scales and 4 aspect ratios. These aspect ratios (height/width) were { 0.25, 0.5, 1.0, 2.0}
  - RPN network has been used to extract area proposal by changing the anchors from 4 scales {0.25, 0.5, 1.0, and 2.0} to 6 scales {0.125, 0.25, 1.0, 1.5, and 2.0} and 6 aspect ratios {0.48, 0.65, 1.09, 1.28, 1.48, and 1.70}. According to the experiment, the appropriate aspect ratios of anchors obtained higher accuracy and lower loss function than the others ones.
  - Some valuable negative samples, such as false positives, were selected from the negative ones. At the time of the prediction, we use a maxima deletion algorithm to filter the multiple boxes per object that can appear.

As the training process progresses, the expectation are that total loss (errors) gets reduced to its possible minimum (That means we have to get less than 1 or lowest one). We ran our training job for 500k steps (took about one day) and stopped at a total Loss (errors) value of 0.07413.

The result of The Function loss of each part after 450 k steps is shown in figure 3a that is the combination of: the proportion of bounding boxes produced by RPN that are correctly classified (as the correct object class) and some distance measure between the predicted and target regression coefficients.

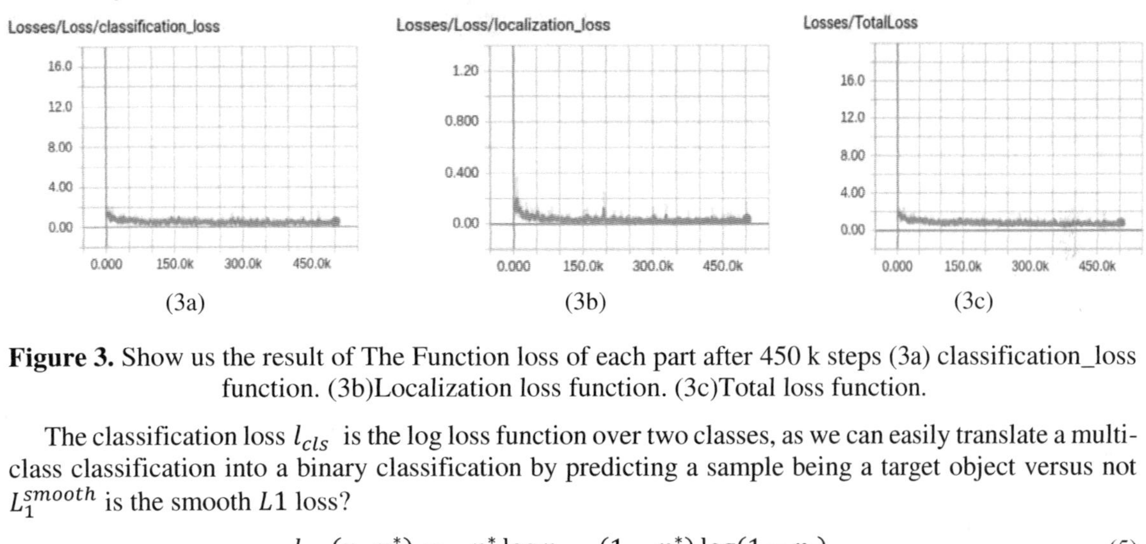

(3a) (3b) (3c)

Figure 3. Show us the result of The Function loss of each part after 450 k steps (3a) classification_loss function. (3b)Localization loss function. (3c)Total loss function.

The classification loss l_{cls} is the log loss function over two classes, as we can easily translate a multi-class classification into a binary classification by predicting a sample being a target object versus not L_1^{smooth} is the smooth $L1$ loss?

$$l_{cls}(p_i, p_i^*) = -p_i^* \log p_i - (1 - p_i^*) \log(1 - p_i) \quad (5)$$

The result of localization loss in figure 3b uses feature maps that belong to the proposed RoIs:

$$l(\{p_i\}, \{t_i\}) = \frac{1}{N_{cls}} \sum_i l_{cls}(p_i, p_i^*) + \frac{\lambda}{N_{box}} \sum_i p_i^*. L_1^{smooth}(t_i - t_i^*) \quad (6)$$

The multi-task loss function combines the losses of classification and bounding or regression, as shown in figure 3c:

$$l = l_{cls} + l_{box} \quad (7)$$

IOP Conf. Series: Journal of Physics: Conf. Series **1284** (2019) 012072 doi:10.1088/1742-6596/1284/1/012072

The main interest here is about our approach with area proposals. In training and in testing, our approach collects raw images and area proposals having important features in the images with the shape of bounding boxes. This stage is about localizing and classifying every bounding box area proposal as a kind of logo or "background". It means that the region in the area proposal is not a logo at all. In case the area proposal includes a logo, it also returns a bounding box regression as output, modifying the area proposal to better highlight the area with the logo. Our area proposals were made from very deep search. After that, these areas on the images are given for training and testing purposes. The AP values per class are shown in Table 1.

Table 1. Experimental result of each classes using FlickrLogos-32. Detection exhibits fairly good detection performance, especially on distinctive logos such as that of Coca-Cola with 98.1%.

method									mAP
	Adidas	Aldi	Apple	Becks	Bmw	Carls	Chim	Cocacola	
	Corona	Dhl	Erdi	Esso	Fedex	Ferrari	Ford	Fosters	
	Google	Guiness	Heineken	Hp	Milka	Nvidia	Paul	Pepsi	
	Ritt	Shell	Sing	Starb	Stel	Texa	Tsin	ups	
Our Approach	89.7	94.9	94	90.7	88.4	93.5	89.4	98.1	92.40
	87.3	96.6	95.1	93.9	94.8	90.2	89.9	91.3	
	97.2	N/A	97.4	92.1	95.3	91.8	92.3	95.7	
	85.3	95.0	89.9	89.8	90.7	94.9	88.9	90.3	

A commonly used metric for performance is mean average precision (mAP) which is single number used to summarize the area under the precision-recall curve. mAP is a measure of how well the model generates a bounding box that has at least @0.5IOU overlap with the groundtruth bonding box in our test dataset and obtained an average accuracy of 92.4% as process in figure 4.

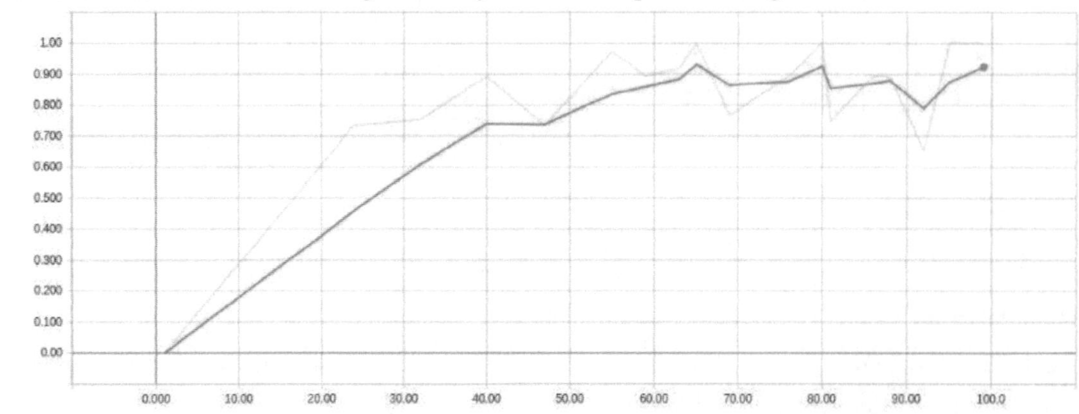

Figure 4. Show us the result of mean average precision (mAP) after 100k steps.

In this work, the experimental environment: NVIDIA GeForce Gtx 1070, CUDA9.0, Ubuntu16.04, memory 8GB, Tensorflow 1.8, Android Sdk 3.2, Android Ndk-r16b and Huawei Mate 10(EMUI 9.0.0181). TensorFlow provides integrated support for calculating the mAP metric via TensorBoard during training and evaluation. By conducting a series of experiments, we collected enough data to evaluate the performance of the models. We have made the process result as shown in figure 5 by conducting a series of experiments, the performance measure of logo detection, we collected FlickrLogo-V2 dataset and added 6400 images to evaluate the performance of the Average Precision (AP) for each individual logo class, and the mean Average Precision (mAP) for all classes. A detection is considered being correct when the Intersection over Union (IoU) between the predicted and groundtruth exceeds 50%. If predicted and true classes match, and the Intersection over Union (IoU) between the predicted bounding box A_2 and the ground truth bounding box A_1 is larger than 50%, the prediction is a True Positive (TP), otherwise is a False Positive (FP).

Figure 5. Shows the logo detection through Huawei mate 10.

$$IoU(A_2, A_1) = \frac{|A_2 \cap A_1|}{|A_2 \cup A_1|} \tag{8}$$

Precision values are computed to calculate the mAP:

$$Precision = \frac{TP}{TP+FP} \tag{9}$$

4. Conclusion

In this paper, we improved the Regional Proposal Network (RPN) by adopting appropriate anchors selection, and integrated Faster R-CNN with MobileNet to get higher-resolution feature maps for mobile devices. At the first, we initialize the network hyper-parameters using the selected appropriate scales and aspect ratios of the anchors. Secondly, we evaluated our approach in detail for the motion phase and the classification phase using FlickrLogo-V2 that the system achieved object accuracy of 92.4%. Throughout experiments performed on different feature maps using MobileNet as backbone which influences higher-resolution feature maps for mobile devices and we notice that the resolution of the feature map has an important role in detecting reliable logo detection. Our approach can usually detect more precisely in earlier feature maps, even if the logo are not so expressive from deeper layers. We have also shown that the choice of anchoring measures for detecting objects is important and specified a criterion for selecting an anchor scale based on the accuracy of the required translation. Finally, we documented our observations as a simple Faster R-CNN extension that can improve the overall performance of a true global corporate dataset for logo detection. In future work, we would like to think about how to increase productivity in classifying objects.

In future work we would like to integrate into our approach using an improved network architecture based on Feature Pyramid Networks by selecting a suitable resolution for generating proposals that may improve the performance of the RPN considerably.

References

[1] Yifei Zhang #1,2, MingMing Zhu#1, Daling Wang #1,2, and Shi Feng #1, *Logo Detection and Recognition Based on Classification* Springer International Publishing Switzerland 2014

[2] Hang Su, Shaogang Gong, Xiatian Zhu Queen Mary University of London Vision Semantics Ltd. *Scalable Deep Learning Logo Detection*

[3] Alireza Alaei, Mathieu Delalandre Laboratoire d'Informatique (LI EA6300, *A Complete Logo*

Detection/Recognition System for Document Images 2014 11th IAPR International Workshop on Document Analysis Systems 12 June 2014

[4] Yuanyuan Li #1, Qiuyue Shi #2 Jiangfan Deng #3, Fei Su #4 ,*A logo is a graphic representation of a brand or company that is used on different communication media*

[5] The Anh Pham#1, Mathieu Delalandre#2, and Sabine Barrat#3. *A Contour-based method for logo detection*. In Document Analysis and Recognition (ICDAR), 2011 International Conference on, pages 718–722. IEEE, 2011

[6] Hongye Wang#1 and Youbin Chen#2. *Logo detection in document images based on boundary extension of feature rectangles*. In Document Analysis and Recognition, 2009. ICDAR'09. 10th International Conference on, pages 1335–1339. IEEE, 2009.

[7] Girshick. In: CoRR abs/1504.08083 (2015). *Fast R-CNN.* ArXiv: 1504.08083

[8] Ross B. Shaoqing Ren, Kaiming He, Ross Girshick, and Jian Sun *Faster R-CNN: Towards Real-Time Object Detection with Region Proposal Networks*. IEEE TRANSACTIONS ON PATTERN ANALYSIS AND MACHINE INTELLIGENCE, VOL. 39, NO. 6, JUNE 2017

[9] Zibo Meng , Xiaochuan Fan†, Xin Chen†, Min Chen‡ and Yan Tong *Detecting Small Signs from Large Images* arXiv:1706.08574v1 [cs.CV] 26 Jun 2017

[10] P. F. Felzenszwalb and D. P. Huttenlocher. *Efficient graph-based image segmentation* Int. J. Comput. Vision, vol. 59, no. 2, pp. 167_181, 2004

[11] P. Arbelaez, M. Maire, C. C. Fowlkes, and J. Malik#4, *From contours to regions: An empirical evaluation* in CVPR, pp. 2294_2301, IEEE, 2009

[12] J.R.R. Uijlings, K.E.A. van de Sande, T. Gevers, and A.W.M. Smeulders#4. *Selective Search for Object Recognition*. International Journal of Computer Vision (2013). DOI: 10.1007/s11263-013-0620-5

[13] Zhuoyao Zhong, Lei Sun , Qiang Huo2 1School of EIE. *An Anchor-Free Region Proposal Network for Faster R-CNN based Text Detection Approaches*

[14] Fei Xia and HuiZhou Li ,*Fast Detection of Airport on Remote Sensing Images with Single Shot MultiBox Detector*

[15] Wei Liu1(B), Dragomir Anguelov2, Dumitru Erhan3, *SSD: Single Shot MultiBox Detector* ECCV 2016: Computer Vision – ECCV 2016 pp 21-37

[16] Karanbir Chahal and Kuntal Dey *A Survey of Modern Object Detection Literature using Deep* Learning arXiv:1808.07256v1 [cs.CV] 22 Aug 2018

Amplitude-Fluctuation Separation Based on BPM Waveform Reconstruction in MIMO Systems

Xinrui Qin[a], Xiaoqi Yang[b] and Weidong Jiang[c]

College of Electronic Science and Engineering, National Univ. of Defense Technology, Changsha 410073, China

[a]1335104481@qq.com; [b]yangxiaoqinudt@sina.com; [c]jwd2232@vip.163.com

Abstract. In Multi-input-multi-output (MIMO) systems, the orthogonality among sub-waves could be realized by binary phase modulation (BPM) encoding. Especially in the case of insufficient matched filters, Hadamard-matrix-based BPM encoding could be employed to separate the transmitted signal from echo effectively, which is verified firstly in this paper by utilizing target location, when introducing fluctuations in amplitude as well as in phase. Hence the amplitude-fluctuation separation method based on waveform reconstruction is put forward. The waveform is reconstructed on basis of the known array structure, and then the results are optimized obeying the minimum mean square error (MSE) criterion to obtain more accurate estimation. Compared with the traditional single-channel blind source separation approach, this method shows significant computation reduction by taking full use of the phase and structure information. Monte-Carlo experiments validate the method application in angle estimation improvement.

1. Introduction

Multi-input-multi-output (MIMO) systems, which are composed of multiple transmitters and receivers, use the orthogonal waveform to separate different transmit-receive channels, achieving the improvement of detection and estimation [1]-[3]. In contrast to traditional phased array radar (PAR), the orthogonal signals transmitted by MIMO radar could achieve waveform diversity through transmitting orthogonal sub-waves. Thus the design and blind source separation for MIMO radar has attracted attentions from researchers in recent years [4] [5].

Appropriate waveform design is a fastidious in practical operation to satisfy complete orthogonality among sub-waves in MIMO systems. To solve this problem, variety of optimization methods for MIMO radars are proposed. [6] has given a brief introduction of MIMO radar waveforms, including four types of orthogonalization schemes and a particular MIMO waveform. The characteristic of orthogonal waveforms is analysed in detail in [7], and also has provided an improved OFDM chirp waveform method. The performance analysis including ambiguity function, correlations, as well as ideal peak side lobe ratio (PLSR) are provided thoroughly by [8]-[11].

Advanced Driver Assistance Systems(ADAS), benefiting from increased angular resolution in radar system, especially in MIMO radar, require massive calculation [12]-[14], causing a conflict between performance and resource. So most systems adopt time division multiple access (TDMA) to realize the orthogonal MIMO waveform so as to reduce filters overhead [15]. However, according to the reference [16], TDMA will increase coupling degrees on account of switching antennas as well as Doppler ambiguity.

This paper has proposed a Hadamard-matrix-based BPM encoding approach aiming at antenna coupling problem caused by TDMA. It is organized as follows. In Section II, the signal model is introduced in ideal condition. Hence, the influence of the fluctuation in amplitude as well as in phase are analyzed. In order to reduce the impaction on result, separation based on waveform reconstruction is put forward in Section III. This method shows computation reduction by utilizing the phase and significant structure information. Monte-Carlo experiments validate its application in angle estimation improvement. At last, conclusions are drawn and the future work is prospected.

2. Signal model and error analysis

2.1 Signal model

A co-located MIMO radar system is shown in Figure 1, including M transmitting and N receiving elements, operated in corresponding far-field uniform arrays. The distance between elements is set as $d=\lambda/2$ The chirp signal in each transmitting antenna element is described as $s(t)=\exp\left(j2\pi\left(f_c t+0.5kt^2\right)\right)$, where f_c denotes the carrier frequency, and k denotes the chirp slope. Thus the echo from a target at (Rm,θ°), transmitted by $Tx1$, and received by $Rx1$ can be expressed as:

$$sr(t)=\exp\left(j2\pi\left(f_c(t-\tau_0)+0.5k(t-\tau_0)^2\right)\right) \tag{1}$$

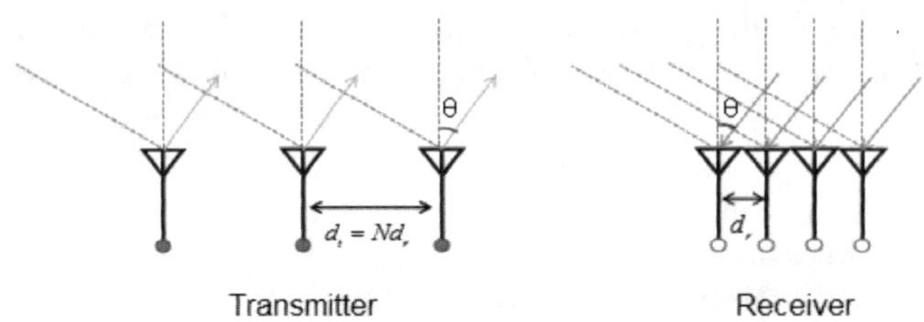

Figure 1. The structure of a MIMO radar.

where $\tau_0=c/2R$ represents the time delay. And the d-chirp signal is:

$$s_{11}(t)=sr(t)\cdot s(t)^*=\exp(-j2\pi k\tau_0 t)\exp(-j2\pi f_c\tau_0)\exp\left(jk\pi\tau_0^2\right) \tag{2}$$

Generally, we regard $Tx1$ and $Rx1$ as the reference antennas. Hence the d-chirp signal from Txm to Rxn is expressed as:

$$s_{mn}(t)=\exp[-j2\pi k(\tau_0+\tau_{mn})t]\exp[-j2\pi f_c(\tau_0+\tau_{mn})]\exp[jk\pi(\tau_0+\tau_{mn})] \tag{3}$$

Where τ_{mn} is the time delay, brought by array structure, and equals to $\tau_{mn}=(d_t+d_r)\sin\theta/c$. Because of a far-field target, which means $\tau_{mn}<<\tau_0$, Equation (3) can be rewritten as:

$$\begin{aligned} s_{mn}(t)&\approx\exp(-j2\pi k\tau_0 t)\exp(-j2\pi f_c\tau_0)\exp\left(jk\pi\tau_0^{\ 2}\right)\exp(-j2\pi f_c\tau_{mn}) \\ &=s_{11}\exp(-j2\pi f_c\tau_{mn})=s_{11}\exp[-j\pi((m-1)N+(n-1))\sin\theta] \end{aligned} \tag{4}$$

Here a brief introduction about Hadamard-matrix-based BPM encoding is provided. For convenience, we divide the system into two parts. On the transmit side, we assume 3 transmit antennas, and each antenna transmits a chirp with same parameters. Each block of frame signals transmitted by 3 antennas are modulated by a 3-by-4 Hadamard matrix. On the receiving side, by multiplying the 4-by-4 Hadamard matrix, the echo can be converted to the signals from 3 transmit antennas. Figure 2 and Figure 3 give explanations about this process.

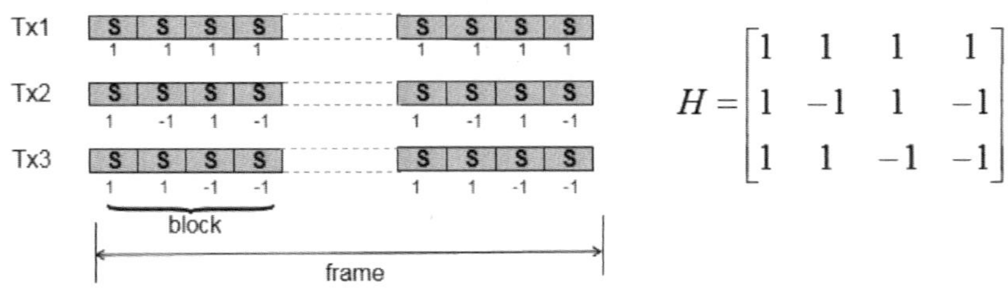

Figure 2. Illustration of transmitting of Hadamard coding MIMO waveform .

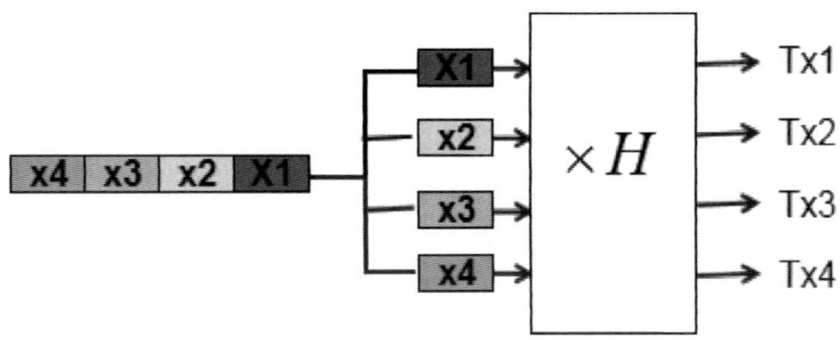

Figure 3. Illustration of receiving of Hadamard coding MIMO waveform .

Therefore the echo signal can be rewritten as:

$$X(t) = AS(t) + N(t) \tag{5}$$

Where A is an array manifold of the virtual array:

$$A = \left[a(\theta_1), a(\theta_2), \cdots, a(\theta_K)\right] \tag{6}$$

$$a(\theta) = [e^{-j\pi 0 \sin\theta}, e^{-j\pi \sin\theta}, \cdots, e^{-j\pi (MN-1)\sin\theta}]^T \tag{7}$$

Considering a 77GHz chirp with 192MHz bandwidth and the maximum unambiguous speed is 120km/h, Figure 4 shows the receive signal amplitude simulation result from Rx1. Figure 5 (a)-(c) illustrates the signals from 3 transmit antennas and (d) shows the redundancy signal. In Figure 6 (a) and (b), the one-dimensional range profile and the DOA estimation of the target are displayed respectively.

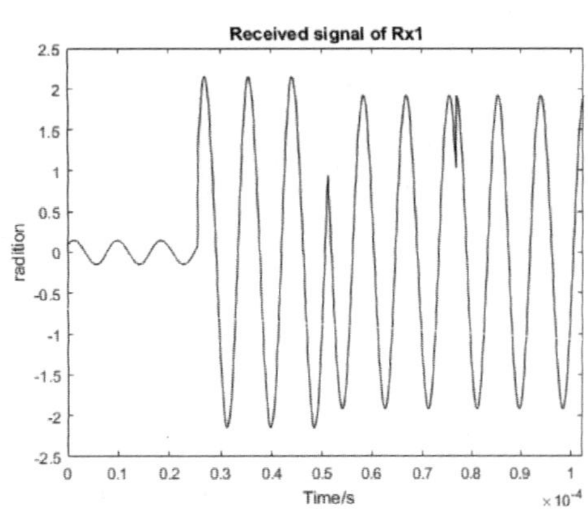

Figure 4. The receive signal amplitude simulation result from Rx1.

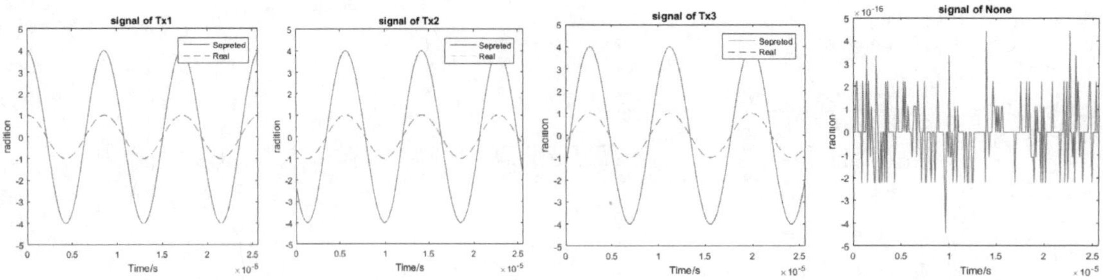

Figure 5. The signals from 3 transmit antennas. (a) The signal from Tx1; (b) The signal from Tx2; (c) The signal form Tx3; (d) The redundancy signal.

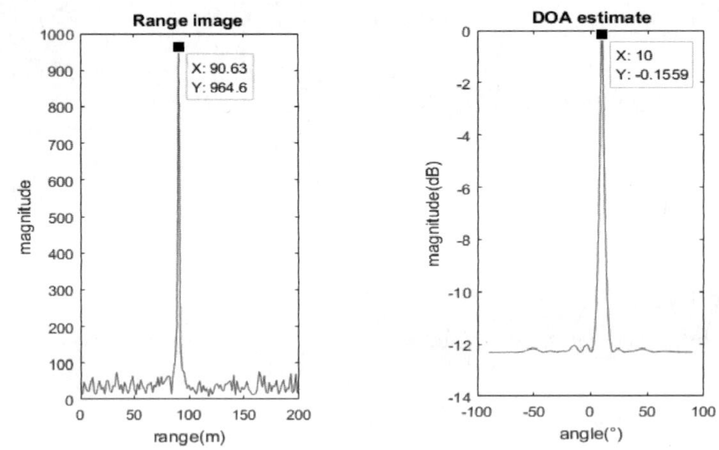

Figure 6. One-dimensional range profile and the DOA estimation. (a) One-dimensional range profile; (b) The DOA estimation

2.2. Error analysis

The estimation results are not right if the sick do not be a better deal which is due to the fluctuations are inevitable in the BPM encoding process. The mathematical models of fluctuations in amplitude as well as in phase are established. On the basis, the effects from the fluctuations on the spatial spectrum estimation are analyzed. The fluctuations can be introduced by error matrix $\mathbf{\Gamma}$. With the assumption that only amplitude fluctuation exists, $\mathbf{\Gamma} \in \mathrm{C}^{12\times12}$ can be expressed as:

$$\mathbf{\Gamma} = \begin{bmatrix} \boldsymbol{\rho}_1 & 0 & 0 & 0 & \boldsymbol{\rho}_2 & 0 & 0 & 0 & \boldsymbol{\rho}_3 & 0 & 0 & 0 \\ 0 & \boldsymbol{\rho}_1 & 0 & 0 & 0 & \boldsymbol{\rho}_2 & 0 & 0 & 0 & \boldsymbol{\rho}_3 & 0 & 0 \\ 0 & 0 & \boldsymbol{\rho}_1 & 0 & 0 & 0 & \boldsymbol{\rho}_2 & 0 & 0 & 0 & \boldsymbol{\rho}_3 & 0 \\ 0 & 0 & 0 & \boldsymbol{\rho}_1 & 0 & 0 & 0 & \boldsymbol{\rho}_2 & 0 & 0 & 0 & \boldsymbol{\rho}_3 \end{bmatrix} \tag{8}$$

where $\boldsymbol{\rho}_1, \boldsymbol{\rho}_2, \boldsymbol{\rho}_3$ are:

$$\begin{bmatrix} \boldsymbol{\rho}_1 & \boldsymbol{\rho}_2 & \boldsymbol{\rho}_3 \end{bmatrix} = \begin{bmatrix} (\rho_{11}+\rho_{21}+\rho_{31}+\rho_{41}) & (\rho_{12}-\rho_{22}+\rho_{32}-\rho_{42}) & (\rho_{13}+\rho_{23}-\rho_{33}-\rho_{43}) \\ (\rho_{11}-\rho_{21}+\rho_{31}-\rho_{41}) & (\rho_{12}+\rho_{22}+\rho_{32}+\rho_{42}) & (\rho_{13}-\rho_{23}+\rho_{33}-\rho_{43}) \\ (\rho_{11}+\rho_{21}-\rho_{31}-\rho_{41}) & (\rho_{12}-\rho_{22}-\rho_{32}+\rho_{42}) & (\rho_{13}+\rho_{23}+\rho_{33}+\rho_{43}) \end{bmatrix} \tag{9}$$

ρ_{ij} means the amplitude fluctuation of the ith sequence from the jth transmitting antenna. According to the reference [6], the covariance matrix can be deduced as:

$$R'_{\mathrm{xx}} = \mathbf{\Gamma} A R_{\mathrm{ss}} A^{\mathrm{H}} \mathbf{\Gamma}^{\mathrm{H}} + \sigma_n^2 I \tag{10}$$

and the spatial spectrum calculated by multiple signal classification (MUSIC) algorithm is:

$$P_{MUSIC}(\theta)=\frac{1}{a^H(\theta)\tilde{U}_N\tilde{U}_N^H a^H(\theta)} \tag{11}$$

$$\tilde{U}_N\tilde{U}_N^H=\sum_{i=M+1}^{N}\tilde{e}_i\tilde{e}_i^H=I-\sum_{i=1}^{M}\tilde{e}_i\tilde{e}_i^H \tag{12}$$

Note that there is only one target which is located at (Rm,θ_1°), the largest eigenvalue λ_1 and corresponding eigenvector $\tilde{e}_1=\Gamma a(\theta_1)$ can be achieved by eigenvalue decomposition. So Equation (12) turns into:

$$\tilde{U}_N\tilde{U}_N^H=I-\sum_{i=1}^{M}\tilde{e}_i\tilde{e}_i^H=I-\frac{\Gamma a(\theta_1)a^H(\theta_1)\Gamma^H}{|\Gamma^2|} \tag{13}$$

Therefore, the denominator of $P_{MUSIC}(\theta)$ equals to:

$$a^H(\theta)\tilde{U}_N\tilde{U}_N^H a^H(\theta)=N-\frac{|a^H(\theta)\mathbf{\Gamma}a(\theta_1)|^2}{|\mathbf{\Gamma}^2|} \tag{14}$$

For convenience, γ and η are employed to replace $(\rho_{1i}+\rho_{2i}+\rho_{3i}+\rho_{4i})$ and $(\rho_{1i}+\rho_{2i}-\rho_{3i}-\rho_{4i})$ so that $a^H(\theta)\mathbf{\Gamma}a(\theta_1)$ are:

$$\begin{aligned}a^H(\theta)\mathbf{\Gamma}a(\theta_1)&=e^{j0\pi\sin\theta}\left(\gamma_1 e^{j0\pi\sin\theta_1}+\eta_1 e^{-j4\pi\sin\theta_1}+\eta_2 e^{-j8\pi\sin\theta_1}\right)+\cdots\\&\cdots+e^{j11\pi\sin\theta}\left(\gamma_{12}e^{-j11\pi\sin\theta_1}+\eta_{23}e^{-j3\pi\sin\theta_1}+\eta_{24}e^{-j7\pi\sin\theta_1}\right)\\&=e^{j0\pi\sin\theta}\beta_1 e^{j\varphi_1}+e^{j\pi\sin\theta}\beta_2 e^{j\varphi_2}+\cdots+e^{j11\pi\sin\theta}\beta_{11}e^{j\varphi_{11}}\end{aligned} \tag{15}$$

On account of fluctuations in amplitude, Equation (14) will not reach the minium when $\theta=\theta_1$, indicating the evaluated errors. Similarly, in the case that only fluctuations existing in phase, $a^H(\theta)\mathbf{\Gamma}a(\theta_1)$ can be shown by:

$$a^H(\theta)\mathbf{\Gamma}'a(\theta_1)=e^{j0\pi\sin\theta}\kappa_1 e^{j\vartheta_1}+e^{j\pi\sin\theta}\kappa_2 e^{j\vartheta_2}+\cdots+e^{j11\pi\sin\theta}\kappa_{11}e^{j\vartheta_{11}} \tag{16}$$

Figure 7 (a) and (b) illustrates the simulations of amplitude and phase fluctuations, and Table 1 displays a part of results on the basis of 100 Monte-Carlo runs.

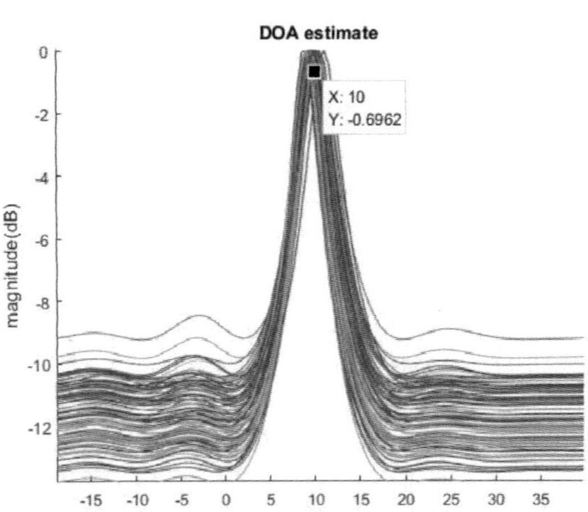

(a)

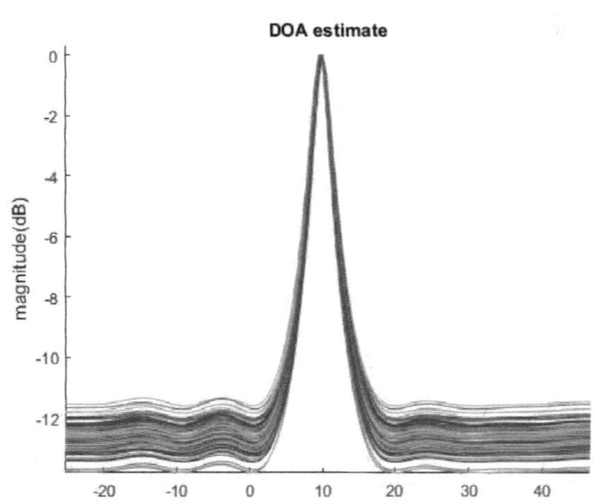

(b)

Figure 7. The amplitude and phase fluctuation simulation results within 100 Monte-Carlo runs. (a) The amplitude fluctuation simulation result; (b) The phase fluctuation simulation result.

Table 1. The partial results of DOA estimation within 100 Monte-Carlo runs.

No.	15	25	30	35	40	45	50	55	60	70	75	80	85
fluctuation in amplitude	9.5	8	9	11	12	10	10.5	9.5	10	11.5	9.5	9.5	10.5
fluctuation in phase	9.5	10	10	9.5	10	9.5	9.5	10	10	9.5	10	10	9.5

Generally speaking, for the error of phase is fluctuating around $\pm 5°$ as well as the cosine is nonlinear, the consequence resulted by phase fluctuation can be ignored reasonablly.

3.Signal separation method based on waveform reconstruction

Supported by Equation (15), it can be deduced that incomplete signal separation mainly results in estimating errors. Considering the random amplitude fluctuation and the known array structure in MIMO radar, a method of waveform reconstruction fitting based on least mean-square error rule is put forward. Thus we provide the processing for fitting in signal separation:

- Initialize the arriving direction of the target echo by MUSIC algorithm, where $\theta = \theta_0$.
- Normalize the signal amplitude of each unit in ULA to get the $A^* = \left[e^{j\phi_1}, e^{j\phi_2}, \cdots, e^{-j11\pi \sin\theta_0}\right]^{\mathrm{T}}$.

For this part, each receiving antennas can separate different signals corresponding to different transmitting antennas. For a better estimation, normalization is necessary.

Set the size and direction of scanning step. The step parameter setting need to consider computation and accuracy.

- Waveform reconstruction supported by the information of the array structure. Reconstruct the manifold $A' = \left[e^{j0\pi \sin\theta_0}, e^{-j1\pi \sin\theta_0}, \cdots, e^{-j11\pi \sin\theta_0}\right]^{\mathrm{T}}$.
- Obeying least mean square error criterion.

Figure 8 shows the initialization of A' and normalized A^*. Figure 9 demonstrates the processing of iteration and get the optimal estimation. For obvious contrast from Table 2, this method can reduce the errors caused by amplitude fluctuation effectively, and the results of Monte Carlo experiments shows the estimation MSEs have been reduced by 0.4 when the amplitude fluctuation obeys Gaussian distribution indicated by $N(1,0.5)$.

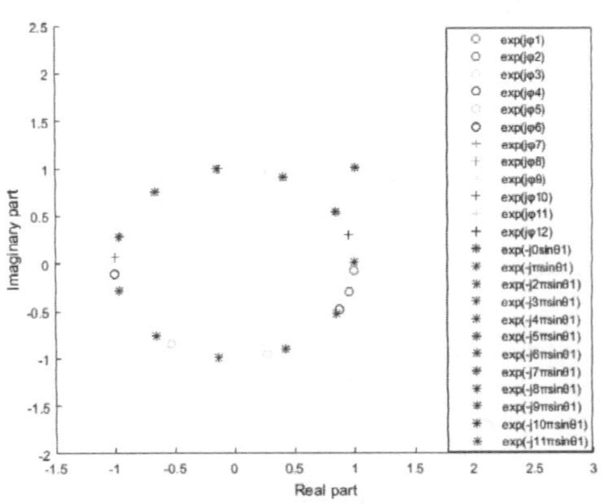

Figure 8. The phase of A' and A^*.

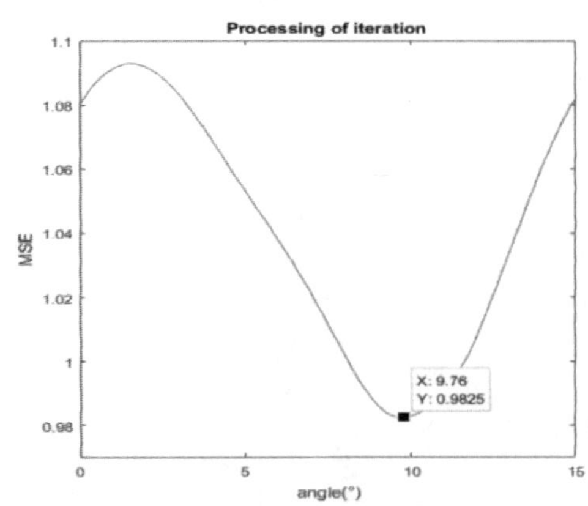

Figure 9. The processing of iteration

Table 2. The partial compare between the original and the proposed method

No.	15	25	30	35	40	45	50	55	60	70	75	80	85
fluctuation in amplitude	9.5	8	9	11	12	10	10.5	9.5	10	11.5	9.5	9.5	10.5
proposed method	10.0	9.8	9.8	10.1	10.2	9.9	9.8	9.9	9.9	10.1	10.0	10	9.9

4. Conclusions

In this paper, a Hadamard-matrix-based BPM encoding method was confirmed is suitable for MIMO radar systems, especially in the case of insufficient matched filters. This paper also had presented results on influence caused by fluctuations in amplitude as well as in phase, which pointed out that amplitude fluctuations have a bigger effect than phase. Within the known MIMO radar architecture, the paper proposed a waveform reconstruction fitting method based on minimum mean-square error rule. Monte-Carlo experiments validate the method application in angle estimation improvement.

References

[1] Zheng, L., Lops, M. , Wang, X. ,& Grossi, E. . 2017. Joint design of overlaid communication systems and pulsed radars. *IEEE Transactions on Signal Processing*, PP(99), 1-1.

[2] Wu, L. , Qi, C. , & Chen, P. . 2016. Antenna placement optimisation for compressed sensing-based distributed mimo radar. *IET Radar, Sonar & Navigation.*

[3] Chen, Peng . 2017.Moving Target Detection Using Colocated MIMO Radar on Multiple Distributed Moving Platforms. *IEEE Transactions on Signal Processing* ,pp(99):1-1.

[4] Pang, Y. C. , & Zhang, C. Z. . 2014. A new method for angle estimation in mimo radar with blind source separation. *Applied Mechanics and Materials*, 543-547, 2500-2504.

[5] Ren, C., Liu, F., & Zhou, J.. 2018. MIMO Radar Waveform Design with Practical Constraints: A Low-Complexity Approach. *2018 IEEE 18th International Conference on Communication Technology (ICCT)*. IEEE.

[6] H Sun, Frédéric Brigui, & M Lesturgie. 2014.Analysis and Comparison of MIMO Radar Waveforms.*2014 International Radar Conference.*IEEE

[7] Wang, & Wen-Qin. 2015. Mimo sar ofdm chirp waveform diversity design with random matrix modulation. *IEEE Transactions on Geoscience and Remote Sensing*, 53(3), 1615-1625.

[8] San Antonio, G. , Fuhrmann, D. R. , & Robey, F. C. . 2007. Mimo radar ambiguity functions. IEEE Journal of Selected Topics in Signal Processing, 1(1), 167-177.

[9] Zhang, X. , Zhang, Y. , Xie, W. , Wang, Y. , & Yan, T. . 2010. Clutter characteristics analysis of airborne MIMO radar under the degradation of transmit waveforms orthogonality. *International Conference on Industrial Mechatronics & Automation*. IEEE.

[10] Hutchison, D. , Kanade, T. , Kittler, J. , Kleinberg, J. M. , & Juristo, N. . 2012. Optimization of orthogonal poly phase coding waveform based on bees algorithm and artificial bee colony for mimo radar. *Lecture Notes in Computer Science,7343*.

[11] Li, Y. , Vorobyov, S. A. , & Koivunen, V. . (2014). Generalized ambiguity function for the MIMO radar with correlated waveforms. *ICASSP 2014 - 2014 IEEE International Conference on Acoustics, Speech and Signal Processing (ICASSP)*. IEEE.

[12] O. Andreas ,P. Clemens,S. Jochen, S. Stefan ,W. Robert, G.H DICE & Co KG,2018,A scalable 77 GHz massive mimo fmcw radar by cascading fully-integrated transceivers,*2018 Asia-Pacific Microwave Conference*.

[13] Meinl, F. , Stolz, M. , Kunert, M. , & Blume, H. . 2017. An experimental high performance radar system for highly automated driving. *2017 IEEE MTT-S International Conference on Microwaves for Intelligent Mobility (ICMIM)*. IEEE.

[14] Dickmann, J. , Klappstein, J. , Hahn, M. , Appenrodt, N. , & Sailer, A. . 2016. “Automotive radar the key technology for autonomous driving: From detection and ranging to environmental understanding”. *2016 IEEE Radar Conference (RadarConf16)*. IEEE.

[15] Chen Bin, ZHU, CS. , LI, W., Wei Jibo. 2016. Original Symbol Phase Rotated Secure Transmission Against Powerful Massive MIMO Eavesdropper. *IEEE ACCESS*.

[16] Xiao-Yu, J. I. , Chen, M. , Hong-Xu, S. , & Wen, W. U. . 2017. Research on signal processing method of 77ghz multi-carrier-frequency mimo radar for car. *Journal of Microwaves*.

IOP Conf. Series: Journal of Physics: Conf. Series **1284** (2019) 012074 doi:10.1088/1742-6596/1284/1/012074

Improved Fast Adaptive IDW Interpolation Algorithm based on the Borehole Data Sample Characteristic and Its Application

Liming Sun[1*], Yingqi Wei[1], Hong Cai[1], Jun Yan[1] and Jianzhang Xiao[1]

[1] China Institute of Water Resources and Hydropower Research, Beijing, China.

[*]E-mail: sunlm@iwhr.com

Abstract. The inverse distance weighted algorithm (IDW) is one of the most widely used algorithms in geoscience calculation based on its university and simple theory. However, in the practical application based on borehole sampling data, the distance between boreholes is further than the sampling distance along the bedding direction of one borehole, and it needs to search all the near points through a large number of distance calculation, and the process is very slow, so this lacks practicality in the spatial interpolation of the large amount and non-uniform distribution data. Based on the special features of the borehole sampling data, this paper presents a fast adaptive and improved inverse distance weighted interpolation method. And firstly the sample point position could be indexed by the adaptive cube grid raster, then determine the index grid in the horizontal and vertical directions of the interpolation point, and combine these grids for the finally IDW calculation, which significantly improves the search speed of the adjacent interpolation calculation points. It can improve the engineering applicability of the IDW method, especially the three-dimensional spatial interpolation and computing has a good application prospect.

1.Introduction

Spatial interpolation is an important approach to solve the shortage of data sampling in the 3D geoscience related issues. The borehole data is one of the important data sources[1-3], and the 2D and 3D data interpolation method based on borehole data is widely used because of the difficulty in subsurface sampling. For example in the 3D geological modeling using borehole sample data, the 2D smooth interpolation of the geological interface can obtain the smooth geological surface and the 3D spatial interpolation of the geological property field can get the heterogeneous distribution of the geological property [4,5]; in the groundwater analysis application, the spatial interpolation method can be used to calculate the 3D distribution field of the chemical elements in groundwater according to the borehole data; property interpolation and grade calculation of metallic minerals are completed in 3D stratigraphic soil modeling, etc.[6], so the 3D spatial interpolation theory based on borehole sample data is an important foundation work for numerical calculation and analysis. The suitable spatial interpolation method can calculate continuous space field data more quickly and accurately, and it could provide basic data for subsurface geological modeling and numerical analysis application. Furthermore, studying a suitable spatial interpolation method based on the borehole data characteristics is the core issues in the geological geometric modeling and calculation analysis [7].

At present, the most commonly used interpolation methods are two categories: one is geometric smooth interpolation based on mathematical and geometric methods, including B-spline[8], NURBS

IOP Conf. Series: Journal of Physics: Conf. Series **1284** (2019) 012074 doi:10.1088/1742-6596/1284/1/012074

surface[9], polynomial and fuzzy mathematics, etc.; and the other is the geoscience interpolation method commonly used in the geology field designed based on the data distribution characteristics of geoscience space, including kriging[10], the inverse distance weighted algorithm (IDW) [11], discrete smooth interpolation, neural network and other interpolation methods, and this kind of interpolation method is more widely used in geospatial interpolation, because the distribution characteristics of geospatial data are usually considered. The IDW interpolation method is suitable for the case of heterogeneous distribution and continuous change, the calculation theory is comprehensive and effective, but it is difficult to deal with the case of uneven and irregular distribution of raw data, and the calculation speed of search neighbor points is slow[12].The kriging interpolation method reflects not only the spatial distance relationship between the original sample points and the interpolation point, but also the directivity and aggregation characteristics of the raw points distribution, which are obtained by creating a semi-variogram function, these characteristic guarantees higher accuracy and better indicators in most cases[13]. If it is assumed that the equations don't converge or have too few raw points, and the kriging calculation results will be worse. The borehole data are often sampled along the borehole line densely, and they do not follow the spherical, cube or exponential distribution model obviously, so the existing general geological interpolation methods (such as IDW and kriging interpolation algorithm) cannot fully reflect the borehole data characteristics. In addition, due to the dense sampling and linear distribution of borehole data, the traditional interpolation method used in the interpolation calculation could not meet the actual needs of the accuracy and calculation time [14].

2. Borehole sample data

The main feature of the borehole data is that it is composed of discrete points, each point has (x, y) coordinates and elevation z coordinates. In addition, the borehole data is spatiality, the vertical stratigraphic point are orderly distributed along the borehole line. The same stratigraphic points of the different boreholes are located at the same stratigraphic interface, which is the basis for 3d stratigraphic modeling with the borehole data. Since the borehole exploration data is a linear structure perpendicular to surface, each geological borehole could be abstracted as a point set element with 3D structure [15, 16], as shown in Figure 1.

The definition of borehole data model is as follows:

(1) Borehole: borehole number, exploration type, hole coordinates (x, y, z), depth h, number of number of strata: n.

(2) Stratum: borehole number, stratum sequence number s, top surface point (x_1, y_1, z_1) and bottom surface point (x_2, y_2, z_2) of the stratum, the corresponding property sample value p, other description information of stratum.

(3) Borehole incline: depth sounding h0, dip Angle α and azimuth β, etc.

(4) Property sample: borehole sample data, which is usually the distributed along the borehole line, sample depth h_0, sample number, sample time, and sample property value.

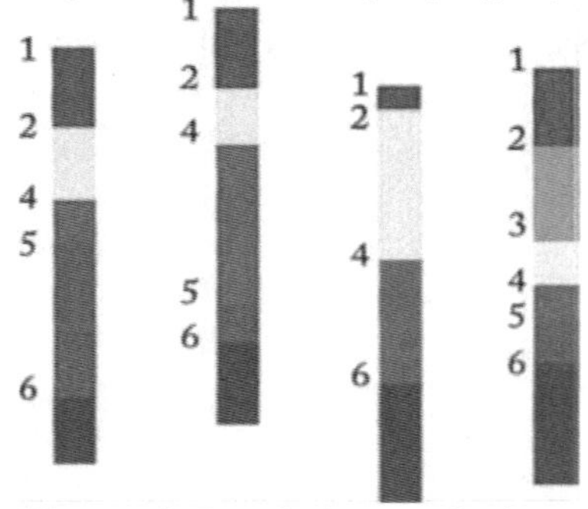

Figure 1. Borehole sample data.

IOP Conf. Series: Journal of Physics: Conf. Series **1284** (2019) 012074 doi:10.1088/1742-6596/1284/1/012074

3. Improved fast adaptive IDW interpolation algorithm

IDW interpolation algorithm is a kind of deterministic algorithm widely used in the geoscience field, for its theoretical principle is simple[17].The IDW principle is to measure distance influence between the interpolation point and the adjacent point. Obviously, the closer the distance is, the greater influence on the interpolation point; and the larger weight on the corresponding point; otherwise, the smaller influence on, so it is called the IDW method. In addition, the interpolation mathematical expression (power exponent), the radius range of adjacent points (number of adjacent points) or the number of adjacent points in different quadrants could be adjusted to improve the interpolation accuracy. Also, IDW is a multidimensional nonlinear data interpolation method and an important tool in scientific visualization of 3d discrete data fields.

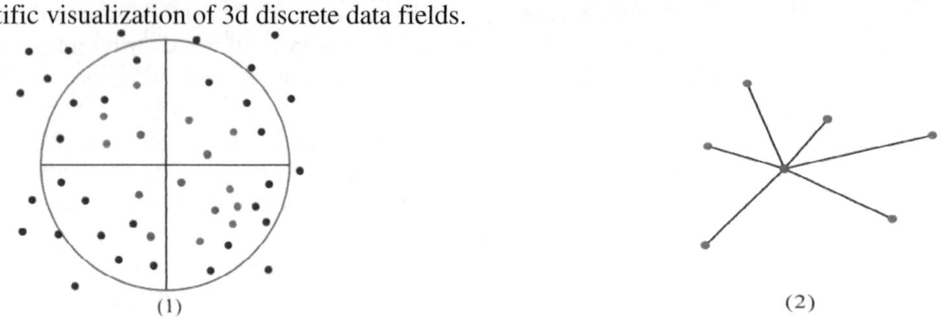

Figure 2. 2D IDW method, the different colour points represent the different weight value, (1) the points inside the interpolation radius range, (2) the distance between the original sample point and the interpolation calculation point.

As shown in Figure.2, the general mathematical expression of IDW is as formula (1):

$$z_j = k_j \sum_i^j \frac{1}{d_{ij}^{\alpha}} z_i \tag{1}$$

Among them, z_i is the value of the i point, k_j is the adjustment value that guarantees that the final weight sum is 1, d_{ij} is the distance from the i point to the current point, and for N-D interpolation, the distance calculation formula between two points is as formula (2):

$$d = \sqrt{(x_1 - x_0)^2 + (x_2 - x_0) + \ldots + (x_n - x_0)} \tag{2}$$

And α is power exponent.

IDW mainly uses power parameters to control the influence of known points on interpolation points. The power exponent is a positive number, when the power exponent is larger, and the effect of distance on the result can be further increased, which varies due to the specific calculation. Borehole sample data is usually sparse in horizontal direction, dense in vertical direction, and the two value is greatly difference. Therefore, the traditional adjacent points searching from spherical surround box are non-uniform distribution along and vertical the borehole line direction, in addition, the available points are not enough, and there are too many similar points, so the accuracy is limited by the traditional sphere bounding box for the search calculation [18], as shown in Figure 3.

Traditionally, the weight correction method is used to correct the points of clustering effect, and assume the correction coefficient *f* [0,1], if *f* is 0, it will not be corrected, and if *f* is 1, it will be the max influence. The *f* could be calculated based on the angle between the connecting line between the current point and the point in the neighborhood and the two points in the neighborhood[19]. This method has a good effect on the local clustering effect, but the borehole sample point is along line, the data clustering effect is not regular, because it is difficult to directly use this method. So it is necessary to study an improved IDW method based on the characteristics of the borehole data[20].

3.1 Improved method flow

In this paper, an improved IDW method based on borehole data features is proposed, and the flowchart is shown in Figure 4 below. Firstly, a 3D search grid is built according to the maximum range of the borehole sample points, and the grid parameters are determined by the interpolation accuracy and the borehole distribution. Secondly, all the borehole sample points are iteratively assigned to the corresponding grids by the surround box coordinates, a 3d grid (as surround box) may contain many points, and big grids is used to replace points to speed up the search process in the improved IDW algorithm. Thirdly, different parameters of the bounding box are set according to the horizontal and vertical directions, and the corresponding points of search grid are merged to be used for the final interpolation calculation. Finally, the parameterized interpolation calculation of the entire subsurface will be completed. The entire interpolation process is introduced as follows, as shown in Figure 4.

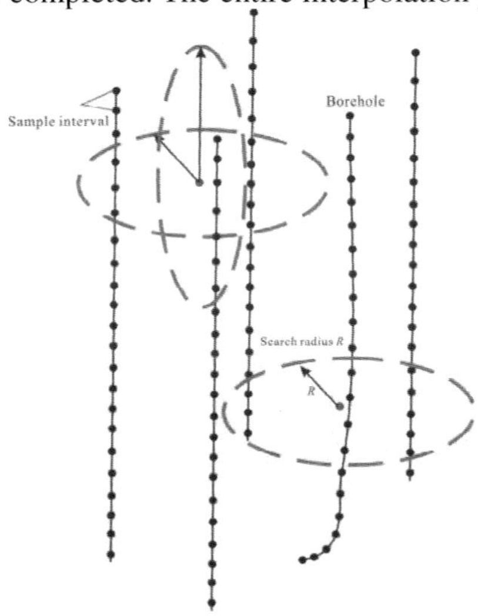

Figure 3. The search extent of borehole sample data in the IDW method.

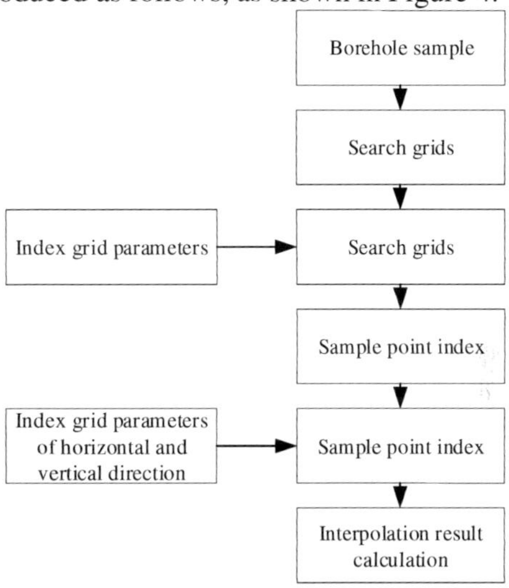

Figure 4. The method process.

3.2 The determination method of interpolation calculation parameters

The cube search grid is used for fast indexing all the original sample points. The whole subsurface is seamlessly filled with adjacent cube grids with the same size, and all the sample points will be assigned to their own grids by position relationship, the grid search will replace the original spherical search in the improved method. First, the side length parameter m of the cube grid needs to be determined. According to the characteristics of the borehole data, two search bounding boxes in the horizontal and vertical directions of the borehole will be introduced to determine the interpolation calculation parameter, each bounding box was composed of a different number of grids, and then the size parameters of the two bounding boxes will be determined.

According to the description in section 2, the borehole data model is firstly built, and the borehole sample data is stored structurally to maintain the topological relationship between the point and the borehole for grid query and calculation. And then, set the minimum number of original sample points involved in the IDW calculation is S_{min}, the maximum search distance is $D_{\max}$, and the maximum search distance ensures that the search range will not be too large when the original sample points are extremely sparse. Generally, the more borehole sample points are used in the calculation, the higher the calculation accuracy will be, but the search speed and calculation speed will slow. Assuming the vertical sampling distance of borehole is d, the distance of the adjacent borehole is l, and $d<<l$, the number of adjacent borehole is B_{num}, $B_{num}>S_{min}$ must be guaranteed, and the size of horizontal search bounding box $B_{horizontal}$ and vertical search bounding box $B_{vertical}$, it is ensure that both S_{min} and D_{max} requirements are met. The calculation rules must be followed:

IOP Conf. Series: Journal of Physics: Conf. Series **1284** (2019) 012074 doi:10.1088/1742-6596/1284/1/012074

(1) The cube grid: points are not too much and uniform distribution, assuming that the search radius is *a*, and the radius *a* is depended on the distribution of the points, $a>d$.

(2) For the horizontal direction box, the bounding box in vertical borehole line direction, its side length is $B_{horizontal}>l*4$, and the number of boreholes is guaranteed that $B_{num} > S_{min}$.

(3) For the vertical direction, the bounding box in the direction of the borehole line, which its length is related to the length of the borehole, the length in the horizontal direction is greater than the vertical direction $B_{vertical}>d*12$.

3.3 Creating search cube grid

According to the section 3.2, the grid length size *m* will be calculated, and the search grid could be built according to the grid size. The size of the search grid is usually large, which is a collection of a points with similar location in a single grid. Firstly, the maximum bounding box of the entire subsurface is calculated according to all the borehole data points, and then the space is divided according to the search grid size *m*. The storage data structure of single grid only needs to store the maximum and minimum XYZ value of the maximum bounding box, and it needs to calculate the number of corresponding grids in the XYZ three directions. In this case, only row and column numbers need to be stored, and all the grid vertices can be calculated according to the maximum and minimum values of row and column numbers and the bounding box. In the end, all sample points need to be iterated once and the point number is recorded in the grid object. Then, when using the IDW interpolation, the corresponding search grids within the specified distance can be directly obtained according to the XY coordinate of the interpolation point, and the interpolation distance is not required to be calculated for many times. In this way, the search speed of the interpolation points involved in the calculation can be greatly accelerated. Since the two squared operation of all sample points is changed to a simple linear sum, so the search time is greatly reduced, which is the core idea of this paper.

3.4 Horizontal and vertical search grid computing

According to the search grid obtained in section 3.3, all the original sampl3 points have been stored in the search grid. The distance between the original sample point and the interpolation point can be obtained using the existing grid directly, and the distance formula needs not to recalculate. On the basis of accelerating the search time and further combining the characteristics of the borehole data, the difference between the horizontal direction and the vertical direction of the search bounding box will be considered, the accuracy is improved on the basis of fast search. In this way, the algorithm optimization based on the borehole sample data can improve the calculation efficiency and accuracy of the IDW algorithm. Since the horizontal and vertical directions need to be calculated separately when searching the grid, it needs to be completed in three steps:

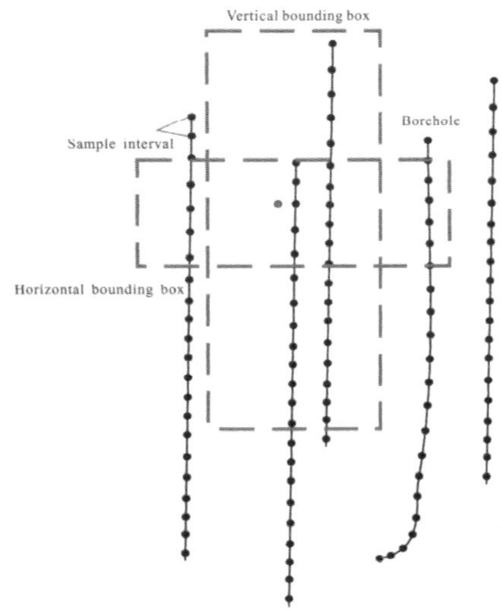

Figure 5. Two search grid surround boxes in horizontal and vertical directions.

(1) Interpolation bounding box of horizontal direction, after determining the position of the interpolation points, the search grids of the interpolation points need to be determined firstly. And then, according to the bounding box side in the horizontal direction calculated in section 3.2, the search grid collection of the interpolation points in the XYZ three direction are obtained directly. Finally, all sample points of the horizontal bounding box can be directly obtained according to the point index in the grid, as shown in Figure. 5.

(2) Interpolation bounding box of vertical direction, it is similar to the horizontal interpolation bounding box, the points in the vertical direction can be directly obtained according to the calculated side length in section 3.2. After deduplication with the horizontal grid, the points in both grids can be merged to obtain all the original sample points involved in the calculation.

(3) Put the bounding box data into the original calculation formula of the IDW interpolation method for calculation, and the interpolation result of any point will be obtained, as shown in Figure 5.

3.5 Application

According to the optimized IDW interpolation method based on the borehole data characteristic in this paper, which will be applied to the grade calculation and analysis of metallic minerals to verify the feasibility of the method. Taking the borehole sample data of a gold mine as an example, there are 421 boreholes with a sample interval of 2m in the direction of the borehole. The boreholes are evenly distributed, most of which are inclined boreholes, and there are infill boreholes in the position with high metal content. Finally, according to the borehole data shown in Figure 6, there are a total of 43,000 sample points, with a east-west span of 1500m and a height span of 1000m, as shown in Figure 6.The single search speed is not acceptable, so it is not practical to directly use the inverse distance interpolation method. By using the IDW method proposed in this paper, the full size of 125m grid was calculated in 0.5 seconds, and the grade interpolation of the whole geological space was completed. The accuracy of cross validation method meets the 5% error requirement, which has advantages over the traditional method in terms of time and accuracy. The result is shown as in Figure.7.

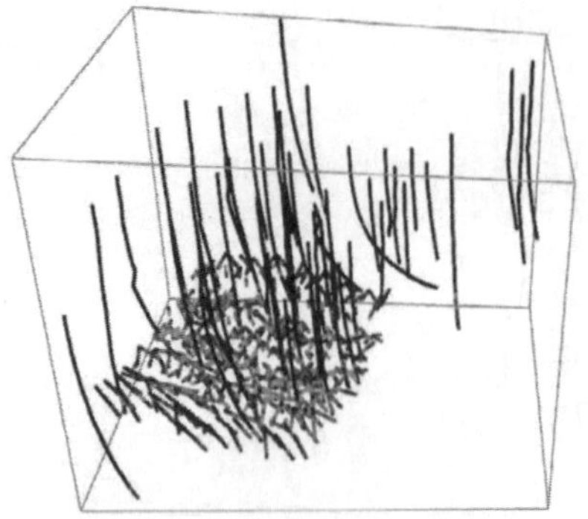

Figure 6. Borehole data of gold mine.

Figure 7. The imporved interpolation method result.

4.Summary

The method proposed in this paper is an improved method of IDW interpolation based on borehole sample data, compared with the traditional IDW interpolation, the spherical search radius in the original IDW interpolation is transformed into the horizontal bounding box and the vertical bounding box. The radius length of both bounding boxes is determined by the borehole data characteristics, which not only considers the borehole data characteristics, the clustering effect of original data, and the distance search efficiency of interpolation points, but also improves the final calculation accuracy by adding reasonable original points. These are important improvements to the IDW algorithm. To sum up, this paper puts forward the following two improvements:

(1) Compared with the traditional IDW method, the distance between each interpolation point and the original sample points will be obtained by iterative calculation, the method indexes original sample points by moving search grid, which significantly accelerates the search speed of adjacent interpolation points. Method could quickly index the distance between two points without calculation.

(2) The method of combining horizontal and vertical bounding boxes is proposed by taking into account the borehole sample data characteristics. The different data distribution density in the two directions is considered, which greatly improves the accuracy of the IDW method. The corresponding parameter quantification criterion is proposed, which provides the feasibility for the practical engineering application of the IDW method, and an example is given to verify the algorithm.

Acknowledgements

The research presented is funded by the National Key Research and Development Program of China (No.2018YFC1508602, 2017YFC0804607).

References

[1] Caumon G, Grosse O, Lepage F. Unstructured Stratigraphic Grids : Construction , Population and Visualization Issues[J]. 2004(1): 94305.

[2] Durlofsky L J. A general gridding, discretization, and coarsening methodology for modeling flow in porous formations with discrete geological features[J]. Advances in Water Resources. 2016. [3] Yamamoto J K, Kikuda A T, Koike K, et al. Determination of volumetric uncertainty for geological bodies detected by boreholes[J]. Measurement. 2015, 66: 45-53.

[4] Wang W, Sun L, Li Q, et al. Representing the geological body heterogeneous property field using the quadratic generalized tri-prism volume function model (QGTPVF)[J]. Arabian Journal of Geosciences. 2017, 10(5): 115.

[5] Zhu L, Li M, Li C, et al. Coupled modeling between geological structure fields and property parameter fields in 3D engineering geological space[J]. Engineering Geology. 2013, 167: 105-116.
[6] Touch S, Likitlersuang S, Pipatpongsa T. 3D geological modelling and geotechnical characteristics of Phnom Penh subsoils in Cambodia[J]. Engineering Geology. 2014, 178: 58-69.
[7] Hwang Y, Clark M, Rajagopalan B, et al. Spatial interpolation schemes of daily precipitation for hydrologic modeling[J]. Stochastic Environmental Research and Risk Assessment. 2012, 26(2): 295-320.
[8] Fu Q, Wu Z, Wang X, et al. An algorithm for finding intersection between ball B-spline curves[J]. Journal of Computational and Applied Mathematics. 2018, 327: 260-273.
[9] Ruiu J, Caumon G, Viseur S, et al. Modeling Channel Forms and Related Sedimentary Objects using a Boundary Representation Based on Non Uniform Rational B-Splines[J]. In Proc. 34th Gocad Meeting. 2014(Umr 7330): 1-23.
[10] Ma Y Z, Royer J J, Wang H, et al. Factorial kriging for multiscale modelling[J]. Journal of the Southern African Institute of Mining and Metallurgy. 2014, 114(8): 651-657.
[11] Setianto A, Triandini T. Comparison of Kriging and Inverse Distance Weighted (IDW) interpolation methods in lineament extraction and analysis[J]. Journal of Southeast Asian Applied Geology. 2013, 5(1): 21-29.
[12] Mei G, Tian H. Impact of data layouts on the efficiency of GPU-accelerated IDW interpolation[J]. SpringerPlus. 2016, 5(1): 104.
[13] Stein M L. Interpolation of spatial data: some theory for kriging[M]. Springer Science & Business Media, 2012.
[14] Setianto A, Triandini T. Comparison of Kriging and Inverse Distance Weighted (IDW) interpolation methods in lineament extraction and analysis[J]. Journal of Southeast Asian Applied Geology. 2013, 5(1): 21-29.
[15] Wang Y, Ge F, Shao Y. The disperse ore bodies rebuilt based on sparse borehole data[J]. 2011 International Conference on Computer Science and Service System, CSSS 2011 - Proceedings. 2011: 2710-2713.
[16] Charifo G, Almeida J A, Ferreira A. Managing borehole samples of unequal lengths to construct a high-resolution mining model of mineral grades zoned by geological units[J]. Journal of Geochemical Exploration. 2013, 132: 209-223.
[17] Woodard R. Interpolation of Spatial Data: Some Theory for Kriging[J]. Technometrics. 2000, 42(4): 436-437.
[18] Cressie N. The origins of kriging[J]. Mathematical Geology. 1990, 22(3): 239-252.
[19] Shepard D. A two-dimensional interpolation function for irregularly-spaced data[J]. 23rd ACM national conference. 1968: 517-524.
[20] Lu G Y, Wong D W. An adaptive inverse-distance weighting spatial interpolation technique[J]. Computers & Geosciences. 2008, 34(9): 1044-1055.